International Conference on Advances in Manufacturing and Materials Engineering (ICAMME 2017)

IOP Conference Series: Materials Science and Engineering
Volume 290

Kuala Lumpur, Malaysia
8 – 9 August 2017

ISBN: 978-1-5108-5748-3
ISSN: 1757-8981

Printed from e-media with permission by:

Curran Associates, Inc.
57 Morehouse Lane
Red Hook, NY 12571

Some format issues inherent in the e-media version may also appear in this print version.

Printed by Curran Associates, Inc. (2018)

For permission requests, please contact the Institute of Physics
at the address below.

Institute of Physics
Dirac House, Temple Back
Bristol BS1 6BE UK

Phone: 44 1 17 929 7481
Fax: 44 1 17 920 0979

techtracking@iop.org

Additional copies of this publication are available from:

Curran Associates, Inc.
57 Morehouse Lane
Red Hook, NY 12571 USA
Phone: 845-758-0400
Fax: 845-758-2633
Email: curran@proceedings.com
Web: www.proceedings.com

Table of contents

Volume 290

International Conference on Advances in Manufacturing and Materials Engineering (ICAMME 2017)

8–9 August 2017, Kuala Lumpur, Malaysia

Accepted papers received: 20 December 2017
Published online: 30 January 2018

Preface

Papers

Nanomaterials

Modeling & Simulation, Manufacturing Systems

Composite Materials

Machining and Welding

Ceramics & Smart Materials

Mechanical Engineering

Green Technology

Manufacturing Engineering & Allied Area

IOP Conf. Series: Materials Science and Engineering **290** (2017) 011001 doi:10.1088/1757-899X/290/1/011001

PREFACE FROM THE CHAIRMAN of the 3rd ICAMME'17

Assoc. Prof. Dr. Mohd Hanafi Ani
Chairman
International Conference on Advances in Manufacturing and Materials Engineering (ICAMME) 2017

Assalamu'alaikum Warahmatullahi Wabarakatuh

It is a great pleasure to welcome you to the "International Conference on Advances in Manufacturing and Materials Engineering 2017 (ICAMME 2017)" that will take place at our International Islamic University Malaysia (IIUM), Gombak campus. The campus covers 700 acres, with elegant Islamic-style buildings surrounded by green-forested limestone hills. It may possibly be the longest crystal quartz outcrop in the world, a geological wonder right next to the city, Kuala Lumpur. The conference is part of our IIUM Engineering Congress, which aims to provide an international platform for academicians, scientists, and researchers to present results of ongoing research in their respective areas.

The call for papers attracted submissions from various backgrounds. There are four keynote lectures, ninety oral presentations and over thirty posters covering the different research areas of the conference. ICAMME 2017 focuses on advances in manufacturing and materials engineering covering wide spectrum of topics. Topics range from green manufacturing, lean manufacturing, subtractive manufacturing, synthesis and application of 2D materials, functional and smart materials, and nanoelectronic among others.

All accepted papers is peer reviewed, and will be published in "*IOP Conference Series: Material Science and Engineering*". The journal offers open access making the proceedings available to a wider scope of readership.

I would like to express my gratitude and appreciation to the keynote speakers, session chairs, staffs and students, and to all participants for their full support and efforts. I am also thankful to the members of the Organizing Committee for their endless efforts and diligence.

Best wishes for an enjoyable and memorable experience.

Wassalam

Assoc. Prof. Dr. Mohd Hanafi Ani

IOP Conf. Series: Materials Science and Engineering **290** (2017) 011002 doi:10.1088/1757-899X/290/1/011002

ORGANIZING COMMITTEE

Chairman
Assoc. Prof. Dr. Mohd Hanafi Ani

Co-Chairman
Dr. Irfan Hilmy

Secretariat
Dr. Yose Fachmi Buys
Dr. Suhaily Mokhtar

Technical Committee
Dr. Muataz Hazza Faizi Al-Hazza (Chairman)
Prof. Dr. Ahsan Ali Khan
Prof. Dr. Md. Abdul Maleque
Prof. Dr. Iis Sopyan
Assoc. Prof. Dr. Mohamed Abdul Rahman
Assoc. Prof. Dr. Souad A. Mohamad
Assoc. Prof. Dr. Suryanto
Dr. 'Atiah Abdullah Sidek
Dr. Noor Azlina Hassan
Dr. Maizatulnisa Othman

Publication Committee
Dr. Farah Diana Mohd Daud (Chairman)
Dr. Noor Azlina Hassan
Dr. Natasha A. Raof
Dr. Norshahida Sarifuddin
Dr. Norita Hassan
Dr. Sharifah Imihezri S. Shahanudin
Nor Amyra Zulianey binti Kahlib
Nur Aishah Ahmad Shahrim
Norhasnidawani Johari
Aida Ashikin Binti Abdullah
Amrina Zakiyuddin
Ainun Shaheera Mohamad
Fatin Bazila Fauzi

Registration and Program Committee
Dr. Alya Naili Rozhan (Chairman)
Dr. Assayidatul Laila Nor Hairin
Dr. Nur Idayu Ayob
Dr. Nur Ayuni Jamal

Invitation, Reception and Ushering Committee
Dr. Noorasikin Samat (Chairman)
Dr. Rubina Bahar

IOP Conf. Series: Materials Science and Engineering **290** (2017) 011002 doi:10.1088/1757-899X/290/1/011002

Sponsorhip Committee
Assoc. Prof. Dr. Hazleen Anuar (Chairman)
Dr. Adibah Amir
Zahir Hussain Syed Meera

Website, Publicity and Promotion Committee
Dr. Irfan Hilmy (Chairman)
Prof. Dr. Mohammad Yeakub Ali
Assoc. Prof. Dr. Mohamed Konneh
Dr. Zahurin Halim

Exhibition Committee
Dr. Abd. Malek Abdul Hamid (Chairman)
Assoc. Prof. Dr. Hadi Purwanto
Dr. Hafizah Hanim Mohd Zaki

Food and Souvenir Committee
Dr. Norhashimah Mohd Shaffiar (Chairman)
Dr. Nor Khairusshima Muhammad Khairussaleh

Transportation, Venue and Logistic Committee
Dr. Ahmad Zahirani Ahmad Azhar (Chairman)
Dr. Mohd Radzi Che Daud
Dr. Afifah Mohd Ali

IOP Conf. Series: Materials Science and Engineering **290** (2017) 011003 doi:10.1088/1757-899X/290/1/011003

Peer review statement

All papers published in this volume of *IOP Conference Series: Materials Science and Engineering* have been peer reviewed through processes administered by the proceedings Editors. Reviews were conducted by expert referees to the professional and scientific standards expected of a proceedings journal published by IOP Publishing.

IOP Conf. Series: Materials Science and Engineering **290** (2017) 012001 doi:10.1088/1757-899X/290/1/012001

In-situ Polymerization of Polyaniline/Polypyrrole Copolymer using Different Techniques

A S Hammad[1,2], H Noby[4], M F Elkady[1,3] and A H El-Shazly[1]

[1] Chemical and Petrochemicals Engineering Department, Egypt-Japan University of Science and Technology, New Borg El-Arab, Alexandria, Egypt.
[2] Chemical Engineering Department, Faculty of Engineering, Port Said University, Port Said, Egypt
[3] Fabrication Technology Department, Advanced Technology and New Materials Research Institute (ATNMRI), City of Scientific Research and Technology Applications, Alexandria 21934, Egypt.
[4] Mechanical Design and Production, Faculty of Energy Engineering, Aswan University, Aswan, Egypt.

Email:ahmed.hammad@ejust.edu.eg

Abstract. The morphology and surface area of the poly(aniline-co-pyrrole) copolymer (PANPY) are important properties which improve the efficiency of the copolymer in various applications. In this investigation, different techniques were employed to produce PANPY in different morphologies. Aniline and pyrrole were used as monomers, and ammonium peroxydisulfate (APS) was used as an oxidizer with uniform molar ratio. Rapid mixing, drop-wise mixing, and supercritical carbon dioxide ($ScCO_2$) polymerization techniques were appointed. The chemical structure, crystallinity, porosity, and morphology of the composite were distinguished by Fourier transform infrared spectroscopy (FT-IR), X-ray diffraction (XRD), Brunauer, Emmett and Teller (BET) analysis, and transmission electron microscopy (TEM) respectively. The characterization tests indicated that the polyaniline/polypyrrole copolymer was successfully prepared with different morphologies. Based on the obtained TEM, hollow nanospheres were formed using rapid mixing technique with acetic acid that have a diameter of 75 nm and thickness 26 nm approximately. Also, according to the XRD, the produced structures have a semi- crystalline structure. The synthesized copolymer with $ScCO_2$-assisted polymerization technique showed improved surface area (38.1 m^2/g) with HCl as dopant.

1. Introduction

Polyaniline (PANI) and polypyrrole (PPy) are extensively studied conductive polymers. Both have unique characteristics such as a high conductivity and environmental stability. In the past, there was a difficulty to make a copolymer of aniline and pyrrole because of synthesis method for each of them [1]-[2]. Now, there are two main ways to synthesize a copolymer of aniline and pyrrole by chemical [3]-[7] or electrochemical [8]-[10]methods. Chemical oxidation process has been checked in previous studies where ammonium peroxydisulfate solution (APS) is used as oxidant [3]-[7]. Modifications to the chemical oxidation method have been conducted to improve the conductivity such as surfactant-assisted chemical oxidative polymerization [11], inverted emulsion [12], pre-polymerization [13] and template synthesis method [14]. Recently, extensive studies were conducted to investigate the idea of using supercritical CO_2 ($ScCO_2$) as an alternative aid in polymer synthesis. Supercritical CO_2 ($ScCO_2$) have some interesting properties due to the intermediate phase between liquid and gas phase such as density like liquids, viscosity and diffusivity like a gas. It is clean, non-toxic, non-flammable, chemically inert, inexpensive and considerable in the atmosphere. By small changes in temperature and pressure, a supercritical CO_2 can be tuned and controlled. It is easy to remove after the polymerization process by reverting to ambient conditions. Moreover, it was stated that the employment of compressed CO_2-assisted polymerization process has a significant effect on the produced morphology of the produced polymer [15],[16].

This investigation deals with the first chemical copolymerization of polyaniline/polypyrrole using three distinct preparation chemical polymerization methods to prepare various copolymers with different morphologies and properties. Conventional chemical oxidation, rapid mixing, supercritical carbon dioxide $ScCO_2$-assisted polymerization techniques were utilized with the same molar ratios of the reactants. The copolymer was characterized by TEM, XRD, FT-IR, and BET surface area analysis for different physical and chemical examinations.

2. Experimental

2.1. Materials

Aniline, Pyrrole, Ammonium peroxydisulfate solution (APS), Acetic acid, Hydrochloric acid, Ethanol, and Methanol were purchased from Sigma-Aldrich. All chemicals were used without further refining. Freshly distilled water was used for the preparation of all aqueous solutions and the washing process.

2.2. Synthesis of copolymer by conventional chemical oxidation

In conventional chemical oxidation process, 25 mL of aniline/pyrrole mixture was prepared by dissolving aniline and pyrrole in aqueous dopant solution (0.2 M acetic acid or 0.2 M HCl) with molar ratio 1:1. At the same time, 25 mL ammonium persulfate solution (APS) was added drop-wise in a flask with stirring. The resultant mixture was left for 24 h seeking the full polymerization. The precipitates were collected and washed with diluted HCl, ethanol, methanol, and finally distilled water. Then, it was centrifuged for 10 min at 6000 rpm. Finally, the copolymer was dried for 24 h at 80 °C.

2.3. Synthesis of PANPY by rapid mixing

Rapid mixing was used to produce the aniline/pyrrole copolymer. The same molar ratio of oxidant: monomer with the same solutions were prepared at the same conditions of previous technique. The oxidizer solution was added at once to the monomers solution followed by continuous stirring at 1000 rpm for 2 h. The produced solution was kept at room temperature for 24 h. The same steps of collecting, filtering, washing, and drying was conducted.

2.4. Synthesis of PANPY by CO_2-assisted method

CO_2-assisted polymerization method in which supercritical CO_2 is used to improve the morphology of the polymers. The same amount of aniline, pyrrole, acid and APS solutions were used. The mixture of aniline and pyrrole were placed in the autoclave which is connected to the CO_2 pump and placed in an oven. The temperature was increased to 40 °C before the injection of CO_2 inside the autoclave until the pressure reaches 9 MPa. Then, APS solution was fed by another high-pressure pump at a uniform flow rate of 0.5 mL/min. Then, the mixture was left to polymerize for 3 h. After polymerization, a valve was opened to release the CO_2 from the vessel in 3 min completely. The prepared polymer was collected, filtered, washed, and dried at 80 °C for 24 h.

2.5. Characterization

The copolymer samples were characterized using TEM, FT-IR, XRD, and BET respectively. The morphology of the samples was investigated by using Transmission electron microscopy (TEM, JEOL Ltd., JEM-1010). Infrared spectra were recorded using a Fourier transform infrared (FT-IR) spectrometer (Vertex 70, Bruker scientific instruments, Germany) to identify the chemical structure of the samples. To measure crystallinity, X-ray diffraction was used (LabX XRD-6100 Shimadzu, Japan). N2 adsorption/desorption isotherms at 77 K and the Brunauer-Emmett Teller (BET) surface area analysis were conducted using (BelsorbminiII, BEI Japan Inc., Japan). All the samples were degassed under vacuum at 150 °C for 2 h.

3. Results and Discussion

3.1. PANPY morphology

Figure 1. TEM for the produced PANPY copolymer (a) CH_3COOH dropwise polymerization, (b) CH_3COOH rapid mixing polymerization, (c) CH_3COOH CO_2-assisted polymerization (d) HCl dropwise polymerization, (e) HCl rapid mixing polymerization, and (f) HCl CO_2-assisted polymerization.

TEM images of the manufactured PANPY copolymers using the two different dopant acids with the employment of the three different polymerization techniques are shown in Figure 1. The results indicate that samples prepared using acetic acid as a dopant acid produced different shapes. When the conventional polymerization process (Oxidizer dropwise adding) were used, a core and shell shape was produced. The shell which may return to PANI had a 25 nm thickness which may be due to the pre-polymerization of the two polymers separately. Also, Figure 1(b) reveals that nanospheres with hollow interiors at the centers was produced when rapid mixing technique was taken into consideration. The interior part has a diameter of 75 nm and thickness 26 nm approximately. Furthermore, the copolymer produced using the $ScCO_2$-assisted polymerization method was more homogeneous and had a one-phase net structure. Therefore, the $ScCO_2$-assisted polymerization process may have a unique mixture which helps to produce the copolymer without separation between PANI and PPY. Using HCl acid (**Error! Reference source not found.**1) produced nanoparticles morphology with an average diameter of 53 nm when dropwise adding polymerization was utilized. Fibrous structure with an average diameter of 80 nm was prepared when rapid mixing polymerization process was used. Finally, using $ScCO_2$-assisted polymerization method produce an aggregated structure copolymer.

3.2. PANPY chemical structure

FT-IR images of PANPY copolymers prepared by different methods are shown in Figure 2. The spectrum of the copolymer prepared by HCl and acetic acid by the three different methods showed characteristic band at 1113 cm^{-1} and 1194 cm^{-1} which are assigned to C=C stretch, C-C in ring stretch, and C-N stretch of polypyrrole and polyaniline [14]. This peak disappeared in the samples prepared using acetic acid as a dopant acid by CO_2 assisted polymerization. Another feature in the spectra is

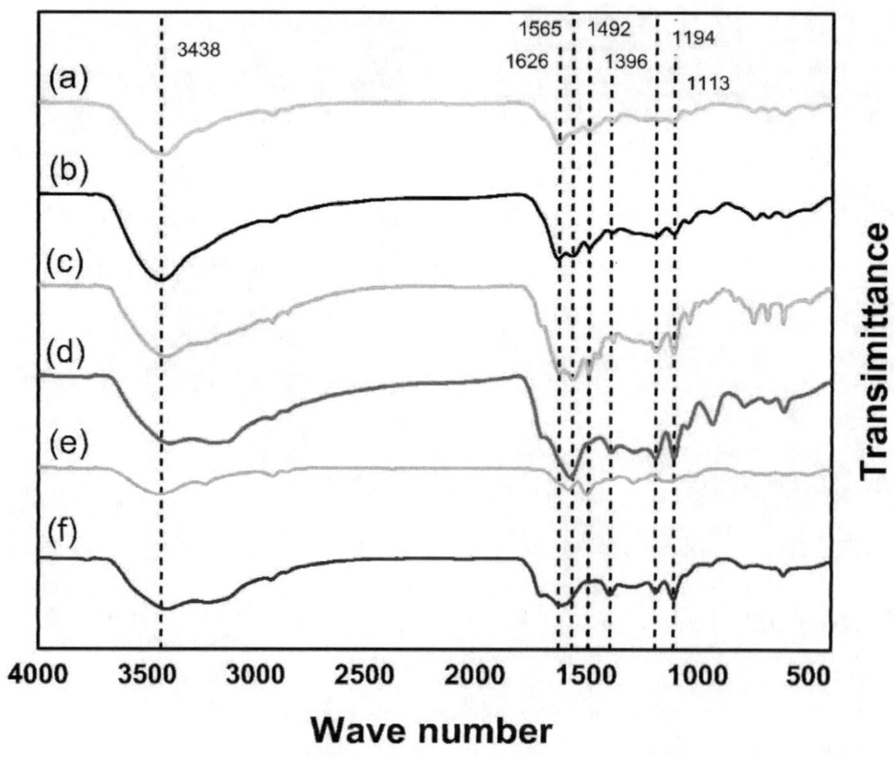

Figure 2. FT-IR curve for PANPY samples (a) CH_3COOH CO_2-assisted polymerization (b) CH_3COOH dropwise (c) CH_3COOH rapid mixing (d) HCl dropwise polymerization (e) HCl CO_2-assisted polymerization (f) HCl rapid mixing polymerization

the band at 3438 cm^{-1}, which corresponds to the N-H stretch of both polyaniline and polypyrrole [5]. The prepared copolymers with acetic acid showed a peak at 1492 cm^{-1} which are assigned to the C=C, N-N, and C-N stretch mode of pyrrole [14]. The semiquinonoid ring (-N-ring) of PANI (1378 cm^{-1}) was shifted to around 1396 cm^{-1} which appeared in the samples synthesized with HCl. A sharp peak at 1565 cm^{-1} which corresponds to C=C stretch showed in the HCl samples, was shifted to around 1626 cm^{-1} in acetic acid samples. This result confirms the successful preparation of the copolymer that includes both polymers.

3.3. PANPY crystallinity

XRD patterns of the six PANPY samples are shown in Figure 3 to investigate its crystallinity. It indicates that the main diffraction peaks centred between $2\theta = 20$ and 24° which can be attributed to the characteristics peaks of PANI ($2\theta = 19.5$ and 25.2°) [16] and PPy ($2\theta = 24^{\circ}$) showing that the two polymers are present in the copolymer and all prepared copolymers have semi-crystalline structures.

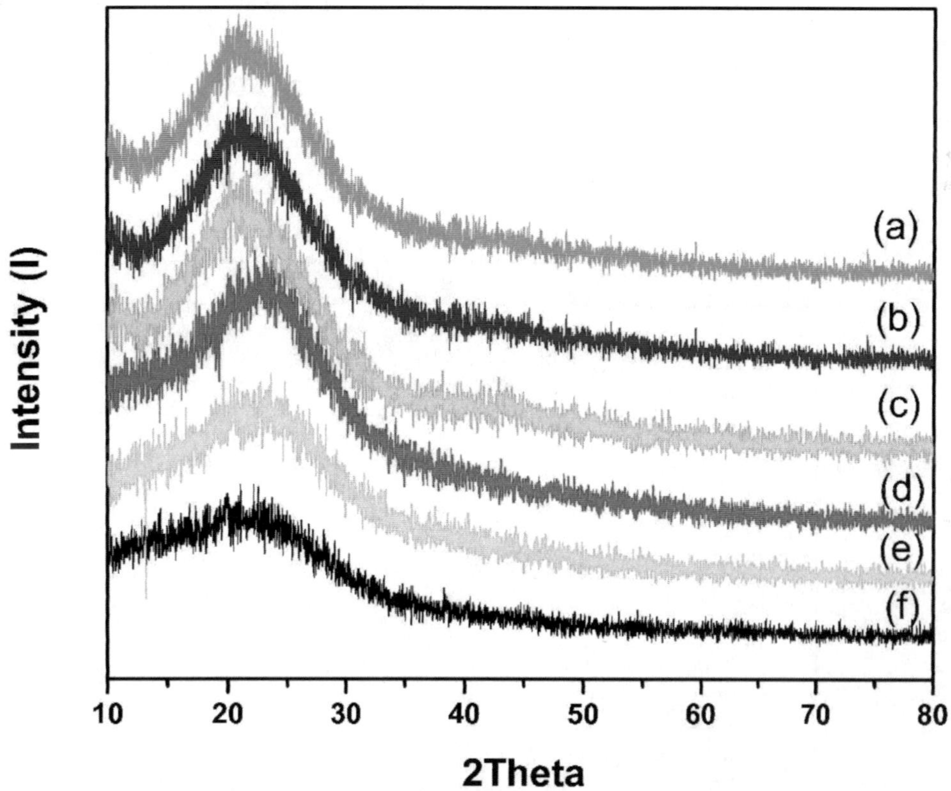

Figure 3. XRD curve for PANPY (a) HCl CO_2-assisted polymerization (b) CH_3COOH CO_2-assisted polymerization (c) CH_3COOH dropwise polymerization (d) CH_3COOH rapid mixing (e) HCl dropwise polymerization (f) HCl rapid mixing polymerization

3.4. PANPY surface area and pore size

Table 1 shows the BET surface area, total pore volume, and average pore diameter of the produced copolymer using the different preparation conditions. For acetic acid, rapid mixing technique gives the best values of surface area (30.73 m^2/g) compared with other preparation techniques which indicate that the $ScCO_2$-assisted polymerization technique has a noticeable effect on the surface area of the prepared copolymer. Also, the hollow structure characteristics produced using the rapid mixing technique (as shown in Figure 1(b)) play a significant role in the material surface area. The dropwise adding technique shows the worst surface area compared with other preparation techniques. On the other hand, the effect of using HCl as a dopant for the preparation of PANPY by the mentioned polymerization processes on the surface area has a different ranking. The surface area produced by CO_2 assisted polymerization is better than the dropwise technique.

The N_2 adsorption/desorption curve and pore size distribution of the produced PANPY structures are shown in Figure 4 and Figure 5. Based on the data obtained, the prepared PANPY structures showed a hysteresis loop of H3 type according to the classification [17]. H3 means that the pores are slit-shaped. Additionally, the prepared PANPY structures have a wide pore size distribution, and it is considered as a macroporous structure according to the IUPAC pore classification.

Table 1. Surface area, total pore volume, and average pore diameter of different PANPY samples.

Parameters	HCl rapid mixing	HCl dropwise	HCl CO_2-assisted	CH_3COOH rapid mixing	CH_3COOH dropwise	CH_3COOH CO_2-assisted
Surface Area (BET) (m^2/g)	10.34	23.33	38.1	30.73	12.02	27.8
Total Pore Volume (BJH) (cm^3/g)	0.050	0.028	0.157	0.182	0.070	0.145
Average Pore Size (BJH) (nm)	2.50	1.96	2.26	2.21	2.21	2.1

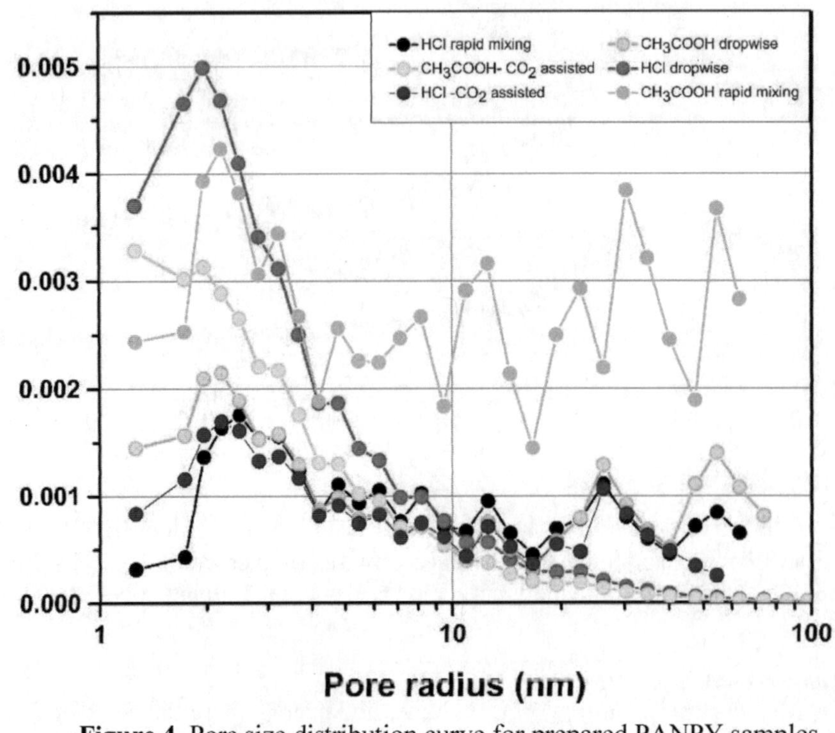

Figure 4. Pore size distribution curve for prepared PANPY samples

IOP Conf. Series: Materials Science and Engineering **290** (2017) 012001 doi:10.1088/1757-899X/290/1/012001

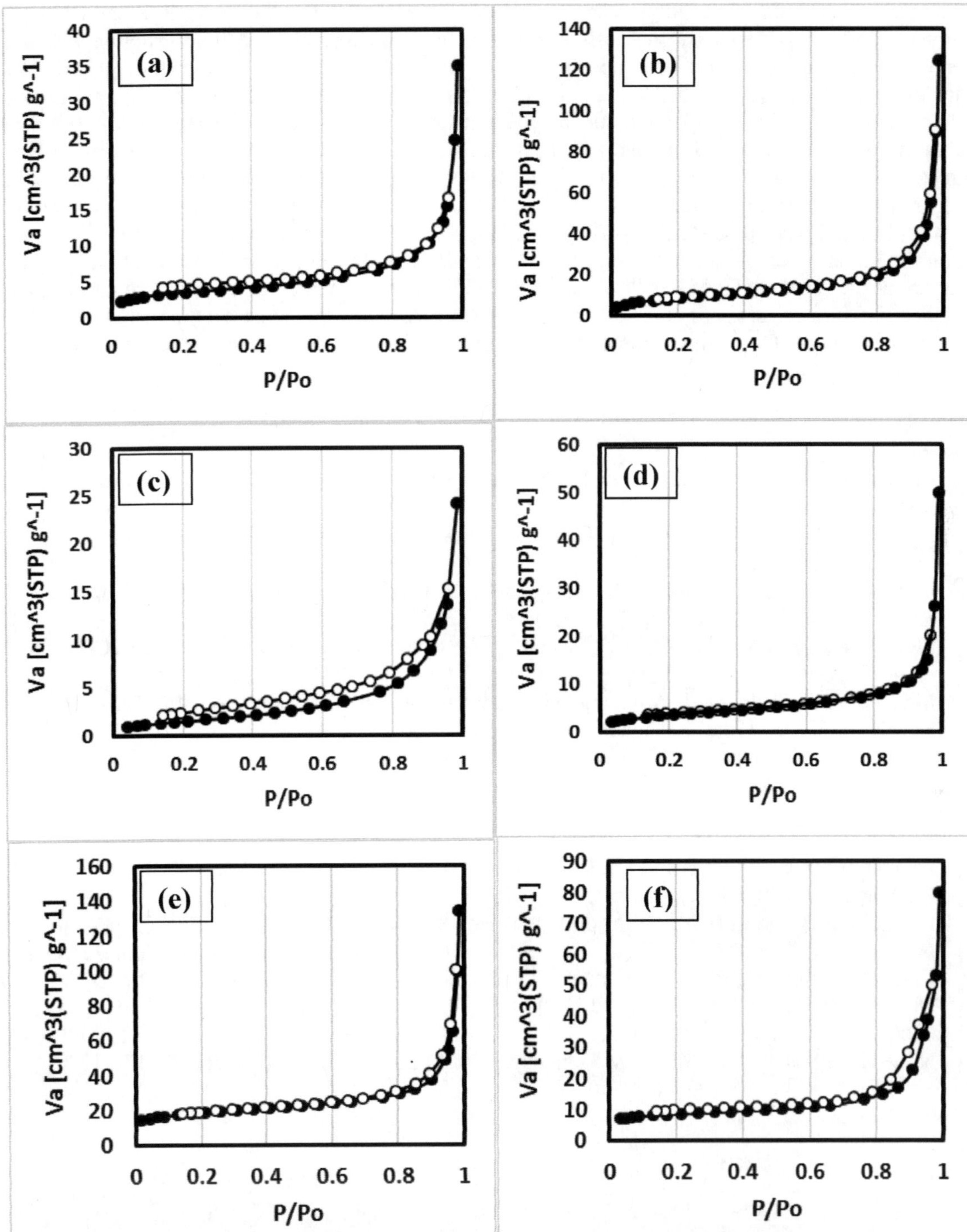

Figure 5. N_2 adsorption/desorption isotherm for (a) HCl rapid mixing polymerization (b) CH_3COOH rapid mixing (c) HCl dropwise polymerization (d) CH_3COOH dropwise polymerization (e) HCl CO_2-assisted polymerization (f) CH_3COOH CO_2-assisted polymerization

4. Conclusion

In this paper, three different polymerization methods were implemented to synthesize PANPY copolymer. The copolymers were successfully prepared and have a semi-crystalline structure with six different morphologies. In case of acetic acid medium with; adding APS dropwisely to the aniline/pyrrole solution produced PANPY with core & shell structure with 25 nm as shell thickness. While adding APS at once with rapid stirring produced PANPY with hollow nanospheres structure with an average diameter of 75 nm and thickness of 26 nm. The $ScCO_2$-assisted polymerization was conducted at 80 °C, CO_2 at 9 MPa was pumped into the aniline/pyrrole solution, resulting in homogenous copolymer structure. In case of HCl acid medium; PANPY nanoparticles with an average diameter of 53 nm were produced by adding APS dropwisely to the aniline/pyrrole solution. Fibrous structure with an average diameter of 80 nm. Finally, the $ScCO_2$-assisted polymerization was conducted at the same conditions as in acetic acid medium produced aggregated structure copolymer. The specific surface area of the different morphologies showed that CO_2-assisted PANPY polymerization with HCl showed improved surface area. Also, macroporous structures were attained from all the studied polymerization methods.

References

[1] Elkady M and Shokry H 2015 Curr. Nanosci. **6** 805-814.
[2] Elkady M and Shokry H 2015 Nanoscale Res. Lett. **10** 474-480.
[3] Stejskal J, Trchová M, Ananieva I A, Janča J, Prokeš J, Fedorova S and Sapurina I 2004 Synthetic Met. **146** 29-36.
[4] Lim V W, Kang E T, Neoh K G, Ma Z H and Tan K L 2001 Appl. Surf. Sci. **181** 317-326.
[5] Kim J W, Cho C H, Liu F, Choi H J and Joo J. 2003 Synthetic Met. **135** 17-18.
[6] Moon D K, Yun J Y, Kanbara T, Osakada K and Yamamoto T 2007 Mol. Cryst. Liq. Cryst. **464** 759-767.
[7] Mavundla S E, Malgas G F, Motaung D E and Iwuoha E I 2010 J. Mater. Sci. **45** 3325-3330.
[8] Elkady M, Shokry H and Hashim A 2016 Mater. **9** 1-24.
[9] Fusalba F and Bélanger D 1999 J. Phys. Chem. B. **103** 9044-9054.
[10] Cakmak G, Küçükyavuz Z and Küçükyavuz S 2005 Synthetic Met. **151** 10-18.
[11] Xu P, Han X, Wang C, Zhang B, Wang X and Wang H L 2008 Macromol. Rapid Comm. **29** 1392-1397.
[12] Palaniappan S, Sydulu S B and Srinivas P 2010 J. Appl. Polym. Sci. **115** 1695-1701.
[13] Ou X and Xu X 2016 RSC Adv. **6** 13780-13785.
[14] Li X, Zhang X and Li H 2001 J. Appl. Polym. Sci. **81** 3002-3007.
[15] Noby H, El-Shazly A H, Elkady M and Ohshima M 2015 International Conference on Chemical, Biological and Environmental Engineering (ICBEE 2015) **14** 48-53.
[16] Noby H, El-Shazly A H, Elkady M F and Ohshima M 2016 MATEC Web of Conferences **69** 04001-04005.
[17] Noby H, El-Shazly A H, Elkady M F and Ohshima M 2017 J. Chem. Eng. Jpn. **50** 170-177.

IOP Conf. Series: Materials Science and Engineering **290** (2017) 012002 doi:10.1088/1757-899X/290/1/012002

Comparison study on biosynthesis of silver nanoparticles using fresh and hot air oven dried *IMPERATA CYLINDRICA* leaf

Noor Najmi Bonnia[1], Afiza Ahmad Fairuzi[1], Rabiatuladawiyah Md. Akhir[1], Sabrina M. Yahya[2], Mohd Azri Ab Rani [2],Suzana Ratim[1] and Norafifah A. Rahman[1], Hazizan Md Akil[3]

[1]School of Physics & Material Sciences, Faculty of Applied Sciences,
[2]School of Chemistry & Environment, Faculty of Applied Sciences,
Universiti Teknologi MARA, 40460 Shah Alam, Selangor, Malaysia
[3]School of Material and Mineral Resources Engineering,Engineering Campus,
Universiti Sains Malaysia
Email:noornajmi@salam.uitm.edu.my

Abstract: The perennial rhizomatous grass; *Imperata cylindrica (I. cylindrica)* has been reported rich in various phytochemicals. In present study, silver nanoparticles were synthesized from aqueous leaf extract of *I. cylindrica* at two different leaf conditions; fresh leaves and hot-air oven dried leaves. Biosynthesized silver nanoparticles were characterized by UV-visible spectroscopy, field emission scanning electron microscopy (FESEM) and Fourier transform infrared spectroscopy (FTIR). Maximum absorption was recorded between 400 nm to 500 nm. FESEM analysis revealed that the silver nanoparticles predominantly form spherical shapes. The particles sizes were ranging from 22-37 nm. The elemental composition of the synthesized silver nanoparticles was confirmed by using energy dispersive X-ray spectroscopy (EDX) analysis. Fourier transform infrared spectroscopy (FTIR) confirmed the reducing and stabilizing actions came from biomolecules associated with *I. cylindrica* leaf extract. Thus in this investigation, an environmentally safe method to synthesized silver nanoparticles using local plant extract was successfully established.

1. Introduction

Cogon grass is a perennial grass that can be found commonly thrives around areas where disturbed by human activities (figure 1). This plant is native in Southeast Asia, the Philippines, China, and Japan [1] and has been listed as one of the most problematic weeds in the world [2]. Unlike other several types grass species, this perennial rhizomatous grass (division of *Magnoliophytai*; class *Liliopsida*, order *Cyperales*, family *Poaceace/Gramieane* and genus *Imperata Cirillo*) is not suitable for forage. This is due to the sharp edge of mature leaves. Apart planted as soil erosion control in some places, this plant was also claimed to have medicinal value. Studied conducted by previous researchers [3] confirm the presence of tannins, alkaloids, saponins and flavonoids in *I. cylindrica* leaf extract. These compounds are claim to have potential for the reduction of ionic into bulk metallic nanoparticles formation [4].

Metallic nanoparticles such as silver have many applications in various areas such as medicine, biology, material science, physics and chemistry [5]. Silver nanoparticles is a fascinating noble due to their unique properties which include good electrical conductivity, chemical stability, catalytic and antibacterial activity [6]. In addition silver possess minimal risk of toxicity towards humans making it a popular choice to be incorporated in health industry, food storage, textile, coatings and several environmental applications. With the emergence of silver nanoparticles in various applications, the development of environmental friendly synthesis is a promising field to be explored in order to cater the increasing demand.

Figure 1. Image of *Imperata cylindrica.*

The use of plant extracts to synthesis nanoparticles is often termed as 'green synthesis' method. This type synthesis procedure reduces or eliminated the use and/or generation of hazardous substances [7]. Silver nanoparticles have been previously synthesized using leaf extract of *Hibiscus cannabius, Pimenta dioica, Anacardium occidentale, Hibiscus rosa sinensis* and *Azadirachta indica* [8]–[12]. Although generally *I. cylindrica* has the power of reducing silver ions to silver nanoparticles, the selection between dry or fresh plant extract can affect the properties of synthesized silver nanoparticles. Hence, the present investigation was carried out to study the effect of oven drying leaf before extraction on optical and morphological properties of biosynthesized nanoparticles.

2. Experimental Procedures

2.1. Plant Material Preparation

Fresh *I. cylindrica* leaves were collected from Universiti Teknologi MARA Shah Alam, Selangor, Malaysia. Taxonomy identification were carried out by Forest Research Institute Malaysia (Sample no. PID 200916-17).Two type of leaves samples were used for this experiment; fresh and oven dried leaves. The green reducing agents were prepared by cutting the leaves of *I. cylindrica* before washing it thoroughly using tap water and followed by distilled water. Leaves specimen were oven dried at 80°C before finely grounded using dry mill of conventional food processor. The powder was screened through 100 mesh sieve to eliminate coarse particles. The aqueous leaves extract of the specimen was prepared by dissolving 5 g of leaf powder in 200 ml distilled water. The mixture was boiled at 60 °C for 30 min. The aqueous leaves extract of fresh leaf specimen was prepared by heating 20 g of fresh leaves together with 100 ml distilled water at 60 °C for 30 min. The extract from both type of leaves were filtered to be use in the experiment.

2.2. Biosynthesis of Silver Nanoparticles

The silver nanoparticles from both types of *I. cylindrica* leaves extract was synthesized by adding10 ml of the aqueous leaves extract into 90 ml of 5 mM aqueous $AgNO_3$. The mixture was stirred for 5 min at 150 rpm before heated using water bath at 60 °C for 10 min. Appearance of brown colour marks the formation of silver nanoparticles.

2.3. Characterization of Biosynthesized Silver Nanoparticles
The formation of silver nanoparticles was monitored by recording SPR bands of both samples using a UV-visible double beam spectrophotometer (Perkin Elmer, Lambda 35 UV/Vis System) operated in 700 -300 nm range at a resolution 1 nm. Morphological properties and elemental composition of the silver nanoparticles were acquired by field emission scanning microscope (FESEM) SUPRA-40 VP together with energy dispersive X-ray (EDX) unit. The EDX analysis was carried out to analyse the elemental composition of the synthesized nanoparticles.

3. Results and Discussion

3.1. Biosynthesis of Silver Nanoparticles
The present study deals with the synthesis of silver nanoparticles using aqueous leaf extract of *I. cylindrica* as green reducing and stabilizing agent. Comparison study between the reducing ability of two variations of leaf extracts (fresh and dried leaves) were observed. The formations of nanoparticles were identified by colour changing from light yellow to dark brown [13].The colour changes in reaction mixture was known due to surface plasmon resonance (SPR) phenomenon [7]. The time taken for leaves and hot-air oven dried leaves to mediate synthesis process was 30 minutes.

3.2. Uv-visible Spectroscopic Studies
Preliminary characterization was carried out using Uv-visible spectrophotometer at absorption wavelength of 300- 700 nm (figure 2). The colloidal solution of silver nanoparticles synthesized from dried leaves extract exhibited strong absorption at 448 nm which confirm the formation of Ag nanoparticles [14]. Silver nanoparticles synthesized from fresh leaves extract gave low intensity and broad absorption spectrum at 437 nm. The appearance of bands at lower wavelength mark smaller particles size[10] which was later confirmed by FESEM images. The SPR bands as reported by previous study [15] are influenced by the size, shape, morphology, composition and dielectric environment of the synthesized nanoparticles.

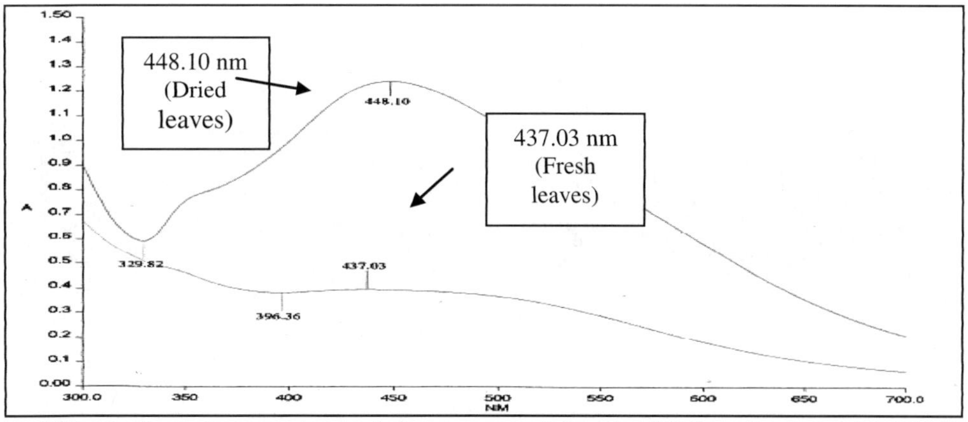

Figure 2. UV- vis absorption spectrum of silver nanoparticles (5 mM) from fresh and dried leaf extract.

3.3. FESEM Studies
Morphology and size of both samples were identified from FESEM images. The particles from both samples were predominately spherical shape (figure 3). Fresh leaf results in nanoparticles size ranging from 22 to 35 nm with least agglomeration. Silver nanoparticles synthesized from oven dried leaf have slightly larger particles with size ranging from 27 to 37 nm. The particles also exhibit some agglomeration

and lack of shape uniformity. Agglomeration of nanoparticles typically took place due to the insufficient capping agent in the leaf extract during synthesis of nanoparticles [16]. According to previous study[17], the quantity and type of flavonoid groups changed according thermo stability during leaf drying and extract preparation. Consequently, the stabilizing and capping behaviour of the plant extract were also altered. These might cause the differences in particles morphology and size of biosynthesized silver nanoparticles.

Figure 3. Micrograph (magnification of 50 k) of silver nanoparticles synthesized from (A) fresh leaves and (B) hot-air oven dried leaves.

3.4. EDX Spectroscopy Study

The presence of silver was confirmed from the Ag peaks obtained from the EDX spectrum (figure 4 (A) & (B)). The signal for silver was obtained at the energy of 3 keV indicates that the silver has been correctly identified [18]. Elemental composition of nanoparticles synthesized from both type of leaf extract was presented in figure 4(C). About 71.42 % and 75 % of silver was detected in both sample of nanoparticles synthesized from fresh leaf and hot-air oven dried leaf respectively.

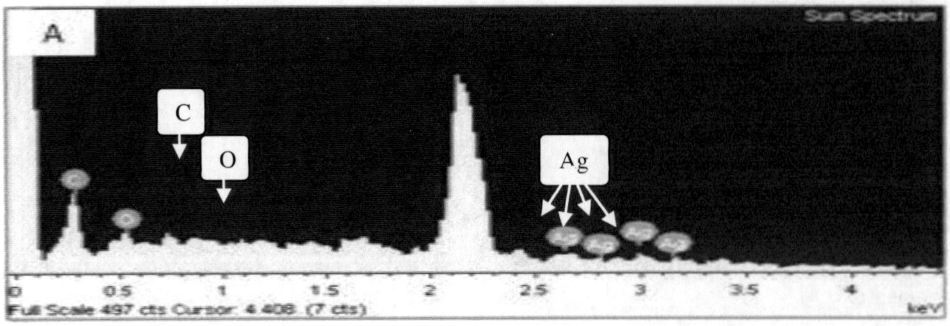

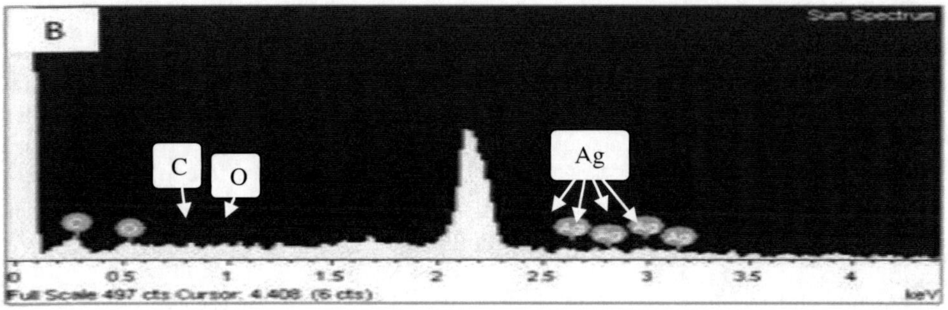

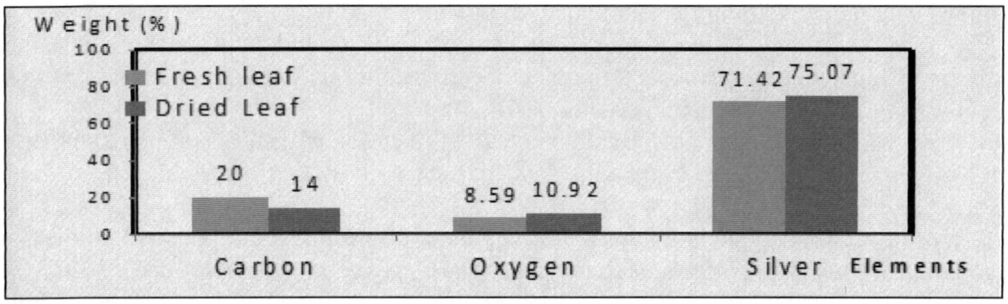

(C)

Figure 4. EDX result showing the presence of silver nanoparticles and bioorganic components of (A) fresh leaves and (B) oven dried leaves. (C) Elemental composition of nanoparticles synthesized from both plant extract.

4. Conclusion

In conclusion, this study successfully demonstrated the use of oven dried and fresh *I. cylindrica* leaf as silver particle size reducing and stabilizing agent. In UV-visible spectroscopy analysis, silver nanoparticles synthesized from hot-air oven dried leaf gave sharper and more prominent peak (448 nm) compared to fresh leaf. FESEM analysis revealed fresh leaf extract produced spherical silver nanoparticles with less agglomeration and smaller particle size range (22 to 35 nm). Silver nanoparticles synthesized from dried leaves showed more agglomeration with particle size ranging from 27 to 38. EDX analysis has correctly identified formation of silver as the result of biosynthesis in both samples.. From this study it can be recognized that drying plant materials for synthesizing nanoparticles may affect the capping abilities and cause particles to agglomerate. A lot work need to be carried out in future to understand the synthesis mechanism hence improving the morphology of biosynthesized nanoparticles.

Acknowledgements
The authors acknowledge Ministry of Higher Education Malaysia for RACE Grant Scheme (600-RMI/RACE 16/6/2(3/2014) and Faculty of Applied Sciences, Universiti Teknologi Mara for research facilities.

References

[1] G. E. MacDonald, “Cogongrass (Imperata cylindrica) - biology, ecology, and management,” *CRC.*

Crit. Rev. Plant Sci., vol. 23, no. 5, pp. 367–380, 2004.

[2] S. K. M. E. R. R. L. J. D. G. S. Hallie Dozier James F. Gaffney, "Cogongrass in the United States: History, Ecology, Impacts, and Management," *Weed Technol.*, vol. 12, no. 4, pp. 737–743, 1998.

[3] D. Krishnaiah, T. Devi, A. Bono, and R. Sarbatly, "Studies on phytochemical constituents of six Malaysian medicinal plants," vol. 3, no. 2, pp. 67–72, 2009.

[4] S. A. Aromal and D. Philip, "Spectrochimica Acta Part A : Molecular and Biomolecular Spectroscopy Green synthesis of gold nanoparticles using Trigonella foenum-graecum and its size-dependent catalytic activity," *Spectrochim. Acta Part A Mol. Biomol. Spectrosc.*, vol. 97, pp. 1–5, 2012.

[5] K. Yokohama and D. Welchons, "The conjugation of amyloid beta protein on the gold colloidal nanoparticles surfaces," *Nanotechnology*, vol. 18, pp. 105101–105107, 2007.

[6] V. K. Sharma, R. A. Yngard, and Y. Lin, "Silver nanoparticles green synthesis their antimicrobial activities," *Adv. Colloid Interface Sci.*, no. 145, pp. 83–96, 2009.

[7] R. Thombre, A. Chitnis, V. Kadam, Y. Bogawat, R. Colaco, and A. Kale, "A facile method for synthesis of biostabilized silver nanoparticles using Eichhornia crassipes (Mart .) Solms (water hyacinth)," vol. 13, no. July, pp. 337–341, 2014.

[8] S. A. Part, M. Spectroscopy, M. Umadevi, and M. Teresa, "Synthesis of monodispersed silver nanoparticles using Hibiscus cannabinus leaf extract and its antimicrobial activity," no. April, 2016.

[9] A. R. Geetha, E. George, A. Srinivasan, and J. Shaik, "Optimization of Green Synthesis of Silver Nanoparticles from Leaf Extracts of Pimenta dioica (Allspice)," vol. 2013, no. i, 2013.

[10] D. S. Sheny, J. Mathew, and D. Philip, "Spectrochimica Acta Part A : Molecular and Biomolecular Spectroscopy Phytosynthesis of Au , Ag and Au – Ag bimetallic nanoparticles using aqueous extract and dried leaf of Anacardium occidentale," *Spectrochim. Acta Part A Mol. Biomol. Spectrosc.*, vol. 79, no. 1, pp. 254–262, 2011.

[11] D. Philip, "Green synthesis of gold and silver nanoparticles using Hibiscus rosa sinensis," vol. 42, pp. 1417–1424, 2010.

[12] S. Ahmed, M. Ahmad, and B. L. Swami, "ScienceDirect Green synthesis of silver nanoparticles using Azadirachta indica aqueous leaf extract," *J. Radiat. Res. Appl. Sci.*, pp. 2–8, 2015.

[13] A. Ahmad, P. Mukherjee, S. Senapati, D. Mandal, M. I. Khan, and M. Sastry, "Extracellular biosynthesis of sil v er nanoparticles using the fungus Fusarium oxysporum," *Colloids Surfaces B Biointerfaces*, vol. 28, pp. 313–318, 2003.

[14] J. Y. Song and B. S. Kim, "Biological synthesis of bimetallic Au / Ag nanoparticles using Persimmon (Diopyros kaki) leaf extract," vol. 25, no. 4, pp. 808–811, 2008.

[15] K. L. Kelly, E. Coronado, L. L. Zhao, and G. C. Schatz, "The optical properties of metal nanoparticles: the influence of size, shape, and dielectric environment," *J. Phys. Chem. B*, vol. 107, no. 3, pp. 668–677, 2003.

[16] M. Vanaja, K. Paulkumar, M. Baburaja, S. Rajeshkumar, G. Gnanajobitha, C. Malarkodi, M. Sivakavinesan, and G. Annadurai, "Degradation of Methylene Blue Using Biologically," vol. 2014, 2014.

[17] B. N. Vaidya and T. A. Brearley, "Antioxidant Capacity of Fresh and Dry Leaf Extracts of Sixteen Scutellaria Species," vol. 2, no. 3, pp. 42–49, 2014.

[18] M. Vanaja, "Phytosynthesis of silver nanoparticles by Cissus quadrangularis : influence of physicochemical factors," no. December 2013, pp. 1–17, 2015.

[19] R. U. Maheswari and V. Euginamala, "Analyzing and Determining the Activity of Antimicrobial , Functional Group and Phytochemicals of Cymbopogon citratus using Well Diffusion , FT-IR and HPLC," vol. 4, no. 3, pp. 2013–2016, 2015.

IOP Conf. Series: Materials Science and Engineering **290** (2017) 012003 doi:10.1088/1757-899X/290/1/012003

Effect of Pineapple Leaf Fibers (PALF) concentration on nanofibers formation by electrospinning

S N Surip[1], F M Abdul Aziz[1a)], N N Bonnia[1] and K A Sekak[1]

[1]Faculty of Applied Sciences, Universiti Teknologi MARA,40450 Shah Alam, Malaysia.

Email:fatimahmuyassarah@gmail.com

Abstract. Electrospinning method has been studied widely in producing nanofibers due to its straightforward and versatile method. In this study, Pineapple Leaf Fibers (PALF) solution were electrospinning to obtain mat of PALF electrospun. PALF were diluted in Trifluoacetic Acid (TFA) into five different concentrations to study the effect of concentration to the nanofibers formation. Raw sample of PALF (PALF $_{raw}$), PALF after dewax (PALF $_{dewax}$) and PALF after dilute with TFA (PALF $_{TFA}$) were analyzed and compared using FTIR to study the structural change occur. TFA solvent has removed and recreated some of the functional group in PALF thus disrupt strong hydrogen bonds that hold hemicellulose, cellulose and lignin together. All the PALF sample has been proceed to electrospinning process. Low concentration of solution cause the solution jet to break up even before reach the collector however high concentration of solution made the solvent volatile faster and the solution dried easily. Therefore, PALF with optimum concentration of 0.02 gml^{-1} had favors the formation of nanofibers and succeed in forming membrane at the collector.

1. Introduction

Nanofiber materials have develops rapidly due to its large surface area-to-volume properties that enhance its compatibility with other materials [1]. Through the years, fibers have become highly demand in market and often use as reinforcing fibers for composite [2]. As the synthetic fibers has critical drawback such as limited supply, toxicity to surrounding and non-biodegradable, comply with the world policy to conduct a green friendly environment, researchers start to put their interest on natural fibers [3]. Natural fiber come from two sources which is plant fibers (lignocellulosic) and animal fibers [4].This study focus on the potential of lignocellulosic fibers as replacement for current petroleum-based fibers.

High specific properties that come with lignocellulosic fibers thus contributed to excellence mechanical properties comparable with synthetic fibers. Saheb et al (1999) in his review papers reported that when specific modulus were considered, natural fibers shows comparable values or even higher than glass fibers [5]. In fact, lignocellulosic fibers consist high cellulose content therefore carried high tensile strength that help improves composite properties during interaction with the polymers [6]. Among this fibers, pineapple leaf were one of the potentially unexplored fibers that could offer more to the industry.

Pineapple plant abundantly available especially in tropical country thus provides unlimited supply. Pineapple plant is referred as secondary plant which by-product wastes were extracts to obtain cellulose [7]. Khengketkit et al (2014) compared properties of several parts in pineapple plant and

PALF part shows the best values in mechanical properties [3]. PALF contain high cellulose content that help the leaf to maintain its turgidity thus give great properties [7]. Besides, Neto et al (2015) has study 12 varieties of pineapple leaf fibers in term of its mechanical and thermal properties. His paper stated that increase in cellulose percentage does increase the tensile strength and specific modulus properties [8]. A previous study has been done by Uma Devi et al (2010) on the PALF polyester composite that agreed with this statement. The tensile strength and young modulus increase with cellulose content however the flexural strength shows opposite [9]. However, even there are papers already been published on PALF, there were only few proceed it to the PALF nanofibers size.
Electrospinning method is a convenient with a multi-purpose method that aims on producing fibers with diameter range from several micrometers to tens of nanometers [10]. Through this method, nanofibers size of PALF could be produced. Recently, Guihe and the co workers succeed in produced mats of electrospun PET/PVA with a diameter range 200 μm to 700 μm with higher mechanical properties when compared to pure PET [11]. Not only enhance the mechanical properties, natural fibers also manage to compete with current synthetic fibers. Bodros et al (2006) in his paper reported that PLLA/flux composite show high tensile strength which is almost close to the polyester/glass composite (12).
Electrospinning lignocellullosic biomass using ionic liquid (solvent) had successfully produced nanoscale ultrathin fibers through studied by Ahn et al (2007) using DMF as solvent and the latest studies was from Santos et al (2015) and it is conducted using TFA. Selection of a suitable solvent system for the polymers plays a big role to ensure electrospinning success [14]. Beatriz et al studied on the most effective solvent for PET and found that high percentage of TFA favour the formation of continues nanofibers of PET [10].

2. Experimental procedures

2.1. Materials
Pineapple Leaf fibers were received from local Malaysian company in fibers form. Polyethylene Terephthalate were obtained from Aldrich and received in granular form. The chemicals which are trifluoroacetic acid, cyclohexane and ethanol all were supplied by Aldrich Merck.

2.2. Soxhlet method (Dewaxing)
PALF were first dewax using soxhlet instrument to remove wax and terpenes thus increase its surface interaction with others chemicals. 10 g of PALF were put into the thimble and rinse with 200ml of ethanol:cyclohexane 1:1 ratio for 6 hours. Then, the remaining sample were rinse using distilled water and put in the oven with temperature 40 ^{0}C for 12 hours until the sample were completely dried.

2.3. Dissolution of PALF
Dried PALF were weight accordingly to table 1. Four different concentrations were made in order to find out the best concentration for the electrospinning process. All samples were dilute in 3mL of Trifluoroacetic acid (TFA) and stirred for 12 hours until homogenous solution were obtained.

Table 1: Concentration of PALF

NO	CONC. (gml^{-1})	FIBERS (WEIGHT) (g)	TFA (VOLUME) (mL)
1	0.01	0.03	3
2	0.02	0.06	3
3	0.03	0.09	3
4	0.04	0.12	3

2.4. Characterization (FTIR)

FTIR spectroscopy was carried out in order to study the interaction that occurred in TFA and PALF. This kind of interaction can be proven through frequency shifts, changes in band intensity, and shape of the FTIR spectra. The spectra were recorded by FTIR (Thermo Fisher Scientific Nicolet iS 10) equipped with Attenuated Total Reflectance (ATR) in the transmittance mode over a frequency range 500-4000 cm^{-1} with 2 cm^{-1} resolution.

2.5. Electrospinning

Electrospinning set up were mainly consists of power supply, syringe pump, syringe, needle and collector (figure 1).

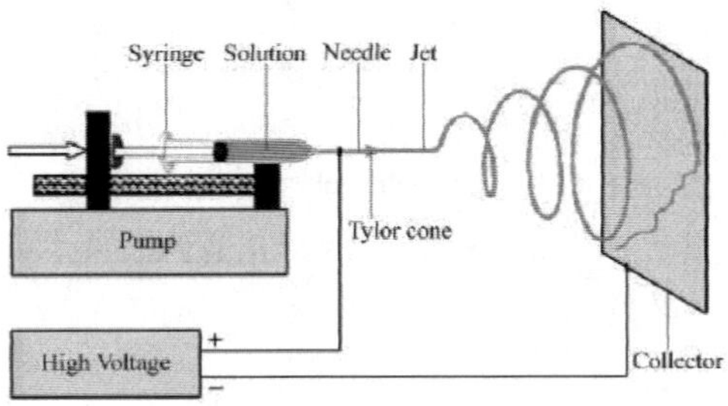

Figure 1: Electrospinning set-up

3. Results and Discussion

3.1. Solvent effect on functional group of PALF

Figure 2 shows FTIR spectra after zoom in at (500-2000) cm^{-1} region. Peak at 1630-1640cm^{-1} suggested from lignin aromatic structure [14]. It can be seen in sample (a) and (b) spectrum however, after addition of TFA the peak were completely absent. At sample (a) and (b) region, a peak approximately at 1730-1740cm^{-1} were observed attribute to the C=O of the acetyl group present in hemicellulose however dissolve at (c) region. Thus, it is suggested that TFA addition favors the elimination of lignin and hemicellulose. At peak 1020-1050 cm^{-1}, a C=O stretching vibration indicated on the cellulose backbone were present in all sample thus suggested the remain of cellulose in the sample. Although a small peak can be observed at peak 1774cm^{-1} in sample (c) that indicated the present of trifluoroacetyl group it is well known that constant exposure to air could lead to the complete hydrolysis of trifluoroacetic group [18]. Disrupt of lignin usually needed to break the high covalent bond that hold the hemicellulose, cellulose and lignin together. Thus, TFA addition favors the elimination of lignin and hemicellulose. The present of C=O in TFA structure make it susceptible for esterification reaction take place as shown in fig 3. Recently, a study on electrospinning from TFA of sisal pulp were conducted by Santos et al (2015) and the FTIR result also showed the possibility of the reactions occur (figure 3).

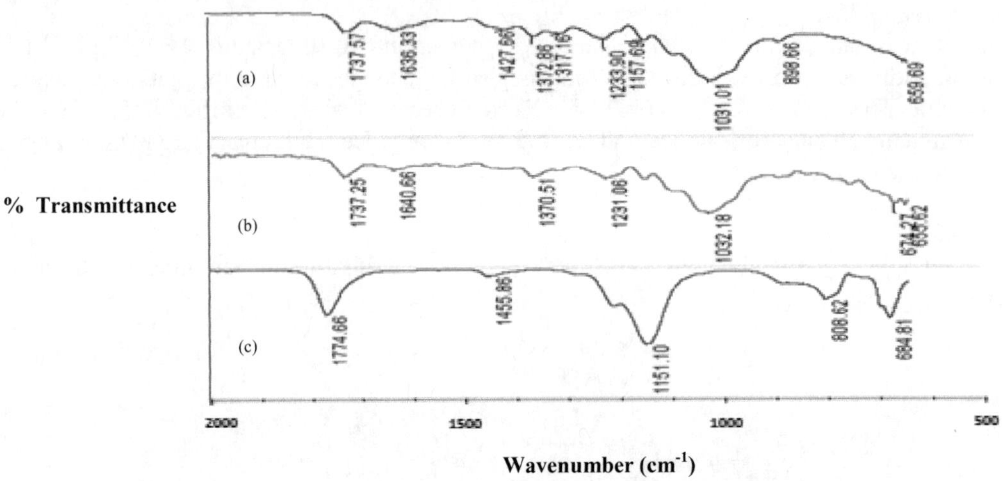

Figure 2: FTIR result for (a) $PALF_{raw}$, (b)$PALF_{dewax}$ and (c)$PALF_{TFA}$ Zoom in for (500-2000) cm^{-1} region

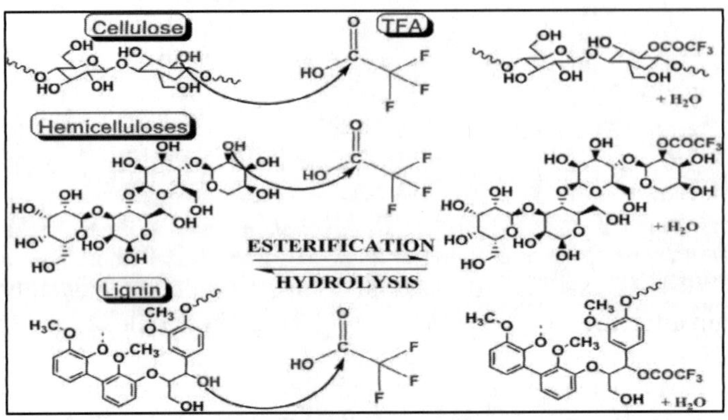

Figure 3 : Main components of a lignocellulosic fiber and their reactive sites susceptible to esterification by TFA (*R.P.O Santos et al., 2015)*

3.2. Concentration effect on nanofibers formation

Table 2 shows the result of PALF nanofibers formation. Nanofibers formation via electrospinning were highly related with the concentration, viscocity and surface tension of the solution. As the concentration increase, the viscocity also increase caused improvement to the polymer entanglement [15]. During 0.01gml^{-1} solution jet, only electrosprayed were formed and no membrane produced at the collector. Low viscocity attributed to the low viscoelastic force thus made the fibers jet easily break up even before reached the collector [16]. Therefore, increasing the concentration helps to prevent the break up due to the improvement in the chain entanglement as can be seen at concentration 0.02gml^{-1}. Membranes were successfully formed at the collector. However, increasing the concentration higher has cause the solution dry faster and did not form any fiber jet as can be seen during 0.04gml^{-1} of solution. TFA has a high volatile values compare to other solvent [17]. Costa et al (2010) suggested high volatile of solvent should set up with low concentration to prevent fast volatilization [10].

Table 2: PALF nanofibers formation by electrospinning

Concentration (gmL^{-1})	Condition	
	Jet formation	**Membrane formation**
0.01	-Solution were electrosprayed	-No membrane formed at collector
0.02	-Solution successfully electrospinning	-Membrane were formed at the collector
0.03	-Solution start dry at 4th second -No solution were eject afterward	-No membrane formed at collector
0.04	-Solution hardly eject and dried at the tip of metallic needle and block the solution	-No membrane formed at collector

4.Conclusion

As the conclusion, TFA used as solvent helps PALF to achieve homogenous solution. FTIR result shows addition of TFA help to disrupt the hydrogen bonding thus eliminate lignin while keep remain the cellulose content. PALF concentration of 0.02 gml^{-1} was the optimum concentration need to electrospin PALF. Low concentrations of PALF produce an electrosprayed rather than electrospinning fibers while high concentration caused solution to dry faster.

Acknowledgement

The authors would like to acknowledge University Technology Mara for providing grant to carry out this research under Nanocomposite Material & Industrial Application Research Group (600-IRMI/DANA 5/3/REI (0001/2016)).

References

[1] Huang Z M, Zhang Y Z, Kotaki M and Ramakrishna S 2003 *Composite Science and Technology* **63** 2223-2253
[2] Kaewpirom S and Worrarat C 2014 *Fibers and Polymers* **15** 1469-1477
[3] Khengketkit N and Amornsakchai T 2014 *Materials and Design* **55** 292-299
[4] Chandramohan D and Marimuthu K 2011 www.arpapress.com/volume/vol8Isuue2/IJJRAS
[5] Saheb D N and Jog J P 1999 *Adv. In Polymer Science* **18** 351-363
[6] Farok O, Bledzki A K, Fink H P and Sain M 2012 *Progress in polymer science* **37** 1552-1596
[7] Asim M, Abdan K, Jawaid M, Nasir M, Dashtizadeh Z, Ishak M R, Hoque M E 2015 *Int. Journal of Polymer Science* ID 950567
[8] Neto A R S, Araujo M A M, Souza F V D, Mattoso L H C and Marconcini J M 2013 *Industrial Crops and Products* **43** 529-537
[9] Dewi L U, Bhagawan S and Thomas S 1997 *Polymer Composite* (Wiley Inter-Science)
[10] Veleirinho B, Rei M F and Lopes J A 2007 *Journal of Polymer Science* (Wiley Inter-Science)
[11] Li G, Zhao Y, Lv M, Shi Y and Cao D 2013 *Colloids and surfaces A :Physicochem. Eng. Aspects* **436** 417-424
[12] Bodros E, Pillin I, Montrelay N and Baley C 2007 *Composites Science and Technology* **67** 462-470
[13] Ahn Y, Lee S H, Kim H J, Yang Y H, Hong J H and Kim Y H 2012 *Carbohydrate Polymers* **88** 395-398
[14] Santos R P O, Rodrigues B V M, Ramires E C, Ruvolo A C and Frollini E 2015 *Industrial*

Crops and Products **72** 69-76

[15] Pillay V, Dott C, Choonara Y E, Tyagi C, Thomar L, Kumar P, Du Toit L C and Ndesendo V M K 2013 *Journal of Nanomaterials* ID 789289

[16] Tarus B, Fadel M, Oufy A A, Messiry M E 2016 *Alexandria Engineering Journal* **55** 2975–2984

[17] Costa L M M, Bretas R E S, Gregorio R and Jr 2010 *Materials Science and Application* **1** 247-252

[18] Rodrigues, B.V., Ramires, E.C., Santos, R.P. and Frollini, E., 2015 *Journal of Applied Polymer Science*, 132(16)

IOP Conf. Series: Materials Science and Engineering **290** (2017) 012004 doi:10.1088/1757-899X/290/1/012004

Optimization of TiO_2 thin film thickness for dye sensitized solar cell applications

S A M Al-Bat'hi[1], N Ahmed[1], R Othman[1] and M Othman[1]

[1]Department of Manufacturing and Materials Engineering, International Islamic University Malaysia (IIUM), Jalan Gombak, 53100 Kuala Lµmpur, Malaysia

Email: su3ad@iium.edu.my

Abstract. Dye sensitized solar cells (DSSCs) rely on the absorption of photons by the dye molecules which are transported to the conduction band of the TiO_2 electrode. The microstructure, energy gap and the absorption spectra of the TiO_2 electrodes highly affects the efficiency of the cell. In this paper, the absorption spectra and energy gap has been studied by varying the thickness of the TiO_2 paste. Nanocrystalline TiO_2 thin films were deposited on ITO glass substrate with three different thickness (4.54µm, 7.12µm and 12.3µm) by using doctor blade method. After deposition all the samples were sintered at 450°C after deposition to enhance the particle bonding and for achieving better adhesion. The samples were characterized by UV-VIS spectra for determining the absorption spectra and Scanning Electron Microscopy (SEM) for investigating the thickness and the surface morphology. Fabricating the electrodes with different thickness showed significant changes in the energy gap and from the results it can be concluded that the energy gap increases with the increased thickness. The highest energy gap of 2.25ev and absorption 3.791 was achieved by 12.3µm thick sample. The absorption spectra also shows better absorption throughout the whole visible light range but the SEM images suggests that 12.3µm thick sample shows cracks all over the deposited region which will cause current leakage when the cell is assembled. Therefore, the optimum result was achieved by 7.12µm thick sample providing 1.9 ev energy gap and 3.91 absorption peak.

1. Introduction

DSSCs has been accepted widely to be used as potential solar cell apart from traditional solid-state cells. It is considered as photo electrochemical solar cell that consists of mesoporous metal oxide semiconductor layer on transparent conductive glass, which is used as working electrode [1]. Other major components are the organic or inorganic dye material, electrolyte and the counter electrode. Oxide semiconductors possess decent stability under irradiation but due to the large bandgap this kind of material cannot absorb visible light directly [2]. In case of dye-sensitized solar cell, TiO_2 is mostly used as the oxide semiconductor having 3.2ev energy gap [3].

To recover the absorption issue organic/inorganic dyes are used to sensitize the TiO_2 based electrode. Dyes help to excite the electrons received from the sunlight to change its state and upon excitation; the dye molecule injects the electron towards the conduction band of the TiO_2 based electrode. The semiconductor electrode then transports the electron to external load. The efficiency of this kind of solar cell highly affected depending on the performance of the TiO_2 coated electrode [4]. The TiO_2 paste formulation is very crucial since it helps to adsorb large amount of dye molecule to the electrode surface and using Ru-complex dye the energy conversion efficiency was achieved 11% with standard measurements [5]. The most important parameters like surface area, roughness, pore size, film thickness, sintering temperature, chemicals greatly affects the result and these properties can be controlled during the film preparation process.

2. Experimental procedures

Degussa P25 TiO_2 powder was used for making the paste. This powder contains 70% anatase phase and 30% rutile phase of TiO_2. The particle size of TiO_2 is around 25nm. To prepare the TiO_2 slurry or paste, 6g TiO_2 powder was grounded in the mortar with 10ml of dilute acetic acid. Each time 1ml of dilute acetic acid was dropped and grounded for 2 minutes. This process was continued for 20 minutes until the solution turned into a milky white paste. To prevent re-aggregation of the particles 0.8ml acetyleaetone was added. Dilute acetic acid helps to modify the surface with H^+ ions and works as binding agent. To lower the surface tension 2 drops of detergent solution was added and stirred for another 5 minutes [6]. The solution was kept in the room temperature for 15 minutes to equilibrate [7].

After preparing the paste ITO glass was masked with tape and the TiO_2 was deposited on the glass by doctor blade method [8]. For deposition, the ITO glass was positioned in such a way that the conductive side of the glass faces up. 2-3 drops of the TiO_2 paste were dropped on the upper portion of the substrate and a glass rod was used to distribute the paste by sliding from up to bottom of the glass. After completing the deposition process, the substrate was kept for drying around 10 minutes. The tape was removed from the glass carefully without damaging the coated side. Then the substrate was placed in the furnace for calcination. The calcination was operated with Carbolite ELF 1/148 furnace and the temperature was set to 450 °C for 30 minutes. Initially coated side was turned into brown color and after 30 minutes, it transformed into white color again. The sample was kept in the furnace until it was slowly cooled [9] down to normal temperature. After the calcination process, the sample was referred for characterization.

3. Results and Discussion

Figure 1 shows the absorption spectra of TiO_2 photo anode with three different thickness respectively 4.54 μm, 7.12 μm and 12.3 μm. For all the three samples the absorption spectra shows good absorption throughout the whole visible light region starting from 400nm to 700nm. The highest peak for 4.54 μm thick paste is 3.314 at 400nm wavelength. After this point, there is a slight decrease in the graph, which is used to calculate the bandgap of the sample from the following equation.

$$Energy\ gap = \frac{hc}{\lambda} \tag{3.1}$$

h= planks constant
c=speed of light
ʎ= corresponding wavelength
Energy gap table corresponding to the absorption spectra

Table 1. Calculated energy gap table from absorption spectra.

Thickness	No of coatings	Highest peak	Wavelength for calculating energy gap	Energy gap
4.54 μm	1	3.314	665nm	1.86ev
7.12 μm	2	3.491	650nm	1.9ev
12.3 μm	3	3.791	550nm	2.25ev

The calculated energy gap of the 4.54 μm thick TiO_2 electrode is 1.86ev, where the corresponding wavelength is 550nm. Energy gap of the 7.12 μm thick TiO_2 electrode is 1.9ev with 650nm wavelength and finally the energy gap of the 12.3 μm TiO_2 electrode is 1.86ev having wavelength of 665nm. From the results, it can be signified that the energy gap of the TiO_2 electrode is increasing with the thickness. The band gap of TiO_2 is 3.2 ev and with the paste formulation the badgap can be

IOP Conf. Series: Materials Science and Engineering **290** (2017) 012004 doi:10.1088/1757-899X/290/1/012004

varied as mentioned in table 1. Reduction in the energy gap increases the conductivity of the material since the electrons will require less energy to travel the Fermi level of the material [10]. When the photon will hit the dye molecule, it will be excited and will jump from the HOMO to LUMO state. After reaching the LUMO state the electron will be to transported to the conduction band of the TiO_2 electrode. Therefore, reducing the bandgap of the paste will result in better transportation of electron to the conduction band of the TiO_2 [11]. From the result, it can be concluded that reducing the thickness tends to achieve smaller bandgap of the TiO_2 electrode.

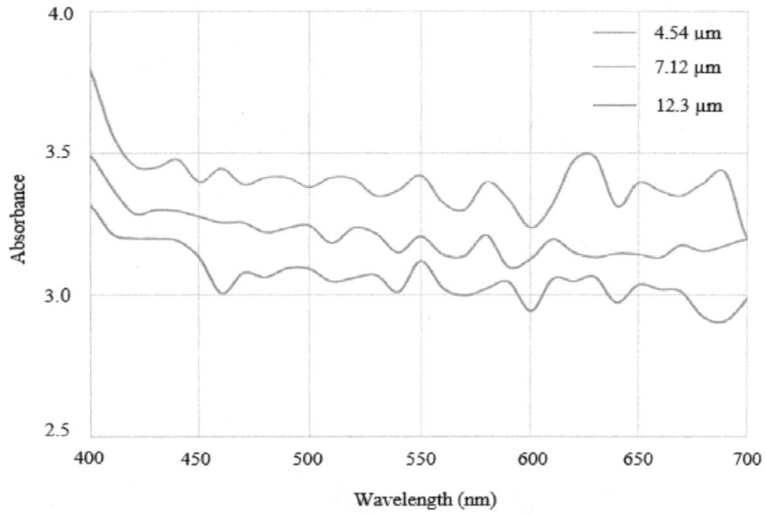

Figure 1. UV-VIS absorption spectra of TiO_2 photoanode with 4.54 µm thickness, 7.12µm thickness and 12.3µm thickness

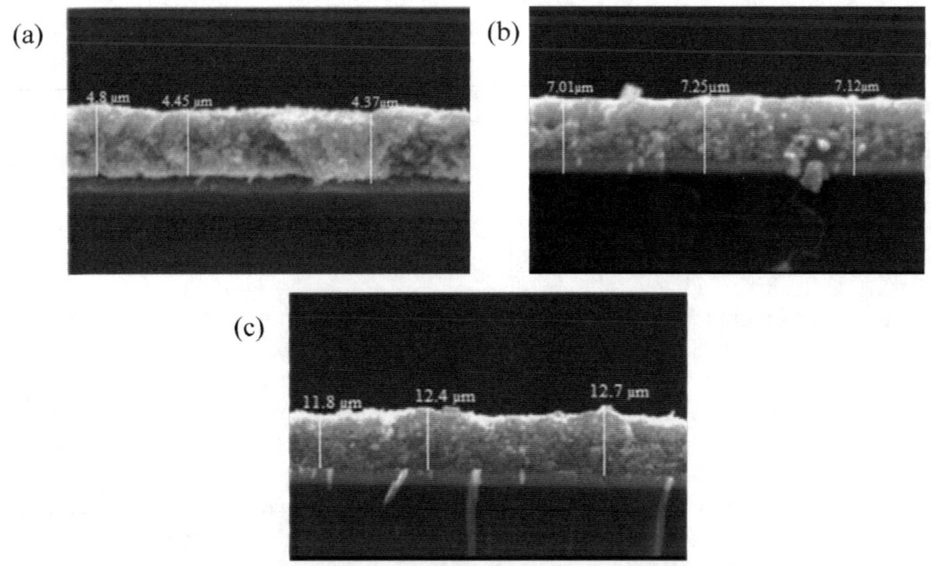

Figure 2. Cross sectional SEM images of the TiO_2 photoanode 4.54µm (a) 7.12µm (b) and 12.3µm (c).

IOP Conf. Series: Materials Science and Engineering **290** (2017) 012004 doi:10.1088/1757-899X/290/1/012004

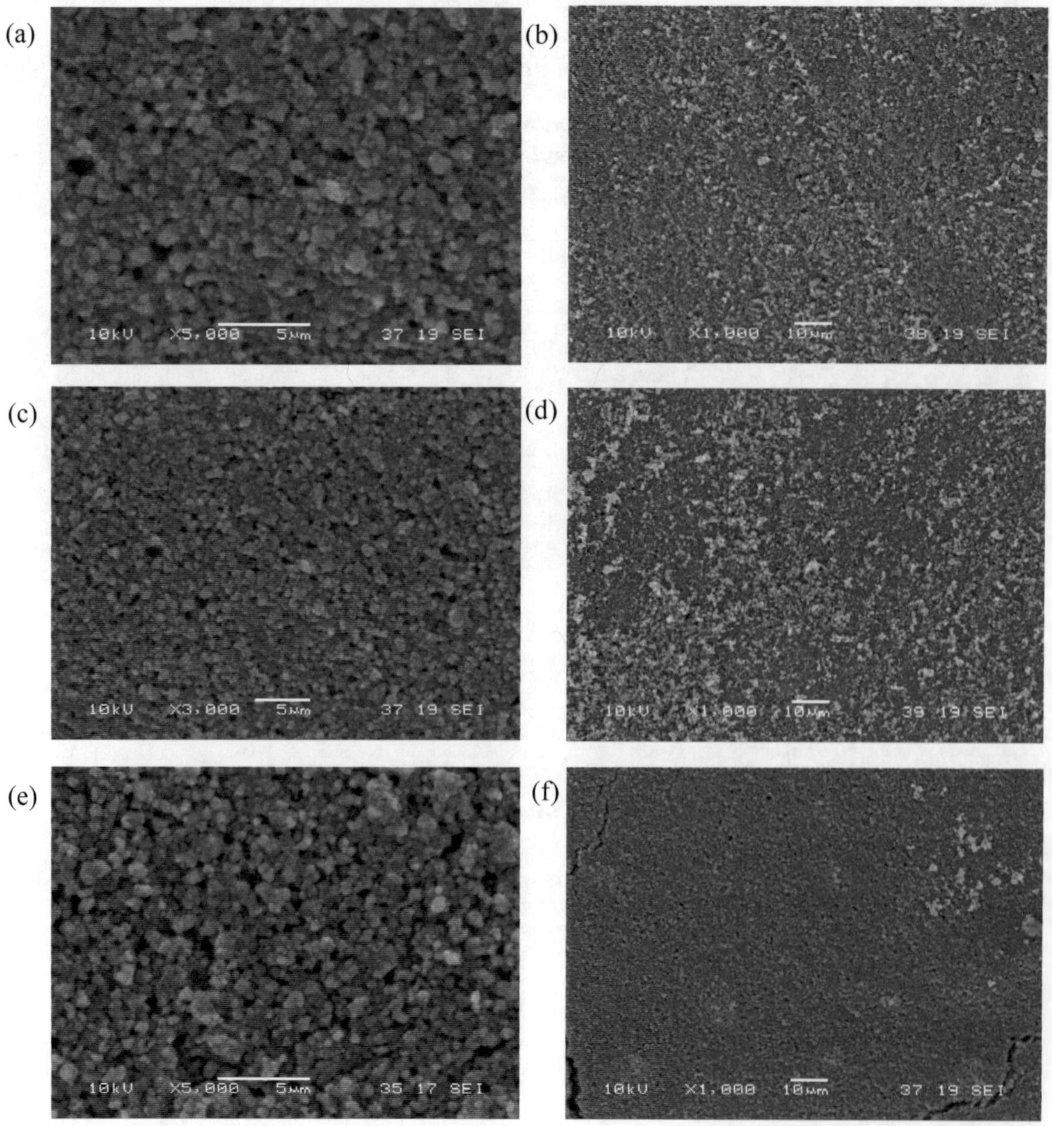

Figure 3. Surface morphology of TiO_2 photo anode with thickness of 4.54 μm (a,b); 7.12μm (c,d) and 12.3 μm (e,f) with x5000 and x1000 magnification.

The SEM images from Figure 3(a), 3(c) and 3(e) suggests that the pores appeared more frequently with the increased thickness of the TiO_2 paste. The pore sizes are random but the film is uniformly distributed with the TiO_2 paste. For all the samples the film did not peeled off from the surface after sintering. However, pores are necessary for the film to absorb the dye molecule and for the electrolyte diffusion [12]. Figure 3(f) shows large cracks all over the region whereas figure 3(b) and 3(d) does not appears with considerable amount of crack. Large cracks occurred due to the rapid evaporation of the materials during the sintering process [7]. For the 12.3μm thick TiO_2 electrode, the paste was formulated with acetic acid and ethanol. Due to the rapid evaporation, the samples were affected for 12.3μm thick electrodes but for 4.54μm and 7.12 μm thick electrodes did not show significant amount of cracks. This kind of cracks reduces the bonding strength of the nano particles. Generally, sintering process makes the particle bonding stronger and improves surface adhesion but if the film is thick, it tends to form cracks with the shrinkage of the film and also for chemical evaporation [13]. Higher shrinkage of the paste causes peeling off the film from the glass substrate after the sintering process or after soaking the electrode in the dye.

During sintering process, 50% of the weight loss occurs between the temperature range of 100°C-300 °C. Acetic acid and acetyl acetone are mostly evaporated within the temperature range and if the thickness of the TiO2 paste is high, the electrode tends to form visible cracks. Increasing the thickness helps to achieve better current density J_{sc} from the electrode but if the thickness is greater than the light penetration depth, it causes charge recombination, which ultimately results in loss of electrons. On the other hand, the open circuit voltage V_{oc} also decreases with thicker electrode paste. Therefore optimizing the thickness is crucial to prevent the electron charge combination loss and also for achieving greater V_{oc} and $J_{sc.}$ However, for all the samples the paste showed good adhesion without peeling off from the surface after sintering and drying process.

4. Conclusion

In summary, this can be concluded that the TiO_2 thin films were uniformly deposited on the ITO glass by controlling thickness using doctor blade method. All the three samples showed strong absorption throughout the whole visible light range. However, increasing the thickness lead towards higher energy gap of the TiO_2 paste. If the energy gap is high, the photons will require higher energy to enter the conduction band of the TiO_2 electrode and if the energy gap is too low that results in lower open circuit voltage for the solar cell. Therefore, the optimum bandgap is achieved with 7.12 μm thick TiO_2 electrode. From the samples, it can be also observed that the absorption and the bandgap is proportionally related with the thickness of the sample. The UV-VIS characterization suggests that the absorption and the band gap can be controlled by varying the thickness of the material. However, the SEM images signifies that if the sample is thicker than 12.3 μm the film starts to form cracks which leads to the leakage of the current when the cell is assembled.

Acknowledgement

I am thankful to my co-authors from the Department of Manufacturing and Materials Engineering, IIUM for the technical assistance and helpful suggestions.

References

[1] Sedghi A and Miankushki H 2015 *Int. J. Electroche. Sci.* **17** 3354– 3362
[2] Dhungel S K and Park J G 2010 *J. Renew. Energy* **35** 2776-2780
[3] Dette C, Miguel A and Pérez-Osorio 2014 *Nano Lett.*. **14** 6533–6538
[4] Ming-Hsiu, Fu-Lin and Shyh S 2010 *Int. J. Mol. Sci.*. **11** 329-353
[5] Peng, Qin Y and Qiang 2012 *Int. J. Photoenergy* **12** 3100-3221
[6] Ito S, Chen P, Comte P, Nazeeruddin M K, Liska P, Péchy P and Grätzel M 2007 *Prog. Photovoltaics Res.Appl.* **15** 603–612
[7] Meen T H, Tsai J K, Tu Y S, Wu T C, Hsu W D and Chang S 2014 *Nanoscale Res. Lett.* **9** 510-523
[8] Jena A, Mohanty S P, Kumar P, Naduvath J, Gondane V, Lekha P, Das J, Narula H K, Mallick S and Bhargava P 2012 *Trans. Indian Ceram. Soc.* **71** 1-16
[9] Qiang Y Q and Saehana S 2011 *Electrochem.Sci.***11** 215-228
[10] Ito S, Murakami T N, Zakeeruddin S M, Yazawa T, Mizuno M, Kayama S and Grätzel M 2014 *Nano* **9** 111-117
[11] Singh M, Mohan M and Ahlawat D S 2015 *nt. J. Human. Soc. Sci. Res.* **5** 35-41
[12] Xu S *et al* 2011 *J. Am. Chem. Soc.* **25** 1168–1172
[13] Gemeiner P and Mikula M 2013 *J. Chem. Sci.*.**6** 29-34

IOP Conf. Series: Materials Science and Engineering **290** (2017) 012005 doi:10.1088/1757-899X/290/1/012005

Mechanical properties of hybrid SiC/CNT filled toughened epoxy nanocomposite

S Ratim[1, 2], S Ahmad[1], N N Bonnia[2], Sabrina M Yahaya[3]

[1]School of Applied Physics, Faculty of Science and Technology, Universiti Kebangsaan Malaysia, Bangi, 43600 Selangor, Malaysia
[2]Materials Technology Programme, Faculty of Applied Sciences, Universiti Teknologi MARA, 40450 Shah Alam, Selangor, Malaysia
[3]Department of Applied Chemistry, Faculty of Applied Sciences, Universiti Teknologi MARA, 40450 Shah Alam, Selangor, Malaysia

Email:suzanar@salam.uitm.edu.my

Abstract. Mechanical properties of epoxy nanocomposites filled single filler have been extensively studied by various researchers. However, there are not much discovery on the behavior of hybrid nanocomposite. In this study, single and hybrid nanocomposites of toughened epoxy filled CNT/SiC nanoparticles were investigated. The hybrid nanocomposites samples were prepared by combining CNT and SiC nanoparticles in toughened epoxy matrix via mechanical stirring method assisted with ultrasonic cavitations. Epoxy resin and liquid epoxidized natural rubber (LENR) mixture were first blend prior to the addition of nanofillers. Then, the curing process of the nanocomposite samples were conducted by compression molding technique at 130°C for 2 hours. The purpose of this study is to investigate the hybridization effect of CNT and SiC nanoparticles on mechanical properties toughened epoxy matrix. The total loading of single and hybrid nanofillers were fixed to 4% volume are 0, 4C, 4S, 3S1C, 2S2C, and 1S3C. Mechanical properties of hybrid composites show that the highest value of tensile strength achieved by 3S1C sample at about 7% increment and falls between their single composite values. Meanwhile, the stiffness of the same sample is significantly increased at about 31% of the matrix. On the other hand, a highest flexural property is obtained by 1S3C sample at about 20% increment dominated by CNT content. However, the impact strength shows reduction trend with the addition of SiC and CNT into the matrix. The hybridization of SiC and CNT show highest value in sample 1S3C at about 3.37 kJ/m^2 of impact energy absorbed. FESEM micrograph have confirmed that better distributions and interaction observed between SiC nanoparticles and matrix compared to CNT, which contributed to higher tensile strength and modulus.

1. Introduction

Nanocomposite materials have gained a great attention to the various fields recently. The used of this material in high performances application such as electronics, telecommunications, defense, aerospace possessed high surface area, therefore polymer nanocomposite offer enhanced mechanical, thermal, barrier and flame retardant properties compared to conventionally filled materials [2]. The discovery of CNT by [3] brought the explosion of nanotechnology in various field. Due to this reason, CNT based product received high demand for bulk production and currently are available in open market [4].

However, there are some problems facing by composite industry in nanocomposite production. According to [5], the major problem in nanocomposite materials production is the dispersion and agglomeration of nanofiller in matrix system. These phenomena will reduce the mechanical properties of nanocomposite and limits their usage in specific applications. The other problem is the use of single nanofiller in a polymer matrix cannot fulfill certain requirement for specific application. Thus, hybrid composite has been proposed in order to solve these problems.

Hybrid nanocomposite is a composite system produced by combining two types or more nanofillers in a matrix system. Hybridization of different nanofiller in one matrix system is a new approached used by several researchers to solve a problem that cannot be attained by single filler [6]. Other researcher [7] claimed that the hybridization of different filler in a matrix system are able to optimized the composite properties and complement the other filler weaknesses. Two or three types of filler present in single matrix not only improved the mechanical properties, but electric, magnetic and microwave absorption as well.

Tensile strength of hybrid composite have similar curve as single nanocomposite but changed from linear to nonlinear due to inconsistent hybrid filler content [8]. Study by [9] on hybrid composite of multiwalled carbon nanotube (MWCNT) and organoclay montmorillonite (OMMT) in epoxy matrix found that impact strength of nanocomposite has been increased and the value is falls between their single filler nanocomposite.

In this study, hybrid composite of CNT and SiC nanoparticle were combined in order to study the effect of their hybridization on mechanical properties of toughened epoxy nanocomposite.

2. Experimental procedures

Toughened epoxy matrix were first prepared by blending epoxy resin with liquid epoxidized natural rubber (LENR) by using mechanical stirrer assisted with ultrasonic cavitations for 1 hour at 1300 rpm. During stirring, the SiC nanoparticles were added and mixed for 15 minutes prior to the addition of CNT addition. Then, the mixture was further mixing for 45 minutes followed by the addition of crosslinking agent at the ratio of 1:3 to the resin amount. After that, the mixture was degassed in vacuum oven for 5 minutes at room temperature for entrapped air removal. The mixture were then poured onto mould and followed by pre-curing at room temperature for 12 hour before proceed to the post cure at 130°C for 2 hours. The mechanical tests were carried out using Testometric M350-10CT Universal Testing Machine. Type I dumbbell shape of tensile sample were prepared according to the ASTM D638. Flexural properties of the hybrid nanocomposite were measures according to ASTM 790 three point bending test while Izod impact tests were performed according to ASTM D256.

3. Results and Discussion

The mechanical properties of single and hybrid nanocomposite were determined by comparing the tensile strength and modulus at different loading condition of CNT and SiC nanoparticles. Measuring the materials behavior under tension load with the present of two different nanofiller were done to study the effects of CNT/SiC nanoparticle loading on mechanical properties of single and hybrid nanocomposites. The tensile properties of toughened matrix, single filled composite and hybrid composite are summarized in figure 1 (a) and (b).

It can be seen that the tensile strength and Young's modulus increased with the addition of 4% single nanofiller of CNT or SiC nanoparticles in toughened epoxy matrix. However, the addition of 4% SiC nanoparticles shows slightly higher value in tensile strength as compared to 4% CNT and similar trend are observed in Young's modulus value which, the modulus of SiC nanoparticle composite received greater value than CNT nanocomposite. This is because higher forced is required to break the sample due to plastic deformation increases during crack deflection and pinning [10]. The matrix chain segment pinning due to the increasing of filler content will increase the stiffness of nanocomposite [11].

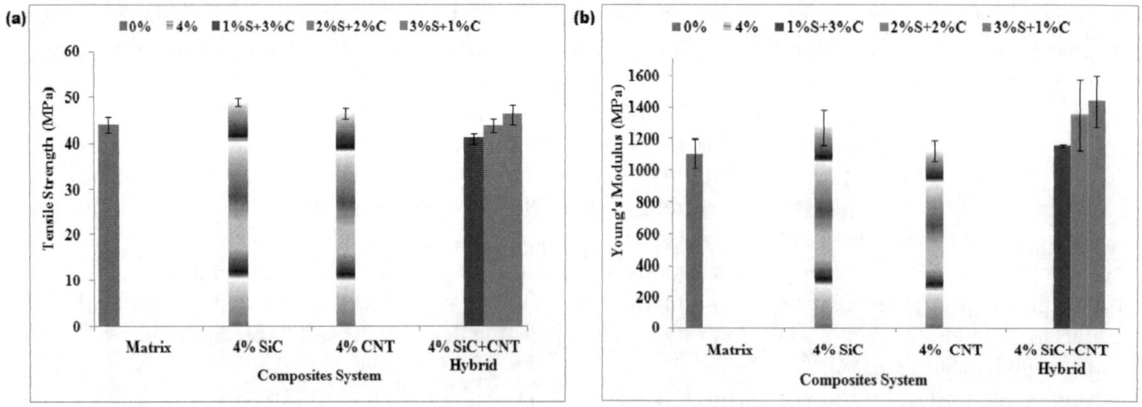

Figure 1. (a) Tensile strength and (b) Modulus elasticity of toughened epoxy matrix, single and hybrid SiC/CNT nanocomposites.

Figure 2 show the flexural (a) strength and (b) modulus comparison between single and hybrid composite of SiC and CNT filled toughened epoxy nanocomposite. It is obvious that the addition of 4% SiC and CNT in toughened epoxy matrix has increased the flexural strength and moudulus. However, CNT nanocomposite of flexural strength and modulus value is higher than SiC nanocomposite.

Hybridization of SiC and CNT in the same toughened epoxy matrix reduces the flexural strength and modulus value but remain higher than matrix. This is due to the domination of CNT content proved that CNT nanotubes provide better flexural properties than SiC nanoparticle. In hybrid nanocomposite sample dominated by SiC content. The presence of large SiC nanoparticle amount restricted the CNT aggregation and increased the tensile strength and modulus. Meanwhile, in hybrid sample dominated by CNT agglomeration acted as a stress concentration sites thus reduced the tensile strength [4].

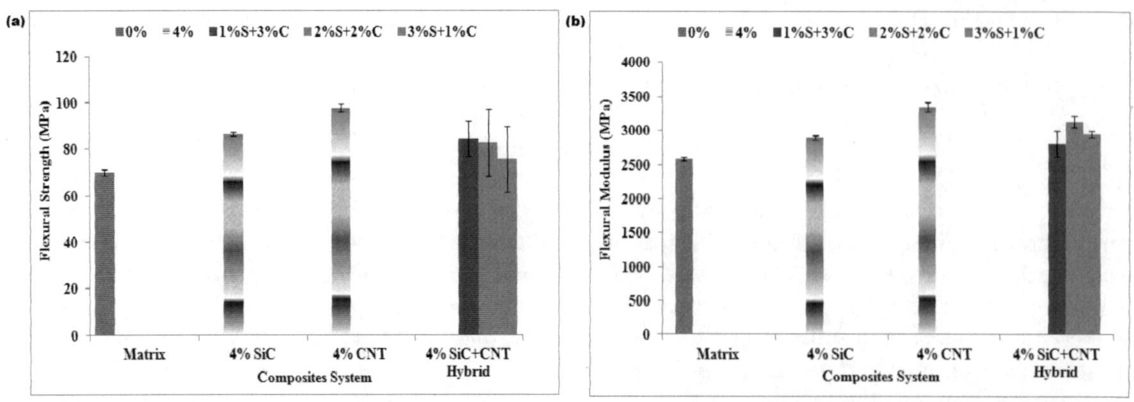

Figure 2. (a) Flexural strength and (b) Flexural modulus of the toughened matrix, single and hybrid SiC/ CNT nanocomposites

The impact strength of single and hybrid nanocomposite of SiC and CNT filled toughened epoxy is shown in Figure 3. There is significant reduction in impact strength with the addition of CNT and SiC nanoparticle in the matrix. However, higher value is observed in CNT compared to SiC nanocomposite. The impact strength value of hybrid nanocomposite falls between single nanocomposite of CNT and SiC nanoparticle. Hybrid sample dominated CNT at 1%S+3%C remain higher than sample dominated by SiC of 3%S+1%C.

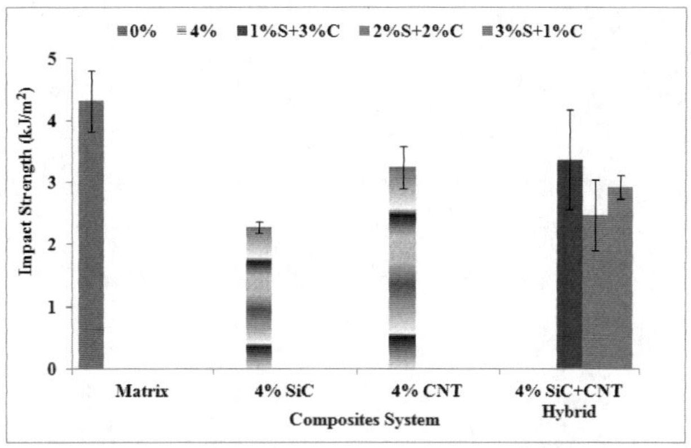

Figure 3. Impact strength of matrix, single and hybrid SiC/CNT nanocomposites

Domination of 3%C+1%S hybrid composite is due to high aspect ratio of CNT requires more energy for crack propagation during impact. Therefore, the energy required to break the sample dominated by CNT nanofiller is higher than the sample dominated by SiC. The stiffness behavior of SiC nanoparticle itself contributed to the stiffness of the composites, thus low energy is needed to break SiC nanocomposite sample.

Figure 4 depicts selected FESEM images of the toughened epoxy matrix and single filled of SiC and CNT nanocomposite fractured surfaces. It clearly can be seen in figure 4(a) that a smooth fractured surface of toughened epoxy matrix with the presence of dispersed rubber particles. Rough surface CNT and SiC nanoparticle show a good distribution on figure 4(b) and (c) support the improvement of tensile strength and Young's Modulus as compared to the matrix. According to [3], rough surface produced by crack propagation changes deviated from original plane. However, there are fiber pullouts of CNT and leave cavity by fiber pullout can be observed samples in figure 4(c) that affected to the reduction of strength in CNT nanocomposite. Meanwhile, the improvement of tensile properties in hybrid nanocomposite sampel dominated by SiC nanoparticle proved complete coverage of toughened epoxy matrix onto the SiC nanoparticle surfaces.

Good distribution of SiC nanoparticle will increased the density and mechanical properties of the hybrid composite. Meanwhile, inhomogeneous distribution of CNT nanofiller leads to the nanoparticles interaction by weak Van Der Walls thus reduces reinforcing behavior of filler in matrix [12].

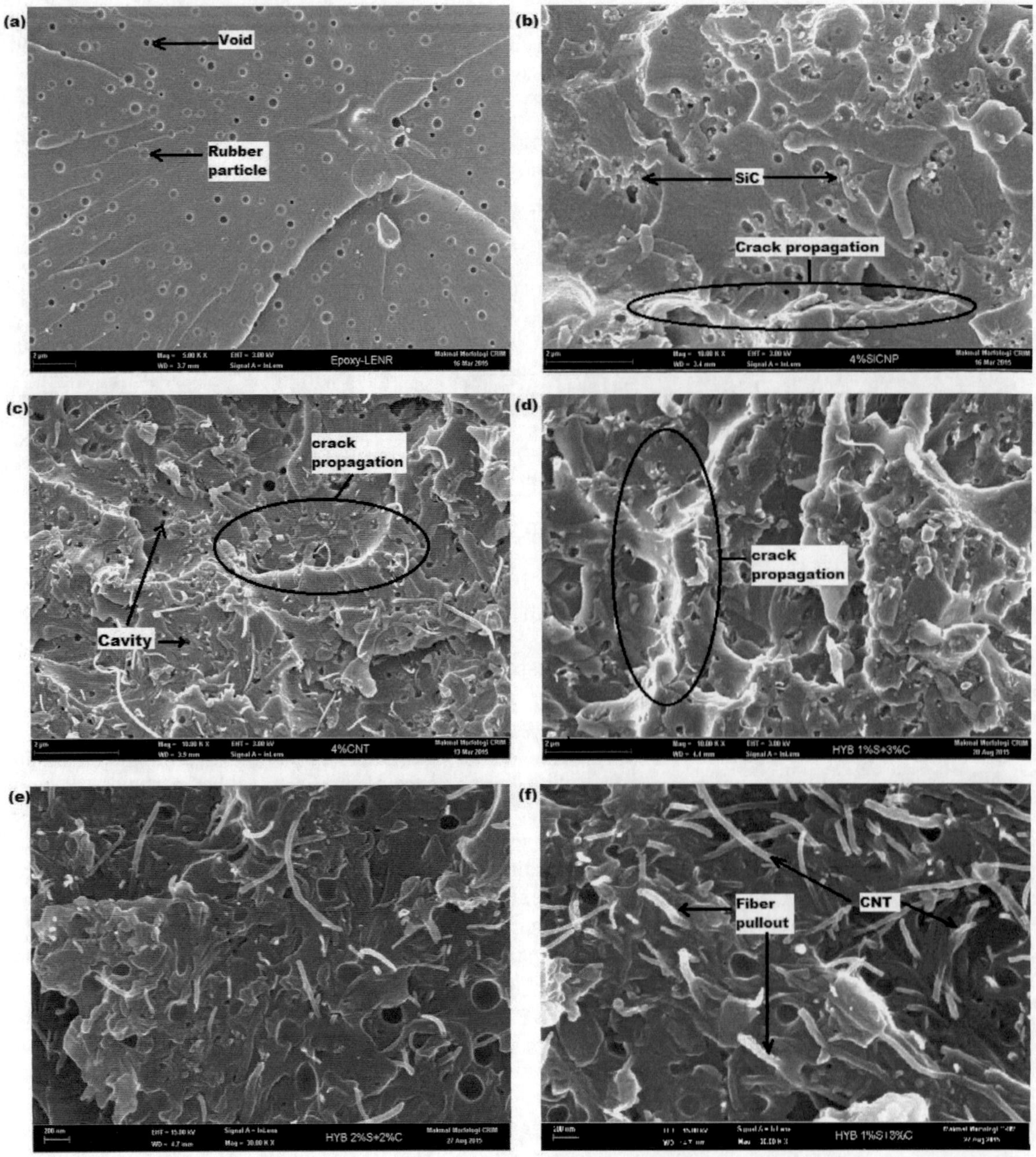

Figure 4. Tensile fracture surface morphology of (a) toughened epoxy matrix (b) 4% S nanocomposite and (c) 4% C (d) 1C3S, (e) 2S2C and (f) 3C1S hybrid nanocomposites.

4. Conclusion

In summary, the presence of CNT and SiC nanoparticles in toughened epoxy matrix has increased 6% of the tensile strength and 12% stiffness respectively. The flexural strength and modulus of SiC and CNT nanocomposite are also increased at about 37% and 25% respectively. However, the impact properties are reduced at about 25% and 47% respectively for both SiC and CNT nanocomposite systems. Meanwhile, the hybridization of SiC with CNT in the toughened epoxy matrix has falls the

IOP Conf. Series: Materials Science and Engineering **290** (2017) 012005 doi:10.1088/1757-899X/290/1/012005

mechanical properties between the value of their single nanocomposite. Domination of SiC nanoparticles content are clearly can be seen in higher tensile strength of 47MPa and Young's modulus of 1442 MPa for 3S1C in hybrid nanocomposite sample. The optimum flexural strength achieved by 1S3C sample at about 85 MPa and the optimum impact strength of hybrid nanocomposite is also observed by 1S3C at about 3.37kJ/m^2.

Acknowledgement
The authors are acknowledge Universiti Teknologi MARA Grant (Lestari) 600-IRMI/MyRA 5/3/ (053/2017), for the financial support and Universiti Kebangsaan Malaysia for the technical support.

References
[1] Simcha S, Dotan A, Kenig S and Dodiuk H 2012 *Nanomaterials* **2** 348-365
[2] Okpala C C 2013 *International Journal of Engineering Research and Development* 8(11) 17-23.
[3] Iijima S 1991 *Nature* (London) pp. 56-58
[4] Belluci S, Micciulla F, Sacco I, De Bellis G and Rinaldi G 2009 *IEEE Int. Symp. on Electromagnetic Compatibility* pp. 89-96
[5] Shokrieh M M and Saeedi A 2013 *J. Nanostruct. Chem.* **3** 1-5
[6] Moosa A A, Ramazani A S A, Kubba F A K and Raad M 2017 *Am. J. Mater. Sci.***7** 1-11
[7] Harris B 1986 *Eng. Comp.Mater.* (London: Institute of Metals)
[8] Zhou X, Depeng S, Wu C and Liu L 2012 *J. Nanomater.* **2012** 1-7
[9] Tarawneh M A, Sahrim H A, Noum S Y E and Ahmad K Z K 2011 *J. Reinf. Plast. Comp.* **30** 1745–1752
[10] Karapappas P, Baltopoulos A, Vavouliotis A, Fiamegko E, Athanasopoulos N, Fotiou I, Kostopoulos V, Borsella E and Fabbri F 2009 *Proceeding of Second International Conference on Smart Materials and Nanotechnology in Engineering* pp. 1-8
[11] Grande D H, Mandell J F and Hong K C C 1988 *J. of Mater.Sci.* **23** 311–328
[12] Ahmad I, Kennedy A and Zhu Y 2009 *Proceeding of International Conference on Composite Materials (ICCM-17)*

IOP Conf. Series: Materials Science and Engineering **290** (2017) 012006 doi:10.1088/1757-899X/290/1/012006

Understanding the significance variables for fabrication of fish gelatin nanoparticles by Plackett-Burman design

Deni Subara[1], Irwandi Jaswir[1,2], Maan Fahmi Rashid Alkhatib[1] and Ibrahim Ali Noorbatcha[1]

[1]Department of Biotechnology Engineering, Faculty of Engineering, International Islamic University Malaysia 53100 Kuala Lumpur, Malaysia
[2]International Institute for Halal Research and Training (INHART), International Islamic University Malaysia Gombak, Kuala Lumpur, Malaysia

Email: deni.subara@yahoo.com

Abstract. The aim of this experiment is to screen and to understand the process variables on the fabrication of fish gelatin nanoparticles by using quality-design approach. The most influencing process variables were screened by using Plackett-Burman design. Mean particles size, size distribution, and zeta potential were found in the range 240±9.76 nm, 0.3, and −9 mV, respectively. Statistical results explained that concentration of acetone, pH of solution during precipitation step and volume of cross linker had a most significant effect on particles size of fish gelatin nanoparticles. It was found that, time and chemical consuming is lower than previous research. This study revealed the potential of quality-by design in understanding the effects of process variables on the fish gelatin nanoparticles production.

1. Introduction

Gelatin is an excellent materials for production nanoparticles that have been described well in the literature [1], [2]. Gelatin is obtained from hydrolysis of collagen. Furthermore gelatin considered as GRAS or generally regarded as safe by FDA (food drug administration) [3]. Gelatin is also the cheap materials because it's abundant, readily available, high nutrition value, biocompatible, and biodegradable [3]. Due to the characteristic of gelatin, it has been successful encapsulated different type of drug, such as antibiotics [4], bioactive extract [5], and anti-cancer [6]. The origin gelatin in the market was from porcine skin and bovine bone. Because of that, almost gelatin nanoparticles have been produced from mammalian gelatin. However, the application of mammalian gelatin was not suitable for some group, such as Jews prefer Kosher and Muslim prefer Halal [7]. Alternative source such as fish gelatin nanoparticles would be fulfill the requirement of some groups.

Mainly, mammalian gelatin nanoparticles was produced by several method, such as emulsion [8], coarcevation [9], self-assembly [10], and two-desolvation method [11]. In this experiment, two-step desolvation method was applied because this method could produce small size nanoparticles with narrow size distribution [3]. However, due to lower gelling point, lower gel strength, and lower melting point of fish gelatin [12], the preparation of fish gelatin nanoparticles will be different compare to mammalian gelatin nanoparticles. In order to fabricate the fish gelatin nanoparticles with small size, the fabrication variables need to be screen and optimize.

In two-step desolvation method, many formulation parameters affect the quality of gelatin nanoparticles. Less chemical consuming and short production time in the process production were became the main purpose, and as well as the ability of the process production brings the product to commercialization. Because of that, quality by design (QbD) was initiated by FDA [13]. QbD requires better understanding on phenomenon in each process production in order to improve their performance. One of the common methods is design of experiment. Design of experiment was the precise technique to know the most significant variables in the pharmaceutical development. Jones et al., [14] explains that the design of experiment have been used to define the objective of investigation, decide the nature of variable and response, and had a minimum number of experimental run. Usually experimental design divides to two step, the first step to screen the significant factors (Plackett-Burman design or Factorial design), the second step to optimized the significant factors (usually by Central composite design or Box-Behnken) [15].

This research focused on the potential of fish gelatin in producing gelatin nanoparticles by two-desolvation method. Plackett-Burman design was applied to screen the effect of various process variables. The selected process variables were pH of solution, concentration of acetone, volume of glutaraldehyde, stirring speed, and stirring time. The effects of the variables on particles size were analyzed. The particles size distribution and shape of fish gelatin nanoparticles was also studied using Zeta sizer and FESEM.

2. Experimental procedures

2.1. Materials

Fish gelatins were fabricated by our research group using suggested method by Jamilah [12]. The fish gelatins were stored in the 4 °C fridge before further use. Others chemicals were purchased from Sigma Aldrich such as Acetone and glutaraldehyde grade 1 (25% vol% aqueous solution). Deionized waters (18.2 mΩ cm) were used trough experiment.

2.2. Preparation of fish gelatin nanoparticles

Fish gelatin nanoparticles were prepared by using two-step desolvation method that introduced by Coester [16]. This method divided on two steps; the first step is fractionation and the second step is precipitation step. Precipitation step was the main focus in this current experiment.

About 0.9 g fish gelatin was dissolved in 10 ml deionized water with constant stirring (600 rpm) and constant heating (45 °C) until clear solutions were received. Fish gelatin solutions then were transferred to falcon tube and were added with 10 ml acetone. The mixed solutions were centrifuged for 5 min in constant speed (12000 g). The supernatants were immediately discarded and the precipitated were re-dissolved in 10 ml deionized water. The pH of precipitate solution was adjusted using 0.1 M HCl until reach certain number of suggested design. The solutions then were added with acetone and were cross-linked using glutaraldehyde solution. The solutions were stirred with specified hour. The fish nanoparticles colloids were centrifuged for 30 min at 14000 rpm. The precipitates were stored at fridge until further use.

2.3. Design of experiment

These experiments were focused on the screening of the significant variables that involve in precipitation step or in second step from two-step desolvation method. In this experiment experimental design (DOE) using Plackett-Burman design was applied. Plackett-Burman (PB) design was applied because less time consuming and less chemical consuming. Plackett-Burman design involved relatively less runs experiment (12 runs) to screen huge number of variables resulting critical and good degree of accuracy compared to factorial design [17]. PB matrix was generated by using Minitab® 16. The expression of linear equation was developed and shown in equation 1 below.

$$Y = b_0 + b_1X_1 + b_2X_2 + b_3X_3 + b_4X_4 + b_5X_5 + \cdots + b_nX_n \qquad (1.0)$$

Where Y is the dependent variable or response, b_0 is the constant and $b_1, b_2 ... b_n$ are the coefficient of the variables $X_1, X_2, X_3, X_4, X_5 .... X_n$ (representing the effect of each factor ordered within −1, +1).

In this PB matrix were involved twelve experimental runs including five independent variables, one dummy variable and one dependent variable. Each experimental run was carried triplicate. Table 1 shows the list of the independent variables and table 2 describes experimental run matric. Before experimental design, the chosen variables (five variables) and levels have been made base on previous researcher and preliminary research [18]–[25]. The selected independent variables were pH of the fish gelatin solution (X_1), acetone concentration (X_2), volume of glutaraldehyde (X_3), stirring speed (X_4), and stirring time (X_5). One dependent variable was fish gelatin nanoparticles size. The results were evaluated by statistical analysis using Analysis of variance (ANOVA). Means variable plot were also establish to analyzed the most significant variables.

Table 1. Experimental variables and level of Plackett-Burman design.

Variables	Symbol	Units	Experimental value	
			Low (−1)	High (+1)
pH	X_1		1.5	5.5
Percentage of Acetone	X_2	%(v/v)	15	65
Volume of Glutaraldehyde	X_3	µl	100	600
Stirring Speed	X_4	rpm	300	900
Stirring Time	X_5	hour	3	21

2.4. *Particles size measurement and FESEM measurement*
The fish gelatin nanoparticles size was measured by dynamic light scattering (DLS) at 90° in 10 mm diameter cells with a Malvern (Zen3600, UK). About 300 µl nanoparticle colloid was mixed to 30 ml of deionized water. The mixtures were sonicated at 60 Amp for 20 min. The size and particles distribution were analyzed. The polydistersity index (PdI) is ranging from 0.0 to 1.0. The surface charge of nanoparticle was measured using Zeta potential Malvern.

Shape, size, and surface morphology of fish gelatin nanoparticles were visualize using Field Emission Scanning Electron Microscope (FESEM) (JEOL, JSM 6700F Model) at an accelerating voltage of 10 kV. Prior to visualization, fish gelatin nanoparticles were mounted on small adapter and was sputter with a thin layer of gold. The adapter was then placed in the chamber.

3. Results and Discussion

3.1. *Impact of fabrication parameters on fish gelatin nanoparticles size*
Table 2 depicts the experimental results of twelve run from five independent variables. The statistical analysis were applied to this results in order to quantify the most influence significant variables on the fabrication fish gelatin nanoparticles by using Minitab® 16 software. The behavior of the individual variables to particles size were also analyzed by alternatively treated at the highest and lowest level. While others variables were kept constant at center levels.

Figure 1 shows the plot of main effect of each variable on the particles size. This figure not only depicts the effect of respective processing variables, but also shows the positive or negative trend of each variable in studied range. It can be seen that pH of solution and percentage of acetone had significant positive effects, while volume of glutaraldehyde had a significant negative effect on the particles size. This statement describes that small size fish nanoparticles could be produced at low pH solution, low percentage of acetone and huge volume of glutaraldehyde. This result agrees with previous works [19]–[21], that small size gelatin nanoparticles were produced bellow isoelectric point, low concentration of acetone and high volume of glutaraldehyde. It is because isoelectric point of fish gelatin had same range of with mammalian gelatins that have been used at previous work. The isoelectric point of fish gelatin is around pH 6-9 [12], [26].

Table 2. Plackett-Burman design matrix.

Run	pH (X_1)	Concentration of Acetone (%, vol v/v) (X_2)	Volume of Glutaraldeyde (µl) (X_3)	Stirring Speed (rpm) (X_4)	Stirring Time (hours) (X_5)	Observed Particles size (nm)	Predicted Particles size
1	5.5	15	600	300	3	275.72±0.07	274.54
2	5.5	65	100	900	3	850.45±6.57	722.38
3	1.5	65	600	300	21	244.49±2.05	264.16
4	5.5	15	600	900	3	245.88±4.09	280.12
5	5.5	65	100	900	21	882.13±3.85	708.95
6	5.5	65	600	300	21	518.19±6.46	364.52
7	1.5	65	600	900	3	227.85±1.07	283.18
8	1.5	15	600	900	21	205.25±5.60	166.34
9	1.5	15	100	900	21	191.84±3.16	182.09
10	5.5	15	100	300	21	233.04±2.10	276.86
11	1.5	65	100	300	3	267.34±2.76	293.35
12	1.5	15	100	300	3	204.71±0.72	189.93

Furthermore, we found fish gelatin nanoparticles could be produced below pH 3.5 while optimum pH to produce mammalian gelatin was at pH 3.5. Producing fish gelatin nanoparticles within isoelectric point would produce large-scale of gelatin nanoparticles, due to the electrostatic attraction [21]. Decreasing the pH solution leads to protonate the side chain amino acid and create the stable suspension [21]. High concentration of acetone creates large aggregate [19]. Moreover, high concentration of glutaraldehyde forced the particles become harder [27]. In other hand, short or long time stirring of gelatin solution had insignificant effect for production small nanoparticles.

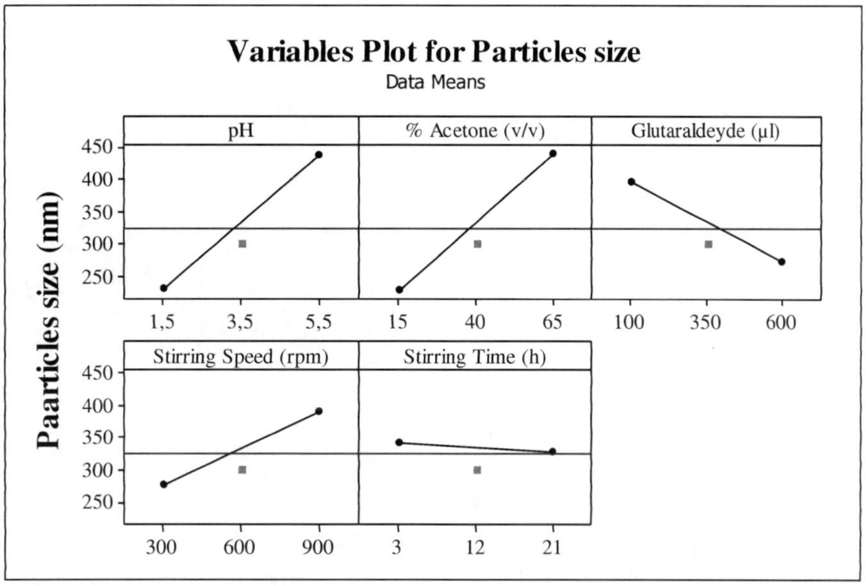

Figure 1. Main effect of variables in Plackett-Burman design of experiment on particles size.

3.2. Statistical analysis of Plackett-Burman design

In this experiment, we performed analysis of variance (ANOVA). The statistical analysis with ANOVA would produce the significance of models and regression coefficient in Plackett-Burman design. This experiment carried out by using 95% of confidence level. Means, the significance variables will be calculated based on p value at 0.05. The variable would be considered as insignificant variables when p value are higher than 0.05. The ANOVA results agree that pH of the solution, concentration of acetone, and volume of glutaraldehyde are the effective variable on producing small size nanoparticles, because they have lower p value than 0.05. The p value of model was obtained to be low value, this result representing that the regression model was also significant with a 95% confidence level. The R square of the results was at 0.86, means good correlation between observed particles size and predicted particles size. This R square also depicts that this model could explain almost 86% of the response variation. However, the p value of curvature was found to be higher than 0.05. That means, the levels of variables is in slope area and need to be changed in order to choose the correct level before optimization.

The polynomial equation between particles size and independent variables was given below.

$$Y = -34.98 + 52.01X_1 + 4.22X_2 - 0.25X_3 + 0.19X_4 - 0.75X_5 - 53.84X_6 \quad (3.0)$$

3.3. Characterization of prepared fish gelatin nanoparticles

The fish gelatin nanoparticle was fabricated using suggested variables such as pH at 1.5, percentage of acetone at 15%, and volume of glutaraldehyde at 600 µl, briefly, while other insignificant variables were treated in the center point. Figure 2 shows the particles size distribution from zeta-sizer analysis. It can be seen that, the average particles size of fish gelatin nanoparticles was about 240±9.76 nm, and the polydispersity index was at 0.3. This fish gelatin nanoparticle had slightly difference in particles size compared with mammalian gelatin nanoparticles was produced by Kumari around 110-257 nm [1]. The morphology of fish gelatin nanoparticles was quantified using FESEM (Figure 2b). FESEM result shows fish gelatin nanoparticles had a round shape. However, some variation in size distribution was depicted in the FESEM images. This phenomenon due to the uncontrolled effect of neutralization the molecule charge during specific pH [19], [28].

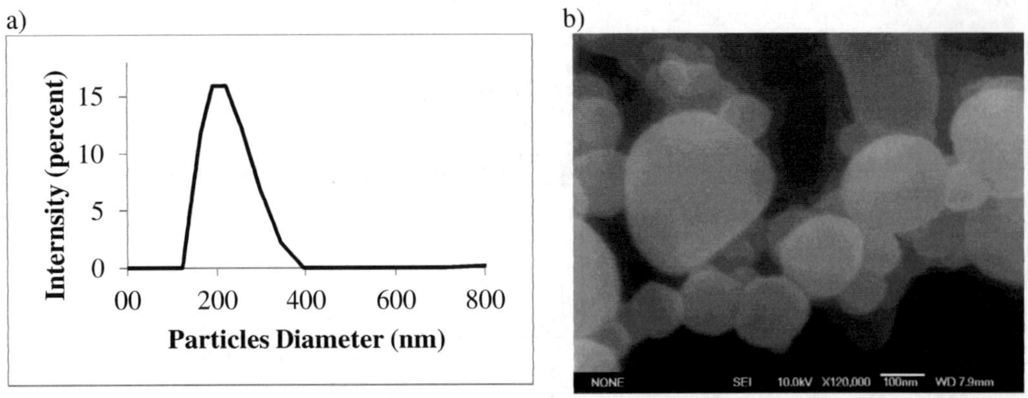

Figure 2. Particles size distribution of fish gelatin nanoparticles (a) and FESEM images of fish gelatin nanoparticles (b).

4. Conclusion

Fish gelatin nanoparticle was successfully developed using two-step desolvation method. Plackett-Burman quality by design approach was used to screen the independent variables and to understand the effect of most significant factors for production of fish gelatin nanoparticles. From the

Plackett-Burman results, three variables such as pH solution, percentage of acetone, and volume glutaraldehyde were found to have significant effects. While other variables like stirring time and stirring speed were insignificant on response. This results means by this method, we can reduce time and energy consuming in producing fish gelatin nanoparticles. To be concluded Plackett-Burman design is an efficient method for screening the most significant factors with minimum experimental run. Additionally, studies will continue to further optimization on production of fish gelatin nanoparticles.

Acknowledgement
This work was financially supported by Postgraduate Research Grant Scheme (PRGS) (Grant no. 13-087-0328) from Malaysian Ministry of Education.

References

[1] Kumari S K A Yadav and Yadav S C 2010 *Colloids Surf. B: Biointerfaces* **75** 1–18
[2] Farokhzad O C and Langer R 2009 *ACS. Nano.* **3** 16–20
[3] Elzoghby A O 2013 *J. Control. Release* **172** 1075–1091
[4] Mahor A, Prajapati S K, Verma A, Gupta R, Iyer A K and Kesharwani P 2010 *J. Colloid Interface Science* **483** 132–138
[5] Hani N, Azarian M H, Torkamani A E, and Kamil Mahmood W A 2016 *Int. J. Food Sci Technol.* **51** 2327–2337
[6] Han S, Li M, Liu X, Gao H, and Wu Y 2013 *Colloids Surf. B:Biointerfaces* **102** 833–841
[7] Karim A A and Bhat R 2008 *Trends Food Sci. Technol.* **19** 644–656
[8] Bajpai A K and Choubey J 2006 *J. App. Polym. Sci.* **101** 2320–2332
[9] Kasankala L, Xue Y, Yao W, Hong S, and He Q 2007 *Bioresour. Technol.* **98** 3338–43
[10] Li Z and Gu L 2011 *J. Agric. Food Chem.* **59** 4225–4231
[11] Dixit N, Vaibhav K, Pandey R S, Jain U K, Katare O P, Katyal A, and Madan J 2015 *Biomed. Pharmacother.* **69** 1–10
[12] Jamilah B and Harvinder K 2002 *Food Chem.* **77** 81–4
[13] Rahman Z, Zidan A S, Habib M J, and Khan M A *Int. J. Pharm.* **389** 186–194
[14] Jones R *Pharm. Eng. product quality* **29** 1–10
[15] Gupta B, Poudel B K, Pathak S, Tak J W, Lee H H, Jeong J, Choi H G, Yong C S and Kim J O 2015 *AAPS PharmSciTech* **15** 1–11
[16] Coester C, Langer K, Von Briesen H and Kreuter 2000 *J. Microencapsul.* **17** 187–193
[17] Plackett R L and Burman J P 1946 *Biometrika***33** 305–325
[18] Ofokansi K, Winter G, Fricker G, Coester C, Winte R G, Fricker G and Coester C 2010 *Eur. J. Pharma. Biopharm.* **76** 1–9
[19] Azarmia S, Huang Y, Chend H, Steve M, Abramse D, Road W, Löbenberga R, Azarmi S, Huang Y and Chen H 2006 *J. Pharm. Sci.* **9** 124–132
[20] Langer K, Balthasar S, Vogel V, Dinauer N, Von Briesen H and Schubert D 2003 *Int. J. Pharm.* **257** 169–180
[21] Saxena A, Sachin K, Bohidar H B B and Verma A K 2005 *Colloids Surf. B: Biointerfaces* **45** 42–48
[22] Taheri E S, Jahanshahi M and Mosavian M T H 2012 *Part. Part. Syst. Charact.* **29** 211–222
[23] Von Storp B, Engel A, Boeker A, Ploeger M and Langer K 2012 *J. Microencapsul.* **29** 138–46
[24] Baharifar H and Amani A 2017 *J. Pharm. Sci.* **106** 411–417
[25] Balavandy S K, Shameli K, Biak D R B A and Abidin Z Z 2014 *Chem. Cent. J.* **8** 11–14
[26] Gudmundsson M 2002 *J Food Sci.* **6** 2172–2176
[27] Nahar M, Mishara D, Dubey V and Jain N K 2008 *Nanomedicine* **4** 252–261
[28] Mazerski J, Grzybowska J and Borowski E 1990 *Eur. Biophys. J* **18** 159–64

IOP Conf. Series: Materials Science and Engineering **290** (2017) 012007 doi:10.1088/1757-899X/290/1/012007

Effects of forming temperature and sintering rate to the final properties of FeCuAl powder compacts formed through uniaxial die compaction process

M M Rahman[1], M A Ismail[1], I Sopyan[2] and H Y Rahman[1]

[1]Department of Mechanical Engineering, Universiti Tenaga Nasional, Putrajaya Campus, Jalan IKRAM-UNITEN, 43000 Kajang, Selangor, Malaysia

[2]Department of Manufacturing and Materials Engineering, Kulliyyah of Engineering, International Islamic University Malaysia, PO Box 10, 50728 Kuala Lumpur, Malaysia

Email:mujibur@uniten.edu.my

Abstract. This paper presents the outcomes of an experimental investigation on the effects of forming temperature and sintering schedule to the final characteristics of FeCuAl powder mass formed at different temperature and sintered at different schedule. A lab-scale uni-axial die compaction rig was designed and fabricated which enabled the compaction of powder mass at room temperature as well as elevated temperature. Iron (Fe) powder ASC 100.29 was mechanically mixed with other elemental powders, namely copper (Cu), and aluminum (Al) for 60 minutes and compacted at three different temperature, i.e., 30°C, 150°C, and 200°C by applying 425 MPa of simultaneous downward and upward axial loading to generate green compacts. The as-pressed samples were inspected visually and the defect-free green compacts were subsequently sintered in an argon gas fired furnace at 800°C for 60 min at three different heating/cooling rates, i.e., 5, 10, and 15°C/min, respectively. The sintered samples were then characterised for their physical, electrical, and mechanical properties. The microstructures of the sintered samples were also analysed. The results revealed that a forming temperature of 150°C and a sintering rate of 10°C/min could produce a product with better characteristics.

1. Introduction

Pure materials are proven to have a lot of disadvantages, i.e., soft, brittle, easy to degrade in hot and humid environment, etc. In order to overcome the disadvantages of pure materials, other materials are used as additives to produce a coherent mass named as alloy which is a homogeneous mixture of two or more materials where the dominant one is called matrix and the other materials are dissolved within it. Alloys have superior properties compared to individual pure materials [1-2]. Alloys are produced either through casting process or mechanical alloying. Casting process requires huge amount of thermal energy to melt the metals and subsequently machining operation to produce end products. On the other hand, mechanical alloying is used to prepare a solid solution of more than one metallic or non-metallic materials and subsequently the solid solution is pressed and sintered to produce end products. The solid solution is prepared through high energy ball milling of the intended powders for a long time. Both methods, i.e., casting and mechanical alloying are found to be time consuming and energy ineffective.

However, alloys could also be produced by shaping a mixture of intended powders and subsequently sintering the powder mass in inert gas fired furnace. This technique is relatively new but offers an approach to produce high quality end products [3]. However, no thorough study found in the literatures regarding this method of product manufacturing from of a mixture of materials. This method is considered as an alternative lower cost process to machining, casting, stamping, forging and other similar metal working technologies. Sintering is a part of a full cycle of powder compaction process, which must be done at controlled environment to avoid contamination of the component by other materials. The basic principle of sintering is the achieving rate of the desired degree of bonding among the particles in powder compacts. The sintering rate plays an important role in controlling the microstructure and porosity that determine the degree of particles bonding. The densification and the strengthening of particles inside the metal powder compacts reflect the properties of the sintered component. The sintered density relates directly to the dimensional and mass changes, which give perfect understanding of the densification process. Most importantly, the relative density of sintered parts determines the porosity in the sintered products. The sintered strength implies on the degree of particles bonding and particles arrangement at the surface of the compact, respectively. Therefore, the objective of this paper is to present the investigation of the effects of compaction temperature and sintering rate to the final properties of FeCuAl powder mass formed through uniaxial die compaction method.

2. Materials and Methods

The study consists of four consecutive steps, i.e, (i) feedstock preparation, (ii) green compact generation by powder forming through uniaxial die compaction method, (iii) pressureless sintering, and (iv) sintered sample characterization. Iron (Fe) powder ASC 100.29 of 20-180 μm particle size range was used as main powder constituent. The composition of copper (Cu) was 7.5 wt%, aluminum (Al) was 0.5% whereas 0.1 wt% of zinc stearate ($Z_n(C_{18}H_{35}O_2)_2$) was added as inter-particle lubricant, and 0.1 wt% of fine carbon powder was used as additive. The main powder constituent and the other elemental powders were mixed mechanically for 60 minutes. The prepared feedstock was filled into the rectangular shape die cavity, then all the die assembled together with the powder mass was heated up to the desired temperature and kept for 30 minutes for uniform heating of the powder mass and the die assembly. The green samples were generated by compacting the powder mass at three different temperature, i.e., 30°C, 150°C, and 200°C through simultaneous upward and downward axial loadings of 425 MPa.

The green compacts were inspected visually and the defect-free ones were subsequently sintered using custom made argon gas fired sintering furnace (Model: HT3-1400-SIC, S/N: LT007). All the samples were sintered at 800°C, at three different sintering rates, i.e., 5°C/min, 10°C/min, and 15°C/min for a holding time of 60 min. Sintered samples were characterised for their physical, electrical, and mechanical properties as well as their microstructure was evaluated. The density of the products were calculated from the dimension and mass of the final products. The dimensional measurement was conducted by using a digital Vernier calliper (Brand: Mitutoyo, Model: CD-6"CS, S/N: 04171546, Accuracy: +/- 0.001 in) whereas the flexure stress was measured through three point bending test machine (Brand: Instron, Model: Instron 3365, S/N: SAA61569) following the standard ASTM E290-09. The electrical resistivity was measured by a digital multi-meter (Brand: Fluke, Model: Fluke 115, S/N: 28341190ws). The image of the fractured surface of each sample was captured through scanning electron microscopy (Brand: JEOL, Model: JSM- 6010PLUS/LA).

3. Results and Discussion

The relative density of final products compacted at different temperatures and sintered at different rates are depicted in figure 1 whereas the volumetric expansion of the same samples during sintering are presented in figure 2. It is clear from figure 1 that samples compacted at 30°C obtained the lowest relative density regardless of sintering rate. The samples compacted at 150°C were found to obtain the highest relative density. The relative density was found to decrease at the sample compacted at 200°C.

These phenomena are inline with volumetric expansion (figure 2) since the density is the ratio between the mass and the volume of the sample. Higher volumetric expansion caused the relative density of the samples to be lower. Deformation of metal powder mass at elevated temperature can be considered as a thermally activated process [5]. The powder particles soften and slide to fill the inter-particle voids hence the powder mass densifies easily compared to the compaction at room temperature [6]. The higher density obtained by the samples formed at 150ºC might also be due to the use of zinc stearate as lubricant. Since the compaction temperature, i.e., 150°C was above the melting temperature of zinc stearate, i.e., 120-130°C, the zinc stearate melted during compaction, which probably reduced the friction among the powder particles [7]. Attainment of lower density for the compaction at 200°C is believed to be due to the thermal expansion of powder particles, which might cause the higher residual stress in green compacts hence volumetric expansion became bigger.

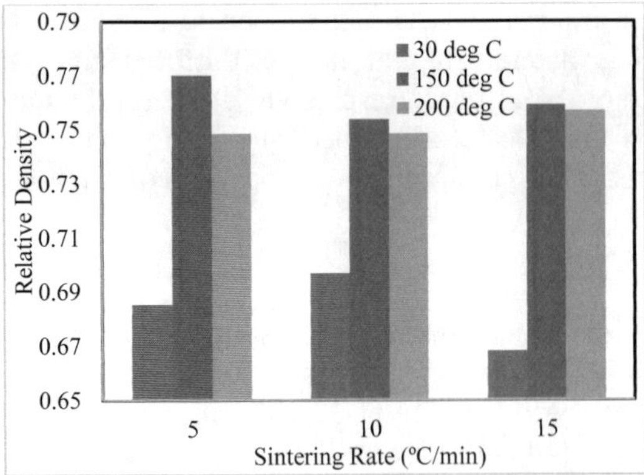

Figure 1. Relative density of sintered sample

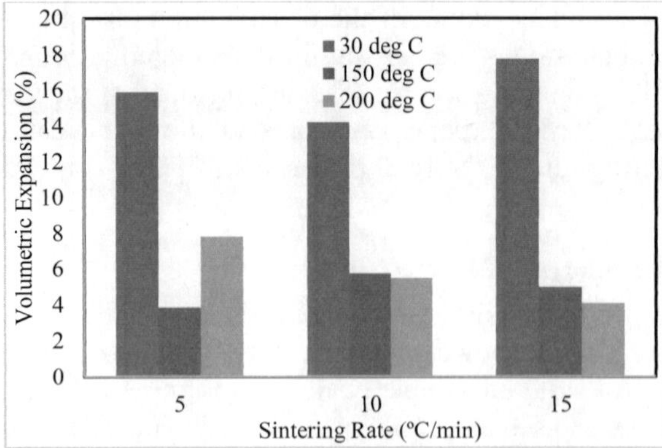

Figure 2. Volumetric expansion of sintered sample

It is also interesting to highlight that the highest density is obtained by the samples sintered at slow rate, i.e., 5ºC. Slow sintering rate caused the powder particles to form perfect necking among them which is manifested by smaller swelling [8] hence density did not drop much, i.e., less volumetric expansion (figure 2).

Electrical resistivity and bending strength are presented in figures 3 and 4. Highest electrical resistivity is found to be obtained by the sample formed at 200ºC and sintered at a rate of 5ºC. Higher electrical resistivity means lower electrical conductivity means it is difficult for the product to pass electrical current. Higher electrical resistivity is mainly due to the presence of interconnected pores or cracks inside the product.

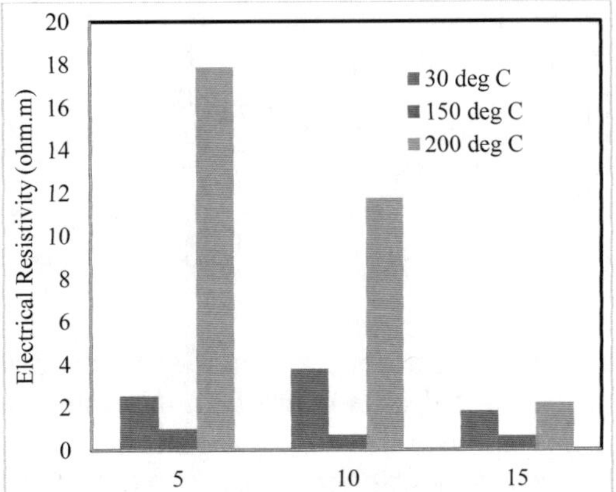

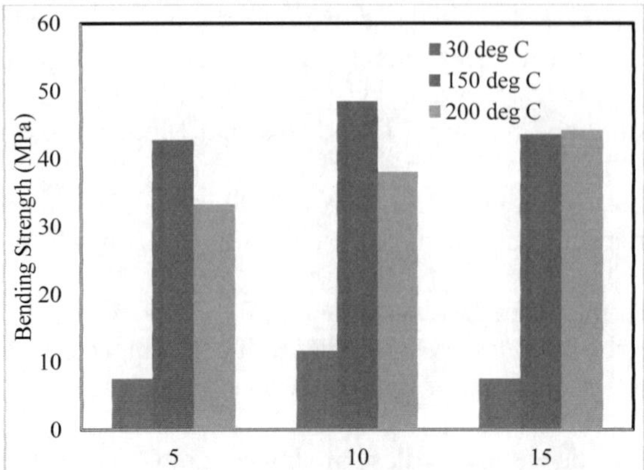

Figure 3. Electrical resistivity of sintered sample

Figure 4. Bending strength of sintered sample

Microstructures of sintered products are depicted in figure 5 where the microstructure of sample formed at 150ºC and sintered at a rate of 10ºC/min was found to be better (figure 5(b)) compared to the other products formed at different temperature and sintered at different rate. Less presence of grain growth is observed at that sample, which is inline with the strength of the sample. Slow sintering rate may cause the grain growth in metal powder mass during sintering which lessens load bearing capability of the products. Faster sintering rate also causes the initiation of micro-cracks inside the products due to the thermal shock hence lower down the strength of the products [9].

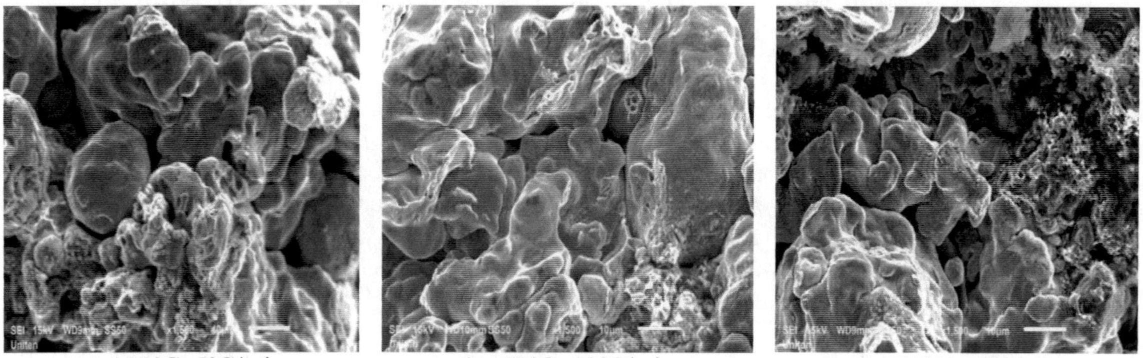

(a) 150ºC, 5ºC/min (b)150ºC, 10ºC/min (c) 150ºC, 15ºC/min

Figure 5. Microstructures of final products formed at 150ºC and sintered at different rates

4. Conclusion

The effects of forming temperature and sintering rate in producing bar shaped products from FeCuAl powder mass was investigated. The results revealed that the suitable forming temperature was 150ºC and the suitable sintering rate was 10ºC/min, which produced sintered product with higher relative density, lower volumetric expansion and electrical resistivity, higher bending strength and better microstructure.

Acknowledgement

This research is financially supported by Ministry of Higher Education (MOHE) Malaysia through the research project 20140117 FRGS.

References

[1] Yoon E Y, Lee D J, Ahn D H, Lee E S and Kim H S 2012 *J. Mater. Sci.* **47** 7770-7776
[2] Kipouros G J, Caley W F and Bishop D P 2006 *Metall Mater Trans A* **37** 3429-3436
[3] Rahman M M, Rosli M H and Namasivayam D 2016 *Int. J. Automot. Mech. Eng.* **13** 3168-3177
[4] Mahdi A S, Mustapa M S, Iajis M A and Rashid M W A 2016 *J. Eng. Appl. Sci.* **11**
[5] Rahman M M, Ariffin A K and Nor S S M 2009 *Appl. Math. Modelling* **33** 4031-4048
[6] Rahman M M, Ariffin A K and Nor S S M, Rahman H Y 2011 *Mater. Des.* **32** 447-452
[7] Rahman M M, Nor S S M, Rahman A T and Rahman H Y 2011 *ASM Sci. J.* **5** 11-18
[8] Cristofolini I, Menapace C, Cazzolli M, Rao A, Pahl W and Molinari A. 2012 *J. Mater. Proc. Technol.* **212** 1513-1519
[9] Fang Z Z, Wang H and Kumar V 2017 *Int. J. Refract. Metals Hard Mater.* **62** 110-117

IOP Conf. Series: Materials Science and Engineering **290** (2017) 012008 doi:10.1088/1757-899X/290/1/012008

The role of tin and magnesium in assisting liquid phase sintering of aluminum (Al)

Nur Ayuni Jamal[1], Farazila Yusof[2], Yusilawati Ahmad Nor[3], Maizatulnisa Othman[1], Khalisanni Khalid[4] and Muhamad Nazarudin Zakaria[5]

[1]Manufacturing and Materials Engineering (MME) Department, Kulliyyah of Engineering, International Islamic University Malaysia, P.O. Box 10, 50728, Kuala Lumpur, Malaysia
[2]Centre of Advanced Manufacturing and Materials Processing (AMMP), Faculty of Engineering, University Malaya, 50603, Kuala Lumpur, Malaysia
[3]Biotechnology Engineering Department, Kulliyyah of Engineering, International Islamic University Malaysia, P.O. Box 10, 50728, Kuala Lumpur, Malaysia
[4]Food and Agriculture Analysis Laboratory Program, Technical Service Centre, Malaysian Agricultural Research and Development Institute (MARDI), 43400 Serdang, Selangor, Malaysia
[5]Bio-composite Technology Programme, School of Industrial Technology, Faculty of Applied Sciences, University Technology MARA, 40450, Shah Alam, Selangor, Malaysia of Manufacturing and Materials Engineering, International Islamic University Malaysia (IIUM), Jalan Gombak, 53100 Kuala Lumpur, Malaysia

Email: ayuni_jamal@iium.edu.my

Abstract. This study aims to investigate the effect of tin (Sn) and magnesium (Mg) on the sintering response of sintered Al. Although this topic has been extensively reported, details on the combined effect of Sn and Mg that function as sintering additives are still limited. The current study discusses the effect of the combined use of Sn and Mg to assist aluminium (Al) in liquid phase sintering via the powder metallurgy technique. The results demonstrated that the densities of sintered Al increased from 2.5397 to 2.575 g/cm^3 as the Sn content increased from 1.5 to 2.5 wt. % respectively. Accordingly, the physical characteristics of sintered Al were transformed from black to silver, which confirmed the reduction in the oxygen content (oxide layer reduction) from 0.58 to 0.44 wt. % respectively. Additionally, the microstructure of the resultant sintered Al demonstrated that effective wetting by Sn addition was obtained at its maximum content of 2.5 wt. % with a greater micro pores reduction and better metallurgical bonding between Al particles. Therefore, the introduction of different Sn content, along with Mg element, was found to further improve the sintering response of the resultant sintered Al that consequently improved its densities and physical characteristics.

1. Introduction

Aluminium (Al) and its alloys have been extensively exploited in the automotive and aerospace industries due several prominent properties including low density, low melting temperature, high corrosion resistance and excellent electrical and thermal properties [1, 2]. Various processing techniques, including casting and powder metallurgy, have been generally utilized to fabricate Al and its alloys [3, 4]. Owing to the efficient materials utilization, greater control of starting materials ratio, and capability to produce complex products with uniform microstructure, the latter process has been widely applied in fabricating Al and its alloys [3-5]. In a recent study by Mahesh et al. [4], titanium nitride (TiN) reinforced Al composites with a maximum compressive strength of 150 MPa have been successfully fabricated using the powder metallurgy technique. However, the processing of Al and its alloys is considered a challenging task, attributable to the existence of thermodynamically stable layers (1-2 nm thickness) of aluminium oxide (Al_2O_3) on the powder surfaces [6, 7]. As such, effective metallurgical bonding among Al particles cannot be obtained during sintering, since they hinder the enhancement of the physical and mechanical properties of the resultant Al and its alloys. Liquid phase sintering by means of sintering additives addition has been commonly introduced to overcome the abovementioned problem [6-8]. Magnesium (Mg) and tin (Sn) elements are broadly employed as sintering additives in order to assist Al in liquid phase sintering [6-8]. MacAskill et al. [8] and Sercombe et al. [6] have been reported that the utilization of sintering additives (Mg and Sn) enhanced the sintering response of sintered Al, which consequently improved its tensile strength up to 118 MPa and 300 MPa respectively.

Despite progressive work on the sintering response of Al and its alloys via the powder metallurgy technique, further investigation is required, especially with different formulations of Al and sintering additives. Therefore, the current study focused on the use of Mg and Sn elements as sintering additives in assisting liquid phase sintering of Al via the powder metallurgy method. More specifically, the effects of various Sn content and fixed Mg content on the physical characteristics, oxygen content levels, density and microstructure of sintered Al were examined. The present work will provide additional knowledge on the existing behaviour of liquid phase sintering of Al and its alloys.

2. Materials and Methods

2.1 Raw materials

Elemental Al powder (99.9% purity, NovaScientific, Malaysia), with an average particle size of 45 μm, was used as the matrix constituent, whereas elemental Mg (99.9% purity, NovaScientific, Malaysia) and Sn powders (99.5% purity, Sigma Aldrich, Malaysia) with average particle sizes of 10 and 45 μm were introduced as sintering additives to initiate the liquid phase sintering of Al. In this study, the content of Mg was fixed at 0.5 wt. %, while the Sn content varied between 1.5, 2 and 2.5 wt. % respectively.

2.2 Preparation of sintered Al compact

The sintered Al compact was prepared using the powder metallurgy method, which comprises mixing, compaction and sintering, as schematically illustrated in figure 1. The elemental powder mixture containing Al, Mg and Sn was initially mixed in a table-top ball mill with the powder-to-ball (Zirconia) ratio of 1:10, for a 12 h period. The elemental mixed powder was then cold compacted in a cylindrical die of 10 mm in diameter and 6 mm in height, at an applied pressure of 250 MPa. Next, the compacted specimen was sintered at 580 °C for 2 h under Argon ambient to obtain the pure sintered Al body. Finally, the sintered Al specimen was washed with acetone and dried in an electrical oven at 90 °C overnight to remove impurities, prior to characterization.

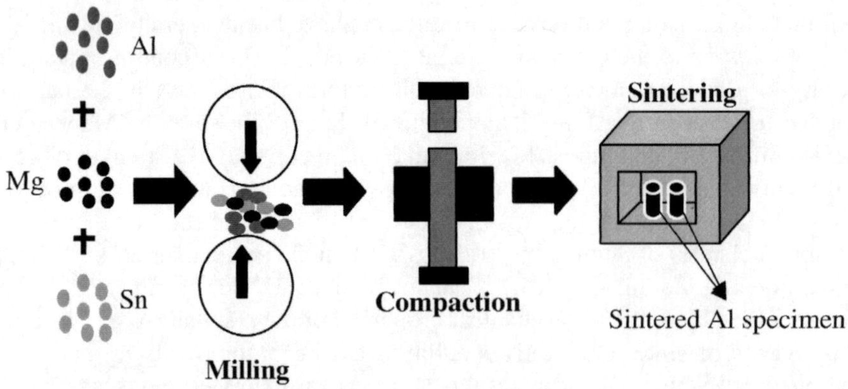

Figure 1. Schematic flow diagram of sintered Al preparation.

2.3 Characterization techniques

The morphology of the as-received starting powders and the elemental powder mixture were observed using a scanning electron microscope (SEM, Jeol JSM6500F, JEOL Ltd., Tokyo, Japan). In terms of the microscopic examination of the sintered specimen, a cross-section of the specimen was prepared. The microstructure of the cross-section was viewed by SEM. In conjunction with SEM, the distribution of Sn element as well as oxygen content analysis using energy dispersive spectroscopy (EDS) was carried out to identify the remaining content of Sn after sintering and oxygen at every stage. The phase transformation of elemental powder mixture and sintered Al was monitored using X-ray diffraction analysis (XRD; PANanalytical empyrean 1032, Eindhoven, Netherlands) in the 2θ range of 20-80°. The density of the sintered Al was calculated using the Archimedes principle, as described in ASTM C830-93.

3. Results and Discussion

3.1 Morphology characterization of starting materials

Figure 2 (a-d) illustrates the SEM images of each starting powder and elemental powder mixture. Al and Mg powders were found to be mostly spherical in shape, with some irregular-shaped Al particles, as seen in figure 2 (a-b). Conversely, Sn powder was predominantly irregular in shape, as shown in figure 2 (c). Since Al is a passive material due to the presence of thermo-chemically stable Al_2O film, it is important to interrupt this film to form a strong metallurgical bonding between the Al particles. Hence, in this study, Sn powder was added to increase the fluidity of Al during sintering, while Mg powder was served to enhance the segregation of Sn particles on the Al surface [6-8]. On the other hand, a uniform distribution of elemental powder mixture without any powder agglomeration was observed after a sufficient mixing time of 12 h, and a compaction pressure of 250 MPa, as shown in figure 2 (d). Subsequent to the mixing process, the particles of the elemental powder mixture were found to possess a lamellar structure with a nearly constant powder particle size and shape. Repeated welding, fracturing and re-welding of the powder particles was the reason for such a phenomenon during the mixing stage using the ball milling technique [9].

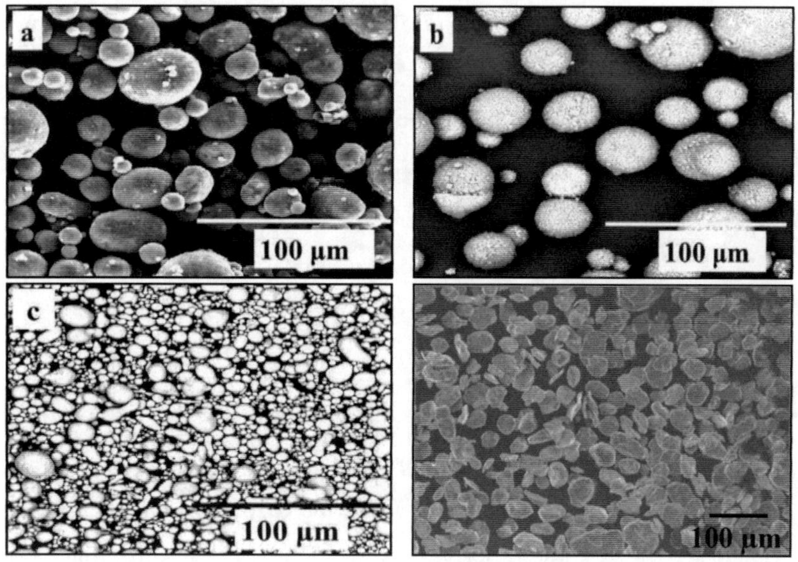

Figure 2. SEM micrographs of a) Al powder, b) Mg powder, c) Sn powder and d) Elemental powder mixture after 12 h mixing.

3.2 Physical characteristic of sintered aluminum specimen

Figure 3 (a-c) shows the physical characteristics of sintered Al specimen at different Sn content. It can be seen that the black colour of the sintered Al body, which is usually attributed to oxidation (i.e. higher coating of Al_2O_3 film on the surface of sintered Al body), was reduced with increasing the Sn content. The highest addition of Sn content (2.5 wt. %) resulted in a silver-like colour that replicated the colour of the starting Al powder. This shows that the film coating of Al_2O_3 was greatly disrupted with the addition of a maximum Sn content of 2.5 wt. %. The EDS analysis tabulated in table 1 further confirmed a greater reduction in oxygen content of the resultant sintered Al from 0.58 to 0.44 wt. % with increasing the Sn content from 1.5 to 2.5 wt. % respectively.

Figure 3. Physical characteristic of sintered Al specimen at different Sn content of a) 1.5 wt. %, b) 2 wt. % and c) 2.5 wt. %.

Table 1. Oxygen content reading of elemental powder mixture and sintered Al at various Sn content values from EDS analysis. Data are presented in mean ± standard deviation.

Tin content (wt. %)	Oxygen Content (wt. %)
Elemental powder mixture	0.22±0.10
1.5 (sintered at 580 °C)	0.58±0.12
2 (sintered at 580 °C)	0.51±0.13
2.5 (sintered at 580 °C)	0.44±0.11

3.3 Scanning electron microscopy (SEM) analysis

Figure 4 (a-c) demonstrates the SEM micrographs of the cross section of sintered Al at 1.5, 2 and 2.5 wt. % of Sn content respectively. At low Sn content of 1.5 wt. %, poor sintering quality was observed due to minimal inter-particle necking, profusion of particle boundaries and lack of micro pores rounding (isolated pores), as shown in figure 4 (a). This indicates that the lowest liquid level was found to be inadequate to fill all micro pores, thus preventing efficient metallurgical bonding between Al particles. As such, incomplete elimination of pores could not be obtained due to the insufficient of liquid formation, poor wettability and limited spreading of the liquid phase [10]. Theoretically, Sn melts during sintering and fills the gaps between Al particles and wetting their surfaces [8, 11]. In order to realize this theory, the addition of 0.5 wt. % of Mg was introduced in the current study to break up the Al_2O_3 layer and promote metal to metal contact between Al particles prior to liquid phase sintering[8, 11]. On the other hand, the sintering response was found to improve with the addition of 2 and 2.5 wt. % of Sn, as evident in figure 4 (b-c). Furthermore, the presence of particle boundaries also became less noticeable with enhanced inter-particle necking and greater micro pores reduction. It is postulated that a higher liquid level is sufficient to fill the pores due to the effect of Sn wetting on the surface of Al particles that improved the metallic contact among the Al particles [7, 8]. In addition, Sn particles (bright colour) were identified to be homogenously distributed within the microstructure, showing the effective wetting effect of Sn particles on the Al phases that enhanced the densification of sintered Al, as illustrated in figure 5 [8-11]. EDS analysis further confirmed the uniform distribution of Sn element throughout the Al structure as observed in figure 5 (a-c). In relation with this, the number of Sn distribution was found to increase with increasing Sn content. Moreover, this criteria also imply that the associated phase was transformed into liquid during sintering that was able to wet the Al surfaces, and concurrently, spread throughout the compact [8].

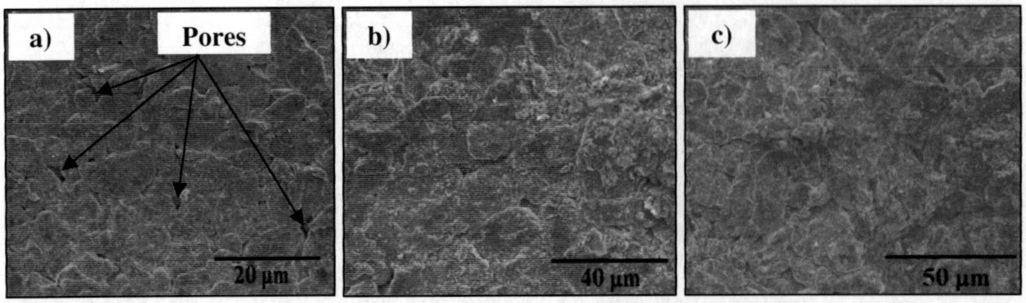

Figure 4. Microstructure of sintered Al structure at Sn content of a) 1.5 wt. %, b) 2 wt. % and c) 2.5 wt. %.

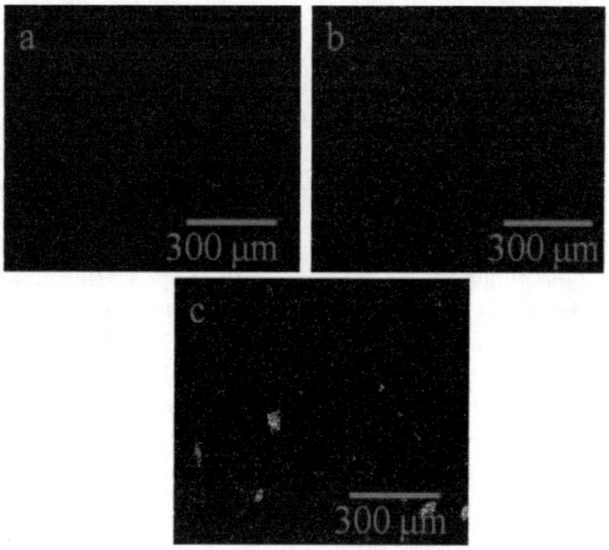

Figure 5. EDS analysis of sintered Al with Sn content of a) 1.5 wt.%, b) 2 wt.% and c) 2.5 wt.%.

3.4 X-ray diffraction (XRD) analysis

The XRD patterns of sintered Al with different Sn content values are revealed in figure 6 (a-c). It is worthy to note that the Al rich phase (JCPDS card number 04-0787) was majorly found in the XRD pattern of all of the samples, characterized by the (111), (200), (220) and (311) diffraction peaks at 38.87°, 45.42°, 67.16° and 78.54° respectively. No additional peaks were seen in the XRD patterns of all the sintered Al, suggesting the absence of intermetallic phases formation during sintering. On the other hand, the intensity of Sn peak increased with Sn contents, confirming a higher Sn content in the sintered Al. As seen in figure 6 (a-c), the peaks intensities for the sintered Al at Sn content of 2 and 2.5 wt. % were found to be higher as compared to the Sn content of 1.5 wt. %, demonstrating the formation of higher crystalline Al during sintering. However, the presence of Mg peak could hardly be detected, possibly due to the small content of Mg powder employed (0.5 wt.%) during the fabrication process [12].

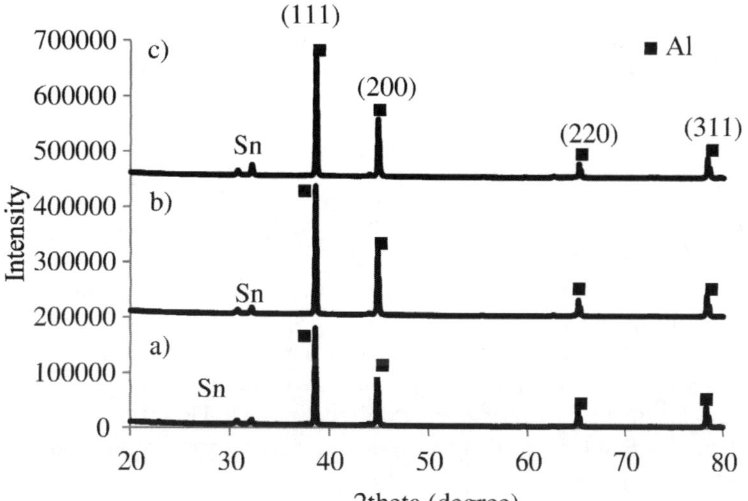

Figure 6. XRD patterns of sintered Al with Sn contents of a) 1.5 wt.%, b) 2 wt.% and c) 2.5 wt.%.

3.5 Sintered density of aluminum

Figure 7 shows the sintered density of compacted Al at various Sn content values. It can be observed that the sintered density of compacted Al increased as the Sn content increased. The sintered densities of compacted Al increased from 2.5397 to 2.575 g/cm^3 when the Sn content was increased from 1.5 wt. % to 2.5 wt. % respectively. It is clear that the Sn addition promoted the densification of sintered Al. This can be attributed to the pores reduction and particle boundaries disappearance at higher Sn content values, as confirmed by the SEM images in Figure 4 (b-c). In other words, an increased in the sintered density of compacted Al can also be attributed to the fact that higher Sn content produced higher liquid content that can fill all pores with sufficient wettability and spreading of the liquid phase. Su et al. (2013) also reported similar finding in which higher formation of liquid phase was observed to increase the density of the sintered Al due to the porosity reduction. Moreover, considering higher vacancy binding energy and diffusivity of Sn in Al, the vacancies (pores) could be bound with Sn elements, thereby enhanced the sintered density of compacted Al as the liquid phase could persist longer time with increasing Sn content [6-8]. Similarly, Sercombe and Schaffer [6] stated that the formation of liquid phase was slightly slower with the addition of Sn element but larger liquid content could be produced especially greater Sn content that tend to remain for longer time.

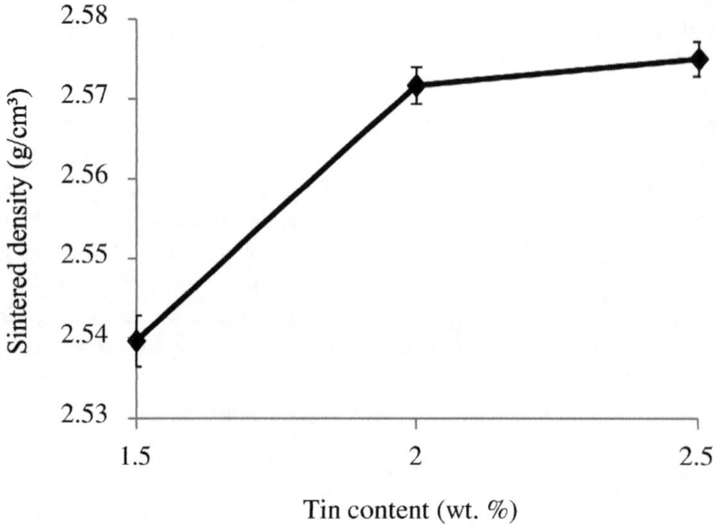

Figure 7. Sintered density of compacted Al as a function of different Sn content.

4. Conclusion

It can be concluded that the sintering response of the resultant sintered Al highly depends on the combined use of Mg and Sn that serve as sintering additives. In line with this, the physical characteristics (black to silver colour transformation), sintered density (2.575 g/cm^3) and oxygen content (0.44 wt. %) of the resultant sintered Al improved greatly particularly at a maximum Sn content of 2.5 wt. %. This can be attributed to the effective wetting effect of Sn which was induced by the Mg addition, resulting in micro pores reduction, less formation of particle boundaries and good metallurgical bonding among Al particles, as revealed by the SEM micrographs. The addition of both Mg and Sn have facilitated the sintering response of the resultant sintered Al, which consequently enhanced its densities, physical characteristics and microstructure. Therefore, maximum content of 2.5 wt. % of Sn could be considered as an ideal sintering additive in promoting liquid phase sintering of compacted Al under the current experimental conditions.

Acknowledgement
The current work was financially supported by a research initiatives grant (RIGS) of International Islamic University Malaysia (IIUM) under project number of 16-091-0255, which is appreciatively acknowledged.

References

[1] Aliyu I K, Saheb N, Hassan S F and Al-Aqeeli N 2015 *Metals* **5** 70-83
[2] Gokce A and Findik F *Scientific Research and Essays* 2011 **6** 4378-4388
[3] Qiu F, Gao X, Tang J, Gao Y Y, Shu S L, Han X, Li Q and Jiang Q C 2017 *Metals* **7** 3-8
[4] Mahesh L and Reddy J S 2017 *Mechanics and Mechanical Engineering* **21** 29-36
[5] Reddy S P, Ramana B and Reddy A C 2010 *International Journal of Materials Science* **5** 777-783
[6] Sercombe T B and Schaffer G B 1999 *Materials Science and Engineering* **A268** 32– 39
[7] Katsuyoshi K, Tachai L, Thotsaphon T and Atsushi K 2007 *Transactions of JWRI* **36** 29-33
[8] MacAskill I A, Hexemer Jr R L, Donaldson I W and Bishop D P 2010 *Journal of Materials Processing Technology* **210** 2252–2260
[9] Suryanarayana C 2001 *Progress in Materials Science* **46** 1-84
[10] Su S S, Changa I T H, Kuo W C H 2013 *Materials Chemistry and Physics Journal* **139** 775-782
[11] Katsuyoshi K, Tachai L, Thotsaphon T and Atsushi K 2007 *Transactions of JWRI* **36** 29-33
[12] Jha N, Mondala D P, Majumdar J D, Badkul A, Jha A K and Khare A K 2013 *Materials and Design* **47** 810–819

IOP Conf. Series: Materials Science and Engineering **290** (2017) 012009 doi:10.1088/1757-899X/290/1/012009

Effects of sintering time and temperature to the characteristics of FeCrAl powder compacts formed at elevated temperature

M M Rahman[1], H Y Rahman[1], M A A Awang[1] and I Sopyan[2]

[1]Department of Mechanical Engineering, Universiti Tenaga Nasional, Putrajaya Campus, Jalan IKRAM-UNITEN, 43000 Kajang, Selangor, Malaysia
[2]Department of Manufacturing and Materials Engineering, Kulliyyah of Engineering, International Islamic University Malaysia, PO Box 10, 50728 Kuala Lumpur, Malaysia

Email:mujibur@uniten.edu.my

Abstract. This paper presents the outcomes of an experimental investigation on the effect of sintering schedule, i.e., holding time and temperature to the final properties of FeCrAl powder compacts prepared through uniaxial die compaction process at above room temperature. The feedstock was prepared by mechanically mixing iron powder ASC 100.29 with chromium (22 wt%) and aluminium (11 wt%) for 30 min at room temperature. A cylindrical shape die was filled with the powder mass and heated for one hour for uniform heating of the die assembly together with the powder mass. Once the temperature reached to the setup temperature, i.e., 150°C, the powder mass was formed by applying an axial pressure of 425 MPa simultaneously from upward and downward directions. The as-pressed green compacts were then cooled to room temperature and subsequently sintered in argon gas fired furnace at a rate of 5°C/min for three different holding times, i.e., 30, 60, and 90 min at three different sintering temperatures, i.e., 800, 900, and 1000°C. The sintered samples were characterized for their density, electrical resistivity, bending strength, and microstructure. The results revealed that the sample sintered at 1000°C for 90 min achieved the better characteristics.

1. Introduction

Pure metals are typically weak and soft at high temperature due to their malleability and ductility. By imparting other elements into a pure metal, a coherent mixture termed as alloy can be formed to enhance its properties [1-3]. FeCrAl alloy is used as engineering materials due to its high strength, improved form stability at high temperature, corrosion oxidation/resistance, and longer high temperature life. This alloy is formed through foundry or mechanical alloying process. Foundry process requires huge amount of heat to melt all the metals whereas mechanical alloying method requires the elements in powder form to be mixed by a high energy ball mill for a long time. Both methods require further processing to produce end products, which are time and energy consuming hence expensive. Another possibility is powder compaction process, which offers an approach to produce end products with minimal processing steps. However, research activities are lacking on the method of alloy forming through this route.

Sintering is a heat treatment process for bonding particles together into a coherent, predominantly solid structure via mass transport events. Such bonding improves the strength and other engineering properties of powder compacts. The standard way of performing the sintering operation is to subject the parts to a high temperature in a controlled atmosphere using a sintering furnace. The process is known as free sintering since no mechanical stress is applied on the parts. The effect of sintering is the change of inter-particle contact, which begins with rapid growth among particles because of the cold welding taken place in the green compact. At this stage, the void volume is reduced and finally as the surface tension and diffusion continue, the spheroidization of isolated pores occurs and it results in a relatively homogenous component. Correct sintering is of paramount importance to the powder metallurgy process to ensure not only the development of the strength needed for the part to fulfil its intended role as an engineering component, but also that the dimensions of the part are correct.

2. Materials and Methods

The experiments consist of four consecutive steps, i.e., (i) feedstock preparation, (ii) green sample generation, (iii) sintering, and (iv) product characterization. Iron powder ASC 100.29 having particle size range of 20 -180 μm was used as main powder constituent since this type of powder has been in use in most of the powder compaction industries [4]. Two other elemental powders, i.e., chromium (Cr) and aluminum (Al) were used as alloying element. The composition of iron powder was 67 (wt%) whereas the chromium was 22 (wt%) and the rest was aluminium powder. All of these elemental powders were mixed mechanically through a low speed (30 rpm) powder mixer for 30 min at room temperature [5]. A lab-scale uniaxial die compaction rig was designed and fabricated to shape metal powder mass at above ambient temperature. The cylindrical shape die cavity of 20 mm diameter was filled with the as-prepared feedstock. The top punch was brought closer to the upper level of the powder mass. The die assembly together with the powder mass was heated to 150ºC and kept for 30 min for the uniform heating of the die assembly and the filled powder mass.

The powder mass was subsequently compacted by applying simultaneous downward and upward loading of 425 MPa. Once the desired loading is achieved, the top punch was kept at that position for a while for the settlement of the powder particles and brought back to the initial position which enabled the spring-back or release of residual stress from the powder compact [6]. The as-pressed powder compact termed as green compact was subsequently ejected from the die cavity by means of bottom punch. All the green samples were inspected visually for any defects or cracks. The defect-free samples were then sintered in a custom made argon gas fired furnace (Model: HT3-1400-SIC, S/N: LT007) at a rate of 5ºC/min for three different holding times, i.e., 30, 60, and 90 min at three different sintering temperatures, i.e., 800, 900, and 1000ºC.

Dimensions of the green as well as the sintered samples were measured to analyse the shape distortion from where the densities (green and sintered) were calculated. The dimensional measurement was conducted through a digital Vernier calliper (Model: CD-6"CS, S/N: 04171546) with an accuracy of +/- 0.01in. Electrical resistivity of the green as well as sintered samples were measured through a digital multimeter (Model: Fluke 115, S/N: 28341190ws). The purpose of electrical resistivity measurement was to detect the presence of any interconnected pores or cracks. The flexure stress was measured through three point bending test machine (Brand: Instron, Model: Instron 3365, S/N: SAA61569) following the standard (ASTM E290-09). Each sample was hold tightly at both ends, force was applied at the middle of the sample until it was broken down. The image of the fractured surface of each sample was captured through scanning electron microscopy (Brand: JEOL, Model: JSM- 6010PLUS/LA).

3. Results and Discussion

The difference in density from green to sintered product is presented in figure 1. The highest density drop was found at the sample sintered at 1000ºC for 30 min, which is 18.19%. It can clearly be observed in figure 1 that the density difference increased when the sample was sintered at high temperature. All the samples sintered at 1000ºC experienced higher volumetric expansion compared to the samples sintered at 900ºC and 800ºC for the sample holding time. Sintering of powder compacts at high temperature might cause the grain growth, which caused the swelling of compacts after sintering [7-8]. Another possible reason is the addition of aluminum, which has a lower melting temperature (667°C) compared to the other elements, i.e., iron and chromium. Mixed sintering might occur, which caused the significant swelling of the compacts after sintering.

Electrical resistivity difference from green to sintered is presented in figure 2 from where it is clear that samples sintered at 800°C and 1000°C experienced the highest electrical resistivity difference. Higher electrical resistivity means electrical current cannot flow easily through the samples. Higher electrical resistivity difference means micro-cracks or interconnected pores might be formed during sintering.

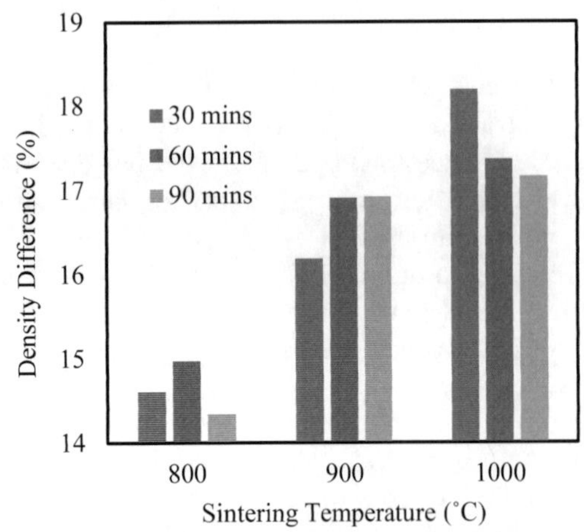

Figure 1. Density difference after sintering.

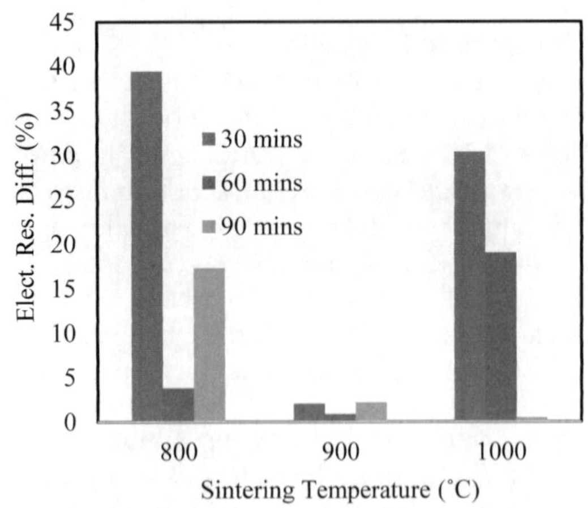

Figure 2. Electrical difference after sintering.

Bending strength or flexure stress of sintered samples is presented in figure 3. It can clearly be observed that samples sintered at 1000°C for 90 min obtained the highest bending strength. This finding is inline with electrical resistivity change (figure 2) where the sample sintered at 1000°C for 90 min experienced the negligible electrical resistivity change. This phenomenon might be due to the longer sintering time and higher sintering temperature caused the perfect sintering [9-10] of the samples. The microstructures of the samples sintered at 800°C, 900°C, and 1000°C for 90 min are shown in figures 4-6. It is clear from these figures (figures 4-6) that sample sintered at 90 contains less interconnected pores compared to the other samples.

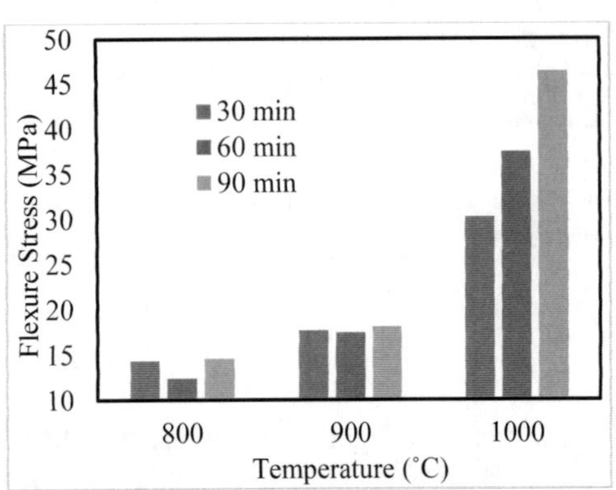

Figure 3. Bending strength of sintered samples.

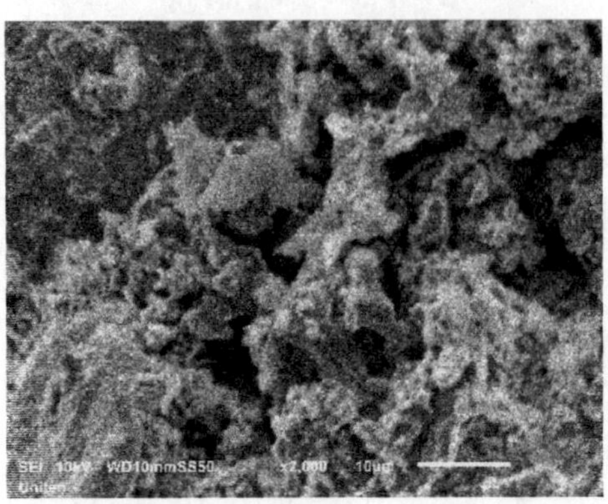

Figure 4. SEM image of sample sintered at 800°C for 90 min.

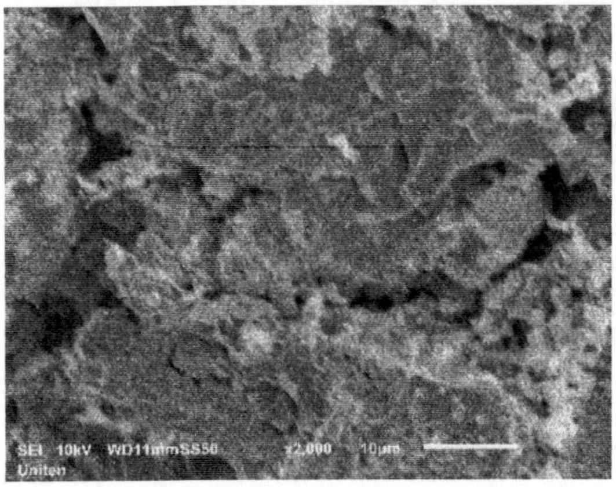

Figure 5. SEM image of sample sintered at 900°C for 90 min.

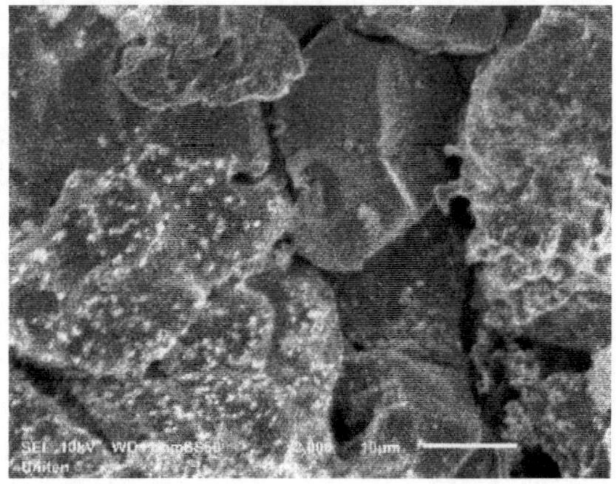

Figure 6. SEM image of sample sintered at 1000°C for 90 min.

4. Conclusion
The effects of sintering temperature and holding time in producing solid cylindrical shape products from FeCrAl powder mass was investigated. The results revealed that the suitable sintering temperature was 1000°C and the holding time was 90 min, which produced stronger sintered product with better microstructure.

Acknowledgement
This research is financially supported by Ministry of Higher Education (MOHE) Malaysia through the research project 20140117 FRGS.

References
[1] Lim X 2016 *Nature* **533** 306-307
[2] Rahman M M, Namasivayam Dand Rahman H Y *2015 Malaysian Journal of Microscopy* **11** 74-78.
[3] Akca E and Gursel A 2015 *Periodicals of Engineering and Natural Sciences* **3** 15-27
[4] Mondol S, Alam T, Banerjee R, Kumar S and Chattopadhyay K 2017 *Materials Science & Engineering A* **687** 221-231
[5] Rahman M M, Ariffin A K and Nor S S M 2009 *Applied Mathematical Modelling* **33** 4031-4048
[6] Demétrio K B, Klein A N, Schaeffer L, Consoni D R, Martinelli A E and Bendo T 2013 *Materials Research* **16** 1030-1038
[7] Rahman M M and Talib A A A 2015 *Materials Today Communications* **4** 42-49
[8] Steinmetz P, Ian GW, Gerry M, Alain G, Bernard P and Renaud P 2004 *Materials Science Forum* **461-464** 455-462
[9] Rahman M M, Rosli M H and Namasivayam D 2016 *International Journal of Automotive and Mechanical Engineering* **13** 3168-3177
[10] Rahman M M and Yogaswerarow S R 2016 *Advanced Materials Research* **1133** 264-268

IOP Conf. Series: Materials Science and Engineering **290** (2017) 012010 doi:10.1088/1757-899X/290/1/012010

Electrodeposition of zinc antimony alloy thermoelectric materials

A L N Hairin[1], M N Romainor[1], R Othman[2], F D M Daud[1]

[1]Department of Manufacturing and Materials Engineering, International Islamic University Malaysia (IIUM), Jalan Gombak, 53100 Kuala Lumpur, Malaysia
[2]Department of Science in Engineering, International Islamic University Malaysia (IIUM), Jalan Gombak, 53100 Kuala Lumpur, Malaysia

Email:assayidatul_laila@iium.edu.my

Abstract. Zinc antimonite, Zn_4Sb_3 is a promising thermoelectric material because of its high thermoelectric performance and abundance of Zn and Sb in nature. Thus, in this study, samples of Zn-Sb alloy were prepared using electrodeposition method because of its simple experimental set-up, which also carried out in the room temperature. From the XRD results, all samples deposited exhibit Zn-Sb alloy compositions. The best results were S1 and S3 as they had dominant peaks that showed the crystal lattice of Zn_4Sb_3. From the SEM images, the surface morphology of Zn-Sb alloy deposited samples showed were all-irregular, course and rough structures. While, the atoms arrangement of the deposited samples were all flowery-like. Based on physical properties characterization, the best samples; S1 (0.1M $ZnCl_2$-0.1M $SbCl_3$, 100mA, 120min) and S3 (0.1M $ZnCl_2$-0.1M $SbCl_3$, 50mA, 120min), were selected and investigated their thermoelectric performances; electrical conductivity and Seebeck coefficient, to determine their power factor, PF. Heat capacity of the samples was also examined to relate it with thermal conductivity of Zn-Sb deposited samples. For thermoelectric performance, S1 obtained power factor of 1.37×10^{-7}V/K. Ω.cm at 102°C with the Seebeck coefficient of 181μV/K. While as for S3, the power factor was 1.58×10^{-7}V/K. Ω.cm with Seebeck coefficient of 113μV/K at 101°C. From DSC analysis, it showed that S3 obtained higher C_p than S1. C_p for S3 was 46.8093mJ/°C while S1 was 38.3722mJ/°C.

1. Introduction

Studies of thermoelectricity were rather active at the time of world wars. Even after the world wars era end, the studies still carried out for use in valuable technologies, which mainly based on cooling and power generation for application in the military and civilian uses. However, due to the beneficial of some individual in the political and economic, the development of thermoelectric devices were very secretive and made it very difficult to advance to another stage of studies. This was because of slow publication and information sharing especially between Eastern European and Western countries. Despite the difficulties of information sharing, the development of thermoelectricity had significant achievement by the 1950's when cooling application demonstrates the ability to cooling from ambient temperature to the temperature below 0°C and the efficiencies of power generator had achieved 5%. These successes in the development of thermoelectricity ultimately lead to some feasible industries.

There were many speculations that assumed thermoelectricity capable of replacing refrigeration and conventional heat engines, which lead to big interest in these technologies.

A good thermoelectric material is the one that has a high Seebeck coefficient, a good electrical conductivity and a low thermal conductivity. These parameters then are related into the dimensionless figure of merit, $ZT=S^2\sigma T/\kappa$ where σ is electrical conductivity, S is Seebeck coefficient, κ is thermal conductivity and T is absolute temperature of the materials. For thermal conductivity, it is the sum of lattice thermal conductivity, κ_l and electronic thermal conductivity, κ_e ($\kappa_l + \kappa_e$). In addition to ZT, the performance of thermoelectric materials can also be characterized by power factor, PF=$S^2\sigma$. Thus, high quality thermoelectric materials will have higher PF and ZT [1–8]. Based on these two factors, among the materials that known as thermoelectric materials; bismuth telluride (Bi_2Te_3), bismuth antimony ($Bi_{1-x}Sb_x$), lead telluride (PbTe), silicon-germanium alloy, skutterudites, clathrates, boron carbide and oxides, bismuth telluride and lead telluride are regarded as high class of thermoelectric materials all the time. Despite their high dimensionless figure of merit, these two materials are toxicity, which are very harmful to the user if have direct contact and regarded as expensive materials [6-7].

Considering those factors of Bi_2Te_3 and PbTe, in this study, zinc antimony was chosen as the thermoelectric materials. Zinc antimony is greatly intoxicant and abundance than Bi_2Te_3 and PbTe. Zinc also has significantly lower in price compared to bismuth and lead [4-8]. In term of thermoelectric properties; dimensionless figure of merit and power factor, zinc antimony has a high performance in the temperature range of 450 – 650 K. Zinc antimony usually exist in two phase; β-Zn_4Sb_3 and ZnSb. β- Zn_4Sb_3 with a complex hexagonal crystal structure has been discovered to be one of the promising candidates for thermoelectric power generation applications. The figure of merit, Z of β-Zn_4Sb_3 is in the range of 450 – 650 K is comparable to both bismuth telluride and lead telluride [4-8]. In this work, zinc antimony was prepared by using electrodeposition method and the effect of deposition parameters; current, time, electrolytic bath formulation, on Zn-Sb alloy composition, and thermoelectricity properties; electrical conductivity, Seebeck coefficient and thermal conductivity, of Zn-Sb alloy were identified and investigated.

2. Experimental procedures

2.1 Materials and apparatus

In the early period of this study, a design of copper substrate which in form of flame retardant 4 (FR4) or also known as printed circuit board (PCB) was fabricated by photolithography process. For holding the substrate at fix distance with the working electrode, Zn plate (99.99% purity), a holder from acrylic board was constructed. To fit Zn plate to the holder, it cut to rectangular form with dimension of 20mm X 50mm. The center of the holder then drilled to form specific depositing area. Electrolytic bath that had been used in this study were 0.1M and 0.2M of zinc chloride, $ZnCl_2$ (EMSURE grade) and antimony chloride, $SbCl_3$ (Merck KgaA grade). $ZnCl_2$ were prepared by dissolving the weighted salt with distilled water. While for $SbCl_3$, the weighted salt was dissolved with citric acid, $C_6H_8O_7$ (HmbG grade) and sodium citrate, $Na_3C_6H_5O_7$ (Bendosen grade) with ratio of 16:49:35 in distilled water. The electrodeposition process was done using AUTOLAB PGSTAT128N machine.

2.2 Electrodeposition process of Zn-Sb alloy

The electrolytic cell consisted of zinc plate as the working electrode and copper substrate as the counter electrode. The copper substrate and Zn plate clamped to acrylic holder. The distance between the two electrodes and deposition area fixed to standardize the process for all experiments. The displacement, D between two electrodes was 20 mm and the diameter, d of the holder's windows was 11.5 mm. For this work, the electrolytic baths used were the mixture of 0.1M $ZnCl_2$-0.1M $SbCl_3$ and 0.2M $ZnCl_2$-0.2M $SbCl_3$. The deposition time varied from 30min to 120min and the current density varied from 50 mA to 100 mA as listed in table 1. The electrolyte volume maintained at 60 ml. The negative terminal of the machine connected to copper substrate and positive terminal connected to Zn plate.

IOP Conf. Series: Materials Science and Engineering **290** (2017) 012010 doi:10.1088/1757-899X/290/1/012010

Table 1. Parameters of electrodeposition of Zn-Sb in $ZnCl_2$-$SbCl_3$.

Sample	**Concentration $ZnCl_2$ (M)**	**Concentration $SbCl_3$ (M)**	**Deposition time, min**	**Working electrode**	**Current, (mA)**
1	0.1	0.1	120	Zn	100
2	0.1	0.1	60	Zn	100
3	0.1	0.1	120	Zn	50
4	0.1	0.1	60	Zn	50
5	0.2	0.2	30	Zn	100

3. Results and Discussion

3.1 XRD analysis of deposited samples

Figure 1 shows the XRD patterns for prepared samples with different molarity, current applied and deposition time during electrodeposition process. The patterns shows that, there are two group of patterns from S1 to S5. The patterns for S1 and S3 show the dominant peaks are located in the range of 36°-44° which indexing plane (223) to (006) of Zn_4Sb_3. Other than that, there is also a peak observed at 54° define the (244) Zn_4Sb_3 plane. There are also significant diffraction peak located at 70°-71° that related to Zn_4Sb_3 (360) to (238) planes as shown in figure 1. As for S2,S4 and S5, the patterns of these three samples show almost identical pattern with the major diffraction peak located around 29° which indexing the reflection from (220) planes of Zn_4Sb_3. Other than that, the peaks are all very weak, thus cannot used for evaluation. The lack of the diffraction intensity produced from these three samples may because of the samples thickness, which are very thin. Besides, scan rate of the XRD machine also may contribute to the lacking in the diffraction intensity. Thus from the diffraction patterns of these five samples, it can be seen that samples prepare for 30-60min deposition time have very weak peaks which can also be related to worst crystal quality. However as the deposition time increase, the samples shows a good patterns with several dominant peaks that indicate the lattice atom of the samples start to crystallize and form Zn_4Sb_3.

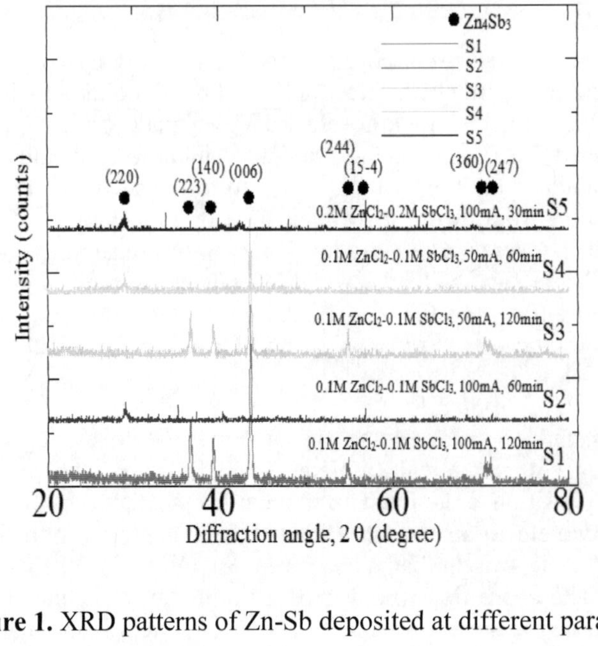

Figure 1. XRD patterns of Zn-Sb deposited at different parameters.

3.2 Surface morphology from SEM

Figure 2 shows the surface morphology of the samples prepared with different deposition parameters, S1 (0.1M $ZnCl_2$-0.1M $SbCl_3$, 100mA, 120min), S2 (0.1M $ZnCl_2$-0.1M $SbCl_3$, 100mA, 60min), S3 (0.1M $ZnCl_2$-0.1M $SbCl_3$, 50mA, 120min), S4 (0.1M $ZnCl_2$-0.1M $SbCl_3$, 50mA, 60min) and S5 (0.2M $ZnCl_2$-0.2M $SbCl_3$, 100mA, 30min) in different magnification. From the figure 3.2, samples S1, S2, S3 and S4 microstructures were very irregular, sparse and flowery-like. Crystalline grain of the samples are fuzzy and surface morphology also very rough. The roughness of the surfaces make crystalline grain almost invisible which indicate the poor crystallization. Figure 2 also indicates that as the current applied during electrodeposition process decrease, the quality of the deposited Zn-Sb alloy also increase. Thus, making S3 and S4 look denser than S1 and S2. However, when the deposition time increase from 60min to 120min, the quantity of deposited ions increase, thus, make the samples become rougher and thicker. The adhesion of the samples also decrease as the Zn-Sb deposited easily wear off from the deposited area.

Based on figure 2, as the molarity of electrolytic bath increase, the structure of Zn-Sb deposited change significantly. Although the surface morphology of S5 is also sparse and rough but flowery-like arrangement of this sample, become significant. The figures also show that the atoms arrange themselves in stacking manner, which then form the flowery-like arrangement. For all the samples, they are to be expected that the surface morphology of Zn-Sb alloy to be course and rough but the flowery-like arrangement is not expected [9]. The complex hexagonal crystal structure of Zn_4Sb_3 as reported also unable to be verified [4-8].

Figure 2. SEM images for samples with different deposition parameters at magnification x100. (a) S1 (0.1M $ZnCl_2$-0.1M $SbCl_3$, 100mA, 120min), (b) S2 (0.1M $ZnCl_2$-0.1M $SbCl_3$, 100mA, 60min), (c) S3 (0.1M $ZnCl_2$-0.1M $SbCl_3$, 50mA, 120min), (d) S4 (0.1M $ZnCl_2$-0.1M $SbCl_3$, 50mA, 60min) and (e) S5 (0.2M $ZnCl_2$-0.2M $SbCl_3$, 100mA, 30min).

3.3 Seebeck coefficient, S of Zn-Sb deposited

The Seebeck coefficient of the Zn-Sb alloy deposited of S1 (0.1M $ZnCl_2$-0.1M $SbCl_3$, 100mA, 120min) and S3 (0.1M $ZnCl_2$-0.1M $SbCl_3$, 50mA, 120min) were measured in temperature range of 30°C to 300°C. As shown in figure 3, Seebeck coefficient of S1 has both negative values and positive values. The values started around zero and then drastically decreased to approximately -100μV/K before up back to be around zero. Then the values showed slight increased and then peaked to approximately 181μV/K at 102°C. The values then drop back and fluctuating in the range of 0μV/K to 44μV/K throughout the rest of temperature range. As for S3, the pattern of the graph was almost identical to S1 but the different it only had positive values. The graph started to have peaks at 38.1°C with Seebeck value of 15.5μV/K and at 46.6°C with Seebeck value of 16.5μV/K. The values then drop to almost zero and the speeded to 113μV/K at 101°C. The graph then had sharp declined and fluctuated in a range of 0μV/K to 27μV/K. When comparing graphs pattern of S1 and S3 with the previous work and Bismuth Telluride thin film, it can be said that approximately at 100°C, the *S* for both samples were higher. However, as the temperature increase, the *S* values of the samples decrease and fluctuate which differed from the result that has been expected [10, 12].

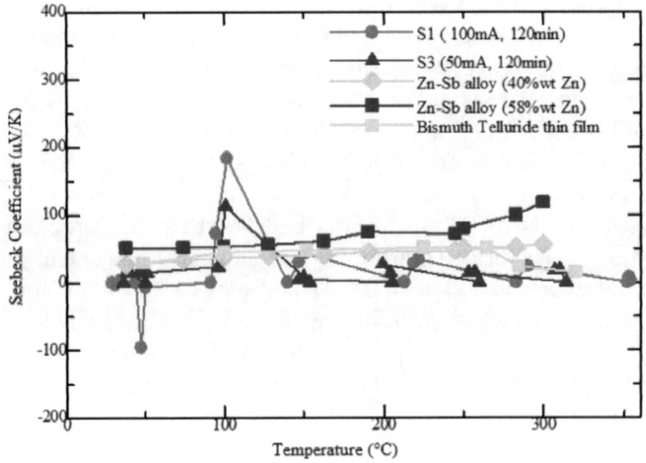

Figure 3. Seebeck coefficient of S1 and S3 deposited sample.

3.4 Electrical conductivity, σ of Zn-Sb deposited

The electrical conductivity of S1 (0.1M $ZnCl_2$-0.1M $SbCl_3$, 100mA, 120min) and S3 (0.1M $ZnCl_2$-0.1M $SbCl_3$, 50mA, 120min) are seen to be constant throughout the temperature range. It can be said that both samples have semiconductors electrical conductivity (around 10^{-6} to $10^{4}\Omega^{-1}.cm^{-1}$) which are less then metals' electrical conductivity (around 10^{7} $\Omega^{-1}.cm^{-1}$) [11]. For S1, the σ is $4.06\Omega^{-1}.cm^{-1}$ and for S3, it has the value of $12.2\Omega^{-1}.cm^{-1}$ as shown in the figure 4. These values considered as very low because from previous work, the σ for Zn-Sb alloy deposited around $100\Omega^{-1}.cm^{-1}$ if the Zn content more than 50% and around $600\Omega^{-1}.cm^{-1}$ if the Zn content around 40%. The reasons as the σ for both samples are very low may because of the poor crystalline structure of the deposited Zn-Sb as shown in the figure 2.

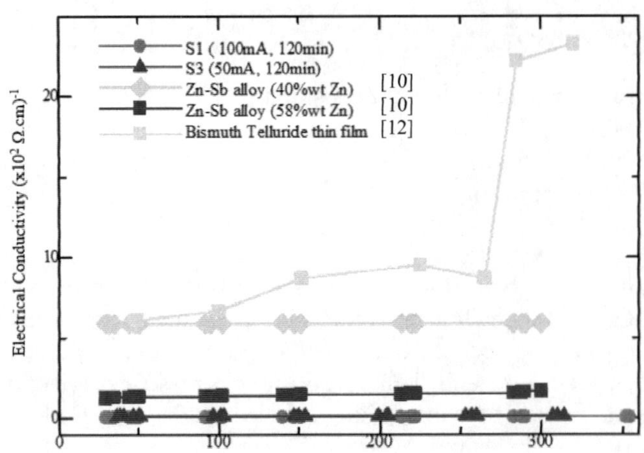

Figure 4. Electrical conductivity of S1 and S3 deposited samples.

3.5 Power factor, PF of Zn-Sb deposited

From Seebeck coefficient, *S* and electrical conductivity, σ, power factor, PF have been calculated. As shown in figure 3.5, S1 (0.1M $ZnCl_2$-0.1M $SbCl_3$, 100mA, 120min) has a significant power factor at 102°C with value 1.37x10^{-7}V/K. Ω.cm. Then the values plummet significantly to be around 0x10^{-9}V/K. Ω.cm to 7.8x10^{-9}V/K. Ω.cm. The peak value of power factor of S3 (0.1M $ZnCl_2$-0.1M $SbCl_3$, 50mA, 120min) is just above 1.5x10^{-7}V/K. Ω.cm which is at 101°C. The rest of the graph for S3 was basically having the same pattern as S1. The reason for a power factor of S3 higher than S1 was expected due to the compactness of the sample. As the S3 is prepared with less current that is 50mA, the ions deposited on the substrate at a slower rate that eventually give enough time to the ions to construct themselves better as shown in figure 5. Due to this, S3 becomes smoother which contributed to the higher electrical conductivity value. On the other hand, because of the rough surface of the samples, the power factor of both samples become inconsistent that in the end reduce the thermoelectric performance of Zn-Sb deposited. These results are not as expected as the previous work [10].

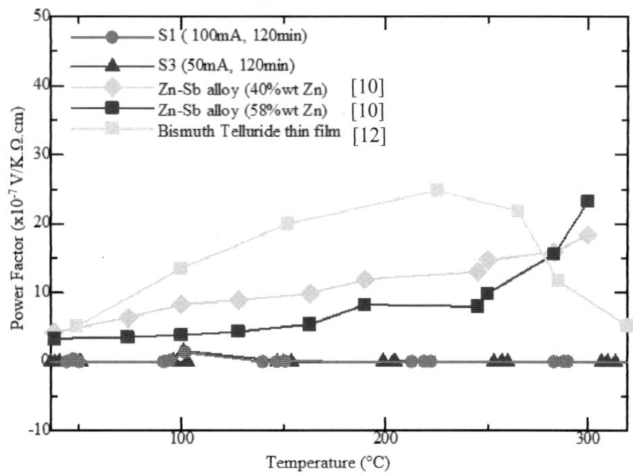

Figure 5. Power factor of S1 and S3 deposited samples.

3.6 Heat capacity, Cp of Zn-Sb deposited

To compensate the thermal conductivity, heat capacity, C_p of S1 (0.1M $ZnCl_2$-0.1M $SbCl_3$, 100mA, 120min) and S3 (0.1M $ZnCl_2$-0.1M $SbCl_3$, 50mA, 120min) have been investigated using DSC. The data from the DSC have been recorded in Table 2. Based on Table 2, it shows that S3 has a higher cap than S1. *Chip* for S3 is 46.8093mJ/°C while S1 is 38.3722mJ/°C with the difference of 8.4371m J/°C between the two samples. This difference could be related to the compactness of S3 compare to S1, which allow the samples to store more heat energy. When the samples have the ability to store more heat energy, it can be concluded that the thermal diffusivity of the samples are low. Thus, according to thermal conductivity equation, when the thermal diffusivity low, the thermal conductivity also low.

Table 2. Heat capacity, C_p of S1 and S3.

S1	Heat Flow (mW)	Heat Capacity (mJ/°C)	S3	Heat Flow (mW)	Heat Capacity (mJ/°C)
a	0.2597	1.5577	a	0.2650	1.5898
	n/a	n/a		1.5429	9.2558
b	0.2501	1.5005	b	17.9410	107.6244
	4.9225	29.5294		n/a	n/a

c	13.3096	79.8418	c	3.6604	21.9579
	0.4479	2.6871		n/a	n/a
Average heat capacity		38.3722	Average heat capacity		46.8093

4. Conclusion

In summary, it can be demonstrated that Zn-Sb alloy was successfully fabricated by using electrodeposition method. Two types of electrolytic bath formulation; the mixture of 0.1M $ZnCl_2$-0.1M $SbCl_3$ and 0.2M $ZnCl_2$-0.2M $SbCl_3$ were prepared for Zn-Sb deposited. From the XRD results, all the samples deposited to form thin films samples exhibit Zn-Sb alloy compositions. The obvious results were S1 and S3 as they had significant peaks that shows the crystal lattice of Zn_4Sb_3. S2, S3 and S5 exhibited the alloy needed, but had a lot of noise in the spectra. The noise was assumed cause by the thickness and roughness of the samples. SEM images showed that the surface morphology of all samples were rough and coarse. When the current lower, the samples become denser. However, when the deposition time increased, quantity of deposited ions increased, thus, made the samples become rougher and thicker. The compactness of the samples affected the thermoelectric performances of the samples significantly. Due to that S3 had higher power factor compared to S1. Although the Seebeck coefficient of S3 (113µV/K) was not as high as S1 (181µV/K), but because of its atomic arrangement that made the sample denser, the electrical conductivity of the S3 surpasses the value of S1. Thus because of this difference only, S3 had produced Zn-Sb alloy with better power factor ($1.58x10^{-7}$V/K. Ω.cm at 101°C) which indicated its efficiency in converting thermal energy to electricity compared to S1 ($1.37x10^{-7}$V/K. Ω.cm at 102°C).

Acknowledgement
This work was financially supported by RIGS16-073-0237, International Islamic University Malaysia (IIUM).

References

[1] Budak S, Guner S, Muntele C. I, and Ila D 2015 *Phys. Procedia* **66** pp 329–335

[2] Carlini R, Marré D, Pallecchi I, Ricciardi R and Zanicchi G 2014 *Intermetallics* **45** pp 60–64

[3] Bjerg L, Iversen B. B, and Madsen G. K. H 2014 *Phys. Rev. B - Condens. Matter Mater. Phys.* **89 (2)** pp 024304

[4] Li J, Li J, Zhang Q, Zhang Z, Yang G, Ma H, Lu Z, Fang W, Xie H, Liang C and Yin F 2016 *Comput. Mater. Sci.* **125** pp 183–7

[5] Okamura C, Ueda T and Hasezaki K 2010 *Materials Transactions* **51(5)** pp 860-862

Pothin R, Ayral R M, Berche A, Granier D, Rouessac F and Jund P 2016 *Chem. Eng. J.* **299** pp 126-134

[6] Rouessac F and Ayral R M 2012 *J. Alloys Compd.* **530** pp. 56-62

[7] Wrona A, Bilewska K, Mazur J, Lis M and Staszewski M 2014 *J. Alloys Compd.* **616** pp 350-355

[8] Fan P, Fan W F, Zheng Z H, Zhang Y, Luo J T, Liang G X and Zhang D P 2014 *Journal of Materials Science: Materials in Electronics* **25** pp 5060-5065

[9] Zheng Z H, Fan P, Liu P J, Luo J T, Liang G X and Zhang D P 2014 *Journal of Alloys and Compounds* **594** pp 122-126

[10] Quirk M and Serda J 2001 Semiconductor manufacturing technology. *Upper Saddle River, NJ: Prentice Hall,* **1**

[11] Kim D H, Byon E, Lee G H and Cho S 2006 *Thin Solid Films* **510** pp 148-153

IOP Conf. Series: Materials Science and Engineering **290** (2017) 012011 doi:10.1088/1757-899X/290/1/012011

Effect of EFB content on reducibility of low grade iron ore composite at 1000°C – 1200°C

H Purwanto[1], H M Salleh[2], A S Mohamad[1], A Zakiyuddin[1], A N Rozhan[1]

[1]Department of Manufacturing and Materials Engineering, International Islamic University Malaysia (IIUM), Jalan Gombak, 53100 Kuala Lumpur, Malaysia
[2] Department of Biotechnology Engineering, International Islamic University Malaysia (IIUM), Jalan Gombak, 53100 Kuala Lumpur, Malaysia

Email:hadi@iium.edu.my

Abstract. This paper focus on reduction of low grade iron at high temperature using palm oil empty fruit bunch as a reductant. The samples were crushed and compacted into composite pellet with EFB before drying at 110°C for 24 hours. The reduction test was then conducted by heating the composite pellet in an electric tube furnace. The parameter used was at temperature range 1000°C to 1200°C and under argon gas atmosphere. The efficiency of the empty fruit bunch in the reduction process was investigated based on the extent of reduction of the iron oxide. The results showed that the extent of reduction obtained was 38.97% at 1200°C. This study demostrates that within the temperature range studied, reducibility of low grade iron ore with an empty fruit bunch tends to increase as the temperature increase.

1. Introduction

Malaysia has been developing on steel making since this industry is one of the strategic industries to promote the country to become a developed country in near future. Steel consumption has significantly increased year by year as reported by Malaysia Steel Industry Federation. Previously, iron ore is imported from other countries as Malaysia has no significant output of good quality iron ore for steel processing. At the same time, in 2014 almost 10 million tonnes are deposited from various mining area in Malaysia. This situation has led to an abundance of low iron ore and ways to utilize it efficiently and effectively is needed [1].

The present process of iron making in Malaysia using coal as the reducing agent in the reduction process of iron ore. However, some problems arising from the previous process of iron making, such as high demand of coal, higher energy consumption, environmental problems [2]. International Energy Statistics estimated about 59 million metric tonnes of CO_2 are given off from the total coal used in Malaysia until 2013.

At the same time, due to the depletion of high-grade resources, properties of natural resources such as iron ore and coal will change drastically [2]. Hence, alternative ways of utilizing low grade iron ore by replacing coal with biomass for iron making process has been taken. Many countries have been used biomass in iron and steel making, for instance in Brazil, biomass have been used as fuels and also as a reductant [3]. The idea of using biomass to replace coal is a promising alternative as biomass is a carbon-neutral and renewable fuel source and contain low amounts of harmful elements such as sulphur and phosphorus [4]. Biomass are categorized as one of the most abundant element in the world, making it a feasible choice in replacing coke in iron and steel making as reductant.

The reduction process with several types biomass for iron making have been done by a previous researcher such as sawdust [2], [5], [6], rice husk [7], wood powder [8], chipped cedar [9], biomass char [3] -[10] and limited research have been done by using palm oil biomass. Hence this present research focused on potential of palm oil empty fruit bunch for the reduction process in reducing low grade iron ore into quality iron materials.

2. Experimental procedures

Raw material, of iron ore and oil palm empty fruit bunch (EFB) was taken from Bahau, Negeri Sembilan mining site and from palm oil mill at Mempaga, Bentong, Pahang. The raw materials were compacted to produce pellet with 10 mm – 20 mm in diameter [11]. Before undergoing reduction test the pellet was dried at 110°C for 24 hours to release moisture and chemical analysis of low grade iron ore used as raw material is provided in table 1.

Table 1. Chemical analysis of iron ore from Bahau, Negeri Sembilan.

Element	Fe_{total}	Al_2O_3	MgO	SiO_2	K_2O	TiO_2	Cr_2O_3	MnO
Content (%)	51.92	9.20	0.44	14.87	0.06	0.52	0.06	0.23

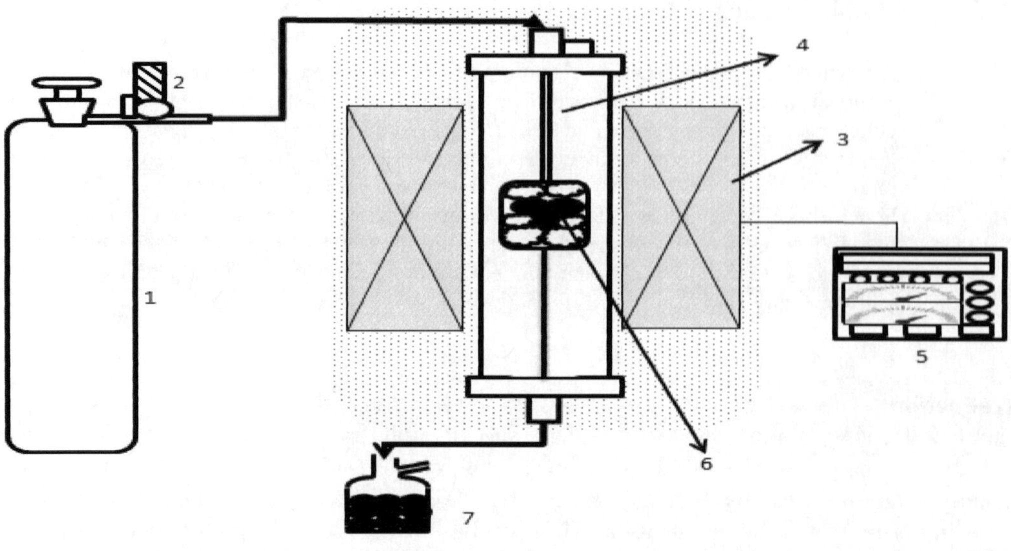

Figure 1. Schematic diagram of experimental apparatus (1- argon gas container, 2- gas controller, 3- electric heater, 4- thermocouple, 5- temperature controller, 6- composite pellet, 7- exhaust gas trap).

Reduction reaction of the pellet was conducted in a vertical electric tube furnace with a quartz tube reactor at the centre as schematically shown in figure 1. Inert condition was established by flowing argon gas into the furnace. For the reduction to be completed, the pellet was held for 30 minutes after heating at the desired temperature between 1000°C to 1200°C. Argon gas was kept flowing into the furnace until the end to ensure no oxidation occurred. The reduced composite pellet was then subjected to characterization to assess the extent of the reduction, magnetization properties, phase change and morphology as well as total iron content.

The complete reduction reaction occurred when hematite phase in the ore completely transformed into metallic iron phase. Transformation takes place when oxygen being removed from the ore and the amount of oxygen removed can be calculated referring to the molecular weights. For every phase transformation involved in reduction reaction the oxygen removed is estimated as shown in table 2. Calculation of extent of reduction and reducibility shown in equation (1) and (2) respectively.

Table 2. Estimation of oxygen removal

Reaction	Reductant	Oxygen removal (%)
$FeO(OH) \rightarrow Fe_2O_3$		2.5
$Fe_2O_3 \rightarrow Fe_3O_4$	CO / H_2	3.33
$Fe_3O_4 \rightarrow FeO$		6.89
$FeO \rightarrow Fe$		22.22

$$Extent\ of\ reduction = W_{pellet} - W_{ore} - W_{EFB} \quad (1)$$

Where; Wpellet : weight loss of the composite pellet after the reduction process
Wbiomass : weight loss of the biomass during the thermal analysis
Wore : weight loss of the iron ore during the thermal analysis

$$Reducibility = \frac{Extent\ of\ reduction}{Theoretical\ oxygen\ removal} \quad (2)$$

3. Results and Discussion

The effect of reduction test at different temperature and composition can be observed in figure 2. As the temperature increases the size of the composite pellet decreases during reduction reaction at implying that the volume of the composite pellet decreases. At all temperatures tested iron with metallic gloss was detected in the composite pellets suggesting that EFB was able to reduce iron oxide into metallic iron.

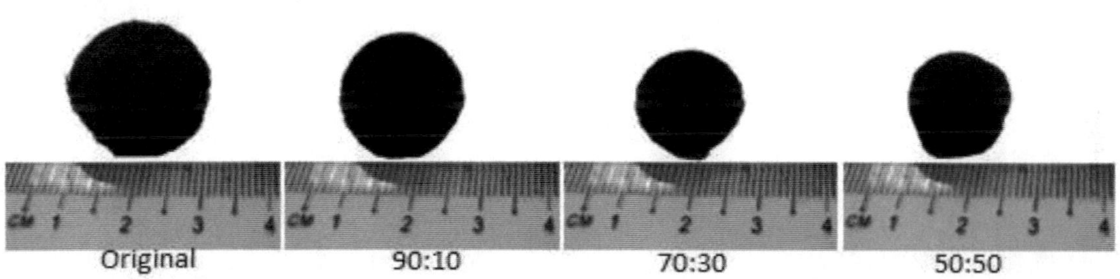

Figure 2. Physical changes of composite pellets of different composition before and after reduction at 1000° C.

Effect of composition on the extent of reduction of composite pellet at different temperatures shown in table 3. From the table it can be seen that extent of reduction reached the highest value of 38.97% at 1200°C. It can be observed that the extent of reduction was increased approximately 15% - 25% as the reaction proceeds from 1000°C to 1200°C. The extent of reduction corresponds to the increasing temperature as there was a reaction between the ore particles and the EFB as the reduction proceed. Based on the oxygen removed from iron ore the extent of reduction was calculated as shown in equation (3.1). Removal of oxygen from the iron oxide caused a phase change as the carbon in the EFB acted as a solid reductant during the reduction reaction.

Table 3. Extent of reduction with different composition at different temperature

Temperature (°C)	Composition (Iron ore: Biomass)	Extent of Reduction (%)
1000°C	90:10	25.42
	70:30	29.12
	50:50	32.87
1200°C	90:10	25.76
	70:30	33.27
	50:50	38.97

Figure 3 shows that the reducibility of the composite pellet is influenced by the reduction temperature and composition of the iron ore: EFB ratio. The reducibility of the reduced composite pellet was calculated using equation (3.2). Increasing the composition of EFB in the composite pellet and reduction temperature increases the reducibility, reflecting improvement in a reduction reaction. Generally, for every 1 tonne of iron ore processed 825 kg of CO_2 will be released when coke is used [12]. It is that indicated that 30 wt.% of EFB will give 84.36% reducibliity and up to 98.80% was achieved when 50wt. % of EFB was used. This reduction proceeds very fast could be caused by the iron reduced by two different reductant which is hydrogen and carbon monoxide gas. Hence, the lab scale data suggest that by replacing 30wt.% of coke with EFB, the theoretical amount of can be reduced to 701.25 kg and up to 775.5 kg of CO_2 can be reduced when 50wt.% of EFB is used.

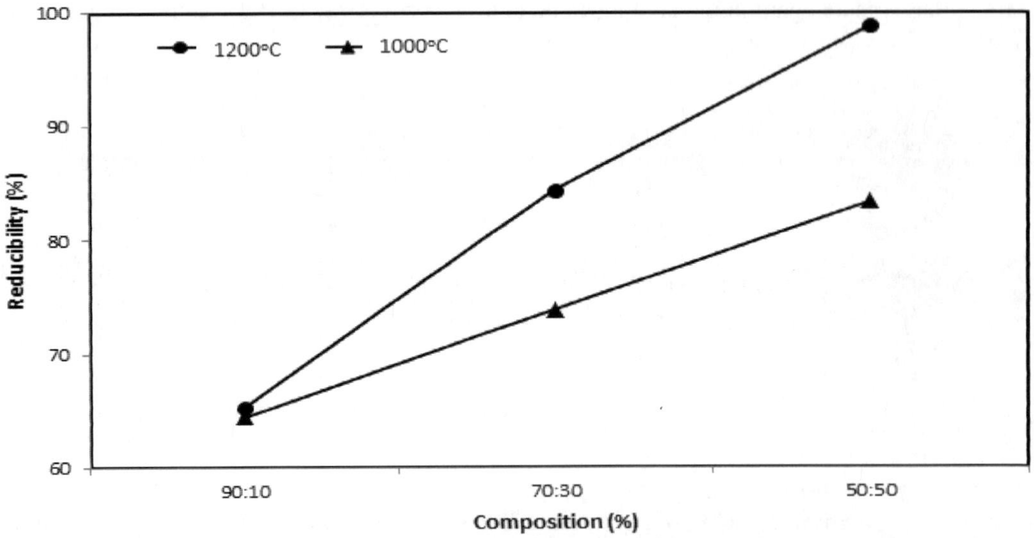

Figure 3. Effect of composition on the reducibility of composite pellets at different temperatures

4. Conclusion

The following conclusions were achieved when reduction test of iron ore with EFB at 1000°C and 1200°C was completed. Both parameters used in this research, temperature and composition were affected the reduction process of the composite pellet. Higher temperature and composition of EFB gives better extent of the reduction. Highest reducibility of 94.22% was achieved strongly at 1200°C compared to other temperature suggests that higher removal of oxygen is occurring at high temperature. Higher extent of reduction proved that the composite pellet was successfully reduced to another phase and metallic iron phase was dominant at 1200°C. In terms of feasibility study of the reduction reaction process using EFB, it was estimated that the theoretical maximum value of 775.5 kg of CO_2 can be reduced for each metric tonne of Bahau ore used.

Acknowledgement
This work was financially supported by the International Islamic University Malaysia through IIUM Research Initiative Grant Scheme (RIGS) 16-353-0517 research fund.

References

[1] Rashid R Z A, Salleh H M, Ani M H, Yunus N A, Akiyama T and Purwanto H 2014 *Renewable energy* **63** pp 617-623
[2] Luo S, Zhou Y and Yi C 2013 *Journal of Renewable and Sustainable Energy* **5** 063114
[3] Norgate T, Haque N, Somerville M and Jahanshahi S 2012 *ISIJ international* **52** pp 1472-1481
[4] Panwar N L, Kothari R and Tyagi V V 2012 *Renewable and Sustainable Energy Reviews* **16** pp 1801-1816
[5] Strezov V 2006 *Renewable energy* **31** pp 1892-1905
[6] Guo D, Hu M, Pu C, Xiao B, Hu Z, Liu S, Zhu X 2015 *International journal of hydrogen energy* **40.14** pp 4733-4740
[7] Purwanto H, Mohd Selamat N F, Pramusanto and Akiyama T 2010 *SEAISI Quarterly Journal* **39** pp 18-22
[8] Ueki Y, Yoshiie R, Naruse I, Ohno K I, Maeda T, Nishioka K and Shimizu M 2013 *Fuel* **104** pp 58-61
[9] Kudo S, Sugiyama K, Norinaga K, Chun Z L and Akiyama T 2013 *Fuel* **103** pp 64-72
[10] Ueda S, Watanabe K, Yanagiya K, Inoue R and Ariyama T 2009 *The Iron and Steel Institute of Japan (ISIJ) International* **49** pp 1505-1512
[11] Rashid R Z A, Purwanto H, Salleh H M, Ani M H, Yunus N A and Akiyama T 2013 *Advanced Materials Research* **701** pp 28-31
[12] Chunbao C X and Cang D Q 2010 *Journal of Iron and Steel Research International* **17** pp 1-7

IOP Conf. Series: Materials Science and Engineering **290** (2017) 012012 doi:10.1088/1757-899X/290/1/012012

Effect of temperature on porosity of iron ore sinter with biochar derived from EFB

H Purwanto[1], A N Rozhan[1], A Zakiyuddin[1], A S Mohamad[1]
[1]Department of Manufacturing and Materials Engineering, International Islamic University Malaysia (IIUM), Jalan Gombak, 53100 Kuala Lumpur, Malaysia

Email: hadi@iium.edu.my

Abstract. In this research, the replacement of fossil fuel energy (coke) with oil palm empty fruit bunch as a potential energy in sintering of iron ore was investigated. Carbon derived biomass has been produced by using oil palm empty fruit bunch by heat treatment process. In the present investigation, sintering process was carried out by heating the mixed iron ore and biochar at various temperatures. The apparent density and porosity for iron sinter show a significant increase and gradual decrement as the temperature increase, respectively. The porosity of iron sinter shows a gradual decrement from 950 °C to 1050 °C but up to 1150 °C it shows a significant decrement about 44%. Inferring to the micrograph, the agglomeration and assimilation of sinter at high temperature is better compared with low sintering temperature.

1. Introduction

Depletion of high grade iron ore happened throughout the world [1] and Malaysia is a renowned resource on low grade iron ore. The usage of low grade iron for sintering process has been continuing due to abundance of low grade iron ore [2]. Sinter is the main iron-bearing burden for the blast furnace and its production is based on coke breeze as a source of fuel.

Currently, coke was used as main fuel in sintering of iron ore but there is some problems arises from this application such as energy consumption during the production of coke. In order to mitigate these problems, biomass can be a probable source of energy to be utilized to replace coke [3]. The potential alternative energy that can substitute fossil fuels is biomass [4]-[8]. In order to highlight the problems in this steel industry, usage of biomass as energy fuel is an interesting solution for iron making, so many different ways of its usage should be tried to reduce the energy consumption and CO_2 emissions [9].A biomass is considered carbon neutral and can be used as a source of heat and as reducing agent for iron reduction process.

In iron making industry, iron sinter is one type of raw materials to be charged into blast furnace. Sintering is the process of converting iron ore fines into coarse agglomerate granules product due to the partial fusion and coalescence bonding for blast furnace usage [10]-[11]. In blast furnace operation, the structure produced from sintering process seize a vital factor that give effect to the strength and reducibility of sinter [12]. Porosity is one of the vital component in physical of property of iron ore. By having high porosity provides large contact area for reducing gas to facilitate into the interior of the lump ore [13].The mechanical properties of final iron sinter could be influenced by the fusion phase produced during sintering process in which the sinter must have a good strength and porosity [14]. In this work, the effects of oil palm empty fruit bunch (OPEFB) addition on the

IOP Conf. Series: Materials Science and Engineering **290** (2017) 012012 doi:10.1088/1757-899X/290/1/012012

sintering of low-grade iron ore were investigated with regards to the density and porosity of iron sinter.

2. Experimental procedures

2.1. Materials

In this experiment, an iron ore taken from mining site in Bahau, Negeri Sembilan was selected as raw material for sintering. The ore then is crushed and ground to get fine particle. The chemical compositions of iron ore is shown in table 1. OPEFB was supplied by local palm oil mill and used as the source of material to produce biochar. Chemical composition of OPEFB is given in table 2. In order to reduce the melting temperature of sinter granules the limestone was added to the mix of biochar and iron ore.

Biochar making was carried out by putting OPEFB in an electric furnace under vacuum atmosphere for 10 minutes to remove the present of oxygen. The temperature was increased to 450°C at a heating rate of 10°C/min, and held for 30 minutes to allow complete production of biochar.

In order to produce a granulated green sample, the iron ore, biochar and limestone were mixed by adding water binder and placed in a ceramic bowl. The granulated green sample was dried at 100°C for 24 hours to remove the water.

2.2 Experimental Method

Sintering process was conducted by putting the granulated green sample in an electric furnace at various temperatures from 950°C to 1150°C with heating rate of 10°C/min and holding time of 30 minutes. The physical properties of iron sinter such as apparent density and porosity was examined. The morphology of iron sinter was also observed by using FESEM. The schematic diagram of furnace for sintering and biochar making are shown in figure 1.

Table 1. Elemental composition of used iron ore (Bahau,Negeri Sembilan).

Compound	**Fe**	**Al**	**Mg**	**Si**	**K**	**Ti**	**V**	**Cr**	**Mn**	**Zr**
Percentage	51.92	9.202	0.436	14.87	0.06	0.516	0.064	0.058	0.226	0.016

Table 2. Elemental composition of OPEFB.

Proximate analysis	**wt%**	**Ultimate analysis**	**wt%**
Moisture	6.8	Carbon	45.64
Volatile matter	83.94	Hydrogen	6.19
Fixed carbon	8.98	Oxygen	48.17
Ash	7.08	Nitrogen	0.35

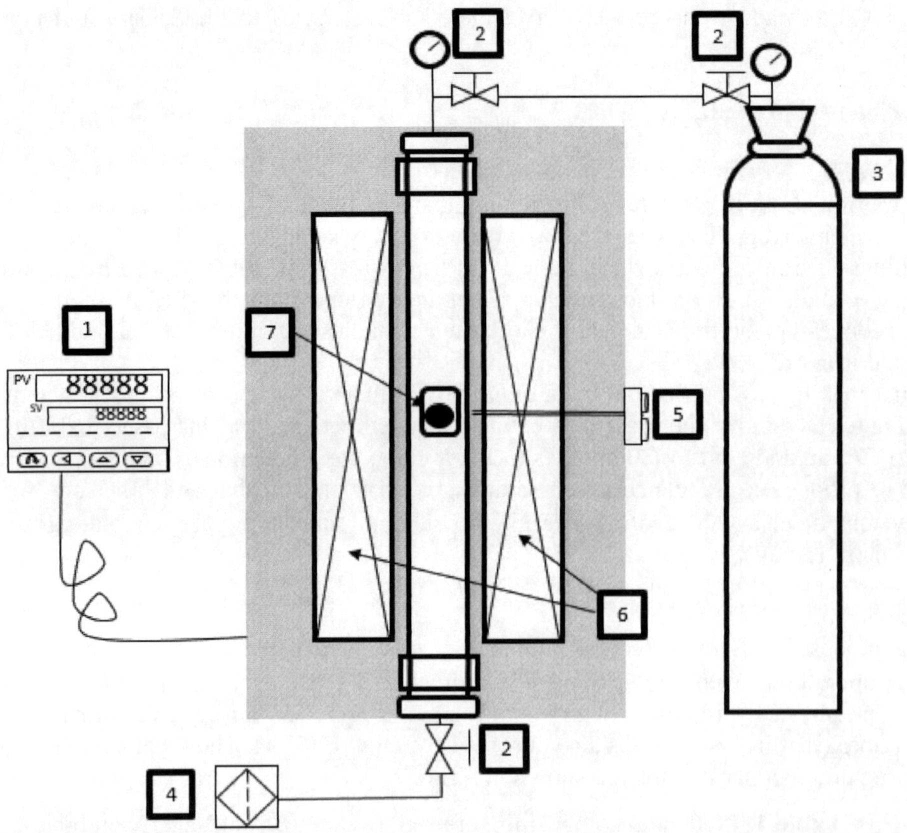

Figure 1. Schematic diagram of furnace for sintering and biochar making (1- Temperature controller, 2- Gas flow controller,3- Gas container,4- Gas filter,5- Thermocouple,6- Electric heater,7- Green sample).

3. Results and Discussion

The figure 2 shows the apparent density and porosity on the iron sinter with 10% content of biochar. In this case, the apparent density for iron sinter shows significant increase as the temperature increase from 950°C to 1150°C. The porosity of iron sinter shows a gradual decrement from 950 °C to 1050 °C but up to 1150 °C it shows a significant decrement about 44%. This result is in line with the porosity shown in figure 2 in which apparent density high could produce dense and compact iron sinter.

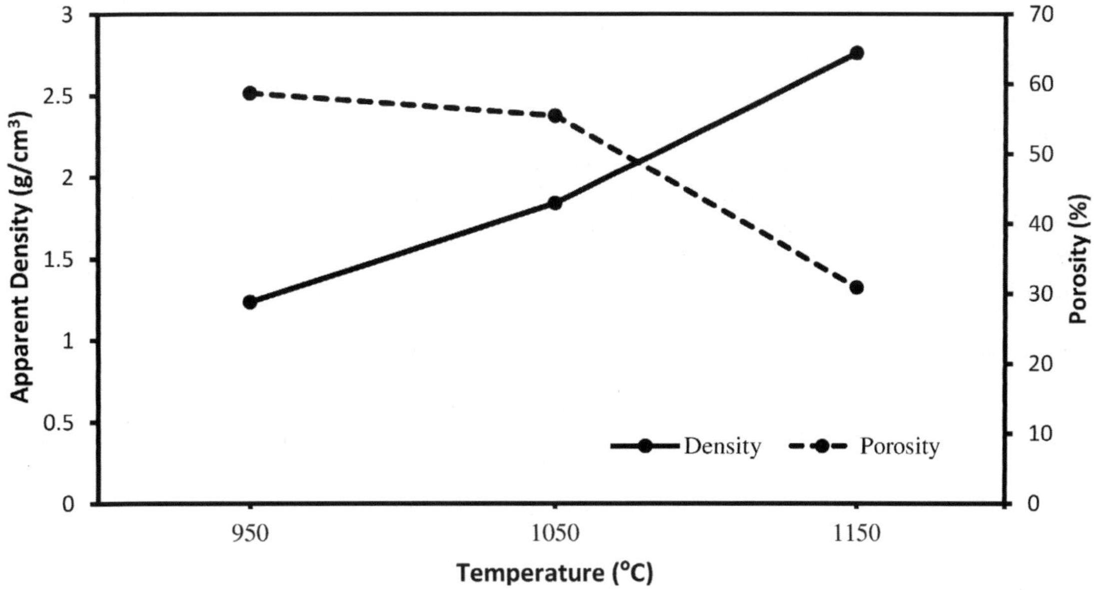

Figure 2. Apparent density and porosity of the iron sinter with 10% content of biochar.

Morphology of iron sinter for 10% biochar content with different sintering temperature (a) 950°C, (b) 1050°C, and (c) 1150°C are shown in figure 2. Noticeable granules have been produced at figure 3(b) and (c) with dense sinter compared with figure 3(a).The micrograph in figure 3 (a), (b) and (c) in line with the porosity and density shown in figure 2. At high temperature shows higher agglomeration and assimilation of sinter mixtures compared to low temperatures. Reshaping of irregular shape pores into more rounded shape indicates the combination of small pores to form a reduced number of larger pores as shown in figure 3(b).

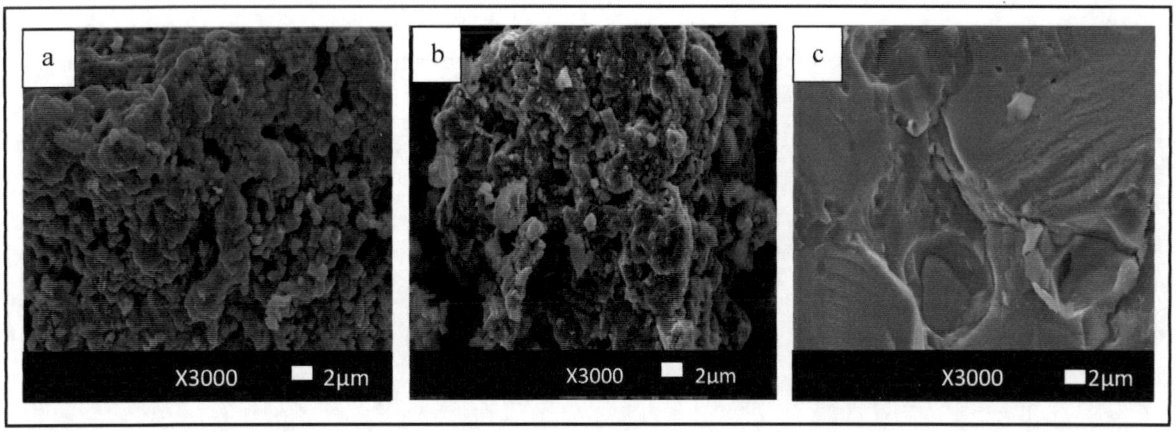

Figure 3. Morphology of iron sinter for 10% biochar content with different sintering temperature (a) 950°C (b) 1050°C (c) 1150°C.

4. Conclusion

In conclusion, the apparent density for iron sinter shows a significant increase as the temperature increase from 950ºC to 1150ºC. Morphology of the sinter revealed that the dense iron sinter has been produced in terms of agglomeration and coalescence of sinter particles at high sintering temperature. In summary, it shows that the usage of carbon derived biomass can produced compact and dense iron sinter. Thus the usage of biochar from oil palm empty fruit bunch could be an alternative energy source for sintering of low grade iron ore for iron making that could lead to the improvement of density and porosity of iron sinter.

Acknowledgement

This work is partially supported by the International Islamic University Malaysia through IIUM Research Initiative Grant Scheme (RIGS 16-353-0517).

References

[1] Lu, L., Ishiyama, O., Lu, L., & Ishiyama, O. 2016 *Mineral Processing And Extractive Metallurgy* **125(3)** 132-139
[2] Murakami, T., & Kasai, E. 2011 *ISIJ International* ***51*(8)** 1220–1226
[3] Gan, M., Fan, X., Ji, Z., Jiang, T., Chen, X., Yu, Z., … Yin, L. 2015 *Ironmaking and Steelmaking* **42(1)** 27–34
[4] Jiang, Z. 2010 *J. Mater. Rev* **24(4)** 66–80
[5] Chen, X., Han, Z., Kong, F., & Xuteng, H. (2007) *J. Proc. Chem*, ***19*(7)**, 1091–1096
[6] Hata, Y., Purwanto, H., Hosokai, S., Hayashi, J., Kashiwaya, Y., & Akiyama, T. 2009 *Energy & Fuels* **23(2)** 1128–1131
[7] Sun, Y., Yuan, Z., & Sun, Z. 2006 *Renewable Energy* **2** 77–82.
[8] Purwanto, Hadi and Selamat, Nurfazilah and ., Pramusanto and Akiyama, T. (2010) *SEAISI Quarterly Journal* ***39*(2)** 18–22
[9] Mašlejová, A. 2013 *In Proceedings of the 22nd international conference on metallurgy and materials (Metal 2013)*
[10] Loo, C. E., & Leung, W. 2003 *ISIJ International***43(9)** 1393–1402.
[11] Umadevi, T., Bandopadhyay, U. K., Mahapatra, P. C., Prabhu, M., & Ranjan, M. 2010 *Steel Research International* **81(6)** 419–425
[12] Ramos, M. V, Kasai, E., Kano, J., & Nakamura, T. 2000 *ISIJ International* **40(5)** 448–454.
[13] Muwanguzi, A. J., Karasev, A. V., Byaruhanga, J. K., & Jönsson, P. G. 2012 *ISRN Materials Science*
[14] Mašlejová, A., Fröhlichová, M., Černík, M., Vlašič, P., & Ivanišin, D. 2014 *Acta Metallurgica Slovaca* **4** 74–81

IOP Conf. Series: Materials Science and Engineering **290** (2017) 012013 doi:10.1088/1757-899X/290/1/012013

Effect of dissolved hydrogen on Schottky barrier height of Fe-Cr alloy heterojunction

A N Berahim, M Z Zaharudin, M H Ani and S K Arifin

Department of Manufacturing and Materials Engineering, International Islamic University Malaysia (IIUM), Jalan Gombak, 53100 Kuala Lumpur, Malaysia

Email:mhanafi@iium.edu.my

Abstract. The presence of water vapour at high temperature oxidation has certain effects on ferritic alloy in comparison to dry environment. It is hypothesized that at high temperature; water vapour provides hydrogen, which will dissolve into ferritic alloy substrate and altering their electronic state at the metal-oxide interface. This work aimed to clarify the change in electronic state of metal-oxide heterojunction with the presence of hydrogen/water vapour. In this study, the Schottky Barrier (SB) was created by sputtering Cr_2O_3 onto prepared samples by using RF Magnetron sputtering machine. The existence of Fe/Cr_2O_3 junction was characterized by using XRD. The surfaces were observed by using Optical Microscope (OM) and Scanning Electron Microscope (SEM). The samples were then exposed in dry and humid condition at temperature of 473 K and 1073 K. In dry condition, 100% Ar is flown inside the furnace, while in wet condition mixture of 95% Ar and 5% H was used. *I-V* measurement of the junction was done to determine the Schottky Barrier Height(SBH) of the samples in the corresponding ambient. The results show that in Fe/Cr_2O_3 junction, with presence of hydrogen at temperature 473 K; the SBH was reduced by the scale factor of 1.054 and at 1073 K in wet ambient by factor of 1.068. Meanwhile, in Fe-Cr/Cr_2O_3 junction with presence of hydrogen, the value of SBH was increased by scale factor of 1.068 at temperature 473 K while at 1073 K, the SBH also increased by factor of 1.009.

1. Introduction

Ferritic alloys are the most common material being used to fabricate boiler tubes due to its properties such as low thermal expansion coefficient and high thermal conductivity [1]. Despite the properties mentioned, ferritic alloys also susceptible to corrosion that will cause the material to scale, exfoliate and undergo changes in physical properties. Many studies have shown ferritic alloys resistance towards corrosion is significantly challenged by the presence of water vapour [2–8]. However, the reason behind the phenomena is still unexplainable.

Oxide layer act as a semiconductor at high temperature [9-10]. The region where metal-oxide contacts will cause rectifying effect which causes the presence of potential difference between the metal and semiconductor itself. This area can be modelled as Schottky Barrier Height (SBH). The SBH indicates the mismatch in the energy position of the majority carrier band edge of the semiconductor and the Fermi level of the metal across the metal-oxide junction. At high temperature; in the case of Fe-Cr alloy, Fe remains its characteristics as a metal.

IOP Conf. Series: Materials Science and Engineering **290** (2017) 012013 doi:10.1088/1757-899X/290/1/012013

It is well known that hydrogen is highly soluble in iron. However, its solubility is one to three magnitude lower in iron oxide [11]. This difference in solubility will create space charge layer as shown in figure 1, where solute hydrogen in metal will contributes to positive charge and attract negative charge from the side of oxide layer [10]. This layer will alter the mass transport in the oxide layer, hence changing the overall oxidation rate during the presence of water vapour [10, 12].

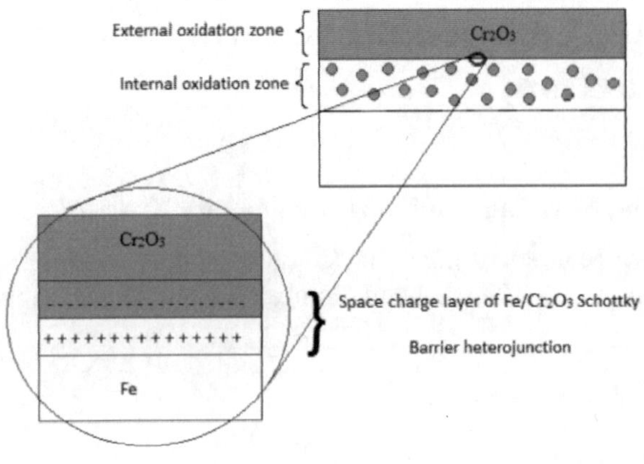

Figure 1. Schottky barrier of Fe/Cr_2O_3 heterojunction. Image are taken from reference [9].

Figure 2 shows the schematic diagram of a Schottky Barrier of a metal-oxide interface. The mismatch in the energy position of the majority carrier band edge at the metal semiconductor (MS) interface and the Fermi level across the MS interface and the surface states of the semiconductor will determine the magnitude of the SBH. This paper intent to clarify the effect of hydrogen on SBH by measuring *I-V* change of metal/oxide samples in dry and humid condition.

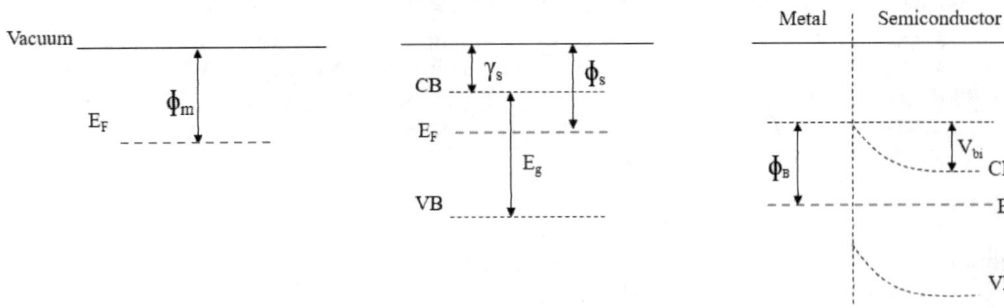

Figure 2. Schematic diagram of Schottky barrier at the contact between the metal and the semiconductor with the different electron affinities.

2. Experimental procedures

SBH are usually deduced from transport measurements of Schottky Barrier Diodes patterned on non-degenerate semiconductors. Current-voltage (*I-V*) method is one of the most common techniques for SBH determination. One tacit assumption in using these measurement techniques is that the SBH under investigation is homogeneous. In an *I-V* experiment, the junction current is measured as a function of the applied bias voltage. From a plot of the logarithmic of the forward-bias current, the saturation current and the ideality factor can be determined. According to the thermionic emission theory, the diode saturation current is related to the SBH by the following equations:

$$I = AA^{*}T^{2}\exp\left(\frac{-q\emptyset_{b}}{kT}\right) \tag{1}$$

$$\emptyset_{b} = \frac{kT}{q}\ln\left(\frac{AA^{*}T^{2}}{I_{0}}\right) \tag{2}$$

Two different samples were prepared; 99.99% Fe and Fe-Cr ferritic alloy (T91). Samples were cut into coupons using automated saw and EDM. The dimension is 10mm x 10mm x 1mm. The sample were then ground with rotating discs of emery paper of grit size P60, P800, P1000, P1500 and P2000.Then the samples were polished with 0.5 μm down to 0.03μm diamond paste. Samples were then sonicated and let dry in autoclave. The polished samples were then sputtered in RF magnetron sputtering machine at room temperature in base pressure less than $5x10^{-8}$ Torr for 1 hour. RF power is 150 W with gas flow set to be 30sccm. The sputtering target was 99.99% Cr. The concept of sputtering Cr_2O_3 onto the samples can be seen in figure 3. Characterization after the sputtering were done using XRD (scanning range, 2θ = 20-80°) for each sample.

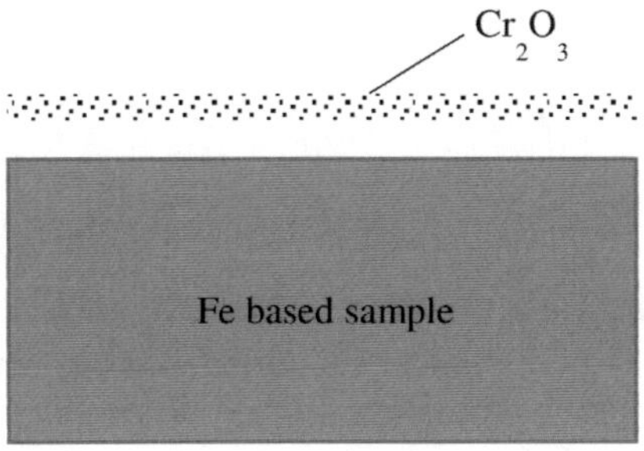

Figure 3. Schematic of Cr_2O_3 to be sputtered on the samples.

After sputtering and XRD were done, samples were exposed separately in different temperature and ambient (dry and humid). Two temperature parameter has been chosen which are 473 K (lower temperature) and 1073 K (higher temperature). The experiment setup is shown in figure 4. The ambient was dry and wet. In dry condition; at lower and higher temperature, 100% Argon was supplied into the isothermal zone in the reaction chamber. For wet condition, 95% Argon was supplied with addition of 5% Hydrogen. Electrical potential was biased into the circuit and the corresponding current were recorded by the data logger. The temperature inside the isothermal zone were monitored using R-type thermocouple.

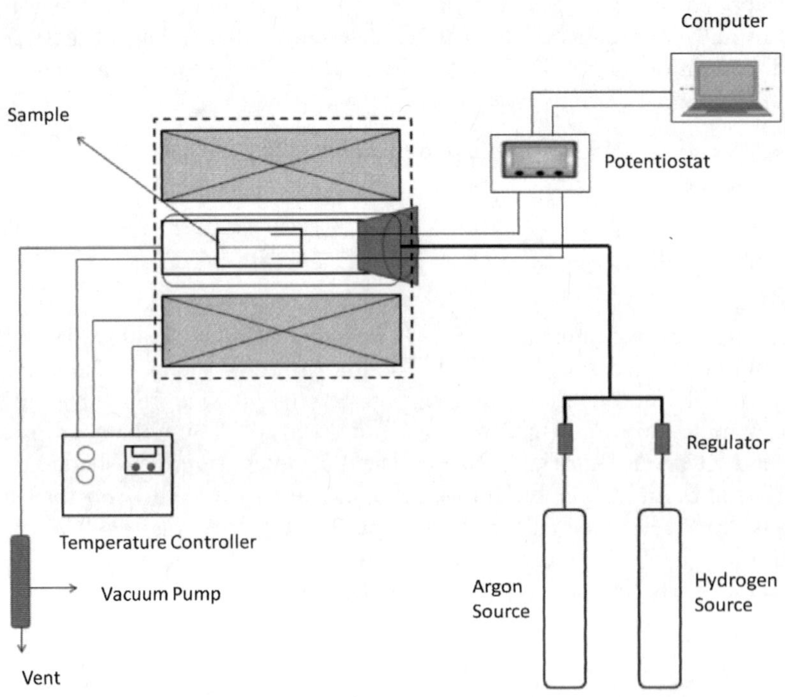

Figure 4. Experiment setup.

3. Results and Discussion

The image of microstructure for 99.99% Fe after sputtering can be seen in figure 5. It shows an almost uniform microstructure of Fe grain size. The cross-section of 99.99%Fe specimen, after sputtering can be clearly seen. The chromia layer is clearly shown and its layer was also being measured by SEM. The layer as observed in the figure was approximately 2.50 µm.

The X-Ray Diffraction (XRD) pattern for 99.99% Fe and T91 alloy after Cr_2O_3 sputtering is presented in figure 6. The result confirms the structural formation of Cr_2O_3 on 99.99%Fe. Fe and Cr_2O_3 is characterized by a prominent (110) and (200) peak at 2θ. The pattern of the spectra matches the value in the database for ferritic alloy. It shows a dominant peak at reflection appearing at an angle of $2\theta = 44.5^o$.

Figure 7 and figure 8 shows *I-V* plot comparison for 99.99% Fe/Cr_2O_3 and Fe-Cr/Cr_2O_3 alloy in 100% Ar; and 95% Ar-5%H_2 environment. There are variations in the value of I_o for each sample in different ambient and temperature. Here, I_o for 99.99% Fe is 3.17 x 10^{-5}A for temperature at 1073 K whilst I_o for Fe-Cr at the same temperature is 1.9068 x 10^{-5} A. For sample 99.99% Fe at 473 K, the value of I_o is 3.42 x 10^{-4} A.

The difference between the values of *I* stated above and the corresponding I_o from linearized curves contributed significantly in the values of SBH for each sample. It can be seen in figure 9 that for 99.99%Fe, the calculated SBH value decreased with the presence of water vapour by scale factor of 1.054 at 473 K, while at 1073 K, the SBH calculated was reduced by the factor of 1.068. For Fe-Cr alloy however; as seen in figure 10 and figure 11, the value of SBH increased in the presence of water vapour. At 473 K, the SBH increased by factor of 1.003 with presence of humidity, and SBH also increased at 1073 K by factor of 1.009 with presence of water vapour. The increase of SBH value for Fe-Cr alloy in humid environment probably due to presence of protective Cr layer in the alloy.

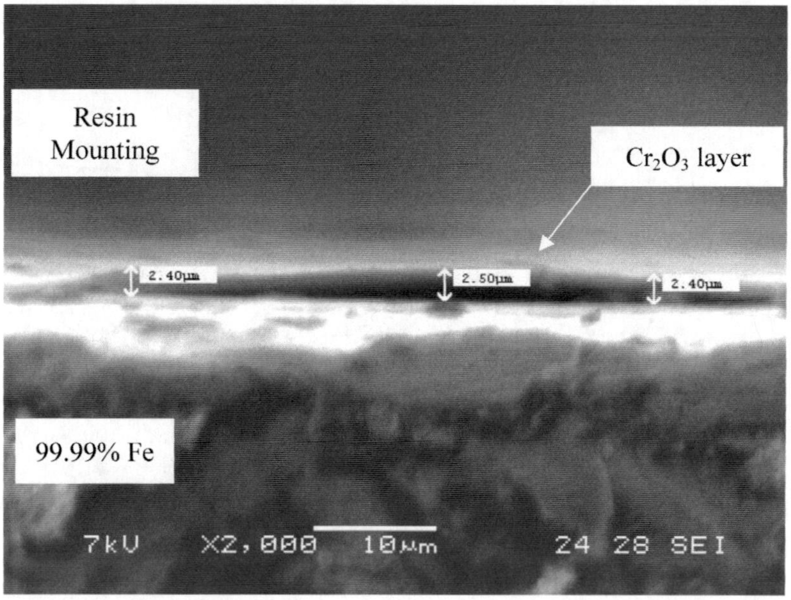

Figure 5. Image of SEM for 99.99% Fe cross section after Cr_2O_3 sputtering. The measurement of Cr_2O_3 layer were 2.40mm, 2.50mm and 2.40mm respectively.

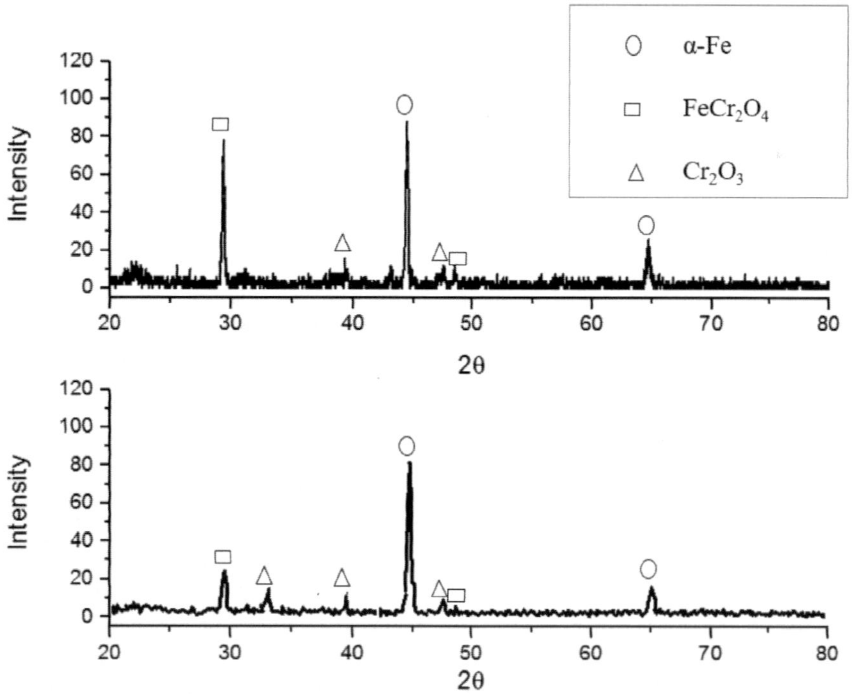

Figure 6. X-RD characteristic of Fe-Cr alloy (top) and 99.99% Fe sputtered with Cr_2O_3 (bottom).

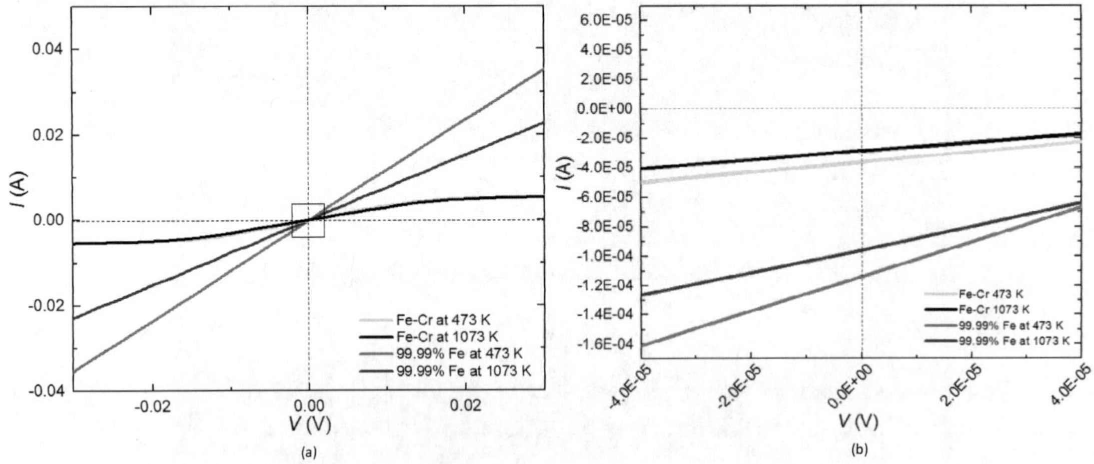

Figure 7. (a) Plot of *I-V* for samples in dry condition (b) the corresponding close up.

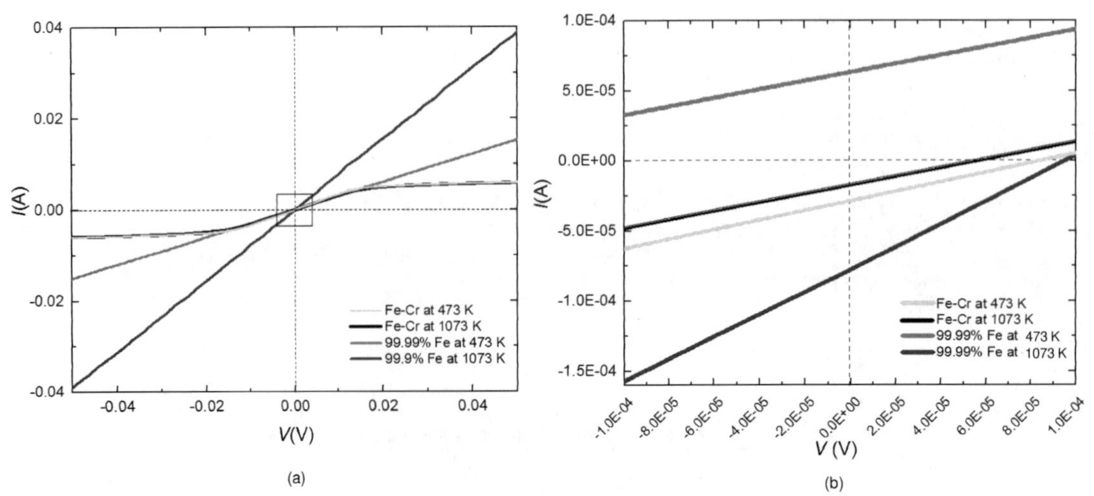

Figure 8. (a) Plot of *I-V* for samples in wet condition (b) the corresponding close up.

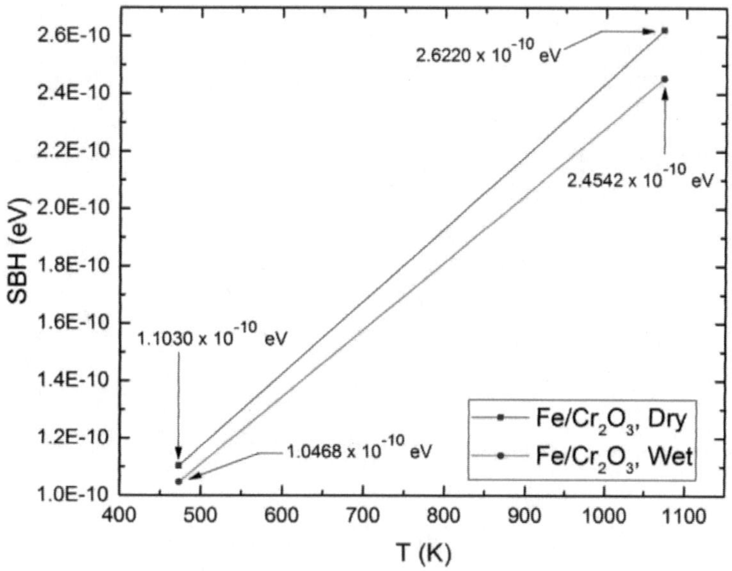

Figure 9. Plot of SBH for 99.99%Fe in dry and humid condition at 473 K and 1073 K.

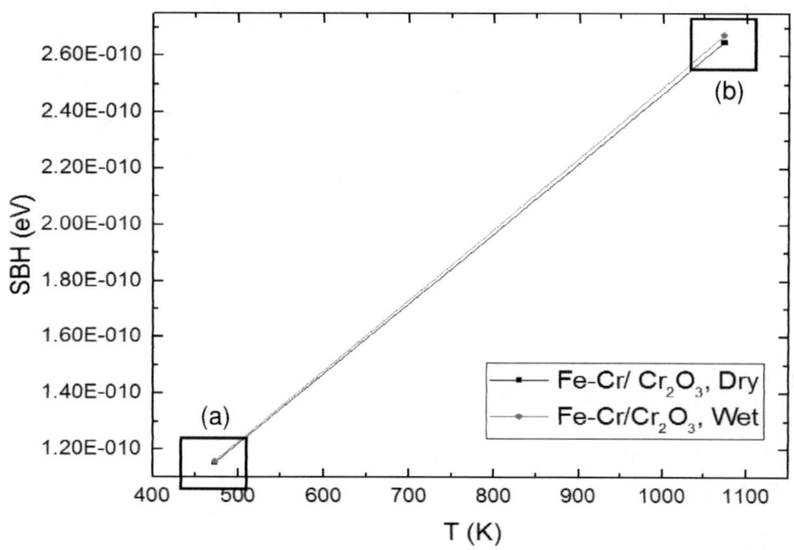

Figure 10. Plot of SBH for Fe-Cr in dry and humid condition at 473 K and 1073 K.

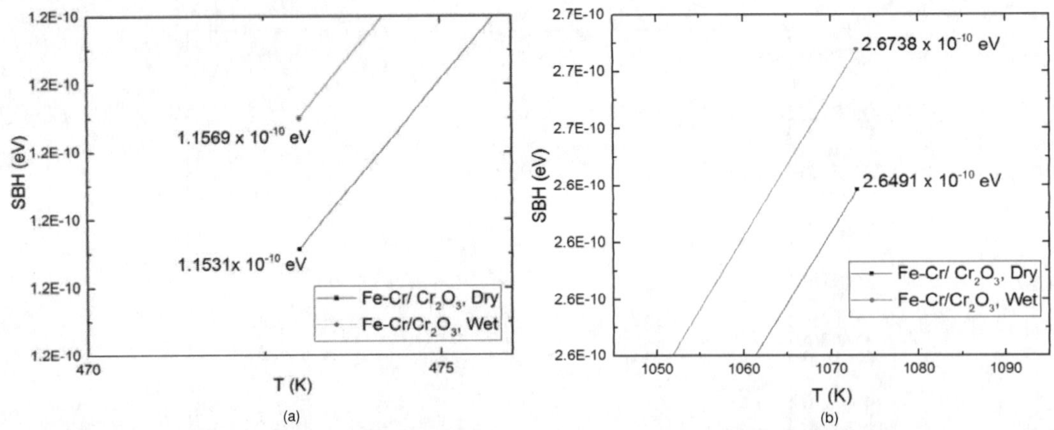

Figure 11. Corresponding close up for values in figure 10. (a) values of SBH at 473 K, (b) values of SBH at 1073 K.

In this work, it is clearly shown that the presence of hydrogen in humid condition may have change the region of space charge layer at the metal oxide junction. Many previous work suggested the presence of dissolved hydrogen have significant effect on high temperature oxidation of Fe-Cr alloy [2, 13–16]. The variation of corresponding *I* with different temperature and environment (dry and humid) proved the space charge layer was altered and supported by the values of the corresponding SBH in the given condition. Therefore, the results obtain from this work are in agreement with other previous work regarding increased in oxidation rate of Fe-Cr alloy in humid condition at high temperature due to effect of dissolved hydrogen in the metal oxide junction.

Conclusion

In conclusion, we have successfully fabricated Fe/Cr_2O_3 and $Fe\text{-}Cr/FeCr_2O_4$ junction. The *I-V* curves obtained from the experiment shown that the presence of water vapour/ hydrogen does affects the electronic state of the metal-oxide junction by altering its Schottky Barrier Height, thus proving the presence of hydrogen altered the electronic state of metal oxide junction. However, the solubility of hydrogen in the metal oxide junction was not measured in this work. Therefore, the relation between hydrogen solubility and alteration of space charge layer will be further investigated in future work.

Acknowledgement

This work was financially supported by International Islamic University Malaysia (IIUM).

References

[1] Viswanathan R, Sarver J, and Tanzosh J M 2005 *Journal of Materials Engineering and Performance* **15** 255–274

[2] Michalik M, Tobing S L, Hänsel M, Shemet V, Quadakkers W J, Young D J 2014 *Materials and Corrosion* **65** 260–266

[3] Othman N K, Othman N, Zhang J and Young D J 2009 *Corrosion Science* **51** 3039–3049

[4] Kaderi A, Ani M H, Herman S H and Othman R 2012 *Advanced Materials Research* **557-559** 100–107

[5] Ani M H 2009 *The Effect of Water Vapor on Transition of Internal/External Oxidation in Ferritic Fe-Cr Alloys at 1073 K* thesis

[6] Issartel C, Buscail H, WangY, Rolland R, Vilasi M and Aranda L 2011 *Oxidation of Metals* **76** 127–147

[7] Othman N K, Zhang J and Young D J 2010 *Oxidation of Metals* **73** 337–352
[8] Lepingle V, Louis G, Petelot D, Lefebvre B and Vaillant J C 2001 *Material Science Forum* **369– 372** 239–246
[9] Zaharudin M Z 2016 *The Schottky Barrier Of Ferritic Alloy/* Cr_2O_3 *Heterojunction* thesis
[10] Zainal M S 2013 *The effect of dissolved hydrogen on Schottky barrier at Fe/*Cr_2O_3 *heterojunction* thesis
[11] Mitsutoshi U, Hideto K, Kenichi K and Toshio M 2004 *Joint International Meeting of The Electrochemical Society and Fall Meeting of the Electrochemical Society of Japan* 127
[12] Ku Saad K M S 2014 *The Effect of Dissolved Hydrogen on Schottky Barrier of Fe/*Cr_2O_3 *Heterojunction at 1073K* thesis
[13] Surman P L 1973 *Corrosion Science* **13** 113-124
[14] Tveten B, Hultquist G and Norby T 1999 *Oxidation of Metal* **52** 221–233
[15] Mu N, Jung K, Yanar N M, Pettit F S, Holcomb G R, Howard B H and Meier G H 2013 *Oxidation of Metal* **79** 461–472
[16] Fukumoto M, Sonobe H, Hara M and Kaneko H 2017 *Oxidation of Metals* 1–8

IOP Conf. Series: Materials Science and Engineering **290** (2017) 012014 doi:10.1088/1757-899X/290/1/012014

Pressure variation of developed lapping tool on surface roughness

A K Hussain[1], K Q Lee[1], L M Aung[2], A Abu[1], L K Tan[1] and H S Kang[3]

[1]Department Malaysia-Japan International Institute of Technology, Universiti Teknologi Malaysia Kuala Lumpur, 54100 Kuala Lumpur, Malaysia
[2]Department of Mechanical Engineering, Myanmar Nobel College, Union of the Republic of Myanmar
[3]Faculty of Mechanical Engineering, Universiti Teknologi Malaysia, 81310 Skudai Johor, Malaysia

Email: lkquen@utm.my

Abstract. Improving the surface roughness is always one of the major concerns in the development of lapping process as high precision machining caters a great demand in manufacturing process. This paper aims to investigate the performance of a newly designed lapping tool in term of surface roughness. Polypropylene is used as the lapping tool head. The lapping tool is tested for different pressure to identify the optimum working pressure for lapping process. The theoretical surface roughness is also calculated using Vickers Hardness. The present study shows that polypropylene is able to produce good quality and smooth surface roughness. The optimum lapping pressure in the present study is found to be 45 MPa. By comparing the theoretical and experimental values, the present study shows that the newly designed lapping tool is capable to produce finer surface roughness.

1. Introduction

Lapping is literally known as one of the oldest machining processes where the work pieces are lapped inside the hole by twisting a stick and sand between them. Lapping is also known as an abrasive machining process. Bronze mirrors, jewels polishing and polishing glass are the products of lapping process. Having the capabilities to produce high quality surfaces, flatness, minimum subsurface damage and high dimensional accuracy make lapping to be known as the modern grinding process since 20th century ago. Metals, non-ferrous metal, ceramics, and some semiconductor are the examples of materials that are possible to be lapped which makes lapping as one of the best method to use for economical machining compared to the other manufacturing method.

According to DIN 8589 part 15 [1], lapping is defined as cutting process with loose abrasive grains dispersed in the paste which is guided by lapping tool with non-directional parts. Nowadays, manufacturing process requires very high machining precision for machine parts and tools. Those critical factors that contribute to current goals such as tolerances, parallelism, flatness, and smooth surface are the finishing processes [2]. According to Klocke [3], characteristics of lapping need to be emphasized especially in the lapping processes. Moe et al. [4] mentioned that lapping was possible to produce excellent surface finishing and higher geometrical form. The author explained various factors that affect the process of mirror-like finishing such as material remover, limit surface roughness and geometrical form improvement by series of experiments. Tam et al. [5] investigated the influences of processing parameters such as rotating speed, abrasive size on the surface roughness of optical devices. Zhao and Chang [6] examined the influence of polishing parameters on material removal rate by the chemical mechanical polishing method where they used silicon as the material to improve surface roughness for their research. Yang and Lee [7], on the other hand, introduced method of lapping to a smaller aspherical lens die by using a spherical tool. They stated that local removal rate was obtained around the contacting points of curvature change of the aspherical surface. According to Yi et al. [8], the typical optic machining consumed longer time cycle and labour intensive, thus there were several computer control being introduced for improving the process. The authors also mentioned that even though there was a precision machining in the industry, but these either had some limitations on materials selection or cycle time. Therefore, polishing system assisted by computer controlled process was introduced. In some other cases, lapping head material is one of the important part in fine finishing process. Lapping head helps in distributing the lapping pressure evenly on the work surface. Thus, mirror-like surface finishing can be achieved and a higher accuracy can be obtained.

The purpose of this research is to investigate the effect of pressure variation on the material surface roughness for a newly designed lapping tool [9]. The specimen material that is used in this research is mild steel, SKD1. Theoretical prediction of surface roughness using Vickers Hardness is implemented as well to have comparison with the outputs of this new lapping tool.

2. Methodology

New lapping tool is designed to achieve the objective of the research. The lapping tool is consists of lapping head, linear guide, soft coil spring, main body, block, and shaft, as shown in figure 1. The main body is machined from a block of mild steel. The total length of the main body is 241.50 mm with width and thickness of 36 mm and 50 mm, respectively. This part was processed with several machining method inclusive of milling as the vital process, drilling, turning by using 4-jaws lathe machine for block, and threading for the bolts and screws holes. The dimensions of linear guide are 122.50 mm x 13.65 mm x 5.20 mm in length, width and thickness respectively. A linear carrier is attached on the rail to keep the carrier always move in the right track. Spring is used to generate lapping pressure on the lapping head. It is located between the holder and collar on the base of lapping head. In order to change the lapping pressure, the spring must be adjustable, changing the relative distance between the holders and the lapping head. The specifications of the spring are shown in table 1. In order to get an optimum improvement of surface roughness, low stiffness of spring is considered to avoid the change in lapping pressure variation with small amount of the spring relative distance.

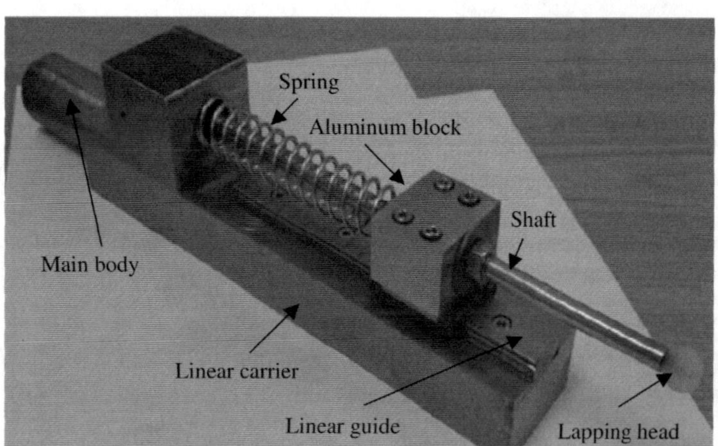

Figure 1. the newly designed lapping tool [9]

Table 1. Specifications of spring

Material	Stainless Steel
Spring Stiffness	1.5 N/m
Inner diameter	2 mm
Length	85.2 mm
Pitch	16 mm

In this research, polypropylene is chosen as the lapping head due its long chain type of the inner molecular. Because of this characteristic, the lapping grains can attach to the lapping head without falling off. The workpiece material is mild steel (SKD1).

The developed lapping tool was mounted on the tool post of the VMC machine (vertical direction towards work surface). Figure 2 shows the lapping tool that is installed in the milling machine. The work piece (SKD1) was fixed on the work table and the lapping slurry was placed on the upper side of work specimen. The function of lapping slurry is to ensure the abrasive grains engage in the lapping process. The CNC program is written on the CNC computer block. Then the simulation is operated to ensure all the movement is under the circumference of work piece. Measurement of surface roughness is conducted using Mitutoyo 178-561-02A Surftest SJ-210 Surface Roughness Tester.

Figure 2. Installation of lapping tool on VMC machine

3. Results and Discussion

Before the experiment, prediction on the critical / limit surface roughness is conducted using Vickers hardness theory with assumption that the work surface is in perfectly plastic condition. The critical / limit surface roughness is the processing surface roughness that cannot be improved even by increasing the friction distance. In this research, the material of lapping head involved is polypropylene. The Vickers Hardness of polypropylene is 9.0. With the same size of mesh grain and work piece, surface roughness limit was examined in order to produce good surface roughness. Table 2 shows the limit surface roughness for polypropylene lapping head. It can be predicted by equation. (1).

$$Rz(\lim) = 0.768 \frac{1}{\tan\alpha} \sqrt{\frac{P}{HV\ work\ x\ Ne}} \quad (1)$$

Where *P* is the lapping force, α is the angle of tip of the grain, *HV work* is the Vickers hardness of the work piece and *Ne* is the (active) effective number of grain on the lapping head. Three grain sizes are predicted in the study, namely #320, #1200 and #2500. Table 2 shows the predicted limit surface roughness at lapping pressure of 25 MPa.

Table 2. Limit surface roughness for Polypropylene ball [9]

Material of work piece	SKD1		
HV of work piece	330		
Material of lapping head	Polypropylene		
HV of lapping head	9.0		
Lapping pressure (MPa)	25		
Grain size	#320	#1200	#2500
Average angle of tip, α (°)	118	117	114
Effective number of grain (Ne)	116	214	490
Predicted limit surface roughness (μm)	0.50	0.35	0.18

Based on the result in table 2, polypropylene shows good quality in producing good surface roughness. This is because polypropylene has small wear effect and can hold the grains strongly. In fact, Yahya et al. [9] have made the comparison between the polypropylene and nylon lapping head and found that polypropylene lapping head gave less wear effect and can strongly hold the grains compared to nylon.

Surface roughness identification based on different lapping pressure is conducted experimentally on the newly designed lapping tool. Although the critical surface roughness can be calculated theoretically, there are other factors that affect the lapping process in reality such as plastic deformation, wearing and chemical reaction. In the present study, four lapping pressures (25, 35, 45 and 55 MPa) are tested for comparison. Table 3 shows the limit surface roughness for grain size of #320 at lapping time of 24 minutes. It is found that higher lapping pressure produces better surface finishing because more grains are matched on the surface of lapping head as larger lapping pressure is applied. Hence, the effective grain number received on the lapping ball increases and the removal rate is more effective. However, at lapping pressure 55 MPa, higher surface roughness is found because it already reaches its pressure limit, thus the limit surface roughness will not decrease any further if the lapping process is continued.

Table 3. Limit surface roughness of grain size #320 at 24 minutes

Lapping speed (mm/min)	Lapping pressure (MPa)	Limit surface roughness (μm)
3500	25	0.87
	35	0.76
	45	0.61
	55	0.73

In the next process, finer grain sizes are used to improve the surface roughness. In the study, #2500 grain is used as the final grain size as it produces finer surface finishing due to the smaller cutting edge. Figure 3 shows the refinement on the surface roughness by lapping the surface using new lapping tool with grain size of #1200 and followed by #2500 at 25 MPa. After the refinement process, the surface roughness achieves the mirror-like finishing surface which is below 0.1 μm. Compare the experimental data with the predicted values in table 2, it is obvious that the new lapping tool produces finer surface roughness. Besides, the reported value is also lower than the commercial lapping tool, which is around 1μm [10]. Therefore, it can be concluded that the new lapping tool is able to produce better surface roughness compare to the conventional tool.

IOP Conf. Series: Materials Science and Engineering **290** (2017) 012014 doi:10.1088/1757-899X/290/1/012014

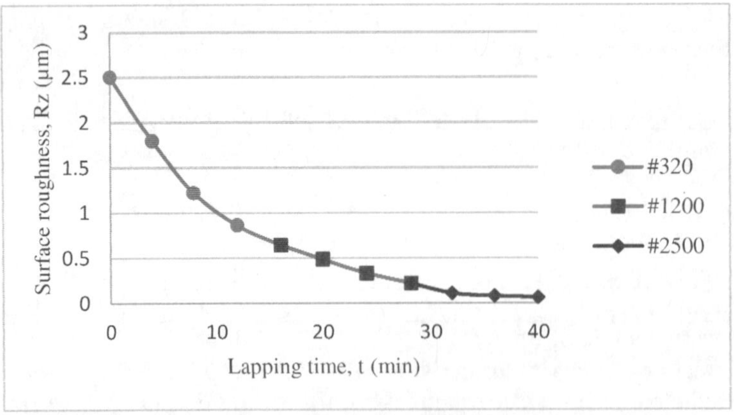

Figure 3. Improvement of surface roughness from grain size of #320 to #2500

4. Conclusion

A new lapping tool is designed for lapping machine where polypropylene is used as the lapping head. The purpose of the present study is to identify the performance of this lapping tool in improving the surface roughness of a block. Four different lapping pressures are tested on this new lapping tool and it is found that pressure of 45 MPa is more appropriate to be used for the lapping process. Polypropylene is found to produce good quality surface with fine surface roughness. Besides, the new lapping tool is found to perform better than the theoretical prediction and the commercial lapping tool.

Acknowledgement

This research was financially supported by Universiti Teknologi Malaysia under research grant Q.K130000.2643.14J65.

References

[1] Deutsches Institut für Normung e. V. 2003 *Manufacturing Processes Chip Removal - Part 15: Lapping; Classification, Subdivision, Terms And Definitions* (Standard: DIN 8589-15)
[2] Kassab S Y and Khoshnaw Y K 2007 *Eng.Technol.* **25** 879-889
[3] Klocke F 1996 *Manufacturing Processes: Grinding, Honing, Lapping* vol 2 (Dusseldorf: VDI-Verlag)
[4] Moe A L, Fukuda K, Abu A and Radzi R 2014 *Appl. Mech. Mater.* **564** 469-474
[5] Tam H Y, Cheng H B and Wang Y W 2007 *J. Mater. Process. Technol.* **192-193** 276-280
[6] Zhao Y W and Chang L 2002 *Wear* **252** 220–226
[7] Yang M Y and Lee H C 2001 *J. Mater. Process. Technol.* **116** 298-304
[8] Yi A Y, Hezlep M and Pol T 2004 *J. Mater. Process. Technol.* **146** 156–162
[9] Yahya M F 2015 *Undergraduate Thesis on Study on Aspherical Lens Mould Lapping* (Universiti Teknologi Malaysia)
[10] Taniguchi N 1983 *CIRP Annals - Manuf. Technol.* **32** 573 - 582

IOP Conf. Series: Materials Science and Engineering **290** (2017) 012015 doi:10.1088/1757-899X/290/1/012015

Nitride alloy layer formation of duplex stainless steel using nitriding process

M A Maleque[1], P H Lailatul [1], A A Fathaen[1], K Norinsan[2] and J. Haider[3]

[1]Department of Manufacturing and Materials Engineering
Kuliyyah of Engineering, International Islamic University Malaysia, P.O. Box 10, 50728 Gombak, Kuala Lumpur, Malaysia.
[2]Materials Science Program, School of Applied Physics, Faculty Science and Technology, National University of Malaysia, 43600 Bangi, Selangor.
[3]Faculty of Science and Engineering, Manchester Metropolitan University, U. K.

Email: maleque@iium.edu.my

Abstract. Duplex stainless steel (DSS) shows a good corrosion resistance as well as the mechanical properties. However, DSS performance decrease as it works under aggressive environment and at high temperature. At the mentioned environment, the DSS become susceptible to wear failure. Surface modification is the favourable technique to widen the application of duplex stainless steel and improve the wear resistance and its hardness properties. Therefore, the main aim of this work is to nitride alloy layer on the surface of duplex stainless steel by the nitriding process temperature of 400°C and 450°C at different time and ammonia composition using a horizontal tube furnace. The scanning electron microscopy and x-ray diffraction analyzer are used to analyse the morphology, composition and the nitrided alloy layer for treated DSS. The micro hardnesss Vickers tester was used to measure hardness on cross-sectional area of nitrided DSS. After nitriding, it was observed that the hardness performance increased until 1100 Hv0.5kgf compared to substrate material of 250 Hv0.5kgf. The thickness layer of nitride alloy also increased from 5μm until 100μm due to diffusion of nitrogen on the surface of DSS. The x-ray diffraction results showed that the nitride layer consists of iron nitride, expanded austenite and chromium nitride. It can be concluded that nitride alloy layer can be produced via nitriding process using tube furnace with significant improvement of microstructural and hardness properties.

1. Introduction

Duplex stainless steel consists of approximately similar amounts of austenite and ferrite phases. DSS is widely used as a paper machinery and paper industry machinery, pressure vessels, pipes and heat exchangers as well as offshore applications and chemical tankers. It is because of the excellent performance with its high strength and stress corrosion resistance. However, duplex stainless steel expressed lower hardness and poor wear resistance during service. Therefore, a new development of surface modification could be a vital solution to produce hard protective layer or coating on duplex stainless steel. Surface modification can be categorized into several types which are melting process [1],[2], thermochemical nitriding [3], and laser cladding [4].

Thermochemical gas nitriding is categorized under surface hardening process, where nitrogen is diffused into the surface of steel using dissociated ammonia and nitrogen gas at a certain composition as the source. This process is mainly applied to steels and consists of enriching the surface of the material to be treated with certain metalloids. The gaseous compound containing the nitrogen element is supplied and deposited to the surface of material via diffusion mechanism. Steels that have undergone this treatment are characterized by good resistance to wear, abrasion and fatigue [5]. However, the process must be controlled and ensure the process temperature is not too high to avoid the reduction of corrosion resistance.

Subbiah et al. [6] demonstrated that low temperature nitrided process of AISI 316LN austenitic stainless steel have been successfully employed to increase the hardness and better wear resistance. The hardness of the untreated sample was found to be 330 Hv, while for treated sample achieved maximum surface hardness ~1269 Hv. The compound layer thickness increased by increasing the nitriding temperature. The formation of S-phase as the nitride layer contributed from the gas nitriding process at 500°C with the gaseous mixture of ammonia and hydrogen. Paijan et al. [7] observed that the thermochemical nitriding process on DSS using tube furnace has produced increment of hardness from 250Hv to 666.3 Hv. This is due to nitride layer formation at temperature treatment of 450°C for 6 hours with 50% of ammonia composition. This nitride layer is the result of the quantity of nitrogen supplied to the surface and its diffusion into the structure surface. However, at temperature 500°C, the dark phases in the nitride layer is associated with chromium nitride has developed and reduce the corrosion resistance of material. In this study, the mixtures of nitrogen and ammonia gas were used for low temperature nitriding techniques and the temperature is restricted at 400°C and 450°C.The purpose of this investigation is to analyse the characteristics of nitride alloy layer which was formed on DSS after nitriding using tube furnace.

2. Experimental procedures

The material used for this study was duplex stainless steel with the grade specification of AISI 2205. The chemical composition of the AISI 2205 duplex stainless steel is 22% chromium, 5.7% nickel, 3.1 moleybdenum, 0.17% nitrogen and 0.02 % carbon. The duplex stainless steel sample was cut into a small size of 12.5 mm x 33 mm x 10 mm. The surface of the duplex stainless steel sample was ground using 250, 350, 500, 800, and 1200 grades of emery paper. Then polished using alumina paste in order to remove the scratch marks or contamination during grinding.

Prior to nitriding process, the samples were immersed in concentrated hydrochloric acid (2M) for 15 minutes with the purpose to remove native oxide film on the surface. This oxide layer is believed to act as a barrier for diffusional nitrogen transport. After immersion, the samples were removed from the beaker. Then, the samples were placed at the alumina specimen holder and put in tube furnace immediately. Prior to heat treatment, the furnace was purged by nitrogen for 30 minutes to remove air containing oxygen whilst heating up to the desired temperature. The nitriding thermal cycle is showed in figure 1.

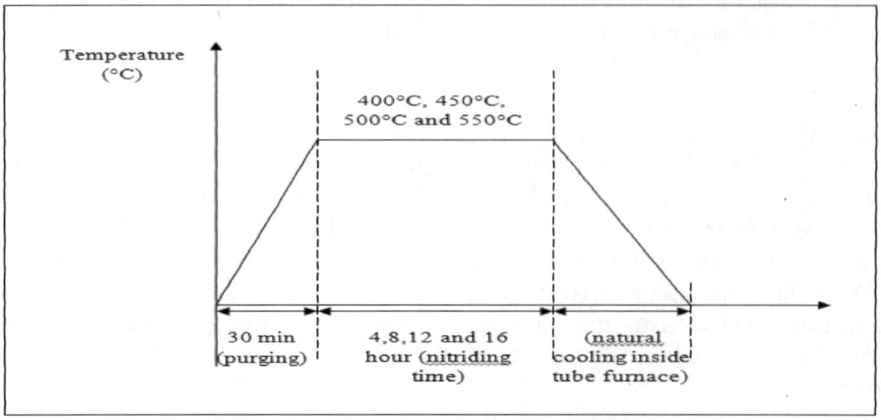

Figure 1. Nitriding thermal cycle for DSS.

The nitrided alloy layer was analysed for their hardness property using Vickers hardness tester on the cross sectional area. The test used load of 500 gm and dwelling time of 10 seconds. The phase analysis was done using x-ray diffraction whereas thickness layer measurement and micrograph using scanning electron microscopy. Table 1 showed the parameters for the experiments.

Table 1. Parameters for the experiment.

No. of run	Temperature (°C)	Time (hrs)	*Composition of ammonia(%)
1	400	4	20
2	400	8	30
3	400	12	40
4	400	16	50
5	450	4	30
6	450	8	20
7	450	12	50
8	450	16	40

*Balance is nitrogen gas

3. Results and discussion

3.1. Thickness layer and microstructure analysis

The nitride thickness layer of the nitrided duplex stainless was measured using scanning electron microscopy. Table 2 has showed the thickness layer of the nitrogen diffused with the respective parameters. Referring to figure 2 and 3, white layer is visibly which was formed on all nitrided samples. The white layer is also known as compound zone is made up of a nitride which is either involving of iron nitride or others complex nitride layer. It is not desirable because of the white layer is very brittle however, it has an extremely hard behavior. Therefore, it is necessary to remove the white layer after nitriding process to maintain the performance of nitrided samples. Besides that, an alloy nitride layer could be seen in all of the nitrided duplex stainless steel except for the samples treated at 400°C for 4, 8 and 12 hours and sample 450°C at 4 hours. However, the mentioned samples exhibited the existance of the diffusion zone as well as the white layer.

The largest thickness layer was produced at 450°C temperature for16 hours of nitriding time with 40% ammonia composition. The longer time and higher temperature of nitriding process has developed high amount of nitrogen element diffused in the nitride alloy layer of DSS [7]. According to the previous research, the thickness layer is dependable on the nitriding temperature [8]. The lowest thickness layer was developed at 400°C for 4 hours with 20% ammonia composition. This is due to insufficient temperature and time for nitrogen to diffuse in the surface layer. However, after 12 and 16 hours of nitriding, the nitrogen was reacted with chromium to form cromium nitride precipitate with dark phases which led to the depletion of chromium in the solid solution phase of the nitrided alloy layer as can be seen in figure 2 (c and d) and figure 3 (c and d). The formation of chromium nitride precipitates leads to the deterioration in corrosion resistance of DSS.

Table 2. Thickness layer and surface hardness of nitride alloy layer with respective parameters.

No. of Run	Temperature (°C)	Time (hrs)	Percentage of Ammonia	Thickness layer (μm)	Maximum surface hardness

					$(Hv_{0.5kgf})$
1	400	4	20	5.00	305.0
2	400	8	30	37.78	450.8
3	400	12	40	55.35	417.4
4	400	16	50	95.52	1000
5	450	4	30	15.6	353.0
6	450	8	20	45.00	513.3
7	450	12	50	88.00	428.8
8	450	16	40	100.00	1100.0

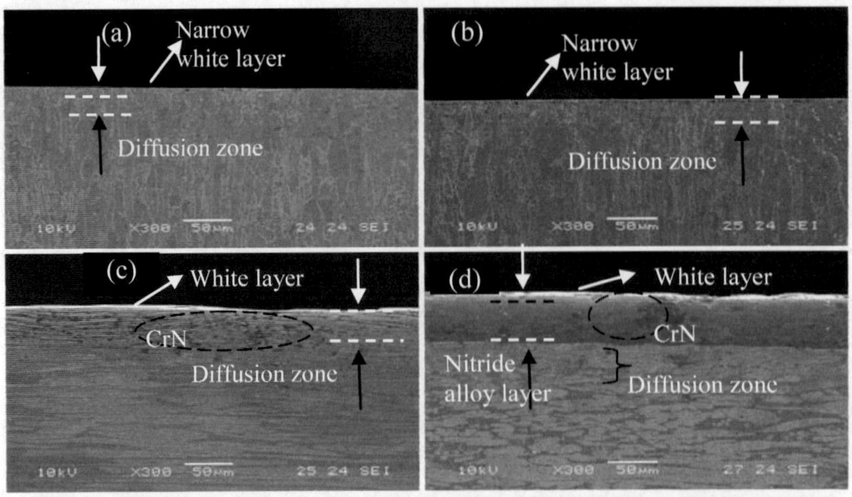

Figure 2. SEM micrographs of cross-sectional area at 400 °C of nitrided alloy layer: (a) 4 hour, (b) 8 hour (c) 12 hour and (d) 16 hour.

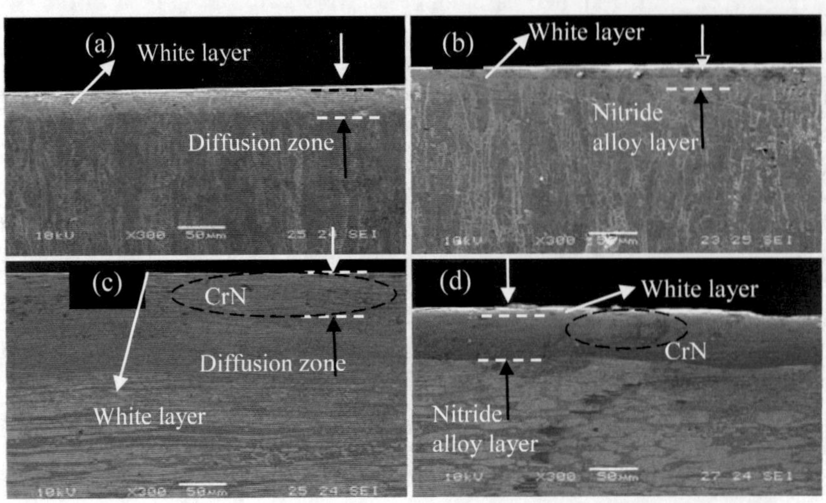

Figure 3. SEM micrographs of cross-sectional area at 450 °C of nitrided alloy layer: (a) 4 hour, (b) 8 hour (c) 12 hour and (d) 16 hour.

3.2 Vickers micro-hardness

The hardness of nitride alloy layer at the surface of DSS was tabulated in Table 2. It can be observed that all nitride alloy layer has improved the hardness compared to untreated hardness of 250Hv. This improvement associated from the formation of expanded austenite and be confirmed from the XRD result in figure 4. The hardening effect is significant due to diffusion of nitrogen into the surface of nitrided specimen to form expanded austenite or S phase layer. This result has a good agreement with previous researcher by Subbiah et al. [6]. The highest hardness produced at 450°C of temperature and 16 hours of time with 40% ammonia composition while the lowest thickness layer was developed at 400°C for 4 hours with 20% ammonia composition. This result depending on the amount of the nitrogen diffusion in the layer at different time and temperature. The trend of hardness is increasing as the temperature and time for nitriding process increase. In addition, the formation of CrN in the nitride layer at a specific range of temperature an time could lead to increasing of hardness.

3.3. Phase analysis of XRD

The phase analysis was conducted on the surface of the untreated and nitride alloy layer duplex stainless steel. In figure 4, the untreated DSS consists of α and γ phases at angle of 35° to 90°. For all treated DSS in figure 4 and 5, both peaks still exist but the peak was slightly shifted to lower angle due to formation of expanded austenite, γ_N. This is the result of the quantity of nitrogen supplied to the surface and its diffusion into the structure. However, after the nitriding time increased to 12 and 16 hours, the chromium nitride precipitate is formed in the nitride alloy layer. The formation of this phase is similar to those observed in plasma nitrided and fluidized bed nitrided samples of austenitic stainless steels, and can be attributed to the decomposition of the expanded austenite and formation of chromium nitrides [3]. The formation of CrN on the nitride layer makes the nitrided duplex stainless steel become harder which increases the hardness properties, but the previous researcher found that it also led to weakening the corrosion resistance [8].

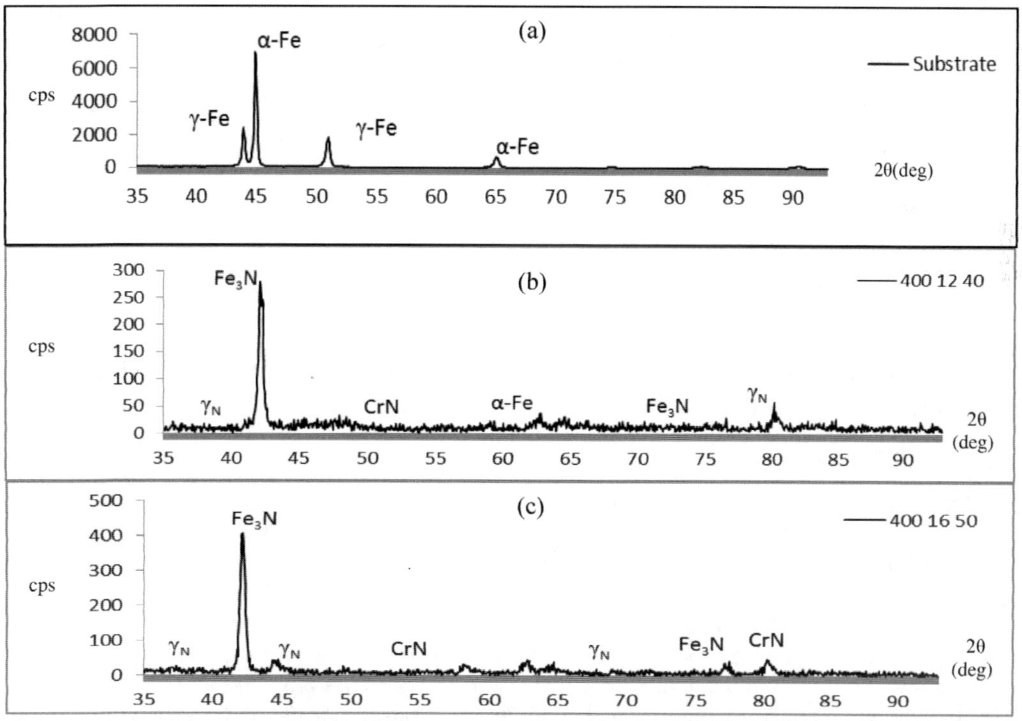

Figure 4. XRD patterns for DSS (a) substrate material (b) nitriding at 400°C + 12 hours + 40%NH_3 (c) nitriding at 400°C + 16 hours + 50%NH_3.

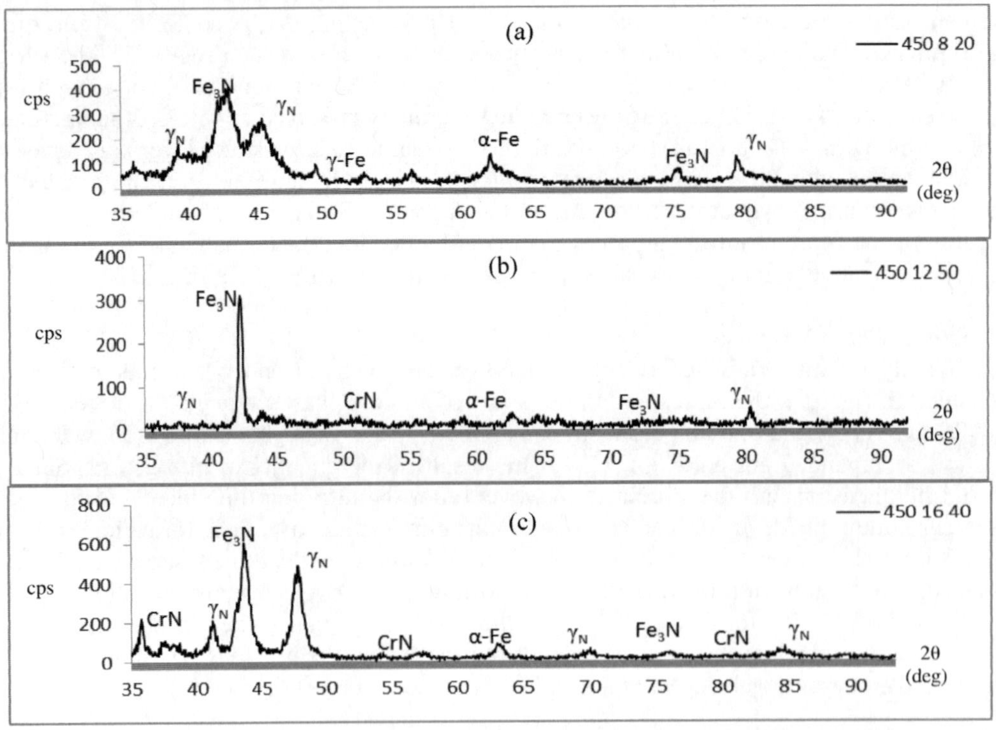

Figure 5. XRD patterns for nitriding DSS at 450°C (a) 8 hours + 20%NH_3 (b) 12 hours + 50%NH_3 (c)16 hours + 50%NH_3.

4. Conclusion

From the results, it can be concluded that the nitriding process is a feasible technique to produce hard nitride alloy layer on DSS. At temperature 400°C and 450°C, the formation of expanded austenite can be developed; however, increased treatment time such as 12 and 16 hours significantly contributes to the formation of chromium nitride which in turn can reduce the corrosion resistance. Therefore, the possible formation of chromium nitride temperature and time range should be avoided.

Acknowledgement

Thanks to the Ministry of Higher Education (MOHE) who has funded this research project under Fundamental Research Grant Scheme of FRGS15-203-0444 and National University of Malaysia.

References

[1] Maleque M A, Ghazal B A, Ali M Y, Hayyan M and Ahmed A S 2015 *Materials Science Forum* **819** 76-80
[2] Lailatul P H and Maleque M A 2017 *Procedia Engineering* **184** 737-742
[3] Haruman E, Sun Y, Malik H, Sutjipto A G E, Mridha S and Widi K 2006 *Solid State Phenomena* **118** 125-130
[4] Yan J, Gao M and Zeng X 2010 *Optics and Lasers in Engineering* **48** 512-517
[5] Takadoum J 2008 *Materials and Surface Engineering in Tribology* (France - Wiley-ISTE)
[6] Subbiah R, Karthick P, Ilavarasan R, Prasant T and Manjunath R 2014 *International Journal of Recent Technology and Engineering (IJRTE)* **03** 139 -141
[7] Paijan L H, Berhan M N, Adenan M S, Yusof N F M and Haruman E 2012 *Advanced Materials Research* **576** 260-263
[8] Yuan X, Zhao Y, Li X, and Chen L 2017 *Metals* **7** 102

IOP Conf. Series: Materials Science and Engineering **290** (2017) 012016 doi:10.1088/1757-899X/290/1/012016

Surface modification of Polycaprolactone (PCL) microcarrier for performance improvement of human skin fibroblast cell culture

N Samsudin[1], Y Z H Hashim[1,2], M A Arifin[3], M Mel[1], H Mohd Salleh[1,2], I Sopyan[4], M Abdul Hamid[5]

[1]Department of Biotechnology Engineering, Kulliyyah of Engineering, International Islamic University Malaysia, P.O. Box 10, 50728 Kuala Lumpur, Malaysia.
[2]International Institute for Halal Research and Training (INHART), Level 3, KICT Building, International Islamic University Malaysia, P.O. Box 10, 50728 Kuala Lumpur, Malaysia.
[3]Faculty of Engineering Technology, University Malaysia Pahang, Lebuhraya Tun Razak, 26300 Gambang, Kuantan, Pahang, Malaysia
[4]Department of Manufacturing and Materials Engineering, Kulliyyah of Engineering, International Islamic University Malaysia, P.O. Box 10, 50728 Kuala Lumpur, Malaysia.
[5]Institute of Bioproduct Development, Department of Bioprocess and Polymer Engineering, University Technology Malaysia, 81310 Skudai, Johor, Malaysia.

Email: yumi@iium.edu.my

Abstract. Polycaprolactone (PCL) has many advantages for use in biomedical engineering field. In the present work PCL microcarriers of 150-200 μm were fabricated using oil-in-water (o/w) emulsification coupled with solvent evaporation method. The surface charge of PCL microcarrier was then been improved by using ultraviolet/ozone treatment to introduce oxygen functional group. Immobilisation of gelatin onto PCL microspheres using zero-length crosslinker provides a stable protein-support complex, with no diffusional barrier which is ideal for mass processing. The optimum concentration of carboxyl group (COOH) absorbed on the surface was 1495.9 nmol/g and the amount of gelatin immobilized was 1797.3 μg/g on UV/O_3 treated microcarriers as compared to the untreated (320 μg/g) microcarriers. The absorption of functional oxygen groups on the surface and the immobilized gelatin was confirmed with Fourier Transformed Infrared spectroscopy and the enhancement of hydrophilicity of the surface was confirmed using water contact angle measurement which decreased (86.93° – 49.34°) after UV/O_3 treatment and subsequently after immobilisation of gelatin. The attachment and growth kinetics for human skin fibroblast cell (HSFC) showed that adhesion occurred much more rapidly for gelatin immobilised surface as compared to untreated PCL and UV/O_3 PCL microcarrier.

1. Introduction

Microcarrier is a term used in reference to microspheres that support cells in mammalian cell culture in which cells grow as monolayer on the surface of microspheres' surface [1],[2]. In a laboratory scale cell culture anchorage-dependent cells are commonly cultivated on the walls of roller bottles or non-agitated vessels such as tissue culture flasks [3].

Moving upscale, particularly to produce large amounts of bioproducts, animal cells, are routinely carried out in a bioreactor. In this condition, microcarriers are required as substrate to the anchorage-dependent cells. Different types of bioreactors such as stirred tank and fluidised bed bioreactor have utilised microcarrier to grow anchorage-dependent cells. Microcarriers are among the most established technological platforms for industrial production to increase productivity [4]. Biomaterials such as biodegradable polymer are particularly suited for the development of microcarrier. The surface can be further improved as desired by introducing functional groups such as hydroxyl and carbonyl onto the surface of polymer microparticle. For instance, a combination of ozone aeration and UV irradiation have been reported to improve immobilisation of gelatin onto the microcarrier, therefore, enhancing the anchorage dependent cell proliferation [5].

This study was set to fabricate biodegradable microcarrier beads using solvent evaporation method and identify the controlling parameters that affect particle size. Ultraviolet ozone (UV/O_3) process conditions were then optimised to improve the surface of the microcarrier beads by introducing functional groups prior to optimisation of gelatin immobilisation on the microcarrier surface. Finally, the fabricated microcarrier beads were evaluated for their cytocompatibility by culturing mammalian cell lines on the microcarrier in spinner vessel culture system.

2. Experimental procedures

2.1Materials

Polycaprolactone (PCL) (MW=45,000) in pellet form, dichloromethane used to dissolve PCL, and poly(vinyl alcohol) (PVA) were supplied by Sigma Aldrich (USA). Pure oxygen (>99%; Linde Malaysia Sdn. Bhd), toluidine blue O (Bendosen Laboratory Chemical, Germany), phosphate buffer (PBS; EMD Chemical Inc, USA), sodium dodecyl sulfate (SDS; Merck, Germany), and absolute ethanol (HmbG Chemicals, Germany) were used. Bovine gelatin was purchased from Halagel (Malaysia). Absolute ethanol from HmbG Chemicals, (Germany) and olive oil was obtained from a local hypermarket. Dulbecco's modification of eagle's medium (DMEM) in powder form, fetal bovine serum (FBS) and antibiotics (100 U/ml penicillin, 0.1 g/l streptomycin) were supplied by Gibco (USA). Sodium bicarbonate, hydrochloric acid, sodium hydroxide and trypan blue were supplied by Sigma-Aldrich. Silicone oil and ethyl acetate were purchased from Merck Millipore (Germany).

2.2 Microcarrier preparation

The preparation of microcarrier was achieved by a solvent evaporation method [6,7] with slight modifications [8]. PCL was dissolved in organic solvent. The dissolved mixture was then added dropwise into a polyvinyl alcohol (PVA). The resulting emulsion was stirred at appropriate speed for 6 hours at room temperature. The microcarrier was collected and washed with distilled water prior drying overnight at 30 °C. Parameters tested to study the effect of operation condition on the size of PCL microcarrier was stirring speed, PVA concentration and polymer matrix ratio. The UV/O_3 system equipped with ozone generator that supply with pure oxygen at flowrate of 0.5 lpm at constant standard working pressure of 20 psig. The generated ozone was supply to the Dresher bottle containing microcarrier that was placed in the UV box for 60 minutes to introduce oxygen functional group [9]. The UV/O_3 surface of PCL microcarrier were activated using 1-ethyl-3-(3-dimethylaminopropyl)-carbodiimide hydrochloride (EDAC) and *N*-hydroxysuccinimide (NHS). The microcarrier was then washed with MES buffer prior to gelatin immobilized in 80 mg/ml gelatin solution. Lastly the microcarrier was washed and dried in an oven overnight.

2.3 Characterization

For characterization, analytical techniques were used to magnify the microcarrier, visualized the structure and investigate the distribution elements on the surface of microcarrier by means of scaning electron microscope (SEM) (Hitachi, S3400N), contact angle (Phoenix, 300), attenuated total reflectant Fourier transform infrared spectroscopy (ATR-FTIR) (Thermo scientific, Nilolet iS50), and gel permeation chromatography (GPC) (Waters, 2690/DAD 996/RID V2410/LSD CD 432). Lastly, in the biological testing component, the microcarriers were tested in mammalian suspension cell culture (human fibroblast cell, HSFC) to determine the biocompatibility of the microarrier.

3. Results and Discussion

3.1 Effect of preparation condition on particle size of PCL microcarrier

The effects of the preparation conditions on the size and morphology of the microspheres were investigated. The result showed that high stirring speed (300 rpm), the emulsion is easily dispersed therefore producing smaller droplets. The size of microspheres was independent of surfactant at low concentrations (0.05-0.1%) as the surfactant is not sufficient to stabilize the droplet formation. However, the size of microspheres decreased from 384.50-190.77 μm as the surfactant concentration is increased. The microspheres size was not only affected by those two parameters, but also by the amount of (PCL) in dichloromethane (DCM) solvent. At high ratio (1:30) of the polymer matrix to the organic solvent, small microspheres size (94.64 μm) was observed and as the ratio decrease (1:10), the microspheres size increase (293.26). Convincingly, the size of microspheres can be control by controlling the parameter's variables by mean of solvent emulsion method.

3.2 Optimised microcarrier

Polycaprolactone (PCL) microcarrier was successfully produced using solvent evaporation method. The microcarrier was then treated with UV/O_3 treatment, which intendedly to introduce the oxygen functional group on the surface of the microcarrier to make it susceptible for gelatin immobilisation. The amount of oxygen functional group increased from 283.4 nmo/g to 1495.9 nmol/g [10]. Subsequently, gelatin was immobilised on the surface of treated also shows increment (Table 1) in amount of gelatin immobilised on the microcarrier surface with optimised condition.

Table 1. Amount of absorbed gelatin on untreated PCL microcarrier and UV/O_3 treated PCL microcarrier.

Microcarrier	Concentration of immobilised gelatin (μg/g)	% increased
Non-optimised PCL microcarrier	320.00±0.9	82.2%
Optimised PCL microcarrier	1797.33±21	

3.3 Contact angle measurement

The wettability of UV/O_3 treated film was compared to the untreated and gelatin coated film. Table 2 shows the water contact angle values and surface energy of untreated PCL film, UV/O_3 treated PCL film and gelatin coated PCL film. Treatment was made by relative measurement under similar condition as the microcarrier.

Table 2. Contact angle values and surface energy of untreated, UV/O_3 treated and UV/O_3 treated and gelatin immobilised PCL microspheres.

Microcarrier	**Contact angle (°)**	**Surface energy (mJ/m²)**
Untreated PCL	86.93	24.94
UV/O_3 treated PCL	69.34	41.12
UV/O_3 treated + gelatin immobilisation (i.e. gelatin coated PCL)	49.34	61.28

The value of contact angle of untreated microcarrier (86.93°) imposed a relatively hydrophobic behaviour. The angle was consequently decreased to 69.34° after UV/O_3 treatment, thus leading to the increased surface energy from 24.94 mJ/m^2 to 41.12 mJ/m^2. This increase in surface energy could be due to the incorporation of oxygen-containing functional groups like O–C=O, C=O, C–O and OH [11] on the surface of the microcarrier. According to Gomathi and Neogi [12], an increase in surface energy is due to the incorporation of the polar components on the surface by the presence of polar groups, electric charges and free radicals.

The introduction of the functional polar components on the PCL microcarrier surface not only improved its hydrophilicity but may also accommodate biomolecules components such as protein and cell growth factors to make the surface more biocompatible for cell growth and proliferation [13]. A drastic decrease in contact angle from 69.34° to 49.34° for gelatin coated PCL microcarrier was observed. This indicates further improvement in hydrophilicity as compared to UV/O_3 treated PCL microcarrier which could be due to the presence of large amount of amino terminal and carboxyl groups [14]. Figure 1 shows the decrement of contact angle of water drops on the raw PCL surface, UV/O_3 treated and gelatin coated PCL surface, respectively.

(a) (b) (c)

Figure 1. Water drop images of contact angle analysis on the surface of: (a) raw PCL, (b) UV/O_3 treated PCL and (c) gelatin coated PCL.

3.4 ATR-FTIR analysis

Figure 2 shows the ATR-FTIR spectra of PCL, UV/O_3 treated PCL and gelatin coated PCL, respectively. The peaks at 1723, 1175 and 1230 cm^{-1} are the signature peaks of polyesters which correspond to C=O, C-O-C and C-C respectively in the IR spectra [15].

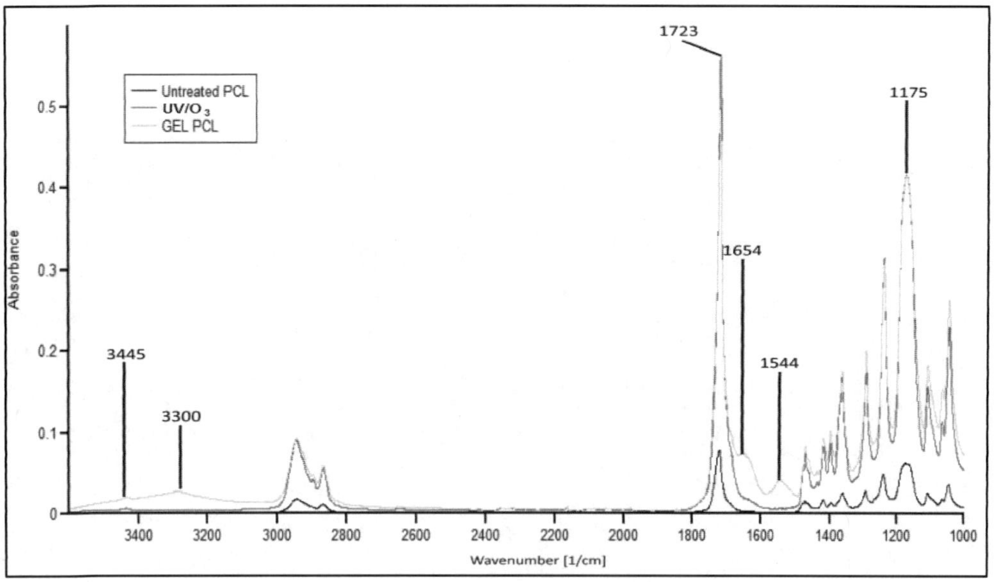

Figure 2. ATR- FTIR spectra of untreated PCL, UV/O_3 treated PCL (UV/O_3 PCL) and gelatin coated PCL (GEL PCL).

Interestingly, figure 2 shows the presence of a new peak in UV/O_3 PCL spectra at 3445 cm^{-1}. This suggests that the oxidation process by UV/O_3 could introduce O-H group in the main chain of PCL [16]. Meanwhile, marked broadening in the region of 1600-1750 cm^{-1} could be attributed to the free acids, modification of chemical environment around the carbonyl group or the formation of vinyl groups [17]. Subsequently, the successful immobilisation of gelatin onto the UV/O_3 treated PCL surface could be deduced by the presence of a broadband at 3300 cm^{-1}, possibly due to the overlapping of a hydroxyl group (O–H) and an amine group (N–H) stretching vibrations. An increase in the relative intensity of amide I band (at 1654 cm^{-1}) and amide II (at 1544 cm^{-1}) [14] also attributed to the successful of gelatin immobilisation.

3.5 SEM analysis

The morphology of the PCL microcarrier was examined by SEM, as displayed in figure 3. The microcarriers were observed to be spherical with a uniform particle size in the range of 100 to 150 μm. The surface of the untreated PCL (figure 3 (a)) appears to be smoother than the surface of the UV/O_3 treated PCL microcarrier (figure 3 (b)). Upon UV/O_3 treatment the surface appears to crease and has plenty of holes which can be clearly seen in figure 3 (e). According to Teare and Bradley [18], the roughness of the PCL surface were due to distortion caused by a series of photo-cleavage reaction involving UV rays and ozone on the polymer surface.

Further, microcarrier coated with gelatin exhibit rather smooth surface as compared to UV/O_3 treated PCL (figure 3 (c) & (f)). The creases were covered by gelatins that were immobilised onto the surface through covalent bonding with intermediate crosslink. The immobilisation was uniform over the microcarrier surface and this may help improve the biocompatibility of the microcarrier [19].

3.6 Gel Permeation Chromatography (GPC)

The effect of PCL microcarrier surface modification on weight-average molecular weight (M_w) were examined by using GPC. HPLC-grade tetrahydrofuran (THF) was used as mobile phase at room temperature with a solvent flow rate of 1 ml/min with 35 bar pressure. From the analysis, the average molecular weight of untreated PCL microcarrier was 23×10^4. After the treatment with UV/O_3, M_w of PCL microcarrier slightly decreased to 22×10^4. The change in M_w of PCL microcarrier was due to the oxidation and chain scissions of polymer through the formation of many free radicals which consequently lowers the molecular weight of the treated sample [20,21].

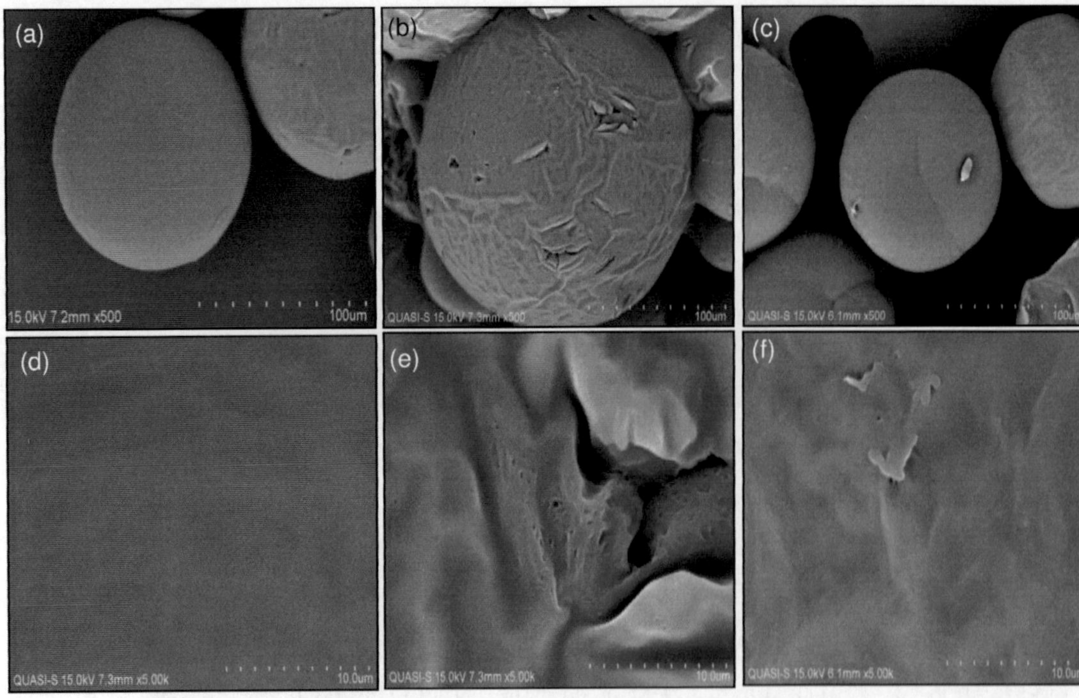

Figure 3. SEM images of the untreated PCL microcarrier (a), UV/O_3 treated PCL microcarrier (b), and gelatin coated PCL microcarrier (c). (d), (e) and (f) are higher magnification of (a), (b) and (c) respectively.

3.7 Cell attachment on gelatin coated PCL microcarrier

The growth of human skin fibroblast cells (HSFC) (1184 (ECACC 90011883) on different PCL microcarrier surfaces was investigated by suspension culture in stirred spinner flasks. Figure 4 shows the growth kinetics of HSFC on gelatin coated PCL microcarrier, UV/O_3 treated PCL microcarrier and untreated PCL microcarrier as a control. Table 3 shows the number of maximum cells attached on microcarrier, growth rate and doubling time of the three different cultures. From the graph, HSFCs were observed to attach and proliferate well on the gelatin coated microcarrier (Gel PCL) within two days after inoculation. The attached cells continued to grow until it reached maximum number of cells of 2.06×10^6 cells/ml (14 fold) as compared to the seeding concentration (1.5×10^5 cell/ml) at 72 hours after cultivation, with the fastest doubling time of 25.4 hours as compared to the cell growth on UV/O_3 PCL microcarrier (maximum cell number of 6.1×10^5 cells/ml, 84 hours after inoculation).

Meanwhile, in the untreated PCL microcarrier culture, slow cell growth with doubling time of 193.37 h and a low maximum number of cells (1.2×10^5 cells/ml) was observed. This low yield can be attributed to the difficulties of cells to attach on the surface of the untreated microcarrier which is less biocompatible as compared to gelatin coated PCL microcarrier.

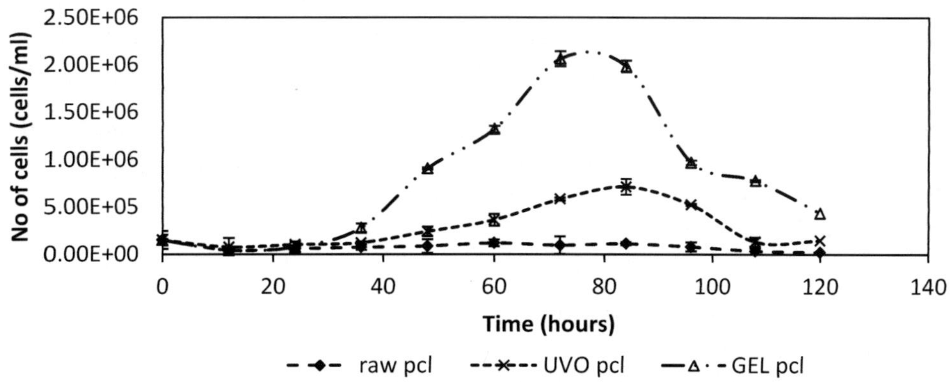

Figure 4. Growth kinetics of human skin fibroblast cell (HSFC) on different microcarriers in stirred spinner flasks: (x) UV/O_3 treated PCL, (▲) gelatin coated PCL, (●) untreated PCL.

Table 3. Values of maximum cell concentration, growth kinetics and doubling time of HSFC on different types of microcarrier.

Microcarrier	Maximum cell concentration ($x10^5$ cell/ml)	Growth rate, μ (h^{-1})	Doubling time, t_d (h)
Untreated PCL	1.2 ± 14.1	0.004	193.37
UV/O_3 treatedPCL	6.1 ± 8.4	0.017	41.51
Gelatin coated PCL	20.6 ± 4.7	0.027	25.40

In the light of the selection of dermal related cells (skin fibroblast) as models in this study, the aim was to demonstrate the usability of the biodegradable PCL microcarrier in tissue engineering applications. Microcarriers offer the advantage of mass production of cells, large scale production of bioproduct, as well as more specific application such as transplantation tools to carry cells in wound healing therapies. Skin fibroblasts have been reported to have great potential for treating skin diseases such as genetic abnormalities, infections and skin cancer [22]. They are also commonly utilized as an *in vitro* cell model for toxicity testing and the discernment of process of chemically induced skin carcinogenesis [23].

4. Conclusion

The characterization of PCL microcarrier indicates the presence of new functional groups on the surface as were evident by ATR-IR spectroscopy. The oxidation process by UV/O_3 treatment introduced O-H group in the main chain of PCL. The successful immobilisation of gelatin onto the UV/O_3 treated PCL surface was also revealed by FTIR measurement. A drastic decrease in contact angle from "69.34°" to "49.34°" for gelatin coated PCL microcarrier was observed, indicating further improvement in hydrophilicity as compared to UV/O_3 treated PCL microcarrier. The ability of gelatin coated PCL microcarriers to support growth and proliferation of cells were assessed using primary cell (human skin fibroblast cell, HSFC) shows that gelatin coated PCL microcarriers were able to support the growth of HSFC which is the primary type of cell that have low plating efficiency. The oxidation of PCL microcarrier surface by UV/O_3 treatment enhances surface wettability and promotes higher gelatin immobilisation and in turn improved cell adhesion and proliferation on the microcarrier.

Acknowledgement
The authors are grateful to the Ministry of Higher Education Malaysia, for financing the research project (PRGS 11-001-0001) under the Prototype Development Research Grant Scheme (PRGS) and to the Department of Biotechnology Engineering, International Islamic University Malaysia, for laboratory facilities.

References

[1] Brun-Graeppi A, Richard C, Michel S, Daniel M and Otto-Wilhelm 2001 *Journal of controlled release* **149** 209-224.
[2] Tan H, Huang D, Lao L, and Gao C 2009 *Journal of Biomedical Materials Research* **91** 228-238.
[3] White L A and Ades E W 1990 *J. Clin. Microbiol* **28 (2)** 283-286.
[4] Chu L and Robinson D K 2001 *Current opinion in biotechnology* **12(2)** 180-187
[5] Ahmad N Y, Mel M, Salleh H M, Ng K H, and Wong C S 2010 *American Journal of Applied Science* **7(6)** 724–731
[6] Heiskanen H, Denifl P, Pitkänen P, and Hurme M. 2012 *Advanced Powder Technology* **23(6)** 779–786
[7] Tan H, Wu J, Huang D, and Gao C. 2010 *Macromolecular Bioscience* **10(2)** 156–63
[8] Arifin M A, Mel M, Samsudin N, Hashim Y Z H-Y, Salleh H M, Sopyan I and Nordin N 2015 *J. Chem. Technol. Biotechnol.* **91** 2607-2619
[9] Murakami T N, Fukushima Y, Hirano Y, Tokuoka Y, Takahashi M, and Kawashima N 2005 *Applied Surface Science* ***249*** 425–432
[10] Samsudin N, Hashim Y Z H-Y, Arifin M A, Mel M, Salleh H M, Sopyan I, Jimat D N 2016 *Cytotechnology*
[11] Darain F, Chan W Y, and Chian K S 2010 *Soft Materials* **9(1)** 64–78
[12] Gomathi N, and Neogi S 2009 *Applied Surface Science* ***255*(17)** 7590–7600
[13] Zhu Y, Gao C, Liu X, and Shen J. 2002 *Biomacromolecules* **3(6)** 1312–9
[14] Yuan S, Xiong G, Roguin A, & Choong C 2012 *Biointerphases* **7** 30
[15] Kemala T, Budianto E and Soegiyono B 2012 *Arabian Journal of Chemistry* **5(1)** 103–108
[16] Sabino M A 2007 *Polymer Degradation and Stability* **92(6)** 986–996
[17] Wu, C 2002 *Journal of Applied Polymer Science* **88** 2888-2895
[18] Teare D O H, & Bradley R H 2000 *Surface and Interface Analysis* ***283*** 276–283
[19] Hong Y, Gao C, Xie Y, Gong Y, and Shen J 2005 *Biomaterials* ***26*(32)** 6305–13
[20] Li L, Zhong Y, Li J, Chen C, Zhang A, Xu J, and Ma Z. 2009 *Journal of Materials Chemistry,* **19(39)** 7222
[21] Yousif E, and Haddad, R 2013 *Springerplus* ***2*(1)** 398
[22] Deyrieux, A F and Wilson, V G 2007 *Cytotechnology* **54(2)** 77–83
[23] Liu, Q 2004 *Biomaterials and Tissue Engineering.* New York., Springer-Verlag Berlin Heidelberg

The influence of machining condition and cutting tool wear on surface roughness of AISI 4340 steel

A R Natasha[1], J A Ghani[2], C H Che Haron[2] and J Syarif[3]

[1]Department of Manufacturing and Materials Engineering, International Islamic University Malaysia (IIUM), Jalan Gombak, 53100 Kuala Lumpur, Malaysia
[2] Department of Mechanical and Materials Engineering, Faculty of Engineering and Built Environment, Universiti Kebangsaan Malaysia, 43600, Bangi, Selangor, Malaysia
[3]Department of Mechanical Engineering, College of Engineering, University of Sharjah, P.O.Box 27272, Sharjah, United Arab Emirates

Email:natashar@iium.edu.my

Abstract. Sustainable machining by using cryogenic coolant as the cutting fluid has been proven to enhance some machining outputs. The main objective of the current work was to investigate the influence of machining conditions; dry and cryogenic, as well as the cutting tool wear on the machined surface roughness of AISI 4340 steel. The experimental tests were performed using chemical vapor deposition (CVD) coated carbide inserts. The value of machined surface roughness were measured at 3 cutting intervals; beginning, middle, and end of the cutting based on the readings of the tool flank wear. The results revealed that cryogenic turning had the greatest influence on surface roughness when machined at lower cutting speed and higher feed rate. Meanwhile, the cutting tool wear was also found to influence the surface roughness, either improving it or deteriorating it, based on the severity and the mechanism of the flank wear.

1. Introduction

AISI 4340 steel is one of the most widely used materials in engineering products. The advantages of AISI 4340 steel such as high toughness and high strength make them dominant in various applications including automotive and machine tools industry. Among its applications include spindles, main shafts, axle shafts, gears and couplings [1]. AISI 4340 is a heat treatable alloy and the material usually undergo a heat treatment process like quenching and tempering in order for them to meet the required level of hardness, strength, and ductility [2]. It is usually machined in hardened state. According to Suresh et al. [1], the main problems associated with machining this type of steel are related to tool wear and surface quality.

Machined surface quality is one of the most important machining outputs and can be evaluated based on two categories; geometrical features, the texture of the surface and the metallurgical characteristics of the subsurface [3]. The texture of the surface includes the machined surface roughness. The value of surface roughness needs to be achieved depends on the requirement of the part's application [4]. During the turning operation, high heat generated during the process is said to improve the chip formation as well as the quality of the machined surface [5]. However, high cutting

IOP Conf. Series: Materials Science and Engineering **290** (2017) 012017 doi:10.1088/1757-899X/290/1/012017

temperature will also increase the tool wear rate during the cutting. Therefore, to control the amount of heat generated, cutting fluid is applied during the machining process.

Cryogenic application as a cutting fluid by using several types of cryogen such as liquid nitrogen (LN), oxygen, helium (LHe), methane, ethane, and argon to reduce the cutting temperature during machining is becoming more popular [6]. Extensive research on the effect of cryogenic application of numerous work materials on the machining outputs such as cutting temperature, tool wear, cutting forces, and surface quality have been conducted. Kaynak et al. [7] had reviewed some of these research in their comprehensive review paper and listed some of the work materials being studied, such as Inconel 718 [8–10], Ti-6Al-4V [10–15], various steels [16–20], shape memory alloys [21–23], magnesium alloys [24] and other engineering materials [25].

As the machining progresses, the cutting tool wear is also progresses. However, the effect of cryogenic coolant and the tool wear on machined surface roughness has not been fully explored yet. In this study, AISI 4340 steel was machined in dry and cryogenic conditions by using multilayer coated carbide cutting tool. The effects of machining conditions and cutting tool wear on machined surface roughness were evaluated.

2. Experimental procedures

AISI 4340 round bar with 100 mm initial diameter was turned in dry and cryogenic machining conditions on a TORNADO CNC lathe machine with the capability of 6000 rpm maximum spindle speed. The experimental tests were carried out using a Sumitomo flat face P10 carbide insert coated with TiC + Al_2O_3. Cylindrical liquid nitrogen (LN) tank was connected to a flexible hose and a copper pipe was used as a nozzle pointing to the cutting zone as shown in figure 1(a) and figure 1(b). The distance between the tip of the nozzle to the cutting point was fixed at 2 cm.

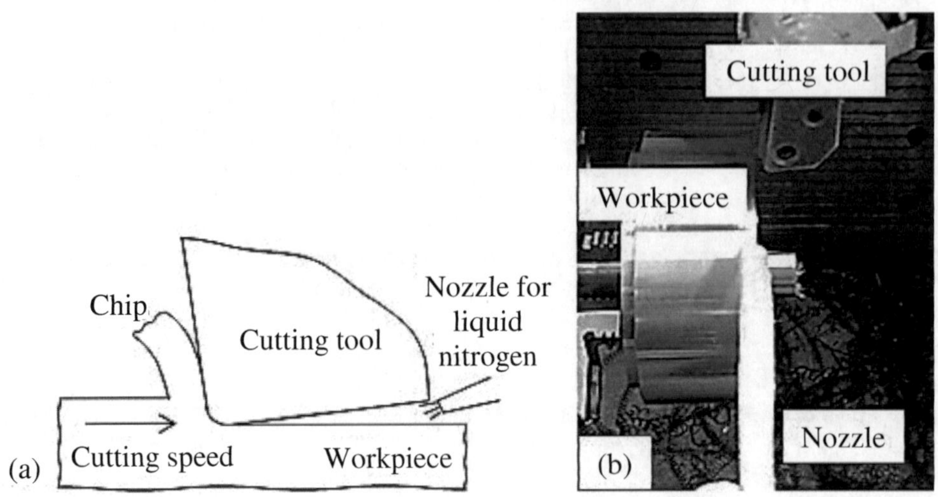

Figure 1. Nozzle orientation at the clearance face of the tool (a) Schematic diagram and (b) Photograph of the nozzle setup at CNC lathe machine.

Machining tests were carried out at cutting speeds of 160, 200 and 240 m/min. The feed was varied at 0.3 and 0.4 mm/rev and the depth of cut employed was 1.0 mm. Machining was interrupted at regular intervals to measure the tool flank wear using an Olympus SZ61 stereo microscope. The machining operation was stopped when the tool flank wear measurement reached one of the tool failure criteria as mentioned in ISO 3685 [26]; that is when average flank wear (V_B) $\geq$ 0.3 mm or the maximum flank wear (V_{Bmax}) $\geq$ 0.6 mm.

The surface roughness of the machined part were measured using a Mitutoyo Surftest SJ-310 portable surface roughness tester. The arithmetic average roughness value (R_a) in micrometer (μm) was taken at every first, middle, and end of the run of each experimental test. This was done as to compare the effect of tool wear on the roughness of the machined surfaces. The measurement was performed five times at five different spots, then the average value was calculated to represent the value of surface roughness.

3. Results and Discussion

The results of machined surface roughness of the work material AISI 4340 alloy steel when cutting using a new multilayer coated carbide tool are presented in figure 2. The value of the surface roughness, R_a reported for each experimental run was the average value of five roughness readings. Figure 2 shows the comparison of R_a of the machined surface using a new cutting tool in dry and cryogenic condition for various values of cutting speed and feed rate.

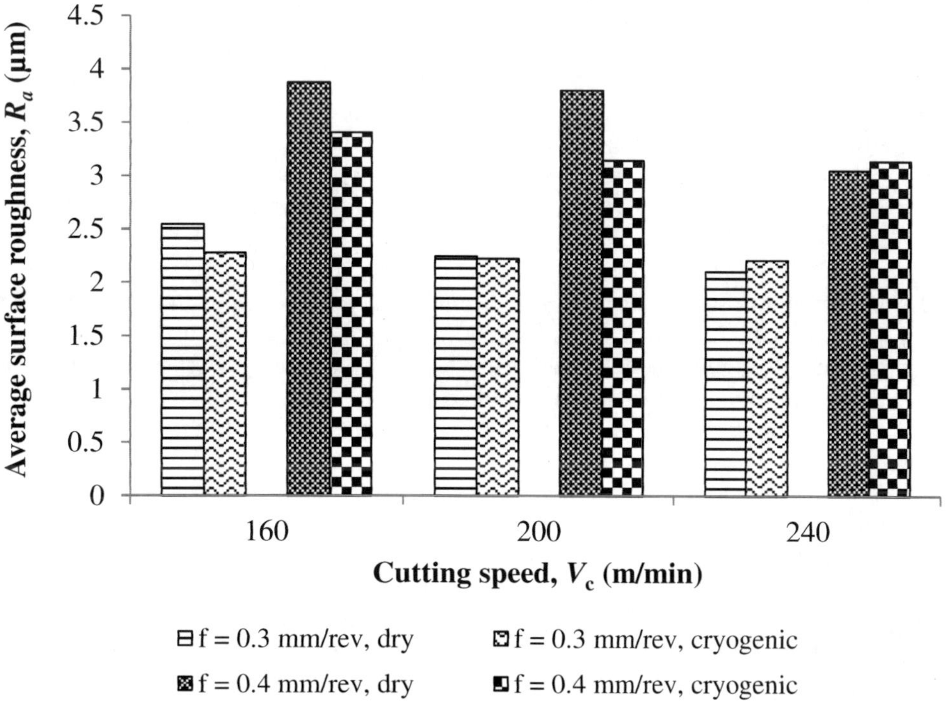

Figure 2. Comparison of average surface roughness (R_a) in experimental test of turning AISI 4340 alloy steel in dry and cryogenic condition for various cutting speeds and feed rate (doc = 1.0 mm).

The experimental result shows that the surface roughness values ranging from 2.10 μm to 3.88 μm were recorded during the turning process of AISI 4340 alloy steel under dry and cryogenic condition using a new cutting tool. The lowest value of machined surface roughness was measured at cutting speed of 240 m/min, feed rate 0.3 mm/rev, machined in dry condition, while the highest value of surface roughness was obtained at cutting speed 160 m/min, feed rate 0.4 mm/rev, machined in dry condition.

It was observed that the value of surface roughness decreased as the cutting speed increases both in dry and cryogenic condition, at both feed rate of 0.3 mm/rev and 0.4 mm/rev. This might be due to the higher temperature generated at higher cutting speed, hence increasing the thermal softening effect of the work material. Therefore, lower forces needed to cut the workpiece hence resulted in better roughness compared to those obtained at lower cutting speed [27]. In addition, the high temperature may also lead to the adhesion of the work material onto the tool nose, consequently smoothening the nose profile, thus contributing to the better surface finish [28]. However, the differences in the roughness values at different cutting speeds were not that significant and apparent in cryogenic turning.

The effect of feed rate on surface roughness was observed to be more significant compared to the cutting speed. As shown in figure 2, the values of machined surface roughness increased tremendously as the feed rate was increased from 0.3 mm/rev to 0.4 mm/rev. This increment pattern was observed in both dry and cryogenic conditions at all values of cutting speed. These results are in agreement with equation (3.1) where high feed rates will produce a rougher machined surface. On the other hand, low feed rate will improve the machined surface finish when cutting using the same value of tool nose radius. With the increase of the feed rate, higher cutting forces are needed for the cutting tool to cut the work material, which resulted in higher cutting temperature. This may lead to a worse tool wear condition and deteriorate the surface finish [29]. However, despite a better surface finish when cutting at the lower feed, the rate of material removed will be decreased too.

$$R_a = \frac{f^2}{32r} \tag{3.1}$$

This result is consistent with some other previous researches [1,28,30,31]. According to Thakur et al. [32], at the higher feed rate, the friction between the tool and the workpiece increases. As the cutting progresses, the temperature will increase and result in reduction of material shear strength. The chips of the work material that stick to the tool edge will cause the surface roughness escalated.

From figure 2 as well, it can be said that the application of liquid nitrogen during the turning process produced slightly better surface roughness compared to dry turning at cutting speed of 160 m/min and 200 m/min. At cutting speed of 240 m/min, the cryogenic application, however resulted in slightly rougher machined surface than dry turning. This might be due to the lower temperature generated in cryogenic turning, hence reducing the tendency of sticky work material at the cutting edge which leads to improved surface finish. Besides, the application of LN during machining reduces the coefficient of friction at the interfaces and this can be another factor that contribute to the better surface roughness [33].

The influence of cutting tool wear on the machined surface roughness was also investigated in this study. Figure 3 shows the readings of average surface roughness as a function of cutting tool condition. The R_a values were measured at flank wear, $V_B = 0$ μm (new tool), $V_B \geq 0.15$ μm (medium wear), and $V_B \geq 0.3$ μm (wear). It was observed that there were a few patterns of the graph relation between surface roughness and tool wear. Several machining conditions produced better surface roughness when cutting using a worn tool. This is most probably due to the effective flattening of the tool nose resulted from the increase in the worn flat on the flank of the tool [34], which increases the tool nose radius as shown in figure 4, thus contributing to better machined surface quality. A significant decrement of the roughness value when cutting using a worn tool at cutting speed of 240 m/min in cryogenic condition might be due to the combined benefits of higher speed and cryogenic application to the surface roughness improvement as discussed previously.

Meanwhile, several other machining conditions produced rougher machined surface when cutting using a worn tool. As machining time increases, tool sharpness deteriorates and leads to a degraded surface roughness [34]. Besides, the adhesion of the work material on the tool edge and the tool flank face may also be a factor contributing to a bad surface finish. A significant increment of the roughness value was observed when cutting using a worn tool at cutting speed 200 m/min in cryogenic condition. Closer observation of the cutting tool used in this machining condition showed that the cutting tool experienced fracture wear on the tool flank face. This might deteriorate the machined surface even more compared to other machining conditions.

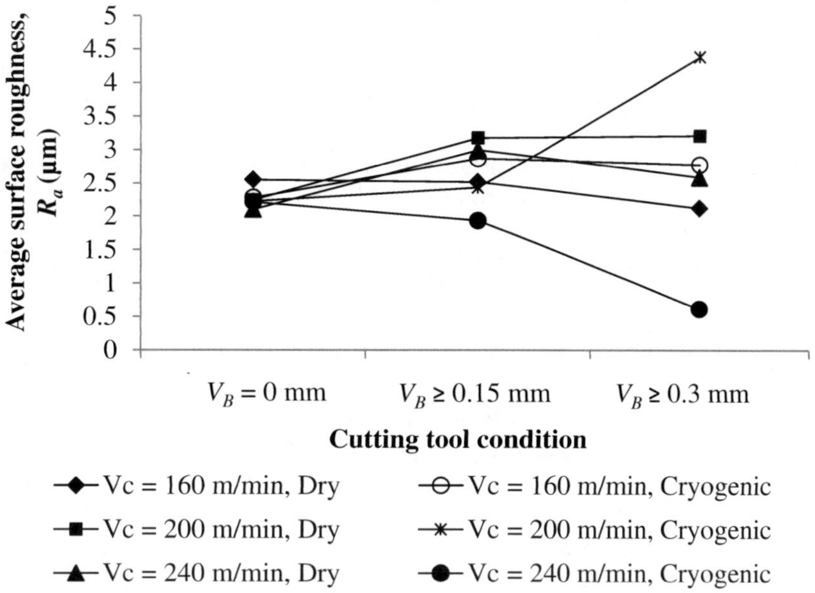

Figure 3. Effect of cutting tool condition on the average surface roughness (R_a) of AISI 4340 alloy steel machined at different cutting speed under dry and cryogenic environment (f = 0.3 mm/rev).

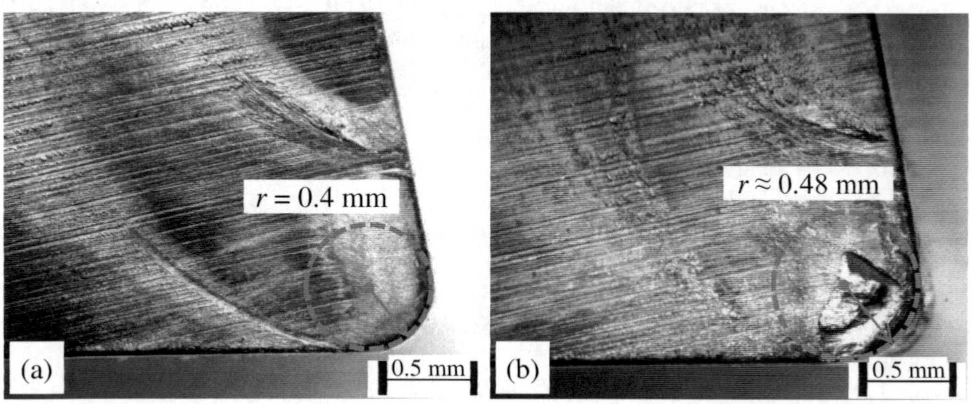

Figure 4. Comparison of tool nose radius in (a) new tool, and (b) worn tool.

4. Conclusion

In this work, AISI 4340 steel was turned using coated carbide insert in order to assess the effect of machining conditions and cutting tool wear on the machined surface roughness. It was found that higher cutting speed and lower feed rate produced better surface finish both in dry and cryogenic cutting conditions. However, the differences between the roughness values were not apparent in cryogenic turning as the cutting speed varied. Cryogenic cutting improved the machined surface roughness, except at cutting speed 240 m/min. For the effect of cutting tool wear on the surface roughness, several machining conditions produced better surface roughness when cutting using a worn tool with cutting at 240 m/min speed in cryogenic condition showed the most significant decrement of roughness value. On the other hand, several other machining conditions produced rougher machined surface when cutting using a worn tool especially when machined at cutting speed of 200 m/min in cryogenic condition.

References

[1] Suresh R, Basavarajappa S, Gaitonde V N and Samuel G L 2012 *Int. J. Refract. Met. Hard Mater.* **33** 75–86
[2] Al-Ghamdi K A, Iqbal A and Hussain G 2014 *Proceedings of the Institution of Mechanical Engineers, Part B: Journal of Engineering Manufacture* 1–21
[3] Davim J P 2010 *Surface Integrity in Machining* vol 53(London: Springer-Verlag London)
[4] Petropoulos G P, Pandazaras C N and Davim J P 2010 *Surface Integrity in Machining* ed J P Davim (Springer) 37–66
[5] Klocke F, Settineri L, Lung D, Claudio Priarone P and Arft M 2013 *Wear* **302** 1136–44
[6] Shokrani A, Dhokia V and Newman S T 2012 *Int. J. Mach. Tools Manuf.* **57** 83–101
[7] Kaynak Y, Lu T and Jawahir I S 2014 *Mach. Sci. Technol.* **18** 149–98
[8] Courbon C, Pusavec F, Dumont F, Rech J and Kopac J 2013 *Tribol. Int.* **66** 72–82
[9] Kaynak Y 2014 *Int. J. Adv. Manuf. Technol.* **72** 919–33
[10] Wang Z Y and Rajurkar K P 2000 *Wear* **239** 168–75
[11] Bermingham M J, Kirsch J, Sun S, Palanisamy S and Dargusch M S 2011 *Int. J. Mach. Tools Manuf.* **51** 500–11
[12] Hong S Y, Ding Y and Jeong W 2001 *Int. J. Mach. Tools Manuf.* **41** 2271–85
[13] Hong S Y, Markus I and Jeong W cheol 2001 *Int. J. Mach. Tools Manuf.* **41** 2245–60
[14] Strano M, Chiappini E, Tirelli S, Albertelli P and Monno M 2013 *Proc. Inst. Mech. Eng. Part B J. Eng. Manuf.* **227** 1403–8
[15] Yasa E, Pilatin S and Çolak O 2012 *J. Prod. Eng.* **2** 1–9
[16] Dhar N R, Paul S and Chattopadhyay A B 2001 *Wear* **249** 932–42
[17] Paul S and Chattopadhyay A B 1996 *Int. J. Mach. Tools Manuf.* **36** 63–72
[18] Venugopal K a., Paul S and Chattopadhyay a. B 2007 *Wear* **262** 1071–8
[19] Pušavec F, Govekar E, Kopač J and Jawahir I S 2011 *CIRP Ann. - Manuf. Technol.* **60** 101–4
[20] Rotella G, Umbrello D, Dillon O W and Jawahir I S 2012 *J. Adv. Mech. Des. Syst. Manuf.* **6** 989–98
[21] Kaynak Y, Karaca H E and Jawahir I S 2011 *6th International Conference and Exhibition on Design and Production of Machines and Dies/Molds* pp 123–8
[22] Kaynak Y, Karaca H E, Noebe R D and Jawahir I S 2013 *Procedia CIRP* **8** 498–503
[23] Kaynak Y, Karaca H E, Noebe R D and Jawahir I S 2013 *Wear* **306** 51–63
[24] Pu Z 2012 *Cryogenic machining and burnishing of AZ31B Magnesium alloy for enhanced surface integrity and functional performance*
[25] Wang Z and Rajurkar K 1997 *Int. J. Mach. Tools Manuf.* **37** 319–26
[26] ISO 3685 1993 *Tool-life testing with single-point turning tools* (Geneva: International Organization for Standardization (ISO))
[27] Pawade R S, Joshi S S, Brahmankar P K and Rahman M 2007 *J. Mater. Process. Technol.* **192–193** 139–46
[28] Dhar N R and Kamruzzaman M 2007 *Int. J. Mach. Tools Manuf.* **47** 754–9
[29] Arunachalam R M, Mannan M A and Spowage A C 2004 *Int. J. Mach. Tools Manuf.* **44** 1481–91
[30] Lima J G, Ávila R F, Abrão a. M, Faustino M and Davim J P 2005 *J. Mater. Process. Technol.* **169** 388–95
[31] Suresh R, Basavarajappa S and Samuel G L 2012 *Measurement* **45** 1872–84
[32] Thakur D G, Ramamoorthy B and Vijayaraghavan L 2009 *Mater. Des.* **30** 1718–25
[33] Natasha A R, Ghani J A, Syarif J, Haron C H C and Hadi M A 2014 *Appl. Mech. Mater.* **554** 7–11
[34] More A S, Jiang W, Brown W D and Malshe A P 2006 *J. Mater. Process. Technol.* **180** 253–262

IOP Conf. Series: Materials Science and Engineering **290** (2017) 012018 doi:10.1088/1757-899X/290/1/012018

Melting of SiC powders preplaced duplex stainless steel using TIG welding

M A Maleque[1] and M Afiq[1]

[1]Department of Manufacturing and Material Engineering, Kulliyyah of Engineering, International Islamic University Malaysia, Kuala Lumpur, Malaysia

Email:maleque@iium.edu.my

Abstract: TIG torch welding technique is a conventional melting technique for the cladding of metallic materials. Duplex stainless steels (DSS) show decrease in performance under aggressive environment which may lead to unanticipated failure due to poor surface properties. In this research, surface modification is done by using TIG torch method where silicon carbide (SiC) particles are fused into DSS substrate in order to form a new intermetallic compound at the surface. The effect of particle size, feed rate of SiC preplacement, energy input and shielding gas flow rate on surface topography, microstructure, microstructure and hardness are investigated. Deepest melt pool (1.237 mm) is produced via TIG torch with highest energy input of 1080 J/mm. Observations of surface topography shows rippling marks which confirms that re-solidification process has taken place. Melt microstructure consist of dendritic and globular carbides precipitate as well as partially melted silicon carbides (SiC) particles. Micro hardness recorded at value ranging from 316 HV0.5 to 1277 HV0.5 which shows increment from base hardness of 260 $HV_{0.5kgf}$. The analyzed result showed that incorporation of silicon carbide particles via TIG Torch method increase the hardness of DSS.

1. Introduction

Duplex stainless steel has been used in various applications such as high pressure valves and heat exchangers due to its high strength and stress corrosion resistance. Despite many of duplex stainless steel advantages, there are also some limitations in which prohibit the usage of duplex stainless steel in certain application. The usage of duplex stainless steel in corrosive environment and aggressive environment will lead to the material's failure due to combined action of wear and corrosion simultaneously [1]. In order to improve properties of duplex stainless steels, surface modification technique can be done taking into account corrosion are phenomenon occurring at the surface of the components. Also, in service components or parts do not solely rely on the inherent properties of their bulk material but also on the properties of their surface. Surface modification is a method involving all processes that are used in order to improve surface properties of engineering materials. Surface modification can be done by transforming surface microstructure or by developing a new intermetallic compound layer on a material's surface in which the process done does not affect properties of bulk material [2].

In recent studies, TIG Torch has gained more interest as this method has been considered as cheaper alternatives of lasers treatments [3, 4]. This method can be used to incorporate ceramic reinforcement materials which enhance mechanical properties of substrate materials via surface melting. Surface melting method is used because of its ability to eliminate interfacial incompatibility by creating thermodynamically stable reinforcements based on their nucleation from the substrate material. Idris et al. [5] investigated the TiC composite coating on AISI 4340 using TIG torch method. Other finding by Adeleke et al. [3] studied the Fe-C-Si coatings on CP-Ti melted under TIG torch techniques. The hardness value increased with the value of 800 Hv compared to 200 Hv of substrate material. Silicon carbide is chemically inert to acids and alkalis in temperature range up to 800 ^{0}C. In the range of 1200℃ and 1900℃ in air, silicon carbide has the ability to form a protective layer of silicon oxide [6]. Silicon carbide also has high chemical and wear resistance due to its ability to resist degradation under physical condition. This material is also one of the semiconductor materials that have the ability to withstand high temperature electronic and electro-optic application. These properties of silicon carbide make them suitable candidates for surface modifications of stainless steel.

Yet, no study has been carried out to melt the SiC powder with DSS using TIG technique for surface modification. Therefore, the purpose of this work is to investigate the microstructure and hardness behavior of DSS surface with SiC powder preplacement using TIG torch technique melted at energy input of 480, 768 and 1728 J/mm.

2. Experimental procedures

S31803 duplex stainless steel was used as a substrate with dimension of 33 mm x 50 mm x 10 mm. The surfaces of the samples are milled and ground in order to obtain flat surfaces for silicon carbide preplacement. SiC particles of size 20 µm, 40 µm and 60 µm were used as reinforcement materials. SiC content of 0.5, 1.0 and 1.5 mg/mm^2 were preplaced on the surfaces by mixing with PVA (polyvinyl acetate) binder. Powder preplaced samples were then dried in Contherm oven at 80 ^{0}C for 1 hour to ensure removal of moisture.

Conventional TIG Torch welding equipment is used to melt surface of powder preplaced samples in order to produce SiC reinforced composite coating samples. Two types of layers were produced viz. single track and overlapped tracks welding. Tungsten electrode was used to generate an arc between the electrode and the powder preplaced samples using Argon as shielding gas at flow rate of 15, 20 and 25 L/min. During the process, current and voltage were set at constant of 80 A and 30 V while traversing speed of 1.0 mm/s, 1.5 mm/s and 2.0 mm/s were used. The parameters used in TIG Torch are shown in table 1.

Table 1. Parameters used in TIG torch technique.

No. of DSS sample	Preplacement Rate (mg/mm^2)	SiC particles size (um)	Traveling speed (mm/s)	Heat Input (J/mm)	Gas Flow Rate (L/min)
Sample 1	0.5	60	1	1080	25
Sample 2	0.5	40	1.5	720	20
Sample 3	0.5	20	2	540	15
Sample 4	1.0	40	1	1080	15
Sample 5	1.0	20	1.5	720	25
Sample 6	1.0	60	2	540	20
Sample 7	1.5	20	1	1080	20
Sample 8	1.5	60	1.5	720	15
Sample 9	1.5	40	2	540	25

A Miller TIG torch source attached with a semi-automatic traversing arm was used for melting purpose. The direct current electrode negative (DCEN) mode was used and a 3.2 mm diameter tungsten electrode was used to strike an arc between the electrode and the workpiece in pure Argon atmosphere to prevent excessive oxidation [3]. Minitab 17 software was used to determine the no. of experiments to be conducted with the predetermined variables and their levels [5]. The optimum setting is the parameter combination, which has the highest S/N ratio [7]. The experiments have been carried out as per Taguchi's L16 orthogonal array (OA) design.

After welding, a complete cross section of the weldments composed of the parent plate and TIG melted surface layer was removed transverse to the welding direction to produce specimens for microstructural examination and hardness measurements. All specimens were ground on silicon carbide "wet" papers from P320 to P1200 grade and polished sequentially on diamond wheels using 3, 1 and 0.25 μm grades. The weld metal microstructure was revealed by using 2% Nital. The microstructure was evaluated using a JMS 5600 field emission scanning electron microscope.

3. Results and Discussion

3.1 Surface Topography

The topographies of the TIG melted surface layer produced were observed under FESEM and shown in figure 1. Rippling marks can be seen at surface of all samples which proved that re-solidification has occurred at region near substrate surface. Ripple marks are produced when there is a difference between re-solidification rates. Area behind the electrode is usually re-solidify first compared to area being melted.

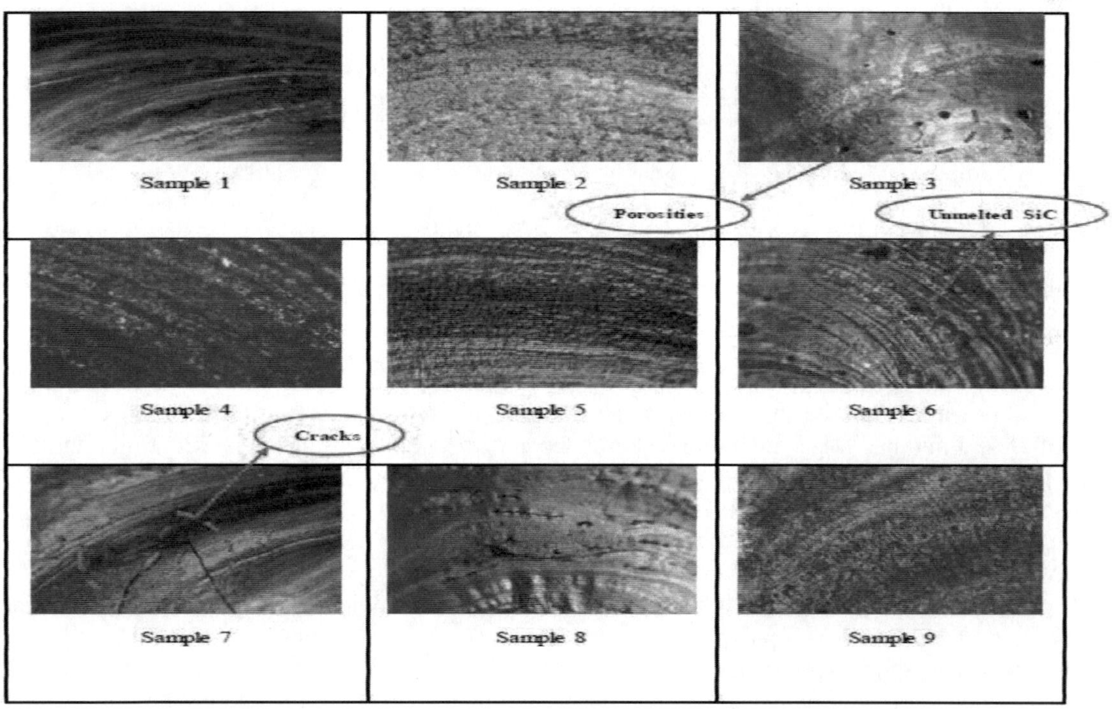

Figure 1. Surface topography of TIG melted layer of DSS under FESEM.

Surface layer produced with heat input of 1080 J/mm^2 at SiC content of 0.5 mg/mm^2 (sample 1) shows relatively smooth surface texture (weld track) and contain more ripple marks compared to other samples. This is because higher heat input and lower preplacement rate increased melt fluidity and prolong solidification time during TIG Torch process [8]. On the other hand, sample produced at heat input of 540 J/mm^2 and preplacement rate of 1.5 mg/mm^2 (sample 9) demonstrate less smooth surface texture (weld track) with less number of ripples. This is due to its high viscosity melt produced during TIG Torch process [3].

Unmelted SiC particles can be observed on surface of sample 3 and sample 6. Unmelted particles is caused by lower heat input used during TIG Torch process due to increased welding speed of glazing. Other features that can be observed are the presence of porosities and cracks. Porosities occurred when entrapped gas escaped from the melt surface during resolidification due to high viscosity of the melt produced. Pores is observed when heat input used is low thus producing viscous melt. Upon, resolidification, the viscous melt is unable to fill the void left by escaping gas thus producing pores on the surface of the samples [6].

3.2 Melt pool dimension

The melt pool width and depth for all samples are shown in table 2. The largest melt pool dimension was observed in sample 1 which is produced with preplacement rate of 0.5 mg/mm^2 and heat input at 1080 J/mm^2. Smallest melt pool was observed in sample 9 which is produced with preplacement rate of 1.5 mg/mm^2 and heat input at 540 J/mm^2.].

Table 2. Melt pool dimension of TIG melted layer of DSS.

No. of DSS sample	Width of surface layer (mm)	Depth of surface layer (mm)
Sample 1	6.14	1.24
Sample 1	5.47	0.99
Sample 3	5.17	0.68
Sample 4	5.83	1.2
Sample 5	5.15	0.85
Sample 6	4.87	0.79
Sample 7	5.46	1.12
Sample 8	5.04	0.76
Sample 9	4.68	0.62

It can be seen that at the same preplacement rate of 0.5 mg/mm^2, surface layer produced at heat input of 1080 J/mm has larger melt dimension compared to samples produced at 720 J/mm and 540 J/mm. Higher heat input produced higher width and depth of the melt pool as can be seen in sample 1 compared to sample 2 and sample 3. This is because of high heat input resulted higher temperature of melt during TIG welding. This higher temperature melt has lower solidification rate which assists the melt to move towards the substrate thus producing larger melt pool [5].

Melt pool of sample produced at 0.5 mg/mm^2 is larger than sample produced at 1.0 mg/mm^2 and 1.5 mg/mm^2 at the same heat input of 1080 J/mm. Sample 1 (which is produced at highest heat lowest preplacement rate) shows largest melt pool dimension when compared to sample 4 and sample 7. It can be seen that increasing preplacement rate during the process at the same energy input affects the dimension of melt pool. This is because lower preplacement rate (i.e. loer SiC) produced content on the surface. Lower content of SiC particles on the surface prolong heat dissipation time from the base metal thus producing larger depth melt pool.

3.3 Hardness measurement

The new surface layer samples were cut along the cross-section for micro hardness measurements and reading was taken along the melt pool region. The hardness of substrate material was 260 $HV_{0.5kgf}$. Hardness profiles for each sample was constructed by plotting Vickers hardness measurements against depth at which measurement was taken. Hardness profiles of the samples are shown in table 3 after measuring maximum hardness for each sample from the average of five readings.

Table 3. Maximum Micro Hardness of the Samples.

No. of DSS sample	Width of surface layer (mm)
Sample 1	316.8
Sample 1	614.3
Sample 3	899.8
Sample 4	864.5
Sample 5	661.1
Sample 6	1126
Sample 7	830.4
Sample 8	1250
Sample 9	1263

From table 3, it can be seen that there is an increment in hardness value of all SiC reinforced composite coating (surface layered) sample compared to hardness value of substrate sample. Sample 8 shows the highest value of hardness recorded in this experiment with the value of 1277 $HV_{0.5}$ while sample 1 shows the lowest value of hardness with the value of 316.8 $HV_{0.5}$. Sample 8 (with preplacement rate of 1.5 mg/mm^2) shows highest hardness value of 1250 $HV_{0.5}$ while sample 2 (with 0.5 mg/mm^2) at the same heat input shows lower hardness value of 614.3 $HV_{0.5}$. Higher content of SiC preplaced on the sample results from the higher amount of SiC which precipitates during re-solidification, hence increase the value of hardness of the DSS material.

4. Conclusion

From the present study, the following conclusions are drawn: Silicon carbide particles incorporation into duplex stainless steel can be done by using TIG Torch method. The usage of different preplacemet rate, size of SiC particles, heat input and gas flow rate affect melt dimension, microstructure and hardness of the produced samples. Melt pool dimension produced with 1.5 mg/mm^2 preplacment rate is smaller compared to those produced with 0.5 and 1.0 mg/mm^2. On the other hand, heat input of 1080 J/mm produced larger melt pool dimension compared to those produced with 720 and 540 J/mm. Highest hardness recorded is 1250 $HV_{0.5}$ from the heat input of 720 J/mm with SiC content of 1.5mg/mm^2. The hardness development depends on the population of dendrite structures formed during TIG Torch process.

Acknowledgement

Special thanks to the Ministry of Higher Education (MOHE) who has funded this research project under Fundamental Research Grant Scheme of FRGS15-203-0444.

References

[1] Chen Q, Xie Z, Chen T and Gong F 2016 *Materials* **9** 963
[2] Shibe V and Chawla V 2014 *International Journal of Research In Mechanical Engineering & Technology* **4** 92-95
[3] Maleque M and Adeleke S 2013 *International Conference on Mechanical, Industrial and Materials Engineering* 668-673
[4] Maleque M A, Bello K A, Adebisi A A and Akma N 2017 *IOP Conf. Series: Materials Science and Engineering* **184** 012035

[5] Idriss A N M, Mridha S and Maleque M A 2012 *Advanced Materials Research* **576** 467-470
[6] Patel P, Mridha S, Baker N 2014 *Materials Science and Technology* **30** 1506-1514
[7] Idriss A N M and Mridha S, Baker T N 2014 *Materials Science and Technology* **31** 1209-1213
[8] Maleque M A, Bello K 2013 *Applied Mechanics and Materials* **378** 259-264

IOP Conf. Series: Materials Science and Engineering **290** (2017) 012019 doi:10.1088/1757-899X/290/1/012019

Influence of Wire Electrical Discharge Machining (WEDM) process parameters on surface roughness

Mohammad Yeakub Ali[1], Asfana Banu[1], Mazilah Abu Bakar[2]

[1]Department of Manufacturing and Materials Engineering, International Islamic University Malaysia, PO Box 10, 50728 Kuala Lumpur, Malaysia
[2]Department of Mechanical Engineering, Politeknik Sultan Haji Ahmad Shah, 25350 Semambu, Kuantan, Pahang, Malaysia

Email: mmyali@iium.edu.my

Abstract. In obtaining the best quality of engineering components, the quality of machined parts surface plays an important role. It improves the fatigue strength, wear resistance, and corrosion of workpiece. This paper investigates the effects of wire electrical discharge machining (WEDM) process parameters on surface roughness of stainless steel using distilled water as dielectric fluid and brass wire as tool electrode. The parameters selected are voltage open, wire speed, wire tension, voltage gap, and off time. Empirical model was developed for the estimation of surface roughness. The analysis revealed that off time has a major influence on surface roughness. The optimum machining parameters for minimum surface roughness were found to be at a 10 V open voltage, 2.84 μs off time, 12 m/min wire speed, 6.3 N wire tension, and 54.91 V voltage gap.

1. Introduction

Electrical discharge wire cutting, more commonly known as wire electrical discharge machining (WEDM) is a specialized thermal machining process capable of accurately machining parts with varying hardness or complex shapes, which have sharp edges that are very difficult to be machined by the main stream machining processes. At present, WEDM is a widespread technique used in industry for high-precision machining of all types of conductive materials such as metals, metallic alloys, graphite, or even some ceramic materials, of any hardness [1], [2]. The degree of accuracy of workpiece dimensions obtainable and the fine surface finishes make WEDM particularly valuable for applications involving manufacture of stamping dies, extrusion dies, and prototype parts. Without WEDM, the fabrication of precision workpiece requires many hours of manual grinding and polishing [3]. Apart from that, WEDM is widely used in the area of production of aerospace parts micro gas turbine blades and electronic components [4].

WEDM is based on electrical discharge machining process, which is also called electro-erosion machining process. When the gap voltage is sufficiently large (i.e reaches the breakdown voltage of dielectric fluid), high power spark is produced, where the temperature increases up to 10,000 °C and allows the removal of metal from the machining area [5], [6]. In WEDM, performance measures are the indicators that are used to observe or to assess the quality of finished products or parts, whereas, the process parameters are the variables that influence the machining process. The most substantial performance measures in WEDM are surface finish, material removal rate, and kerf width. Surface roughness is used to control the quality of the finished part while kerf width is used to determine the accuracy of dimension and material removal rate for economic purposes. According to previous research, the surface roughness improves when the pulse duration and the discharge current decreases [7]. All of these performance measures are usually affected by the process parameters such as pulse on-time, open voltage, servo voltage, wire feed, dielectric pressure, wire tension, and etc. These parameters have the ability to improve the machined surface quality with less cracks and surface damages [8]. Hence, the objective of this paper is to investigate the influence of WEDM process parameters on surface roughness of stainless steel using distilled water as the dielectric fluid and brass wire as the tool electrode.

2. Experimental procedures

The experimental investigation was performed using the Mitsubishi FX10K CNC Wire EDM. The workpiece material, which was stainless steel S304, was prepared in desired dimension of 150 mm × 8 mm ×10 mm with flat surface finish. Stainless steel is widely used almost in all industrial applications, is accounted for approximately 50% of the world's stainless steel production and consumption. Because of its aesthetic view in architectures, resistance against corrosion and chemicals, high hardenability, and well mechanical property, it becomes the most preferred material [7], [9]. Brass wire with 0.2 mm diameter was used as the tool electrode whereas distilled water as the dielectric fluid. The experimental parameters are listed in table 1. The parameters were selected based on the type of machining material, tool material, height of the workpiece, and the capability of the machine. The experiments were designed using the Taguchi's L_{16} orthogonal array statistical model. The controlled parameters were voltage open, wire speed, wire tension, voltage gap, and off time. Mitutoyo Surftest (SV-514) was use to measure the surface roughness (R_a). The tester used a cut-off length of 2 mm and evaluation length of 0.8 mm. The tester uses Surfpak V4.10 (2) software with a resolution of 0.01 μm and stylus speed of 0.10 mm/sec. Measurements of R_a were repeated three times, and the average of the R_a was calculated. The experimental results are tabulated in table 2. The results are analysed using the signal to noise (S/N) ratio and analysis of variance (ANOVA) approach.

3. Results and Discussion

3.1 Signal to Noise (S/N) ratio

The S/N ratio according to Taguchi method is the ratio of signal to noise where signal represents the desirable value and noise represents the undesirable value. The S/N ratio for the experiments conducted is shown in table 2 meanwhile the mean of S/N ratio for R_a is presented for four level are tabulated in table 3. According to Taguchi method, minimizing average R_a would be better for precision manufacturing. Hence, "smaller-the better" type problems would give a smaller values for average R_a in order to get a better quality of finish parts [10]. Based on table 3, it was found that S/N ratio for R_a decreases when the wire speed, wire tension, and voltage gap are at level 4. The wire vibration tends to reduce when the wire tension increases which helps the surface roughness to improve [11]. Meanwhile, the S/N ratio for R_a increases when the off time is at level 4 and level 3 for voltage open. Voltage open should be kept as low as possible in order to get a better surface roughness. It is to avoid powerful explosion to occur which can cause deep crater on the machined surface and resulting with poor surface quality at the machined area [11-12].

3.2 Analysis of variance (ANOVA)

An empirical model as expressed in Eq. 1 is developed by ANOVA. Based on table 4, the model F-value of 102.38 implies that the model is significant. There is only a 0.14% chance of noise that could occur in the model F-value. The prob>F values less than 0.0500 indicates that the model terms A (voltage open), B (off time), E (voltage gap), AB (voltage open and off time), AC (voltage open and wire speed), BD (off time and wire tension), BE (off time and voltage gap), CD (wire speed and wire tension), CE (wire speed and voltage gap), and DE (wire tension and voltage gap) are significant. The most significant factor that affects the R_a is B with 77.57 F-value. However, values that are greater

IOP Conf. Series: Materials Science and Engineering **290** (2017) 012019 doi:10.1088/1757-899X/290/1/012019

than 0.1000 indicates the model terms are not significant. If there are many insignificant model terms (not counting those required to support hierarchy), then the model reduction may improve the model. Factors C and D are the most uninfluenced factors since their prob>F are 0.5508 and 0.9052 respectively. Nevertheless, when factor C interact with factor A, D, and E, the model terms becomes significant. Hence, it means that factor C cannot be removed from the hierarchy, even though it is not significant if it is stand-alone. It goes same with factor D, where the model terms become significant when it interacts with factor B, C, and E. The Predicted R^2 of 0.8133 is in reasonable agreement with the Adjusted R^2 of 0.9878. Adequate precision measures the signal to noise ratio, where the ratio greater than 4 is desirable. The ratio of 44.649 indicates that the signal is adequate. This model can be used to navigate the design space.

Table 1. Experimental parameters

Control Parameters	**Factors**	Level **I**	**II**	**III**	**IV**
Voltage open (V)	A	10	12	14	16
Off time (μs)	B	1	2	3	4
Wire speed (m/min)	C	6	8	10	12
Wire tension (N)	D	6	7	8	9
Voltage gap (V)	E	40	45	50	55
Fixed Parameters:					
Workpiece material	Stainless steel S304				
Tool electrode	Brass wire (Ø 0.2 mm)				
Dielectric fluid	Distilled water				

Table 2. Experimental result

Run	Parameters: Voltage open (V)	OFF time (μs)	Wire speed (m/min)	Wire tension (N)	Voltage gap (V)	Response: Average surface roughness R_a (μm)	S/N value Average surface roughness
1	14.00	3.00	6.00	7.00	55.00	3.22	-10.1571
2	10.00	3.00	10.00	8.00	50.00	3.13	-9.9109
3	14.00	4.00	8.00	6.00	50.00	2.05	-6.2351
4	16.00	3.00	8.00	9.00	40.00	2.79	-8.9121
5	12.00	3.00	12.00	6.00	45.00	3.45	-10.7564
6	12.00	2.00	6.00	9.00	50.00	3.11	-9.8552
7	14.00	2.00	12.00	8.00	40.00	3.44	-10.7312
8	16.00	4.00	6.00	8.00	45.00	2.96	-9.4258
9	16.00	2.00	10.00	6.00	55.00	2.73	-8.7233
10	10.00	2.00	8.00	7.00	45.00	2.72	-8.6914
11	10.00	4.00	12.00	9.00	55.00	3.86	-11.7317
12	16.00	1.00	12.00	7.00	50.00	3.51	-10.9061
13	12.00	4.00	10.00	7.00	40.00	2.97	-9.4551
14	14.00	1.00	10.00	9.00	45.00	3.01	-9.5713
15	10.00	1.00	6.00	6.00	40.00	2.82	-9.0050
16	12.00	1.00	8.00	8.00	55.00	3.09	-9.7992

$$R_a = 3.755 + 0.831A - 1.063B + 1.949C - 2.314D - 0.199E - 0.096AB - 0.058AC + 0.206BD + 0.011BE - 0.035CD - 0.019CE + 0.044DE \quad (1)$$

Where, R_a = surface roughness (μm), A = voltage open (V), B = off time (μs), C = wire speed (m/min), D = wire tension (N), and E = voltage gap (V).

Table 3. S/N ratio means for R_a

Process parameters	S/N Ratio mean			
	Level 1	Level 2	Level 3	Level 4
Voltage open (V)	-9.8347	-9.9665	-9.1737	-9.4918
OFF time (μs)	-9.8204	-9.5003	-9.9341	-9.2119
Wire speed (m/min)	-9.6108	-8.4094	-9.4151	-11.0314
Wire tension (N)	-8.6799	-9.8024	-9.9668	-10.0176
Voltage gap (V)	-9.5258	-9.6112	-9.2268	-10.1028

Table 4. ANOVA for average R_a

Source	Sum of Square	Degree of Freedom (DF)	Mean Square	F Value	Prob>F
Model	2.56	12	0.21	102.38	0.0014
A	0.12	1	0.12	55.36	0.0050
B	0.16	1	0.16	77.57	0.0031
C	9.364E-004	1	9.364E-004	0.45	0.5508
D	3.491E-005	1	3.491E-005	0.017	0.9052
E	0.026	1	0.026	12.62	0.0380
AB	0.098	1	0.098	47.14	0.0063
AC	0.14	1	0.14	69.43	0.0036
BD	0.54	1	0.54	259.29	0.0005
BE	0.040	1	0.040	19.05	0.0222
CD	0.061	1	0.061	29.29	0.0124
CE	0.48	1	0.48	231.98	0.0006
DE	0.39	1	0.39	185.59	0.0009
Residual	6.256E-003	3	2.085E-003		
Cor Total	2.57	15			

Standard deviation	0.046	R^2	0.9976
Mean	3.05	Adjusted R^2	0.9878
Coefficient of variation	1.50	Predicted R^2	0.8133
Predicted residual error of sum of square (PRESS)	0.48	Adequate precision	44.649

3.3 Optimization and Verification

The ANOVA-based optimization was done in order to get the optimum values of the process parameters for minimum R_a. Minimum R_a (2.56 μm) can be achieved at 10 V voltage open, 2.84 μs off time, 12 m/min wire speed, 6.3 N wire tension, and 54.91 V voltage gap. Experiments were conducted to validate the results obtained from the optimization. Based on the experiment, the actual R_a (2.76 μm) was higher compared to the optimize R_a (2.56 μm) with maximum error of 7.81%. The percentage error for minimum Ra is relatively small which shows the empirical Eq. 1 is valid.

4. Conclusion

In this research, the influence of WEDM process parameters on the surface roughness of stainless steel (S304) using brass wire as the tool electrode and distilled water as the dielectric fluid has been investigated. Following conclusions are drawn from the experimental study:

1. From S/N ratio, it can be inferred that wire speed, wire tension, and voltage gap should be at the highest level, which is level 4 while voltage open and off time should be at level 3 and 4 respectively to obtain low surface roughness.
2. Based on ANOVA, the most significant factor that influences the surface roughness is off time.
3. The minimum surface roughness is found to be 2.56 μm at 10 V voltage open, 2.84 μs off time, 12 m/min wire speed, 6.3 N wire tension, and 54.91 V voltage gap which are considered to be the optimum process parameters for WEDM of stainless steel.
4. The predicted value and experimental R_a value is within 7.81% error.

Acknowledgement

This research was funded by MOSTI, under Research Grant SF15-016-0066. The authors are thankful to the staff of Tool and Die Laboratory from the Faculty of Engineering, International Islamic University Malaysia (IIUM) where the experimental studies were conducted.

References

[1] Singh H and Garg R 2009 *Journal of Achievements in Materials and Manufacturing Engineering* **32** 70-74
[2] Ali M Y, Khamarruzaman S N Z, and Banu A 2017 *IOP Conference Series: Materials Science and Engineering* **184** p 012033
[3] Nihat T, Can C, and Gul T 2004 *Journal of Materials Processing Technology* **152** 316-322
[4] Kumar K and Ravikumar R 2013 *International Journal of Modern Engineering Research* **3** 1645-1648
[5] Durairaj M, Sudharsun D, and Swamynathan N 2013 *International Conference on Design and Manufacturing , IConDM 2013* p 868-877
[6] Ali M Y, Karim A M, Adesta E Y T, Ismail A F, Abdullah A A, and Idris M N 2010 *International Journal of Precision Engineering and Manufacturing* **11** 779-784
[7] Huang J, Chen Y, and Liao Y 2004 *Journal of Materials Processing Technology* **149** 165-171
[8] Rizauddin R, Roazam A, and Jaharah A G 2012 *Journal Teknologi UTM* **59** 209-213
[9] Akkurt A 2009 *Indian Journak of Engineering & Materials Sciences* **16** 373-384
[10] Ali M Y, Mohamed A R, Khan A A, Asfana B, Lutfi M, and Fahmi M I 2013 *World Applied Sciences Journal (Mathematical Applications in Engineering)* **21** 73-78
[11] Ikram A, Mufti N A, Saleem M Q, and Khan A R 2013 *Journal of Mechanical and Science Technology* **27** 2133-2141
[12] Banu, A and Ali, M Y 2016 International Journal of Engineering Materials and Manufacture **1** 3-10.

IOP Conf. Series: Materials Science and Engineering **290** (2017) 012020 doi:10.1088/1757-899X/290/1/012020

A comparison between destructive and non- destructive techniques in determining coating thickness

F I Haider, Suryanto, M H Ani and M H Mahmood

Department of Manufacturing and Materials Engineering, International Islamic University Malaysia (IIUM), Jalan Gombak, 53100 Kuala Lumpur, Malaysia

Email: surya@iium.edu.my

Abstract. Measuring coating thickness is an important part in research works related to coating applications. In general, techniques for measuring coating thickness may be divided into destructive and non-destructive methods which are commonly used depending on the applications. The objective of this study is to compare two methods measuring the coating thickness of electroplating copper coating on the austenitic stainless-steel substrate. The electroplating was carried out in a solution containing 200 g/L $CuSO_4$, 100 g/L H_2SO_4 at room temperature and current of 40mA/cm^2 during 20,40,60,80 and 100 mins as coating periods. And the coating thickness was measured by two methods, cross sectional analysis as a destructive technique and weight gain as a non-destructive technique. The results show that at 20 mins coating time interval, the thickness measured by cross sectional method was 16.67 µm and by weight gain method was 17.37 µm, with difference of 0.7 µm and percentage error of 4.11%. This error increased to 5.27% at 100mins time interval, where the values of the thickness measured by cross sectional and weight gain were 86.33 µm and 81.9 µm respectively, and the difference was 4.43 µm. Moreover, though the weight gain method is fast and gives the indication for the termination of a coating process, information regarding the uniformity, porosity and the presence of cracks cannot be obtained. On the other hand, determining the coating thickness using destructive method will damage the sample.

1. Introduction

Coatings applied on a material substrate provides properties not inherent in the material that include corrosion and wear resistance, conductivity, color, and solderability. Also, the final product's cost depends not only on the appropriate selection of coating material, but also on the coating amount applied and the final surface texture. In this sense, the amount of coating applied is the coating thickness which is essential to the product's final use and price. Measuring the coating thickness is a significant part in researches involving coating applications. Generally, techniques for measuring coating thickness may be divided into non-destructive and destructive techniques. For the non-destructive case, there are five commonly described methods namely; X-ray fluorescence, beta backscatter, eddy-current, magnetic induction and weight gain which arranged from highest to lowest cost respectively.

First of the non-destructive techniques is the X-ray fluorescence, which is considered as the most precise measurement method. It is used to measure the thickness of small-diameter parts or dual coatings such as gold and nickel over copper. When a material is subjected to x-ray bombardment, some of its electrons will gain energy and leave the atom, releasing a photon of x-ray energy known as x-ray fluorescence. The energy level or wavelength of fluorescent x-rays is proportional to the atomic number of the molecule and is characteristic for a peculiar fabric. The quantity of energy released will be dependent upon the thickness of the material being measured [1-2].

The second method is beta backscatter that measures many typical thickness applications, including gold on nickel, copper on epoxy, silver on copper, titanium nitride on steel, and tin-lead alloys. This method is similar to the X-ray fluorescent in that the area tested is the target of radiation, and the energy emitted from the surface is measured. Beta rays are electrons emitted from unstable radioisotopes. The electrons penetrate the plating material and reflected back (back scatter) toward the source. They are collected and counted with a Geiger-Mueller tube for conversion to coating thickness [3].

The third method is eddy current which is used to measure the thickness of nonconductive coatings on non-ferrous metal substrates. Eddy current inspection is based upon the principles of electromagnetic induction, and in turn has many similarities to the electromagnetic induction test method. Magnetically induced eddy currents generate an opposing magnetic field, which changes the circuit impedance and the output voltage. The change in output voltage is applied to calculate the coating thickness [4-5].

Fourth is magnetic induction, which is employed to quantify the thickness of a non-magnetic coating (zinc, cadmium, paint, powder coating, etc.) over a steel substrate. The execution of this test method is straightforward; a measurement probe is placed directly on the surface of the test specimen. Upon placement of the probe, the linear distance between the contact point of' the probe (with the coating surface) and the substrate is then measured [6].

The last non-destructive method is the weight gain, which is used only if the coating distribution is uniform. The thickness can be determined by weight of the sample before and after coating to measure the mass of coating and calculate the surface area of the sample and use the following equation [7].

$$Thickness\ (\mu m) = \frac{M\ (g) * 10^4}{A\ (cm^2) * \rho\ (\frac{g}{cm^3})} \qquad Eq\ (1)$$

Where M is the mass of coating, A is the surface area and ρ is the density of the coating material.

On the other hand, the most commonly used destructive method to measure the coating thickness is the cross-sectioning based method. In this case, coating thickness can be determined by cutting the coated sample and then view the cut microscopically by optical microscopy or scanning electron microscopy. It can also be measured by making a geometrically designed incision through the coating and viewing its cross-section with a scaled magnifier. A special cutting tool is used to make a small precise V-groove through the coating and into the substrate. This destructive method allows direct observation and is often the only means to identify multiple coatings economically, provided that alternating colors have been used. Gauges are available that come complete with cutting tips and illuminated scaled magnifier. ASTM D4 I38 outlines a standard method of this measurement system [7].

In general, a non-destructive technique is a process of inspecting, testing, or evaluating the coating, components or assemblies for discontinuities, or differences in characteristics without destroying the serviceability of the part. In other words, when the inspection or test is completed the part can still be used. In contrast to non-destructive, other tests are destructive in nature and are therefore done on a limited number of samples, rather than on the materials, components or assemblies being put into service.

This paper presents a comparison study between weight gain method as non- destructive technique and cross sectional based method as destructive technique in determining coating thickness of copper electroplating.

2. Experimental procedures

Austenitic stainless-steel grade 316L and 99.91 % pure copper samples with dimensions of 20 mm x 20 mm x 2 mm, were used as cathode and the anode respectively. The electroplating was carried out in a solution containing 200 g/L $CuSO_4$, 100 g/L H_2SO_4 at room temperature and current density of 40 mA/cm^2 during experimental processing times of 20, 40, 60, 80 and 100 mins. All the stainless-steel samples were polished up to 1000 SiC paper, and ultrasonically cleaned in acetone for 5 mins before being coated [8-9]

Coating thickness was determined using two techniques, weight gain as non-destructive and cross sectional based as destructive method. For weight gain, the sample was weighed by using 4-digit electric balance before and after coating to calculate the mass of coating with the equation:

$$M\,(g) = W_2 - W_1 \qquad \text{Eq (2)}$$

Where M is the mass of coating, W_2 and W_1 are the weights of the sample after and before coating respectively. The coating thickness was measured by the equation:

$$Thickness\ (\mu m) = \frac{M\ (g) * 10^4}{A\ (cm^2) * \ \rho\ (\frac{g}{cm^3})} \qquad \text{Eq (3)}$$

Where the area of each sample was (9.6 cm^2), and the density as $\rho = 8.96$ g/cm^3. All measured thicknesses are listed in table 1.

In the cross-sectional method, coating thickness was determined by cutting the coated sample using the EDM wire cutting machine and viewing the cut microscopically by optical microscopy which is equipped with scale magnifier. The cross-sectional image is shown in figure 1. Measurements were taken in 10 locations across the cross-section of the sample on every side for every sample, and the average value was calculated. All thicknesses measured are listed in table 1.

Figure 1. Optical microscopy image of cross-section copper coating on 316L substrate.

IOP Conf. Series: Materials Science and Engineering **290** (2017) 012020 doi:10.1088/1757-899X/290/1/012020

Table 1. Thickness measured by weight gain and cross-sectional methods.

Coating Time (mins)	Thickness by weight gain (μm)	Thickness by Cross sectional (μm)	Difference (μm)	Average Thickness (μm)	Error in %
20	17.37	16.67	0.7	17.02	4.11
40	33.77	35.24	1.47	34.505	4.26
60	51.42	54.06	2.64	52.74	5.01
80	69.5	66.09	3.41	67.795	5.03
100	81.9	86.33	4.43	84.115	5.27

The percentage error between the thicknesses in the two methods was calculated by using the following equation:

$$Error\ (\%) = \frac{The\ difference}{Average\ thickness} \text{ x } 100 \qquad \text{Eq (4)}$$

3. Discussion

The values of the difference between the coating thicknesses measured by the two methods as depicted in table 1, are too insignificant compared with the average coating thickness value. But the process of the two methods is totally different, the weight gain method is very simple and just needs to weigh the sample before and after coating. In contrast, the cross-sectional based method is a complicated technique since it needs more processing and equipment such as cutting machine and optical microscopy as well as time. Additionally, the most important thing is the resulting damage to the sample which prevents it from further usage for other processes.

Figure 1 shows that, the electroplating copper coating distribution is uniform; therefore, the weight gain method can be used to measure the thickness. The graph in figure 2 illustrates the comparison between the thickness values which were determined by weight gain as non-destructive and cross-sectional as destructive method. The thickness values are in the same range, but the difference between the values measured by the two methods increases with the increase in the time of coating, which indicates that, the difference increases with increasing thickness of the. In the same way, figure 3 explains the increasing of percentage error with increasing of time as well as the thickness/ increasing time as well as thickness, where the maximum error in percentage is 5.27% after 100mins and the minimum error in percentage is 4.11% at 20 mins interval.

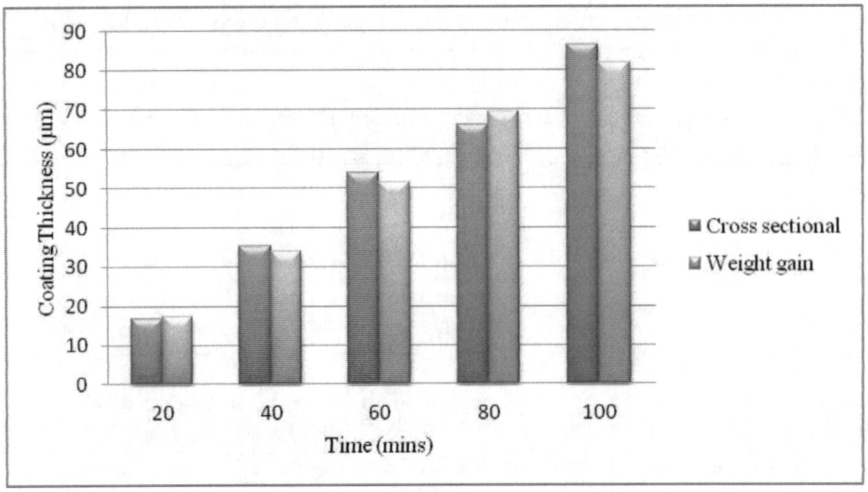

Figure 2: Comparison between the thickness values which determined by weight gain and Cross-sectional methods.

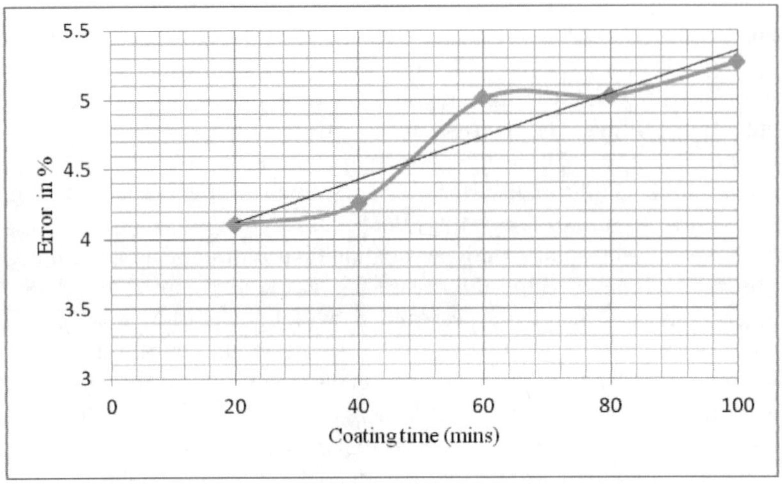

Figure 3: The error percentage between the two methods.

4. Conclusion

The results show that the thickness values measured by the weight gain method are as accurate as a cross-sectional method, with a small error of around 4.11% at coating time interval of 20 mins with a difference of 0.7 μm, this error increases with the increase in coating time and due to increasing coating mass. However, this error is still non-significant since its values is around 5.27% and the difference between the two methods values is 4.43 μm at 100 mins coating time. For this reason, the weight gain as non-destructive technique is more interesting than the cross-sectional method as destructive one since all samples can be tested without any damage. On the other hand, there are some limitations of weight gain method like, it can be used only with small parts and it is difficult to relate the mass of the coating to thickness if the substrate is rough or the coating distribution is not uniform, also information regarding the uniformity, porosity and the presence of cracks cannot be obtained. Likewise, the weight gain as non-destructive method is the simplest method that can be used to determine the coating thickness. In contrast, cross sectional based technique is more complicated due to the time and process it takes.

IOP Conf. Series: Materials Science and Engineering **290** (2017) 012020 doi:10.1088/1757-899X/290/1/012020

References

[1] Mantler M and Schreiner M 2000 *X-ray Spectrometry* **29** 3-17

[2] Russe I S, Brock D, Knop K, Kleinebudde P, and Zeitler J A 2012 *Molecular Pharmaceutics* **9** 3551-3559

[3] Gongora G J and Vail M A 2015 U.S. Patent No. 8,969,833, Washington, DC: U.S. Patent.

[4] Garcia-Martín J, Gómez-Gil J and Vázquez-Sánchez E 2011 *Sensors* **11** 2525-2565

[5] Zhang D, Wu M, Li H, Chen Z, Li Y and Xie S 2016 *International Journal of Applied Electromagnetics and Mechanics* **52** 1409-1415

[6] Petrilli C 2001 *Metal Finishing* **99** 810-813

[7] Beamish D 1999 *Metal Finishing* **97** 548-550.

[8] Suryanto, Haider F, Ani M and Mahmood M 2017 *Materials Science and Engineering* **204** 12-17

[9] Haider F , Suryanto, Ani M and Mahmood M 2017 *Applied Mechanics and Materials* **864** 121-126

IOP Conf. Series: Materials Science and Engineering **290** (2017) 012021 doi:10.1088/1757-899X/290/1/012021

An integrated approach for facilities planning by ELECTRE method

E M Y Elbishari[1], M H F Al Hazza[1], E Y T Adesta[1], Nur Salihah Binti Abdul Rahman[1]

[1]Department of Manufacturing and Materials Engineering, International Islamic University Malaysia (IIUM), Jalan Gombak, 53100 Kuala Lumpur, Malaysia

Email: mauataz@iium.edu.my

Abstract. Facility planning is concerned with the design, layout, and accommodation of people, machines and activities of a system. Most of the researchers try to investigate the production area layout and the related facilities. However, few of them try to investigate the relationship between the production space and its relationship with service departments. The aim of this research to is to integrate different approaches in order to evaluate, analyse and select the best facilities planning method that able to explain the relationship between the production area and other supporting departments and its effect on human efforts. To achieve the objective of this research two different approaches have been integrated: Apple's layout procedure as one of the effective tools in planning factories, ELECTRE method as one of the Multi Criteria Decision Making methods (MCDM) to minimize the risk of getting poor facilities planning. Dalia industries have been selected as a case study to implement our integration the factory have been divided two main different area: the whole facility (layout A), and the manufacturing area (layout B). This article will be concerned with the manufacturing area layout (Layout B). After analysing the data gathered, the manufacturing area was divided into 10 activities. There are five factors that the alternative were compared upon which are: Inter department satisfactory level, total distance travelled for workers, total distance travelled for the product, total time travelled for the workers, and total time travelled for the product. Three different layout alternatives have been developed in addition to the original layouts. Apple's layout procedure was used to study and evaluate the different alternatives layouts, the study and evaluation of the layouts was done by calculating scores for each of the factors. After obtaining the scores from evaluating the layouts, ELECTRE method was used to compare the proposed alternatives with each other and with the existing layout; ELECTRE compares the alternatives based on their concordance and discordance indices. The alternatives were ranked from best to worst where regarding to the layouts concerned with the manufacturing area B.4 is the best alternative.

1. Introduction

In today's modern age, technology made the process of developing new products from the initial stages of designing until the final stages of assembly more efficient. This paved the way for personalized, custom products to satisfy the various demands. In order to remain competitive, firms had to increase the variety while maintaining high volume to meet the changing demands of the markets. A facility layout problem (FLP) usually compromises with the placing of machines, departments, transportation flow in a layout area objective to maximize the production rate of the system. The majority of the FLP use flow relationship between machines to reduce the material handling costs and increasing the machines utilization. According to Tompkins [1], material handling costs can account for 20–30% of total production costs. A FLP can be defined as one which aims to place the resources or departments in optimal locations which are essential for running a successful production system within the available space. The relationship between the facility layout in production area have been analysed by different researchers [2-5]. Many of them used the artificial intelligent methods such as genetic algorithm [6-7], simulation annealing method [8-9], and others used particle swarm [11-12]. However, very few researchers tried to investigate and analyse the effect of service and supporting departments on the work efficiency and the human efforts. Maximizing the efficiency of the factory layout usually is evaluated by the material handling system and the human efforts to have a smooth flow of the raw material, work in progress and final products and the needs of the human and manpower. This will increase the cost dependency on the distances between the different locations of machines, cells and other departments in the manufacturing systems. Therefore, the most used factor in facility layout design is the flow–distance metric based on weighted average between the material transferred and the distance travelled. However, material handling process for materials such as the raw material, semi-finished products and final products are affected directly by many factors such as manufacturing flexibility, work force capacity, lead-time, handling cost and efficiency of material flow. Therefore, facility layout design based on the flow–distance is insufficient. Selecting of the handling method and equipment based on factory layout during the design phase of the manufacturing system is one of the main complicated tasks that may affect the production rate and shortening the time to market of the products.

2. Methodology

To achieve the goal of this article a ELECTRE method [13] as one of multi criteria decision-making (MCDM) approach have been used to compare a number of alternatives for manufacturing facility layout. ELECTRE has supported by using apple's procedure [14]. In this research, apple's procedure has been customised to fit with the requirement of the case study: Dalia Industries, the steps of apple's procedure are shown in figure 1.

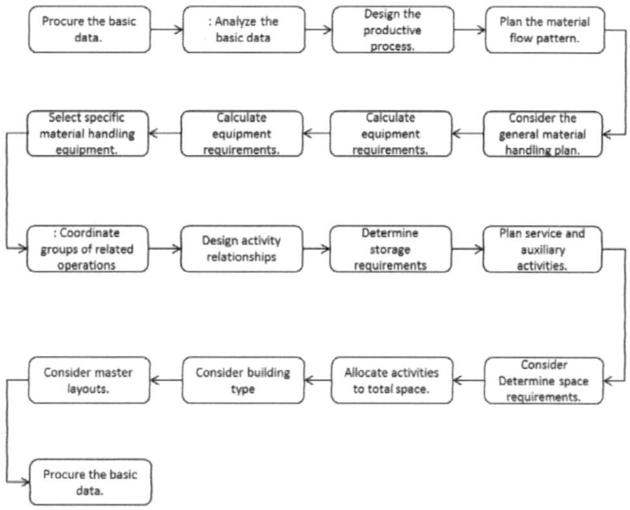

Figure 1. Apple's layout procedure steps

ELECTRE is a method for multi criteria decision making which was developed in France ELECI'RE is an acronym for elimination and choice translating algorithm. Given a set of alternatives on K objectives, this algorithm seeks to reduce the size of the non-dominated set. The k objectives all have a set of weights that is determined by the decision maker. The fundamental idea of the ELECI'RE technique is that alternatives can be eliminated which are dominated by other alternatives to a specified degree. There are two indices used in this method to determine advantages and disadvantages of one alternative to the other ones, which are:

1. Concordance index: This measures relative advantages.
2. Discordance index: This measures relative disadvantages.

Figure 2 shows the steps that are followed to complete the ELECRE method and obtain the ranking result which is the goal of using this multi criteria decision making method. [15]

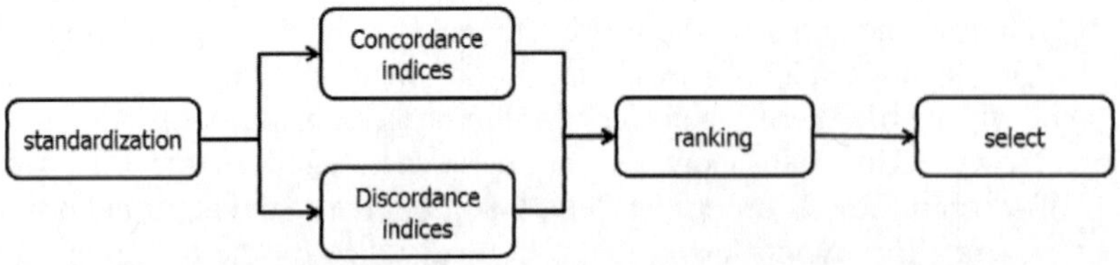

Figure 2. ELECTRE method steps

3. Results and Discussion

This section presents the results achieved from the integration of apples procedure and ELECTRE method. From data collected in Dalia industry the facility was divided into two layouts the first layout is considered with the whole facility which include management offices and utilities and the rest of the departments in Dalia Industries, this layout will be known as "Layout A", the second layout is considered with the manufacturing area itself and the utilities considered in manufacturing and fabricating the product and is labelled as "Layout B". The layouts for the facility will be known as "Layout A", and the manufacturing area layout will be known as "Layout B" with "B.1" being the original layout and "Layout B.2,B.3,B.4" will be the alternative layouts. Layouts A.1 and B.1 are shown in figure 3.

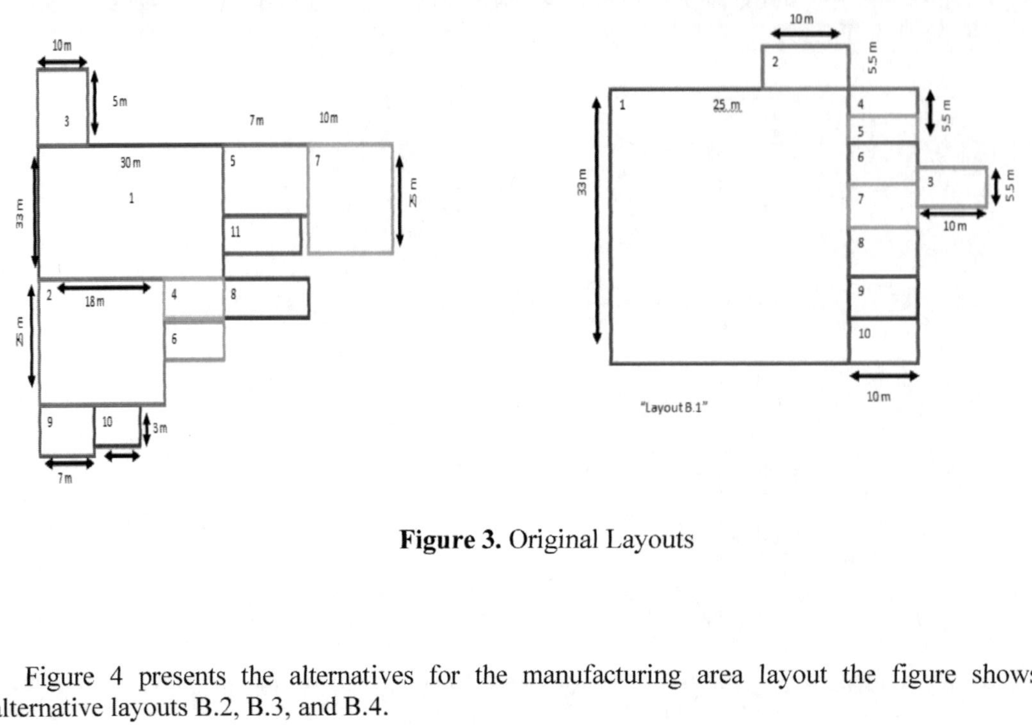

Figure 3. Original Layouts

Figure 4 presents the alternatives for the manufacturing area layout the figure shows there alternative layouts B.2, B.3, and B.4.

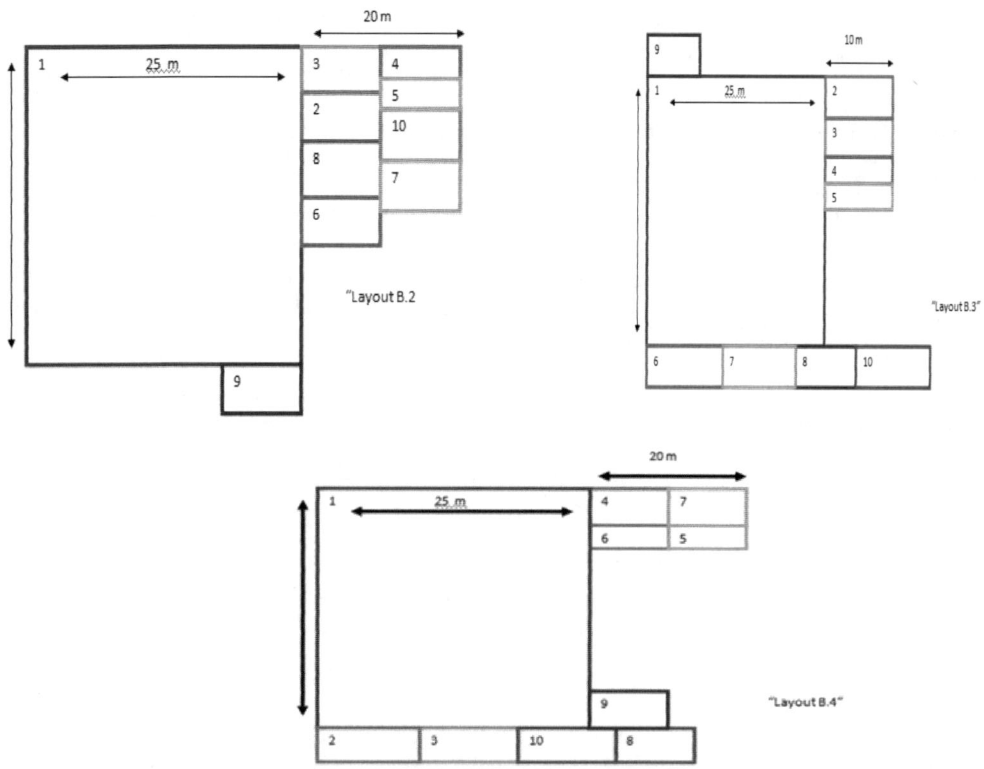

Figure 4. Alternative Layouts

Steps of the apples procedure will be inducted to analyze existing layouts and there produced alternatives. The evaluation of the alternative layouts will be compared based on each of the criteria's which are (Inter department satisfactory level, distance and time travelled for both the product and the workers), the result of these chapters will then be used as inputs ELECTRE method which will be used to compare the alternatives factoring in all the criteria's and coming up with a hierarchy evaluating the alternatives. Finally, a simulation have been conducted using delmia quest software to analyses the production rate and utilization for the manufacturing processes. Layouts were compared based on a number of factors starting with the inter department satisfactory level, which in it the layouts design were evaluated and checked for how many of the important relations were satisfied which are (A relations, and E relations), and checked if any of the undesirable relations (X relations) were found in the layouts. Table 1 summarizes the score of each layout in the Inter department relation satisfactory Level factor.

Table 1. Score for each layout in the inter department satisfactory level "Layout B"

	B.1	B.2	B.3	B.4
A Relations	9	9	6	6
E Relations	3	7	5	5
X Relations	-2	0	-2	0
Score	10	16	9	11

The second factor is the total distance travelled for the workers, this was checked by calculating a specified route travelled by workers, and since the layouts are different but conclude the same departments the total distance could be calculated for each layout. Table 2 summarizes the score of each layout in the distance traveled for workers.

Table 2. Score of each layout in distance traveled for workers "Layout B"

Layout	B.1	B.2	B.3	B.4
Score	133.5 m	145 m	172 m	163.5 m

The next factor was total time travelled for workers, which used the total distance travelled and divided by the average worker speed to conclude the time travelled by workers in each layout. Table 3 summarizes the score of each layout in the time traveled for workers.

Table 3. Score of each layout in time traveled for workers "Layout B"

Layout	B.1	B.2	B.3	B.4
Score	47.7 sec	51.8 sec	61.4 sec	58.4 sec

The forth factor was the total distance travelled for the product this was checked by calculating a specified route that the product travels in, and since the layouts are different but conclude the same departments the total distance could be calculated for each layout. Table 4 summarizes the score of each layout in the distance travelled for product.

Table 4. Score of each layout in distance traveled for the product "Layout B"

Layout	B.1	B.2	B.3	B.4
Score	140 m	140 m	135 m	100 m

The final factor considered was total time travelled for the product, which used the total distance travelled and divided by the average product transfer speed to conclude the time travelled by the product in each layout. Table 5 summarizes the score of each layout in the time traveled for product

Table 5. Score of each layout in time traveled for the product

Layout	B.1	B.2	B.3	B.4
Score	84 sec	84 sec	81 sec	60 sec

After comparing between them based on each criteria we will proceed with the multi-criteria decision making tool that was chosen which is ELECTRE I. Table 6 shows the score of each layout based on each score.

Table 6. Summarized scores for the manufacturing area layouts "Layout B"

Layout	Inter department relation satisfactory Level	Distance traveled "Product"	Time traveled "Product"	Distance traveled "Workers"	Time traveled "Workers"
B.1	11	133.5	47.7	140	84
B.2	16	145	51.8	140	84
B.3	9	172	61.4	135	81
B.4	11	163.5	58.4	100	60

From table 6, it can be conclude the best layout based on each of the factors, in regards for the first factory, which is, inter department satisfactory level layout B.2. The layout obtained the highest score, for the total distance traveled and total time traveled for the workers layout B.1 has the lowest measurements which indicates it is the best layout based on those two factors. Finally, for the total distance and total time traveled for the workers layout B.2 obtained the lowest measurement, which indicates it has the best score in these two factors.

The standardized scores for the manufacturing area layout B is concluded in table 7.

Table 7. Standardized scores for the manufacturing area layouts "Layout B"

Layout	S1	S2	S3	S4	S5
B.1	0.43	1	1	0	0
B.2	1	0.3	0.3	0.5	0.5
B.3	0	0	0	0.875	0.875
B.4	0.29	0.78	0.78	1	1

Figure 5 compares between the different standardized scores as bar chart for the manufacturing area facility layouts "Layout B

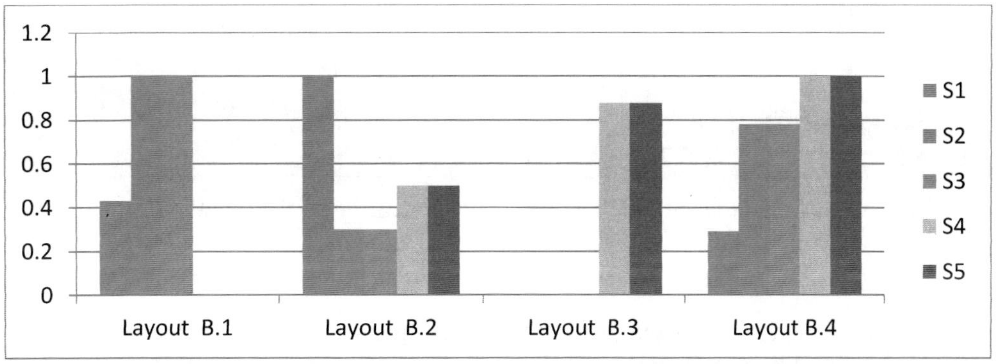

Figure 5. Bar chart for the manufacturing area facility layouts "Layout B"

After standardizing the numbers to a range between 0-1 you proceed to determine the concordance and discordance indices [16] by following the equation (1) and equation (2), respectively:

Concordance index C(A,B) = $\sum$ (W= + .5 W-)/ $\sum$(W= + W+ W-) (1)

Discordance index D(A,B) = Max {[ZBk - ZAt] / [Z"k - Z-k]} (2)

Table 8 and 9 show the concordance and discordance indices for both layouts layout A, and layout B.

Table 8. Concordance indices for the manufacturing area layouts "Layout B"

C	B.1	B.2	B.3	B.4
B.1	-	0.4	0.6	0.6
B.2	0.6	-	0.6	0.2
B.3	0.4	0.4	-	0
B.4	0.4	0.8	1	-

Table 9. Discordance indices for the manufacturing area layouts "Layout B"

D	B.1	B.2	B.3	B.4
B.1	-	0.57	0.875	1
B.2	0.7	-	0.375	0.5
B.3	1	1	-	0.78
B.4	0.22	0.71	0	-

For "p" and "q" the values which have been used are as follows (0.25, 0.75, 0.9). In the drawing the arrows go from the layout that is the dominant to the dominated layout.

P = 0.75, q = 0.9

Concordance Index
B.1: B.3, B.4
B.2: B.1, B.3
B.3: /
B.4: B.2, B.3

Discordance Index
B.1: B.2,B.3
B.2: B.1,B.3
B.3: B.4
B.4: B.1,B.2, B.3

B.1 > {B.3}
B.2 > {B.1, B.3}
B.4 >{B.2, B.3}

Set Kernel: {B.4},
Set not Kernel: {B.1, B.2, B.3}

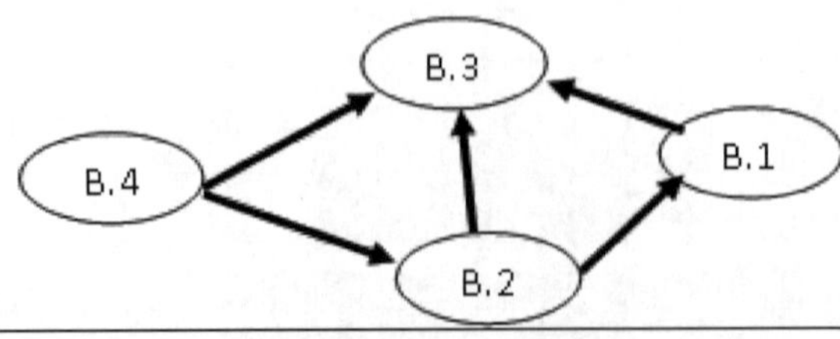

Figure 6. Concordance Index, Discordance Index for p =0.75 and q=0.9

As shown in figure 6 above, we can see that layout B.1 dominates layout B.3 while B.2 dominates layouts B.1 and B.3, and B.4 dominates B.2 and B.3.

P = 0.25, q = 0.75

Concordance Index

B.1: B.2,B.3, B.4
B.2: B.1, B.3
B.3: B.1, B.2
B.4: B.1, B.2,B.3

Discordance Index

B.1:B.2
B.2: B.1, B.3, B.4
B.3: /
B.4:B.1, B.3

B.1 > {B.2}
B.2 > {B.1, B.3}
B.4 >{B.1, B.3}

Set Kernel: {B.4 },
Set not Kernel: {B.1, B.2, B.3}

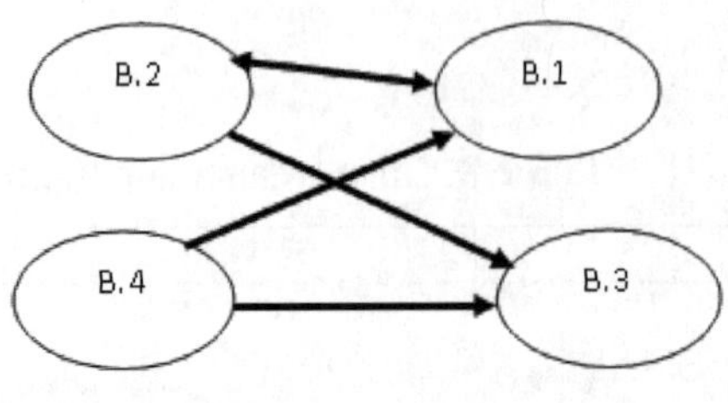

Figure 7. Concordance Index, Discordance Index for P = 0.25, q = 0.75

As shown in figure 7 above, we can see that layout B.1 dominates B.2 while B.2 itself dominates layouts B.1 and B.3, and B.4 dominates B.1 and B.3.

P = 0.25, q = 0.9

Concordance Index	Discordance Index
B.1: B.2, B.3, B.4	B.1: B.2, B.3
B.2: B.1,B.3	B.2: B.1, B.3
B.3: B.1, B.3	B.3: B.4
B.4: B.1, B.2, B.3	B.4: B.1, B.2, B.3

B.1 > {B.2, B.3}
B.2 > {B.1, B.3}
B.4 >{B.1, B.2, B.3}
Set Kernel: {B.4},
Set not Kernel: {B.1, B.2, B.3}

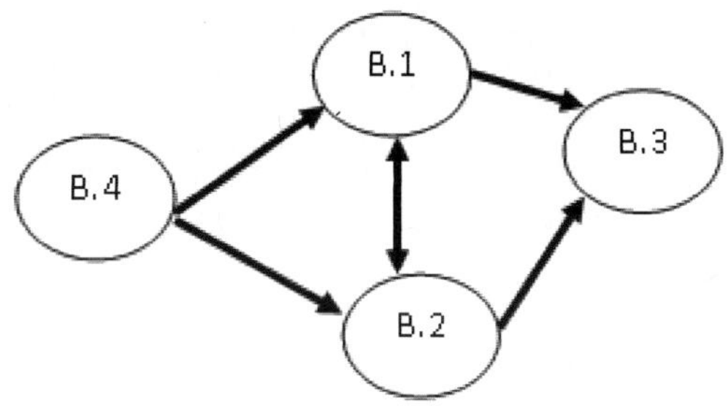

Figure 8. Concordance Index, Discordance Index for **P = 0.25, q = 0.9**

As shown in figure 8 above, we can see that layout B.1 dominates both layouts B.2 and B.3 while B.2 dominates layouts B.1 and B.3, and B.4 dominates B.1, B.2 and B.3.

4. Conclusion

As a result of apple's layout procedure a number of block layouts were developed for the whole facility (layout A) and for the manufacturing area (Layout B), the layouts were evaluated based on five different factors separately which were (inter department satisfactory level, total distance and time travelled for both the workers and the products). After comparing the layouts based on each factor ELECTRE was used as a multi-criteria decision making method to compare the layouts using all of the factors, the result of comparisons is shown below from the best layout to the worst. To conclude, the dominant and dominated layouts depends on the values of "p" and "q" different values were used for both limits, and as a result the layout that dominated the most, and was least dominated by will be at the top of the ranking. As conclusion, the best layout alternative for manufacturing area was layout "B.4 as shown in Table 10.

Table 10. Layouts ranking

	Layout B
1.	B.4
2.	B.2
3.	B.1
4.	B.3

References

[1] Tompkins J A 2003 *Facilities planning* (New York: John)
[2] Hungerländer P and Rendl F 2013 *Comput Optim Appl.* **55** 1-20.
[3] Kothari R and Ghosh D 2014 *J. Heuristics* **20** 125-142
[4] Anjos M F, Fischer A and Hungerländer, P. 2016 *Operations Research Proc.* (Springer

International Publishing) p. 17-23
[5] Allahyari M Z and Azab A 2017 *Computational Optimization in Engineering-Paradigms and Applications* (InTech)
[6] Mak K L, Wong Y S and Chan F T S 1998 *Computer Integrated Manufacturing Systems* **11** 113-127
[8] McKendall A R, Shang J and Kuppusamy S 2006 *Computers and Operations Research* **33** 2431-2444.
[9] Mavridou T D and Pardalos P M 1997 *Computational optimization and Applications* **7** 111-126
[11] Samarghandi H, Taabayan P and Jahantigh F F 2010 *Computers and Industrial Engineering* **58** 529-534.
[12] Kulturel-Konak S and Konak A 2011 *Engineering Optimization* **43** 1263-1287
[13] Roy B. 1991 *Theory and decision* **31** 49-73
14] Meyers F E 2005 *Manufacturing Facilities Design and Material Handling* (Pearson Education)
[15] Hugonnard J C and Roy B 1982 *Cahiers Scientifiques de la Revue Transports*
[16] Benayoun R, Roy B and Sussman B 1966 ELECTRE: Une méthode pour guider le choix en présence de points de vue multiples. Note de travail 49, SEMA-METRA International, Direction Scientifique, 1966.

IOP Conf. Series: Materials Science and Engineering **290** (2017) 012022 doi:10.1088/1757-899X/290/1/012022

Finite element analysis on deformation of stretchable electronic interconnect substrate using polydimethylsiloxanes (PDMS)

M F Roslan, N M Shaffiar, M K N Khairusshima and I S S Sharifah

Department of Manufacturing and Materials Engineering, International Islamic University Malaysia (IIUM), Jalan Gombak, 53100 Kuala Lumpur, Malaysia

Email: shishaff@iium.edu.my

Abstract. Over the years, the technology of electronic industry has growth tremendously. Open ended research on how to make a better concept of electronic circuit is ongoing especially on the stretchable electronic devices. There are many designs to achieve stretchability in electronic circuits. The problem occurs when deformation applied to the stretchable electronic circuit, it cannot maintain its functionality. Fracture may happen on the conductor. In this research, the study on deformation of stretchable electronic interconnects substrate using Polydimethlysiloxanes is carried out. The purpose of this research are to study the axial deformation occur, to determine the optimum shape of the conductor designs (horseshoe, rectangular and u-shape design) for the stretchable electronic interconnect and to compare the mechanical properties of Polydimethlysiloxanes (PDMS) with Polyurethane (PU) using Finite Element Analysis (FEA). The simulation was done on the FE model of the stretchable circuit with dimension of 2.4 X 2.4 X 0.5 mm. The stretching of the FE model was simulated with the range of elongation at 10, 20 and 30 percent from its original length in order to find the strain value for all three of the conductor designs. The best conductor design is used to simulate with different types of substrate (PDMS and PU). From the simulation result, Horseshoe design record the lowest strain value for each elongation, followed by rectangular and U-shape design. Thus, Horseshoe is considered as the optimum design for the conductor compared to the other two designs. From the result also, it shows that PDMS substrate will offer more maximum allowable stretchability compared to PU substrates. Thus PDMS is considered as a better substrate compare to PU. PDMS is a good material to replace PU since it can perform under tension much better mechanically.

1. Introduction

Recently the research into the design and fabrication of flexible and stretchable circuits and systems has received a lot of attention. Throughout the years, the electronic circuits have driven advanced in technology. Electronic circuit has become smaller, smarter and faster. Due to this advance technology, stretchable interconnecting circuits have been created. With the existent of stretchable interconnecting circuit, it can offer application that is impossible for rigid circuit board of today. Stretchable circuit can experience stress and strain from the deformations that occur and still deliver its function. It has great potential in application of wearable electronics, flexible display, biomedical, healthcare and military [1]. In electronic industry, it is crucial to make sure the stretchable electronic circuit works perfectly

IOP Conf. Series: Materials Science and Engineering **290** (2017) 012022 doi:10.1088/1757-899X/290/1/012022

fine when deformation such as stretching and bending occur. At a certain point when elongation happens toward the circuit, it may cause damage toward the circuit and cause failure. The metal line layouts in the circuit can be configured in many different shapes that can accommodate of mechanical stress, while the elastomeric substrate guarantees that the metal line will revert to its original shape [2].

In the recent year, several designs of conductor layout have been created such as U-shape, rectangular shape and horseshoe shape [3]. Each design can vary it performance and functionality of the stretchable electronic circuit. For now, designing the most optimum metal line layout that can deliver its function is still challenging. The stress and strain of the stretchable electronic interconnect substrate PDMS need to be determined when the deformations occur toward it. By doing this, the best shape for the conductor can be determined. The material that is being used in stretchable electronics as the substrates are polymer. Among the characteristics that are needed for any material to be used as the substrate are having a low-modulus with stress/strain responses that can be tailored precisely to match the non-linear properties of biological tissues and it must not be too rigid, too thick or too heavy.

2. Experimental procedures

2.1 Conductor designs

The design of an appropriate shape is crucial to allow stretchability of the conductor [3]. The three design are horseshoe, u-shape and rectangular. All the three designs are chosen because it has the similar shape as serpentine. This is to minimize strains in materials under large deformation, metallic films can be patterned into serpentine-shaped ribbons [4]. Figure 1 illustrates the horseshoe shape, rectangular shape and U-shape formed by copper lines configured in PDMS substrates.

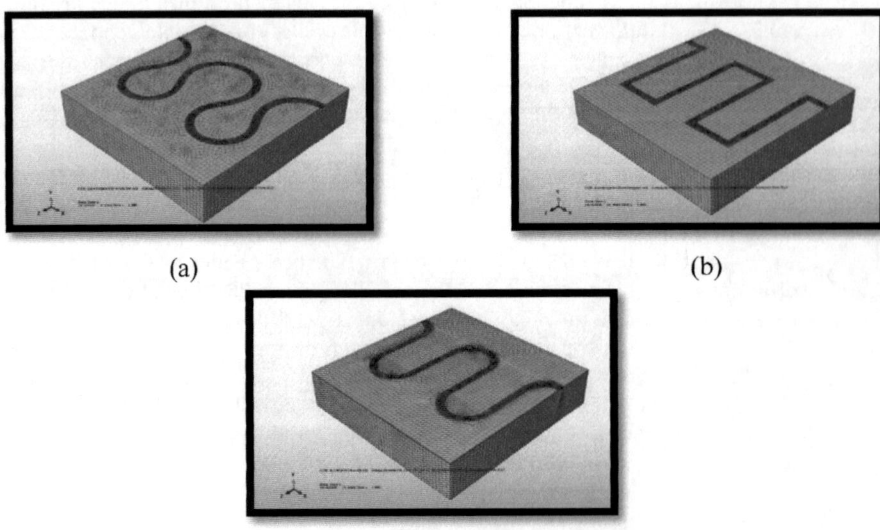

(a) (b) (c)

Figure 1: FE modeling of stretchable electronics. (a) Horseshoe, (b) Rectangular, (c) U-shape.

2.2 Finite element modelling

In this simulation, Abaqus software was used for the modeling and analyzing of the stretchable electronic model. The simulation was done on the FE model of the stretchable circuit with dimension of 2.4x2.4x0.5mm. Three shapes of conductors are designed and embedded into the substrate as shown in figure 1. The thickness of the conductor thin films is 0.018mm with a width of 0.1mm. The stretching of the FE model was simulated with the range of elongation at 10, 20 and 30 percent of its original length for each design.

Two parts involved in the simulation are conductor metal thin films and substrate. The parts are modelled first using solidwork software and imported to Abaqus software for Finite Element Modelling. In Abaqus software, it consist numbers of module that is parts, property, assembly, step, interaction, load, mesh, optimization, job, visual and sketch. Few steps must be done before the simulation can be run. In this research, all finite element models interface between the conductor and substrate are considered as perfectly bonded.

The material properties assign for conductor is copper (Cu) while for the substrates is polydimethysiloxanes (PDMS) and polyurethane (PU) respectively as shown in table 1. Copper is used as metal thin films because it is a good conductor of electricity and it is widely used in stretchable interconnect circuit.

Table 1. Material Properties of conductor and substrates use.

Material	Properties
Copper	Density = 8.96 g/cm^3 Yielding Point = 200 Mpa Poisson's Ratio = 0.3 Young Modulus,E = 85 Gpa
Polydimethlysiloxanes	Density = 0.965 g/cm^3 Yielding Point = 2.24 Mpa Poisson's Ratio = 0.47 Young Modulus,E = 1.2 Mpa
Polyurethanes	Density = 1.2 g/cm^3 Yielding Point = 138 Mpa Poisson's Ratio = 0.5 Young Modulus,E = 69 Mpa

Meshing is used to manage material deformation during crushes on simulation. A finer mesh will result in more accurate result. In Abaqus software, there are several types of mesh generation techniques. In certain condition, mesh cannot be done because of the complex geometry of the model. Meshing technique needs to be change or smaller partition of the model need to be created. The Element shape that can be use is Hex, Hex-dominate, Tet and wedge. The Hex element shape is chosen. For the substrate part, partitioning need to be done because meshing it without partitioning will cause a poor element being created and will affect the accuracy of the result.

For the loading, set one side of the FE model at fixed position using the boundary condition while the other end of the model is set as a region to be pull or elongated on the positive X-axis direction. For the fixed side, ENCASTRE (U1=U2=U3=UR1=UR2=UR3=0). It means that the region is restrained to move along all axes. Next boundary condition is to insert the value of elongation at U1 for the movement in X-axis direction.

3. Results and Discussion

3.1 Result of three different conductor designs

During axial stretching, the copper conductor is under tension when the line is parallel to the stretching direction and is under compression when the copper conductor is perpendicular to the stretching direction [3]. Locally, the outer part of the copper line is in compression and the inner part in tension. In order to make a comparison of results, several simulations were done for each design of copper conductor with an elongation of 10, 20 and 30 percent from original length being applied. Figure 2 shows the strain distribution for each design being applied with 30% elongation from its original length and in the table 2 shows the maximum strain value for each elongation being applied.

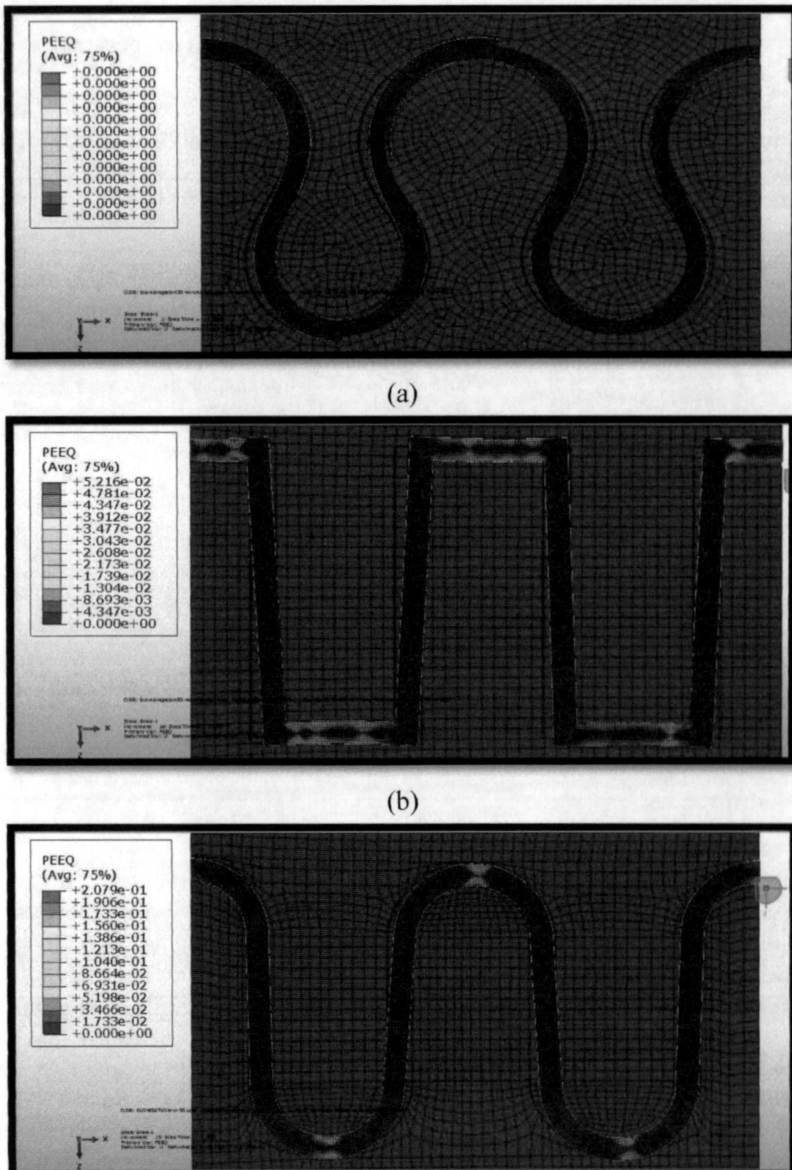

(a)

(b)

(c)

Figure 2: Strain distribution in copper conductor line for three different conductor shapes at 30% elongation (a) Horseshoe, (b) Rectangular, (c) U-shape.

Table 2. Maximum strain value for Horseshoe ,U-shape and Rectangular design.

	Strain		
Elongation %	**Horseshoe**	**U-Shape**	**Rectangular**
10	0.00121984	0.0037091	0.00178798
20	0.00243967	0.0174855	0.00520232
30	0.00364829	0.0390473	0.00835903

Based from the strain values in table 2, a graph was plotted and shown in figure 3.

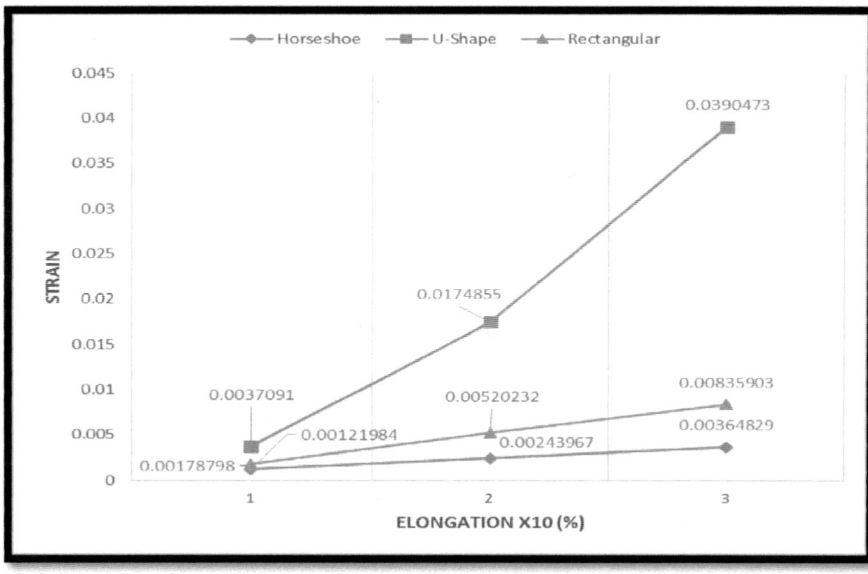

Figure 3. Strain versus Elongation for each designs.

From the graph, it can be seen that as the elongation being applied toward the stretchable circuit increase, the strain value also will be increased. It shows an agreement on the increasing trend for the graph of total deformation Vs plastic strain made by other researcher [3]. Strain comparison is enough for choosing the optimum conductor design, predominant failures are expected in regions where the highest plastic strain concentration located [3]. Thus the region where the copper experiences the highest strain is most likely to fail. This is because as the elongation increase, the amount of deformation and strain will increase thus increasing the stress that copper conductor experience. As the stress increase, it will eventually reach the ultimate stress point where it is the point corresponding to the maximum stress that a material can handle before it failure. The failure of the conductor will cause the circuit to stop conducting the electricity or signal. Thus the lowest strain value will give a better stretchable electronic circuit. From the graph in figure 3, it shows that Horseshoe design record the lowest strain value for each elongation, followed by rectangular and U-shape design. Thus, Horseshoe is considered as the optimum design for the conductor compared to the other two designs.

3.2 Result of different substrates

Simulation being done for two types of substrate materials that is Polyurethanes and Polydimethlysiloxanes to show the comparison between these two materials as a substrate. Horseshoe design conductor was used for the copper conductor. Each model undergoes elongation at 10%, 20% and 30%. Table 3 shows the maximum strain and stress value for each elongation being applied.

Table 3. Maximum strain & stress for PDMS and PU substrates at different elongation.

	Strain		Stress	
Elongation %	PU	PDMS	PU	PDMS
10	0.11338	0.112455	7.82E+06	1.35E+05
20	0.226761	0.224911	1.56E+07	2.70E+05
30	0.340141	0.337365	2.35E+07	4.05E+05

Based from the strain values in table 3, a graph was plotted and shown in figure 4.

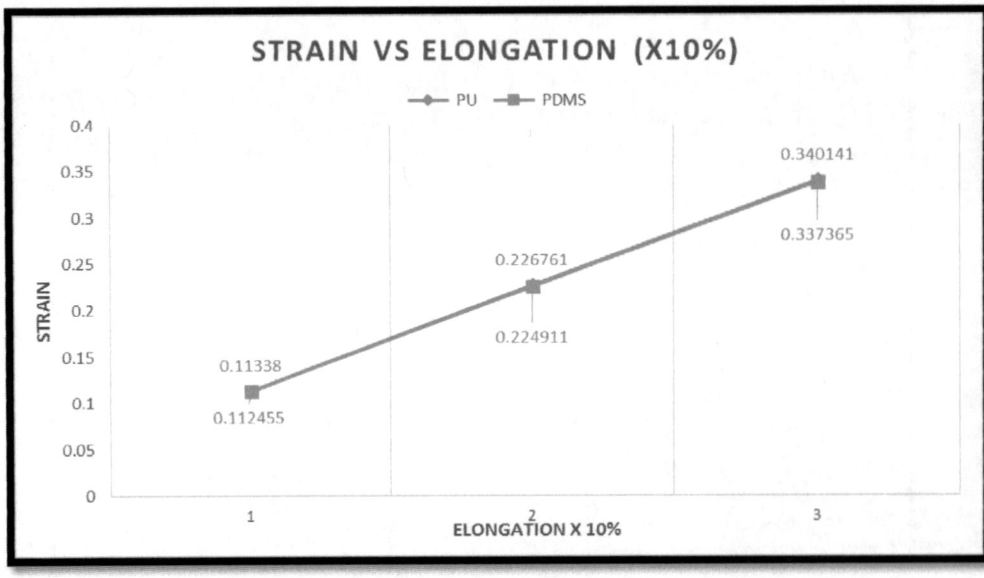

Figure 4. Strain for PU and PDMS versus Elongation.

From the graph, PDMS substrate record a lower strain compare to PU substrate. Thus PDMS is considered as a better substrate to be used for the stretchable electronic circuit. This is because PDMS has a lower Young Modulus compared to PU. Increasing the stiffness of the substrate will decrease the maximum allowable stretchability of the structure [3]. As the Young Modulus increase, the stiffness of the structure or materials will also increase. From the result, it shows that PDMS substrate will offer more maximum allowable stretchability compared to PU substrates. Based from the stress values in table 3, a graph was plotted and shown in figure 5.

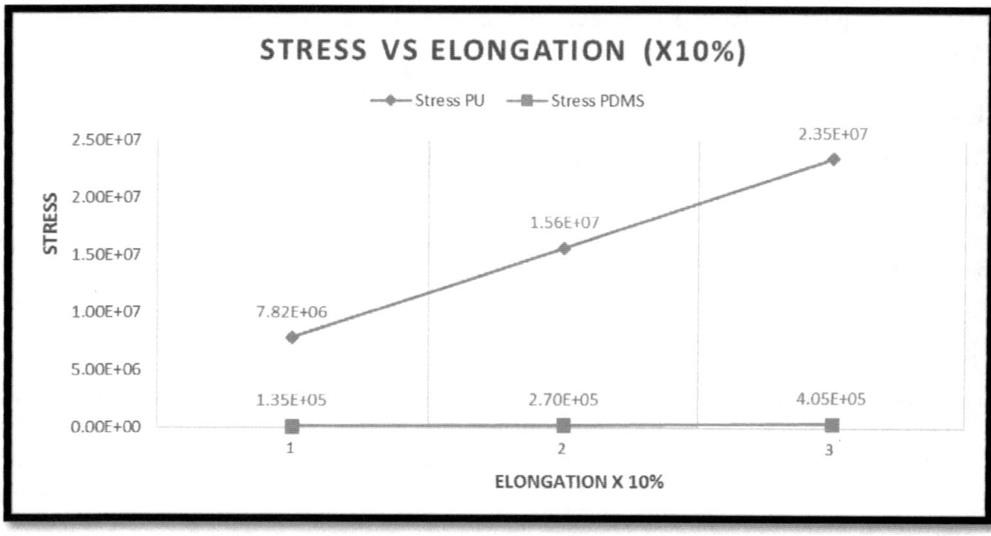

Figure 5. Stress for PU and PDMS vs Elongation.

From the graph, PDMS substrate record a lower stress (Pa) compare to PU substrate at each elongation. This indicates that in order for the stretchable circuit to stretch to a certain elongation or length, PU substrate will give higher stress compare to PDMS substrate. It shows that PDMS is more stretchable than PU substrate. Thus PDMS is considered as a better substrate to be used for the stretchable electronic circuit. This is also because PDMS has a lower Young Modulus compared to PU. According to Verstraete.H et al [5], by increasing the Young's modulus, the entire device has become stiffer and the maximum stress has increased. As the Young Modulus increase, the stiffness of the structure or materials will also increase. From the result, it shows that PDMS substrate will offer more maximum allowable stretchability compared to PU substrates. Thus PDMS is considered as a better substrate compared to PU.

PDMS is a good material to replace PU since it can perform under tension much better mechanically. By using PDMS also it will offer a much softer substrate material compared to PU. The PDMS have a hardness unit of 40-80 Shore A[6]. The PU has a hardness unit of 90 Shore A[7]. Shore A unit was used to measure the hardness of flexible rubbers that range in hardness from very soft and flexible, to medium and somewhat flexible to hard with almost no flexibility at all. Extremely soft and viscous materials were suggested as a good candidate for the encapsulation layer of stretchable electronics [8]. Thus, PDMS is a better material to be used as substrate because it is softer compared to PU.

4. Conclusion

To summarize, in this research, the study on deformation of stretchable electronic circuit was done. In order to get a deeper knowledge on how the deformation affects the stretchable electronic circuit, Abaqus Explicit software is used. Axial deformation was applied for the simulation of the FE model. Based on the several runs of simulation from the three different designs of conductor that are Rectangular, Horseshoe and U-shape at different elongations, the strain value can be observed. As the elongation increase, the strain value will also increase. As strain rate increase it will leads to failure of the copper conductor. In conclusions:

- Among the three designs, Horseshoe recorded the lowest strain value. Thus, it is considered as the optimum design for the copper conductor.
- PDMS recorded a lower strain and stress compare to PU at elongation 10%, 20%, and 30%. It offered more maximum allowable stretchability compared to PU substrates.

- PDMS is a good material to replace PU since it can perform under tension much better mechanically.

Acknowledgement
This work was supported by the International Islamic University Malaysia (IIUM).

References
[1] Harrop D P 2013 *Stretchable electronics comes to market* IDTechEx
[2] Hua Lin 2015 *Analysis of stress concentration phenomenon in stretchable interconnects* University of Electronic Technology Guilin, China
[3] Gonzalez M, Axisa F, Bulcke M V, Brosteaux D, Vandevelde B and Vanfleteren J 2008 *Reliability* **48** 825–832
[4] ThomasWidlund, ShixuanYang, Yung-Yu Hsu and Nanshu Lu 2014 *International Journal of Solid and Structures* **51** 23-24
[5] Verstraete H, Pakazad S and Dekker 2016 *Design of stress-deflecting objects and shapes for metal interconnects in stretchable electronic circuits using finite element analysis* (Eindhoven, Netherlands)
[6] Cidra Precision 2017 Polydimethylsiloxane (PDMS) http://www.cidraprecisionservices.com/life-sciences-materials-pdms.html
[7] Pleiger 2016 Polyurethane Material Propeties http://pleiger.com/polyurethane-materialproperties
[8] Yuan J H, Pharr M, Feng X, Rogers JA and Huang Y 2016 *Design of Stretchable Electronic Against Impact*

IOP Conf. Series: Materials Science and Engineering **290** (2017) 012023 doi:10.1088/1757-899X/290/1/012023

Product Development and Cost Analysis of Fabricating the Prototype of Roller Clamp in Intravenous (I.V) Tubing Medical Devices using Fused Deposition Modeling (FDM) Technology

WAY Yusoff[1,2]

[1]Kulliyyah of Engineering, Manufacturing and Materials Department International Islamic University, Malaysia
[2]College of Engineering, Industrial Engineering Department, University, Malaysia, University of Hail, Kingdom of Saudi Arabia.

Email: yusmawiza@iium.edu.my

Abstract. The main aim of this research is to develop a new prototype and to conduct cost analysis of the existing roller clamp which is one of parts attached to Intravenous (I.V) Tubing used in Intravenous therapy medical device. Before proceed with the process to manufacture the final product using Fused Deposition Modeling (FDM) Technology, the data collected from survey were analyzed using Product Design Specifications approach. Selected concept has been proven to have better quality, functions and criteria compared to the existing roller clamp and the cost analysis of fabricating the roller clamp prototype was calculated.

1. Introduction

Rapid Prototyping technology (RP) offers the user the ability to optimize part design in order to meet customer requirements with few manufacturing restrictions. One of the most common RP processes is Fused Deposition Modeling (FDM). The majority of commercial rapid prototyping system build object by adding one layer after another. RP was mainly used for product development in the manufacturing industry. It was used to speed up design process for household products, toys, tools and mechanical components [1]. The mechanism of rapid prototyping technology basically just same as the mechanism of normal printer that we use to print images. The different is on the product where the product produced from rapid prototyping machines is in 3D form as well as solid physical model while the product produced from normal printer is in 2D form [2]. As shown in Figure 1, FDM machine was developed by StratasysInc. FDM creates 3D models out of heated thermoplastic material, extruded through a nozzle positioned over a computer controlled X-Y table. The table is moved to accept the material until a single thin slice is formed. The next slice is built on top of it until the object is completed. FDM utilizes a variety of build materials, such as polycarbonate, polypropylene and various polyesters which are more robust than the SLA models [3].

Figure 1. FDM machine

Previous works conducted by Petzold et.al [4] and Ed [5] in medical sector mentioned that rapid prototyping can be used as an important tool to improve sub-sectors in medical such are orthopedics, surgery and implant. Compare to other traditional making process of prototype, rapid prototyping is better because it can speed up the process and manufacture the prototype in minutes. Besides, the complex parts of human anatomy are not considered as problem anymore because rapid prototyping can easily manufacture complex fracture without any obstacles. In traditional making process such as foundry, the complexity of product becomes one of the disadvantages of the process Sanghera et.al [6]. FDM models can also be made in wax, enabling custom-made implants to be investment cast for individual patients. Gibson et al. [3], Sanghera et.al [6] and Ed [5] agreed that recent improvements in technology are leading to the increased use of rapid prototyping to make implant human body parts. Jarnieson et al. [7] suggested that the implant human body parts must be customized for each patient for instant nearly all hearing-aids today are made using either Stereolithography or Selective Laser Sintering.

According to Webb [8] and Aspden et.al [9], intravenous therapy is a treatment that experienced by mostly people in the world. It is because the therapy is the most efficient way to solve or operate several common problems such are maintain imbalance electrolyte in our body, blood transfusion, deliver fluids and medications. Even for blood donation process, the treatment used is intravenous therapy. Matthew [10] defines the word 'intravenous' simply means 'within a vein'. I.V. therapy is the infusion of liquid substances into vein. The term vein is taken from Latin phrase which is 'vena' which means blood vessels. Blood vessels are considered as the route for I.V. therapy. Indications for I.V. therapy include achieving or maintaining fluid and electrolyte balance, replacing or supplementing needed blood components, providing nutrients, and administering medications. Jarnieson et al. [7] compared with other routes of administration, the intravenous route is the fastest way to deliver fluids and medications throughout the body. Because of this situation, it is very important to maintain the safety of the I.V. therapy. Aspden et.al [9] noted that medication errors were among the most common medical mistakes, it is among the top 10 leading causes of death in the U.S harming at least 1.5 million people each year and costing more than $3.5 billion annually for preventable drug-related injuries in hospitals alone.

. One of the popular treatments in medical sector is intravenous (I.V.) therapy. It use the device known as Intravenous (I.V) Tubing which is a combination of several parts as shown in Figure 2. Barbara [11] found that the idea Intravenous (I.V) Tubing may be a macrodrip solution administration set that delivers 10 or 15 drops/ml, or a microdrip set of the burette type that delivers 60 drops/ml. Macrodrip sets are used for routine infusions. Microdrip sets are used in pediatric and neonatal care and when relatively small amounts of fluids are to be administered over a long period of time. The infusion tubing drip chamber will achieve maximum flow rate when suspended approximately three feet above the site of injection due to the force of gravity. The higher of the roller clamp height, the greater force flow due to the weight of the fluid column.

Among these parts, roller clamp is one of the parts that have led to many problems as well as medication errors. Roller clamp has an important and critical function for the intravenous therapy which is to control the flow rate of the medication supplied. By referring to the previous research by Barbara [11]; Aspden et.al [9], Webb [8] reported the possibilities for medication errors to occur still are very high. Given the uprising number of medications error and medical mistakes, there will be high risk for the cost of losses to increase and at the same time harming more patients Jarnieson et.al [7]; Matthew [10]. This situation also can lead to dropping of the reputation of medical institutes because the duty of medical institutes is to provide medications or treatments, not to harm their patients.

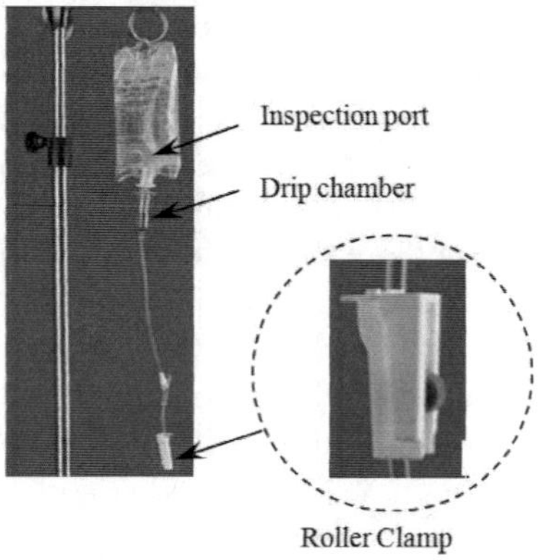

Figure 2. Basic Intravenous IV tubing Setup

In response to this predicted possible national concern, this research was set up in order to improve medical sector by redesign roller clamp to be used in Intravenous therapy. The existing roller clamp still has some weaknesses at the design and need to be improved to avoid medication errors and increase the effectiveness of the treatment. The purpose of the research is to redesign and to the optimum the effectiveness and its functions by considering several factors such as ergonomics, dimension, features and functionality while at the same time reduce the fabrication cost.

2. Methodology

In order to understand the current problem of Roller Clamp in Intravenous (I.V) tubing a survey has been conducted. The purpose of this survey is to identify the common problem of the Roller Clamp in Intravenous (I.V) tubing and to gain some ideas for redesigning task of new Roller Clamp in Intravenous (I.V) tubing. In addition, several interviews have been done with medical assistants thoroughly. The target participant for this survey and the interview were medical students, medical assistants, nurse, surgeons and also experienced physicians in local hospitals. After problems of the current designs of Roller Clamp in Intravenous (I.V) tubing have been obtained, and the customers' voices and needs about the ETT have been known, the basic specifications of the new design of Roller Clamp in Intravenous (I.V) tubing parts were determined. The new features of Roller Clamp in Intravenous (I.V) tubing that has been proposed by participants were analyzed and instilled together into the new design. Several design concepts were generated, which will include all the basic specifications and semi-finished designs of the new product. These visible design concepts were analyzed before the final product is selected. This final product has been tested to ensure that it is conform to the basic specifications, which are already determined at the early stage of the product development and process. The test ensures the new product met the customers' needs. If the test result is negative, the whole process will be redone until the right and most suitable final product is chosen. Then, the specifications of the final design concept will be revised if necessary in order to merge all the positive features that might be good to be added together with the final product.

3. Results and Discussion

Table 1 shows the major problems faced by the existing design of roller clamp that is used as a reference concept to be compared with each of the generated concept. Analysis has been carried out on the existing design of roller clamp. It is found that, the linear bottom part of existing roller clamp should be modified to non-linear to improve the flow rate. Figure 3 shows the selection of criteria of existing roller clamp in Intravenous (I.V) tubing. It shows the selection of criteria demanded by respondents it also provides rough information regarding the customer needs but not specifically and accurately. It is found that only four criterion above clearly rated as important by the respondents. Therefore, to get better analysis and accurate result, the data was analyzed using Product Design Specifications (PDS) by constructing a table of customer need at the beginning of process. From the previous analysis on the data gathered from questionnaire and PDS, the research procedure is preceded and 5 concept designs have been successfully created. Basically, the design of roller clamp consists of two parts, roller and body. These parts will be combined to form roller clamp and below is the table of 5 concept designs that have been successfully produced. Each of the design has its own special features and differences.

As shown in Table 2, the metrics listed are the technical data that roller clamp must have and the metrics will be matched to each of the metrics. This comparison is very important to find marginal value and ideal value that will be used to generate five concept designs Table 3 shows how the analysis conducted, it is found design E obtained highest final score compare to other designs. It is found that concept design E has been successfully chosen as the best design and will be developed. Figure 4 shows the engineering drawing and its main components. The final prototype of roller clamp in Intravenous (I.V) tubing fabricated using FDM machine is shown in Figure 5.

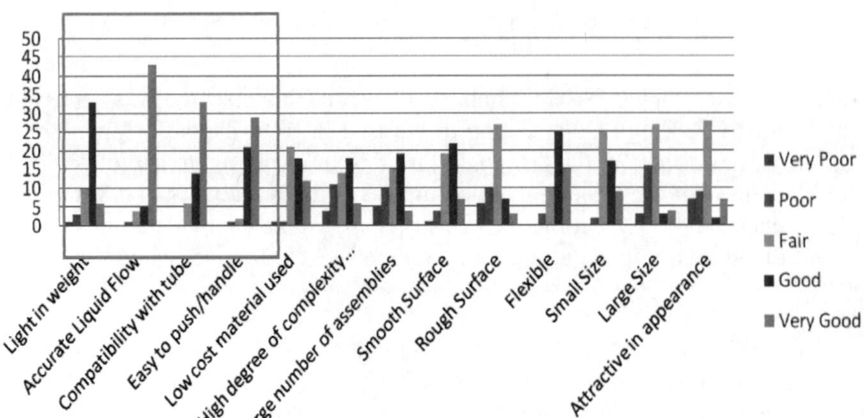

Figure 3. Criteria selection of roller clamp

Table 1. Problems faced by the existing design of roller clamp

No	Problem	Causes	Parts involve
1	Lack of accuracy of the flow rate.	Linear shape of the bottom part of roller clamp.	
2	Pressure or movement from patient disturb the roller clamp and modify the location of the roller.	Lack of gripping force from the roller to the IV tube.	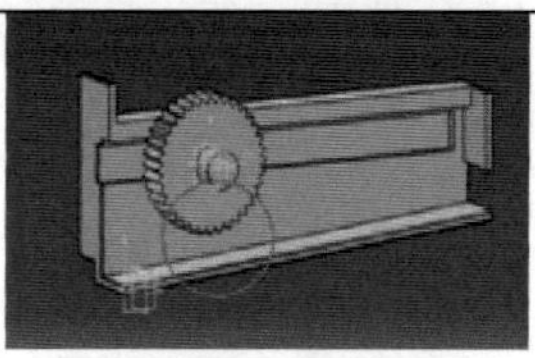
3	Lack of gripping force at the finger to push the roller.	Smooth surface of roller.	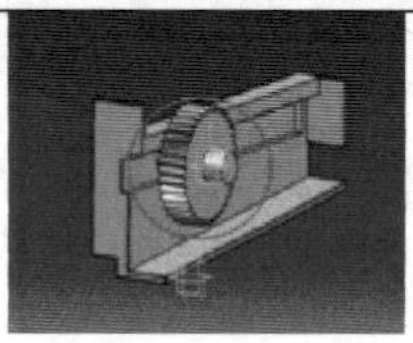

Table 2. Target specification

Need #'s	Metric	Imp.	Units	Marginal Value	Ideal Value
1	Weight	4	kg	>0.010	>0.010
2	Flowrate	5	drops/mL	>20	>30
5	Cost	4	RM	5-50	<20
11,12	Width	4	mm	20-30	<20
11,12	Length	4	mm	30-100	>40
8,9	Surface roughness	4	subj	Good	Good
7	Number of assemblies	4	integer	2	2
5	Biocompatible material for roller clamp	4	list	ABS/PP	ABS
5	Thermal resistance temperature	4	'C	100-110	>110
4,7	Ergonomic	5	Subj	Good	Good
7	Manufacturing time per each	4	min	10-50	<20
13	Color	3	Subj	Black/White	Black/White
3,6,10	Part complexity	3	Subj	Reduce	Reduce

Medical sector is one of the best platforms for rapid prototyping to develop. One of the popular treatments in medical sector is intravenous (I.V.) therapy. In rapid prototyping technology, there are two most popular materials used in the fabrication process. The first one is ABS and the second one is Polypropylene (PP). There are pro and contras between these two materials in term of price, strength, and resistivity and so on. In the fabrication of roller clamp, it is clearly stated that ABS is used as the fabrication material. There is not much different about the price between these two materials and of course ABS material has higher price compared to PP, however both prices are affordable for any fabrication process [2]. In addition, ABS has been proven to have better quality and characteristics compared to PP material. Below is the calculation of the total fabrication cost of the roller clamp [1].

Table 3. Selection matrix for concept screening

Selection Criteria	For the following criteria, please indicate for each of the design concept below. 1 - poor 2 – fair 3 – good 4 - very good 5 – excellent				
	A	B	C	D	E
1.Functionability					
Compatible with tube	0	0	+	0	+
Able to control flow rate of fluid	+	+	+	+	+
2.Convenience					
Easy/comfortable to grip	0	0	+	-	+
Convenience to hold/place tube	0	0	+	0	+
3.Ergonomic					
Easy to push	0	0	0	-	+
Easy to hold with one hand	0	-	-	+	+
4.Appearance					
Attractive	-	-	0	0	+
Non-complicated	+	+	0	0	0
5.Surface texture					
Smooth surface	0	0	0	+	0
Rough surface	-	-	+	-	+
Gripping surface area	0	0	+	-	+
6.Manufacturing ease					
Low number of assembly steps	+	+	0	0	0
Low cost of design complexity	+	+	-	0	-
Low complexity of parts	+	+	0	0	0
7.Size/dimension					
Small	-	-	-	-	-
Medium	+	+	0	+	0
Large	0	0	+	0	+
TOTAL '+'	6	6	7	4	10
TOTAL '0'	8	7	7	8	5
TOTAL '-'	3	3	3	5	2
NET SCORE	3	3	4	-1	8
RANK	3	3	2	4	1
Continue?	Combine	Combine	Yes	No	Yes

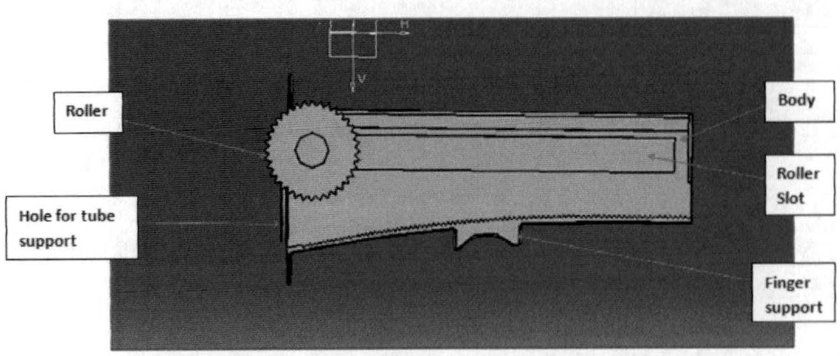

Figure 4. Best design selected (Design E)

Figure 5. Prototype of roller clamp in Intravenous (I.V) tubing

Below is the calculation of the total fabrication cost of the roller clamp.
Cost per cartridge: RM3300
Volume per cartridge: 1700

$$\text{Total cost} = \frac{\text{Model Material (cm}^3\text{) + Support Material (cm}^3\text{)} \times \text{Cost per cartridge}}{\text{Volume per cartridge (cm}^3\text{)}}$$

$$= \frac{1.467\ \text{cm}^3 + 1.615\ \text{cm}^3 \times \text{RM3300}}{1700\ \text{cm}^3}$$

$$= \textbf{RM5.98} \approx \textbf{RM6.00 per roller clamp}$$

$$\text{Total of roller clamp can be produced per cartridge} = \frac{1700\ \text{cm}^3}{1.467\ \text{cm}^3 + 1.615\ \text{cm}^3}$$

$$= \textbf{551 units}$$

$$\text{Total of roller clamp can be produced per hour} = \frac{\text{60 minutes}}{\text{Estimated build time per roller clamp}}$$

$$= \frac{60}{19}$$

$$= 3.15 \approx \textbf{3 units of roller clamp per hour}$$

Some improvements that have been to design E compared to reference concept especially on the non-linear bottom part of body, surface made rougher to increase gripping force, hole is created to support tube and finger rest part created at the bottom to support the finger while pushing the roller.

4. Conclusion

The overall findings in this study showed that the respondents which taken among medical students, doctors and nurses give positive feedback and good supports to the research and most of them clearly want the existing design to be improved. It can be clearly seen that the existing design still do not satisfy the users which are among nurses, doctors and medical trainers. In other words, the existing roller clamp still have some weaknesses and this is the purpose of this research to be done which is to complete the weaknesses and then improves the process to deliver blood and medications through Intravenous Therapy. Moreover, the selected concept has been proven to have better quality, functions and criteria compared to the existing roller clamp and this research successfully achieved the mission. As the prototype of improved design has been fabricated, it will be introduced to doctors and medical officers for some advises regarding the steps should be taken to develop the prototype further.

Acknowledgements

This research is partially funded by the Research Endowment Fund Type B, RMC IIUM, We would like to give special thanks to Dr Saufi Awang, Dr Abdul Hadi, Dr Nurakidah Mohd Nor, medical doctors and nurses at General Hospital Kuala Lumpur, Tengku Ampuan Afzan Hospital Kuantan, for their endless help in the design, development and implementation of this research .

References

[1] Chua C K and Leong K F 1997 *Rapid Prototyping: Principles and Applications in Manufacturing.* John Wiley & Sons, Inc., (Asia).
[2] Rafiq N 2005 *Rapid Prototyping: Principles and applications.* John Wiley & Sons, Inc. (Hoboken, New Jersey).
[3] Gibson I, Cheung L K, Chow S P, Cheung W L, Beh S L, Savalani M, and Lee S H 2006 *Rapid Prototyping J.* **12** 53–58.
[4] Petzold R, Zeilhofer H-F and Kalender W A 1999 *Comput. Med. Imag. Grap.* **23** 277-284.
[5] Ed G 2007 *Medical Applications of Rapid Prototyping.* Retrieved from http://www.additive3d.com/med_lks.htm
[6] Bal Sanghera, Satyajit Naique, Yannis Papaharilaou, Andrew Amis2001 *Preliminary study of rapid prototype medical models, Rapid Prototyping J* **7** 275-284

[7] Jarnieson R, Holmer B and Ashby A 1995 *Rapid Prototyping J.* **1** 38-41.
[8] Webb P A 2000 *J. Med. Eng. Tech.* **24** 149-153.
[9] Aspden P, Wolcott J, Bootman L, Cronenwett L R 2006 *Preventing Medication Errors* Washington, D.C: National Academies Press.
[10] Matthew G 2011 *Capping Intravenous (I.V) Tubing and Disinfecting Intravenous Ports Reduce Risks of Infection*. Retrieved from http://www.ncbi.nlm.nih.gov/pmc/articles/PMC3086084/Sanghera
[11] Barbara L C 2009 *Introduction to Biomedical Instrumentation*. United States of America. Cambridge University Press.

IOP Conf. Series: Materials Science and Engineering **290** (2017) 012024 doi:10.1088/1757-899X/290/1/012024

Evaluating 8 pillars of Total Productive Maintenance (TPM) implementation and their contribution to manufacturing performance

E Y T Adesta[1], H A Prabowo[2], D Agusman[3]

[1]Department of Manufacturing and Materials Engineering, International Islamic University Malaysia (IIUM), Jalan Gombak, 53100 Kuala Lumpur, Malaysia.
[2]PhD Student at Manufacturing and Materials Engineering Department, International Islamic University Malaysia (IIUM), Gombak, Malaysia and a Lecturer staff of Industrial Eng. Dept. Mercu Buana University Jakarta, Indonesia.
[3]Department of Mechanical Engineering, Universitas Muhammadiyah Prof. Dr. HAMKA (UHAMKA), Jalan Limau 2 Kebayoran Baru, Jakarta Selatan, Indonesia

Email: herryagung@gmail.com

Abstract. TPM is one method to improve manufacturing performance through an emphasis on maintenance that involves everyone in the organization. Research on the application of TPM and its relevance to the manufacturing performance has been performed quite a lot. However, to the best of our knowledge, a study that deliberates how the application of 8 pillars TPM (especially in developing countries) is still hard to find. This paper attempts to evaluate in more detail about how the 8 pillars of TPM are applied in Indonesia and their impact on manufacturing performance. This research is a pilot study with a target of 50 companies. From the results of data collection, only 22 companies (44%) are eligible to process. Data processing was performed using SPSS and Smart PLS tools. From the validity and reliability tests, it can be seen that all items/indicators for TPM pillars are valid and reliable with correlation value (R) of 0.614 - 0.914 and with Cronbach's alpha equal to 0.753. As for the Manufacturing Performance construct, the Delivery indicator was not valid. In overall, the model is reliable with Cronbach's alpha of 0.710. From the results of Confirmatory Factors Analysis (CFA) for TPM, it can be seen that four indicators (pillars) are highly significant while four other indicators are less significant. For MP, three indicators are significant, and two are not significant. In general, the structural model of the relationship between TPM and MP is relatively strong and positive with values $R = 0.791$, and R squared $= 0.626$. This means that the TPM Pillars can explain 62.6% MP variability construct variable, while the other 37.4% can be explained by unrelated variables.

1. Introduction

With the challenges of tight competition worldwide, manufacturing industry is under pressure to provide high levels of performance and commitment. To face the ever-changing customer demands, manufacturing companies opted to adopt strategic changes in management approaches, production process and technologies, supplier attitudes and customer behavior. Lean manufacturing principles have been widely used by manufacturing companies to achieve these changes and gain competitive advantage. It also has emphasized the re-examination of the role of improved maintenance management towards enhancing the organization's competitiveness [1]. One approach to improving the performance of the maintenance activities is by implementing Total Productive Maintenance (TPM) strategy [1]. It is required in modern manufacturing for companies to be successful organizations, and it has to be supported by both effective and efficient maintenance program, which is TPM [2].

The basic practices of TPM are often called the "pillars" or "elements" of TPM. The entire edifice of TPM is built and stands on eight pillars [3]; [4]; [1] and [5]. The eight pillars of TPM is a system for maximizing production effectiveness and efficiency. Many researchers have investigated and explained the application of those eight pillars in the industry that are mostly located in developed countries, but very few had examined their implementation in developing countries (such as Indonesia) and their effect on the operating performance. Some Indonesian researches in TPM focused mostly on the impact of TPM implementation on Six Big Losses and Overall Equipment Effectiveness (OEE) regardless of how the execution of the 8 pillars was implemented. For example [6], uses OEE to measure the effectiveness of production machinery and uses 5S methodology as a tool to progress existing maintenance processes and suggested to implement TPM Pillars [7]. Many researchers [8], [9], [10], [11] also only examine Six Big Losses and OEE for measuring performance machine regardless of how TPM pillars are running. Though Six Big Losses and OEE are only 'Results' or "Effects" of TPM implementation. A little progress is made by [12] with examining one of the TPM pillars: Autonomous Maintenance and its impact on OEE machines. From this background, this study aims to evaluate the employment level of 8 TPM pillars and their impact on manufacturing performance in the Indonesian manufacturing industries.

2. Total Productive Maintenance (TPM)

Total Productive Maintenance (TPM) is a productive maintenance concept designed to achieve a comprehensive effectiveness of the production system by involving everyone within the organization [3] and [13]. In more detail, TPM is divided into three important concepts:

1. Total, which means the involvement of all personnel/employees of the company.
2. Productive, which means TPM activities/activities is executed as much as possible, does not interfere with the productivity of the company,
3. Maintenance, which means the selection of the most appropriate/effective method of maintenance.

The eight pillars of TPM is a system for maximizing production effectiveness of any industry. The summary of eight pillars is given in table 1.

Table 1. Brief Summary of Eight Pillars [1]; [5]

TPM Pillars	Description	Advantages
Autonomous Maintenance	Hands operators of equipment responsible for carrying out basic maintenance of equipment	Operators feel responsible for their machines; equipment becomes more reliable
Planned Maintenance (Keikaku-Hozen)	Maintenance scheduled using the historical failure rate of equipment	Maintenance can be scheduled when production activities are few
Quality Maintenance	Quality ingrained in the equipment to reduce defects	Defect reduction & consequent profit improvement
Continuous Improvement (Kaizen)	Use of cross-functional teams for improvement activities	Improves problem-solving capabilities of the workers
Early Equipment Management	Design of new equipment using lesson learned from previous TPM activities	New equipment achieves full the potential in a shorter period
Education & Training	Bridging of the skills and knowledge gap through training of all workers	Employees gain the necessary skills to enable them to solve the problems within the organization
Health. Safety & Environment	Providing of an ideal working environment devoid of accidents and injuries	Elimination of harmful conditions & healthy workforce

TPM in the Office	Spread of the principles to administrative functions within an organization	Support functions understand the benefits of these improvements

The definition of the 8 pillars of TPM above is the basis of this research which is about the evaluation of the application of 8 pillars of TPM in Indonesian manufacturing industries and its effect on their performance.

3. Manufacturing Performance (MP)

In this research, only 5 MP indicators are used, 4 indicators were taken from [14] namely: 1) Quality (Q), 2) Cost (C), 3) Delivery (D), and 4) Flexibility (F) performances and 1 (one) indicator taken from [1] and [15] which is OEE (Overall Equipment Effectiveness) value.

4. Research Methodology

To achieve the objectives of this study and to obtain a valid result, a series of research activities were systematically constructed. These activities can be described as follow:

i. Identifying TPM Pillars and MP indicators according to previous studies (based on literature reviews).
ii. Generating a questionnaire to measure variables and conducting pilot study (n=50)
iii. Performing Validity and Reliability Test.
iv. Formulating the models (CFA models and Structural Model) using SEM tools.
v. Data processing using PLS/Amos software.
vi. Analysing and comparing the results with SEM standard values.
vii. Making conclusion, documentation, and publication.

5. Results and Discussion

Out of the 50 companies targeted for the delivery of questionnaires, only 25 companies responded, and only 22 companies (44%) were eligible for processing which considered to be sufficient. The collected data is then processed using SPSS to test the validity and reliability of the questionnaire. Only valid and reliable data are then processed using Smart-PLS 3.0 program.

Validity and Reliability Tests.

- **Results for TPM Pillars:**

Table 2. TPM pillars validity test

Indicators	Correlation (R)	Conclusion
1st Pillar	0.707	Valid
2nd Pillar	0.658	Valid
3rd Pillar	0.644	Valid
4th Pillar	0.867	Valid
5th Pillar	0.914	Valid
6th Pillar	0.614	Valid
7th Pillar	0.668	Valid
8th Pillar	0.708	Valid

From the above validity test, all TPM pillars indicators have correlation values (R) between 0.614 to 0.914 which is greater than 0.5, so that all question items in the questionnaire are considered valid [16] and can extract the desired latent variable (TPM pillars).

Table 3. TPM Pillars reliability test

Cronbach's Alpha	N of Items
.753	25

From the reliability test, it can be seen that the questionnaire used is reliable because it has a value of Cronbach's alpha greater than 0.7 [16], so it can be continued to the next stage of data processing using Smart PLS to get the indicators that form TPM Pillars in Indonesia.

- **Manufacturing Performance (MP) Test Results:**

Table 4. MP Validity test

Indicators	Correlation (R)	Conclusion
Quality	**0.587**	Valid
Cost	**0.616**	Valid
Delivery	**0.253**	Not Valid
Flexibility	**0.509**	Valid
OEE	**0.595**	Valid

From the result of the validity test above, not all of manufacturing performance indicators are valid as the delivery indicator has R value count less than 0.5 [16], so it was revised for further data retrieval process.

Table 5. MP Reliability test

Cronbach's Alpha	N of Items
.710	5

From the reliability test, it was observed that Cronbach's alpha for MP is greater than 0.7 which revealed the reliability [16]. Then all indicators were further processed using Smart PLS program to determine the significance level of each indicator in forming the latent variable.

Smart PLS Analysis Results

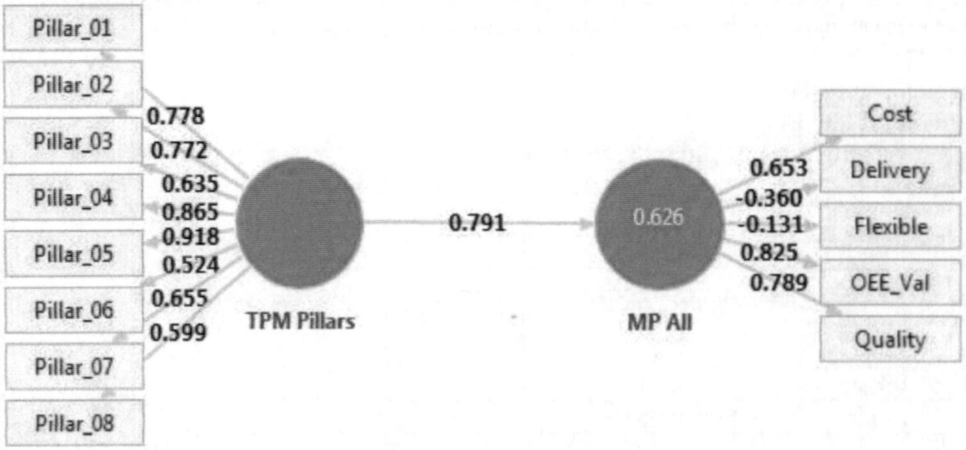

Figure 1. CFA and SEM (Structural Equation Modelling) for TPM Pillars and MP

Confirmatory Factors Analysis (CFA) for TPM Pillars (Outer model)

From the calculation results using Smart-PLS program (Figure.1), it is observed that 4 indicators have been applied and run very well. Those are Pillar 01, Pillar 02, Pillar 04 and Pillar 05 which had loading factor values greater than 0.7 [16]. While Pillar 03, 06, 07 and 08 have been applied but not running maximally as the loading factor values were less than 0.7. However, according to [16], loading factor on the level of development values of 0.5-0.6 can still be considered adequate.

Table 6. TPM Indicators/Pillars

Indicators	Loading factor	Conclusion
1st Pillar: Autonomous Maintenance	0.778	Significant
2nd Pillar: Continuous Improvement	0.772	Significant
3rd Pillar: Planned Maintenance	0.635	Less Significant
4th Pillar: Quality Maintenance	0.865	Significant
5th Pillar: Education and Training	0.918	Significant
6th Pillar: Safety, Health, Environment	0.524	Less Significant
7th Pillar: Office TPM (Supporting)	0.655	Less Significant
8th Pillar: Development Management	0.599	Less Significant

From table 6, it can be explained in detail the activities that have been executed as part of the implementation of 8 pillars TPM as follows:

- 1st Pillar: Autonomous Maintenance, that the operator has done simple machine maintenance activities include cleaning, lubrication, adjustment and tightening and inspection. The operator also has a sense of ownership of the machines/equipment they operate.
- 2nd Pillar: Continuous Improvement (Kaizen), the PDCA process (Plan, Do Check, Act) has been running well, various types of losses have been identified to be systematically eliminated, and the efficiency and effectiveness of the system are continually enhanced.
- 3rd Pillar: Planned Maintenance, preventive, predictive and corrective maintenance have been planned. All maintenance activities have been carried out regularly. The maintenance program tries to optimize the MTBF (Mean Time Between Failures) and MTTR (Mean Time To Repair) engines but still needs improvement.
- 4th Pillar: Quality Maintenance, zero defects targets have been applied, causes of quality issues have been identified appropriately. Machinery, materials, operators have been prepared to achieve the best performance.
- 5th Pillar: Education and Training, human resource competencies have been adapted to organizational goals, multi-skilled workers have been developed as needed. Human resources also have been evaluated, and it updates the skill of employees on a regular basis
- 6th Pillar: Safety, Health, Environment (SHE). SHE standard operating procedures (SOPs), safe and healthy working environment, and the availability of adequate sewage treatment facilities are still being developed. It has not run perfectly because it requires many investment costs.
- 7th Pillar: Office TPM (Supporting), 5S program has been implemented in the office area, work procedure/bureaucracy have been minimized, and synergy between departments have been built, but still, needs improvement.
- 8th Pillar: Development Management, problems in the installation of new equipment have been minimized, the experience of previous machine/system has been utilized for repair, Machine maintenance system has been developed in a better way. This pillar still needs enhancement.

Confirmatory Factors Analysis (CFA) for Manufacturing Performance (Outer model)

For MP indicators, only three indicators that significantly represent them, namely: Quality, Cost, and OEE, which were significant; whereas the other two indicators, Delivery and Flexibility, were not significant because their value of loading factor was below 0.5 [16] as can be seen in table 7.

Table 7. MP Indicators/Pillars

Indicators	Loading factor	Conclusion
Quality	0.789	Significant
Cost	0.653	Significant
Delivery	-0.360	Not Significant
Flexibility	-0.131	Not Significant
OEE	0.825	Significant

Since the study is still in the pilot project stage, the insignificant variables would not be abandoned but might be revised through the detailed questionnaires. In addition to the loading factors, to assess the validity of reflective indicators, it is also necessary to see the discriminant validity and average variance extracted (AVE) values.

Table 8. Discriminant Validity

Fornell-Larcker Criterium		
	MP All	**TPM Pillars**
MP All	0.612	
TPM Pillars	0.791	0.730

Table 9. Average Variance Extracted (AVE)

	AVE
MP All	0.375
TPM Pillars	0.532

Discriminant validity shows that the relationship between the construct (latent variable) of TPM Pillars with the construct of MP with a value of 0.791 is slightly higher than the correlation between the construct indicator (TPM Pillars = 0.730 and MP indicators = 0.612). As a result, the correlation between constructs is greater than the correlation in the construct. From the AVE value, it appears that AVE for MP has a value below 0.5 which is not valid. This can be explained by the discriminant value where only 3 indicators are valid with values above 0.5. For further analysis, the related questionnaire needs some improvements/revisions.

Table 10. Model's Reliability

	Composite Reliability (CR)
MP All	0.502
TPM Pillars	0.898
	Cronbach's Alpha (CA)
MP All	0.323
TPM Pillars	0.872

From the above values (CR and CA), it appears that TPM Pillar construct variables have excellent reliability because the value is above 0.70, while MP construct variables are less reliable.

Structural Equation Model (Inner Model) for TPM and MP

Table 11. Path coefficient

	MP All
TPM Pillars	0.791
	R Square
MP All	0.626

The relationship between TPM Pillars with MP appear to be STRONG POSITIVE based on Path Coefficient (R value) of 0.791. This is in line with previous studies such as [13], [17] and [18] which reported that there is a strong and positive relationship between TPM implementation and manufacturing performance. From the R squared value, 62.6% MP variability can be explained by the TPM Pillars construct variables, while unrelated variables can describe the other 37.4%.

6. Conclusion

In summary, it can be concluded that the implementation of 8 TPM's pillars in Indonesia manufacturing industries was relatively good. Where the 4 pillars, namely: Autonomous Maintenance, Continuous Improvement, Quality Maintenance, and Education and Training, are running very well. The other 4 pillars, namely: Planned Maintenance, Safety-Health & Environment (SHE), Office TPM (Supporting), and Development Management, need to be further improved, particularly SHE pillar which has the smallest loading factor. The insignificant variables in Manufacturing Performance should also be revised through related questionnaires. From the structural model, it can also be concluded that the model of TPM Pillars and MP relationship is reliable, positive and strong, as TPM pillars can explain 62.6% variability of MP, while other factors influenced the rest (37.4%).

References

[1] Ahuja, I.Khamba "TPM: Literature review and directions," *International Journal of Quality and REliability Management,* pp. 709-756, 2008.

[2] Hashim. e. S., "TPM and Innovative Performance," *International Journal of Engineering Research and Development,* vol. 3, no. 11, pp. 62-67, 2012.

[3] Nakajima. S., Introduction to TPM, Cambridge: Productivity Press, 1988.

[4] Ireland. B. D. F., "A study of TPM implementation," *Journal of Quality in Maintenance Engineering,* vol. 7, no. 3, pp. 183-191, 2001.

[5] Pandey. R. N. D.S., "Implementing TPM by doing RCA," *International journal of advanced research in science, engineering and technology,* vol. 3, no. 2, 2016.

[6] J. S. Hery Kristanto Sinurat, "Analisa Efektivitas Mesin menggunakan OEE dan 5S sebagai usulan penjadwalan mesin," *Rekavasi,* vol. 3, no. 2, pp. 86-93, 2015.

[7] A. R. R. M. Baharuddin Yusuf, "Analisa OEE untuk memperbaiki perawatan mesin berbasis TPM," *Jurnal Rekayasa dan Manajemen Sistem Industri,* vol. 3, no. 1, 2015.

[8] G. H. Badik Y. Asgara, "Analisa efektivitas mesin chrane dengan metode OEE," *INASEA,* vol. 15, no. 1, pp. 62-70, 2014.

[9] T. O. G.Adrian, "Perhitungan OEE di PT ABC," *Jurnal Titra,* vol. 3, no. 2, pp. 253-256, 2015.

[10] E. A. N. Sunaryo, "Kalkulasi OEE untuk mengetahui efektivitas mesin Komatsu 80T," *Teknoin,* vol. 21, no. 4, pp. 225-233, 2016.

[11] B. S. Tofiq D. Darmawan, "Analisa OEE untuk Meminimalkan Six big losses di PT Semen Indonesia," in *KOnferensi Nasional IDEC 2017,* Surakarta, 2017.

[12] I. M. A. W. E.R. Supriatna, "Autonomous Maintenance pada Plant 2 PT Ingress," *Journal Teknik Industri,* 2017.

[13] Bakri., "Boosting LM via TPM," *Social and Behavioral Sciences,* vol. 65, pp. 485-491, 2012.

[14] I. Belekoukias, "The Impact of Lean Methods and Toolson The Operasional Performance of Manuf. Org.," *International Journal of Production Research,* pp. 5346-5366, 2014.

[15] K. Tomar S Arun, "Analysis of OEE for TPM Implementation," *International journal of business quantitative economics,* vol. 2, no. 8, 2016.

[16] H. L. Imam Ghozali, PLS Konsep, Teknik dan Aplikasi dengan SmartPLS 3.0, Semarang: Penerbit UNdip, 2015.

[17] Mishra. e. Y., "Total Productive Maintenance," in *Emerging Trends in Engineering and Mangement*, India, 2016.

[18] Perera. GLD, "Effect of TPM practices on Manufacturing Performance," *Journal of Manufacturing Technology Management,* vol. 27, no. 5, 2016.

IOP Conf. Series: Materials Science and Engineering **290** (2017) 012025 doi:10.1088/1757-899X/290/1/012025

Productivity improvement using discrete events simulation

M H F Al Hazza[1], E M Y Elbishari[2], M Y Bin Ismail[3], E Y T Adesta[4], Nur Salihah Binti Abdul Rahman[5]

[1]Department of Manufacturing and Materials Engineering, International Islamic University Malaysia (IIUM), Jalan Gombak, 53100 Kuala Lumpur, Malaysia

Email: mautaz@iium.edu.my

Abstract. The increasing in complexity of the manufacturing systems has increased the cost of investment in many industries. Furthermore, the theoretical feasibility studies are not enough to take the decision in investing for that particular area. Therefore, the development of the new advanced software is protecting the manufacturer from investing money in production lines that may not be sufficient and effective with their requirement in terms of machine utilization and productivity issue. By conducting a simulation, using accurate model will reduce and eliminate the risk associated with their new investment. The aim of this research is to prove and highlight the importance of simulation in decision-making process. Delmia quest software was used as a simulation program to run a simulation for the production line. A simulation was first done for the existing production line and show that the estimated production rate is 261 units/day. The results have been analysed based on utilization percentage and idle time. Two different scenarios have been proposed based on different objectives. The first scenario is by focusing on low utilization machines and their idle time, this was resulted in minimizing the number of machines used by three with the addition of the works who maintain them without having an effect on the production rate. The second scenario is to increase the production rate by upgrading the curing machine which lead to the increase in the daily productivity by 7% from 261 units to 281units.

1. Introduction

The high competition in the market forced the manufacturer to search for new methods and procedures to speed up the production rate and reduce the time to market. Traditional methods of planning and development are no longer sufficient to reach targets and survive in the global market. The increasing in complexity of the new technology was increased the cost of investment as well. In many cases, the theoretical calculation is not consistent with the real environment. However, the development of the new advanced software is protecting the manufacturer from investing money in production lines that may not be sufficient and effective with their requirement in terms of machine utilization and productivity issue. By conducting a simulation, using accurate model will reduce and eliminate the risk associated with their new investment. Many of researchers [1-6] try to use different simulation approaches to support this claim. Most of the industries based on discrete events nature. In discrete events processes, each event occurs at a particular instant in time and adds value to the product. Therefore, discrete event simulation (DES) arises as one of the effective tools in production planning. DES is the process of codifying the behavior of a complex system as an ordered sequence of well-defined events [7-8]. However, discrete event simulation technologies have been extensively used by industry and academia to deal with various industrial problems [9]. Different softwares have been developed recently such as promodel, Sim event, delmia quest and many others. In this research, Delmia Quest will be used for real case study to highlight the importance of simulation approach in protecting manufacturers in unutilized equipment and reduce the risk of wrong decisions.

2. Methodology

Delmia Quest software has been used to analyses the production line of concrete poles in local industry. Most of stages in the production line are shared for the three different sizes of poles produced in the factory. A standard operating procedure (SOP) have been developed as basis to develop the simulation model. To develop the model and then simulating the model, few steps have been conducted.

1. Analyze the product design to determine the main components of the final product to create main sources in the system.
2. Seven main sources have been defined: Caging Part, Spiral Wire, Concrete source, PC wire, Mould A, Mould B and Mould C.
3. The processing time for each individual process has been estimated using direct measurement as well as company resources.

Finally, the model of the production line have been created and ready for simulation as shown in figure 1.

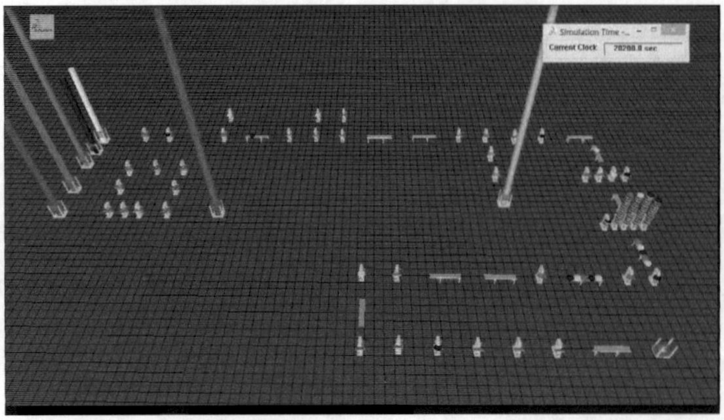

Figure 1. Layout of the manufacturing facility as designed in Delmia Quest (original design)

Based on original design, two different scenarios have been proposed: the first scenario is focusing on improving productivity and the second scenario is to improve the production rate. However, selecting one of these designs depends on the company objectives. In the following sections, the result of the two scenarios is presented.

2.1 Proposed Design

2.1.1 Scenario 1

The first scenario focuses on the utilization percentage and unused resources. Based on the results in original design, the following assumptions are realised in scenario 1:

1. Stretch and Plug Set Ring process in the original design used two machines with utilization of 44% for each. Therefore, by logic and simple calculation, the company can replace one machine to cut the cost and still can achieve the production rate of 261 parts per day
2. Tighten Bolt Accessories process as in the previous process used as well two machines with utilization of about 12% for each machine. The same calculation can be done and expected that the production rate will not be changed but with advantage of reducing the operational cost of workers and capital.

3. Spinning process have four equipment with utilization of about 71%, thus by assuming full utilization of three machines instead of four machines, the production rate will be remain the same.
4. Curing process in the original design have five machines with utilization is between 44% up to 65% based on simulation results. Therefore, the researcher assumes that four curing machines can fulfil the production rate of the company.
5. The Straighten PC Wire process have almost 100% of utilization. This may cause a bottleneck and may affect the whole production line. However, the researcher assumed an additional machine for this process to improve the flow of the production line.
6. Finally, additional handling equipment has been added to improve the flow of the products.

2.1.2 Scenario 2

The second scenario is focuses on improving the production rate per day. However, after different try by focusing on the bottleneck points with high percentage of utilization with different assumption, it was found that the production rate cannot be change by this approach. Therefore, the researches tried another approach by focusing on the longest processing time and try to improve. The research assume that the company upgrade the curing rooms and reduce the total processing time by 10% even though that the curing is not the highest utilization with the average of 52%.

3. Result

The results of the simulation for original design are concluded in figure 2.

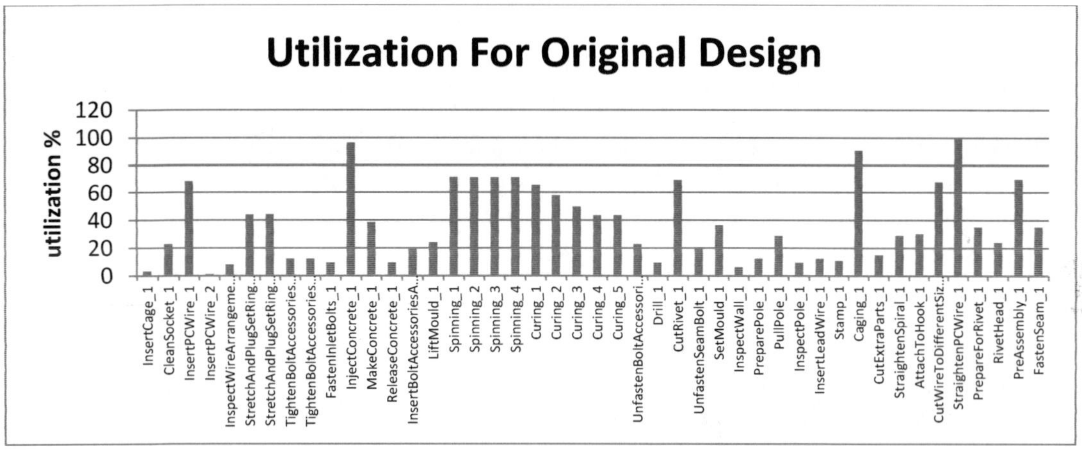

Figure 2. Utilization bar chart for original design

To get the whole picture, the analysis of the handling equipment utilization is concluded in figure 3.

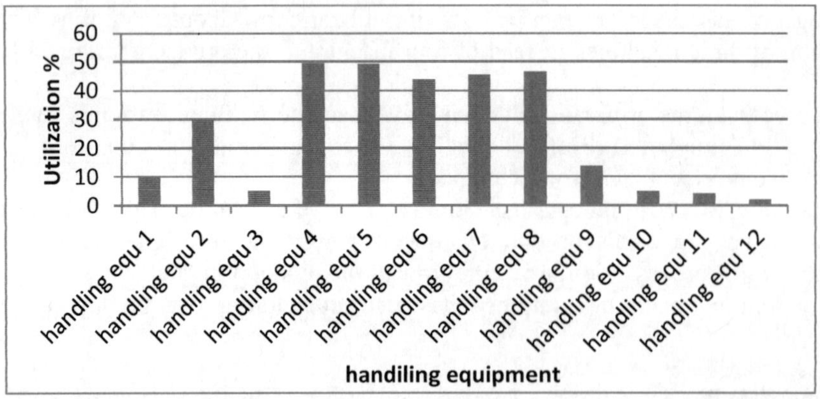

Figure 3. Analysis of the handling equipment utilization of original design

One of the most important factors to detect the bottlenecks is the utilization for each machine and handling equipment. The simulation was done for 2 days (8 hour/day) in order to determine the accurate number of products daily after subtracting the products produced in the first day: The has been done to the include the work in progress (WIP) parts that is in reality available in the production system. However, the results show there are 99 products. In the first day, while the second day the average products was 262 products. The second day shows the accurate number of product that should be produced if the system runs smoothly without any interruption or any unexpected breakdown.

3.1 Result on the proposed design

3.1.1 Scenario 1

This scenario was focused on maintain the production rate, on the other hand to cut the cost by reducing the capital investment and the labour cost. The utilization for the different machine was concluded in Figure 4 and 5.

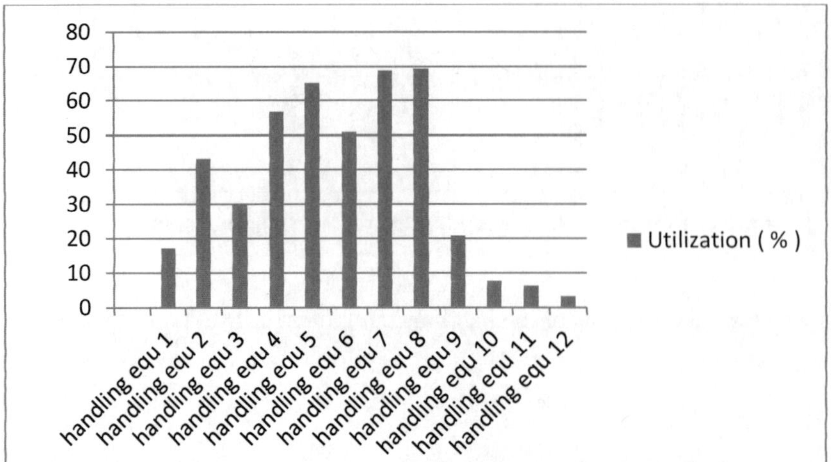

Figure 4: Utilization for the different handling equipment for scenario 1.

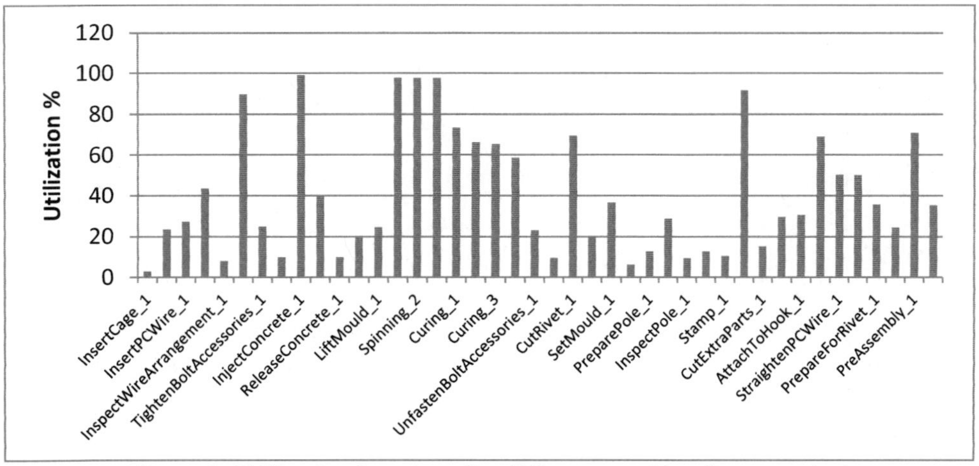

Figure 5: Utilization bar chart for different machine for scenario 1.

A comparison between the original and proposed design based on utilization percentage for selected machine is concluded in figure 6.

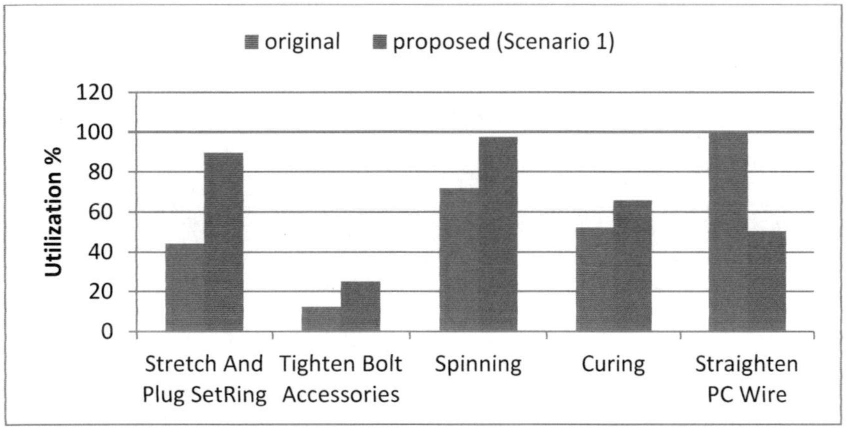

Figure 6. The machine utilization for original and propose design

Based on the simulation results, it can be concluded that the production rate will be the same of 261 parts per day. On the other hand, the advantages of this design are by saving in capital investments as well as in the workers that should be assigned to each machine. With simple calculations the company can save four works with maintaining the same production rate which means that the productivity is increased based on the basic productivity equation: Total output/total input = productivity.

3.1.2 Scenario 2

The second scenario is focuses on improving the production rate per day. Therefore, below is the result of the simulation after the processing time of the curing room has been minimized.

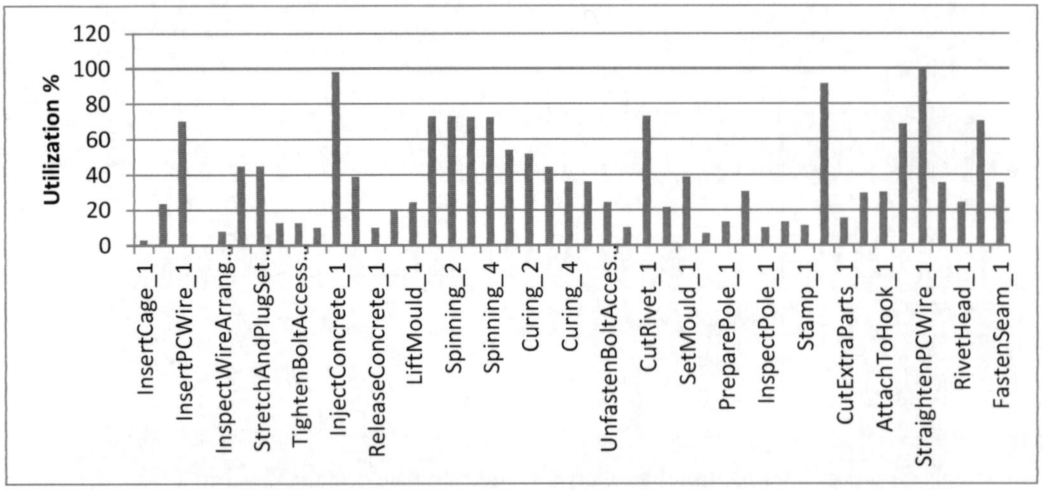

Figure 7. Utilization bar chart production line 2

The results show that the production rate was increased by 20 parts per day. However, there is no much difference in the utilization of the machines compared to the original design as shown in figure 8.

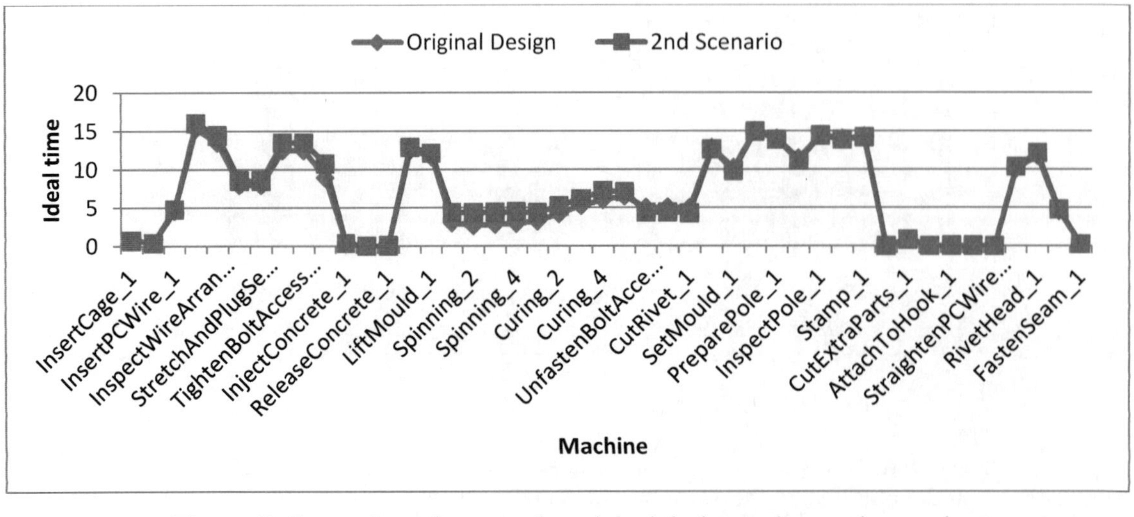

Figure 8. Comparisons between the original design and second scenario

As a conclusion, the second scenario shows an improvement in the production rate with need to invest to upgrade the curing system in order to reduce the processing time base on improving the heating system or the capacity of accommodate more than the 40 piece in each curing room. Finally, the researcher analysed the effect of the three designs on the handling system in term of utilization percentage and the idle time without processing. Figure 9 shows a comparison between the handling system for the original and the second scenario in terms of utilization percentage. The figure shows an increasing of the utilization in parts of the handling system. This is due to the increasing of the production rate.

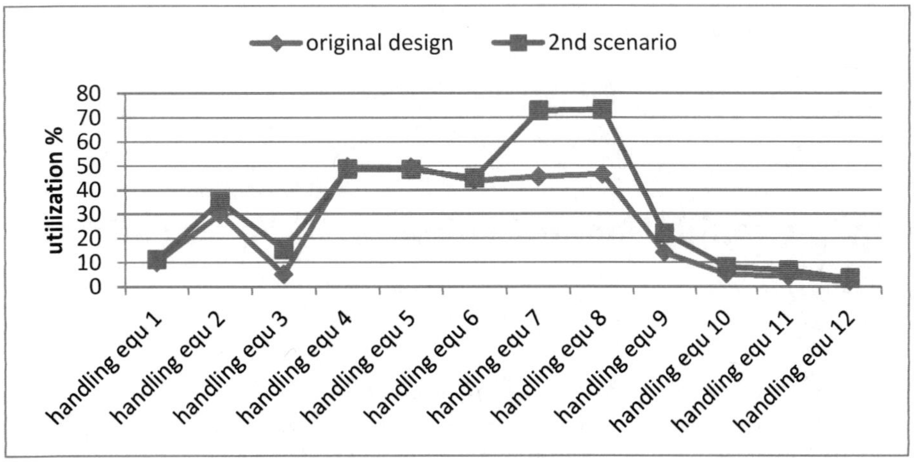

Figure 9. Comparisons between handling system in the original design and second scenario

From the results above, it can be concluded that the company have two alternatives based on their strategy and objectives. The first alternative can improve the productivity by selecting scenario 1. The productivity will improve by reducing the number of machines and workers with maintain the production rate per day on the level of 261 products. On the other hand, the company can select the second alternative by focusing on the production rate per day. However, this alternative needs an additional investment and technology experts to upgrade and improve the curing system in the company

References

[1] Huang S H, Dismukes J P, Shi J, Su Q I., Razzak M A, Bodhale R and Robinson D E 2003 *International Journal of Production Research* **41** 513-527
[2] Spedding T A and Sun G Q 1999 *International journal of production economics* **58** 289-301
[3] Detty R B and Yingling J C 2000 *International Journal of Production Research* **38** 429-445
[4] Windisch J. Väätäinen K, Anttila P, Nivala M, Laitila J, Asikainen A and Sikanen L 2015 *Applied energy* **149** 315-325.
[5] Ab Rashid M F F, Mohamed N N, Rose A M and Kor K Y 2015 *Journal of Mechanical Engineering and Sciences* **8** 1283-1292.
[6] Samanta S and Philip D 2015 In *Industrial Engineering and Operations Management (IEOM), 2015 International Conference on* (pp. 1-8) IEEE
[7] Mandalaki G and Manesis S 2013 In *Computer Modelling and Simulation (UKSim) 2013 UKSim 15th International Conference on* (pp. 554-558) IEEE
[8] Deb S, Chowdhury N F A and Claudio D In *IIE Annual Conference. Proceedings* (p. 1817) Institute of Industrial and Systems Engineers (IISE)
[9] Babulak E Wang M 2008 *International Journal of Online Engineering* **4**

IOP Conf. Series: Materials Science and Engineering **290** (2017) 012026 doi:10.1088/1757-899X/290/1/012026

Modelling and validation of Proton exchange membrane fuel cell (PEMFC)

A K M Mohiuddin[1], N Basran[1], A A Khan[2]
[1]Department of Mechanical Engineering, International Islamic University Malaysia (IIUM), Jalan Gombak, 53100 Kuala Lumpur, Malaysia
[2]Department of Manufacturing and Materials Engineering, International Islamic University Malaysia (IIUM), Jalan Gombak, 53100 Kuala Lumpur, Malaysia

Email: mohiuddin@iium.edu.my

Abstract. This paper is the outcome of a small scale fuel cell project. Fuel cell is an electrochemical device that converts energy from chemical reaction to electrical work. Proton Exchange Membrane Fuel Cell (PEMFC) is one of the different types of fuel cell, which is more efficient, having low operational temperature and fast start up capability results in high energy density. In this study, a mathematical model of 1.2 W PEMFC is developed and simulated using MATLAB software. This model describes the PEMFC behaviour under steady-state condition. This mathematical modeling of PEMFC determines the polarization curve, power generated, and the efficiency of the fuel cell. Simulation results were validated by comparing with experimental results obtained from the test of a single PEMFC with a 3 V motor. The performance of experimental PEMFC is little lower compared to simulated PEMFC, however both results were found in good agreement. Experiments on hydrogen flow rate also been conducted to obtain the amount of hydrogen consumed to produce electrical work on PEMFC.

1. Introduction

Fuel cell is an electrochemical device, which consists of negative charge electrode (anode), a positive charged electrode (cathode), and an electrolyte membrane. A fuel cell converts the chemical energy into electrical energy generally using oxygen and compressed hydrogen gas and produces only heat and water as byproduct [1]. Therefore, the fuel cell vehicles (FCV) are considered as zero-emission vehicle. Hydrogen fuel cell is identified as the most promising technologies as a power source, especially for transportation and stationary power application [2].

In a fuel cell, there are two electrodes, anode and cathode, and in the middle of it is an electrolyte membrane. A single fuel cell will generate about 0.7 volts [3]. In order to achieve higher amount of volts, multiple cells are stacked in series and it is called fuel cell stack system.

Hydrogen gas entering the fuel cell from the anode side and will split into proton and electron. Protons flow through the electrolyte to the cathode side forcing electrons to flow through external circuit producing electric current (figure 1). Oxygen gas fed into the cathode where oxygen, protons and electrons combine to from water which later given off from the exhaust as steam or vapors [4].

IOP Conf. Series: Materials Science and Engineering **290** (2017) 012026 doi:10.1088/1757-899X/290/1/012026

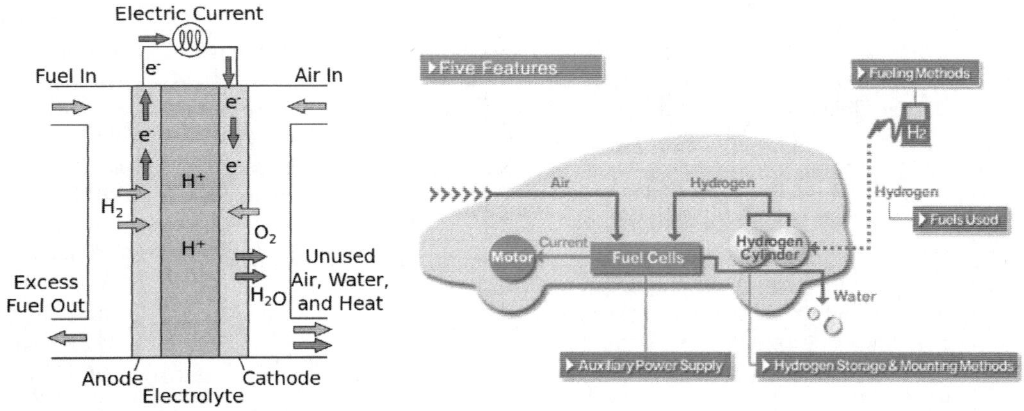

Figure 1. Schematic representation of electrochemical processes in PEM fuel cells and schematic representation of fuel cell vehicles (FCV)

2. Fuel cell

Presently, there are various types of fuel cells that are under research and development. Each type has its particular properties, advantages, and disadvantage which makes them different in possible application. Table 1 below shows the characteristics of six main types of fuel cells that are already being developed and commercially used [5] [6].

Table 1. Types of fuel cell

Fuel Cell Type	Polymer Electrolyte Membrane (PEMFC)	Direct Methanol (DMFC)	Alkaline (AFC)	Phosphoric Acid (PAFC)	Solid Oxide (SOFC)
Fuel	H_2	$CH_3OH + H_2O$	H_2	H_2/CO/reformat	H_2
Oxidizer	O_2, air	O_2, air	O_2, air	O_2, air	O_2, air
Electrolyte	Solid organic polymer poly-perfluoro-sulfonic acid	Solid organic polymer poly-perfluoro-sulfonic acid	Aqueous solution of potassium hydroxide soaked in a matrix	Liquid phosphoric acid soaked in a matrix	Solid zirconium oxide to which a small the amount of yttria is added
Operating Temperature	40-100 °C	60-130 °C	65-220 °C	205 °C	600-1000 °C
Efficiency	50-60%	40%	50-70%	40 %	50-60 %
Applications	-Electric equipment -Portable devices -Transportation -lower power CHP system -Specialty vehicles	-Electric equipment -Portable devices -Specialty vehicles	-Space -Military	-Stationary power generator -Large vehicles	-All sizes of stationary power generation. -Electric utility -Transportation
Advantages	-Solid	-Inexpensive	-Cathode	-Low cost	- Non-precious

	electrolyte reduces corrosion and management problem -Very strong bond, therefore, can be made into very thin film -Low operation temperature -Quick start-up -High power density -Compactness and lightness -Can absorb a lot of water	-High power density -Easily transport and stored -Solid electrolyte advantages (see PEMFC)	reaction faster in alkaline -High performance -Low cost material	electrolyte -Tolerant to impure H_2 -Long time performance -commercial	metals can be used as catalyst -High efficiency - Solid electrolyte advantages (see PEMFC) -Tolerance to CO impurities -Able to use fossil fuel gas -Do not require external reformer
Disadvantages	-Expensive catalyst -Sensitive to CO impurities -Crucial water management -All parts must be sufficiently hydrated	-Low efficiency -High crossover of methanol through electrolyte -Produce intermediate hydrocarbon which poisons the electrode	-Intolerant CO_2 in impure H_2 and air -Corrosion -Expensive catalyst -Water management	-Require more expensive catalyst -Corrosive and sulfur poisoning of electrolyte -Large and heavy	-Slow start-up -Requires significant thermal shielding to retain heat -Corrosive electrolyte causes decreasing in cell life.

2.1. PEM Fuel cell for automotive application

PEM fuel cell is the most used technology in fuel cell vehicles. Many large automotive companies are focusing more on PEM fuel cell development. The main reason is the high power density of PEM fuel cell, which meets the requirement of space constraint in vehicles.

This is also because of PEM fuel cell's lower operating temperature which allows a rapid start-up. The efficiency varies usually between 40-60% and output power can be changed to meet quickly demanded load [5]. Another characteristic of PEM fuel cell is the compactness and lightness. Thus, it is the best candidate for mobile application [7]. However, PEM fuel cell is sensitive to fuel CO impurities and requires precious metal as a catalyst which is expensive. Other than that, the electrolyte must be sufficiently hydrated. Thus, water management on stack becomes crucial [8]. In transportation, PEM fuel cell is used in hybrid configuration with electricity storage devices, such as batteries or super capacitor.

3. Modelling of PEMFC system

3.1. Nernst equation

In PEM fuel cell, the cell potential that obtained in open circuit thermodynamic balance is the reversible voltage of PEM fuel cell (E_Nernst). It can be defined mathematically by the following equation:

$$E_{Nernst} = E_r + \left(\frac{RT}{2F}\right) \ln\left(\frac{P_{H_2O}\,P_{O_2}^{0.5}}{P_{H_2O}}\right) \quad (1)$$

E_{Nernst} can be calculated by modified version of the equation by taking the standard temperature as 25 °C [9]. Using the standard pressure and temperature values, the Nernst equation can be rewritten as:

$$E_{Nernst} = 1.229 + 0.85 \times 10^{-3}(T - 298.15) + 4.3085 \times 10^{-5}\left[ln\left(\frac{P_{H_2}P_{O_2}^{0.5}}{P_{H_2O}}\right)\right] \quad (2)$$

When the byproduct is steam instead of water, 1.229 V should be replaced by 1.18 V. T is the cell operating temperature in Kelvin, P_{H_2} and P_{O_2} are hydrogen and oxygen partial pressure in atm.

3.2. Voltage losses

When the electrical energy is drawn from the fuel cell, the actual voltage decreases due to the several irreversible loss mechanisms. It can be defined as:

$$V(i) = V_{rev} - V_{irrev} \quad (3)$$

Three major classification of losses that cause the voltage drop from the circuit are as follows:

3.2.1. Activation polarization loss. Activation polarization loss is the voltage over potential required to overcome the activation energy of the electrochemical reaction on the catalytic surface. This type of losses dominates at low current density. Activation polarization loss can be used to measure the catalyst effectiveness at a given temperature [10]. The activation losses can be expressed simply by the Tafel equation:

$$v_{act} = -\frac{RT}{\alpha nF} ln\frac{i}{i_o} \quad (4)$$

Where i is the current density, and i_o is the reaction exchange current density. The reaction rate represents by the exchange current density. F is Farady's constant, n is the number exchange protons per mole of reactant, and α is the charge transfer coefficient used to describe the amount of electrical energy applied to change the rate of the electrochemical reaction.

3.2.2. Ohmic polarization loss. Ohmic polarization loss is due to natural resistance upon charge flow in conductors which occurs because of electrical resistance in the cell components [11].

$$v_{ohmic} = iR_{ohmic} = i(R_{ionic} + R_{elec}) \quad (5)$$

Where the current of the fuel cell is i (A), while R_{elec} is the total electric resistance of all other conductive components (Ω), and R_{ionic} is the ionic resistance of the electrolyte. Since the ohmic overpotential for the fuel cell is mainly due to ionic resistance in the electrolyte, this can be expressed as:

$$v_{ohmic} = iR_{ohmic} = jA_{cell}\left(\frac{\delta_{thick}}{\sigma A_{cell}}\right) = \frac{j\delta_{thick}}{\sigma} \quad (6)$$

3.2.3. Concentration polarization loss. The fuel cell concentration losses (or mass transport losses) can be express by the following equation:

$$v_{conc} = cln\left(\frac{i_L}{i_L - i}\right) \quad (7)$$

Where the limiting current density is i_L, while i is the current density of the fuel cell, and c is a constant. The value of c is approximately:

$$c = \frac{RT}{2F} = \frac{(8.314)(373.5)}{2(96485)} = 0.016 \quad (8)$$

The absolute temperature is assumed at 373.5 K (100 °C) which is the boiling point of water [9]. This is due to voltage drop occurs at the high-temperature condition as well as high current density. Therefore, the voltage loss by concentration equation can be expressed as:

$$v_{conc} = 0.016 \ln\left(\frac{i_L}{i_L - i}\right) \quad (9)$$

3.3.4. Reactant gasses diffusion in the electrode. To find Nernst voltage and voltage losses, the partial pressure of water must be calculated first and after that the partial pressure of hydrogen and oxygen. To calculate the partial pressure, the assumptions are as follows [5]:

i. Ideal and uniform gas distribution
ii. The fuel cell gas flow channel at constant pressure
iii. One-dimensional treatment
iv. H_2 is humidified and oxidant is humidified by air. The effective anode water vapor is at 50% of the saturated vapor pressure while the effective cathode water pressure is 100%.

Thus, the saturation pressure of water can be express as:

$$log_{10}(P_{H_2O}) = -2.1794 + (0.02953 * T_c) - (9.1837 \times 10^{-5} * T_c^2) + (1.4454 \times 10^{-7} * T_c^3) \quad (10)$$

Where the operating temperature at degree Celcius is T_c, the partial pressure of hydrogen and oxygen are:

$$P_{H_2} = 0.5 * \left[\frac{P_a}{\exp\left(\frac{1.653*i}{T_K^{1.334}}\right)}\right] - P_{H_2O} \quad (11)$$

$$P_{O_2} = \left[\frac{P_c}{\exp\left(\frac{4.192*i}{T_K^{1.334}}\right)}\right] - P_{H_2O} \quad (12)$$

Where the saturated vapor pressure in atm is P_{H_2O}, while T is the absolute temperature (K), P_c is the partial pressure at the cathode, and P_a is the partial pressure at anode.

4. Results and discussions

4.1 Experiment on performance of PEM fuel cell

An experiment has been conducted using Horizon FCSU-12 PEM Mini fuel cell to determine its performance. To express the performance of a fuel cell, current voltage dependency is the most important property due to kinetics of electrochemical reaction of oxygen and hydrogen [12]. Figure 3 represents polarization curve of fuel cell used. The polarization curve shows the electrochemical efficiency of the fuel cell at operating current. At first, the output voltage was recorded directly from the fuel cell without connecting it to the motor which will determine its open circuit voltage. As can be seen on figure 4, both experimental and simulation data are in good agreement. The open circuit (no load) voltage of the experimental PEM fuel cell is almost similar with the simulation data. But after that, the activation losses that occurred on the experimental PEM fuel cell are greater compared to simulation. This shows that the efficiency of the experimental fuel cell seems to be much lesser compared with the simulation. This may be due to the circuit of the motor that contain more resistance since there are two LED light attached to the motor circuit. The voltage reading only can be take maximum at 0.5 A due to the motor specification. Larger than 0.5 A of load will damage the motor.

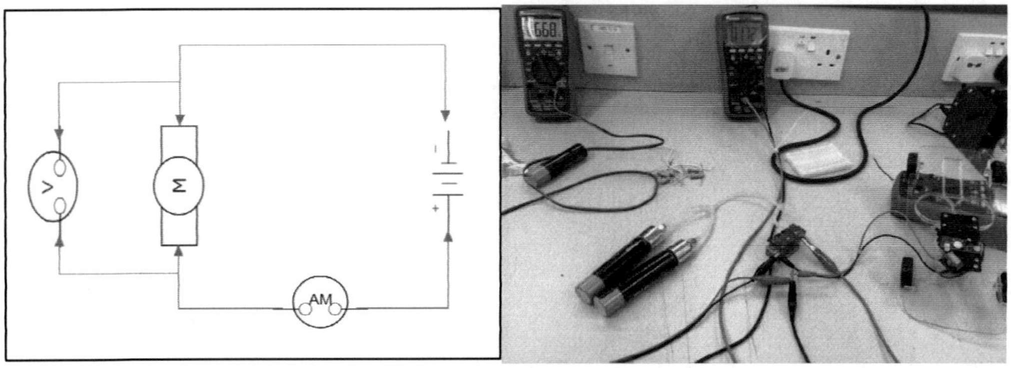

Figure 2. Experiment setup

An experiment has been done using Horizon FCSU-12 PEM Mini fuel cell to determine its performance. To express the performance of a fuel cell, current voltage dependency is the most important property due to kinetics of electrochemical reaction of oxygen and hydrogen [12]. Figure 3 represents polarization curve of fuel cell used. The polarization curve shows the electrochemical efficiency of the fuel cell at operating current. At first, the output voltage was recorded on directly from the fuel cell without connecting it to the motor which will determine its open circuit voltage. As can be seen on figure 4, both experimental and simulation data are in good agreement. The open circuit (no load) voltage of the experimental PEM fuel cell is almost similar with the simulation data. But after that, the activation losses that occurred on the experimental PEM fuel cell are greater compared to simulation. This shows that the efficiency of the experimental fuel cell seems to be much lesser compared with the simulation. This may be due to the circuit of the motor that contains more resistance since there are two LED light attached to the motor circuit. The voltage reading only can be taken at maximum 0.5 A due to the motor specification. Larger than 0.5 A of load will damage the motor.

For further validation, the experimental and simulation data are compared with the results from other studies as can be seen on figure 5. All results are in good agreement. The result from S. Haji [10] is conducted at 25°C which is similar to the condition of this experiment, while T. Selyari et al. [13] fuel cell is conducted at 50°C. Both are using a fuel cell with an active area of 25 cm^2. However, T. Selyari et al. fuel cell performance are much higher due to greater fuel cell operating temperature. This can validate that higher temperature does improve the performance of the fuel cell. Meanwhile, L. Patularu et al. [14] conducted the study at 23°C with 92 cm^2 active area. Although, the operating temperature is much lower, the greater active area improves the performance of the cell. More hydrogen and oxygen are utilized to produce higher electrical work.

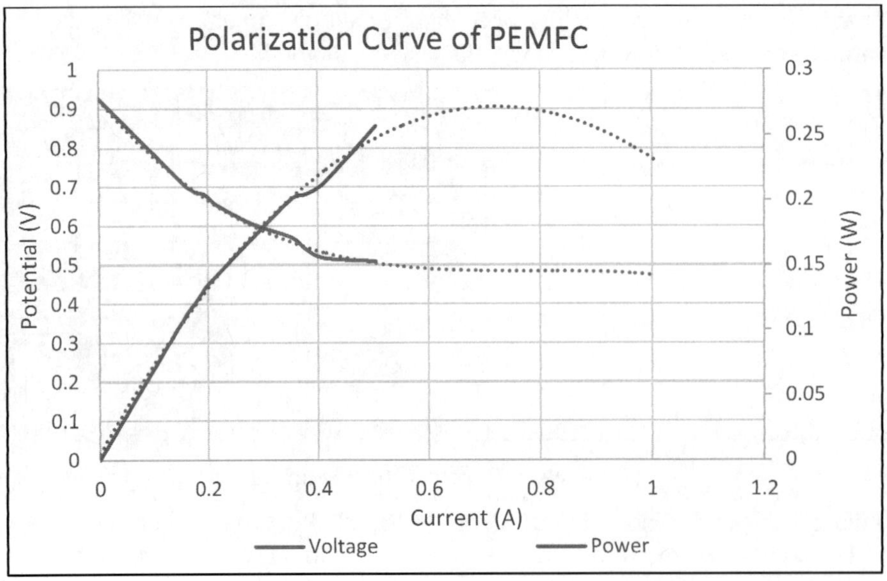

Figure 3. Polarization curve of experimental PEM fuel cell

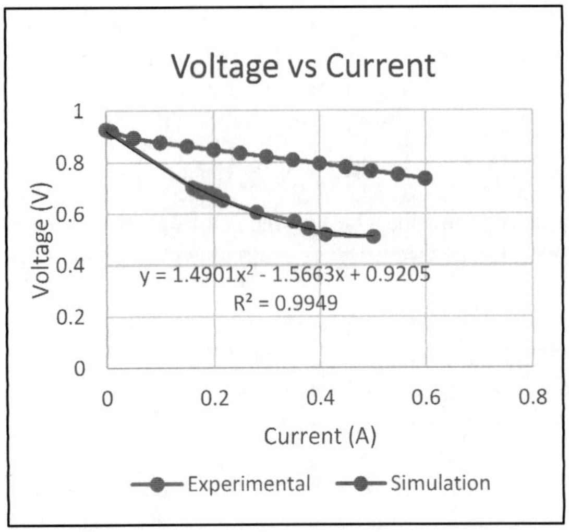

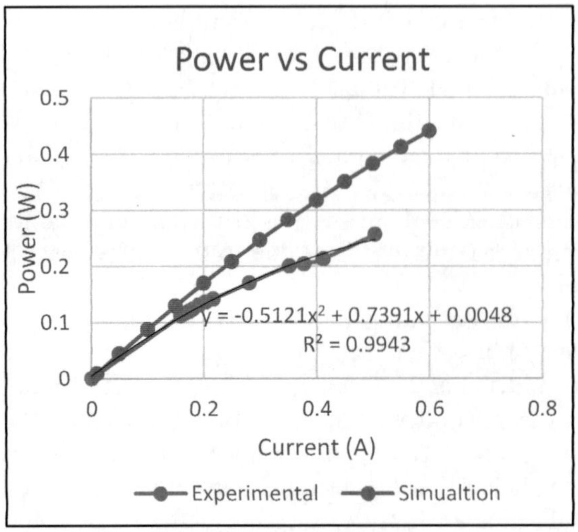

Figure 4. Comparisons between experimental and simulation value of PEM fuel cell polarization curve

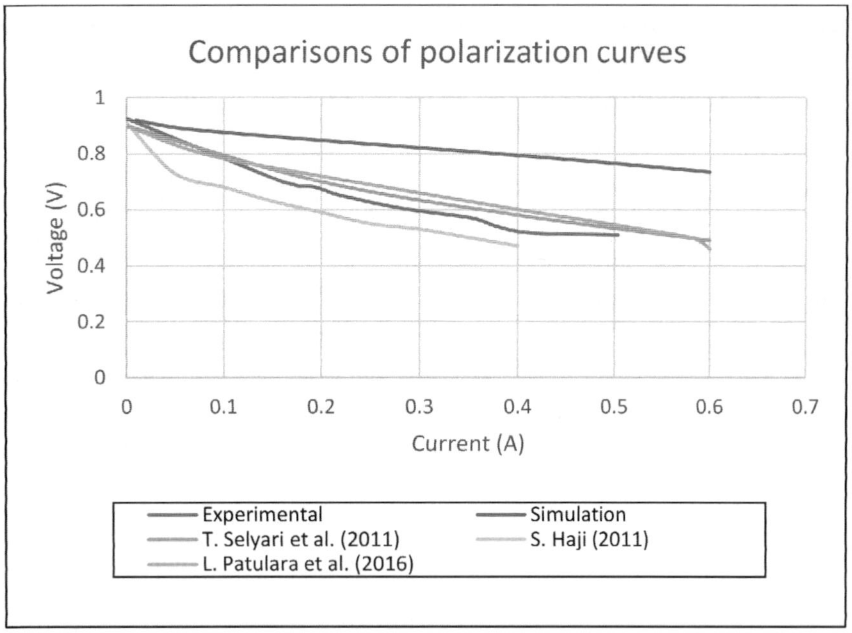

Figure 5. Comparisons of polarization curves with other studies

4.2 Experiment on hydrogen flow rate

Figure 6 shows the configuration of the hydrogen fuel cell flow rate measurement system. The input of flowmeter is connected to the hydrogen supply system while the output flowmeter is connected to the fuel cell. The reading on the flowmeter is taken with one minute interval.

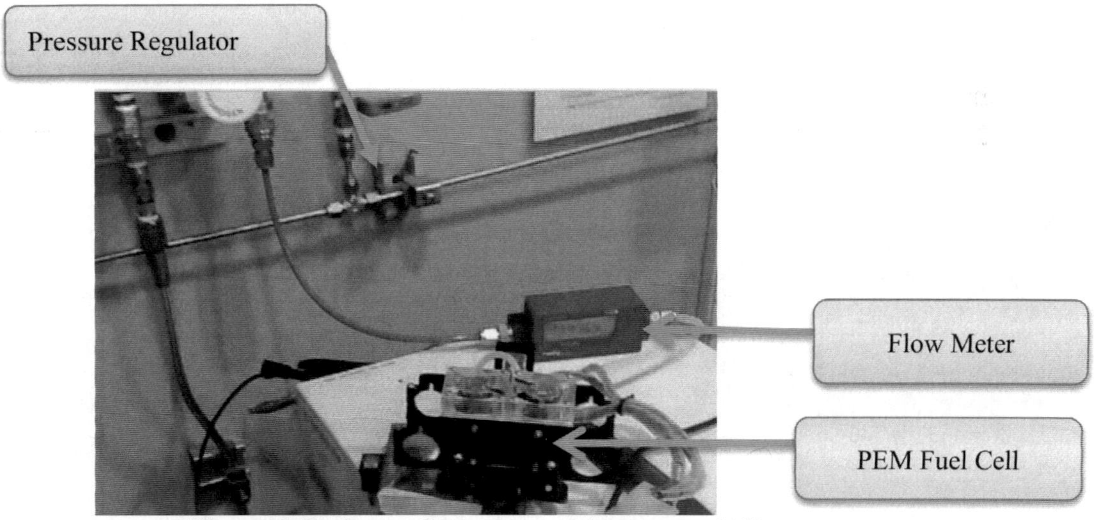

Figure 6. Hydrogen flow rate experiment setup [15]

From both graph on figure 7 shows the variation of hydrogen flow rate based on the change of voltage and time. Both graphs show that there is no rapid change on the amount of hydrogen flow rate. From figure 7, the graph shows a little increasing trendline which proves that higher hydrogen flow rate produce higher voltage. This result also achieved by Khazaee et al. [16] where the experiment was done on different hydrogen and oxygen flow rate to determine the performance of the cell. It is observed that the increasing of oxygen flow rate from 0.5 L/min to 0.9 L/min and hydrogen flow rate from 0.3 L/min to 0.7 L/min enhances the performance

From the equation of the trendline on the graph in figure 7 which is,

$$y = 0.0297x + 0.1519 \tag{13}$$

It can be assumed that the slope of the trendline is m = 0.0297. This value also can represent that 0.0297 l/min of hydrogen is needed to produce 1 Volt.

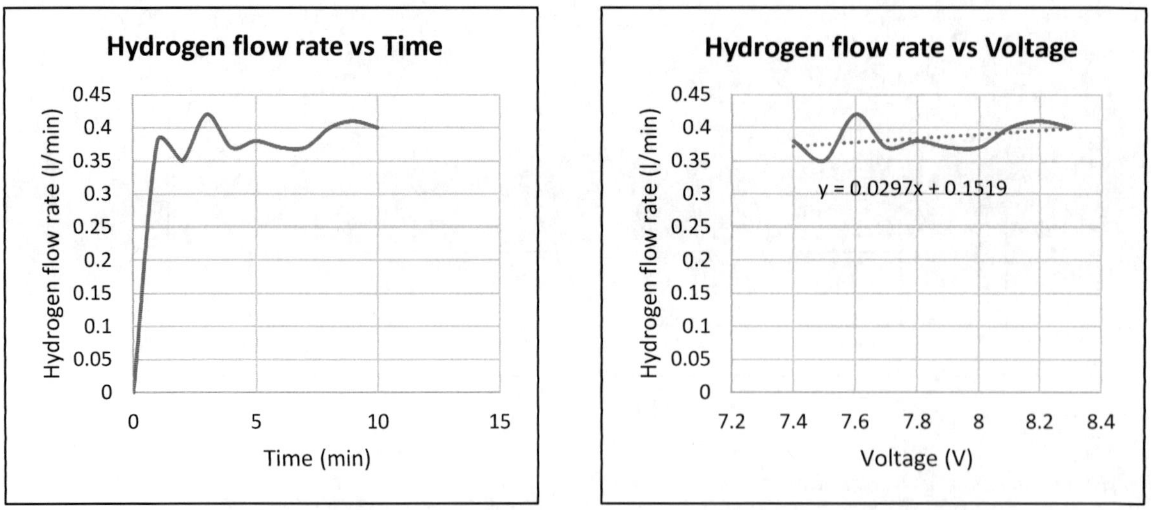

Figure 7. Hydrogen flow rate versus time and voltage

4.3 Simulation of PEM fuel cell using MATLAB Simulink

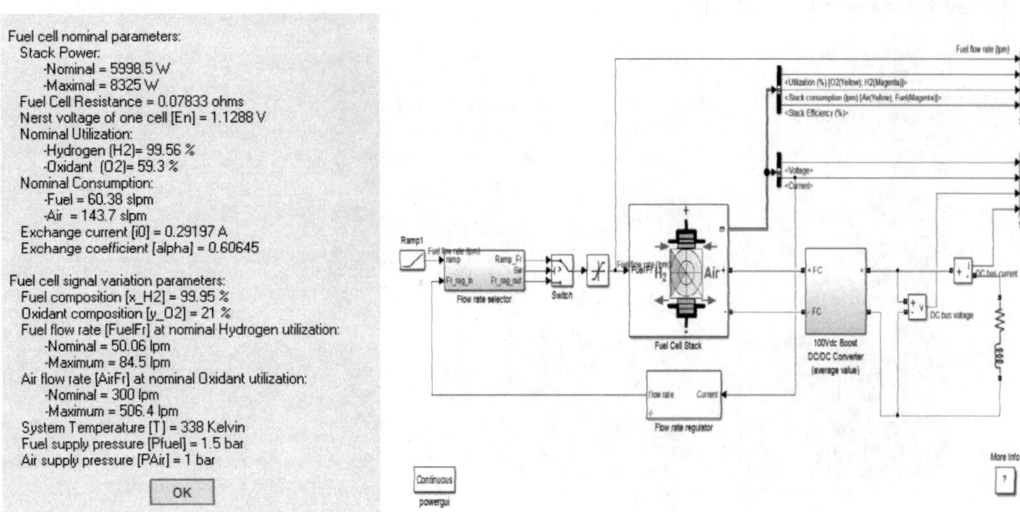

Figure 8. PEM fuel cell stack simulation parameters and setup [17]

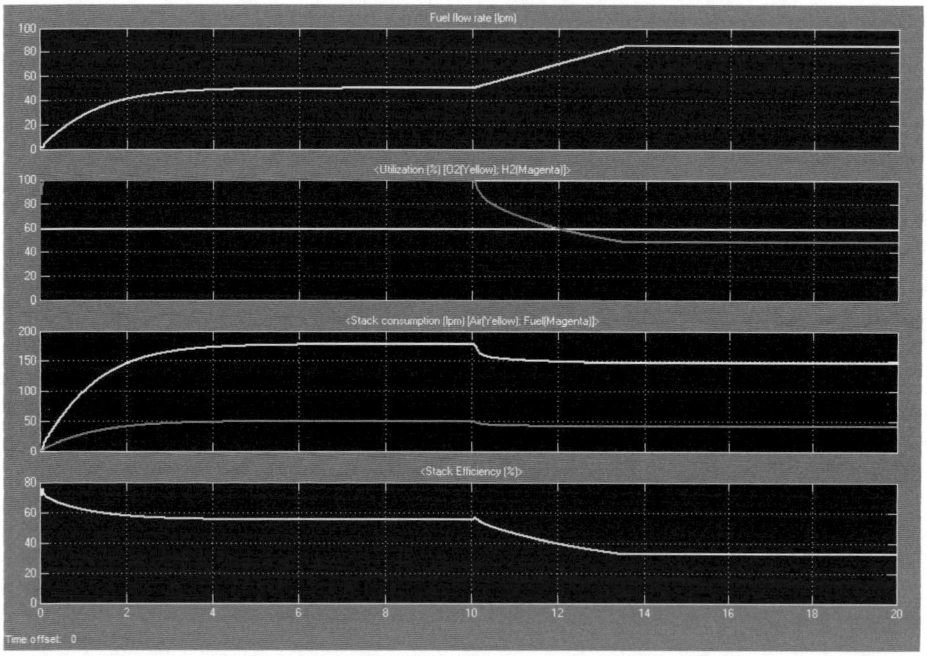

Figure 9. PEM fuel cell fuel flow rate utilization, stack consumption and stack efficiency

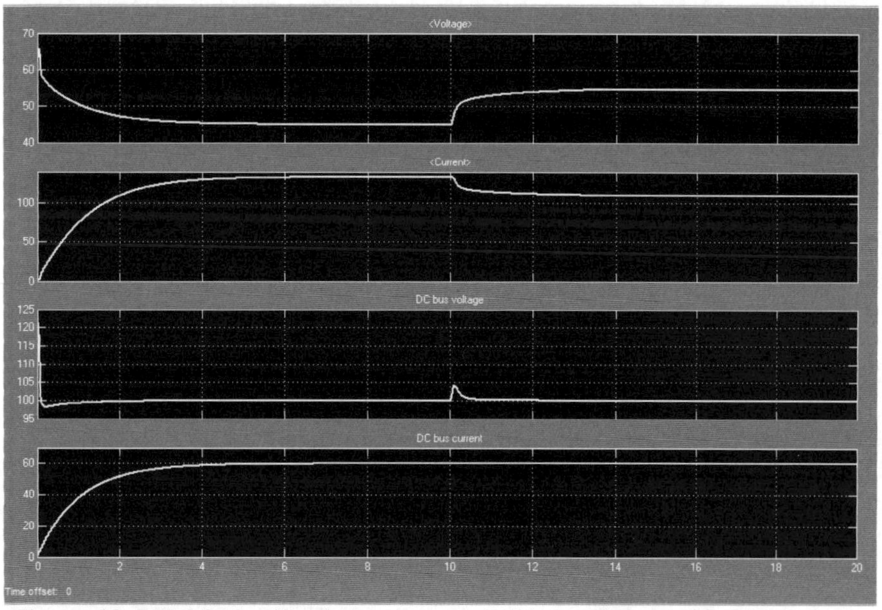

Figure 10. PEM fuel cell voltage, current, DC bus voltage and DC bus current

In this simulation, we want to observe how the fuel flow rate may affect the fuel cell stack voltage which will affect the stack efficiency, fuel consumption and air consumption. For the first 10 secs, the utilization of hydrogen is set to constant (uf_H2 = 99.56%) by via a fuel flow rate regulator. After 10 secs, the flow rate regulator is bypassed and the fuel flow rate is increased from 50 lpm to maximum 85 lpm. The increase of fuel flow rate has resulted in the increase in stack voltage and decrease in stack current as shown in figure 10. From figure 9, stack efficiency is at its highest at 0.7 sec which nearly 76%, and after that it starts to decrease as the load increases. The stack efficiency once again starts to drop after 10 secs from 55% to 33% at 13.5 second. This drop of stack efficiency after 10 second seems to be affected by the decreasing amount of hydrogen utilization, stack consumption and current.

5. Conclusion

Fuel cell are a new promising technology for generation of electrical energy, not just for automotive application but for other applications as well. This technology involves in using hydrogen and oxygen to produce electrical work through electrochemical process within the fuel cell. The experiment and simulation of fuel cell have been conducted using mini PEM fuel cell to measure the output voltage on certain amount of load which can prove the efficiency of the fuel cell. An experiment and simulation on hydrogen flow rate on PEM fuel cell also have been conducted to determine the effect of hydrogen flow rate on the fuel cell voltage and to measure the amount of hydrogen needed for the fuel cell to produce electricity. From the study it can be concluded that:

i. Both experimental and simulation results are in good agreement although the voltage losses on experimental result is little greater.
ii. It is observed that the increasing fuel flow rate will increase the voltage produced by the fuel cell.
iii. It is also observed that higher operating temperature and pressure of the cell improve the performance of the fuel cell.

Acknowledgements
The authors would like to acknowledge the RIGS of International Islamic University Malaysia for the support to carry out this work.

References
[1] Barbir F 2005 *PEM Fuel Cells: Theory and Practice* (Amsterdam: Elsevier)
[2] Briguglio N, Ferraro M and Antonucci V 2011 *Fuel cell hybrid electrical vehicles* (Italy: InTech.)
[3] Larminie J and Dicks A 2003 *Fuel Cell System Explained* (West Sussex: John Wiley & Sons Ltd.)
[4] Proton Exchange Membrane Fuel Cell. Retrieved May 2016, http://en.wikipedia.org/wiki/Proton_exchange_membrane_fuel_cell
[5] Spiegel C 2007 *Designing and Building Fuel Cells* (New York: McGraw Hill)
[6] Barbir F and Yazici S 2008 *Int. J. Energy Re.* **32** 369-378.
[7] Thomas C E 2009 *Hydrogen Energy* **34** 6005-6020
[8] Martin J S, Zamora I, Martin J S, Aperribay V and Eguia P 2010 *Int. Conf. on Renewable Energies and Power* vol 1 p. 735-740.
[9] Spiegel C 2008 *PEM Fuel Cell Modelling and Simulation using MATLAB* (Burlington: Acedemic Press)
[10] Haji S 2011 *Renewable Energy* **36** 451-458
[11] Peighambardoust S J, Rowshanzamir S and Amjadi M 2010 *Int. J. Hydrogen Energy* **35** 9349-9384.
[12] Kabza A 2016 Just another Fuel Cell Formulary. Retrieved February 2017 http://www.pemfc.de/FCF_Smart.pdf
[13] Selyari T, Ghoreyshi A, Shakeri M, Najafpour G and Jafary T 2011 *Chem. Ind. Chem. Eng. Quarterly* **17** 207-214
[14] Laurentiu P, Stanica E and Aurelian C 2016 *U.P.B. Sci. Bull. C* **78** 205-218
[15] Mohiuddin A K M, Chemani M F and Zakaria M B 2014 *Int. J. Vehicle Systems Modelling and Testing* **9** 321-333.

[16] Khazaee I and Ghazikhani M 2012 *Int. J. Energy Environ.* **3** 461-470
[17] Mathworks documentation, MathWorks. Retrieved February 2017, https://www.mathworks.com/help/physmod/sps/powersys/ref/fuelcellstack.html;jsessionid=6cc09e2cc664e4ede0f0cd4eafd8

IOP Conf. Series: Materials Science and Engineering **290** (2017) 012027 doi:10.1088/1757-899X/290/1/012027

Implementing lean in Malaysian universities: Lean awareness level in an engineering faculty of a local university

M. Azim Khairi[1], Mohamed Abd Rahman[1]

[1]Department of Manufacturing and Materials Engineering, International Islamic University Malaysia (IIUM), Jalan Gombak, 53100 Kuala Lumpur, Malaysia

Email: mrahman@iium.edu.my

Abstract. Many academic articles were published in Malaysia promoting the goodness of lean in manufacturing and industrial sectors but less attention was apparently given to the possibility of obtaining the same universal benefits when applying lean in non-manufacturing sectors especially higher education. This study aims to determine the level of lean awareness among a local university's community taking its Faculty of Engineering (FoE) as the case study. It also seeks to identify typical FoE's staff perception on lean regarding its benefits and the obstacles in implementing it. A web-based survey using questionnaires was carried out for 215 respondents consisting of academic and administrative staff of the faculty. Statistical Package for the Social Science (SPSS) was used to analyze the survey data collected. A total of 13.95% of respondents returned the forms. Slightly more than half of those responded (56.7%) have encountered some of the lean terms with mean 1.43 and standard deviation 0.504. However, the large amount of standard deviation somewhat indicates that the real level of lean awareness of FoE as a group was low. In terms of lean benefits, reduction of waste was favored (93.3%) by the respondents with mean 0.93 and standard deviation 0.254. For obstacles in implementing lean, lack of knowledge was selected by most respondents (86.7%) to be the major factor with mean 0.87 and standard deviation 0.346. Through the analysis done, the study may conclude that level of lean awareness among the university's community was low thus may hinder implementation of lean concept.

1. Introduction

Manufacturers always strive to overcome challenges such as soaring customers' expectation for the speed of service, better product quality, and flexibility – all at a lower cost to cope with unstable demand and intense competition in markets. Non-manufacturing sector is similar with manufacturing sector where they still need to serve customers. This paper focuses on lean implementation in non-manufacturing sector namely, higher education (HE). Emiliani [1] revealed HEs are facing the obstacle to giving worthwhile educational service while maintaining everything at affordable cost. Lean concept was developed in Japan to surpass the limitation of mass production and to shorten customers' waiting time by means of waste elimination and application of continuous improvement [2]. Compared to mass production, lean consumes less in everything [3].

To the best of our knowledge, studies on lean implementation in Malaysian HE were relatively scarce. In order to evaluate the suitability of implementing lean in local HE, a survey was administered to an engineering faculty of a local university to gauge how well its community was aware of lean, and to assess the typical perception of the community on the benefits of lean and the challenges of implementing it.

2. Experimental procedures

In this study, the survey tool was adapted from Salem *et al.* [4] and research subject was Faculty of Engineering (FoE) at the chosen local university. Since the data were needed for the calculation of the awareness index, open-ended type of questions was avoided. Section A focused on respondents' demographic while Sections B, C, and D focused on lean awareness and understanding in the faculty. Some questions were of multiple-response types. Section E attempted to identify respondents' thought on lean implementation in HE with special reference to their faculty.

Sample size was calculated using a method developed by Krejcie and Morgan [5]. A total of 215 respondents were targeted consisting of academic and non-academic staff and questionnaires were distributed among them using online platform. No students were involved in this survey due to several shortcomings e.g. manpower shortage, time limitation, and budget constraint.

To ensure the reliability of the data and internal consistency of the test score, Cronbach's alpha test was run. The survey attempted to measure three aspects (i.e. lean awareness, opinion on lean implementation advantages and opinion on lean implementation obstacles). According to Table 1 the value of Cronbach's α is above 0.6 (i.e. 0.911, 0.617 and 0.6897). Pursuant to Nakamura *et al.*, having 0.6 of Cronbach's α value is rather satisfying but Nunnaly had a stance to say 0.7 is adequate for data reliability [8].

Table 1. Cronbach's Alpha.

Factors and indicators	Mean	Std. Dev.	Cronbach's α
Fundamental on lean	1.43	.504	.911
Opinion on lean implementation advantages	1.03	.183	.617
Opinion on lean implementation obstacles	.83	.379	.689

3. Results and Discussion

30 sets of the questionnaire were returned and found reasonable for data analysis representing a response rate of 13.95%. The response rate is low because only one technique of survey distribution i.e. web-based survey was used. However, a study by Nasuha Lee Abdullah *et al.* [6], using almost all possible methods i.e. hard copy mail, paid-service online survey and a lucky draw innitiative, turned out to have only 7% of response rate. Wong *et al.* [3] reported a response rate of 12.6% using postal survey. A similar rate of 11.5% by Ahmed and Hassan [7] was also reported. Hence, the low feedback rate in survey research in Malaysia was not unusual.

Section A: Demographic

Table 2 displays the background of the respondents. Learning respondents characteristic helped in deciding how close the sample in replicating the total population [9]. It can be seen that only 13.3% were from non-academic staff and the remainder was academic staff.

Manufacturing and Material Eng. Dept formed the highest respondent with 30%. This may be attributed to the lean topic which is originally from Manufacturing Engineering syllabus. The only department not participating in this survey was Professional Engineering & Continuing Education (PEaCE) and for 'other', the respondent claims him/herself from the administration. 50% of the respondents had more than 10 years of working experience in the FoE while the rest are as in the table.

This survey investigates the number of years the respondent have been involved in engineering to indicate their maturity in the field. As forecasted, 20 out of 30 respondents are having more than 10 years of experience. With this, this study can somehow differentiate between different sub-groups.

Table 2. Profile of the repondents.

a) Role in the FoE	No. of Respondent	Percent
Academic staff	26	86.7
Non-academic staff	4	13.3
b) Department in the FoE		
Biotechnology Engineering	5	16.7
Manufacturing & Material Eng.	9	30.0
Mechatronics Engineering	4	13.3
Civil Engineering	1	3.3
Mechanical Engineering	1	3.3
Department of Science in Engineering	3	10.0
Electrical & Computer Engineering	6	20.0
Other	1	3.3
c) Experience in the FoE		
0 - 4 years	13	43.3
5 – 10 years	2	6.7
more than 10 years	15	50.0
d) Experience with engineering		
0 - 4 years	6	20.0
5 – 10 years	4	13.3
more than 10 years	20	66.7

Section B: Overview of Lean

The first question in Section B was asking whether the respondents have ever heard about 'lean' or anything related to lean e.g. lean manufacturing or lean concept. Here, this study tries to adopt the principle of Fricke and Buehlmann [10] where respondents will be considered aware of lean if they happened to encounter any of lean vocabulary before. The result shows that 56.67% of the respondents claimed that they have heard about the term lean. Hence, according to [10], the community of the FoE should be aware of lean. However going through the analysis, the large standard deviation, 1.43 (Table 1) indicates that the real level of lean awareness in this study may still be low.

For those who answered the first question negatively were later asked in the next question whether they were interested to know more about lean and from the feedback, it seems that 11 (84.6%) out of 13 are interested to discover the knowledge of lean. This is a good sign since lean implementation in any organization normally begin with lean training or education first. Undergoing this 'orientation phase', the partakers get familiarization with lean thinking, equipped with lean tools, and most importantly understand the benefits of practicing lean [1].

Meanwhile, for those who have heard about lean were asked about their general understanding of the lean concept. Figure 1 shows the outcome of this inquiry.

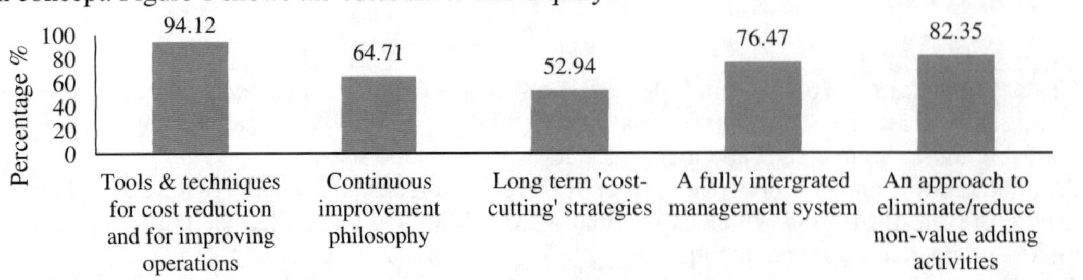

Figure 1. Lean Concept.

For this question, the respondents were allowed to choose more than one answer. The result obtained shows the same trend with [4], where the highest percentage is "tools and techniques for cost reduction and for improving operations". While the least chosen answer is "long term 'cost cutting' strategies". It is likely that "long-term 'cost cutting' strategies" are not a major attraction among the

FoE's staffs although it is a significant feature of a successful lean implementation. As suggested in [4], training is crucial on stressing the importance of this 'element' for preparing an institution for any lean transformations projects. Different finding was shown in [3] where 'waste reduction' was chosen the closest to lean concept. Still, both elements score the highest in this paper which leads to the deduction that this paper's result may be practical.

Next, these respondents were asked about the type of lean principle they are familiar with (the answer could be more than one). Figure 2 shows the result. From the observation, the principle that scores the highest is "continuous improvement" followed by "elimination of waste" and "JIT". The least is "zero defect". From the result, it is important to note that, among the two principles with the highest scores, are the principles with naturally understood. People tend to choose words that are far more widely known than the one which are more 'lean-specific' [10].

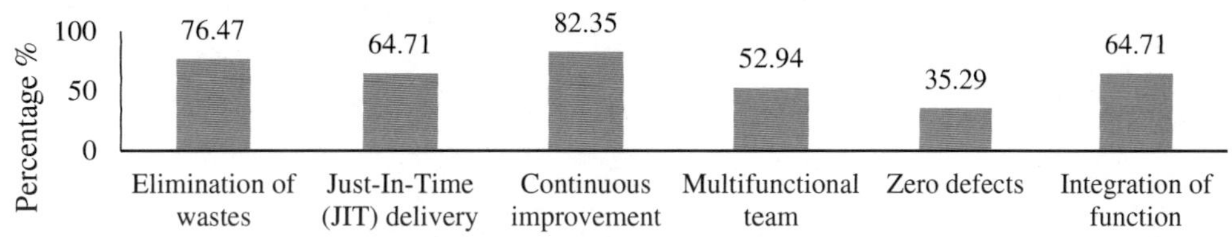

Figure 2. Lean Principle.

Section C: Lean Waste

This section concentrates on respondents' reaction toward lean waste. It is the screening type question where the first question was aimed to divide the respondents into categories, which is the one that aware about the term "lean deadly waste" and another one is the group of people who do not. To summarize, out of 30 responses, 24 (80 %) of them were never heard about lean waste. This ratio indicates the shortage in the aspect of lean thinking, concept, and culture in majority of the respondents in the FoE. This result is rather normal if we compare with [4] as they also have 75% of the respondents which are an industrial worker in Qatar who answer 'no' to the same type of question.

The 2nd and the 3rd question in this section both asked the respondents about their own perspective on what are those that contribute to waste in term of 'lean'. Figure 3 concluded the response by those who claimed they never heard about "lean deadly waste". This question can be answered by more than one answer.

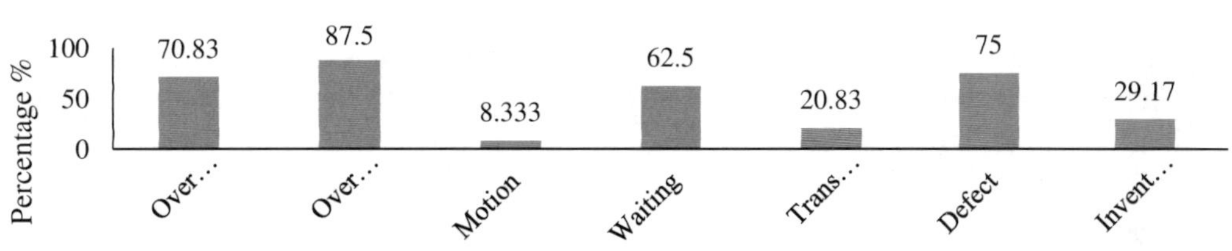

Figure 3. Response of respondents who are not aware of the lean deadly wastes.

"Over processing" was selected as the most waste (87.5%) above all by 21 respondents, followed by "Defect" (75%) and "Over production" (70%). It can be concluded that, although these 30 respondents claimed that they are never aware of what lean deadly waste is, they, however, have a fairly good idea of "over processing" and "defect" as wastes. But, to be fair, that outcome may be due to the phrasing of the terms itself e.g. 'over' here play such a vital psychological expression which describing an excessiveness, or to process to too great a degree, where logically tells it is a waste. Looking at figure 3, two of the lean waste that has the word 'over' in it got among the highest score. And "defect" is definitely a 'direct' word to describe something is broken, where people usually avoiding their money over it – so it is definitely a waste. All these types of the waste term are more natural to people unlike the rest of the terms where they have slightly a deeper meaning, a more specific phrase to distinguish general waste and lean waste. They scored the lowest. Similar results are shown in [4] where "defect", "over-production" and "over-processing" score 73%, 62.2% and 51.4% respectively which makes them the top three.

For those who answered Question 1 of this section affirmatively were asked to identify the type of lean waste that they are familiar with. The scores were all above 80% as in figure 4.

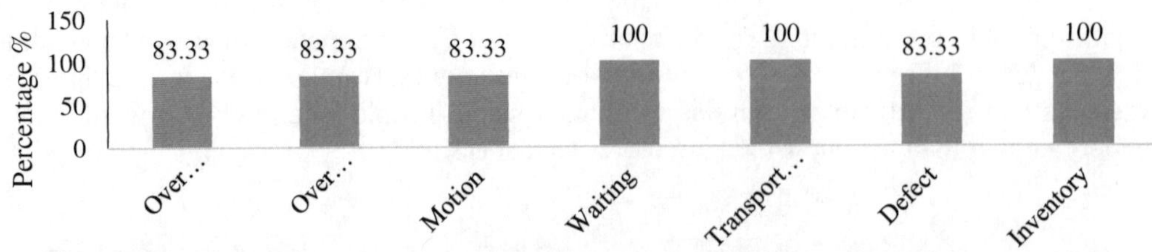

Figure 4. Response of respondents are familiar of lean deadly wastes.

With a high percentage of the vote for every element, this might give an insight that they actually have a deep understanding regarding lean. This claim might be supported by the fact that "motion", "transportation", and "inventory" have a really low vote in the previous section. Because, only the people that have a good understanding about lean know that all these three waste are as important as the others. In [12], where these researchers asked their respondents to rank which kinds of waste that lean eliminates in their respective organizations shows that "motion" and "inventory" comes second and third after "extra processing". To see that lean practice is eliminating those elements mostly in their industries gave us a view that those wastes are worth to give an attention to. As said before only people with a good understanding with lean knows how much "motion", "transportation" and "inventory" means as wastes. Tompkins *et al.* informed that transportation has a cost attribute to material handling about 15% to 18% (as cited in [8]). Likewise, Liker [13] said that excess inventories can be described as "evil" that hide problem. To conclude, with this level of result, it might be a good indicator for the faculty to kickstart the lean implementation.

Section D: Lean Tools & Techniques

The questions in this section intended to collect information specifically related to how much people in the FoE are aware of the lean tools and techniques that existed today. 66.7% of the respondents expressed the not knowing of the existence of any lean tools. While another 33.3% of the respondents were directed to 'tick' on any of the choices in the next question asking what type of lean tools they are familiar with. The outcome is illustrated in the following figure 5.

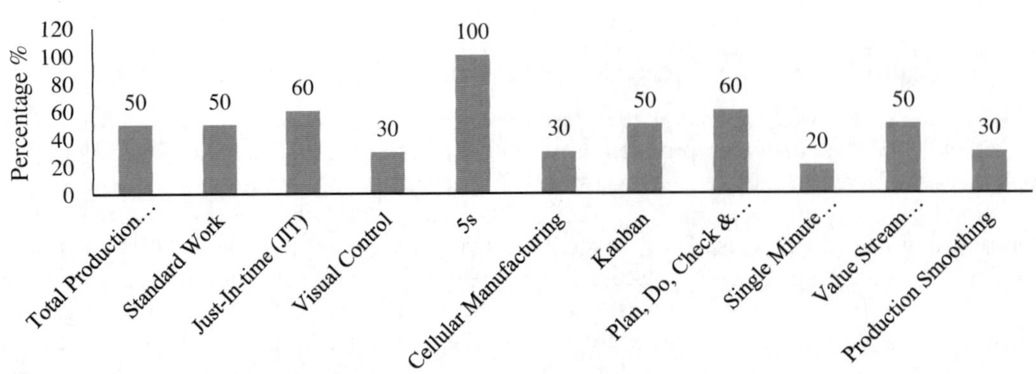

Figure 5. Lean tools and techniques awareness.

Based on the figure, the most popular tool is "5S" having 100% vote from all the 10 respondents. This result may be affected by the popularity of the 5S itself as it is used in most companies in Malaysia. Data in [3] and [4] also have similar finding where 5S dominates first. This proves that many companies in Malaysia are popular with 5S. In fact, 5S is the basic of lean manufacturing tool that you will use first to begin the lean implementation, where no one should fail practicing it [14]. The second most known tools based on the graph should be "PDCA cycle", and "JIT", followed by "TPM", "Standard work" and "VSM". This trend shows that the respondents have a basic or general knowledge about lean tools [4]. Noticed that the percentage are low for such tool like "SMED", visual control and "cellular manufacturing". There is an uncertainty regarding this situation either the area of research i.e. non-manufacturing, hence very little activities - nearly none, to expose the respondents about manufacturing sector activities give effect in this result.

Section E: Lean implementation in HE

The last section in the set of the survey questionnaire was conducted to collects information about respondents' opinion on applicability in HE. Question 1 asked whether the respondents have faith in the possibility of implementing lean in non-manufacturing sector. 96.7% of the respondents seem to have the faith. Moving to the next question, it is to investigate either any of the respondents were ever been involved (as a part of the team) in the lean project. Interestingly, there are a few. Having 10% from the respondent that actually had been involved and have an actual real life experience is such an excitement because the application of lean in HE is still at the beginning level, especially in Malaysia (considering only few publications that actually touched about lean implementation in Malaysian HE has been found). It is supported by Langer [15] that said lean in university is still relatively new. Also, Ziskovsky, B. and Ziskovsky, J. [16] mentioned that the concept of lean management is not new except for education industry. Having some of the experienced people in the team is beneficial because according to University of Central Oklahoma, their first step to lean was outsourcing facilitator from a professional consultant agency to guide and facilitate their progress (as cited in [17]).

The remaining two questions were asking about the benefits that lean would bring to the organization and the challenges of implementing it with reference to their FoE.

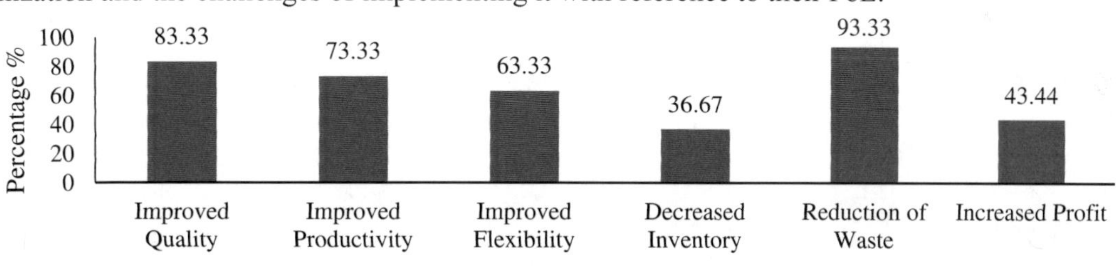

Figure 6. Benefits of lean.

From figure 6, 93% recognizes lean will reduce waste and 83% on "improved quality" (product and/or service) in the FoE. Whilst 73% of the responses indicate that the FoE would improve in productivity. The rest of the corresponding answers are shown in figure 6. What seems to be just a theory to these respondents, are actually the most practical in real cases. In [3], when they asked their respondents what has lean brings (i.e. benefits) to their company, "reduction in waste" and "improved productivity" got placed among the top. Waste minimization is still the major concern regardless the sector (i.e. manufacturing firms and service sector). Noted that since the subject of research is university, "decreased inventory" has the lowest vote. Different to the findings of other cases where the subject of study is 'ndustry i.e. [3], [4] and [8], "decreased inventory" get the proper attention as it should be.

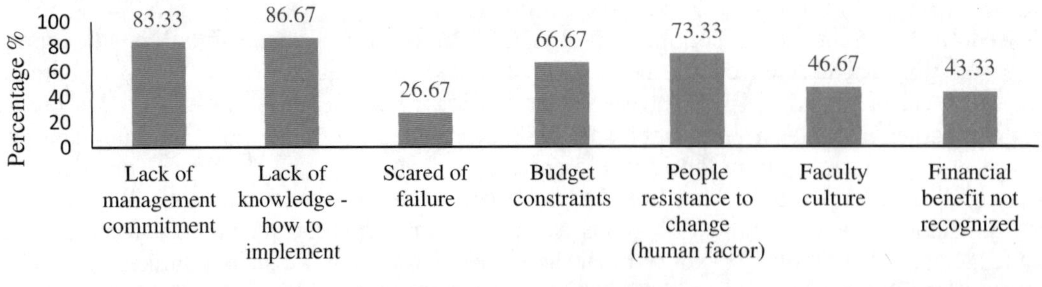

Figure 7. Challenges of lean implementation.

Figure 7 is the most likable challenges to be in the way of implementing lean based on respondent's thought. "Lack of knowledge" and "lack of management commitment" are the two most voted factors, 86%, and 83% respectively that 'cross beam' the successfulness of lean implementation. Since lean is a methodology that requires a long-term supervision, the fright of giving a commitment to it is always the factor blocking an organization from initializes lean [18]. Also, Scherrer-Rathje *et al.*; Worley & Doolen mentioned one of the success factors for the implementation is senior and middle management commitment (as cited in [16]).

The third highest factor (73.3%) acknowledge by respondents is "human factor." This factor is supported in [3] where the major obstacles of applying lean in companies seem to be 'people' factor. People tend to work in the old ways probably because of the perception that lean may burden them with extra chores [3]. Furthermore noticed that "budget constraints" scored quite high in the data, this should contribute such problem if the FoE wants to implements lean later on in fact that the country is having such an economic issues right now. It is also one of the elements that contain high percentage in [3] and [8] where both types of research were done in Malaysia. "Scared of failure" seems not be an issue here since it scored the least vote by the respondents. Maybe this is due to the environment of an educational institute that is full of knowledgeable people.

4. Conclusion

From the data collected and analysis made real awareness of lean concept is somewhat low in the community of the engineering faculty of the chosen local university. Respondents have a general idea on what lean waste is, thus, 'reduction of waste' was perceived as the most important lean benefit. In terms of barriers to successful lean implementation, 'lack of knowledge' was perceived as the major factor.

To capitalize the benefits of lean in Malaysian higher education, achieving strong awareness and real understanding of the concept must be given the highest priority together with real commitment by the management. A success story of lean implementation in a local Malaysian higher institution of higher learning would really serve as an excellent approach to achieve this aim.

References

[1] Emiliani B 2005. Lean in higher education. *Center for Lean Business Management*

[2]Abioye T E and Bello E I 2012. A Review of Awareness and Implementation of Lean Manufacturing within Nigerian Small-Scale Manufacturing Companies. *Journal of Mechanics Engineering and Automation,* **2**, 6

[3]Wong, Y C, Wong K Yand Ali A 2009. A study on lean manufacturing implementation in the Malaysian electrical and electronics industry. *European Journal of Scientific Research,* **38**(4), 521-535

[4]Salem R, Musharavati F, Hamouda A M and Al-Khalifa K N 2015. An empirical study on lean awareness and potential for lean implementations in Qatar industries. *The International*

Journal of Advanced Manufacturing Technology, **82**(9-12), 1607-1625. doi:10.1007/s00170-015-7421-7

[5] Krejcie R V and Morgan D W 1970. Determining sample size for research activities. *Educ Psychol Meas*

[6] Nasuha Lee Abdullah, Khairur Rijal Jamaludin, and Hayati Habibah Abdul Talib 2013. *Insights from Data Collection in Malaysia's Electrical & Electronics Manufacturing Industry*. Proceedings 2nd International Conference on Engineering Business Management, (ICBM2013)

[7] Shamsuddin Ahmed and Masjuki Hassan 2003. Survey and case investigations on application of quality management tools and techniques in SMIs. *International Journal of Quality & Reliability Management,* **20**(7), 795-826. doi:http://dx.doi.org/10.1108/02656710310491221

[8] Hafiz J M, Fakhri S A and Rahman M A 2012. The Implementation of Lean Manufacturing in Malaysian Automotive Industry. *Advanced Materials Research*, **576**, 731-734

[9] *The importance of socio-demographics in online surveys* n.d.. Retrieved from CheckMarket: https://www.checkmarket.com/blog/socio-demographics-online-surveys/

[10] Fricke C F and Buehlmann U 2012. Lean and Virginia's Wood Industry – Part I: Awareness and Implementation. *BioResources*, **7**, 5074-5093.

[11] Taninecz, G. 2005. Lean beyond production. *Lean Enterprise Institute. Accessed*, **12**(27), 7

[12] Mehta E R K, Mehta D and Mehta N K 2012. An Exploratory Study on Implementation of Lean Manufacturing Practices (With Special Reference to Automobile Sector Industry). *Yönetim ve Ekonomi*, **19**(2), 289-299

[13] Liker J K 2004. *The Toyota Way - 14 Management Principles from the World's Greatest Manufacturer*. New York: McGraw-Hill

[14] Earley T 2014. *Lean Manufacturing Tools, Principles, Implementation*. Retrieved from Lean Manufacturing Tools: http://leanmanufacturingtools.org/

[15] Langer T 2011. The application of Lean thinking for improving process in higher education institutions: Evidence from three UK case studies. *Queen's University Belfast*

[16] Ziskovsky B and Ziskovsky J 2007. Doing more with less – Going Lean in education. *A White Paper on Process Improvement in Education*, 1-19

[17] Cristina D and Felicia S 2013. Implementing Lean in Higher Education University. *Constanta Maritime University's Annals*, **18**, 279-282

[18] Salewski A and Klein V 2013. *How to launch lean in a university, ASQ, 2013*

IOP Conf. Series: Materials Science and Engineering **290** (2017) 012028 doi:10.1088/1757-899X/290/1/012028

Lean energy analysis of CNC lathe

N A Liana[1], N Amsyar[1], I Hilmy[1], MD Yusof[1]

[1]Manufacturing and Material Eng. Dept., Kulliyyah of Engineering IIUM, Malaysia

Email: mdyusof@iium.edu.my

Abstract. The industrial sector in Malaysia is one of the main sectors that have high percentage of energy demand compared to other sector and this problem may lead to the future power shortage and increasing the production cost of a company. Suitable initiatives should be implemented by the industrial sectors to solve the issues such as by improving the machining system. In the past, the majority of the energy consumption in industry focus on lighting, HVAC and office section usage. Future trend, manufacturing process is also considered to be included in the energy analysis. A study on Lean Energy Analysis in a machining process is presented. Improving the energy efficiency in a lathe machine by enhancing the cutting parameters of turning process is discussed. Energy consumption of a lathe machine was analyzed in order to identify the effect of cutting parameters towards energy consumption. It was found that the combination of parameters for third run (spindle speed: 1065 rpm, depth of cut: 1.5 mm, feed rate: 0.3 mm/rev) was the most preferred and ideal to be used during the turning machining process as it consumed less energy usage.

1. Background

Malaysia's electricity demand continues to increase rapidly throughout the years and excessive use of the demand may leads to the future power shortage. According to the international energy agency (IEA), the energy demand in Malaysia has doubled in the past decade [1]. In year 2012, the energy consumption in Malaysia has reached up to 134 billion kilowatt-hours and are expected to rise by more than 3% throughout 2020. Peninsular Malaysia as a high demand center is currently facing fuel shortages especially in natural gas while the need for generation capacity is getting higher. Malaysia try to overcome the problems by modifying the portfolio of power generation fuels and reducing the use of expensive fuel sources.

As stated by the Energy Commission of Malaysia, the main source of power demand in Malaysia is controlled by industrial sector by 45% of the total in year 2012. The percentages for commercial and residential demand was 33% and 21%. The percentage of energy demand for transportation sector and agriculture sector is less than 1%. These conclude that major contribution of energy consumption is from the production area.

Fujishima, Mori and Oda [2] stated that highest rate of energy consumption was coming from the usage of machine tool compared to other manufacturing equipment. The implemented methods to reduce the energy consumption were by reducing the energy required by the machines, reduce the cycle time and reduce the idle time of machine. Energy consumption of a factory would become uncontrolled if daily usage of the energy was not being observe carefully. The waste of energy would leads to the loss of a production and company. As stated by Donelly, Kummer and Drees [3], opportunities for energy efficiency can be identified by assessing the portfolio of a building and this method will lead to cost savings and improved building performance. Instead of that, energy analysis was one of the effective method to keep track on the energy consumption of the factory. Kuo, Fu & Cho [4] had stated several fundamental factors such as orientation, building area, total annual electricity consumption, and number of customer which are identified to affect the energy consumption of convenience store. Pramadona and Akhbar Adhiutama [5] mentioned that lean manufacturing system has been applied in industries for cost reduction and quality improvement and productivity by reducing variance and deficiency of product. Based on the Kissock and Seryak [6], the lean energy method could be used to predict energy cost for every energy consumption, measures savings, and to determine cost structures. The energy consumption could be analyzed by using lean energy analysis based on several factor which affect the analysis such as buildings layout and weather. The standby period is when the machine is halted and not performing any machining process. All activities such as coolant operation, servo motors, chip conveyor and displays are also stopped during the period. According to research made by Fujishima et al. [7], the average power consumption time for 10,000 units of machines was 268 hour/month and the average operation time was 150 hour/month. The difference was 118 hour/month and this was when the machines were in idle and halted.

This research was conducted to discuss about the approach of improving energy efficiency in a company by optimizing cutting conditions of the machine tools. The type of machine tool used in this research was an altered Mini Lathe Machine and the cutting parameters that were spindle speed (rpm), feed rate (mm/rev) and depth of cut (mm).

2. Problem Statement and Objectives

Energy sources has become one of the major factors that contribute to the increasing of the cost production of a company. The example of the energy consumption of a company are from facilities, space-conditioning and production-related components. These are among of the crucial aspect that usually being evaluated to improve a performance of a company. However, a focus is given to the production equipment such as machine tools as it stated the highest rate of energy consumption compared to other aspects in a production area. The improvement process of machine tools can be proposed in order to reduce the energy consumption of the machine tool to the minimum level.

Energy consumed the most during machining process of machine tool. However, the rate of energy consumption can be controlled if suitable cutting parameters are selected during the machining process. One of the effective ways to reduce energy usage during machining is by optimizing certain cutting parameters such as spindle speed (rpm), feed rate (mm/rev) and depth of cut (mm). By implementing this technique, energy efficiency of machine tool can be increased.

High amount of energy consumption could affect the production cost and this will lead to the company loss. Cost spent on waste will decrease the profit margin of a company. An applicable approach by enhancing the cutting parameters of machine tool is proposed to reduce the energy consumption and spending cost on the waste can be saved and thus will give benefit to the company in future.

The objective of this research: (1) to identify the techniques used for minimizing the energy consumption of turning machine. (2) to analyze the impact of the cutting parameters on energy consumption of turning machine tools, (3) to select the suitable cutting parameters which affect the energy consumption of turning machine and (4) to minimize the energy consumption of turning machine tools by enhancing the cutting parameters.

3. Methodology

The energy consumption of the machine tool can be analyzed by conducting several experiments on different aspects. In this research paper, the experiment is focusing on increasing the energy efficiency of the machine tool by optimizing the cutting conditions.

Optimization of the cutting conditions refer to the several factors that highly affect the energy efficiency of machine tool are generally caused by cutting parameter used during the machining process. The example of the cutting parameter includes feed rate, depth of cut, cutting speed, cutting condition and tool geometry. By optimizing certain cutting parameter, critical cutting condition that consumed high energy consumption can be identified and eliminated.

The type of machine used in the experiment is CNC Mini Lathe Machine. The main function of the machine is to machine or to cut small size of workpiece. The machine has two axis of the movement of tool and constant position of workpiece at spindle which are x-axis and z-axis. The movement of the stepper motors for the X-axis and Z-axis are controlled by using coding generated in Arduino software. The mini CNC turning machine is shown in the figure 1.

The type of workpiece used in the experiment is an acrylic material which is in a form of cylindrical rod. The acrylic materials are widely used in industrial applications as it provides good electrical properties, good chemical resistance, lightweight, good weatherability and high durability. Figure 2 shows the measurement of the workpiece used in the turning process. The diameter of the workpiece is 20 mm and the length is 60 mm. The main aim of the machining is to reduce the workpiece diameter to 14 mm and workpiece length to 57 mm.

Figure 1. Mini CNC Lathe Machine

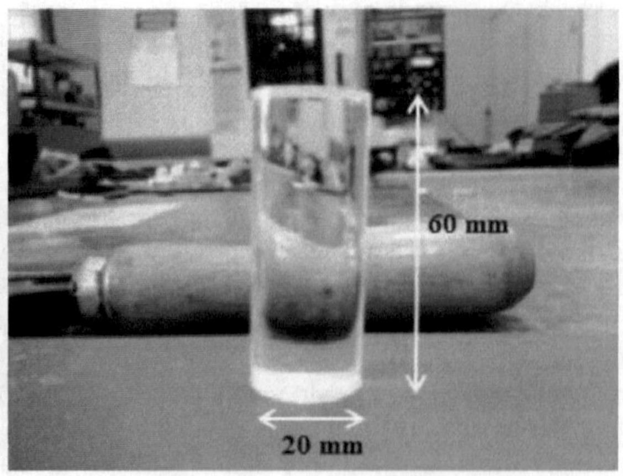

Figure 2. Work piece Measurement

A voltage and current detector device is built to detect a current and voltage flow of lathe machine during machining process. The device consists of several main components such as Arduino board, screen display, current sensor and voltage divider. The device is connected to the motor to detect the amount of current and voltage flow through it. The current and voltage value of the motor are later shown on the display screen. In addition to that, power consumption and cycle time are also displayed on the screen.

The type of the Arduino board used in the device is UNO R3. The board consists of several features as shown in table 1. Another main component of the device is a current sensor. The type of the current sensor used in the circuit is a 50A Current Sensor (AD/DC). The specification of the current sensor is also shown in table 1.

Table 1. UNO board and sensor specification

Board Specification	**Sensor Specifications:**
6 Analog Inputs	Peak Measuring Voltage:3000V(AC),500V(DC)
32k Flash Memory	Sensitivity:40 mV/A
16Mhz Clock Speed	Operating Temperature: -40~150°C
14 Digital I/O Pins (6 PWM outputs)	Dimension:34x34mm

Input voltage 7 12V	Operating Voltage(analog): 5V
ATmega328 microcontroller	Current Measuring Range:-50~50A

The current sensor used in the circuit gives an ability to the user to monitor current flow in the circuit especially for energy saving purposes. The example of the current sensors are shown in figure 3 and figure 4

Figure 3. Arduino Board and Current Sensor

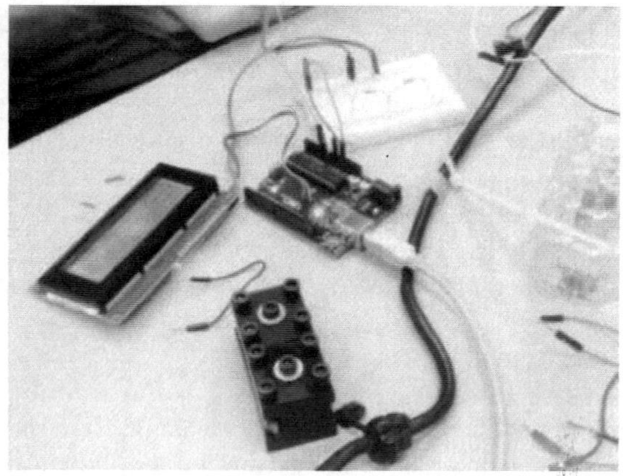

Figure 4. Voltage and Current Sensor

Furthermore, voltage divider also holds an important function in the device. The function of the voltage divider used in the circuit is to convert large amount of voltage into a small value of voltage. It consists of two different values of resistors which are R_1 and R_2. The value of R_1 is set as 0.46 kΩ and R_2 is unknown. The voltage divider formula is used to find the value of the R_2:

$$\text{Vout} = \frac{\text{Vin}\times\text{R2}}{\text{R1}+\text{R2}} \tag{1}$$

The given values of V_{out} = 5V, V_{in}= 12.6 V and R_1 = 0.46 kΩ. R_2 is solved by substituting the given values in the formula. R_2 is calculated by using Equation (2):

$$\text{R2} = -\frac{\text{Vout}\times\text{R1}}{\text{Vout}-\text{Vin}} - \text{R1} \tag{2}$$

The value for R_2 is 0.303 kΩ and as the nearest value of resistor available is 0.21kΩ, the value of R_2 is set as 0.21 kΩ. Both of the transistors (R_1 and R_2) are connected in a series connection in the circuit.

Figure 5 shows the block diagram of the turning process. The noise factors consists of the absent of coolant during machining and the hardness of the material work piece. The control factors represent the cutting parameters such as depth of cut (mm), feed rate (mm/rev) and spindle speed (rpm).

Table 2 shows the process parameter and the levels selected for the experiment. The type of operation performing by the mini lathe machine is turning operation. This process is a process of metal removal from the work piece to produce flat surface. All possible combinations of operation need to be carried out during the experiment to analyze the effects on power and energy consumption. Taguchi's orthogonal array technique is implemented to compose the operation combinations.

IOP Conf. Series: Materials Science and Engineering **290** (2017) 012028 doi:10.1088/1757-899X/290/1/012028

Table 2. Process Parameter and Experiment Level

Factors	Process parameter	Level 1	Level 2	Level 3
A	Spindle speed (rpm)	1065	1400	1700
B	Feed rate (mm/rev)	0.1	0.2	0.3
C	Depth of cut (mm)	0.5	1.0	1.5

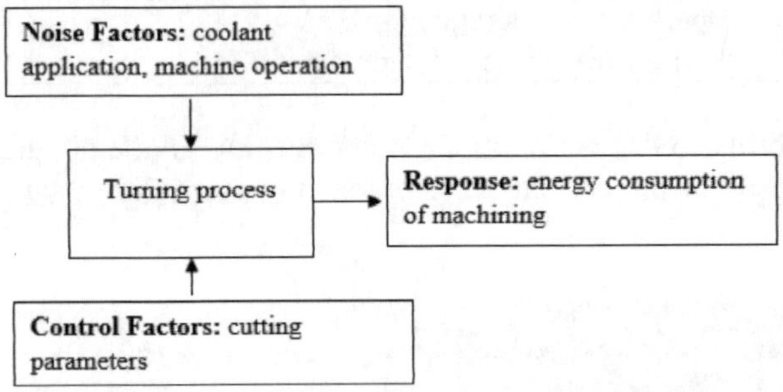

Figure 5. Block Diagram of Turning Process

Design-Expert 9.0.6 software is used to generate the experimental design and L9 Taguchi's orthogonal array technique is selected to compose the operation combinations. The technique consists of nine operation combinations and three experiment trials.

4. Power measurement system

The data for average current and voltage during the turning process are monitored by using two multimeters which are directly connected to the motor circuit of the machine. The first multimeter is connected in series to detect the amount of current used during process. The second multimeter is connected in parallel to detect the amount of voltage used during the process. This is one of the method used to measure the voltage and current flow of the motor besides using the detector device. This method is suggested when the measured voltage is very high. The average power can be obtained by calculating the value of voltage and current using the Power formula as shown in Equation (3):

$$Power\ (W) = Voltage\ (V) \times Current\ (A) \tag{3}$$

The cycle time for each operation is also recorded in order to obtain the rate of energy consumed by the lathe machine for each operation.

5. Result and Discussion

The data collected for the experiment is tabulated in the table 3. The table contains four main columns which are Parameter, Power Demand (Watt), Cycle Time (Sec) and Energy (Joules). The parameter column consists of three cutting parameters which are A, B and C. The parameter A represent spindle speed (rpm), the parameter B represent depth of cut (mm) and the parameter C represent feed rate (mm/rev).

The Power Demand column is divided into P1, P2, and P3 which consists of the amount of power consumption for three trial of experiments. P (Average) is an average amount of power consumed by the machine for three trials of experiments. The Cycle Time column is referring to the total time required for each machining process to complete. The cycle time is recorded and energy consumption is calculated.

Table 3. Experimental Data of Power Consumption during Machining

	Parameter			Power Demand (Watt)				Cycle time (Sec)	Energy (Watt-Sec)
Run	A	B	C	P1	P2	P3	P(AVG)		
1	1065	0.50	0.10	41.04	38.33	39.23	39.53	37.00	1462.73
2	1065	1.00	0.20	56.14	55.66	50.89	54.23	17.67	958.24
3	1065	1.50	0.30	70.99	70.93	73.41	71.78	11.00	789.55
4	1400	0.50	0.30	48.76	50.51	48.03	49.10	35.33	1734.70
5	1400	1.00	0.10	64.39	66.13	64.02	64.85	17.00	1102.39
6	1400	1.50	0.20	77.40	79.68	77.69	78.26	11.00	860.82
7	1700	0.50	0.20	53.38	55.74	67.20	58.77	36.00	2115.84
8	1700	1.00	0.30	68.47	70.84	67.12	68.81	16.67	1147.06
9	1700	1.50	0.10	90.06	88.59	88.00	88.88	10.00	888.83

The energy consumption for each of the machining process is calculated by multiplying the average of power consumption and cycle time (Sec). This calculation is simplified by using Energy Formula as shown in Equation (4).

$$Energy\ (Joules) = Power\ (Watt) \times Cycle\ Time\ (Sec) \quad (4)$$

The information about the machining process can be analyzed by plotting the data in form of graph as shown in figure 6. The type of graph plotted is main effects graph which is used to examine the trend pattern of each of the parameters selected. The relationship of graph is between the cutting parameter and power consumption.

According to the slope of the graphs, it shows that all of the selected cutting parameters give effect towards the energy consumption. However, the cutting parameter of depth of cut and feed rate are highly influence the amount of energy usage during machining process. The statement agreed with the research done by Camposco-Negrete et al. [8] which mentioned that the depth of cut and feed rate give the most impact towards the response.

As referring to the graph for spindle speed, as the value for spindle speed increased, the value for power consumption is also increased. This is because high force is needed to move the motor spindle faster. Hence, the energy consumption can be reduced by using the low level of spindle speed during the machining.

In addition to that, the parameter of cutting depth also affects the amount of power consumed during the machining. Based on the figure 6, the trend for power consumption increased as the cutting depth increased. This is cause of high force is needed to remove the part of material. This correspond with the research done by Camposco-Negrete et al. [8] which mentioned that the system required to consume more power as high force is needed to cut the material because of the increment of cutting depth value. As to reduce the power usage during machining, minimum value of depth of cut is recommended to cut the workpiece.

Another cutting parameter that influences the rate of power consumption during machining process is feed rate. Referring to the figure 6, when the value of the feed rate increased, the amount of power consumption decreased. This relationship happened caused by the reduction of cycle time. High amount of material being cut per revolution will reduce the time taken, thus less amount of power required to complete the process.

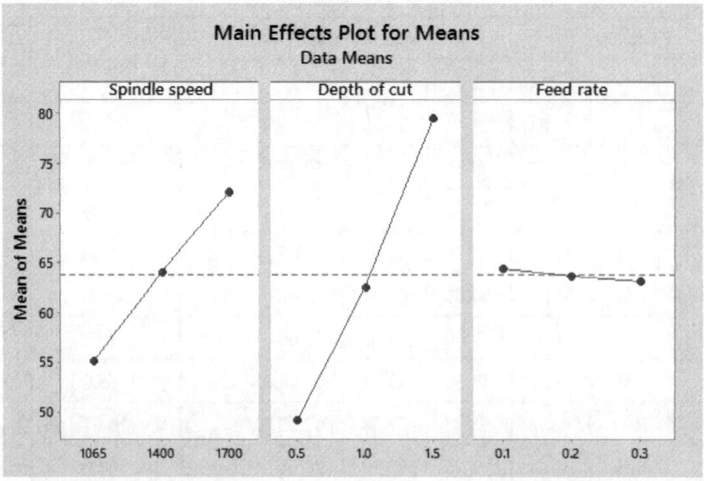

Figure 6. Main Effects Graph (Power vs Cutting Parameter)

The SN ratio graphs were used to examine the variation of cutting parameters towards power consumption. The type of signal-to-noise ratio selected in this experiment is Smaller-The-Better characteristic as the main aim of this experiment is to minimize the energy consumption in the machining process. The levels minimize the power consumption and ensure that the process stay within the target value. The final equation in terms of actual factors for the model is as shown in Equation (5). The equation can be used to predict the value of the energy usage for the each level of factors.

$$Energy = -538.87533 + 2.07080 * spindle\ speed + 6.38900 * depth\ of\ cut + 5378.51633 * ate - 0.76517 * spindle\ speed * depth\ of\ cut - 4.05866 * spindle\ speed * feed\ rate \quad (5)$$

6. Conclusion

Industrial sector is one of the major sectors that contribute to the high power demand in the world. It is important for the sector as electrical energy is required to move the machine tools in manufacturing a product. However, uncontrolled amount of power demand may lead to the power shortage and environmental impact in future. Thus, another alternative need to be implemented in order to minimize the energy consumption.

The main purpose of the research is to reduce the energy consumption of the machine tools by enhancing the cutting parameters. The type of cutting parameters selected in the experiment is spindle speed (rpm), depth of cut (mm) and feed rate (mm/rev). The amount of energy consumption for each combination parameters may not be the same although it has the same amount of material removed.
This research concluded that the third run consumed the least amount of energy usage while machining the workpiece. The run consists of the parameter combination of:

- Spindle speed : 1065 rpm
- Depth of cut : 1.5 mm
- Cutting depth : 0.3 mm

By implementing the selected parameter combination, the energy efficiency of machining process is improved and thus, energy usage can be saved.
Nonetheless, there are also some limitation occurred during conducting the experiments. The type of workpiece used for this experiment is different to the past research which may cause some variation in

IOP Conf. Series: Materials Science and Engineering **290** (2017) 012028 doi:10.1088/1757-899X/290/1/012028

results. Moreover, there are also some difficulties in handling the lathe machine as it has to be operated by generating the Arduino code. The movement of machine motor is may not be as accurate as the CNC software as the rotation of motor has to be manually calculated.

References

[1] Walton et al., *South East Asia Energy Outlook*, World Energy Outlook Special Report, IEA, 2014

[2] Fujishima M, Mori M, & Oda Y 2013 *Energy-efficient manufacturing on machine tools by machining process improvement, Production Engineering* **8** 217–224

[3] Donnelly, M, Kummer J, & Drees, K 2013. *LEAN Energy Analysis Using Regression Analysis to Assess Building Energy Performance Issue Brief*, (March) **2**

[4] Kuo, P Y, Fu, J C, & Cho, C H 2013 *Analysis on the energy consumption and usage characteristics of convenience stores* 687–691

[5] Adhiutama, A, & Pramadona 2013 *The Application of Lean Manufacturing for Operation Improvement : A Case Study of Black Cough Medicine Production in Indonesia*, **6**(1) 56–64

[6] Seryak, J, & Kissock, K 2004 *Lean Energy Analysis: Identifying, Discovering and Tracking Energy Savings Potential* 1–11

[7] Fujishima, M, Mori, M, & Oda, Y 2013 *Energy-efficient manufacturing on machine tools by machining process improvement. Production Engineering*, **8** 217–224

[8] Camposeco-Negrete, C, de Dios Calderón Nájera, J, & Miranda-Valenzuela, J C 2015 Optimization of cutting parameters to minimize energy consumption during turning of AISI 1018 steel at constant material removal rate using robust design. *The International Journal of Advanced Manufacturing Technology*. doi:10.1007/s00170-015-7679-9

Influence of fiber treatment on dimensional stabilities of rattan waste composite boards

Zuraida A[1], Insyirah Y[1], Maisarah T[1], Zahurin H[1]

[1]Department of Manufacturing and Materials Engineering, Kuliyyah of Engineering, International Islamic University Malaysia (IIUM), P.O. Box 10, 50728 Kuala Lumpur, Malaysia

Email: zuraidaa@iium.edu.my

Abstract. The main drawback of using natural fibers in composite boards is its hydrophilic properties which absorb a high volume of moisture. This results in low dimensional stability of the produced composite boards. Hence, the purpose of this study is to investigate the effects of fibers' treatment processes of the rattan waste fibers on the dimensional stabilities of composite boards. The collected fibers underwent two types of retting processes, namely a water treatment and alkaline treatment retting processes; where the fibers were soaked in water and a 1% sodium hydroxide (NaOH) solution, respectively. The fibers were dried and mixed with poly(lactic) acid (PLA) pellets with ratio of 30% fibers: 70% matrix; before being fabricated into composite boards via a hot-pressing process and were labelled as RF/PLA, WRF/PLA, CRF/PLA for untreated rattan, rattan treated by water retting, rattan treated by chemical retting, respectively. The produced composite boards were cut and soaked in water for 24 hours for dimensional stability in terms of water absorption and thickness swelling tests. The results showed that WRF/PLA has the lowest water absorption (3.2%), and the CRF/PLA had the highest water absorption (23.2%). The thickness swelling showed a similar trend as water absorption. The presence of void contents and fibers damaged the insides of the boards, which contributed to low dimensional stabilities of the composite boards. It can be concluded that water retting facilitated in improving dimensional stability of the produced composite board.

1. Introduction

The plastic industry dominates is a leading industry in the world, with almost 234 million tons of plastic produced in 2013. It contributes towards problems of dumping plastic waste due to the abundant usage of petroleum synthesized polymers in plastic production. This leads to a cost burden and a crisis for waste management [1]. The high awareness towards sustainability of the environment, along with transformation of government regulations for green byproducts, have motivated researchers to find new substitutes for present composite materials with eco-friendlier and biodegradable materials. Biopolymer has a potential to function as bioplastics, where one of the most preferable candidate of biopolymer is poly(lactic acid) (PLA) [2]. PLA is derived from renewable resources such as corn starch or sugar cane, where it works as a thermoplastic which provides environmentally-friendly characteristics, biocompatibility, sustainability, as well as potentially useful physical and mechanical properties. The superior performance of synthetic fibers such as glass fibers or carbon fibers are advantageous for composite materials, yet the safety of these fibers become a threat towards the environment and health. Therefore, researchers have actively explored natural fibers for example sisal, banana and kenaf, where these fibers are non-toxic, biodegradable, lightweight, have superior performance and are cheaper due to their availability [3]. One of the fibers that has high potential as reinforcement in the bio-composites is rattan waste fibers. Rattan waste fibers comprise of cellulose, hemicelluloses and lignin; where these fibers feature good properties such as low density, flexibility and versatility [4].

The rattan waste fibers and PLA matrices are, however, incompatible with poor interfacial bonding, due to the impurities and hydrophilic properties in rattan compositions, which makes it unsuitable with the hydrophobic nature of PLA. It results in reduction in poor adhesion between fibers and matrix, causing ineffective stress transfer during interfere of composite produced [5]-[6]. The significant of interfacial bonding between fiber and matrices in composite was illustrated in figure 1, that shows the treatment (or environments) on the fiber surface have great effects onto the adhesion between the fibers and matrices.

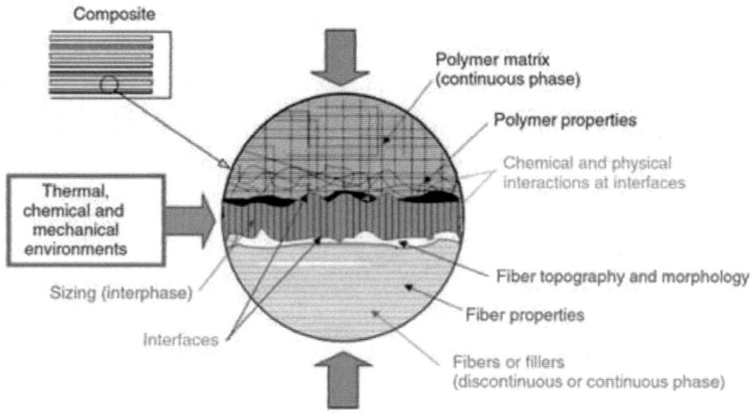

Figure 1. Adapted illustration of schematic diagram shows the importance of interfacial bonding in a composite [7].

Based on those factors, the performance of the produced composite can be improved by surface modification of the rattan fibers, through proposed retting processes. Retting is a microbial process that breaks the chemical bonds that hold the stem together and allows separation of the bast fibers from the woody core [8]. The retting process has a good potential to improve the structure of fibers by washing out the impurities on the fibers such pectin, lignin, celluloses and wax. These impurities are the main factors that reduce the good properties of rattan fibers. There are several common retting processes which are tabulated in table 1, with the advantages and disadvantages of each processes. this study only focused on two processes – water retting and chemical retting. Thus, this study was executed to assess the effects of fiber treatments on rattan waste fibers, effecting dimensional stability of the produced composite boards.

2. Experimental Procedures

The rattan waste fibers was supplied by Rinaat Cane Sdn. Bhd., Perak, Malaysia, in the size of 500μm. The matrix chosen was poly(lactic acid) (PLA) in white crystalline form, purchased from Fuzan Enterprise, Malaysia. The fibers were prepared for two retting processes, which were water retting and chemical retting [9]. The execution of water retting was carried out by fully immersing the fibers in a beaker using tap water at room temperature for 8 days. While, the chemical retting process was conducted by preparing an alkaline solution (1% of NaOH) and the fibers were then soaked in the alkaline solution for 60 minutes before taken out to be washed with tap water to remove the alkali solution from the fiber surface. Both treated fibers were dried in an oven at the temperature of 35°C for 48 hours, and were labelled as WRF and CRF for fibers that underwent water retting and chemical retting processes, respectively.

The Scanning Electron Microscopy, SEM (Jeol JSM-6700F) analysis was done to these treated fibers under 1000x magnification. The fibers were mixed with PLA using an internal mixer (Haake

Polylab System Thermo) machine at a heating temperature of 165°C, rotor speed of 50 rpm and mixing time of 12 minutes, with fiber:matrix ratio of 30:70. The compound materials were labelled as RF/PLA, WRF/PLA, CRF/PLA, and were grinded using crusher into small pieces before being hot-pressed at pressing parameters of 160°C, 3 minutes and 147.5kPa. The water absorption and thickness swelling test was performed as per ASTM D7031-04 method. The values of water absorption and thickness swelling were calculated using the following formula of 3.1 and 3.2.

$$\text{Percentage of water absorption (\%)} = \frac{w_2 - w_1}{w_1} x\ 100 \quad (3.1)$$

where w_1 is initial weigh of specimens and w_2 is final weight of specimens after being immersed

$$\text{Percentage of thickness swelling (\%)} = \frac{t_2 - t_1}{t_1} x\ 100 \quad (3.2)$$

where t_1 is initial thickness of specimens and t_2 is final thickness of specimens after being immersed

Table 1. Types of common retting, adapted from [9].

Retting Types	Advantages	Disadvantages	Retting Duration
Dew retting	- Bacteria easily can removed pectin in the fiber	- Reduced strength - Low and inconsistency quality - Restriction to certain climatic changes - Product contaminated with soil	2 – 3 weeks
Water retting	- Yields greater uniformity and high quality	- High cost - Putrid odour, environmental problems - Requires high water treatment maintenance - Low-grade fiber	7 – 14 days
Enzymatic retting	- Easier refining particularly for pulping purposes - Cause a partial degradation of the components separating the cellulosic fiber from non-fiber tissues - Faster and cleaner process	- Lower fiber strength	12 – 24 hours
Chemical retting	- Efficient - Can produce clean with consistent long and smooth surface bast fiber within a short time.	- Tensile strength of fiber decreases if retted using 1% NaOH - Unfavorable color - High processing cost.	1 – 1.5 hours

Mechanical retting	- Produces massive quantities of short fiber in short time	- High cost - Low fiber quality

3. Results and Discussion

Figure 2 illustrates the percentage of water absorption and thickness swelling of raw PLA, raw rattan fibers, RF/PLA, WRF/PLA and CRF/PLA after 24 hours of water immersion. The bar chart in figure 2 demonstrates that the raw PLA, with no addition of any fiber content, has the lowest water absorption at 0.8%. This confirms the hydrophobic nature of PLA exhibited the water from absorbed into the polymer chains [2]. On contrary, the raw rattan waste fibers (RF) have the highest water absorption at 47.6%, owing to the hydrophilic characteristic of natural fibers. It is believed that hemicelluloses are responsible for moisture absorption [10], where the removal of hemicellulose produces a less dense and less rigid interfibrillar region.

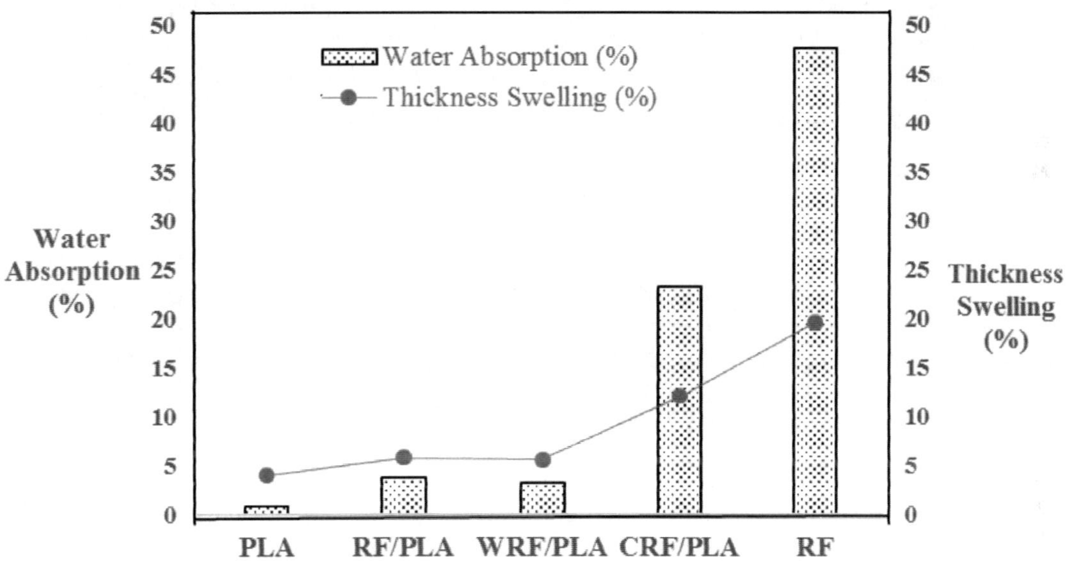

Figure 2. The percentage of water absorption (bar graph) and thickness swelling (line graph) of the samples.

The RF/PLA composite produced water uptake at 3.7%, resulting from the well-dispersed and mixing between rattan waste fibers and PLA that helped to increase the interfacial bonding of fiber-matrix. The WRF/PLA composite had 3.2% of water uptake, proving that the water retting process had slightly improved the composite compared to other composites. This resulted from the modification of hydrophilicity of rattan fiber filling the voids of cellulosic structure with polymers during the formation of the composite. The CRF/PLA composite had 23.2% of water uptake, due to the presence of void contents inside the composite. The void contents in the composite encouraged the water to diffuse into void spaces, which formed during compounding and the presence of micro-voids between fiber-matrix interfaces [11]. The free hydroxyl groups occurred in the natural fibers attracted water molecules and were absorbed into the composite [12]. Thus, the water absorption led to cell expansion and affected the dimensional change of the composite such as thickness swelling. The thickness swelling of PLA, rattan fibers and composites produced in the line inside figure 2, shows a similar trend as the water absorption graph.

Both retting processes have removed the impurities and surface debris such as lignin, hemicelluloses, pectin from the fiber surface resulting in cleaner vascular bundles and rougher surface

[6], [10], compared to untreated fibers, as can be seen in figure 3. This rough surface facilitates both mechanical interlocking and bonding reaction due to the exposure of the hydroxyl groups to the matrix, thereby increasing the fiber–matrix adhesion, increasing the effective surface area available for contact with the matrix. The processes also disrupted the hydrogen bonding in fiber surfaces. Chemical retting has better efficiency in removing the impurities in fiber surfaces compared to other types of retting, within short duration that changes fiber nature from hydrophilic to hydrophobic [9]. On top of that, this treatment has caused diminished strength of fiber, due to certain factors such as extended treatment time or disproportionate concentration of alkali used [9], [13]. Meanwhile, water treatment did not require any additional chemical for its treatment, except for large treatment scale as mentioned in the table 1. It is easy to handle which required less skills and more environmental-friendly [14].

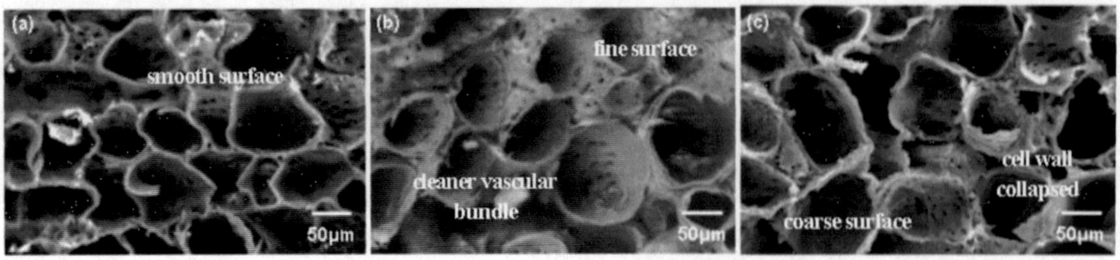

Figure 3. Micrographs of fiber surfaces of (a) RF, (b) WRF, (c) CRF.

The micrographs in figure 4 shows the composite board surfaces after a mechanical test, thereby proving that the WRF/PLA boards (in figure 4(b)) had smaller voids compared to RF/PLA and CRF/PLA boards. Similar result obtained by previous study where fiber underwent water retting process gave the highest tensile strength [9], compared to fiber underwent sodium hydroxide and sodium benzoate treatment. A few void spots were noticed on the surfaces of all composite boards indicate that parts of rattan fibers have poor bonding adhesion with the PLA matrix, due to incompatibility of hydrophobic polymer matrix and hydrophilic of natural fiber [7], [15]. figure 4(c) presence greater void occurrence of pulled-out fibers from matrix, which resulted for incoherent in board strength with high water absorption [11], [13], [16]. There are numerous factors that affects boards properties such as fiber types and orientation, processing parts and many more [15].

Figure 4. Micrographs of surface structures of composite boards (a) RF/PLA, (b) WRF/PLA, (c) CRF/PLA.

Based on the results discussed, water retting has good potential as a surface modification method for rattan fibers which promotes the lowest water absorption and thickness swelling among the samples. It is also recommended to add coupling agents such as maleic anhydride grafted polymers (MAPP) to increase the compatibility between matrix and fibers.

4. Conclusion

The retting process has positive effects on the properties of rattan waste composites. Based on the performance evaluation on the produced composite boards, the water retting process improves dimensional stability of boards by having lowest water uptake. The CRF/PLA has the highest water uptake due to void presence in composite structures. Meanwhile, the thickness swelling test follows a similar trend with the water absorption test which shows that rattan waste that undergo a chemical retting process are not stable in terms of thickness swelling due to high water uptake. Based on that, the water retting process has improved the performance of composite in terms of its mechanical and physical properties.

Acknowledgement
The authors wish to thank the Ministry of Higher Education (Malaysia) for financial support of sponsorship scheme of Grant No. FRGS15-206-0447, as well as the lecturers and technicians in International Islamic University Malaysia (IIUM) for the help and facilities provided.

References
[1] Thompson R C, Charles J M, Frederick S v S, Shanna H S 2009 *Philosophical Transactions R Soc Lond B Biol Sci.* **364** 2153–2166
[2] Nurul Fazita M R, Jayaraman K, Bhattacharyya D, Mohamad Haafiz M K, Saurabh C K, Hussin M H, Abdul Khalil H P S 2016 *Materials.* **9** 435-464
[3] Pickering K L, Aruan Efendy M G, Le T M 2016 *Composite Part A: Applied Science and Manufacturing.* **83** 98-112
[4] Maisarah T, Zuraida A, Roszalia A, Luqman M A 2015 *Advanced Materials Research.* **1115** 317-320
[5] Viviana P C, Claudia V, Kenny J M, Analia V 2004 *Journal of Composite Materials.* **38** 1387-1399
[6] Benyahia A, Merrouche A, Rahmouni Z E A, Rokbi M, Serge W, Kouadri Z 2014 *Mechanics & Industry.* **15** 69-73
[7] Drzal L T, Rich M J, Lloyd P F 1983 *Journal Adhesion.* ***16*** 1–30
[8] Kabir M M, Wang H, Aravinthan T, Cardona F, Lau K-T 2011 *EDDBE2011 Proceedings* 94-99
[9] Paridah M T, Basher A B, Saiful Azry S, Ahmed Z 2011 *BioResources.* **6** 5260-5281
[10] Geng X, Zhang S Y, Deng J 2006 *Journal of Wood Chemistry and Technology.* **26** 313–324
[11] Tao Y U, Yan L I, Jie R E N 2009 *Transactions of Nonferrous Metals Society of China.* **19** 651-655
[12] Tee Y B, Talib A R, Abdan K, Chin N L, Basha R K, Yunos M, Faezah K 2015 *Pertanika Journal of Tropical Agricultural Science.* **38** 563-573
[13] Mohanty A K, Misra M, Drzal LT 2001 Compos. Interfaces. **8** 313–343
[14] Garrote G, Dominguez H, Parajo J 1999 *Holz als Roh- und Werkstoff.* **57** 191-202
[15] Donghwan C, Hyun-Joong K, Lawrence T D 2014 *Polymer Composite.* **3** 133-177
[16] Vaxman A, Narkis M, Siegmann A, Kenig S 2004 *Polymer Composite.* **10** 449 – 453

IOP Conf. Series: Materials Science and Engineering **290** (2017) 012030 doi:10.1088/1757-899X/290/1/012030

Effect of Alkali treatments on physical and Mechanical strength of Pineapple leaf fibres

M Asim[1], M. Jawaid[1], K Abdan[2] and M Nasir[3]

[1]Laboratory of Biocomposite Technology, Institute of Tropical Forestry and Forest Products (INTROP), Universiti Putra Malaysia, 43400 Serdang, Selangor, Malaysia
[2]Department of Biological and Agricultural Engineering, Faculty of Engineering, Universiti Putra Malaysia, 43400 Serdang, Selangor, Malaysia
[3]Composite wood, Forest Product Division, Forest Research Institute, 248006 Dehradun, India

E-mail: khanfatehvi@gmail.com

Abstract. Pineapple leaf fibre (PALF) is a waste material of pineapple plants. PALF is abundant in amount for industrial purpose, cheap, easily available, high specific strength and stiffness. PALF is contributing a sustainable development in bio-composites as reinforcement material. However, natural fibres are not fully compatible with matrix due to hydrophilic in nature. To enhance the compatibility with matrix, fibres are modified its surface to make good interfacial bonding with matrix. In this research, PALF is treated with 3% and 6% concentration of NaOH for 3h, 6h 9h, and 12h soaking time. Surface modification of fibres was investigated by using scanning electron microscopy. Single fibre test and diameter of PALF fibres were evaluated the effects of NaOH treatments.

1. Introduction

There are many eco-friendly natural fibres, wood fibre and non-wood fibres being used to develop bio composites. Natural fibres are eco-friendly, minimum cost, less density, good mechanical properties, sound absorber and carbon dioxide sequestration over synthetic fibres[1]. Natural fibres extract from various sources such as leaves, fruit, grass and bast, which provide mechanical strength and stiffness of biocomposites in various applications [2]. Several industrial crop are being grown to collect fibres, it strengthen the supply of natural fibre for renewable materials and bio-composites[3]. There are some common natural fibres like flex, hemp, jute, kenaf, bamboo, sisal, and cotton being use for biocomposite. Pineapple leaf fibres (PALF) are waste material from pineapple cultivation. Annually, these waste material are available as a renewable cheap and abundant in amount for industries and researches purpose. PALF is the composition of high content of cellulose (70–82%) which is very higher than other natural fibres like oil palm frond, coir, and banana stem fibres while arrangement of PALF is the same as in cotton (82.7%) [4]. Higher cellulosic content in PALF and its lower microfibrillar angle enhance the physical and mechanical strength of matrix reinforced biocomposites [5]. Many researchers [6-8] reported that some limitations of natural fibre reinforced thermoset/thermoplastics are natural fibre-matrix interfacial adhesion and hydrophilic nature of natural fibres. Lower fibre-matrix incompatibility effect the mechanical and physical strength of matrix reinforced composites and incapable to load transfer between fibre and matrix. There are various treatments to increase interfacial adhesion capacity and less hydrophilic nature of natural fibre to enhance to mechanical properties of biocomposites [9]. In this study, alkali, treatments have been used on PALF fibres with various concentrations and immersion of time to modify its surface for good interfacial bonding with matrix. The aim of these treatments is to optimise chemical treatment of these fibres to make affective matrix reinforced composites.

2. Materials and methods

Pineapple leaf fibres (PALF) was collected and cleaned. The chemical used for treatments are NaOH (Sodium Hydroxide 6% w/v soln, R&M), from Jasa Sejiwa Enterprise.

2.1. Chemical treatment

PALF were immersed into distilled water with different chemical concentration with various period of soaking of time. NaOH concentration of 3% and 6% were used with four different soaking times having 3h interval (3h, 6h, 9h and 12h). After treatment, the fibres were thoroughly washed with running water several times until pH values are neutralized and dried in oven at 80 C temperatures for 48 h.

2.2 Scanning Electron Microscopy (SEM)

Morphology of untreated and treated PALF with different chemical concentrations and its various soaking time were investigated with Scanning Electron Microscope (SEM) machine Model HITACHI S-3400N.

2.3 Diameter Measurement

Diameter test was performed on untreated and treated PALF by using image analyzer. Accuracy of diameter measurement of natural fibres is very difficult to calculate due to irregular shape.

2.4 Tensile test

Untreated and treated PALF with different chemicals concentrations and immersion time were evaluated by Instron universal testing machine (UTM). According to the standard test method for single fibre tensile test ASTM D 3379 with speed of 1 mm/min. Before testing, fibres were examined to remove crack, partially break fibres by image analyser.

$$T = F / A \qquad \text{Eq. (1).}$$

Where, T tensile strength in Pa, F force to failure in N, A average fibre area in m^2.

3. Result and discussion

3.1 SEM

SEM analysis observed effect the treatment of NaOH with concentration of 3% and 6% on PALF fibres. Untreated fibres carried many impurities on its surface shown in figure 1(a). PALF fibre treated with 3% concentration of NaOH with various type of period of soaking time showed different types of surface morphology. PALF fibres were soaked for 3 h in figure 1(b) revealed small impurity and smooth surface. Figure 1(c) showed rough surface with very less impurities. While figure 1(d) and 1(e) rough deeply and exposed cellulosic materials in comparison of untreated and figure 1(b) fibres. Fibres treated with 3% NaOH are still not effective enough to remove impurities from the fibre surface[10].

IOP Conf. Series: Materials Science and Engineering **290** (2017) 012030 doi:10.1088/1757-899X/290/1/012030

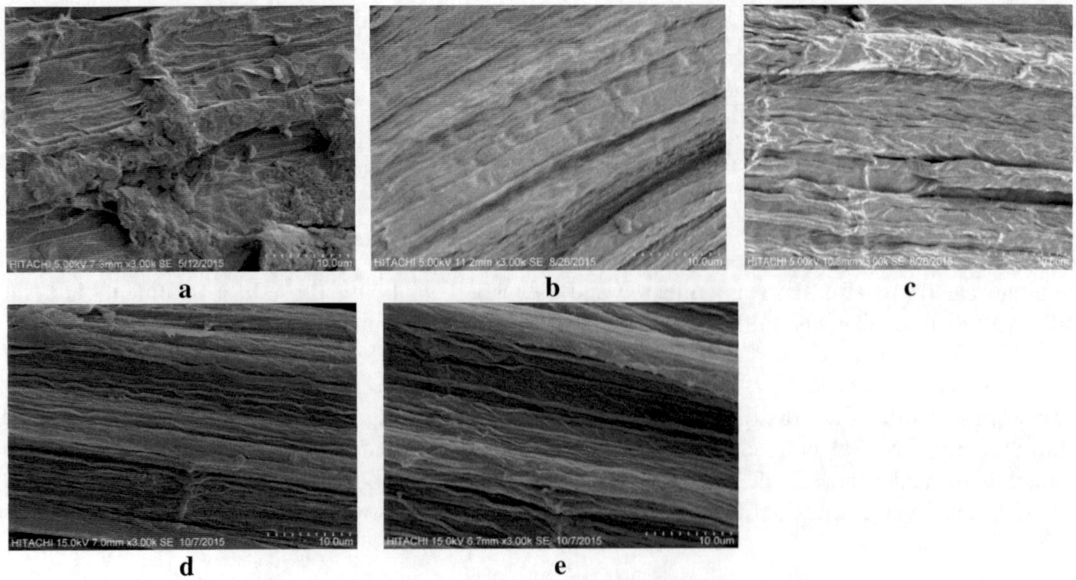

Figure 1. SEM of untreated (a) and 3% NaOH treated PALF fibre for (b) 3h, (c) 6h, (d) 9h and (d) 12h.

PALF fibre treated with 6% concentration of NaOH showed in figure 2(a) showed clean surface with some impurities on fibre surface and corrosive surface area. Soaked Fibre for 6h showed its clear surface with rough surface shown in figure 2(b). Figure 2(c) and 2(d) are very clean and clearly showed effect of chemicals on fibre surface. Higher concentration of NaOH can break hemicelluloses web of treated fibres, and separate fibres from fibre bundle [11]. After the separation of fibre, surface area of fibres increase and interaction of fibre/matrix also[12]. In figure 2 (a-d), there are many grooves like structures there because of removal of cemantic material of fibre and its impurities are removed from the surface. These exposed areas enhance the surface of fibre which comes into contact with polymers and resulting better bonding [13, 14].

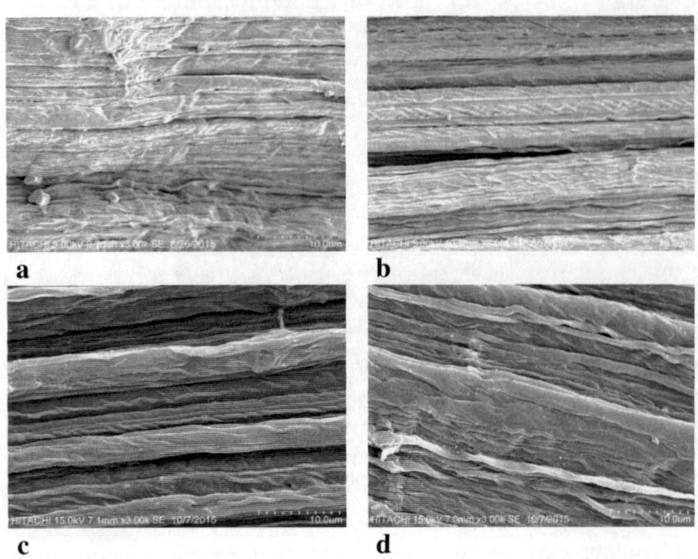

Figure 2. SEM of 6% NaOH treated PALF fibre for (a) 3h, (b) 6h, (c) 9h and (d) 12h.

Fibre treated with NaOH enhance the adhesion between fibre and polymer and mechanical properties of composites[15, 16]. The fibre treated with 6% NaOH is shown absence of impurities on the surface. While fibre treated with 9% NaOH is also clear with jagged and physically rough surface which make fibres very weak and brittle [10].

3.2 Diameter Measurement

Natural fibres are non uniform, structurally heterogeneous and may irregularities in shape [17]. It can be observed that treated fibre deduct its diameter in compare to untreated fibres. This reduction is expected because of NaOH treatment attack on amorphous surface of natural fibre and it removed impurities hemicelluloses and lignin [18].

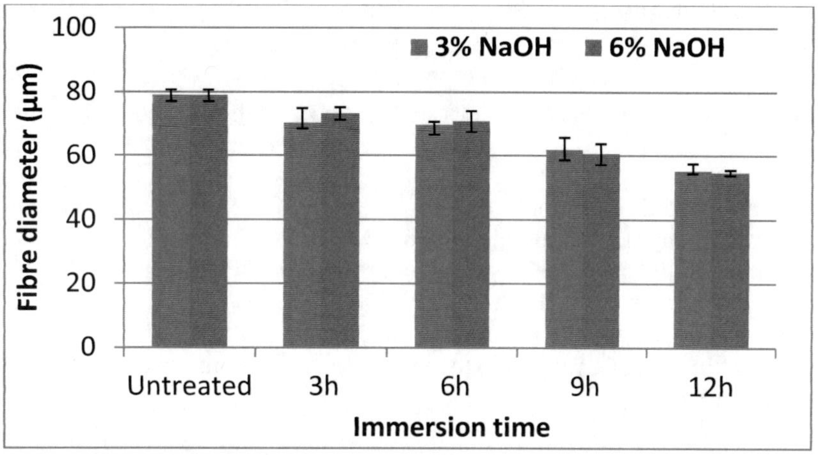

Figure 3. diameter of untreated and treated PALF fibres.

Untreated fibres showed highest diameter than other treated fibre shown in figure.3. PALF fibre treated with 3% NaOH and 6% NaOH reduced its diameter. Among all the treatment, 6% NaOH treatment with soaking time 12 h showed smallest diameter. 3%NaOH treatment also decreased its diameter but fibre treated 6% NaOH showed better result. This study is clearly showed the effect of NaOH treatment on PALF fibres. Both treatments affect the fibre but 6% NaOH treatment for 12h showed lowest diameter.

3.3. Tensile strength

Figure 4 Showed effect of untreated and 6% NaOH treated fibres for 6h showed highest result among all the treatments while 3% NaOH treated fibres for 6h also showed a very good result. The treatments for 3 h are insignificant for both 3% and 6% NaOH treatments. 6% NaOH treatment for 6h is able to remove hemicelluloses and lignin and enhance surface area for fibre/matrix adhesion [19]. 3% NaOH treated fibre did not much affected with time but 6%NaOH treated fibre showed huge fall in tensile strength. From 6h to 12h, 3% NaOH treatment did not show much fluctuation though 6% NaOH treatment fall drastically and reached on its lowest point. Higher chemicals concentration may remove impurity efficiently but may decrease tensile strength due to lignocelluloses degradation and rupture of fibre surface [20].

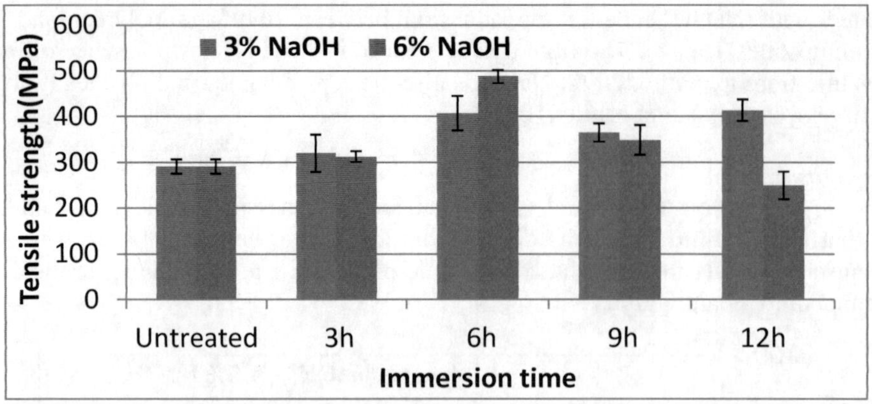

Figure 4. Tensile properties of untreated and chemically treated PALF fibres

4. Conclusion

The 3% concentration of NaOH treatment did not affect the fibre and unable to remove impurities from the surface. However, at the soaking time 6h, 6% NaOH concentration removed the impurity and made the surface rough. Both 3% and 6% NaOH concentration effect diameter of fibre and after 12 h fibre soaking, the diameter was least.. Both 3% and 6% NaOH concentration revealed higher tensile strength at 6 h soaking time. Beyond 6 h soaking time tensile strength of 6% NaOH were declined which showed negative impact of treatment. However, 3% NaOH sample was not affected very much due to lower concentration.

Acknowledgement

The authors are thankful to PUTRA GRAN Vote no 9448600 for supporting this research work.

References

[1] Cordeiro E P, Pita V J and Soares B G 2017 *J Polym Environ* 1-12
[2] Sanyang M, Sapuan S, Jawaid M, Ishak M and Sahari J 2016 *Renew. Sustainable Energy Rev.* **54** 533-549
[3] Khalil H A, Aprilia N S, Bhat A, Jawaid M, Paridah M and Rudi D 2013 *Renew. Sustainable Energy Rev.* **22** 667-685
[4] Asim M, Abdan K, Jawaid M, Nasir M, Dashtizadeh Z, Ishak M and Hoque ME 2015 *Int J Polym Sci.* **1** 1-19
[5] Lopattananon N, Panawarangkul K, Sahakaro K and Ellis B 2006 *J. Appl. Polym. Sci.***102** 1974-1984
[6] Jawaid M, Abdul Khalil H P S and Abu Bakar A 2010 *Mater. Sci. Eng.* A **527** 7944-7949.
[7] Jawaid M, Abdul Khalil H P S, Hassan A, Dungani R and Hadiyane A 2013 *Composites Part B.* **45** 619-624
[8] Arrakhiz F Z, Elachaby M, Bouhfid R, Vaudreuil S, Essassi M and Qaiss A 2012 *Mater. Des.* **35** 318-322
[9] Asim M, Jawaid M, Abdan K and Ishak M 2017 *Fiber Polym* **18** 940-947
[10] Edeerozey A M, Akil H M, Azhar A and Ariffin MZ 2007 *Mater. Lett.* **61** 2023-2025
[11] Asim M, Jawaid M, Abdan K and Ishak M R 2016 *J Bionic Eng.* **13** 426-435
[12] Asim M, Jawaid M, Abdan K and Ishak 2017 M. *J Polym Environ* 1-8
[13] Ariawan D, Mohd Ishak Z A, Salim MS, Mat Taib R and Ahmad Thirmizir M Z 2015 *Polym. Compos*
[14] Shaha S, Dyuti S, Ahsan Q, and Hasan M 2011 *Adv. Mater.sci.* **264** 1922-1927

[15] Hashim M Y, Roslan M N, Amin AM, Zaidi A A, and Ariffin S 2012 *World Acad Sci Eng Technol.* **68** 1638-1644
[16] John M J and Anandjiwala R D 2008 *Polym. Compos.* **29** 187-207
[17] Mohanty A, Misra M and Hinrichsen G 2000 *Macromol. mater. eng.* **276** 1-24
[18] Kabir M, Wang H, Lau K, and Cardona F 2013 *Composites Part B* **53** 362-368
[19] Meon M S, Othman M F, Husain H, Remeli M F, and Syawal M S M 2012 *Procedia Eng.* **41** 1587-1592
[20] Atiqah A, Jawaid M, Ishak M and Sapuan S 2017 *J Nat Fibers* 1-11

IOP Conf. Series: Materials Science and Engineering **290** (2017) 012031 doi:10.1088/1757-899X/290/1/012031

Physical properties of coir and pineapple leaf fibre reinforced polylactic acid hybrid composites

R Siakeng[1], M Jawaid[1,2], H Ariffin[3] and S M Sapuan[1]

[1]Laboratory of Biocomposite Technology, Institute of Tropical Forestry and Forest Products (INTROP), Universiti Putra Malaysia, 43400 Serdang, Selangor, Malaysia
[2]Department of Chemical Engineering, College of Engineering, King Saud University, Riyadh, Saudi Arabia
[3]Laboratory of Biopolymer and Derivatives, Institute of Tropical Forestry and Forest Products (INTROP), Universiti Putra Malaysia, 43400 UPM Serdang, Selangor, Malaysia

Email: ramengmawii8@gmail.com

Abstract. This study examined the physical behaviour of Coir fibres (CF) /Pineapple leaf fibres (PALF) /Poly lactic acid (PLA) composites. In this research, coir and PALF reinforced PLA hybrid composites were fabricated by hand lay-up process and hot press. The aim of this work is to do comparative study on density, water absorption (WA) and thickness swelling (TS) of untreated CF/PALF reinforced PLA composites and hybrid composites. The effect of different fibre ratios in hybridization on density, WA and TS of CF/PALF hybrid composites were also analyzed and C7P3 showed highest density while P30 had lowest. The results indicated that the density varies on different fibre ratio. WA and TS of CF/PALF composites and hybrid composites vary with fibres ratio and soaking duration. WA and TS of untreated CF/PALF hybrid composites were increased by increasing coir fibre ratio so, C30 showed highest WA and TS whereas P30 and C1P1 showed least WA and TS respectively apart from neat PLA.

1. Introduction

Recently, natural materials have attracted substantial interest due to the rising need for lowering the environmental load and to ensure sustainability [1, 2]. This has also led to a considerable change in the research direction of fibre reinforced polymer composites; which is a biodegradable composite composed of a biodegradable polymer and natural cellulosic materials which are expected to be environmentally friendly [3-6]. Polylactic acid (PLA), made from renewable raw materials has been considered to be a type of commercially available thermoplastic polymer and fully biodegradable [7, 8]. PLA is now one of the most promising biodegradable polymers for industrial plastic applications, as an alternative to conventional synthetic polymers due to its high mechanical properties and good processability [8, 9]. Nonetheless, it has some limitations such as brittleness, low impact strength, and low thermal resistance [8, 10]. In order to make them suitable in various applications, the mechanical properties of PLA needs to be enhanced by using reinforcements such as natural fibres [1, 11]. Natural fibres such as coconut coir, jute, hemp, kenaf, pineapple leaf fibres can be combined with a polymer matrix to produce biodegradable composites with specific properties comparable to synthetic fibre based polymer composites [12-18].

Coconut coir is a multipurpose lignocellulosic fibre extracted from the husk of coconut and is one of the stiffest among the tough natural fibres and draw large attention due to their potential to reinforce polymer and bulk density to yield cost-effective composite products [19]. Coir fibres have lower cellulose (36–43%) and hemicellulose (15.17%) contents, higher lignin content (32.25%) and microfibrillar angle (30–45°) higher than other natural fibres [20]which results in their relatively low tensile strength and good impact as well as highest elongation at break among natural fibres [2] .Coir drawn much attention due to their good thermal conductivity and biodegradability properties [13]. Likewise, PALF also has the potential for use as reinforcing filler biocomposites with gratifying physical and mechanical properties [21]. PALF is a smooth thin yet strong natural fibre with high cellulose content (70-82 %), lignin (5-12 %) and ash (1.1 %) [22].

Natural fibre reinforced polymer composites are sensitive to moisture and humidity. Water absorption affects the mechanical strength and fibre-matrix interfacial bonds leading towards poor stress transfer efficiencies from matrix to fibres [17]. Study of water absorption in natural fibre composite is necessary for various applications such as outdoor component, packaging and building industry [23]. In natural fibre reinforced composites, fibre absorb water due to many reason such as temperature, fibre loading and orientation, fibre characteristics, area of exposed surfaces, diffusivity, and surface protection [24]. Mechanism of water absorption in fibre composites is based on three ways; first method is diffusion method, water molecules flow into micro-cracks of polymer chains. Second method is capillary flow; water flow along with the interface if interfacial bonding of fibre and matrix is weak. Third is moisture content; storage of water in micro-cracks present in both polymer and natural fibre [25]. Various research [23, 26, 27] have been conducted on the water absorption and thickness swelling of different types natural fibre composites and discussed the problem associated with water absorption in composites. In this research water absorption and thickness swelling properties of Coir/PALF/PLA composites have been studied on various fibre ratios and time span. The objective of this work, fibre mixing ratio of Coir/PALF/PLA composite are optimised by water absorption and thickness swelling testing, it will be helpful for further study of mechanical, thermal and biodegradability properties of the same biocomposites.

2. Experimental procedures

2.1. Materials

Polylactic acid (PLA) pellets with a molecular weight (M_w) of approximately 74,000 g/mol, density of 1.25 g/mol, glass transition temperature (T_g) of 55 to 60°C and melting point (T_m) of 160 to 180°C was purchased from TT Biotechnologies Sdn. Bhd., Malaysia. Coir fibres (*Cocus nucifera*) were harvested in west Malaysia while pineapple leaf fibres (PALF) (*Ananas comosus*) were harvested in India and was supplied by Innovative Pultrusion Sdn Bhd., Malaysia.

2.2. Preparation of composites

Coir and PALF used as reinforcements in fabrication of composites were chopped into short fibres (1-2 mm) by using ring flakers machine; fibres were maintained at 6-8% moisture content. Coir and PALF composites were fabricated by using 15x15x3 mm stainless metal plate mould. The mould was placed into hydraulic pressure hot press at 180° temperature and was removed from the press after 15 minutes pre-heat and 5 minutes press kept for cooling in cold press for 3 minutes, and then samples were cut for testing according to ASTM standard. The formulations of hybrid composites were described in table 1.

Table 1. Density and formulation of Coir/ PALF/PLA hybrid composites.

Composite Code	PLA (Weight %)	Coir (Weight %)	PALF (Weight %)	Density (g/cm^3)
Pure PLA	100	_	_	1.2
C-30	70	30	_	1.13
P-30	70	_	30	1.02
C1P1	70	15	15	1.07
C3P7	70	9	21	1.05
C7P3	70	21	9	1.09

2.3. *Composite density determination*

Density was measured by using ASTM D 1895 standard. The density of the samples was calculated by using following equation (1):

$$\rho = m/v \tag{1}$$

where, *m* represents mass of fibre/matrix and *v* represents volume of fibre/matrix.

2.4. *Water absorption test*

Prepared samples were immersed into distilled water at room temperature. Water absorption was calculated by the equation (2) for various period of time. All the samples were weighed before and after soaking and recorded carefully. At the three intervals, water absorption was examined such as 3 day, 5 days and 7 days. After taking out samples from water, all the samples were wipe dry thoroughly to remove excess amount of water on the surfaces. The percentage equilibrium water absorption was calculated as an average value of several measurements. The percentage of water absorption was calculated from equation (2) using ASTM D570 [28].

$$\text{Water absorption (\%)} = \frac{Wn - Wd}{Wd} \times 100 \tag{2}$$

where, *Wn* is the weight of composites samples after immersion and *Wd* is the weight of the composite samples before immersion.

2.5. *Thickness swelling test*

Six samples of pure PLA and each different fibre ratio combination composites were prepared for the experiment of thickness swelling. Thickness swelling was calculated according to ASTM D570 by using Equation (3) [28]. The thickness swelling experiment was conducted by measuring the swelling of the each sample. Measurement of the composites was taken before and after the soaking. Samples were measured three times, first measurement was taken on third day of immersion second was fifth days of immersion and last was seven days of immersion.

$$\text{Thickness Swelling (\%)} = \frac{\text{T1} - \text{T0}}{\text{T0}} \times 100 \tag{3}$$

where *T1* is the thickness after soaking and *T0* is the thickness before soaking.

3. Result and discussion

3.1. Density
The experimental densities of untreated coir/PALF reinforced PLA composites and hybrid composites were calculated shown in table 1. The density of P30 composites is lowest (1.02 g/cm^3) among all composites. However C30 composites showed highest experimental value (1.13 g/cm^3). The experimental density of C7P3 is highest among the hybrid composites which is 1.09 g/cm^3 which is lower than the coir/PLA composite and higher than PALF/PLA composite. The differences among composites are mainly due to manufacturing process and fibre ratio or loading. In some case, composites have void space due to air trap inside composites [29].

3.2. Water absorption
Water absorption test was conducted to study the amount of water absorbed in a definite period of time. Generally, the natural fibre composites absorb moisture due to several reasons such as fibre fraction, voids, viscosity of matrix, humidity and temperature [29]. As shown in the **Error! Reference source not found.** 1 all untreated composites showed high WA than neat PLA sample. Among the all composites, P30 composites absorb lowest moisture contents, while C30 composites showed highest WA which showed 10% WA increment. This study indicated the presence of void content and porosity on the surface of untreated composites. Due to the presence of voids and pores on the surface of composite, the weight of composite increases by trapping the water inside the voids [29]. On the other hand, the WA for the pure neat PLA was lowest among all. In this result, it is shown that all hybrid composites showed quite similar pattern of WA rate having the pattern of increased WA with increasing coir fibres ratio.

There are many factors such as porosity, void content, lumen size, and fibre-matrix adhesion to effect water absorption behaviour of the composites besides the hydroxyl groups present in cellulose and hemicelluloses of natural fibres which provides easy access to moisture in to composites [30]. The immersed composites into the water absorb water molecules through capillary action and fills the voids and cracks in the composites [28, 31]. The higher WA rates in coir reinforced composites compared to PALF could be due to the higher porosity and hollowness of the coir fibres.

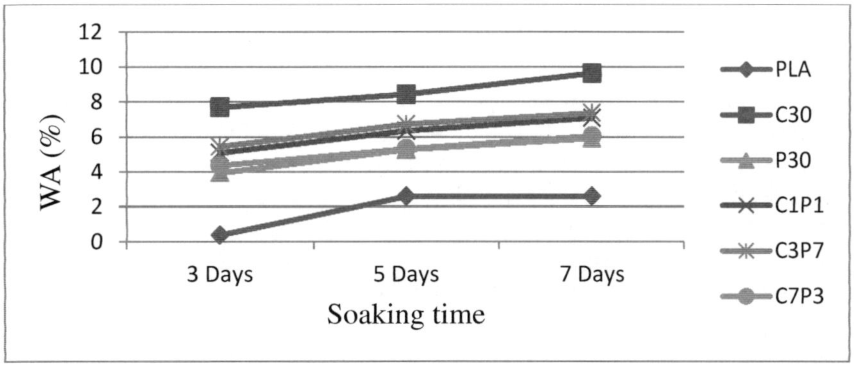

Figure 1. Water absorption of untreated coir/PALF/PLA composites.

3.3. Thickness swelling
Thickness swelling test was conducted to study the affect of water molecules on dimension stability of coir/PALF reinforced PLA composites. The TS is the exposure of the natural fibre on the surface of composite. The hydrophilic nature of lignocellulosic materials and the mechanism called capillary action caused the intake of moisture when the samples were soaked in water which leads to the swelling of fibres and thus increase the dimension of composite [29]. Figure 2 shows that neat PLA

sample showed high dimensional stability, however, composites and hybrid composites revealed higher thickness swelling. Among the all composites, C1P1 hybrid composite showed highest dimension stability and C30 composites showed highest thickness swelling approximately 8%.

Higher TS indicated the presence of void on the surface or high porosity in the composites. It is responsible for the changes in the dimension of lignocellulosic fibre-based composites, particularly in the thickness, and the linear expansion due to reversible and irreversible swelling of the composites [31]. C30 composites have high content of coir fibres which is a hollow and porous lignocellulosic fibre [19], which may be the cause of thickness swelling.

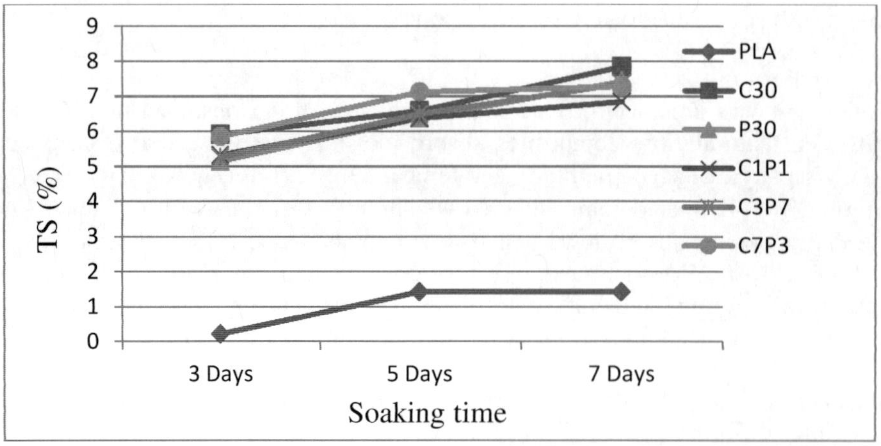

Figure 2. Thickness swelling of untreated coir/PALF/PLA composites.

4. Conclusion

In this work, density, water absorption and thickness swelling were examined. 30% coir fibre reinforced PLA composite showed highest water absorption and thickness swelling than all other tested composites and hybrid composites. Inclination of fibre composites were obvious, higher coir fibre loading showed higher moisture content and swelling in the composites. Neat PLA showed lowest water absorption which indicates presence of less void contents. Thickness swelling is directly related to the water absorption, therefore, the trends of water absorption and thickness swelling of all composites are very similar. High water absorption and thickness swelling affect the interfacial bonding between fibre and matrix. In this study we found that water absorption and thickness swelling did not much affect the physical properties of composites due to very less percentage of water absorption and thickness swelling observed. However, natural fibres are hydrophilic due to the presence of hydroxyl groups from cellulose and lignin. So, surface modification of these fibres would improve the fibre –matrix bonding and hence gives batter composites products. The results of the present study show that a useful composite with good strength could be successfully developed using coir fibres as a reinforcing agent for the polylactic acid matrix.

Acknowledgement
This work was financially supported by Putra Grant (9520200) Universiti Putra Malaysia

References

[1] Summerscales J, Dissanayake NP, Virk AS, and Hall W 2010 *Compos Part A.* **41** 1329-1335
[2] Nam TH, 2011 *Compos Part B* **42** 1648-1656
[3] Yu T, 2010 *Compos Part A* **41** 499-505.
[4] Tao Y, Yan L, and Jie R. 2009 *Transactions of Nonferrous Metals Society of China* **19** 651-655
[5] Song YS, 2012 *Compos Part B* **43** 856-860.
[6] Jawaid M, 2014 *International Journal of Polymer Analysis and Characterization* **19** 62-

69.
[7] Koronis G, Silva A, and Fontul M 2013 *Compos Part B* **44** 120-127.
[8] Pang X, 2010 *Biotechnol journal* **5** 1125-1136.
[9] Rasal RM, Janorkar AV, and Hirt DE. 2010 *Prog. polym Sc* **35** 338-356.
[10] Shukor F, 2014 *Mater. Des* **54** 425-429.
[11] Lu T, 2011 *Compos Part B* **62** 191-197.
[12] Ku H, 2011 *Compos Part B* **42** 856-873.
[13] Jang JY, 2012 *Compos Part B* **43** 2434-2438.
[14] Graupner N and Müssig J 2011 *Compos Part A* **42** 2010-2019.
[15] Sawpan MA, Pickering KL, and Fernyhough A. 2011 *Compos Part A* **42** 1189-1196.
[16] Smitthipong W, Tantatherdtam R, and Chollakup R. 2015 *Journal of Therm Compos Mater* **28** 717-729.
[17] Asim M, 2017 *Fibers and Polymers* **18** 940-947.
[18] Asim M, 2016 *Journal of Bionic Engineering* **13** 426-435.
[19] Nam TH, Ogihara S, and Kobayashi S. 2012 *Adv. Compos.Mater* **21** 103-122.
[20] Dong Y, 2014 *Compos Part A* **63** 76-84.
[21] Kaewpirom S and Worrarat C. 2014 *Fibers and Polymers* **15** 1469-1477.
[22] Asim M, 2015 *Int. J. Polym. Sc.*2015
[23] Venkateshwaran N, 2011 *Mater. Des* **32** 4017-4021.
[24] Devi LU, 2011 *Polym. Compos.* **32** 335-346.
[25] Shirangi M and Michel B. 2010 *Moisture Sensitivity of Plastic Packages of IC Devices:* Springer, 29-69.
[26] Najafi SK and Kordkheili HY 2011. *European Journal of Wood and Wood Products* **69** 553-556.
[27] Kord B and Kiaeifar A. 2010 J*ournal of Reinforced Plastics and Composites* **29** 3480-3485.
[28] Khalil HA, Jawaid M, and Bakar AA. 2011 *BioResources* **6** 1043-1052.
[29] Jawaid M, 2011 *Journal Polym*. Env **19** 106-109.
[30] Mariatti M, 2008. *Journal of composite materials* **42** 931-941.
[31] Ahmed K S and Vijayarangan S. 2007 J. Appl. Polym. Sc.**104** 2650-2662.

IOP Conf. Series: Materials Science and Engineering **290** (2017) 012032 doi:10.1088/1757-899X/290/1/012032

Effect of coupling agent on durian skin fibre nanocomposite reinforced polypropylene

M A Siti Nur E'zzati, H Anuar and A R Siti Munirah Salimah[1]

[1]Department of Manufacturing and Materials Engineering, International Islamic University Malaysia (IIUM), Jalan Gombak, 53100 Kuala Lumpur, Malaysia

Email:snezzatiapandi@gmail.com

Abstract. This paper reports on the development of a composite-based natural fiber to reduce the reliance on petroleum-based product in order to amplify environmental awareness. The production of Durian Skin Nanofiber (DSNF) was conducted using biological fermentation method via rhizopus oryzae in order to obtain the nano dimension of the particle size. Polypropylene (PP) and DSNF were produced using Haake internal mixer via melt blending technique. The significant effect of maleic anhydride grafted polypropylene (MAPP) on the properties of PP/DSNF nanocomposite was investigated to study its mechanical properties which are tensile strength and thermal stability using thermogravimetric (TGA) and differential scanning analysis (DSC). The tensile property of PP nanocomposites increased from 33 MPa to 38 MPa with the presence of MAPP. The addition of MAPP also increased the thermal stability of PP/DSNF nanocomposite where the char residue increased by 52%. Besides that, the thermal degradation of PP/DSNF and PP/DSNF-MAPP were higher than PP where they exerted higher amount of weight loss at an elevated temperature. The percentage of crystallinity, %Xc, of PP nanocomposites improved with the addition of MAPP by 35% based on the differential scanning calorimetry (DSC) result. The SEM analysis showed that the PP/DSNF-MAPP exerts ductile fracture while PP/DSNF exerts brittle fracture.

1. Introduction

A lot of research has been done to investigate the effect of using natural fibers such as jute and kenaf in producing environmental-friendly composites such that there will be improvement on the degradability. Thus, efforts have been made to replace these types of materials with those that are more environmentally friendly. The prospect of using renewable resources gives a huge impact towards the global sustainability of composite materials and natural fiber development [1]. The interest of reinforcing polymer composites with natural fiber offers lower environmental effect since they give similar attributes as artificial fibers such as carbon and glass [2]. The application of natural fiber composites is limited because of its lower mechanical properties but it can still be used for household products such as plastic bottles, plates, cups, and food containers. As an effort to manufacture eco-friendly products, studies on strengthening thermoplastic and thermoset materials with natural fibers such as coir, bamboo, sisal, and pineapple leaves have been done [3].

One of the most popular seasonal fruits in South East Asia is Durian. Durian is a climacteric fruit belonging to the Bombacaceae family [4]. Durian is considered the king of fruits by the locals from South East Asia countries. Durio zibethinus is a type of durian that is commonly used for

IOP Conf. Series: Materials Science and Engineering **290** (2017) 012032 doi:10.1088/1757-899X/290/1/012032

commercial and economic importance. The skin of a durian contains 15.45% lignin, 13.09% hemicelluloses, and 60.45% celluloses [5],[6]. It was discovered that we can make fiber from durian peel and this has the potential to encourage new studies in natural fiber around the world. The target is to value-add the durian skin that so in the next few years, the durian trashes will be decreased and together it will reduce accumulated trashes in the country.

In this paper, the usage of agricultural waste as reinforcement for polymeric nanocomposites is suitable by combining petroleum-derived polymer matrix with natural fiber waste according to their potential performance in the mechanical properties [7]. Since durian can be found in Malaysia, durian waste can be developed as reinforcement for polymer biocomposite. This will also aid in reducing the durian waste and improve the degradability of polymer biocomposites [8]. The addition of maleic anhydride grafted polypropylene (MAPP) coupling agent can be used to modify the compatibility of nanocomposite [9],[10]. Thus, this study is important to analyze the effect of DSNF on the mechanical and thermal properties of polypropylene nanocomposites. Comparison is also made between PP/DSNF nanocomposite with and without the presence of the coupling agent.

2. Experimental Procedure

Durian skin waste was obtained in Kuala Lumpur, Malaysia. Polypropylene grade 6331 was supplied by Titan Pro Polymers (M) Sdn. Bhd. and maleic anhydride polypropylene (MAPP) was provided by Sigma-Aldrich.

The preparation of durian skin nanofiber (DSNF) started with washing, chopping, grinding, and sieving processes until DSF in the size of over 250 μm was attained. The production of DSNF was conducted using biological fermentation method via Rhizopus Oryzae in order to obtain the nano dimension of the particle size. The Rhizopus oryzae was obtained from expired bread. From there, it was transferred to a potato dextrose agar (PDA) medium. Rhizopus oryzae was allowed to grow within the PDA medium on a petri dish at 37 °C inside the oven. The first step of the fermentation was done by adding 50 ml of PDB solution and 50 ml of distilled water into the flask. The PDB medium contained 20 g per liter of dextrose and 4 g per liter of potato starch. Next, 2 g of DSF was added into the flask. After that, the grown *Rhizopus oryzae* was added into the flask and covered with an aluminium foil. Then, the flask was shaken thoroughly using an incubator shaker at the speed of 250 rpm for 7 days [11]. After 7 days, the fermented fiber was filtered, washed, and then dehydrated in the oven for a day at 60 °C.

The PP/ DSNF and PP/ DSNF-MAPP nanocomposite material was done by mixing the sample at 180 °C using Internal Mixer model Haake Polylab System Thermo Scientific. The rotor speed was 50 rpm and mixed for 12 min.

The tensile tests were done to obtain the tensile properties of PP/DSNF and PP/DSNF-MAPP biocomposite. The test was done by following the ASTM D638 standard by using LLOYD universal testing machine where the load was 10 kN and the test speed used was 50 mm/min.

Morphological observation of the tensile fracture specimen was conducted under scanning electron microscope (SEM) Jeol Jsm-5600.

Thermal gravimetry analysis (TGA) was done to analyze the thermal degradation of PP/DSNF and PP/DSNF-MAPP nanocomposite by following the ASTM D3850 standard. TA instruments Q500 were used in this analysis where the samples were heated on a platinum pan under inert atmosphere of nitrogen gas. The samples were heated from 28 °C to 800 °C at a rate of 20 °C /min. Differential scanning calorimetry (DSC) test was done to determine the thermal properties of PP/DSNF and PP/DSNF-MAPP biocomposite such as the crystallization enthalpy and degree of crystallinity. The machine used was Perkin Elmer Pyris Diamond. The samples were heated at a rate of 20 °C/min with the temperature ranging between 40 °C to 300 °C. The percentage of crystallinity, $\%X_c$, was calculated by obtaining the crystallization enthalpy of composite ΔH and the crystallization enthalpy of fully crystalline polypropylene ΔH_m. The value for melting enthalpy of 100% crystalline PP was fixed at 207.1 J/g [12].

$$\%Xc = (\Delta H \div \Delta Hm) \times 100\% \text{..............................Eq. 1}$$

3. Results and Discussion

The tensile strength of neat PP and PP/DSNF-MAPP 2.5% is shown in figure 1. The graph contains the comparison between PP/DSNF without MAPP and PP/DSNF with MAPP. Before the addition of MAPP, neat PP had the highest tensile strength of 32.4 MPa as compared to PP reinforced with 1 wt%, 3 wt%, and 5 wt% DSNF. The tensile strength of PP/DSNF decreased with the increasing amount of fiber loading. This was due to poor interaction of interfacial bonding between the PP matrix and DSNF. This poor interaction is mainly because of the hydrophobicity of PP and hydrophilicity of DSNF. PP has an even electrons distribution which makes it non-polar, thus it cannot attract water [13]. DSNF is a form of cellulose which means that it contains a long chain of hydrocarbon that consists of glucose, thus it has polarity and absorbs water [11]. The effect of MAPP towards the tensile strength of PP reinforced with 1 wt%, 3 wt%, and 5 wt% DSNF were also analysed in figure 1. The addition of 2.5 wt% MAPP increased the tensile strength of PP/DSNF composites with 1 wt% DSNF to 5 wt% DSNF.

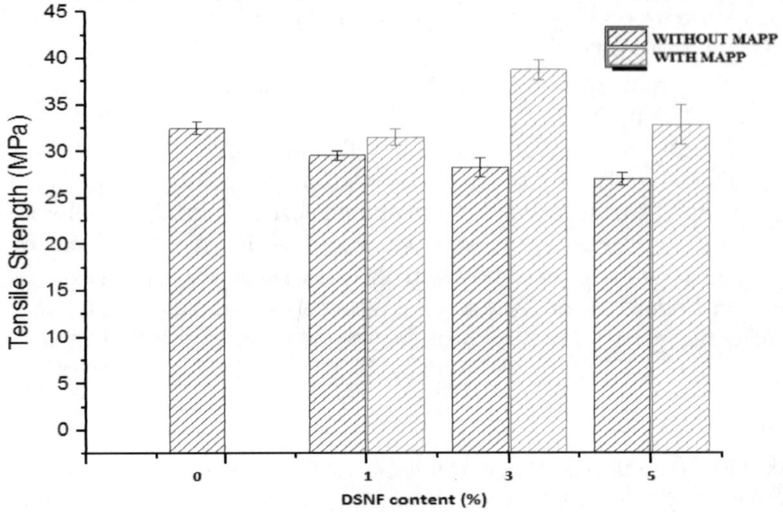

Figure 1. Effect of DSNF content on PP nanocomposites.

The addition of MAPP improved the interfacial bonding between PP and DSNF which was confirmed by SEM (figure 2). The observation under SEM in figure 2(a) and 2(b) reveals the fracture surface of PP and PP nanocomposite without the added MAPP in which it was ductile and had a few spots of brittle fracture surface. The lack of wettability adhesion in composites caused poor interlocking between the fibre and the matrix [14]. Meanwhile, figure 2(c) shows the tensile fracture surface for PP/DSNF-MAPP. The fracture surface mainly consisted of ductile fracture and proves that the addition of MAPP coupling agents aids to improve the tensile strength of the PP nanocomposites. This also suggests that the interfacial adhesion between polymer and fiber reinforcement improves with the addition of MAPP coupling agent [15].

Figure 2. Tensile fracture surface for (a) neat PP, (b) PP/DSNF, (c) PP/DSNF-MAPP.

The tensile modulus of PP/DSNF without MAPP and PP/DSNF with MAPP are shown in figure 3. The tensile modulus of PP/DSNF without MAPP decreased as the fiber loading increased. This result is similar to the tensile strength shown in figure 1. It also shows that at a higher fiber loading, the bonding between fiber and PP became poorer [13]. On the other hand, the addition of 2.5 wt% MAPP towards the PP/DSNF 1% nanocomposite resulted in a slight drop of tensile modulus by 4.4%. In contrast, the tensile modulus of PP/DSNF 3% and PP/DSNF 5% increased after the addition of maleic anhydride coupling agent. This proves that MAPP enhances the interfacial adhesion of filler-matrix. The tensile modulus of PP/MAPP-DSNF 5% was lower than that of PP/MAPP-DSNF 3%. This is because 2.5 wt% MAPP was not sufficient to fully improve the matrix-reinforcement bonding which caused some of the fiber to experience entanglement [15].

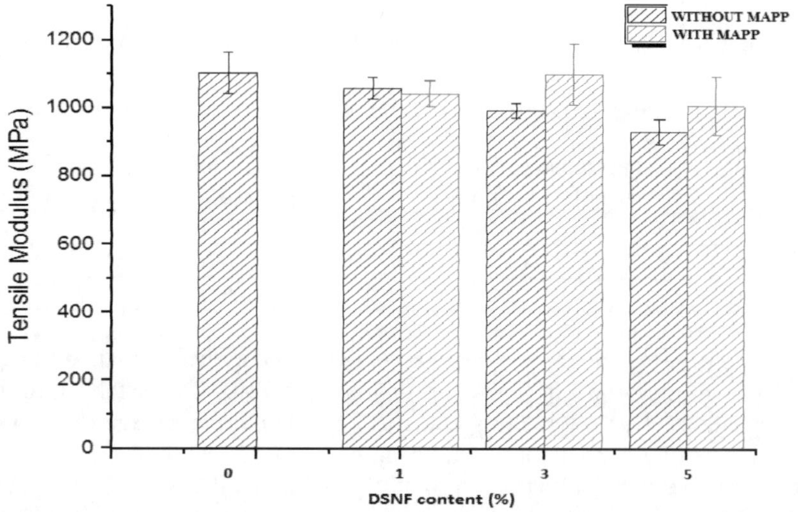

Figure 3. Effect of DSNF content on PP nanocomposites.

The TGA curve shown in figure 4 describes the weight loss of the PP, PP/DSNF, and PP/DSNF-MAPP nanocomposites with the respective temperatures. Table 1 shows the thermal properties of PP nanocomposite obtained from the TGA. The presence of MAPP in the PP/DSNF had a slight effect in terms of the thermal stability. This is because the sample underwent a weight loss of 15% where the weight of the sample reduced with the presence of DSNF and MAPP. Besides that, the degradation temperature increased with the addition of DSNF. This was due to the addition of DSNF that contained hydroxyl group (OH) bonds which allowed it to degrade faster [16]. However, the addition of MAPP coupling agent improved the degradation temperature because at 990 °C, the composite still had not degraded fully to 0% weight loss. The same trend was observed in the study by Zampaloni et al. [17], whereby the addition of MAPP improved the degradation temperature of kenaf polypropylene composites.

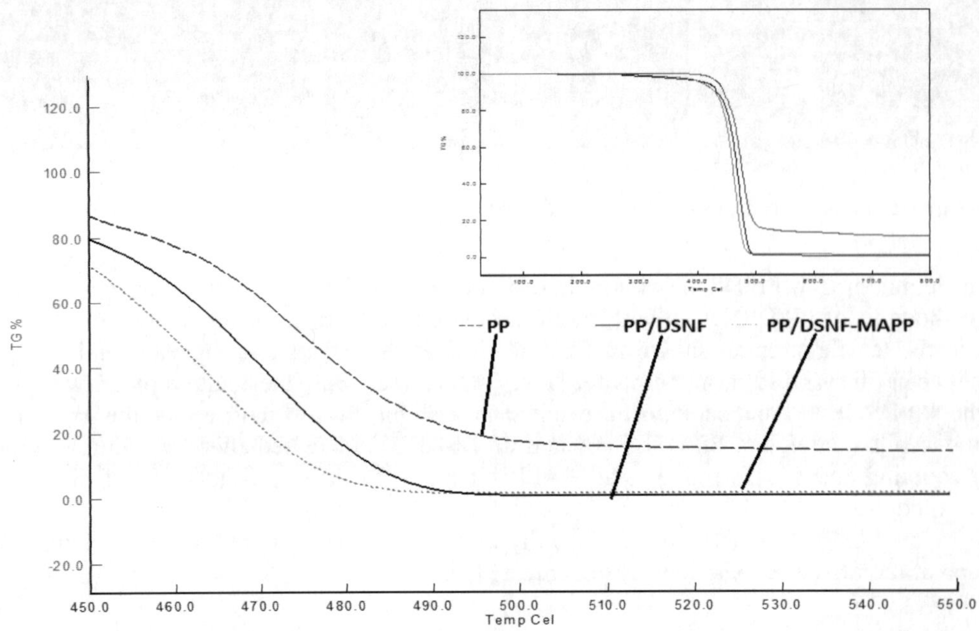

Figure 4. TG curves for PP, PP/DSNF and PP/DSNF-MAPP biocomposites.

Table 1: TG data of PP/DSNF biocomposites.

Sample	Temperature at 15% weight loss (°C)	Final temperature (°C)	Char residue at 600°C, (%)
PP	521	650	13.3
PP/DSNF	480	580	0.6
PP/DSNF-MAPP	472	990	1.25

The melting behaviour of the PP composites was analysed using differential scanning calorimetry (DSC). The DSC data analysis shown in figure 5 is tabulated in Table 2. The table describes the onset melting, melting temperature T_m (°C), melting enthalpy (ΔH_f), and percentage of crystallinity ($\%X_c$) of the composites. The melting temperature of all 2 samples had no significant change. The neat PP's melting temperature was the highest compared to PP/DSNF and PP/DSNF-MAPP. The addition of reinforcement reduced the crystallinity of the composite. According to Lee et al. [18], the addition of

MAPP into polypropylene reinforced with wood flour fiber increases the crystallinity and tensile strength of the composites.

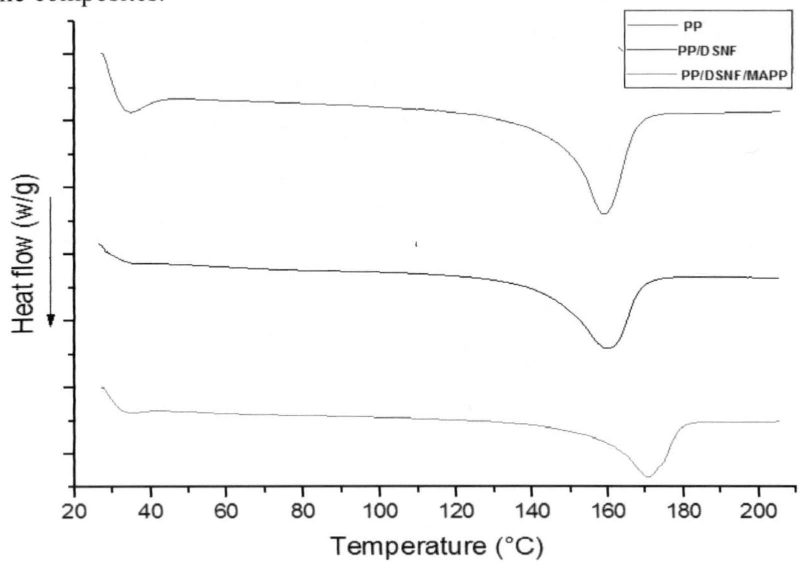

Figure 5. Differential Scanning Calorimetry (DSC) analysis of PP/DSNF biocomposites.

Table 2. Data obtained from DSC analysis of PP/DSNF Nanocomposites.

Sample	T_m (°C)	Δ H_f (J/g)	Crystallinity (%X_c)
PP	160.1	9.86	4.7
PP/DSNF	162.4	8.23	3.9
PP/DSNF-MAPP	173.6	12.37	6.0

The percentage of crystallinity shows that PP/DSNF-MAPP had the highest value compared to PP and PP/DSNF nanocomposite. The addition of MAPP increased the chain scission and reduced the number of crosslinking chain which gave better higher crystallinity and better mechanical strength [19].

4. Conclusion

From the analysis conducted, it can be suggested that the use of durian skin nanofiber (DSNF) as reinforcement improves the mechanical and thermal properties of PP nanocomposite with the addition of MAPP. The tensile strength and tensile modulus improve with the addition of a MAPP coupling agent. Thus, it can be defined concluded that the interfacial adhesion between the matrix and reinforcement improved due to the introduction of the MAPP coupling agent. The crystallinity of PP/DSNF nanocomposites also increased after the addition of the MAPP coupling agent based on the DSC analysis. Research on the suitability of DSNF as a reinforcement is worth exercising to produce polymer composites using waste material for a greener environment.

Acknowledgement

This work was financially supported by the Fundamental Research Grant Scheme (FRGS14-108-0349) from the Ministry of Education Malaysia. The authors would like to thank International Islamic University Malaysia for the research facilities in making this study a success.

References

[1] Cheung H Y, Ho M P, Lau K T, Francisco C, David Hui 2009 *Composites: Part B* **40** 655–663
[2] Anuar H and Ahmad Z 2011 *Composites: Part B* **42** 462-465
[3] Prasad P and Kochhar A 2014 *Journal of Environmental Science, Toxicology and Food Technology* **5** 1-7
[4] Manshor M R, Anuar H, Nur Aimi M N, Ahmad Fitrie M I, Wan Nazri W B, Sapuan S M, El-Shekeil Y A, Wahit M U 2014 *Material and Design* **59** 279-286
[5] Khedari J N, Nankongnab N, Hirunlabh J and Teekasap S 2004 *Building and Environment* **39** 59–65
[6] Nur Aimi M N, Anuar H, Manshor M R,Wan Nazri W B and Sapuan S M 2014 *Industrial Crops and Products* **54** 291-295
[7] Fuqua M A and Ulven C A 2008 *The Canadian Society for Bioengineering* 084364
[8] Charoenvai S 2014 *Composite Science and Technology* 20-28
[9] Deng H, Reynolds C T, Cabrera N O, Barkoula N M, Alcock B and Peijs T 2010 *Journal Applied Polymer Science* **118** 30-41
[10] Bajwa D, Bajwa S and Holt G 2015 *Polymer Degradation Stability* **120** 212-219
[11] Nur Aimi M N, Anuar H, Mairzirwan M, Sapuan S M, Wahit M U and Zakaria S 2015 *Sains Malaysiana* **44** 1551–1559
[12] Dessouky H M, Mahmoudi M R, Lawrence C A, Yassien K M, Sokkar T Z N and Hamza A A 2009 *Polymer Engineering Science* **49** 2116–2124
[13] Smith E R, Howlin B J and Hamerton I 2013 *Journal of Materials Chemistry* A 12971-12980
[14] Joseph P V, Rabello M S, Mattoso L H S, Joseph K and Thomas S 2002 *Composites Science Technology* **62** 1357-1372
[15] Chun KS, Husseinsyah S, Osman H 2013 *Advanced Materials Research* **747** 645-648
[16] Patpen P, Abdul Rahman Rusly, A. Talib R and Abdan K 2015 *International Journal on Advanced Science Engineering Information Technology* **5** 343-349
[17] Zampaloni M, Pourboghrat F, Yankovich S, Rodgers B, Moore J and Drzal L 2007 *Composites Part A Applied Science and Manufacturing* **38** 1569-1580
[18] Lee K Y, Blaker J J and Bismarck A 2009 *Composites Science and Technology* **69** 2724-2733
[19] Tugiman N, Mohamad Z and Wan Abdul Rahman W A 2010 *Universiti Teknologi Malaysia* 1-4

Analysis and modeling of delamination factor in drilling of woven kenaf fiber reinforced epoxy using Box Behnken experimental design

M Suhaily[1], C H Che Hassan[2], A G Jaharah[2], M A Afifah[1], M K Nor Khairusshima[1]

[1]Department of Manufacturing and Materials Engineering, International Islamic University Malaysia (IIUM), Jalan Gombak, 53100 Kuala Lumpur, Malaysia
[2] Department of Mechanical and Material, Faculty of Engineering and Architecture, UniversitiKebangsaan Malaysia, 43600 UKM-Bangi, Selangor, Malaysia

Email:suhailymokhtar@iium.edu.my

Abstract. In this research study, it presents a comprehensive mathematical model for correlating the influences of drilling parameters on the delamination factor during the drilling of woven kenaf fiber reinforced epoxy composite laminates using the Box Behnken experimental design. The purpose of this study is to investigate the influence of drilling parameters such as cutting speed, feed rate and drill sizes on the delamination produced when drilling woven kenaf reinforced epoxy composite using the non-coated HSS drill bits. The damage generated on the woven kenaf reinforced epoxy composite laminates were observed both at the entrance and exit surface during the drilling operation. The experiments were conducted according to the Box Behnken experimental designs.

1. Introduction

The composite are advanced engineering materials that provide requisite properties that unattainable by the conventional materials through proper selection and combination of appropriate reinforcement and matrix materials [1]. The composite is defined as a combination of two or more constituents which are polymer, metal or ceramic (known as matrix) with a reinforcing agent such as glass, carbon, aramid, or other reinforcing material. Globally, the use of natural fibres as the primary reinforcement material in composite materials is gaining popularity over the past years particularly in the market of automotive, aircraft and constructions in replacing of the fibreglass in reinforced composite plastics [2]. The need for incorporation of natural fibre based materials into automotive components has become one of the interests among the research community and the automotive industry as to produce compostable based composites parts that can be safely disposed of after its intended use without polluting the atmosphere [3]. In simple definition, natural fibers are some kind of fibers which are not produce synthetically or human made fibers and can be sourced from plants or animals [4]. Examples of natural fibers that have been the interest of researchers are such as flax, cotton, hemp, jute, sisal, kenaf, pineapple leaf fiber (PLAF), ramie, bamboo, banana leaf fiber and others which are often used as a source of lignocellulosic fibers and applied as the reinforcement materials for developing thermoplastic or thermoset composites[5].

Conventional drilling operation is an essential process in manufacturing industries despite of the existence of advanced manufacturing process in producing a near-net shape parts. It is still necessary to facilitate drilling machining of producing holes for joining purposes (rivets, bolts, fasteners) and assembly of structural component. With the increasing need of the natural fiber reinforced composite usage in several applications, machining of these materials is facing various problems come up owing to deficient data and limited amount of study in cutting and machining of natural fiber-reinforced plastics composite materials. As for today, most of the researchers were unable to attain satisfactory machining information (technique, cutting tools, and machining parameters) for machining this difficult-to-cut material, natural fibre reinforced composite. During composite materials machining, several challenges arises and impedes the performance of the composite parts in the application field. Delamination of the produced composites part was one of the main reasons for failure produced parts particularly in the aviation industry [6] and is often regards as the limiting factor in the use of structural applications. According to Chen [7], delamination in composite materials occurs mainly owing to the localized bending in the zone which is situated at the point of attack of the drill bit. It drastically reduces the assembly tolerances and strength against fatigue which later reducing and degrading the long term performance of the composite parts. Babu et al. [8] stated that the natural fiber reinforced composites materials are recognized as better materials for structural applications, however, machining of these composites may causes a number of problems. Several damages for instance delamination, fiber pull-out, edge chipping, uncut fibers and others are some of the sources occurs in drilling of composite materials [9]. According to Gay et al. [10], in all mechanical components, damaged drilled holes induces weakening of the fracture resistance in comparison with the region without holes by a factor of 40 to 60% in tension and 15% in compression. It then affects poor assembly tolerances, reduces the structural integrity of material and deterioration of the potential machined parts for the long term performances [11]. Delamination is often reckoned as a resin or matrix dominated failure behaviour that passes in the inter-ply rigging. The delamination may occur due to insufficient funding and ease during the machining process in which the composites are more flexible than metals, hence are more potent to bend and deform under loading. The presences of such damages have a highly damaging effect on damage tolerance, survivability and reliability of the fabricated parts.

Thus, this research work aims to look into the operation of drilling operation in the machining of woven kenaf/epoxy composite laminates in order to create a high quality holes with minimal damage on the machined part's surface. Several machining parameters (cutting speeds, feeds, diameter of drill bits, material of drill bit) are investigated in order to place the most suitable machining parameters for the drilling of woven kenaf/epoxy composite laminates material.

2. Experimental procedures

The woven kenaf fiber reinforced composite laminates were produced by hand lay-up process with epoxy as the resin polymer. A backing plate was placed underneath the specimen of woven kenaf fiber reinforced composite laminates. The drilling operation was performed on 3 mm thick specimen of composite laminates using SPINNER VC 450 machining center. Standard two flute straight shanks of non-coated high speed steel (HSS) drill bits of 6, 9 and 12 mm in diameter with 118°point angle were employed. During the drilling woven kenaf fibre reinforced epoxy composites, burrs or delaminations were created at the top and bottom side of the investigated work materials. The induced entry and exit surface delamination were measured using the Carl Zeiss Stemi 2000 stereo microscope to determine the damage quality of the drilled surface by placing the investigated sample underneath of the microscope. Once the image is captured, measurement is carried out by using the tool measurements function provided by the software. Circle was drawn using the draw tool available in the Carl Zeiss Stemi 2000 stereo microscope software for both maximum diameter and the nominal diameter. From the values of the Dmax and D delamination factor (shown in figure 3.9) was calculated using the below equation. The ratio of maximum delaminated area and the nominal area of the drill size is taken as an index for comparing the delamination extent produced by the drills.

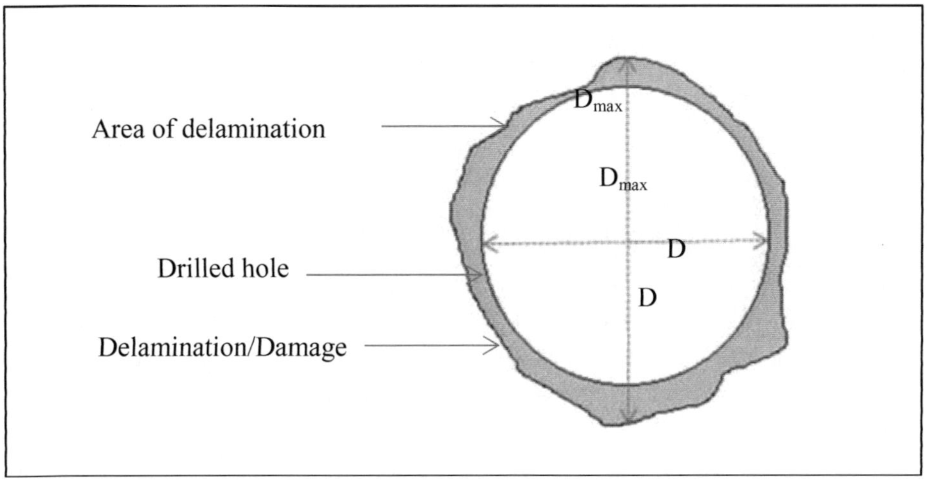

Figure 1. Scheme of delamination factor

$$F_d = \frac{D_{max}}{D} \quad (1)$$

Table 1. Cutting tools and geometric details for non-coated HSS bits.

	Non-coated HSS
Drill type	Standard two flutes
Drill diameter (mm)	6, 9 and 12
Point angle (°)	118
Helix angle (°)	30

Table 2. The details of the experimental settings in drilling of woven kenaf fiber reinforced composite laminates.

Machine tool/ Equipment	*Spinner VC 450 CNC*
Workpiece material	Woven kenaf/epoxy composites
Drilling conditions: Dry	Cutting speed (m/min): 20 ~ 70 Feed (mm/rev): 0.1 ~ 0.3

The woven kenaf fiber reinforced composite laminates used in this study is a woven type with a ply orientation of 0° and 90° was manufactured through a hand lay-up process. The woven sample is 3 mm in thickness. The details of the specimen of the work material fabrication are as listed in the Table 3, Table 4 and Table 5.

Table 3. Details of the fabrication method adopted for manufacturing of woven kenaf iber reinforced composite laminates.

Items	*Descriptions*
Fabrication method	Hand lay-up method
Fiber material	Kenaf fibers
Resin used	Epoxy resin

IOP Conf. Series: Materials Science and Engineering **290** (2017) 012033 doi:10.1088/1757-899X/290/1/012033

Laminate thickness (mm)	3
Weight Volume fraction (%)	Fiber (60%) Matrix (40%)
Types of fibers	Woven mat type
Fiber orientation	0 and 90° woven

Table 4. Mechanical properties of kenaf fibre.

Kenaf properties	*Unit*
Diameter of fibre (microm)	55.27 (avg)
Density (g/cm3)	1.222
Tensile modulus (GPa)	51.98
Tensile strength (MPa)	504.78
Specific modulus $(m/s)^2$	42.5×10^6
Specific strength (kNm/kg)	413.1
% elongation at break (%)	9.8

Table 5. Mechanical properties of epoxy resin.

Epoxy properties	*unit*
Viscosity at 20 degree (mPa.s)	1200
Density (g/cm3)	1.13
Tensile modulus (GPa)	3.60
Tensile strength (MPa)	67
Specific modulus $(m/s)^2$	3.18
Specific strength (kNm/kg)	59.3
% elongation at break (%)	6.0

3. Results and Discussion

3.1 Results of Delamination Factor Using Non-coated HSS Drill Bits

The obtained results from the conducted experiments are presented in Table 6.

Table 6 Box Behnken design for the experimental runs using non-coated HSS drills.

Run	**Cutting speed (m/min)**	**Feed (mm/rev)**	**Drill bit size (mm)**	**Upper delamination, F_{dU}**	**Bottom delamination, F_{dB}**
1	20	0.2	6	1.148	1.117
2	70	0.2	6	1.133	1.111
3	45	0.1	6	1.145	1.087
4	45	0.3	6	1.157	1.123
5	70	0.3	9	1.102	1.050
6	20	0.3	9	1.098	1.072
7	70	0.1	9	1.074	1.028
8	20	0.1	9	1.101	1.043
9	45	0.2	9	1.103	1.031

10	45	0.2	9	1.101	1.028
11	45	0.2	9	1.100	1.024
12	45	0.2	9	1.102	1.032
13	45	0.2	9	1.099	1.020
14	20	0.2	12	1.089	1.032
15	70	0.2	12	1.083	1.018
16	45	0.1	12	1.087	1.014
17	45	0.3	12	1.101	1.036

Runs of number 10, 11, 12 and 13 are repeats of Run 9

3.2 Analysis of Variances for Upper Delamination Factor Using Non-coated HSS Drill Bits

Table 7 shows the ANOVA model for the upper delamination factor using the non-coated HSS drill. The F-value of 612.87 for the model indicates that the model is significant with a probability of $F < 0.0001$. There is only a 0.01% chance that the F model is wrong due to noise. The value of Prob > F was less than 0.05 indicates the significance of the model terms. Owing to the ANOVA data, most of the factors have almost similar P value (less than 0.05).

Table 7. ANOVA result for upper delamination factor using the non-coated HSS drill bit of the RSM model.

Source	Sum of Squares	DF	Mean Square	F value	Prob > F	
Model	$9.038E^{-03}$	9	$1.291E^{-03}$	612.87	< 0.0001	significant
A	$2.420E^{-04}$	1	$2.420E^{-04}$	114.87	< 0.0001	
B	$3.251E^{-04}$	1	$3.251E^{-04}$	154.33	< 0.0001	
C	$6.216E^{-03}$	1	$6.216E^{-03}$	2950.61	< 0.0001	
A^2	$2.837E^{-04}$	1	$2.837E^{-04}$	134.67	< 0.0001	
C^2	$1.784E^{-03}$	1	$1.784E^{-03}$	846.58	< 0.0001	
AB	$2.402E^{-04}$	1	$2.402E^{-04}$	114.04	< 0.0001	
AC	$2.025E^{-05}$	1	$2.025E^{-05}$	9.61	0.00127	
Residual	$1.896E^{-05}$	8	$2.107E^{-06}$			
Lack of Fit	$8.961E^{-06}$	3	$1.792E^{-05}$	0.72	0.6440	not significant
Pure Error	$1.000E^{-05}$	4	$2.500E^{-06}$			
Cor Total	$9.057E^{-03}$	16				

R-Squared = 0.9979
R-Squared (adj) = 0.9963
Pred R-Squared = 0.9922
Adeq Precision = 82.983

3.3 Mathematical Model Development of Upper Delamination Factor Using Non-coated HSS Drill Bits

According to the experimental results, the Design Expert Version 6.0.10, to estimate the upper delamination factor according to the experimental significant factors, generated a quadratic mathematical model equation. The recommended transformation is normal as proposed by Box-Cox plot.

$$F_{dU} = 1.38066 + 7.04211E^{-04}V_c - 0.075750f - 0.051747d - 1.31158E^{-05}V_c^2 + 2.28363E^{-03} + 3.10E^{-03}V_cf + 3.0E^{-05}V_cd \quad (2)$$

Figure 2 shows the normal probability plot of the studentised residuals which is used to check for the normality of the residuals. In this way, the residuals can be checked to determine how well the model satisfies the assumptions of ANOVA. It is used to measure the standard deviation separating of the actual and predicted values. Referring to figure 2, the plot reveals that the points on the plot lie reasonably close to the straight line, which indicates that the errors are distributed normally. The straight line also indicates that there is no response transformation was required and that there was no apparent problem with normality.

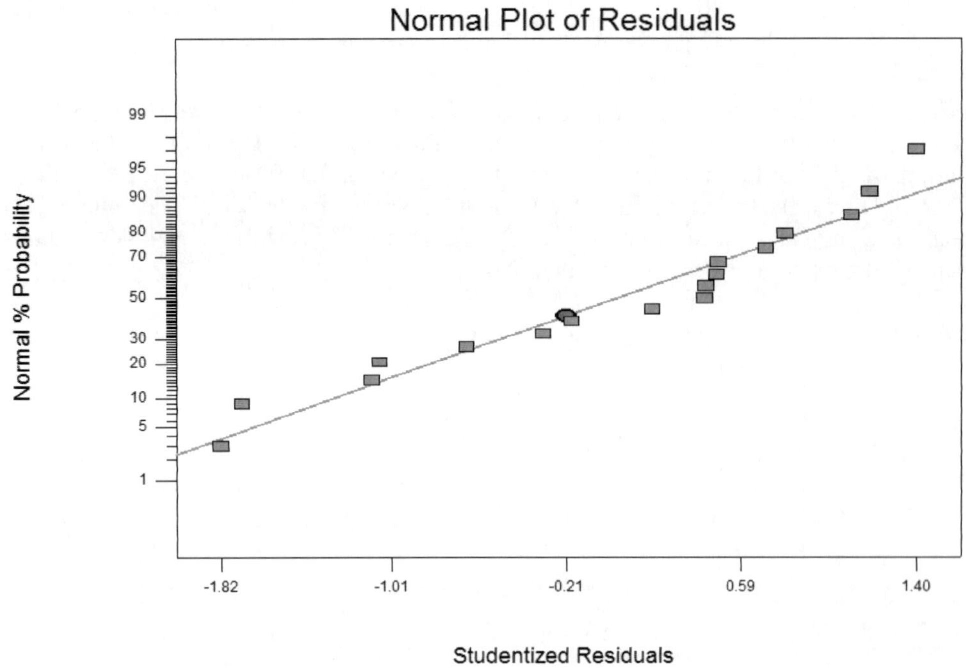

Figure 2. Normal probabilities of residuals for upper delamination factor using non-coated HSS drill.

3.4Effects of process parameters on upper delamination factor

In figure 3, the three-dimensional graph exhibits the interaction effects of cutting speed and feed rate on upper delamination factor with drill bit of 9 mm is held at constant. From the figure, it is observed that the upper delamination factor of woven kenaf fibre reinforced polymer composite increases with increase in feed rate and decreases with increase in cutting speed. This is in agreement with the finding observed by Ramesh et al. [9] in drilling of hybrid glass-sisal-jute reinforced composites. According to the conducted research work, the results indicated that the maximum drilling induced damage (delamination) was observed at higher feed rates. Vinayagamoorthy and Rajeswari [12] stated that at higher feed rates, the tool tend to penetrates the work material more quickly than during at the low feed rates. The faster penetration of drill bit during the drilling process are prompts to an expansion in the thrust force and results in higher delamination rates [13].

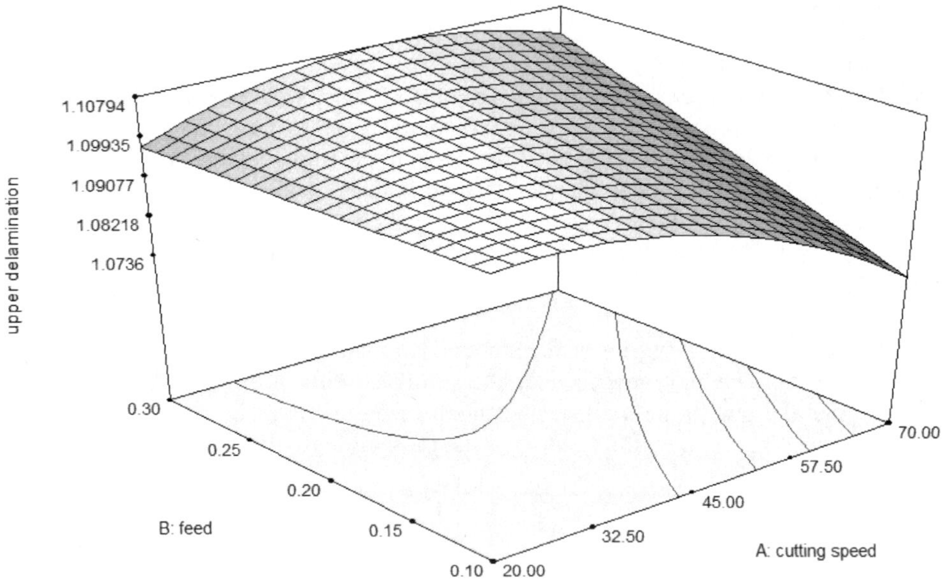

Figure 3. 3D response surface plot interaction effects of cutting speed and feed rate on upper delamination factor for non-coated HSS drill bit.

3.5Analysis of variances of bottom delamination factor for non-coated HSS drill bits

In this part, the backward elimination procedure was selected to automatically eliminate the non-significant model terms that resulted in the ANOVA table for the response surface reduced quadratic model [14]. The resulting ANOVA table for the reduced quadratic model for bottom delamination factor is shown in Table 8. The F-value of 135.50 for the model indicates that the model is significant with a probability of $F < 0.0001$. There is only a 0.01% chance that the F model is wrong due to noise. The value of Prob > F was less than 0.05 indicates the significance of the model terms. From the ANOVA data, most of the factors have almost similar P value (less than 0.05).

Table 8 ANOVA result for bottom delamination factor using the non-coated HSS drill bit of the RSM model.

Source	Sum of Squares	DF	Mean Square	F Value	Prob > F	
Model	0.021	6	$3.530E^{-03}$	135.50	< 0.0001	significant
A	$4.061E^{-04}$	1	$4.061E^{-04}$	15.59	0.0027	
B	$1.540E^{-03}$	1	$1.540E^{-03}$	59.12	< 0.0001	
C	0.014	1	0.014	541.73	< 0.0001	
A^2	$6.711E^{-04}$	1	$6.711E^{-04}$	25.76	0.0005	
B^2	$3.132E^{-04}$	1	$3.132E^{-04}$	12.02	0.0060	
C^2	$3.758E^{-03}$	1	$3.758E^{-03}$	144.26	< 0.0001	
Residual	$2.605E^{-04}$	10	$2.605E^{-05}$			
Lack of Fit	$1.605E^{-04}$	6	$2.675E^{-05}$	1.07	0.4968	not significant
Pure Error	$1.000E^{-04}$	4	$2.500E^{-05}$			
Cor Total	0.021	16				

R-Squared = 0.9878

R-Squared (adj) = 0.9806
Pred R-Squared = 0.9628
Adeq Precision = 34.121

3.6 Mathematical model development of bottom delamination factor for non-coated HSS drill bits

According to the experimental results, a quadratic mathematical model equation was generated using the Design Expert Version 6.0.10, to estimate the bottom delamination factor according to the experimental significant factors. The recommended transformation is normal as proposed by Box-Cox plot.

$$F_{dB} = 1.48235 - 2.10E^{-03} V_c - 0.20625f - 0.07375d + 2.020E^{05} V_c^2 + 0.8625f^2 + 3.31944E^{03} d^2 \quad (3)$$

Figure 4 shows the normal probability plot of the studentised residuals which is used to check for the normality of the residuals. The plot reveals that the points follow a straight line with no major deviations from the line along with the random placement of the points. The points are not following any pattern which will require any transformation of data, thus showing the normality of residuals.

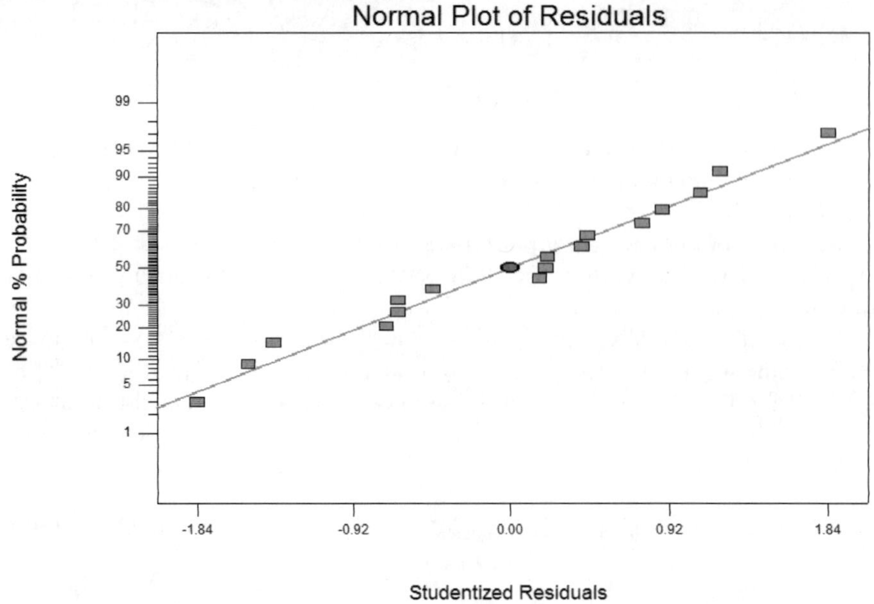

Figure 4 Normal probabilities of residuals for bottom delamination factor using non-coated HSS drill.

3.7 Effects of process parameters on bottom delamination factor

In figure 5, the three-dimensional graph exhibits the interaction effects of cutting speed and feed rate on bottom delamination factor with drill bit of 9 mm is held at constant. From the figure, it is observed that the bottom delamination factor of woven kenaf fibre reinforced polymer composite increases with increase in feed rate and decreases with increase in cutting speed. This is in agreement with the finding observed by Santhanam and Chandrasekaran [15] observed that the value of delamination factor at the exit surface of woven fibre (banana-glass) composite materials increasing with an increase in feed rate.

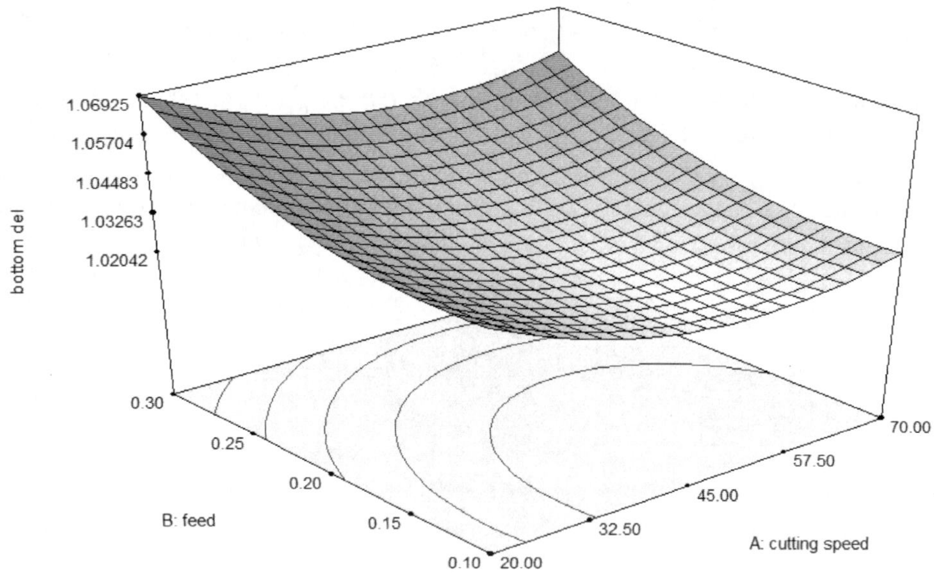

Figure 5. 3D response surface plot interaction effects of cutting speed and feed rate on bottom delamination factor for non-coated HSS drill bit.

4. Conclusion

Machining of fibre reinforced composite materials differs significantly in many aspects from machining of metals and alloys. In the machining of fibre reinforced composites, the materials behaviour is not only non-homogeneous, but it also depends in diverse reinforcement and matrix properties, types of machining processes, machining parameter and machining conditions. In the present study, the performance of non-coated HSS of drill bit during drilling of kenaf fiber reinforced polymer composites have been performed under dry machining condition. It was observed that the interaction of drill size and feed rate were the most significant factors that influence the delamination factor in the drilling of woven kenaf fiber reinforced epoxy composite laminates.

Acknowledgement

Authors would like to thank the Universiti Kebangsaan Malaysia and Malaysian Ministry of Higher Education for support of this work under the Long Term Research Grant Scheme (LRGS/TD/2012/USM-UKM/PT/05).

References

[1] N S Mohan, A Ramachandra and S Kulkarni *Compos Struct* **vol. 71** pp 407–413
[2] G Koronis A Silva and M Fontul *Compos Part B Eng* **vol. 44** pp 120–127
[3] A ElayaPerumal and N Venkateshwaran *IJAEA* **vol. 1** pp 68–74
[4] M. Ho *et al* 2012 *Compos. Part B Part B* **vol. 43** pp 3549–3562
[5] D Chandramohan and S Rajesh 2014 *Acad. J. Manuf. Eng.* **vol. 12** no. 3 pp. 72–77
[6] I El-Sonbaty, U a Khashaba and T Machaly 2004 *Compos. Struct.* **vol. 63** no. 3–4 pp 329–338
[7] W C Chen 1997 *Int. J. Mach. Tools Manuf.* **vol. 37** no. 8 pp 1097–1108
[8] G D Babu, K S Babu and B U M Gowd 2013 *J. Adv. Mech. Eng.* **vol. 1** pp. 1–12
[9] M Ramesh K Palanikumar and K H Reddy 2014 no Aimtdr pp. 2–7
[10] D Gay, S V Hoa and S. W. Tsai, "Composite Materials Composite Materials," *New York*, 2003.

[11] N Koboević, M Jurjević, and Ž Koboević 2012 *Teh. Vjesn.* pp. 391–398

[12] R Vinayagamoorthy and N Rajeswari 2014 *Jordan J. Mech. Ind. Eng.*, **vol. 8**, no. 6, pp. 385–392

[13] R Vinayagamoorthy and N Rajeswari *Jordan J. Mech. Ind. Eng.* **vol. 8** no. 6 pp. 385–392

[14] R. Vinayagamoorthy and N Rajeswari *Jordan J. Mech. Ind. Eng.*, vol. 8, no. 6, pp. 385–392, 2014.

[15] V Santhanam and M Chandrasekaran 2014 "A Study on Drilling Behavior of Banana-Glass Fibre Reinforced Epoxy Composite," no. November, pp. 194–199

Effects of electron beam radiation dose on the compatibilization behaviour in recycled polypropylene/microcrystalline cellulose composites

N Samat[1], S N R Motsidi[1] and N H M Lazim[1]

[1]Department of Manufacturing and Materials Engineering, International Islamic University Malaysia (IIUM), Jalan Gombak, 53100 Kuala Lumpur, Malaysia

Email: noorasikin@iium.edu.my

Abstract. The purpose of this research was to evaluate the influence of dose level of electron beam on the compatibilization behavior of recycled polypropylene (rPP) in rPP/microcrystalline cellulose (MCC) composites. Initially, the rPP was irradiated with various dose of electron beam (5 kGy up to 250 kGy) which then mixed with unirradiated rPP (u-rPP) at a ratio of 30:70 respectively. The composites were prepared by incorporating a series wt% of MCC fibers into rPP (u-rPP : i-rPP) using extruder and finally moulded with an injection moulding machine. The compatibility behavior of irradiated rPP (i-rPP) were analysed with mechanical tensile and thermal methods. The results of mechanical analysis showed great improvement in tensile modulus but an increase in radiation dosage gradually decreased this property. Nevertheless, the tensile strength exhibited a minor effect. The thermal stability of composites is lowered with increase in the absorbed dose, more significantly at higher content of MCC. Fracture surface observations reveal adhesion between the cellulose and rPP matrix.

1. Introduction

Radiation processing is a technique that can be used to modify the chemical and physical structure of polymeric materials. Therefore, the mechanical and thermal properties of polymers will change. In this technique, the polymers will be exposed to the irradiation and the common sources of irradiation are electron beams, gamma rays, X-ray etc. The application of irradiation to polymers results in several reactions such as crosslinking and chain scissions [1]. Both reactions can occur concurrently. However, the occurrence of crosslinking is preferred as it generally promotes the improvement of polymer's property. Moreover, this technique offers several advantages such as simpler, cheaper and easier to use.

According to Luguo et al. [2], radiation dosage is an important parameter to change the properties of polymers. Khalid et. al [3] reported that the high irradiation dose up to 200kGy increased the tensile strength and hardness of natural rubber nanocomposites. A study by Li et. al [4] in silk fibroin (SF) fiber-reinforced poly(e-caprolactone) (PCL) composite also found that sample added with 45% fiber and irradiated with 150 kGy resulted in the optimum tensile strength and modulus. On the contrary an improvement at low dose (<50 kGy) was reported by Ndiaye et. al [5] for polypropylene/wood composites. But in this work the sample was irradiated with gamma beam.

In polymer composites, the compatibility between matrix and filler is necessary to the enhancement of their properties. Incompatibility between these constituents will hinder efficient stress transfer from the matrix to filler. Most previous studies reported that polymer composites added with compatibilizer or coupling agents exhibited higher property [6-8]. As mentioned earlier, besides the chemical-based compatibilizer or chemical reagents, the ionizing radiation process also is able to modify the properties of a material. Studies by Samat et. al [9] on the compatibility effects of irradiated recycled polypropylene (i-rPP) in rPP/microcrystalline cellulose (MCC) composites, found that the tensile modulus of rPP/MCC was comparable to the chemical-based coupling agent. The improvement was associated with the existence of i-rPP in the rPP matrix with a ratio of 50:50 and the radiation dose was below 50 kGy.

Thus, the main objective of this study was to evaluate the influence of variation in radiation dose on the compatibility effect i-rPP in the rPP/MCC composites. The irradiation dose was varied from low to high dose and the ratio of u-rPP: i-rPP was 70:30. The compatibility behaviour associated with the changes in irradiation dose was determined from the mechanical and thermal stability properties. Field emission scanning microscopy analysis was also carried out.

2. Experimental procedures

Recycled polypropylene (rPP) was purchased from Top Flow Sdn. Bhd., Shah Alam, Selangor. The reinforcement material is microcrystalline cellulose (MCC). It was obtained from Sigma-Aldrich Co. (M) Sdn. Bhd. and has a density of 0.6 g/cm^3. The polymer and the MCC were dried in an oven at 40-60°C for at least 7 hours before use.

The rPP pellets were irradiated using a 3MeV electron beam (EB) using the EB model EPS-3000 with an accelerator voltage of 0.5-3.0 MeV at room temperature. The radiation dosages were varied from 5, 10, 50, 100, 150, 200 and 250 kGy. The unirradiated (u-rPP) and irradiated rPP (i-rPP) pellets were blended at a ratio of 70/30 wt% by weight prior to the addition of various MCC loadings (5, 20, and 40 wt%). Several compositions, radiation dose and designation of composite in present study is given in table 1.These materials were compounded in a Brabender twin-screw extruder machine at temperatures ranging from 180-190°C with a screw speed of 70-90 rpm. The compounded materials were then crushed into small pellets and then injected moulded into standard specimens (ASTM D638) at a temperature 180°C.

Table 1. Composition, radiation dose and designation of blends

Sample designation	u-rPP (%)	i-rPP (%)	MCC (wt%)	Irradiation dose (kGy)
rPP	100	-	-	-
rPP/5D/5M	70	30	5	5
rPP/5D/20M	70	30	20	5
rPP/5D/40M	70	30	40	5
rPP/50D/5M	70	30	5	50
rPP/50D/20M	70	30	20	50
rPP/50D/40M	70	30	40	50
rPP/100D/5M	70	30	5	100
rPP/100D/20M	70	30	20	100
rPP/250D/5M	70	30	5	250
rPP/250D/20M	70	30	20	250
rPP/250D/40M	70	30	40	250

The tensile were determined according to ASTM D638 standards by using a Universal Machine (LLYOD Instruments LR10K Plus) at a cross speed of 50 mm/min. Thermogravimetric analysis was conducted on a STA7300. The samples were scanned from 50 to 700°C with a heating rate of 10°C/min in the presence of nitrogen gas. The weight loss and its derivative were measured and recorded as a function of temperature. The fracture surface morphology's of the composites were observed using field emission scanning microscopy (JEOL-JSM 5600) at 10 kV. The sample surfaces were sputter coated with gold coating before analysis to eliminate any electron charging effect

3. Results and Discussion

The effect of radiation dose and MCC loading on the tensile properties of rPP composites are shown in figure 1(a). In comparison with the neat rPP, the tensile strength of composites decreased slightly as the content of MCC was increased from 5 to 40wt%. The lower tensile strength at 40wt% of MCC is presumed due to agglomeration that resulted from the poor interfacial adhesion between the MCC and the matrix. A study by Zulkifli et. al [10] found that the presence of agglomerated MCC or filler with weak bonding strength would induced the formation of voids and led to fast crack propagation rates. Nevertheless, irrespective of the MCC content, no noticeable differences in the tensile strengths were observed for all irradiation dose. It seems that the tensile strength of rPP/MCC composites is insensitive to varying irradiation dose. Similar findings were reported by Chaudhari et al. [11] for PP-PE blends. These blends were irradiated with 100-500kGy.

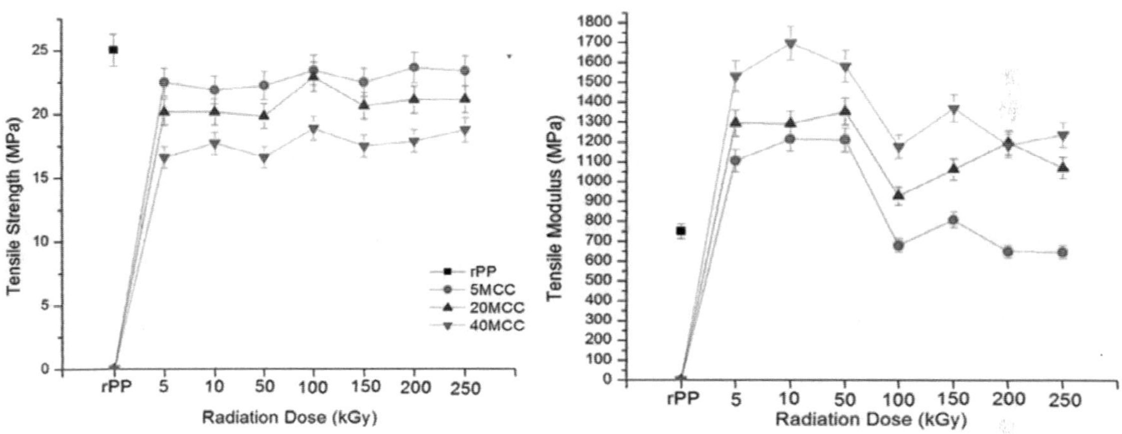

(a) (b)

Figure 1. Tensile properties (a) tensile strength and (b) tensile modulus of rPP/MCC composites with various MCC loadings and radiation dose.

From figure 1(b), it is evident that the Young's modulus of composites is affected by variation in the radiation dose and MCC content. Clearly, the presence of MCC fibers has caused an increase in the tensile modulus, including in samples added with 5wt% of MCC. In other word, though the amount of MCC was low but the improvement in Young's modulus is still significant. Therefore, this observation signifies that the enhancement in the Young's modulus is not merely caused by the stiffness property of MCC fibers but also by the existence of 30% of i-rPP, which acts as a compatibilizer. The i-rPP possesses higher degree of crosslinking that resulted from irradiation process with electron beam. Thus, the presence of high crosslink level also assisted to improve the modulus or stiffness of fabricated composites [12].

Moreover, changes in the modulus values were noted along with the increase in the radiation dose level beyond 50 kGy. Results from the electron spin resonance analysis by Lazim et. al [13] demonstrated that higher radiation dose produced more oxidation products (carbonyl and C-O compounds) and free radicals which associated with the chain scissions. Hence, it is suggested that irradiation of rPP at high dose diminishes the compatibility effects which resulted from the occurrence of chain scission phenomenon and formation of oxidation compounds.

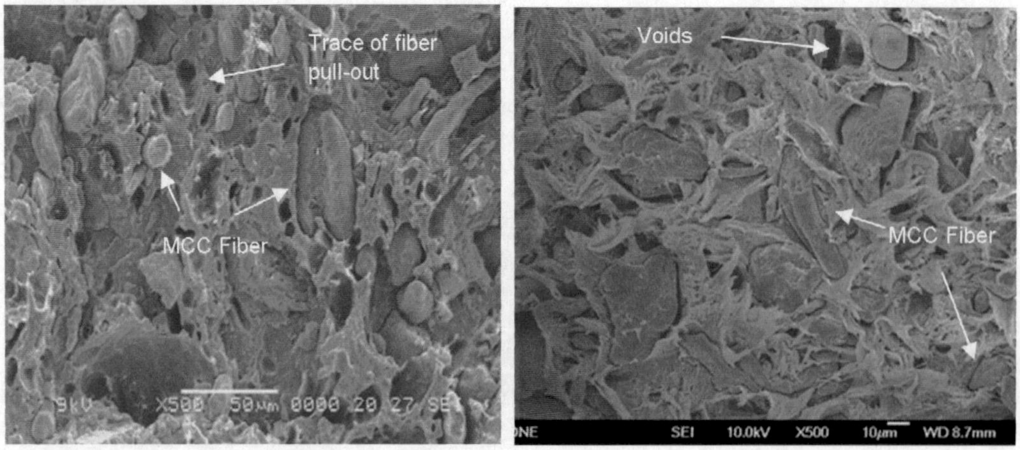

(a) (b)

Figure 2. FESEM micrograph of rPP/MCC composites for (a)rPP/10D/5MCC and (b)rPP/150D/5MCC

FESEM investigations on the tensile fracture surface of rPP composites which contained i-rPP at low and high dose is depicted in figures 2a and b, respectively. The fracture surface exhibits several features: traces of fiber pull-out, dissimilar size of MCC and voids. Interestingly, the cellulose-rPP matrix adhesion is also evident as observed from figure 2. This indicates that the presence of i-rPP improves the compatibility between the fiber and matrix. In this study although the ratio of i-rPP to u-rPP was only 30%, which is lower than in a work by [9](i-rPP = 50%), but the compatibility effects of i-rPP is still present.

The thermogravimetric analysis (TGA) was carried out to investigate the thermal properties of the produced composites. The result is presented as residual weight vs. temperature (Figure 3). Dissimilar to neat rPP, the composite samples exhibit the two-mass loss step feature. The first degradation temperature occurs at a range of 250˚C to 350˚; which attributed to the presence of MCC fibers [9] or natural fiber [14]. Meanwhile, the second degradation corresponds to the neat rPP.

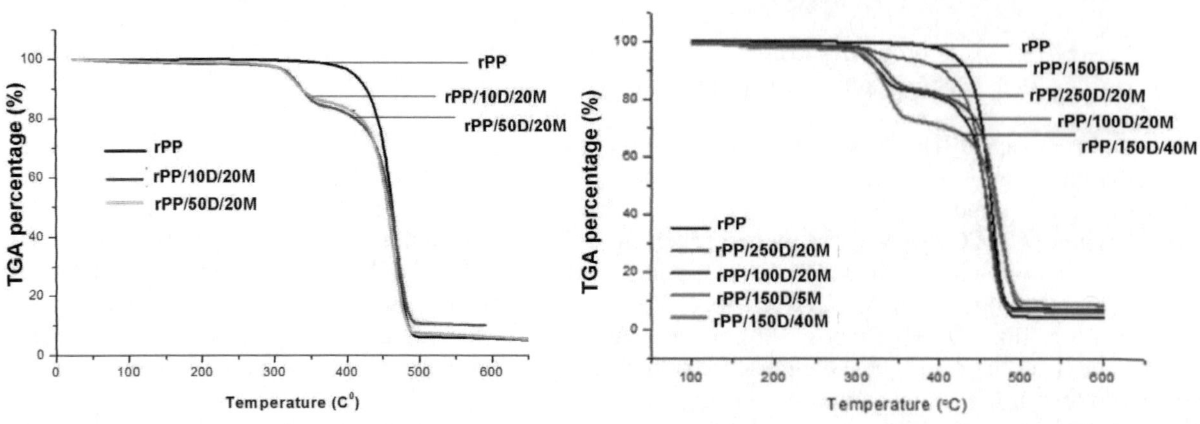

Figure 3.Thermogravimetric mass loss curves of rPP/MCC composites with various MCC loadings at (a)low radiation dose and (b)high radiation dose

Table 2 shows the analysis of thermal degradation temperature for weight loss at 5% (T_5) and 50% (T_{50}). From this table, it is noticed that independent of the MCC content, the presence of i-rPP compatibilizer with different radiation dose level affects the thermal stability of composites. For instance, the irradiation at 250kGy has caused the degradation to occur at 10^{0}C faster than the neat rPP. The degradation at higher irradiation dose is in accordance with previous observation in Young's modulus and a work by [5]. The highest char residue implies greater thermal stability in composites with lower radiation dose. Thus, it can be emphasized that for polymer based-cellulose/natural fibers; the high irradiation dose would lead to chain scission mechanism [1,13] and resulted in lower thermal stability.

Besides the radiation dose, the thermal stability is also dominated by the presence of MCC fibers. Comparison of samples rPP/150/40M and rPP/150/5M shows that the thermal stability deteriorated with the increasing loadings of MCC fibers (Table 2). The incorporation of MCC fibers from 5 to 40wt% lower increased the decomposition rate, which is attributed to their lower thermal stability. Similar findings were reported by Kiziltas et. al [15] for PET-PTT blends/MCC composites.

Table 2: The degradation temperature of rPP/MCC composites

MCC(wt%)	Dose (kGy)	Degradation Temperature (^{0}C)		Char residue at 590^{0}C.
		T_5	T_{50}	
0	0	411.2	461.3	4.18
20	10	321.1	461.1	10.17
20	50	318.7	457.0	7.91
20	100	313.4	460.9	6.74
20	250	311.5	458.8	6.85
5	150	341.6	458.5	5.98
40	150	304.3	458.7	8.45

4. Conclusion

The following conclusions can be drawn from the results obtained:

(1) The presence of i-rPP at 30% enhanced the adhesion between the cellulose fiber and rPP matrix.

(2) The compatibilizing effect of i-rPP in polymer based-cellulose is efficient at irradiation below 50 kGy.

(3) The high radiation dose along with high concentration of MCC fibers reduced the thermal stability of rPP/MCC composites

References

[1] Drobny J G 2013 *Ionizing radiation and polymers: principle, technology and applications* Elsevier USA

[2] Lugao A B, Cardoso E C L, Lima L F C P, Hustzler B and Tokumoto S 2003 *Nuclear Ins. and Methods in Phy. Research* **208** 252–255

[3] Khalid M, Ismail A F, Ratnam C T, Faridah Y, Rashmi W and AlKhatib M F 2010 *Radiation Phy. and Chem.* **79** 1279–1285

[4] Li W, Qiao X, Sun K and Che X 2009 *J of Applied Poly Sci* **113** 1063–1069

[5] Ndiaye D, Badji AM and Tidjani A 2014 *J of Comp Mat* **48** 3063–3071

[6] Awanis J, Anis Sofia S and Samat N 2012 *Adv. Mat. Research* **576** 390-393

[7] Samat N, Marini C D, Maritho M A and Sabaruddin F A 2013 *Comp Interfaces* **20** 497-506

[8] Samat N, Zulkapli N, Halim Z, Ahmad Z and Habibah A I H D 2015 *Adv. Mat. Research* **1115** 402-405
[9] Samat N, Lazim N H M, Motsidi S N R and Azlina H N 2017 *Mat. Sci. Forum* **894** 62-65
[10] Zulkifli N I, Samat N, Anuar H and Zainuddin N 2015 *Mats and Design* **69** 114-123
[11] Chaudhari C, Dubey K, Bhardwaj Y, Naxane G, Sarma K and Sabharwal S 2007 *Nuclear Ins. and Method in Phy. Research Sect B: Beam Int. with Mat. and Atoms* **263** 451-457
[12] Lazim N H and Samat N 2017 *Procedia Eng*. **184** 538 – 543
[13] Lazim N H and Samat N 2017 *Polymer Comp*
[14] Kaiser M R, Anuar H, Samat N and Shamsul AR 2013 *Iranian Poly J* **22** 123-131
[15] Kiziltas A, Gardner D J, Han Y, and Yang H S 2010 *J Therm Anal Calorim*

IOP Conf. Series: Materials Science and Engineering **290** (2017) 012035 doi:10.1088/1757-899X/290/1/012035

Experimental Investigation of Minimum Quantity Lubrication in Meso-scale Milling with Varying Tool Diameter

M Q M Yusof[1], H N S B Harun[1] and R Bahar[1]

[1]Department of Manufacturing and Materials Engineering, International Islamic University Malaysia (IIUM), Jalan Gombak, 53100 Kuala Lumpur, Malaysia

Email:rbahar@iium.edu.my

Abstract. Minimum quantity lubrication (MQL) is a method that uses a very small amount of liquid to reduce friction between cutting tool and work piece during machining. The implementation of MQL machining has become a viable alternative to flood cooling machining and dry machining. The overall performance has been evaluated during meso-scale milling of mild steel using different diameter milling cutters. Experiments have been conducted under two different lubrication condition: dry and MQL with variable cutting parameters. The tool wear and its surface roughness, machined surfaces microstructure and surface roughness were observed for both conditions. It was found from the results that MQL produced better results compared to dry machining. The 0.5 mm tool has been selected as the most optimum tool diameter to be used with the lowest surface roughness as well as the least flank wear generation. For the workpiece, it was observed that the cutting temperature possesses crucial effect on the microstructure and the surface roughness of the machined surface and bigger diameter tool actually resulted in higher surface roughness. The poor conductivity of the cutting tool may be one of reasons behind.

1. Introduction

Minimum quantity lubrication (MQL) has progressively emerged into the area of metal machining and, has already been established as a viable alternative to dry and flood cooling machining. The amount that should be applied in MQL is subjective and depends on the material, process and tools. As a general rule of thumb, MQL uses 5 to 80 ml/hour (0.2 to 2.5 oz./hour) on the tools which has less than 40 mm diameter[1]. Some major advantages of MQL include improved environment, better surface finish, increased tool life, reduced wear and damage at tool tips; reduce health hazards, and overall low costs in various micromachining processes. Machining processes using MQL have been studied for all conventional meso or micro machining processes like micro and meso scale drilling[2],[3] micro milling[4], [5], turning[6],[7] and grinding[8]. Implementing MQL in machining process helps to increase the tool life and produce better surface finish. It is also the best remedy for promoting eco-friendly industry due to reduction in waste. In this paper, the performance of meso-milling under MQL has been studied and compared with the same machining experience under dry condition. There have been numerous studies on micro or meso milling using MQL and dry machining. For example, [9] studied the micro-milling using MQL with varying the cutting parameters and analyzed the tool performance and machined surface condition. The tool performance was analyzed by [5] and it showed that MQL increased the tool life by 100 times compared to dry machining. MQL Assisted Meso-scale milling has been studied by [10] and it was shown how that the use of compressed chilly air and nanofluid MQL is effective for improving surface finish and reducing machining force. However, very limited research article are available where the machining performance has been studied for using different diameters of milling tool. In our work, the diameter of the tool bit has been varied to observe its effect on the overall heat transfer by measuring the machining temperature. The surface condition of the work piece and tool has been investigated after machining with two lubrication methods.

2. Experiments

The experiments were conducted using the MIKROTOOLS Integrated Multi-purpose modelled DT-110. This machine has three travel axes (X, Y, and Z axes) to move the tools with ultra-precision milling capabilities with submicron accuracies. Bluebe MQL system was used with lubricant Accu-Lube LB 2000 (LB-1) which is manufactured from biodegradable ingredients, extracted and blended on the basis of vegetable oils. With its high lubricity and heat transfer properties, it is capable to increase heat dissipation and prolongs tool life. Figure 1(a) and 1(b) shows the milling machine and the lubrication system respectively.

a) b)

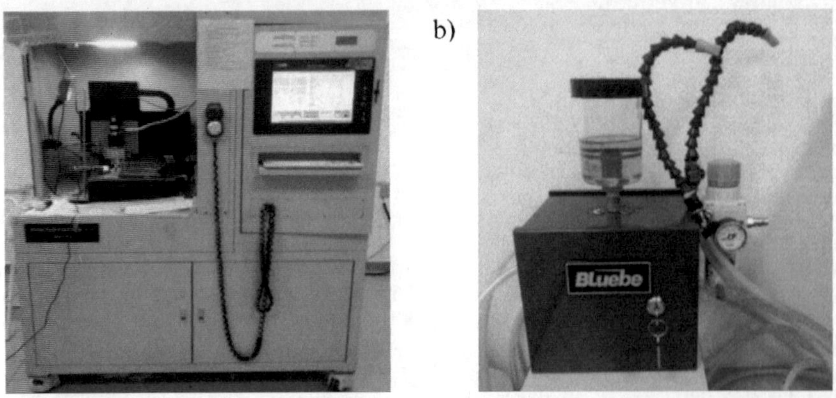

Figure 1. MICROTOOLS Integrated Machine DT-110(a) Bluebe MQL system(b).

The milling tools selected were made of carbide but two different diameters were chosen to observe the effect of cutting tool's increased diameter on the heat transfer during machining. The work piece metal chosen was mild steel for its stable mechanical and chemical properties as well as specific energy for cutting and compatibility with carbide tools. Fluke Ti20 thermal imaging system was employed to measure the machining temperatures at different condition. The other machining parameters are listed in **Table 1**.

Table 1. Machining Condition and parameters.

Parameters	Description
Work piece Material	Mild steel (35mm x 10mm x 10mm)
Cutting tools	Two flute carbide end mill tools (diameter 0.5 mm and 0.8 mm)
Depth of cut	0.15 mm
Machining length	12 mm
Spindle speed	2300 rpm
Feed rate	2 mm/s
Lubrication conditions	Dry and MQL

3. Results and Discussion

The heat generated during micro milling can be divided in two portions; one portion is absorbed by the work metal while the other part is absorbed by the MQL coolant system. Hence, the energy balance equation can be written as Equation 1.0

$$H_{total}=H_{metal}+H_{MQL} \tag{1.0}$$

Where H is the heat energy. H_{total} is the amount of heat dissipated during the machining process. H_{metal} is the energy absorbed by the metal which can be expressed as =kA[dT/dx]

Here, k=thermal conductivity (w/m°C), A=area,

dT/dx=temperature gradient over the metal thickness

While H_{MQL}=the portion of the heat energy absorbed by the MQL mist=$h_{MQL}A(T_m-T_a)$

Here h_{MQL}=Heat transfer coefficient for the mist created during MQL,

T_m=Metal temperature on the machined surface(°C)

T_a=ambient temperature (°C)

Therefore, (1.0) can be re-written as Equation 2.0

$$H_{total}= kA[dT/dx]+ h_{MQL}A(T_m-T_a) \tag{2.0}$$

As can be seen form the equation, if the heat carried by the metal itself is higher, the MQL cooling becomes less dominant. However, it was observed that the machining temperature was lower when MQL was used in the micro milling on the mild steel since it carried away some of the heat generated during the machining compared with the machining temperature during dry machining. The cutting temperature measured in the present work refers mainly to average chip tool interface temperature, as seen in **Table 2**. It is also seen that the machining temperature was increasing with increase in diameter of the cutting tool. The trend may be attributed to the [dT/dx] term in Equation 2.0, as the diameter was increasing, the heat conduction rate by the cutting tool was dropping.

Table 2. Machining temperature for two different diameter tool with MQL and dry condition.

Cutting Tool Diameter	**MQL condition**		**Dry condition**	
	0.5mm	0.8mm	0.5mm	0.8mm
Cutting Temperature (°C)	26.9	27.9	27.6	29.0

The surface conditions including the microstructure and surface roughness were examined for the cutting tool and work piece.

The tools‘ surfaces were observed using Field Emission Scanning Electron Microscope (FESEM) and surface roughness tester while Olympus DP26 Microscope and surface roughness tester were used to observed the microstructure of the mild steel

Table 3 shows the condition of the tool after machining. The figures in Table 3 were developed using field emission scanning electron microscope (FESEM) under magnification of 200 and 5kV. The views are generated after micro milling operations on mild steel using the different diameter tools. Under all environment, abrasive scratch marks appeared on the flanks but the most obvious one was observed for the tools that underwent dry machining. Besides, chipping wear was also be identified especially on 0.8mm tool diameter under dry machining. This can be atrributed to the sharp ragged edges and multiple irregular wear pattern along the edge of the tool. This condition may lead to catastrophic failure in the early life of the tool which concealed the failure mode.

The flank wear formed were uniform over localized area and it was accelerated with higher temperature. This condition appear as rough surface mainly on the flank of the tool and to a lesser extent towards the face of the tool. According to the results obtained, the highest temperature generated (29.0 °C) and the most apparent flank wear occur on tool with 0.8mm diameter under dry machining condition.

Table 3. Observation of tool wear for different lubrication condition and tool diameter.

Dry machining	**MQL**

Tool Diameter: 0.5 mm	Tool Diameter: 0.5 mm
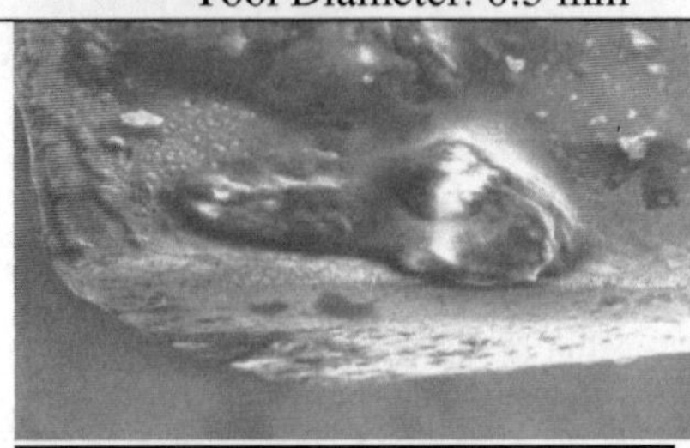	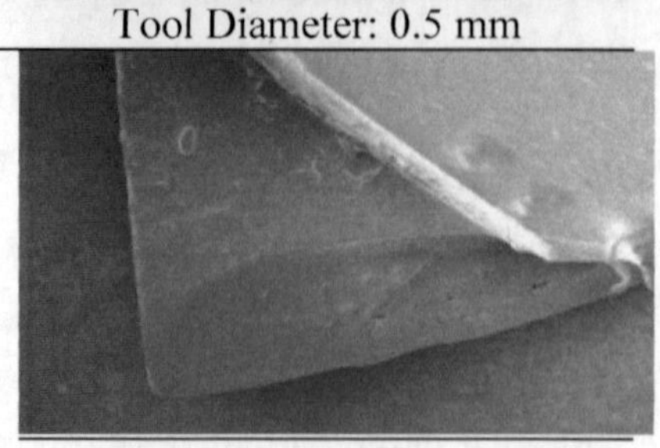
Tool Diameter: 0.8 mm	Tool Diameter: 0.8 mm

The Surface roughness test was done using Veeco WYKO NT1100 optical profiler. The result for both dimeter cutting tools are shown in Figure 2.

It is observed that surface roughness of the cutting tools was higher for dry machining. The highest Ra value obtained was 1.86 for tool diameter 0.8mm at dry machining condition. The trend also shows that with increased diameter, the surface roughness deteriorated regardless of the lubrication method used. The reason may be attributed to the larger torque experienced by the bigger diameter tool. Higher machining temperature may also be one of the reason.

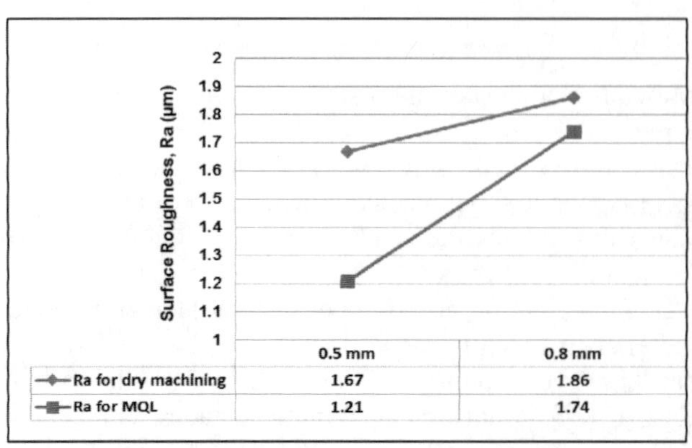

Figure 2. Surface Roughness vs Tool Diameter graph.

For investigation of the machined work-piece, the surface roughness (R_a) and microstructure of machined surface were examined. It is found that the surface roughness increased linearly with increase in tool diameter. The value of Ra for 0.8mm diameter of cutting tool is the highest for both lubrication conditions while the surface roughness under MQL condition is found to be better than dry machining condition, which are 560.92μm under MQL and 792.79μm under dry machining condition respectively. This proved that the presence of MQL will give better surface roughness due to lower cutting temperature provided by the lubrication that reduce the heat generated while machining process. The results are summarized in Table 4.

Table 4. Surface roughness of machined surface under different lubrication and tool diameter.

Tool Diameter	**MQL condition**		**Dry condition**	
	0.5mm	0.8mm	0.5mm	0.8mm
Surface Roughness of machined surface	461.73	560.96	611.46	792.79

Microstructure captured by the Olympus DP26 Microscope, both for the condition under MQL and Dry machining for using two different diameter tool is displayed in **Table 5**.

Table 5. Microstructure of the machined surface under different lubrication and tool diameter.

Tool Diameter → / **Lubricant condition ↓**	**0.5 mm↓**	**0.8 mm↓**
Dry condition	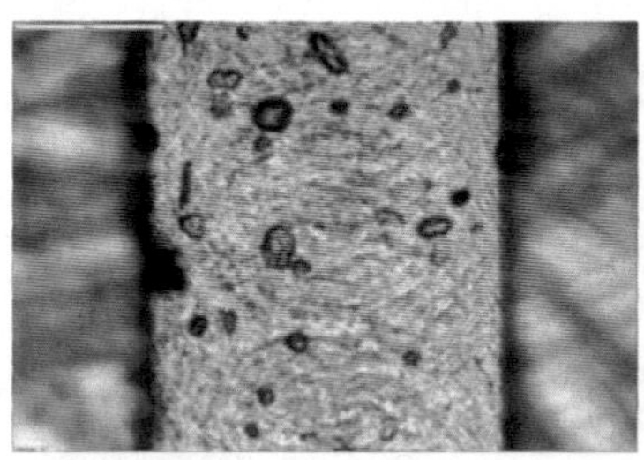	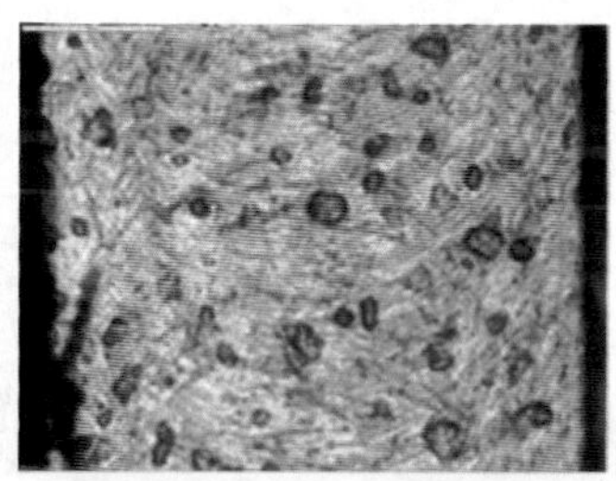
MQL condition	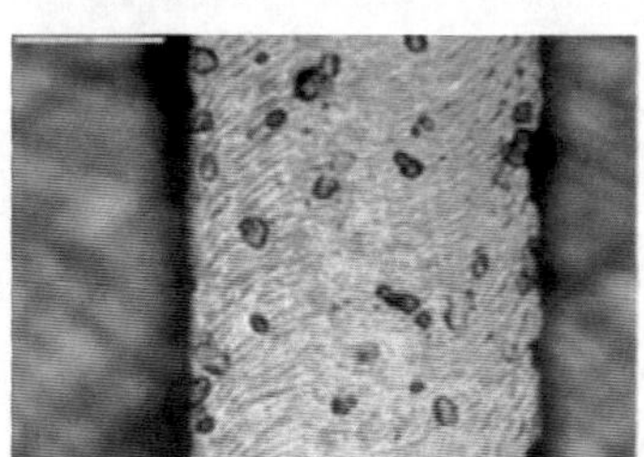	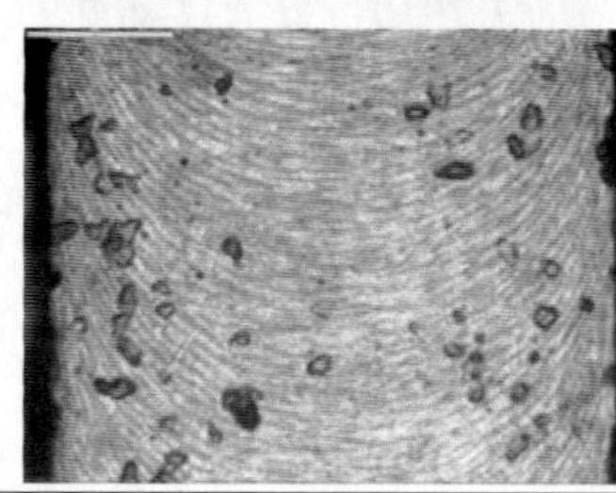

Based on mild steel surface properties, the microstructure of the machined surface is seen to contain ferrite and pearlite structure. As both of ferrite and pearlite, they have very poor hardness. This in turn lead to different physical characteristics (such as variable friction coefficients) causing vibrations to increase and highly fragmented chips to occur. Presence of MQL cannot reduce this chip–tool interface temperature totally because the fluid can hardly penetrate into that the interface where the chip–tool contact is mostly plastic in nature. That is why there are still surface defects like dimples on the machined surface of mild steel, surface dimples do occur as a result of the dual phase structure of the work piece material and always occur at a hard to-soft grain transition. They do not occur when cutting moves across a soft-to-hard grain boundary. The presence of MQL provided at least better dissipation of the heat compared to dry machining and help to move the chips away better that dry machining. The effect of diameter of the tool bit was not a dominating factor for this case.

4. Conclusion

In this study, experimental investigation was carried out to observe the effect of MQL and tool diameter on the machined surface and tool wear. It has been found that the presence of MQL benefits in reducing the cutting temperature, which improves the chip tool interaction and maintain sharpness of the cutting edges. The cutting temperature had a significant effect on the microstructure and the surface roughness on mid steel that is why presence of MQL provided better microstructure and better surface roughness.In terms of tool wear and surface roughness, 0.5mm tool performed better compared to 0.8mm tool with lesser wear under MQL contion.

Acknowledgement

Kulliyyah of Engineering, International Islamic University Malaysia, has funded this work.

References

[1] T Walker 2013 *The HANDBOOK A guide to machining with Minimum Quantity Lubrication*. Unist, Inc

[2] A E D Durval, U Bragaa, N L C Gilberto and W A Miranda 2002 *J. Mater. Process. Technol.*, **vol. 122** pp. 127–138

[3] M N Bahar 2016 *Int. J. Eng. Mater. Manuf.* **vol. 1** no. 2 pp 65–70

[4] A Syahmi, A Bin, R Bahar, and T F Ariff 2016 in *Proceedings of 6th International Conference on Industrial Engineering and Operations Management*, pp. 1782–1787

[5] S Kajaria, S Chittipolu, S Adera and W N Hung 2012 *Mach. Sci. Technol.* **vol. 16** no. 4 pp 524–546

[6] V N Gaitonde, S R Karnik, and J P Davim 2012 *Mach. Sci. Technol.* **vol. 16** no. 2 pp. 304–320

[7] M Hadad and B. Sadeghi 2013 *J. Clean. Prod.* **vol. 54** pp. 332–343

[8] M Hadad and B Sadeghi 2012 *Int. J. Mach. Tools Manuf.* **vol. 63** pp. 1–15

[9] V Shaikh, N Boubekri and T W Scharf 2014 *Int. J. Manuf. Eng.* **vol. 2014** pp 1–13

[10] P H Lee, T S Nam, C Li and S W Lee 2010 *Proc. 2010 Int. Conf. Mech. Ind. Manuf. Technol. (MIMT 2010)*, pp. 457–462,

IOP Conf. Series: Materials Science and Engineering **290** (2017) 012036 doi:10.1088/1757-899X/290/1/012036

Machinability of Soda Lime Glass in High Speed End Milling

Mohamed Konneh, Mst. Nasima Bagum, Tasnim Firdaus Bt. Mohamed Arif and Mohammad Yeakub Ali

Department of Manufacturing and Materials Engineering, Kulliyyah of Engineering, International Islamic University Malaysia, Jalan Gombak, 53100 Kuala Lumpur, Malaysia

Email: mkonneh@iium.edu.my

Abstract. Ductile regime machining can produce crack free surface on brittle material under certain controlled cutting conditions. Although end milling is a versatile process, it has not been frequently applied for machining soda lime glass. The brittle nature of the soda lime glass makes the machining difficult. Soda lime glass is a strain rate and temperature sensitive material. Therefore, high speed machining can be beneficiary to achieve ductile surface with low roughness. Hence, proper combination of cutting parameters needs to identify. In this paper, the effects of cutting parameters such as spindle speed, feed rate and depth cut on the surface roughness parameters R_a, R_q and R_t are investigated to assess machinability of soda lime glass during high-speed end milling. Regression models are generated and process parameters are optimized using the Central Composite Design (CCD) of Response Surface Methodology (RSM). The uncoated tungsten carbide tool was used to conduct milling operation. Experimental verification of optimal condition, spindle speed 40,000 rpm, feed rate 10 mm/min and depth of cut 34 μm confirmed that ductile surface with R_a, R_q and R_t of 0.33μm, 0.49μm and 6.59 μm respectively possible to achieve.

1. Introduction

As of late research were focused on achievement of ductile mode material removal of soda lime glass; an important brittle material [1-3] and critical parameters identification for ductile brittle transition [4] in machining. Roughness is an important performance predictor of mechanical component; a part of the blue print of manufacturing processes [5]. In order to assess machinability of soda lime glass, the executions of ductile mode material removal, as well as the roughness of the generated surface, due to machining need to be consider. In case of ductile metal milling process, achievement of R_a value less than 0.8 μm is less frequent [6]. Although average roughness parameters (R_a) is widely used but different surface texture can have same R_a value, as R_a does not differentiate between peaks and valleys. The complimentary of average roughness R_a is root-mean-square (RMS) deviation of a profile i.e. R_q. In optical application, R_q is still applicable as it directly linked with the optical characteristics of the surface. The extreme parameter, R_t signposts the existence of a sharp spike or burr on the surface as results of poor material processing [7]. Hence, in this study, effects of high-speed end milling parameters on all these three roughness parameters (R_a, R_q and R_t) are investigated and analyzed to assess the machinability of soda lime glass using uncoated carbide end mill. The multi-response optimization was carried out using desirability function of RSM.

2. Experimental procedures

4 mm diameter tungsten carbide tool having 2 flutes was chosen to perform milling on soda lime glass work piece using the upgraded vertical axis CNC milling machine. The cutting conditions of spindle speed ranges from 20,000 to 40,000 rpm, cutting depth ranges from 30 to 50 µm and feed rate ranges from 10 to 30 mm/min were employed in dry condition. The CCD of RSM was employed for designing and analyzing the experimental results. The machined surface roughness R_a, R_q and R_t were measured using Veeco Wyko Profiling system Microscope model NT110. Scanning electron microscope (SEM) was used to examine surface integrity.

3. Results and Discussion

The outcome of the experiment is represented in Table 1, where spindle speed (A), feed rate (B) and depth of cut (C) were taken as input parameters and roughness parameters such as R_a, R_q and R_t as responses. In most of the cases, R_q is 20% larger than R_a and this relation is similar in case of standard machining process [5]. Again, R_t/R_a ratio is about 7 to 16. Hence, high-speed end milling of soda lime glass can be considered as a standard machining process for glass.

Table 1 Central Composite Design using RSM along with the response value

Run	Type	A: Spindle speed (rpm)	B: Feed rate (mm/min)	C: Depth of Cut (µm)	R_a (µm)	R_q (µm)	R_t (µm)
1	Axial	40000	20	40	0.71	0.94	8.37
2	Axial	20000	20	40	0.58	0.71	5.24
3	Fact	40000	30	30	1.47	1.82	13.15
4	Center	30000	20	40	0.94	1.09	9.13
5	Fact	20000	10	30	0.42	0.53	6.18
6	Fact	40000	10	50	0.50	0.6	5.7
7	Fact	20000	30	30	0.45	0.57	7.16
8	Axial	30000	10	40	0.39	0.47	9.52
9	Fact	40000	30	50	1.09	1.33	8.28
10	Fact	40000	10	30	0.49	0.6	5.7
11	Center	30000	20	40	1.06	1.31	10.23
12	Fact	20000	10	50	0.52	0.66	8.65
13	Axial	30000	20	50	1.25	1.6	12.04
14	Center	30000	20	40	0.91	1.11	7.73
15	Center	30000	20	40	0.82	1.04	9.92
16	Axial	30000	20	30	1.66	2.11	13.89
17	Axial	30000	30	40	1.00	1.32	12.78
18	Fact	20000	30	50	0.41	0.51	5.17

The development of response surface model includes problem formulation: transformation checking and selection of the model that provides the finest relationship between dependent responses and independent input variables according to the instruction of fit summary tests; analysis of variance (ANOVA); model diagnostics. In the fit summary statistics, the Sequential Model Sum of Square (SMSS) results and Model Summary Statistics endorsed quadratic relationship between the independent variables and the responses (R_a, R_q and R_t). The "Lack- of- fit" value showed insignificant. The response R_q and R_t necessitate transform to natural log and inverse square root as suggested by Box-Cox diagnostic of CCD.

The analyses of variance (ANOVA) for R_a, R_q and R_t models with all possible model terms implied that all the models are significant with large F-value. Although the R_q model required log transformation suggested by Box-Cox plot, both of the R_a and R_q models have similar significant model terms A, B, AB, A^2, B^2 and C^2. The significant model terms of R_t model are B, AB, BC and A^2. The "Lack of Fit F-value" of the models illustrate that it was not significant relative to the pure error.

However, the predicted R^2 and the adjusted R^2 value of the models implied that each of the model predicted R^2 value was not as close to the adjusted R^2 value as required by the design expert 7.0.0.

After manually removing the non-significant terms (BC and CA for R_a and R_q model, CA, B^2, C^2 for R_t model) that are not contributing to model's hierarchy, using back ward regression method the model showed improvement. For instant, the improved models ANOVA for R_a is given in Table 2.The model equations are therefore obtained based on the reduced model. The equation (3.1), (3.2) and (3.3) perceive the relative influence of each process parameter on response R_a, R_q and R_t respectively based on coefficient of each factors. The regression model in equation (3.4), (3.5) and (3.6) is used to recreate the results of the experiment for each of the responses.

Table 2 ANOVA for R_a (Improved model term)

Source	Sum of Squares	df	Mean Square	F Value	p-value Prob > F	
Model	2.29	7	0.33	14.48	0.0002	significant
A-spindle speed	0.35	1	0.35	15.63	0.0027	
B-Feed rate	0.44	1	0.44	19.50	0.0013	
C-depth of cut	0.052	1	0.052	2.29	0.1609	
AB	0.34	1	0.34	15.05	0.0031	
A^2	0.41	1	0.41	17.95	0.0017	
B^2	0.31	1	0.31	13.61	0.0042	
C^2	0.48	1	0.48	21.44	0.0009	
Residual	0.23	10	0.023			
Lack of Fit	0.20	7	0.028	2.86	0.2091	not significant
Pure Error	0.029	3	9.825E-003			
Cor Total	2.52	17				

$$R_a = +0.98 + 0.19A + 0.21B - 0.072\,C + 0.21\,AB - 0.39\,A^2 - 0.34B^2 + 0.42C^2 \quad (3.1)$$

$$\ln(R_q) = +0.17 + 0.24A + 0.28B - 0.048C + 0.26\,AB - 0.41A^2 - 0.45B^2 + 0.40\,C^2 \quad (3.2)$$

$$1.0/\mathrm{Sqrt}(R_t) = +0.31 - 0.019A - 0.019\,B + 9.547\times10^{-3}\,C - 0.036AB + 0.025BC + 0.070 \quad (3.3)$$

$$R_a = +3.460 + 2.097\times10^{-4}\,A + 0.093\,B - 0.345\,C + 2.062\times10^{-6}\,AB - 3.870\times10^{-9}(A)^2 - 3.370\times10^{-3}(B)^2 + 4.229\times10^{-3}(C)^2 \quad (3.4)$$

$$\ln(R_q) = +1.548 + 2.187\times10^{-4}A + 0.128\,B - 0.324\,C + 2.613\times10^{-6}\,AB - 4.111\times10^{-9}(A)^2 - 4.476\times10^{-3}(B)^2 + 3.993\times10^{-3}(C)^2 \quad (3.5)$$

$$1.0/\mathrm{Sqrt}(R_t) = +0.979 - 3.659\times10^{-5}\,A - 1.167\times^{-3}\,B - 4.047\times10^{-3}\,C - 3.570\times10^{-7}\,AB + 2.501\times10^{-4}\,BC + 6.978\times10^{-10}\,(A)^2 \quad (3.6)$$

In case of R_a and R_q, it is clear from equation (3.1) and (3.2) that the most influencing process parameters are depicted as feed rate followed by spindle speed. A part from that cutting speed, depth of cut and feed rate showed quadratic effect and spindle speed and feed rate showed interaction effect on these responses (equation 3.1 and 3.2). According each factor coefficient as in equation (3.3), quadratic effect of spindle speed showed most influential effect on R_t followed by interaction effect of spindle speed and feed rate, interaction effect of feed rate and depth of cut. Depth of cut confirmed little contribution and its effect is insignificant for all models (According to p-value in ANOVA, for R_a shown in table-2). The adequacy of the models was verified through the statistical features associated with it. For all model R^2 were greater than 0.8, the difference of predicted R^2 with adjusted R^2 was less than 0.2, adequate precision were greater than 4. In addition, the residual of the data were normally scattered, constant variance assumption was satisfied, and residuals were not correlated. Hence, all models are adequate for application in navigating the design space within the experimental limit. The interaction effect between spindle speed and feed rate is significant in all models and their 3D interaction plots presented in Figure 1. When the spindle speed is at upper level, the feed rate showed much stronger effect on these responses. This is possibly due to molten chip contamination [8] or possible thermal degradation of cutting tool [9].

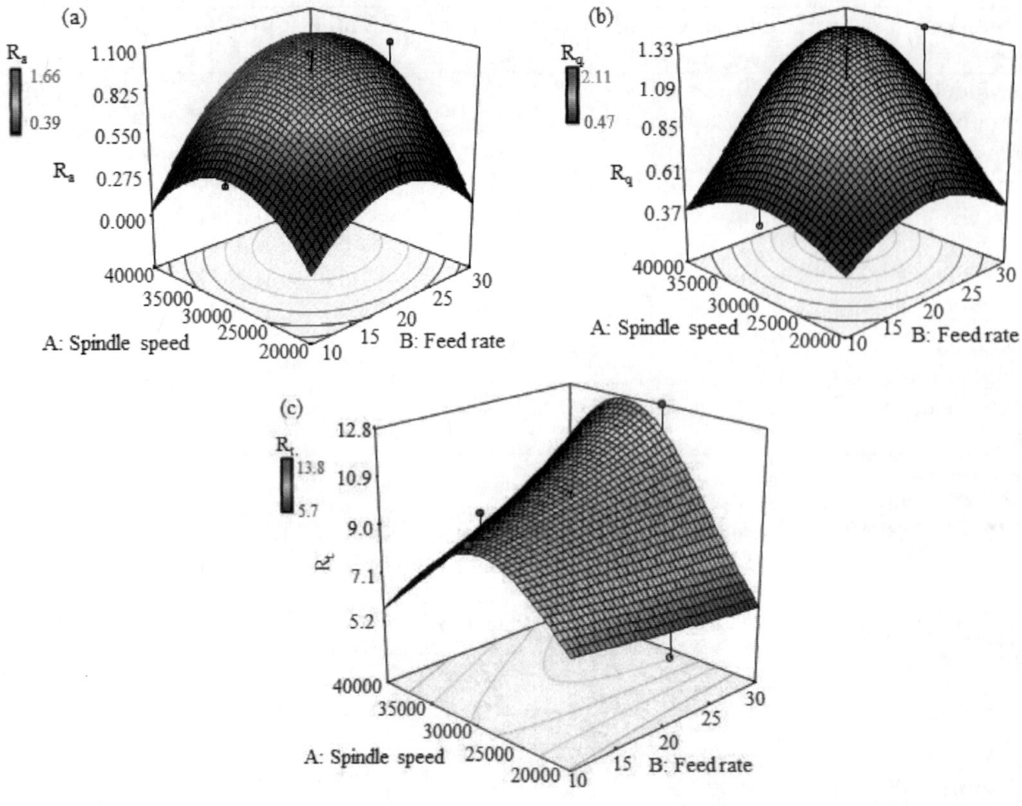

Figure 1. 3D interaction plots of spindle speed and feed rate (a) R_a (b) R_q and (c) R_t.

According to equation (3.6) and as indicated in Figure 1 (c) the extreme parameter, R_t can be used as a signpost of the existence of a sharp spike or burr on the surface as results of chip contamination at upper level of spindle speed. At lower level of spindle speed, roughness parameters R_a and R_q increased quadratically up to an optimal point followed by a decrease in these responses with the increase in feed rate. According to ideal rules, low feed rate and high cutting speed produced smaller roughness [5], this relation also holds here. However, at lower level of spindle speed, R_t dropped sharply with an increase in feed rate. Interaction effects between feed rate and depth of cut are not significant for R_a and R_q, however, are significant for R_t. Similarly, the interaction effect between depth of cut and spindle speed is not significant for all models. However, irrespective of the level of depth of cut, the responses increased up to an optimal point with the increase in spindle speed followed by a gradual decreased in its value as the spindle speed further increases. This might be because at high speed and depth of cut combination, thermal softening effect is prominent [10].

The statistical error associated with experimental and predicted roughness data, R_a, R_q and R_t illustrated that the error limit was within 10% in almost all the run. According to desirability function of RSM, roughness parameters R_a, R_q and R_t were optimized with in the ranges of input parameters. Optimization provides 22 solutions with more than 87.7% combined desirability. Among them the roughness R_a 0.27 μm, R_q 0.47 μm and R_t 5.57 μm was predicted at spindle speed 40,000 rpm, feed rate 10 mm/min and depth of cut 34 μm with 97 % desirability. Confirmation test showed that at this combination R_a, R_q and R_t was 0.33, 0.49 and 6.59 μm respectively shown in Figure 2.

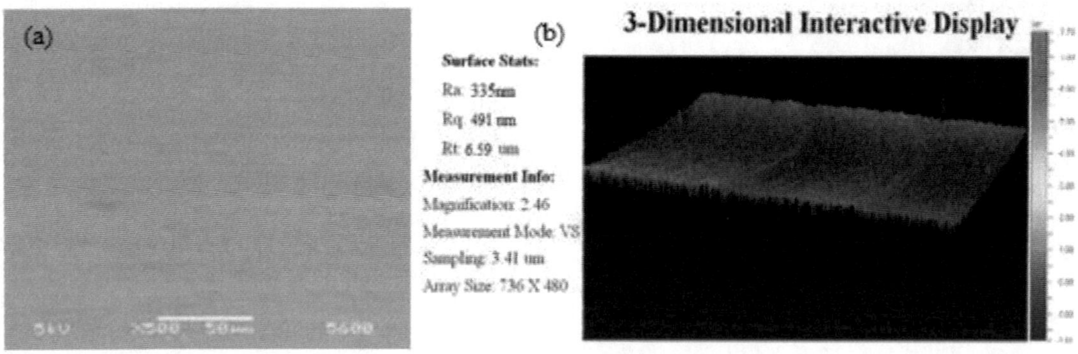

Figure 2. Confirmatory test of optimal condition (a) SEM view (b) Wyko view

4. Conclusion

It can be concluded that high speed end milling is capable to produce necessary profile on soda lime glass. The quadratic models generated for roughness parameters, R_a, R_q and R_t can be used as a performance predictor in soda lime glass milling using uncoated carbide tool. In addition, R_a can be used as a supplementary of R_q as both model shows quite similarity. The extreme parameter, R_t can be used as a signpost of the existence of chip contamination. In case of R_a and R_q, the most influencing process parameters are feed rate followed by spindle speed. Quadratic effect of spindle speed showed most influential effect on R_t. Depth of cut confirmed little contribution and its effect is insignificant for all models. The optimum roughness, R_a 0.27 μm, R_q 0.47 μm and R_t 5.57 μm was predicted to achieve at spindle speed 40,000 rpm, feed rate 10 mm/min and depth of cut 34 μm with 97 % desirability.

References

[1] Arif M et al 2011 *The Inter. J. of Adv. Manuf. Tech.* **53(9-12)** 1063-1073.
[2] Mostofa M. et al 2013 *The Inter. J. of Adv. Manuf. Tech.* **67(5-8)** 1139-1150.
[3] Zhao Y, Zhang Y, Huo D, Teng X and Shi Y 2014 *20th IEEE International Conference on Automation and Computing (ICAC).*
[4] Amin A N, Bagum M N, Fathiah N, Konneh M and Ariff T F B M 2016 *Arabian J. for Science and Eng.* **41(11)** 4553-4562
[5] Whitehouse D J 2004 *Surfaces and their Measurement* Elsevier.
[6] Standard A 1995 ASME B46. 1-1985. *Surface Texture (Surface Roughness, Waviness and Lay)*, New York.
[7] Leach R 2001 *The measurement of surface texture using stylus instruments* National Physical Laboratory.
[8] Onozuka H et al 2015 *J. of manuf. Proces.* **19** 95-101.
[9] Sreejith P 2005 *J.of Mat. Proces. Tech.* **169(3)** 414-417.
[10] Ajjarapu S K et al 2004 *AIP Conf. Proc.*

IOP Conf. Series: Materials Science and Engineering **290** (2017) 012037 doi:10.1088/1757-899X/290/1/012037

Chip morphology as a performance predictor during high speed end milling of soda lime glass

M N Bagum [1], M Konneh [1], K A Abdullah[2] and M Y Ali [1]

[1]Department of Manufacturing and Materials Engineering, Kulliyyah of Engineering, International Islamic University Malaysia, Jalan Gombak, 53100 Kuala Lumpur, Malaysia
[2]Department of Mechanical Engineering, Kulliyyah of Engineering, International Islamic University Malaysia, Jalan Gombak, 53100 Kuala Lumpur, Malaysia

Email: nasimaanam@gmail.com

Abstract. Soda lime glass has application in DNA arrays and lab on chip manufacturing. Although investigation revealed that machining of such brittle material is possible using ductile mode under controlled cutting parameters and tool geometry, it remains a challenging task. Furthermore, ability of ductile machining is usually assed through machined surface texture examination. Soda lime glass is a strain rate and temperature sensitive material. Hence, influence on attainment of ductile surface due to adiabatic heat generated during high speed end milling using uncoated tungsten carbide tool is investigated in this research. Experimental runs were designed using central composite design (CCD), taking spindle speed, feed rate and depth of cut as input variable and tool-chip contact point temperature (T_{tc}) and the surface roughness (R_t) as responses. Along with machined surface texture, R_t and chip morphology was examined to assess machinability of soda lime glass. The relation between T_{tc} and chip morphology was examined. Investigation showed that around glass transition temperature (T_g) ductile chip produced and subsequently clean and ductile final machined surface produced.

1. Introduction

Glass belongs to the hard and brittle ceramic materials. It is a challenging task to achieve ductile machining of brittle material, which is capable to attain fracture free surface. The difficulties in glass machining are mainly due to its low fracture toughness. Soda lime glass is a temperature sensitive material [1]. Researcher stated that with the increase of temperature starting at 300°C, which is around half of the glass softening temperature (T_g), fracture toughness of soda lime glass increased [2]. Hence, improved ductility was obtained near the range of T_g [3]. Again, brittle ductile transition temperature increased with the increase of strain rate. Furthermore, fracture toughness increased around 600°C, and a sharp decrease was detected above 620°C [4]. Vicker's hardness decreased as temperature increased from 20°C to 500 °C [5]. Therefore, at elevated strain rate and temperature material's brittleness index is decreased by reducing its hardness to toughness ratio.

As of late, researcher focused on achievement of ductile mode material removal due to designated cutting condition and tool geometry, typically at low speed machining [6-8]. Also, identification of critical parameters for ductile brittle transition [9]. High speed machining was reported have positive influence on achievement of ductile machining. Reddy et al. [10] developed an analytical model for the surface roughness of machinable glass ceramic under the condition of 10000 rpm to 20000 rpm, feed rate of 30 mm/min, and found that the increase of spindle speed caused the decrease of surface roughness. In the case of peripheral milling of BK-7 glass, Arif et al. [11] identified that by increasing cutting speed the critical chip thickness value could be increased. At high cutting speed and high depth of cut, adequate temperatures were generated within the work piece that make thermal softening as the prominent phase transformation phenomenon [12]. Schinker et al. [13] stated that at high speed, the generated heat in the cutting zone owing to adiabatic micro shearing gives rise to continuous chip formation and produce smooth surfaces on optical glass. Conversely, it might also modify the properties of glass in the processing zone [13]. In case of material those glass transition temperature is comparatively low, this thermal effects might be prominent [14]. However, quantitative research on thermal softening effect is not done yet. In order to assess machinability, along with ductile material removal confirmation from machined surface texture analysis, the extreme surface parameter, R_t which signpost the existence of a sharp spike or burr on the surface [15] is also important to justify process capability. Hence, in this study, along with machined surface texture, R_t was examined to assess machinability of soda lime glass during high speed end milling using uncoated carbide tool. In addition, the relation between the temperature generated due to high-speed end milling, T_{tc} and chip morphology was established.

2. Experimental procedures

4 mm diameter tungsten carbide tool having 2 flutes was chosen to perform milling on soda lime glass work piece using the upgraded vertical axis CNC milling machine. The cutting conditions of spindle speed varied from 20,000 to 40,000 rpm, cutting depth from 30 to 50 μm and feed rate from 10 to 30 mm/min were employed in dry condition. The CCD of RSM was employed for designing the experimental run. The image analysis of surface texture and chip morphology was studied using Scanning Electron Microscope (SEM), JSM-5610 and JBM-6700F, respectively. The tool chip contact point temperatures were recorded using the thermal camera Thermo Pro TP8. The surface roughness R_t, was measured using Veeco Wyko Profiling system Microscope model NT110.

3. Results and Discussion

The experimental run along with roughness R_t and temperature T_{tc} are given in table 1. Figure 1 shows the corresponding SEM image of machined surface and chip in relation to T_{tc} and R_t values. The temperature recorded during experimental run, according to table 1 confirmed that high temperatures were generated in the tool-chip contact point. Subsequently, almost in all run, ductile chips produced ,figure 1. The ductile chip formation and removal has significant effect on final machined surface. The extreme parameter, R_t signpost the existence of a sharp spike or burr on the surfaces, because of poor material processing. Considering only this parameter as an indicator of whether the ductile chips are removed from or contaminated with the machined surface, then, R_t value is equal at both 540°C and 588°C. In addition, surface textures and chip morphology shown in figure 1(b) and (c) are quite similar in these two cases. At 650°C, R_t is 5.17, also SEM image of surface and chip in figure 1(d), illustrated that ductile chip produced and removed. At 710°C, 756°C and 786°C SEM image of surface in figure 1 (g), (h) and (i), confirmed chip contamination and R_t value is high in these three case.

For clear understanding of chip removal process apart from SEM image and R_t value of machined surface, the chip morphology plays an important role. For instance, at 756°C and 786°C SEM image in figure 1 (h) and (i), are quite similar and R_t value is nearby. The chip morphology at these two conditions is different. Within the glass transition temperature Tg, (520-600°C) of soda lime glass and also up to 650°C the chip rolled uniformly and removed from the surface as cutting progress, hence clean final machined surface is produced as shown in figure 1 (a), (b), (c) and (d). Below the T_g (342°C), surface is not clean as both ductile and partial ductile chips are produced shown in figure 1 (e). As temperature increased beyond T_g chip started to molten.

IOP Conf. Series: Materials Science and Engineering **290** (2017) 012037 doi:10.1088/1757-899X/290/1/012037

Table 1 Experimental run along with roughness and temperature.

Run	Type	Spindle speed rpm	Cutting speed m/sec	Feed rate mm/min	Feed/tooth μm	C: Depth of Cut μm	Rt μm	Temp °C
1	Axial	40000	8.37	20	0.25	40	8.37	760
2	Axial	20000	4.19	20	0.50	40	5.24	748
3	Fact	40000	8.37	30	0.38	30	13.15	800
4	Center	30000	6.28	20	0.33	40	9.13	736
5	Fact	20000	4.19	10	0.25	30	6.18	342
6	Fact	40000	8.37	10	0.13	50	5.7	588
7	Fact	20000	4.19	30	0.75	30	7.16	654
8	Axial	30000	6.28	10	0.17	40	9.52	520
9	Fact	40000	8.37	30	0.38	50	8.28	790
10	Fact	40000	8.37	10	0.13	30	5.7	540
11	Center	30000	6.28	20	0.33	40	10.23	722
12	Fact	20000	4.19	10	0.25	50	8.65	422
13	Axial	30000	6.28	20	0.33	50	12.04	800
14	Center	30000	6.28	20	0.33	40	7.73	726
15	Center	30000	6.28	20	0.33	40	9.92	780
16	Axial	30000	6.28	20	0.33	30	13.89	756
17	Axial	30000	6.28	30	0.50	40	12.78	786
18	Fact	20000	4.19	30	0.75	50	5.17	650

In the temperature ranges from 710-800°C two types of surface - clean and chip contaminated are identified. Cutting speed 4.19 m/sec, feed/teeth 0.50μm and axial depth 40 μm generated T_{tc} of 748°C. In this case soften chip are removed away before contaminated on the machined surface, hence almost clean surface is generated as shown in figure 1 (f). When the parameters speed and feed per tooth both are high such as at cutting speed 6.28 m/sec, feed/tooth 0.50 μm and axial depth 40μm generated T_{tc} is 786 °C, in this case as ductile-molten chips are formed and surface is contaminated shown in figure 1 (i). At 6.28 m/sec, combination of moderate feed/tooth (0.33μm) and depth of cut (30μm) produced spiral helical ductile chip in figure 1 (h). Here, low and high strain regions formed in the chip. Hence, thermal softening is not homogeneous throughout the removable layers of material; therefore, molten chips contaminated on the machined surface.

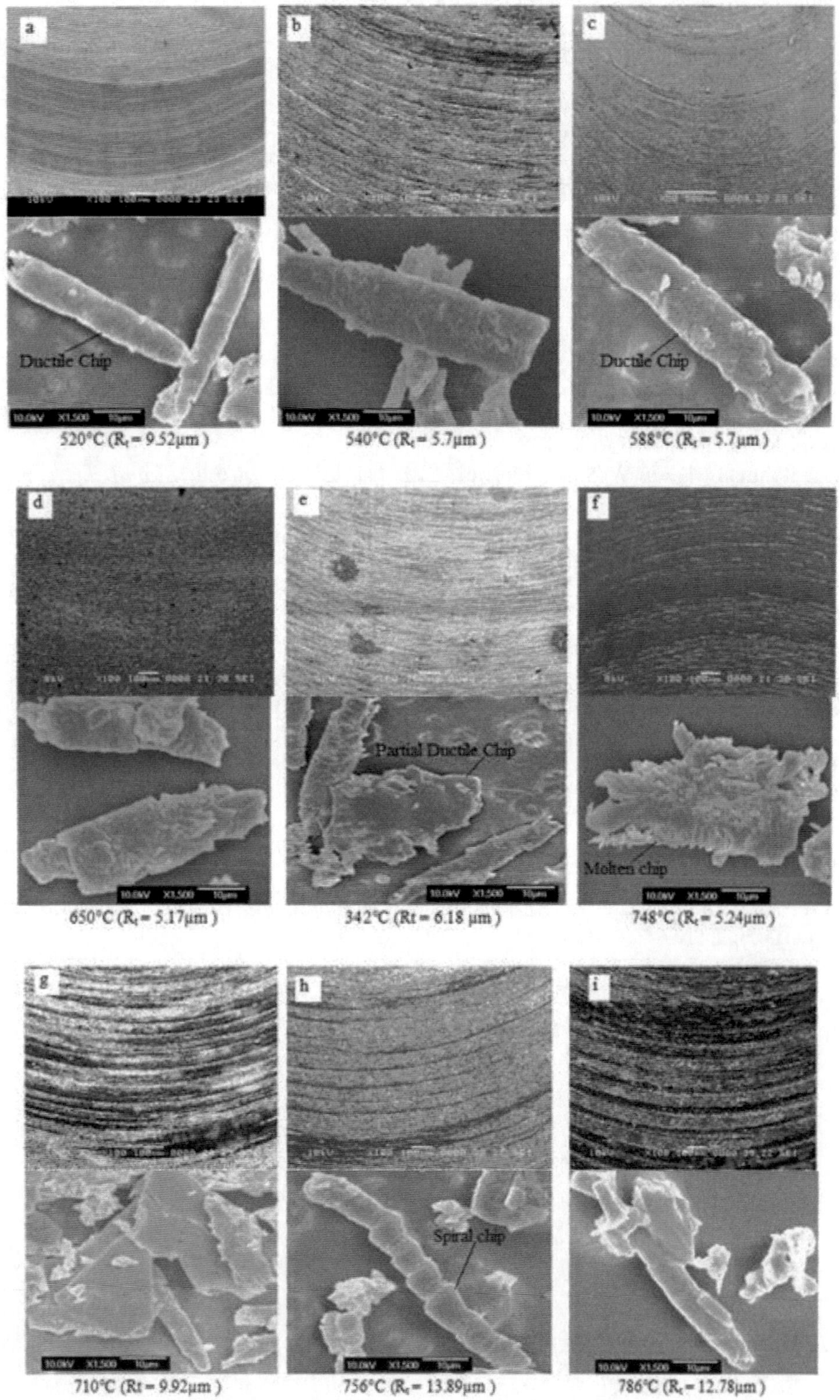

Figure 1. Tool-Chip contact point temperature, surface roughness (R_t) and chip image at different cutting condition.

4. Conclusion

In summary, it can be conclude that High-speed end milling is capable of adiabatic heating of soda lime glass and ductile chip removal. The roughness parameter, R_t can predict existence of chip contamination on the surface. Chip morphology can indicate mode of material removal in ductile machining. Around glass transition temperature, clean, ductile surface with low R_t value is generated as ductile chip removed before contaminated on machined surface. Above $T_{g,}$ both clean or chip contaminated surface produce depending on cutting parameters combinations.

References

[1] Callister W D and Rethwisch D G 2007 *Materials Science and Engineering: An Introduction* (New York: Wiley)
[2] Bourhis E L and Metayer D 2000 *J. of non-crystalline solids* **272** pp 34-38
[3] Rouxel T and Buisson M 1999 *Key Eng. Mater*
[4] Li D, Li W, Wang R and Shen X 2015 *J. of Non-Crystalline Solids* **409** pp 126-130
[5] Michel M D, Mikowski A, Lepienski C M, Foerster C E and Serbena F C 2004 *J. of non-crystalline solids* **348** pp 131-138
[6] Arif M, Rahman M, San W Y and Doshi N 2011*The Inter. J. of Adv. Manuf. Tech.* **53** pp 1063-1073
[7] Mostofa M G, Malekian M, Park S S, Jun M B and Aarts M 2013 *The Inter. J. of Adv. Manuf. Tech.* **67** pp 1139-1150
[8] Zhao Y, Zhang Y, Huo D, Teng X and Shi Y 2014 *20th IEEE International Conference on Automation and Computing (ICAC)*
[9] Amin A N, Bagum M N, Fathiah N, Konneh M and Ariff T F B M 2016 *Arabian J. for Science and Eng.* **41** pp 4553-4562
[10] Reddy M M, Gorin A and Abou-El-Hossein K 2011 *IOP Conference Series: Mater Science and Eng.* IOP Publishing
[11] Arif M, Rahman M and San, W Y 2012 The Inter. J. of Adv. Manuf. Tech. **60** pp 487-495
[12] Ajjarapu S K, Fesperman R R, Patten J A, Cherukuri H P and Brand C 2004 *AIP Conf. Proc*
[13] Schinker M G 1991 Preci Eng. **13** pp 208-218
[14] Owen J D, Davies M A, Schmidt D and Urruti E H 2015 *CIRP Annals-Manuf. Tech.* **64** pp113-116
[15] Leach R 2001 The measurement of surface texture using stylus instruments *National Physical Laboratory*

IOP Conf. Series: Materials Science and Engineering **290** (2017) 012038 doi:10.1088/1757-899X/290/1/012038

Establishing Relationship between Process Parameters and Temperature during High Speed End Milling of Soda Lime Glass

Mst. Nasima Bagum, Mohamed Konneh and Mohammad Yeakub Ali

Department of Manufacturing and Materials Engineering, Kulliyyah of Engineering, International Islamic University Malaysia, Jalan Gombak, 53100 Kuala Lumpur, Malaysia

Email: nasimaanam@gmail.com

Abstract. In glass machining crack free surface is required in biomedical and optical industry. Ductile mode machining allows materials removal from brittle materials in a ductile manner rather than by brittle fracture. Although end milling is a versatile process, it has not been applied frequently for machining soda lime glass. Soda lime glass is a strain rate and temperature sensitive material; especially around glass transition temperature T_g, ductility increased and strength decreased. Hence, it is envisaged that the generated temperature by high-speed end milling (HSEM) could be brought close to the glass transition temperature, which promote ductile machining. In this research, the objective is to investigate the effect of high speed machining parameters on generated temperature. The cutting parameters were optimized to generate temperature around glass transition temperature of soda lime using response surface methodology (RSM). Result showed that the most influencing process parameter is feed rate followed by spindle speed and depth of cut to generate temperature. Confirmation test showed that combination of spindle speed 30,173 rpm, feed rate 13.2 mm/min and depth of cut 37.68 μm generate 635°C, hence ductile chip removal with machined surface R_a 0.358 μm was possible to achieve.

1. Introduction

Glass belongs to the hard and brittle ceramic materials. Under compression, glass exhibits softening of bulk modulus [1]. Glass appears to be strain rate sensitive, as loading rate increase tensile strength become high .Compression test at larger strain rate reported increase in its ultimate tensile strength UTS [2]. In addition, dynamic tensile is reported higher than static tensile strength [3]. At elevated temperature under indentation loading soda lime glass undergoes material softening, which showed effect on its mechanical properties such as fracture toughness, hardness and Young's modulus [4]. The fracture toughness of soda lime glass increased strongly with temperature starting at a temperature 300°C, which is around half of the glass softening temperature [4] and improved ductility was obtained near the range of T_g [5]. The fracture toughness increased around 600 °C, and a sharp decrease was detected above 620 °C [6]. Vicker's hardness decreased as temperature increased from 20°C to 500 °C [7]. Therefore, at elevated strain rate and temperature material's brittleness index decreases by reducing its hardness to toughness ratio.

IOP Conf. Series: Materials Science and Engineering **290** (2017) 012038 doi:10.1088/1757-899X/290/1/012038

High-speed cutting impose high amount of strain rate on materials and temperature generated at chip formation zone allows more plastic deformation than the low speed cutting. Reddy et al. [8] found that the increase of spindle speed caused the decrease of surface roughness. In the case of peripheral milling of BK-7 glass, Arif, et al. [9] identified that the critical chip thickness value could be increased by increasing cutting speed. Adiabatic melting and instant annealing of inorganic glass at higher cutting speed was stated [10] favorable to achieve stress free surface. Schinker et al. [11] stated that at high speed, the generated heat in the cutting zone owing to adiabatic micro shearing gives rise to continuous chip formation and produce smooth surfaces on optical glass. Conversely, it might also modify the properties of glass in the processing zone [11]. At high cutting speed and high depth of cut, adequate temperatures were generated within the work piece that make thermal softening as the prominent phase transformation phenomenon [12]. In case of material those glass transition temperature is low this thermal effects might be prominent [13]. However, quantitative research on thermal softening effect is not done yet. Hence, in this study investigation of the effect of high speed cutting parameters on tool-chip contact point temperature is carried out. Response surface model was generated using central composite design of RSM. The parameters were optimized at glass transition temperature range (520-600) °C to justify machinability of soda lime glass at T_g.

2. Experimental procedures

4 mm diameter tungsten carbide tool having 2 flutes was chosen to perform milling on soda lime glass work piece using the upgraded vertical axis CNC milling machine. The cutting conditions of spindle speed from 20,000 to 40,000 rpm, feed rate from 10 to 30 mm/min and depth of cut from 30 to 50 μm were employed in dry condition. The tool chip contact point temperatures were measured using IR thermal camera. The CCD of RSM was employed for designing and analyzing the experimental results.

3. Results and Discussion

The outcome of the experiment is represented in Table 1. Response model development followed the steps such as problem formulation; transformation checking and selection of the model that provides the finest relationship between dependent responses and independent input variables according to the instruction of fit summary tests; Analysis of variance (ANOVA); Model diagnostics.

Table 1 Central Composite Design using RSM along with the response value.

Run	Type	A: Spindle speed Rpm	B: Feed rate mm/min	C: Depth of Cut μm	Temp °C
1	Axial	40000	20	40	760
2	Axial	20000	20	40	748
3	Fact	40000	30	30	800
4	Center	30000	20	40	736
5	Fact	20000	10	30	342
6	Fact	40000	10	50	588
7	Fact	20000	30	30	654
8	Axial	30000	10	40	520
9	Fact	40000	30	50	790
10	Fact	40000	10	30	540
11	Center	30000	20	40	722
12	Fact	20000	10	50	422
13	Axial	30000	20	50	800
14	Center	30000	20	40	726
15	Center	30000	20	40	710
16	Axial	30000	20	30	756
17	Axial	30000	30	40	786
18	Fact	20000	30	50	650

In the fit summary statistics, the Sequential Model Sum of Square (SMSS) as well as Model Summary Statistics endorsed quadratic relationship between the independent variables such as spindle speed (A), feed rate (B) and depth of cut (C) and the response temperature (T). Lack of fit value showed insignificant. The analyses of variance (ANOVA) with all possible model terms implied that the model is significant with large F-value. The significant model terms are A, B and B^2. The "Lack of Fit F-value" of the model implies that it was not significant relative to the pure error. Also predicted R^2 was in reasonable agreement with the adjusted R^2 value as required by the design expert 7.0.0. Using back ward regression method, the non-significant terms that are not contributing to model's hierarchy were removed to improve model. The model equations are therefore obtained based on the reduced model ANOVA shown in Table 2 as in equation (3.1) in terms of coded factors and equation (3.2) in terms of actual factors.

$$\text{Temperature} = +753.50 + 66.20\,A + 126.80\,B + 15.80\,C - 144.30\,B^2 \qquad (3.1)$$

$$\text{Temperature} = -339.10 + 6.62\times10^{-3}\,A + 70.40\,B + 1.58\,C - 1.44\,B^2 \qquad (3.2)$$

Table 2 Analysis of variance and descriptive statistics of reduced model T.

	Sum of Squares	df	Mean Square	F Value	p-value Prob > F	
Model	2.996E+005	4	74911.90	45.19	< 0.0001	significant
A-Spindle speed	43824.40	1	43824.40	26.43	0.0002	
B-feed rate	1.608E+005	1	1.608E+005	96.98	< 0.0001	
C-Depth of cut	2496.40	1	2496.40	1.51	0.2415	
B^2	92544.40	1	92544.40	55.82	< 0.0001	
Residual	21552.40	13	1657.88			
Lack of Fit	19420.40	10	1942.04	2.73	0.2208	not significant
Pure Error	2132.00	3	710.67			
Cor Total	3.212E+005	17				
R-Squared	0.9329					
Adj R-Squared = 0.91, Pred R-Squared =0.87, Adeq Precision=19.54, C.V. %=6.05						

In equation (3.1) the coefficients of each factor represents the effect of this particular factors on temperature. According to equation (3.1) as well as perturbation plot in Figure 1, it is clear that each of the factors showed positive influence on response and the most influencing process parameters are depicted as feed rate followed by spindle speed and depth of cut. Feed rate showed negative quadratic effect and hence, there is a continuous increase in T up to an optimal point, than decreased with further increase in this response. Depth of cut confirmed little contribution and its effect is insignificant in this model. The adequacy of each of the model was verified through the statistical features associated with it. Here R^2 was greater than 0.8, the difference of predicted R^2 with adjusted R^2 was less than 0.2; adequate precision was greater than 4. The residuals of the data were normally scattered (Figure 2), similarly constant variance assumption was satisfied, and residuals were not correlated.

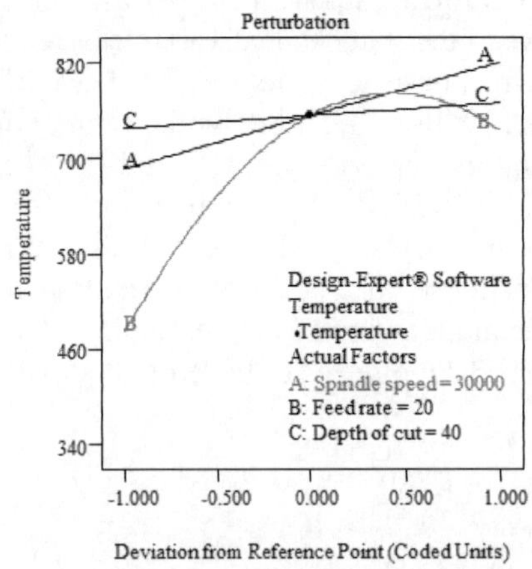

Figure 1. Perturbation plots.

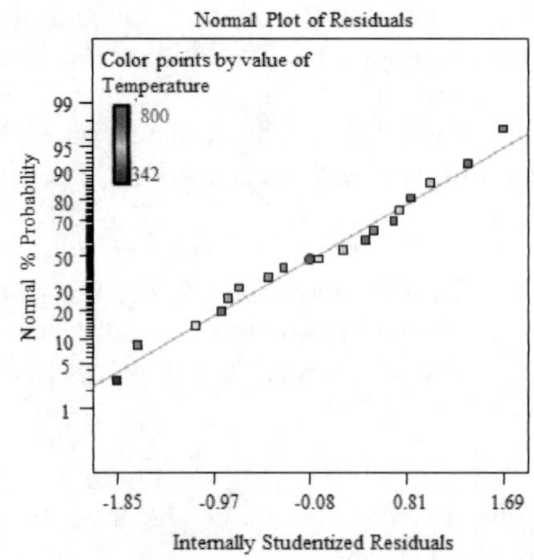

Figure 2. Normal plot of residuals.

The relationship between theoretical and experimental value of T is depicted in Figure 3 also suggests that predicted value is closer to experimental value. Hence, it was concluded that model equation (3.2) was adequate for application in navigating the design space within the experimental limit.

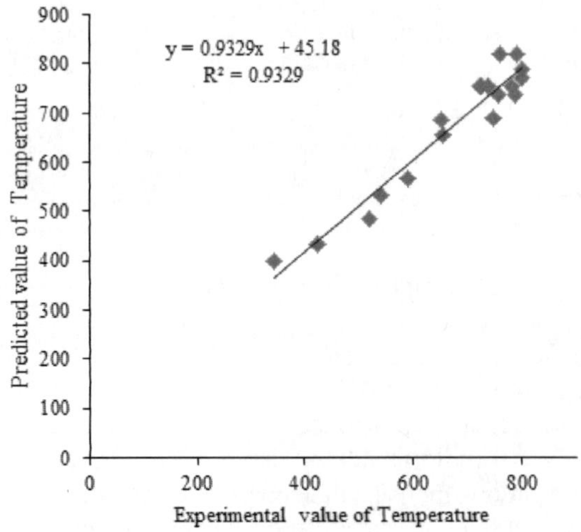

Figure 3. Relationship between predicted and experimental T.

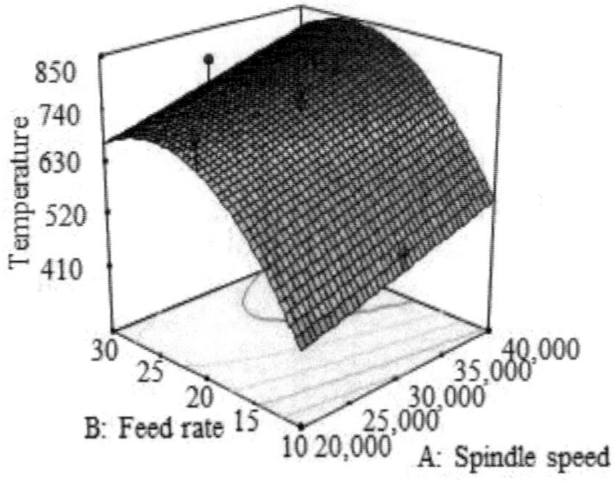

Figure 4. Spindle speed and feed rate interaction.

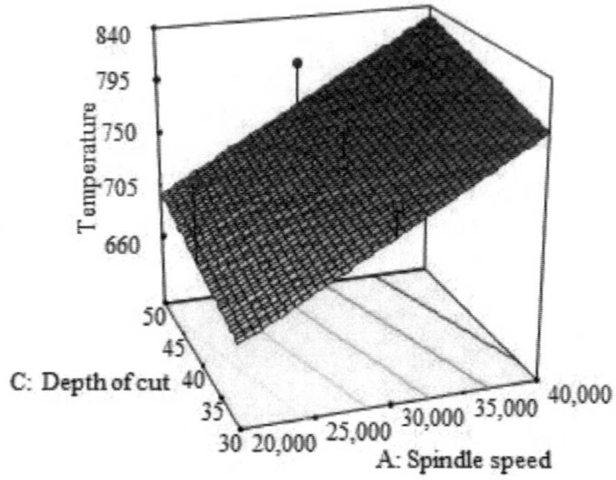

Figure 5. Spindle speed and depth of cut interaction.

The 3D plot of temperature for feed rate and spindle speed interaction demonstrated in Figure 4 shows that when both feed rate and spindle speed increased temperature rises sharply. On the other hand, the depth of cut and spindle speed interaction plot shown in Figure 5 illustrated that effect of depth of cut is almost constant. According to desirability function of RSM, the optimal combinations of parameters to achieve glass transition temperature (520- 600°C) provide 27 alternative solutions with 100% desirability. The optimal combination of parameters spindle speed 30,173 rpm, feed rate 13.2 mm/min and depth of cut 37.68 μm is predicted to achieve 597.94°C. Confirmation test showed that this optimal combination produced 635°C. Due to adiabatic heating ductile chip removal was possible, with surface R_a 0.358 μm shown in Figure 6.

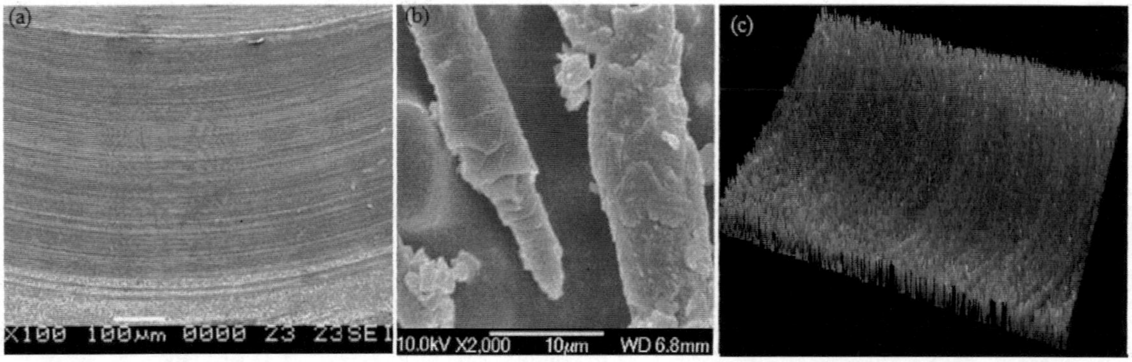

Figure 6. Optimal condition (a) Surface integrity, (b) Chip morphology (SEM view), (c) Surface roughness Wyko 3D display.

4. Conclusion

High speed end milling is capable to achieve ductile removal of soda lime glass around glass transition temperature. Feed rate is the most influencing parameter followed by spindle speed and depth of cut. The combination of 30173 rpm, 13.2 mm/min feed rate and 37.68 μm depth is predicted to achieve 597.94°C with 100% desirability. Confirmation test at optimal combination produced 635°C, where ductile chip removal along with machined surface R_a, 0.358 μm was achieved.

IOP Conf. Series: Materials Science and Engineering **290** (2017) 012038 doi:10.1088/1757-899X/290/1/012038

Reference

[1] Cagnoux J et al 1982 *AIP Conference Proceedings*.
[2] Peroni M et al 2011 *Applied Mechanics and Mater*. Trans Tech Publ.
[3] Zhang X et al 2012 *Inter. J.of Protective Structures* **3(4)** 407-430.
[4] Bourhis, E L and Metayer D 2000 *J. of non-crystalline solids* **272(1)** 34-38.
[5] Rouxel T and Buisson M 1999 *Key Eng. Mater*.
[6] Li D et al 2015 *J. of Non-Crystalline Solids* **409** 126-130.
[7] Michel M et al 2004 *J. of Non-Crystalline Solids* **348** 131-138.
[8] Reddy M M, Gorin A and Abou-El-Hossein K 2011 *IOP Conference Series: Mater Science and Eng*. IOP Publishing.
[9] Arif M, Rahman M and San, W Y 2012 The Inter. J. of Adv. Manuf. Tech. **60(5)** 487-495.
[10] Schinker M and Döll W 1982 *Le Journal de Physique Colloques* **43**(C9) C9-603-C9-606.
[11] Schinker M G 1991 *Preci. Eng.* **13(3)** 208-218.
[12] Ajjarapu S K et al 2004 *AIP Conf. Proc*.
[13] Owen J et al 2015 CIRP Annals-Manuf. Tech. **64(1)** 113-116.

IOP Conf. Series: Materials Science and Engineering **290** (2017) 012039 doi:10.1088/1757-899X/290/1/012039

Tool Wear Mechanisms during Cutting of Soda Lime Glass

Mohamed Konneh, Mst. Nasima Bagum, Mohammad Yeakub Ali, Tasnim Firdaus Bt. Mohamed Arif

Department of Manufacturing and Materials Engineering, Kulliyyah of Engineering, International Islamic University Malaysia, Jalan Gombak, 53100 Kuala Lumpur, Malaysia

Email: mkonneh@iium.edu.my

Abstract. Soda lime glass milling has high performance application. It is a challenging task to achieve fracture free surface on this material due to its brittle nature. High-speed end milling is capable to achieve ductile mode in an enhance flexibility. In this research, end milling of soda lime using uncoated carbide tool was performed where spindle speed varied from 20,000 to 40,000 rpm, cutting depth from 10 to 30 μm and feed rate from 5 to 20 mm/min in dry condition. The effects of cutting parameters (cutting speed, feed per tooth and depth of cut) on tool flank wear as well as wear mechanisms of tool flank investigated. Investigation showed that feed per edge has most influencing effect followed by cutting speed and depth of cut on flank wear and the main wear mechanism is abrasion wear. In some cases, oxidation, thermal diffusion and recast layer on tool flank also observed.

1. Introduction

As of late, researcher focused on achievement of ductile mode material removal due to designated cutting condition and tool geometry, typically at low speed machining. Negative rake angle [1] and micro level radius of cutting edge [2] have proven favorable to achieve ductile mode. In soda lime glass milling, uncoated tungsten carbide tools were used by researcher to utilized adiabatic heating benefits [3], as the temperatures generated during machining using the uncoated tool are higher than those by the coated tools. However, one or more combinations of the criteria such as cutting tool wear, cutting tool life, power consumption, chip formation, machined surface integrity, geometrical accuracy, etc. [4] usually define machining performance of a work material. Tool wear is a complicated issue and still not clear even in case of metal machining and generally depends on the work piece, tool materials, tool geometry, machining parameters, lubricating condition, machine tools characteristics. For wear, monitoring flank wear is commonly used [5]. Contradictory illustrations are found regarding applicability of uncoated carbide tool, mostly in machining of brittle materials. It has proven better performance in terms of cutting force and surface roughness in milling of carbon fiber reinforced composite materials [6]. At high cutting speed, uncoated carbide tools fail mainly due to flank wear because of abrasion, notching, fracturing, cracking. Although coating believed to enhance the tool life, but once coating removed, the tool failure is similar to that of the uncoated tool. In milling process because of intermittent cutting action cyclic stress developed, hence once coating removed micro chipping of cutting edge also observed [7]. In case of turning of composites, wear morphologies of carbide was corner wear due to abrasion, adhesion and micro breakage in the cutting edge [5]. In diamond cutting of soda lime glass main wear mechanism were mechanical, thermo-chemical and abrasion wear [8]. The mechanism of wear was abrasion dominant while TiAlN coated tool used for glass milling and diffusion of tool material was not present [9] because of low cutting speed.

In this research, end milling of soda lime glass performed using uncoated carbide tool at high speed. The purpose of the study is to investigate the effect of cutting parameters (cutting speed, feed per tooth and depth of cut) on uncoated carbide tool flank wear as well as wear (flank) mechanisms during high speed milling of soda lime glass.

2. Experimental procedures

The Box-Behnken Design was employed for designing the experimental run where the spindle speed varied from 20,000 to 40,000 rpm, cutting depth from 10 to 30 μm and feed rate from 5 to 20 mm/min in dry condition. The uncoated solid carbide flat end-milling tools having 4mm diameter, two flutes, edge radius 6 μm, rake angle -5° and grain size 0.6 μm were used to perform milling on soda lime glass work piece, in an upgraded vertical axis CNC milling machine. In each experimental run, tool flank wear after performing 75 mm milling length were measured using optical microscope. Tool wear mechanism were observed using Scanning Electron microscope (SEM) JSM-5610.

3. Results and Discussion

According to IS0-8688-2-1982, the milling time at which the tool flank wear V_b reached to 0.3mm is consider as tool life. In the cutting process of soda lime glass, lot of chips in powder form generated on the face of the tool and brittle break occurred at the tool corner. In real machining, surface integrity and dimensional accuracy are of primary concern. Therefore, in this research at the lowest value of cutting parameters i.e., spindle speed 20,000 rpm, feed rate 5 mm/min and depth of cut 10 μm, both of the tool flank wear and surface integrity of work piece were examined after certain milling path. The result showed that after 75 mm milling length, indentation marks were visible on the machined glass surface as shown in Figure 1, due to edge chipping of the cutting tool. Hence, instead of 0.3mm V_b, considering the surface integrity, at each run the tool flank wear after 75 mm milling length was consider as the limiting value of tool wear. The shape of the wear area of the tool looks like an inverted triangle below the corner of the tool (Figure 2). Only maximum wear V_c was observed and recorded as flank wear.

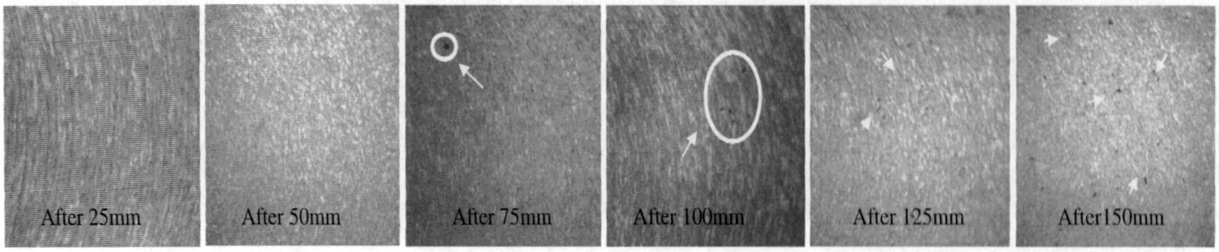

Figure 1. The machined surface integrity at different milling length.

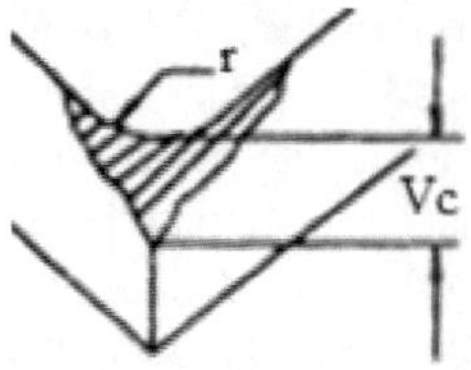

Figure 2. Shape of wear.

Table 1 shows the experimental results of the flank wear as designed by Box-Behnken design. In case of milling process, combination of spindle speed and feed rate along with no of cutting flute, determine the value of feed per tooth. The interaction plot in Figure 3 (a) illustrates that feed rate has a much stronger effect on V_b when spindle speed is at lower level. At lower level of spindle speed with the increase of feed rate tool wear increased sharply. This is because low cutting speed and high feed rate resulted in higher feed/tooth, which caused increased tool wear due to mechanical loading. At upper level of spindle speed, with the increase in feed rate tool wear increased up to an optimal point followed by little or no change in V_b as this parameter further increased. This relation more clearly depicted in the contour plot shown in Figure 3 (b).

Table 1 The Box-Behnken design matrix with response for flank wear.

Run	Type	Factor 1 Spindle speed(rpm)	Cutting speed m/min	Factor2 Feed rate (mm/min)	Feed /tooth(μm)	Factor3 Depth of Cut (μm)	Response 1 Flank wear (V_b)(μm)
1	Center	30000	376.8	12.5	0.21	20	189
2	Center	30000	376.8	12.5	0.21	20	200
3	Ibfact	40000	502.4	5	0.06	20	182
4	Ibfact	20000	251.2	20	0.50	20	228
5	Ibfact	40000	502.4	20	0.25	20	219
6	Center	30000	376.8	12.5	0.21	20	193
7	Ibfact	30000	376.8	20	0.33	10	193
8	Ibfact	20000	251.2	12.5	0.31	10	172
9	Center	30000	376.8	12.5	0.21	20	197
10	Ibfact	20000	251.2	12.5	0.31	30	200
11	Ibfact	30000	376.8	5	0.08	30	165
12	Ibfact	40000	502.4	12.5	0.16	30	210
13	Center	30000	376.8	12.5	0.21	20	191
14	Ibfact	30000	376.8	5	0.08	10	143
15	Ibfact	20000	251.2	5	0.13	20	161
16	Ibfact	30000	376.8	20	0.33	30	217
17	Ibfact	40000	502.4	12.5	0.16	10	202

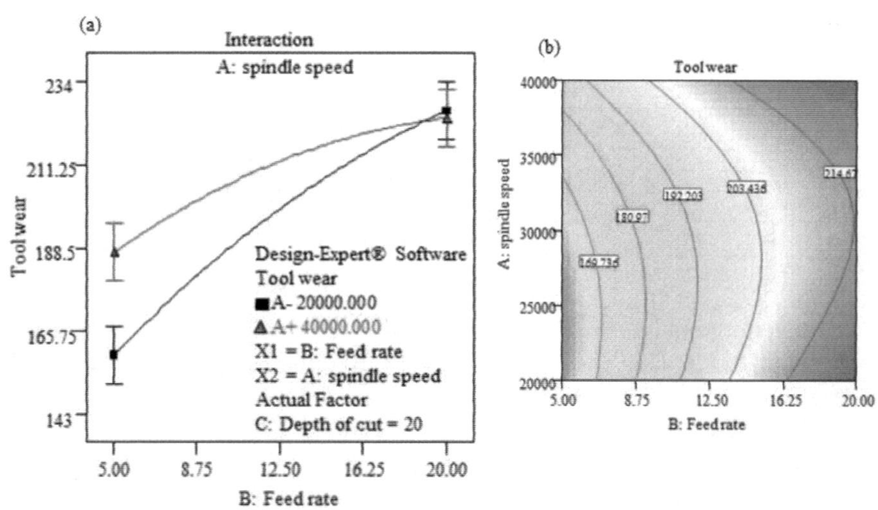

Figure 3. Interaction plot and contour plot of V_b with feed rate and spindle speed (a) Interaction plot and (b) contour plot.

In case of milling process, feed per tooth is reported as the most influencing parameter for ductile machining of brittle materials [2]. This is also true for tool flank wear. The effects of feed/tooth, cutting speed and depth of cut on tool flank wear further illustrated in Figure 4. As depicted in Figure 4 (a), feed / tooth showed dominating influence on tool wear among the cutting parameters. Among the three identified zone, Zone-1 revealed that even though feed/tooth increased, there is a decrease in wear, explained by Figure 4 (b). At 5 mm/min feed rate, flank wear at 251 m/min cutting speed is lower than that of 502 m/min, because at faster cutting speed hard particle from work piece scraped the tool more violently. Zone-2, started at 0.13µm feed/tooth where tool wear is minimum, later as the feed / tooth and cutting speed increased wear increased up to 0.25 µm feed/tooth. Subsequently Zone-3 started, where the tool wear increased at 251 m/min cutting speed due high value of feed/tooth. Furthermore, Figure 4 (b) clarifies that at constant depth of cut, 20 µm and feed rate 20 mm/min, tool wear (high) is almost constant with respect to cutting speed. Also at high cutting speed, the depth of cut showed insignificant effect on tool wear (Figure 4 c).

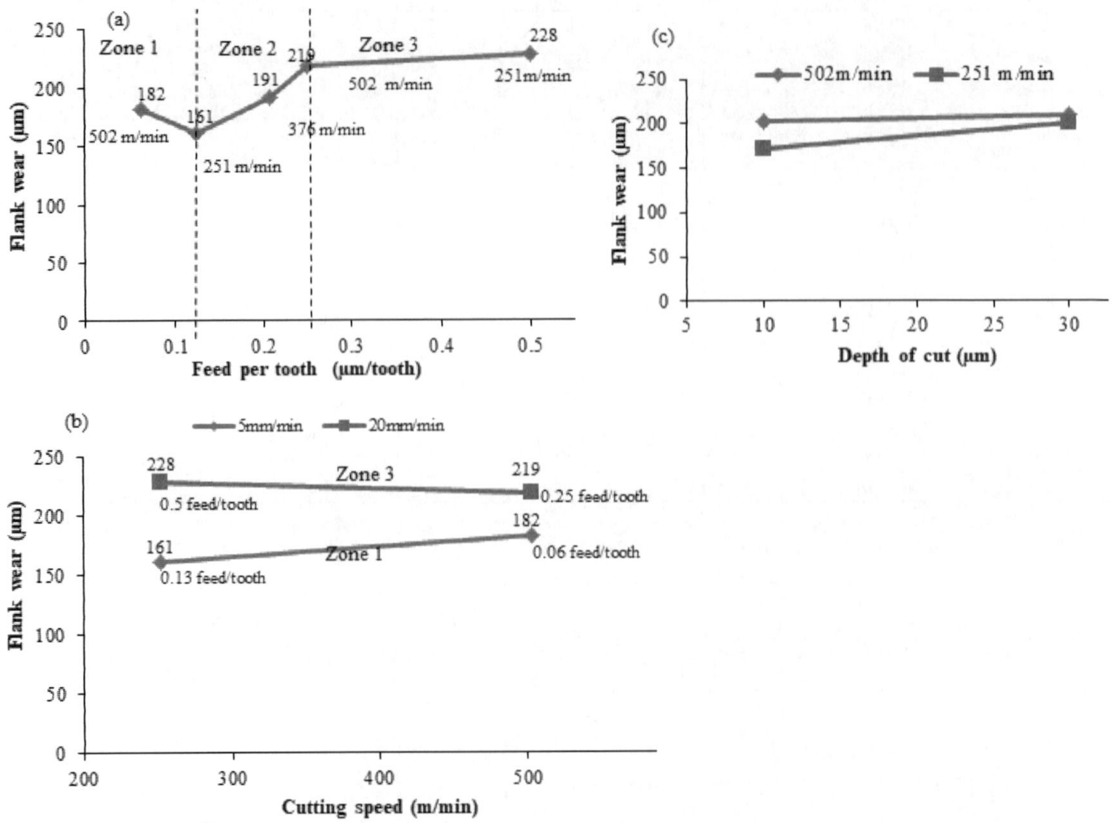

Figure 4. Effect of cutting parameters on tool wear (a) Feed per tooth at constant depth of cut 20µm (b) Cutting speed at different feed rate at constant depth 20 µm (c) Depth of cut at different speed.

At the beginning of the milling process, a brittle break occurred at the corner of the carbide tool in all run. After that edge become round and gradual flank wear occurred. By means of microscopic examination, the major wear form of the tool found was abrasive wear. However, buildup edge also formed some time since there is aluminum exists in soda lime. At extreme combination of cutting parameters, (at cutting seed 502.4 m/min, feed per tooth 0.25µm and depth of cut 30 µm.) oxidation, recasting and thermal deformation occurred due to high temperature generated because of friction shown in Figure 5.

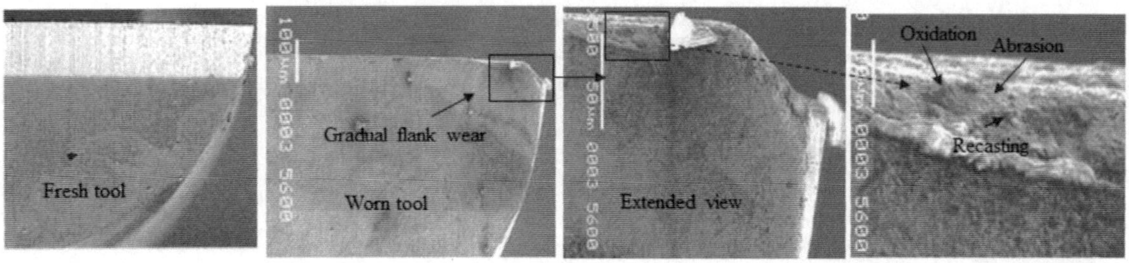

Figure 5. The tool wears mechanisms at high spindle speed, feed rate and depth of cut condition.

At high speed, moderate feed/tooth and low depth cut [502.5 m/min, feed/tooth 0.16 µm and depth of cut 10 µm] along with abrasion wear, thermal deformation and oxidation occurred at tool flank shown in Figure 6. At 20 µm depth of cut, low cutting speed (251.2 m/min) and high feed/tooth (0.5 µm) combination caused high tool wear due to mechanical loading, on the other hand high cutting speed (502.2 m/min) and high feed/tooth (0.25 µm) combination formed recast layer on tool flank, as shown in Figure 7. At this condition, the temperature generated was nearby dissolution temperature of carbide tool (1100°C) captured by thermal camera.

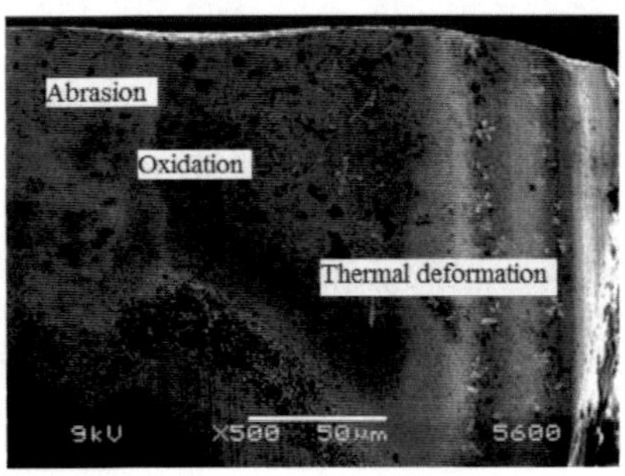

Figure 6. Thermal deformation at cutting speed 502.5 m/min, feed per tooth 0.16 µm and depth of cut10 µm (run 17).

Figure 7. Recast layer at cutting speed 502.2 m/min, feed per tooth 0.25µm and depth of cut 20µm (run 5).

4. Conclusion

From the experimental results it is can be concluded that among the cutting parameters feed per tooth has the most dominant influence on tool flank wear followed by cutting speed and depth of cut. However, at high level of cutting speed (502 m/min), the depth of cut showed no significant effect on tool wear. From the perspective of the wear mechanism, the major wear form of the tool flank is abrasive wear. Furthermore, oxidation, recasting and thermal deformation of tool flank wear occurred due to high temperature generated at extreme cutting condition (at cutting seed 502.4 m/min, feed per tooth 0.25µm and depth of cut 30 µm).

References

[1] Yan J, Zhao H, Kuriyagawa T 2009 Semi. Sci and Tech. **24** 075018
[2] Arif M, Rahman M and San W Y 2011 *Int. J. of Mach. Tools and Manuf.* **51(3)** 170-181.
[3] Amin A N, Bagum M N, Fathiah N, Konneh M and Ariff T F B M 2016 *Arabian J. for Sci. and Eng.* **41(11)** 4553-4562
[4] Cheng K 2008 Machining dynamics: fundamentals, applications and practices, *Springer Science & Business Media.*
[5] Wei X et al 2016 *The Int. J. of Adv. Manuf. Tech.* **86(1-4)** 613-619.
[6] Materiala R K P 2016 I *Materiali in Tehnologije* **50(4)** 591-600.

[7] Arif M, Rahman M and San W Y 2011 *J. of Manuf. Processes* **13(1)** 50-59.
[8] Jia P and Zhou M 2012 *Tribology* **1** 05.
[9] Arif M, Rahman M, San W Y and Doshi N 2011 *The Int. J. of Adv. Manuf. Tech* **53(9)** 1063-1073.

IOP Conf. Series: Materials Science and Engineering **290** (2017) 012040 doi:10.1088/1757-899X/290/1/012040

Toolpath strategy for cutter life improvement in plunge milling of AISI H13 tool steel

E Y T Adesta[1], Avicenna[1*], I hilmy[1], M R H C Daud[1]

[1]Department of Manufacturing and Materials Engineering, International Islamic University Malaysia (IIUM), Jalan Gombak, 53100 Kuala Lumpur, Malaysia

Email: aan.cnive@gmail.com

Abstract. Machinability of AISI H13 tool steel is a prominent issue since the material has the characteristics of high hardenability, excellent wear resistance, and hot toughness. A method of improving cutter life of AISI H13 tool steel plunge milling by alternating the toolpath and cutting conditions is proposed. Taguchi orthogonal array with L9 (3^4) resolution will be employed with one categorical factor of toolpath strategy (TS) and three numeric factors of cutting speed (*Vc*), radial depth of cut (a_e), and chip load (f_z). It is expected that there are significant differences for each application of toolpath strategy and each cutting condition factor toward the cutting force and tool wear mechanism of the machining process, and medial axis transform toolpath could provide a better tool life improvement by a reduction of cutting force during machining.

1. Introduction

Plunge milling is a milling process that differs from traditional milling methods where the cutting process performs at the end of the tool instead of the periphery, making it advantageous due to the change in the direction of the cutting forces from predominantly radial to axial. Plunge milling can be also being categorized as an alternate method when side milling is not possible due to a strong vibration. The process also called as Z-axis milling which includes three phases: plunging phase, rising phase, and offset phase [1]. Material removal occurs on the first phase of plunging, where then the cutter retracts during the rising phase and moves to the next position during the offset phase [2].

The advantage of the plunge milling process is that the process tends to be more vibration-free than conventional end-milling [3]. This is due to the feed axis coincides with the most rigid spindle axis direction. As a result, plunge milling could reduce the deformation of a workpiece and improve the machining efficiency. For that reason, plunge milling recently became popular in roughing of cavities in the die, mold, and aerospace industries.

Plunge milling is generally used for hard materials with significant gains in productivity, particularly in the case of deep workpieces. Machinability of AISI H13 tool steel is a prominent issue since the material has the characteristics of high hardenability, excellent wear resistance, and hot toughness. This is because AISI H13 tool steel is a hardened steel with an excellent mechanical property with some wide applications in the field of machining hot work molds. The material is popular in the manufacturing process of plastic injection molds, pressure casting dies, forging, extrusion tools, and many more [4]. With the tough material, the setup of cutting parameters and conditions should be set up appropriately, since in plunge milling a significant increase of radial depth and friction in rising phase have great influences on cutter life [2].

There have been many significant researches on the plunge milling utilization for hard materials machining. Zhuang et al. [5] conducted a research on the milling process for nickel-base alloy Inconel 718, a material with high strength, good ductility and anti-fatigue. The study was conducted to predict the cutting force which affects greatly on choosing milling parameters, machining time and cutting tools. A later study on the milling conditions by Sun et al. [6] on the plunge milling for damage tolerant titanium alloy, gave a result that the cutting temperature had more influence than cutting force under the same parameters, and higher cutting heat was the main reason of the tool failure.

Toolpath is another critical factor in a milling process regardless the cutting type and conditions. A poorly-defined toolpath can cause material damage, damage to the tool, and can potentially cause bodily harm. There are very real physical limits to how the toolpath should be defined. For each tool, the manner and amount of material to be removed is controlled through the speed of the tool and the path it follows. Since toolpath plays a significant role in the plunge milling process of hardened material, a method of improving cutter life of AISI H13 tool steel plunge milling by alternating the toolpath and cutting conditions is proposed.

2. Cutting toolpath

Toolpath can be simply defined as the movement of the tool during the milling operation. The best toolpath for one operation is dependent on geometry of the part, the machine characteristic, and cutting condition [7]. However, different from any other traditional milling process, a plunge milling operation has some distinctive characteristics on the process thus requiring its own strategy. In basic, for each plunging cycle, the tool should retract away from the wall during the overall retract to reduce chatter and increases tool life when machining hard materials. Retracting the tool from the wall may give a slight improvement to the tool life by 10 to 15 percent [8].

Several studies that relate toolpath strategy to the milling process of hardened materials can be found in the recent years. Liang et al. [9] performed a study in plunge milling by addressing a novel approach to optimize tool orientation and tool location determination of an open blisks. The result shows that the application of modified particle swarm optimization method (PSO) optimized the tool orientation and corresponding non-interferences locations, proving the approach to be efficient and reliable to plan toolpaths for four-axis plunge milling.

Sun et al. [10] in their study proposed for a new plunge milling tool path generation method using medial axis transform to control the radial depth to improve the cutting efficiency and cutter life. The method was based on the inscribed circle of the offset contour generated by medial axis transform, and the result showed that radial depth can be easily controlled in this method. Experiments showed that both the cutting efficiency and the cutter life were improved.

Another study by Pralea & Nagit [11] put the effort to construct a type of milling based on the plunge milling and raster milling toolpath strategy, only this time it was not being used for roughing but only for finishing vertical and angled surfaces of components of injection molds, with the help of a three-axis vertical machining center. This toolpath strategy designed, with a commercial CAM program, for finishing the vertical and angled walls, offers a better surface roughness, and a better dimensional control (accuracy), both for the milling tool and for the work piece.

Almost all the studies mentioned indicate that the most critical factor in the outcome of the milled model is the strategy for removing the material. This strategy will vary depending on the geometry of the workpiece and the material being used. There are several types of cutting toolpath strategy that have been used in milling process, and some are directly adapted to the plunge milling operation.

3. Toolpath strategies in plunge milling

The idea on the selection of the toolpath strategy return to the basic criterion that a pattern should results in less machining time or the short path length [7]. There are many popular software in the market equipped with CAM supporting plunge milling toolpath, and most of them are adapted from the conventional milling strategy. But even if the software does not support the function, a little bit of g-code programming and some decent adjustments with any CAD software may do the work.

Several types of commercial strategy for moving the tool that may commonly be found are the rough parallel (direction parallel), which is also known as zigzag toolpath, and constant scallop (contour parallel) which is also called as spiral toolpath [7, 12]. These conventional strategies are appealing and widely used for machining in many CAD/CAM software, e.g. Mastercam, Cimatron, etc. This is because these toolpath strategies are both computationally tractable and geometrically appealing [13]. For the sake of the introduction to the experimental stage of the research, a recent method of medial axis transform toolpath is also introduced in the description at section 3.3.

3.1 Rough parallel

Rough parallel or zigzag toolpath is typically the first toolpath used in a model to removes bulk of material from all surfaces. It is the path segments which correspond to back and forth motion in a fixed direction within the boundary of a 2D cross-section [7]. Kim and Choi [14] divided the toolpath into three types of directions: 1) unidirectional path, 2) pure zig-zag path, and 3) smooth zig-zag path with contour parallel path. Their study for the proposal of machining time model that takes the effects of tool acceleration and deceleration showed that the smooth zigzag path was the most efficient regardless of the feed rate and the path interval.

The rough parallel toolpath moves the tool in equally spaced parallel passes in the XY plane across the surface. Like all rough toolpaths, it cuts the surface in several Z steps. Rough toolpaths are done with large diameter tools and coarse settings to remove material efficiently before cutting a finish pass with finer settings.

According to Held et al. [15], it will be more advantageous to use the rough parallel strategy for rather simple shapes or when utilizing small tools. This is because the more complicated the shape of the pocket area and the larger the tool will lead to the better offset curve milling seems to be.

3.2 Constant scallop

Constant scallop or spiral toolpath comprises of a series of contours that are parallel to the boundary of a 2D cross-section [7]. The constant scallop height toolpath moves the tool over the surface in a spiral motion, from the inside-out or the outside-in. The step-over is a fixed amount, but it is calculated parallel to the surface at the location being milled (rather than in the XY plane as with the parallel toolpath). It will make small pockets in some places, if necessary. This toolpath is an excellent toolpath for rolling terrain. It gives a very uniform, consistent result on all surfaces when used with a ball endmill. Usage with one large blanket terrain surface with smaller road surfaces will provide a better resulting surface.

There has been a growing interest for the utilization of spiral toolpath for pocket machining [16]. However, Held [15] mentioned that the shape of the contour affects the suitability of spiral toolpath for high speed machining. If a pocket is very long and narrow, one spiral toolpath may not be efficient to cover the entire contour. Patel [17] also stated that the longest toolpath length and longer cutting time were obtained for spiral-look toolpath strategy rather than the zigzag type.

3.3 Medial axis transform

Medial axis, also referred as the topological skeleton, is the set of all points having more than one closest point on the object's boundary. The medial axis transform is a complete shape descriptor which comprises of medial axis together with the associated radius function of the maximally inscribed discs. It is a technique that first proposed in 1967 as a mean to describe figure [18]. Due to its shape, a medial axis of a figure is also called the skeleton or the symmetric axis of the figure.

Medial axis transform has been used in plunge milling process by Sun et al. [10] to obtain a new milling tool path to control the radial depth. The toolpath generation process by this method comprised of six steps: The calculation of offset contour (1), the calculation of medial axis (2), the calculation of inscribed circles along medial axis (3), the tool path generation (4), cutter locations generation (5), and link of cutter locations (6).

The medial axis transform toolpath generation was claimed to be able to control the radial depth during machining process, thus avoiding an excessive amount of cutting force applied at some locations on the toolpath [10].

The medial axis transform toolpath differentiates in a way the cutter moves its location from one point to another. Rather than moving the cutter subsequently along the path like in the conventional end-milling toolpath, the cutter moves along the path of the inscribed circles generated from the medial axis (skeleton) of the cutting profile. Figure 1 provides a simplified example on the comparison between the conventional milling toolpath and medial axis transform toolpath for an isosceles trapezoid milling profile [10].

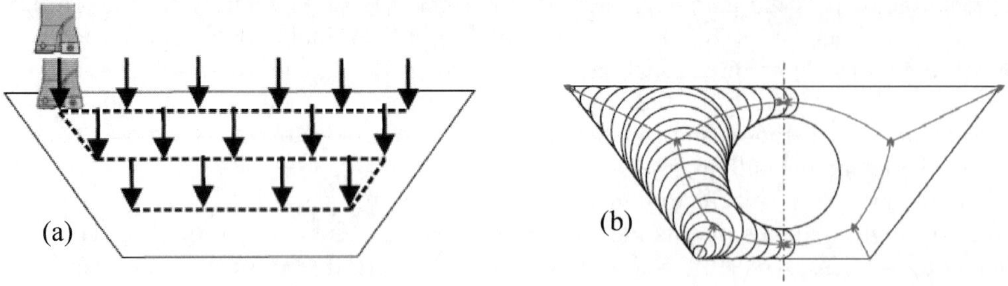

Figure 1. Conventional milling toolpath (a) and medial axis transform skeleton (b) [10]

4. Experiments

The investigation covers the observation of tool life within different cutting conditions and toolpath strategies within the frame of Taguchi Orthogonal Array.

4.1 Materials and Equipment

AISI H13 tool steel has been widely used in hot forging, extrusion, and pressure die casting due to its good wear resistance and large high-temperature strength. A block-shaped with the size of 130 mm x 100 mm x 60 mm will be used in all milling tests. Table 1 shows the chemical properties of the steel under consideration.

Table 1. Chemical Composition of AISI H13 Tool Steel

C	Si	P	S	Cr	Mo	V	MN
0.32 -0.40	0.80 – 1.20	0.011	0.002	4.55 – 5.50	1.00 – 1.50	0.80 – 1.20	Bal.

All machining tests will be conducted on Mazak Nexus 410A-II (Mazatrol Matrix Nexus CNC) three-axis vertical machining center, equipped with maximum spindle rotation speed of 12,000 rpm and motor output (30 minute rating) 16.0 hp/11 kw. All processes will be conducted under dry condition.

Sandvik CoroMill 210 roughing cutter with diameter 25 mm will be used in the milling tests. The insert (ISO Standard R210-09 04 12M-MM 2030) is a double coated consisting of TiN and TiAlN at the outer and inner layer, with 0° rake angle and 7° clearance angle. The details of the insert geometry

IOP Conf. Series: Materials Science and Engineering **290** (2017) 012040 doi:10.1088/1757-899X/290/1/012040

are shown in Figure 1. Only one of the teeth will be used to avoid the effects induced by small difference between the teeth, and to keep the cutting condition constant during tool wear analysis.

4.2 Experimental Procedures

Taguchi orthogonal array with L9 (3^4) resolution will be employed for toolpath strategy and cutting conditions effects toward cutting forces and tool life. The L9 orthogonal array is meant for understanding the effect of 4 independent factors with each having 3 factor level values (UMass Amherst, 2011). It is assumed that the array shows no interaction between any two factors.

One categorical factor of toolpath strategy (TS) and three numeric factors of cutting speed (*Vc*), radial depth of cut (a_e), and chip load (f_z), are employed into the design. Other parameters will be held invariable. Cutting side step will be set to 1.0 mm, axial depth cut at 3 mm, and tool overhang at 4 x D of tool diameter (100 mm). The set of 9 runs orthogonal array will be replicated three times, resulting a total of 27 runs experiment. The layout of the L9 orthogonal array experiment is provided in table 2. The profile of the machined area will have the shape of round-corners rectangle with the dimension of 110 mm x 80 mm as illustrated at figure 2.

Table 2. Taguchi L9 Orthogonal Array layout

Std	Run	Toolpath	Cutting Speed (m/min)	Radial Depth of Cut (mm)	Chip Load (mm/tooth)
1	1	Rough Parallel	100	0.75	0.05
9	2	Medial Axis Transform	200	1.25	0.05
8	3	Medial Axis Transform	150	0.75	0.15
2	4	Rough Parallel	150	1.25	0.10
6	5	Constant Scallop	200	0.75	0.10
4	6	Constant Scallop	100	1.25	0.15
7	7	Medial Axis Transform	100	1.75	0.10
5	8	Constant Scallop	150	1.75	0.05
3	9	Rough Parallel	200	1.75	0.15

Three type of toolpath strategies will be utilized for the first phase of machining study. Rough Parallel (*RP*), Constant Scallop (*CS*), and Medial Axis Transform (*MA*) toolpath. The simplified illustration of the three toolpath layouts are presented in figure 3.

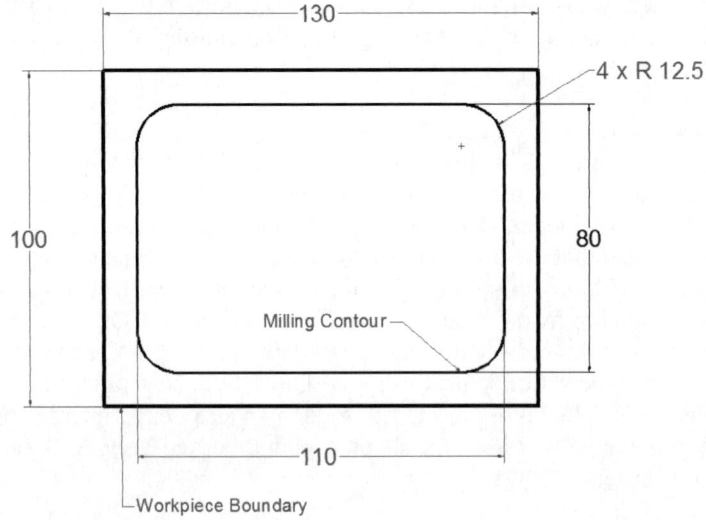

Figure 2. Workpiece profile for machined area and dimensions

Numerical computing software Matlab R2014b will be used for the skeletonization of the profile area skeleton for medial axis transform toolpath generation.

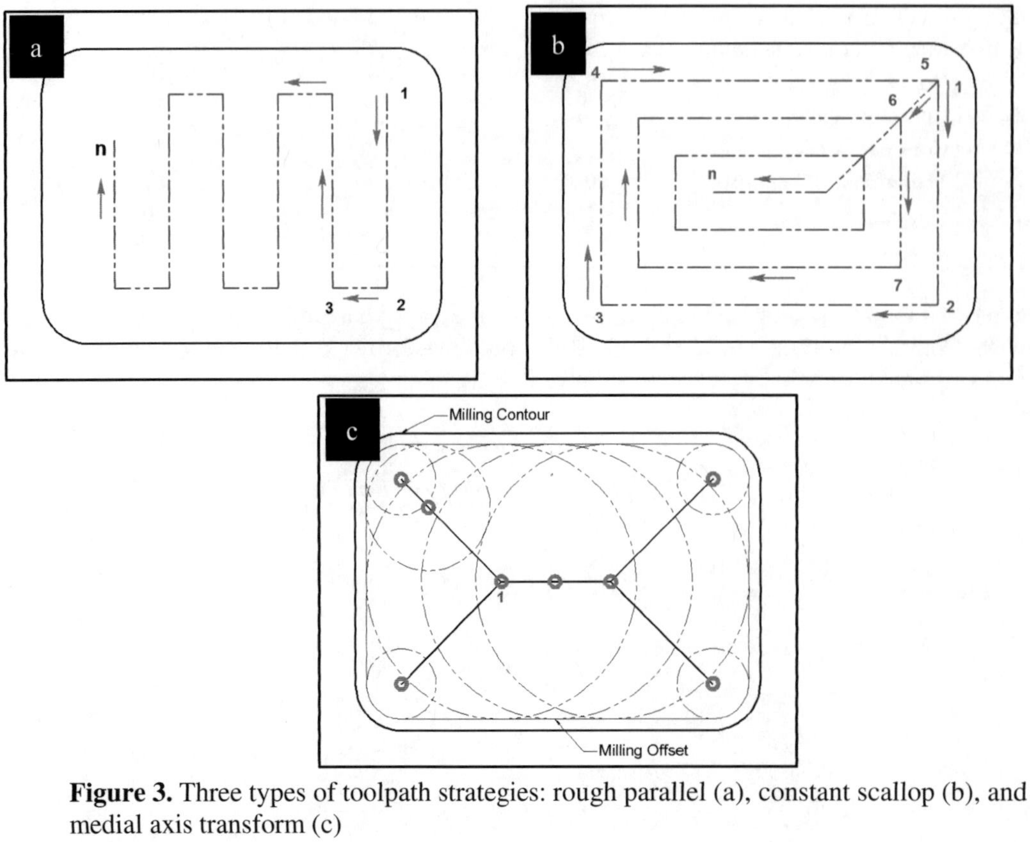

Figure 3. Three types of toolpath strategies: rough parallel (a), constant scallop (b), and medial axis transform (c)

IOP Conf. Series: Materials Science and Engineering **290** (2017) 012040 doi:10.1088/1757-899X/290/1/012040

The measured responses will be focused on the cutting force induced and the tool wear. Cutting forces will be measured by a Kistler 9257B three-piezoelectric dynamometer, paired with Kistler 5019A charge amplifier and the corresponding data acquisition and processing system.

The tool wear will be observed periodically with Union Optical Hisomet-II Measuring Microscope (510D Stages). When the tool flank wear reach or increase by the value of 0.3 mm, the tool life will be recorded. The tool life will be represented by the volume V of the removed metal. The worn tools are also expected to be examined by means of scanning electron microscopy (SEM) and energy dispersive X-ray spectroscopy.

Additional data of cutting temperature will also be taken by TP8 ThermoPro infrared thermal imaging camera. The data will be helpful to understand the effect of machining temperature for the resulted cutting forces and tool wear. All data will be processed for analysis of variance by Stat-Ease Design Expert v7.

5. Expected & Preliminary results

5.1 Expected results

There are several expected results based on the established theories of the topics contribute to the research. But in general, it is expected that:

1. There are significant differences for each application of toolpath strategy and each cutting condition factor toward the cutting force and tool wear mechanism of the machining process.
2. Medial axis transform toolpath as an alternate toolpath is expected to provide a better tool life improvement by a reduction of cutting force during machining.
3. Optimization of significant factors is expected to improve the machinability in terms of shorten machining time by providing a lesser number of cutter locations.

5.2 Preliminary test

A preliminary test was conducted to test the initial weariness of the tool inserts prior deciding the depth of the plunge milling operation. Three different variables were employed within Taguchi L9 framework. Cutting speed (A) was set to 120 – 240 m/min, feed rate (B) at 0.05 – 0.1 mm/tooth, and depth of cut (C) at 0.1 – 0.2 mm. The layout and the result of the tool wear is presented at table 3.

Table 3. Taguchi L9 Orthogonal Array Layout & Result

Run	Block	A Cutting speed (m/min)	B Feed rate (mm/tooth)	C Depth of cut (mm)	Tool wear (mm)
1	Block 1	120	0.05	0.2	0.301
2	Block 1	120	0.1	0.1	0.203
3	Block 1	240	0.05	0.2	0.114
4	Block 1	240	0.1	0.1	0.337

Based on the result, the model of F-values imply that the model is insignificant relative to the noise. The highest rate for tool wear happened during the cutting speed of 240 m/min and feed rate of 0.1 mm/tooth (Run 4) at 0.337 mm, while the lowest one occurred at 0.203 mm when the cutting speed set to 120 m/min and feed rate at 0.1 mm/tooth. The representative graph is presented on figure 4.

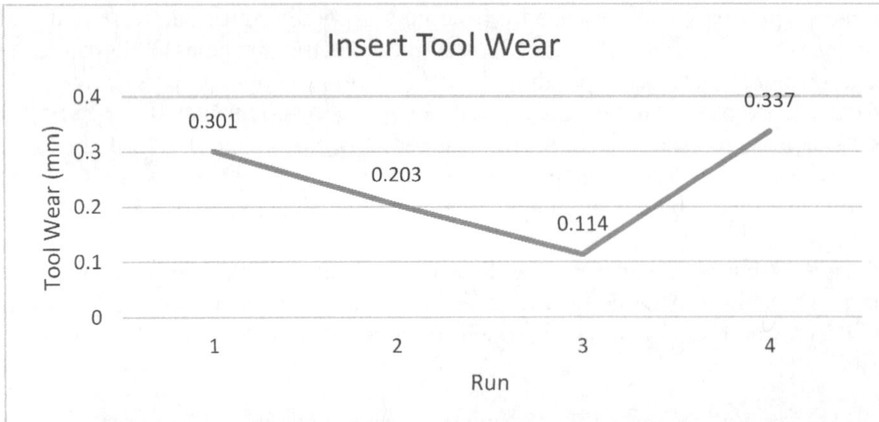

Figure 4. Insert tool wear on the preliminary test of AISI H13 milling

Further calculation and application of inserts was then set based on the preliminary results of the tool weariness displayed during the milling of AISI H13.

6. Conclusion

The main contributions of this work include the machinability in practical plunge milling process for the hard material AISI H13 which has been widely employed in many industrial processes. Aside of that, the application of the non-conventional milling toolpath for plunge milling in the attempt of evading the large radial depth effects is also a significant issue, since a large radial depth may lead to unacceptable large cutting force and make the cutter wear quickly.

References

[1] Rauch M and Hascoet, J Y 2012 *Int. J. Adv. Manuf. Technol.* **59** 47-54
[2] Sun C, Bi Q, Wang Y and Huang N 2015. *Int. J. Adv. Manuf. Technol.* **77** 965-972
[3] Ko J H and Altintas Y 2007 *Int.J. Mach. Tools and Manufacture* **47** 1351-1361
[4] Hioki D, Diniz, A E and Sinatora A 2013 *J. Braz. Soc. Mech. Sci.Eng.* **35** 537-553
[5] Zhuang K, Zhang X, Zhang X and Ding H 2012 *Int. Conf. on Intelligent Robotics and Applications* 255-263
[6] Sun T, Fu Y C, He L, Chen, X M, Zhang, W G, Chen W and Su X B 2016 *Int. J. Adv. Manuf. Technol.* **85** 1315-1323
[7] El-Midany T T, Elkeran A and Tawfik H. 1993 *Geometric Modeling and Imaging* (IEEE) pp. 77-82
[8] Warfield, B 2016 Retrieved on 2017, July 12, http://blog.cnccookbook.com/2016/10/05/toolpath-secret-weapon-complete- guide-to-plunge-milling-roughing/
[9] Liang Y, Zhang D, Chen Z C, Ren J and Li, X 2014 *Int. J. Adv. Manuf. Technol.* **70** 2249-2261
[10] Sun C, Wang, Y and Huang N 2015 *Int. J. Adv. Manuf. Technol.* **76** 1575-1582
[11] Pralea B and Nagit G H 2016 *IOP Conference Series: Materials Science and Engineering* vol 161 p. 012004
[12] Romero P E, Dorado R, Díaz F A and Rubio E M 2013 *Procedia Engineering* **63** 523-531
[13] Xu J, Sun Y and Zhang X 2013 *Int. J. Adv. Manuf. Technol.* **67** 2489–2500
[14] Kim B H and Choi B K 2002 *Computer-Aided Design* **34** 89-95

[15] Li H, Dong Z and Vickers G W 1998 *International IFIP TC5/WG5.3 Conference SSM '98 Sculptured Surface Machining Conference* (Michigan)
[16] Xiong Z, Zhuang C and Ding H 2011 *Proc. Institut. Mech. Eng. Part B: J. Eng. Manuf.* vol 225 pp.483–495.
[17] Patel D D, Dodiya H R and Lalwani, D I 2016 *Int. J. Exp. Design and Process Optimisation* **5** 94-116
[18] Lee D T 1982 *IEEE Transactions on pattern analysis and machine intelligence* **4** 363-369

IOP Conf. Series: Materials Science and Engineering **290** (2017) 012041 doi:10.1088/1757-899X/290/1/012041

An Investigation of TIG welding parameters on microhardness and microstructure of heat affected zone of HSLA steel

M H A Musa [1,a] *, M A Maleque[1,b] and M Y Ali[1,c]

[1]Department of Manufacturing and Material Engineering
Kulliyyah of Engineering, International Islamic University, Malaysia
P.O.Box 10, 50728 Kuala Lumpur, Malaysia

Email: [a]mosahossin@yahoo.com, [b]maleque@iium.edu.my, [c]mmyali@iium.edu.my

Abstract: Nowadays a wide variety of metal joining methods are used in fabrication industries. In this study, the effect of various welding parameters of the TIG welding process on microhardness, depth, and microstructure of the heat-affected zone (HAZ) of L450 HSLA steel and optimizing these process parameters following Taguchi experimental design was investigated. The microhardness tended to increase significantly with the increase of welding speed from 1.0 to 2.5 mm/s whereas the width of HAZ decreased. The current and arc voltage was found to be less significant in relative comparison. Microstructures of the welded samples were also studied to analyze the changes in the microstructure of the material in terms of ferrite, pearlite, bainite, and martensite formations. Welding speed was found to be the most significant factors leading to changes in microhardness and metallurgical properties. The increase of welding heat input caused an increase in width (depth) of HAZ and the growth of prior austenite grains and then enlarged the grain size of coarse grain heat affected zone (CGHAZ). However, the amount of martensite in the HAZ decreased accompanied by an opposite change of paint. It was observed that the hardness properties and the microstructural feature of HAZ area was strongly affected by the welding parameters.

1. Introduction

When high strength low alloy steel is welded, non-uniform heating and cooling in weld metal and base metal create harder heat affected zone (HAZ), cold crack susceptibility and residual stress in weldments [1]. The properties of the welded metal, HAZ, and base metal have been considered with varying parameters. But there is only a little work on welded high strength low alloy steels. Hence HSLA steels become an indispensable steel category for different applications. For instance, ships, railroad equipment, building construction, boilers, offshore oil drilling platforms, pipelines, pressure vessels, aircraft, automobiles [1, 2]. However, These critical zones which created through the welding process playing an important role in the morphology of material properties of welded HSLA steels. The heat affected zone (HAZ) is one of an undesirable zone, but the inevitable region of a fusion welded joint. This region undergoes additional thermal cycle(s) that may have a detrimental effect on the microstructure and mechanical properties of the welding joints. For this reason, the properties of the HAZ, particularly the toughness, and strength, still remains a major concern for many users of these materials [2]. Also, hardness, depth of heat affected zone and microstructure of heat affect zone

(HAZ) are extremely important for HSLA steel structural consideration in welding. These are very much important in the case of welding method, cooling rate, grain structure, and structural integrity. The process parameters like welding current, voltage, welding speed and arc length are the primary variables which control the fusion, depth of penetration, and depth of HAZ. The HAZ was more classified into three regions, i.e. grain growth region, grain refined region and transition region [3]. The hardness during weld joint is not uniform. Moreover, HSLA Steel is sensitive to thermal cycle and the metal of weld joint was having highest temperature, i.e. above the melting point and the parent metal was having temperature very less below the lower critical line. Because of the high-temperature difference between these two regions, the cooling rate was very high and the solidification of weld metal was under non-equilibrium conditions. Due to this, the transformation from austenite to pearlite micro-constituent has not occurred and austenite to martensite or bainitic lathe has happened. So, the hardness of this region was very high and HAZ has become more susceptible to catastrophic cracking [4]. Therefore, a variety of microstructures in welded HSLA steel is obtained depending on the deformation temperature and cooling rate. The proper control of these process parameters and the characteristics will give better results and fewer risks in order to guarantee the reliability and durability of the welding zones [5, 6].

To predict the optimum input welding parameters without consuming more time and raw materials in experimentation trials, there are different methods of optimization for obtaining the required output values through the development of models. A design of experiments (DOE) approach is the most significant and efficient way for optimization of process variables. The use of DOE methods over the last two decades has grown significantly and has been adapted in science and engineering experiments to optimize process parameters [7]. No research work has been done using DOE approach for optimization of TIG torch welding parameters on HSLA steel. Therefore, the main aim of this work is to study the influence of welding parameters (current, arc voltage, travel speed and heat input) on microhardness, depth, and microstructure of the heat-affected zone (HAZ) and to optimize the process parameters after accomplishment of the TIG welding process on X65 HSLA steel following Taguchi experimental design.

2. Experimental procedures

The high strength low alloy steels employed in the present work, designated as L450- HSLA steel, was procured from China in the form of plates with 10 mm in thickness. The chemical composition of this steel can be found elsewhere [8].

A Miller TIG torch source attached with a semi-automatic traversing arm was used for melting purpose. The direct current electrode negative (DCEN) mode was used and A 3.2 mm diameter tungsten electrode was used to strike an arc between the electrode and the workpiece in pure Argon atmosphere to prevent excessive oxidation [8].

Minitab 17 software was used to determine the no. of experiments to be conducted with the predetermined variables and their levels [9]. Taguchi method provides a set of well minimum experimental runs. This method uses a statistical measure of performance called signal-to-noise ratio. The S/N ratio takes both the mean and the variability into account. The optimum setting is the parameter combination, which has the highest S/N ratio [10]. Based on Taguchi's orthogonal array design, experiments have been conducted with four different levels of process parameters as listed in table 1: welding current, welding voltage, welding speed and shielded gas obtain bead-on-plate weldment on L450 HSLA steel plates (60×40×10) mm^3. The experiments have been carried out as per Taguchi's L16 orthogonal array (OA) design.

Table 1. Process parameters and their limits.

No.	Parameters	Notation	Unit	Level 1	Level 2	Level 3	Level 4
1	Current	C	A	70	80	90	100
2	Voltage	V	V	25	30	35	40
3	Speed	S	mm/sec	1	1.5	2	2.5
4	Shielding Gas	SG	L/min	15	20	25	30

Taguchi technique was used for the analysis of results obtained after testing and to find out the main effect of the process parameters and the percentage contribution of each parameter on the output. After welding, a complete cross section of the weldments composed of the parent plate and weld metal was removed transverse to the welding direction to produce specimens for microstructural examination and hardness measurements. All specimens were ground on silicon carbide "wet" papers from P320 to P1200 grade and polished sequentially on diamond wheels using 3, 1 and 0.25 μm grades. The weld metal microstructure was revealed by using 2% Nital solution. An average of three readings was taken to determine the hardness and the width of HAZ. The microstructure was evaluated using a standard optical microscope and JMS 5600 scanning electron microscope.

Table 2. Design matrix with experimental results for Hardness and width of HAZ

	Parameters				Responses	
Run	Current (A)	Voltage (V)	Speed (mm/sec)	Shielded Gas flow rate(L/min)	Hardness of HAZ (Hv)	Width of HAZ (mm)
1	70	25	1	15	249.30	1.39
2	70	30	1.5	20	280.20	1.20
3	70	35	2	25	289.50	0.75
4	70	40	2.5	30	333.90	0.70
5	80	25	1.5	25	283.70	1.52
6	80	30	1	30	243.60	1.19
7	80	35	2.5	15	354.90	0.85
8	80	40	2	20	256.30	1.15
9	90	25	2	30	282.00	1.30
10	90	30	2.5	25	339.60	0.84
11	90	35	1	20	218.50	1.67
12	90	40	1.5	15	219.20	1.84
13	100	25	2.5	20	350.70	1.12
14	100	30	2	15	251.28	1.57
15	100	35	1.5	30	225.90	1.41
16	100	40	1	25	209.00	2.04

IOP Conf. Series: Materials Science and Engineering **290** (2017) 012041 doi:10.1088/1757-899X/290/1/012041

3. Results and Discussion

Based on the DOE runs the experiments are conducted and the values of the responses are shown in table 2. The objective of the parameter design is to optimize the settings of the process parameter values for improving HAZ characteristics and From the output variables, the effect of the process parameters on these variables can be identified. [11]. In this paper, the characteristic values are selected by the hardness of HAZ and the width of HAZ, since a good result is obtained by the smaller HAZ width and lower hardness, the lower the better (LB) is preferred.

3.1 Effect of parameters on Hardness of HAZ

Hardness is one of the most important critical factors indicating the quality of the welding and its performance in service. From the Main effect plot for means of the hardness of HAZ of welded L450 HSLA steel, figure 1 established that there was a significant increase in the hardness of HAZ with the increase of the welding speed, while a little decrease with increasing voltage and current, respectively. This is due to the fact that as the welding speed increases the amount of the heat input will decrease causing high cooling rates lead to increase of hardness levels in the HAZ [12, 13].

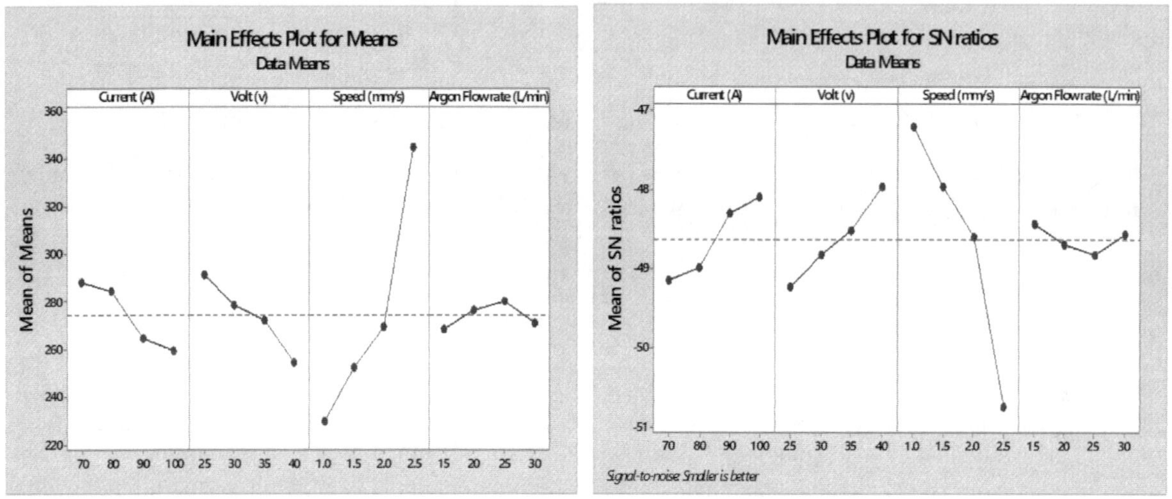

Figure 1. Main effect plots a) for the means and b) for S/N ratio for Hardness of HAZ.

3.2 Effect of parameters on width of HAZ

The width of HAZ as shown in Figure 2 was influenced by the speed and current factors, the width of HAZ is smaller as the current decreased and the speed increased. The high speed causes low heat input and this causes the formation of a smaller HAZ region, the literature survey supports this result [4,14]. Whereas the results of the welding voltage and shielded gas flow rate have a less significant effect on the HAZ width. The welding speed has significantly influenced the HAZ width, this was due to the amount of molten on the base metal influences the HAZ area. It can be said that an increase in welding speed and a decrease in welding current decreases the HAZ area compared to voltage or shielded gas flow rate and by controlling them the size of the HAZ can be controlled.

3.3 S/N Ratio Analysis

Based on S/N ratio analysis, the process parameters which significantly affect the hardness and width of HAZ are shown in table 3 and table 4. The level of a factor with the highest S/N ratio was the optimum level of responses measured. From the Table 4 it is clear that the optimum value levels for lower HAZ hardness are at a Current (100 A), Voltage (40 V), speed (1 mm/sec) and Gas flow rate (15 L/min), and from Table 4 the optimum value levels for lower HAZ width are at a Current (70 A), Voltage (35 V), speed (2.5 mm/sec) and Gas flow rate (30 L/min). The response tables include ranks based on Delta statistics, which compare the relative magnitude of effects. The Delta statistic is the

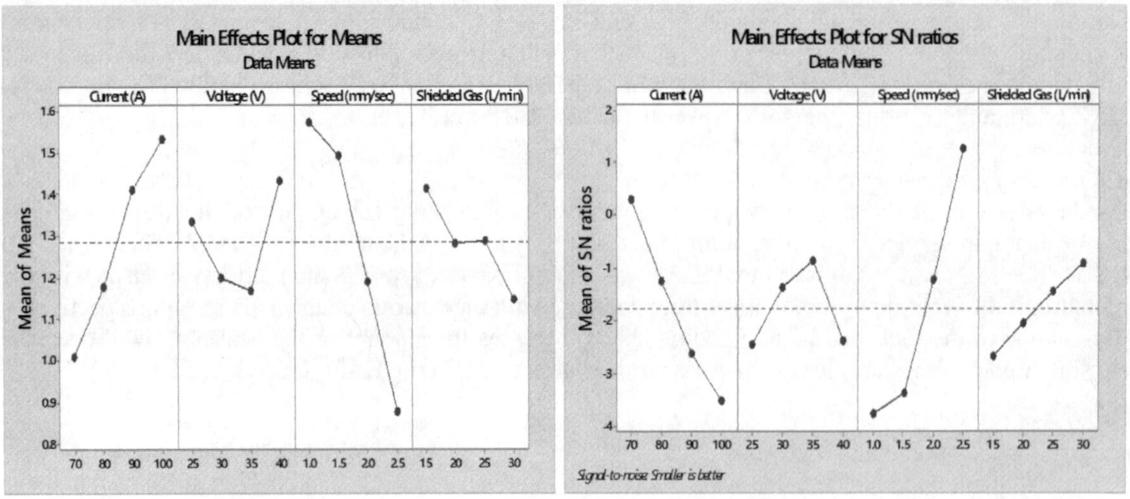

Figure 2. Main effect plots a) for the means and b) for S/N ratio for width of HAZ.

highest minus the lowest average for each factor [11]. Minitab indicates ranks based on Delta values; rank one to the highest Delta value, rank two to the second highest, and so on.

Table 3. Hardness of HAZ Response Table for S/N Ratios.

Level	Current (A)	Voltage (V)	Speed (mm/sec)	Shielded gas flow rate (L/min)
1	-49.15	-49.22	-47.21	-48.44
2	-48.99	-48.83	-47.98	-48.7
3	-48.31	-48.53	-48.6	-48.83
4	-48.1	-47.97	-50.75	-48.57
Delta	1.05	1.26	3.53	0.39
Rank	3	2	1	4

Table 4. Width of HAZ Response Table for S/N Ratios.

Level	Current (A)	Voltage (V)	Speed (mm/sec)	Shielded gas flow rate (L/min)
1	0.3	-2.45	-3.76	-2.67
2	-1.24	-1.37	-3.38	-2.05
3	-2.62	-0.87	-1.21	-1.45
4	-3.52	-2.39	1.27	-0.91
Delta	3.82	1.57	5.03	1.76
Rank	2	4	1	3

3.4 The microstructure of the HAZ

The microstructure of the HAZ depends on the chemical composition and the peak welding temperature. Higher heat-input welding produces wider coarse grain HAZ (CGHAZ). The extent of grain growth in the HAZ depends on the local thermal cycle, which can be controlled by adjusting the welding parameters [6]. The prior austenite grains corresponding to different locations of the HAZ are displayed in figure 3(a–d), and are characterized by distinct grain boundaries. When a welding speed

IOP Conf. Series: Materials Science and Engineering **290** (2017) 012041 doi:10.1088/1757-899X/290/1/012041

is high (low heat input), the microstructure of weld metal and HAZ changes, it will cause the martensite phase transformation and the microhardness increased as shown in figure 3b.

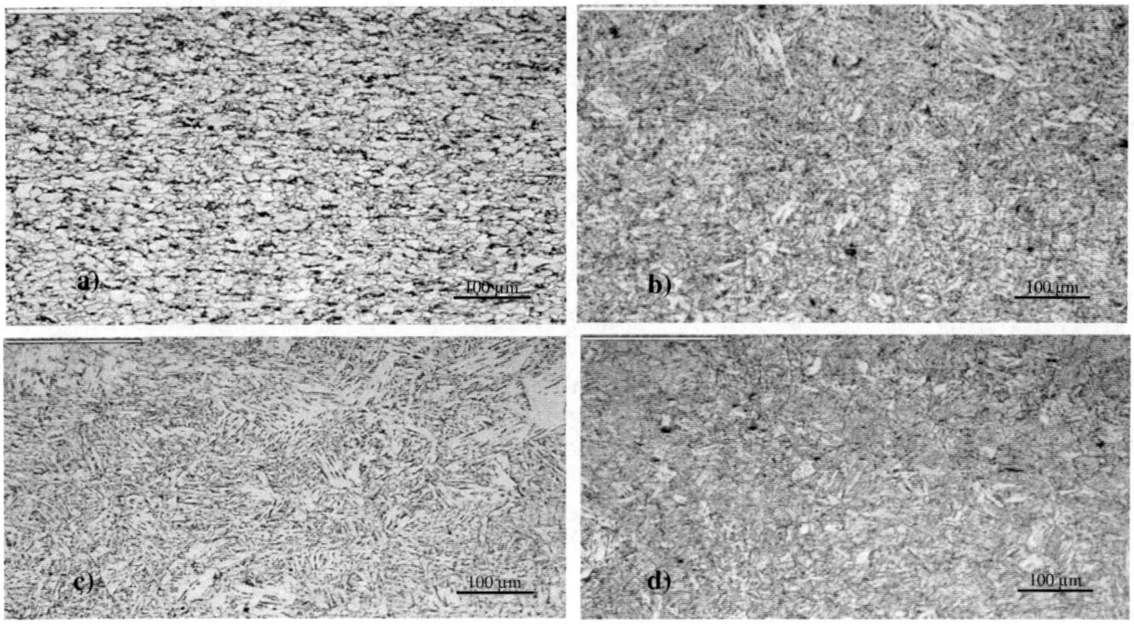

Figure 3. Optical microstructure of various HAZ sections in welded HSLA-X65 steel at different heat input: (a) Base metal, (b) 840 J/mm, (c) 1152 J/mm and (d) 1512 J/mm.

On the contrast, increasing the welding heat input (lower welding speed and higher welding current), this corresponded low cooling rate and beneficial for the formation of bainite and the ferrite was coarsened in HAZ region leading to decreased the microhardness of HAZ phases as shown in Figure 3c. Also, the increasing of welding heat input caused an increase in width (depth) of HAZ and the growth of prior austenite grains and then enlarged the grain size of coarse grain heat affected zone (CGHAZ). However, the amount of martensite in HAZ decreased accompanied by an opposite change of bainite as indicated in figure 3d.

Conclusion

From the present study, the HAZ area was strongly affected by the welding parameters and the welding speed being the most significant parameter on the hardness of HAZ. The increasing in welding speed cause a significant increase in the hardness of HAZ, while a slight decrease with increasing current and voltage. Moreover, the width of HAZ is influenced by the current and the welding speed. It means that an increase in welding speed and a decrease in welding current cause smaller HAZ area compared to voltage or shielded gas flow rate.

The application of Taguchi approach was successful in this study, and from the Taguchi S/N ratio the optimal parameters setting for lower HAZ hardness are at a Current (100 A), Voltage (40 V), speed (1 mm/sec) and Gas flow rate (15 L/min), and the optimal value levels for lower HAZ width are at a Current (70 A), Voltage (35 V), speed (2.5 mm/sec) and Gas flow rate (30 L/min).

When the welding heat input is low, the martensitic phases transformation occurred. Whereas increasing the welding heat input the formation of bainite and the ferrite was coarsened in HAZ region.

Acknowledgement
The authors would like to acknowledge the Department of Manufacturing & Materials Engineering of International Islamic University Malaysia for the support to carry out this work.

References
[1] Tewari, S. P., Gupta, A., & Prakash, J. 2010 *International Journal of Engineering Science and Technology* **2** pp 512-516
[2] Musa, M. H. A., Maleque, M. A., Ali, M. Y., & Hasan, M. H. 2015 *Advanced Materials Research*, pp *1115*
[3] Kumar, R., & Kumar, S. 2014 *International Journal of Research in Engineering and Technology* **3** pp 751-756
[4] Gharibshahiyan, E., Raouf, A. H., Parvin, N., & Rahimian, M. 2011 *Materials & Design* **32** 2042-2048
[5] Maleque, M. A., Bello, K. A., Adebisi, A. A., & Dube, A. 2017 *IOP Conference Series: Materials Science and Engineering* **184** pp 012018
[6] Baptista, C. A. R. P., Versuto, B. C., de Oliveira, F. P., Torres, M. A., & Costa, D. H. 2011 *Proc. of COBEM* **2011** pp 24-28
[7] Pamnani, R., Vasudevan, M., Vasantharaja, P., & Jayakumar, T. 2017 *Proceedings of the Institution of Mechanical Engineers, Part L: Journal of Materials: Design and Applications* **231** pp 320-331.
[8] Musa, M. H. A., Maleque, M. A., & Ali, M. Y. 2017 *Conference Series: Materials Science and Engineering* **184** pp 012053)
[9] Maleque, M. A., Bello, K. A., Adebisi, A. A., & Akma, N. 2017 *Conference Series: Materials Science and Engineering* **184** pp 012035)
[10] Datta, S., Bandyopadhyay, A., & Pal, P. K. 2008 *The International Journal of Advanced Manufacturing Technology* **39** pp 1136-1143
[11] Maleque, M. A., Bello, K. A., Adebisi, A. A., & Dube, A. 2017 *Conference Series: Materials Science and Engineering* **184** pp 012018
[12] Shome, M. 2007 *Materials Science and Engineering: A* **445** pp 454-460
[13] Bodude, M. A., & Momohjimoh, I. 2015 *Journal of Minerals and Materials Characterization and Engineering* **3** pp 142
[14] Mohammed, G. R., Ishak, M., Aqida, S. N., & Abdulhadi, H. A. 2017 *Metals* **7** pp 39

Fabrication of silica ceramic membrane via sol-gel dip-coating method at different nitric acid amount

N A Z Kahlib[1], F D M Daud[1], M Mel[2], A L N Hairin[1], A Z A Azhar[1] and N A Hassan[1]

[1]Department of Manufacturing and Materials Engineering, Kulliyyah of Engineering, International Islamic University Malaysia (IIUM), Jalan Gombak, 53100 Kuala Lumpur, Malaysia
[2]Department of Biotechnology Engineering, Kulliyyah of Engineering, International Islamic University Malaysia (IIUM), Jalan Gombak, 53100 Kuala Lumpur, Malaysia

Email: farah_diana@iium.edu.my

Abstract. Fabrication of silica ceramics via the sol-gel method has offered more advantages over other methods in the fabrication of ceramic membrane, such as simple operation, high purity homogeneous, well defined-structure and complex shapes of end products. This work presents the fabrication of silica ceramic membrane via sol-gel dip-coating methods by varying nitric acid amount. The nitric acid plays an important role as catalyst in fabrication reaction which involved hydrolysis and condensation process. The tubular ceramic support, used as the substrate, was dipped into the sol of Tetrethylorthosilicate (TEOS), distilled water and ethanol with the addition of nitric acid. The fabricated silica membrane was then characterized by (Field Emission Scanning Electron Microscope) FESEM and (Fourier transform infrared spectroscopy) FTIR to determine structural and chemical properties at different amount of acids. From the XRD analysis, the fabricated silica ceramic membrane showed the existence of silicate hydrate in the final product. FESEM images indicated that the silica ceramic membrane has been deposited on the tubular ceramic support as a substrate and penetrate into the pore walls. The intensity peak of FTIR decreased with increasing of amount of acids. Hence, the 8 ml of acid has demonstrated the appropriate amount of catalyst in fabricating good physical and chemical characteristic of silica ceramic membrane.

1. Introduction

Sol-gel technique can be described as a process where the solution or sol experiences sol-gel transition, where at this transition, the solution turn into rigid non-fluid mass [1]. This method comprises forming of mineral phase starting from soluble molecular precursors, resulting an inorganic polymerization reaction. The reproduction of new porous nanomaterials with well-defined structures and complex shape has been manufactured by the sol - gel process [2]. Fabrication of ceramics and glasses by using sol-gel methods have been used for a long time [1]. Sol-gel techniques offer some advantages over other methods in the fabrication of ceramics, such as low temperature needed for synthesis to be carried out, high purity materials is produced, and xerogels porous texture such as pore size and structure can be modified [4]. Fabrication of silica by using the sol – gel technique is started by silicon alkoxides hydrolyzation through condensation, creating the silica nanoparticles. These silica particles then grow and form a sol, which are influenced by the reaction condition. Silica films can be formed by depositing the sol onto a substrate. Parameters such as precursors, reactant molar ratio, pH, synthesis temperature, modifying agent, composition of the solvent; and aging and drying condition affected the textural and structural properties of the fabricated materials [2].

In sol-gel technique, the pH of the solution control the hydrolysis and condensation reaction, where it will affect the metal oxide microstructure. The pH of the solution can be modified by adding catalyst. Normally, silicon alkoxides reaction with water is slow. Hence, by adding the acid or base catalyst, the reaction of the process can be increased [5]. For example, after the addition of HCl into the solution of TEOS and ethanol, the gelation time was reduced to 92 hours from 1000 hours [6]. In the films fabricated by sol-gel methods, acid catalysts are normally used with a small amount of water (R=4) [7]. Previous work by Coltrain et al. [8] shows that under the acidic condition, the condensation of TEOS achieved maximum at pH=7 and minimum an pH=2, where the molar ration of the water, R=4. A study by Fardad [5] indicate that type of catalyst used while preparing precursor solution affected the thickness, shrinkage, porosity and optical quality of the film fabricated.

Catalyst concentration in silica sol is quite important as with only a small change in the concentration may affect the reaction greatly [9]. In acid-catalyzed reaction, hydrolysis is faster than condensation, where normally started after the completion of hydrolysis [10] [11] [12]. On the other hand, condensation step is faster than hydrolysis under base-catalyzed condition, which producing a highly condensed species that may agglomerate into fine particles [10]. Besides that, dense microporous (pore size $<$ 2nm) networks is obtained, while mesoporous (2nm $<$ pore size $<$ 50nm) gels was produced under alkaline condition [13]. The pH also affected silica dissolution and precipitation [14]. More porous structures are produced at a high pH value as the solubility of the particulate in the sol may be high. However, at low pH condition, due to the low dissolution-reprecipitation rate, material with dense structure and fine pore network are obtained.
Previous work by Tuzun et al. [15], the silica sol was prepared by using the molar ratio of TEOS: water: ethanol is 1:4.7:3.8, with small amount of acid. However, the amount of the acid was not specified. In this paper, the effect of the amount of the nitric acid as a catalyst on the characteristics of the fabricated silica ceramic membrane via sol-gel dip-coating method is investigated.

2. Experimental procedures

2.1 Silica Sol Preparation.
Silica sols was prepared by mixing TEOS, water and ethanol together at 298 K with vigorous stirring. The molar ratio of TEOS: water: ethanol is fixed at 1:4.7:3.8 [15]. Then, the nitric acid was added to the solution as a catalyst. The amounts of nitric acid are varied at 8, 10, and 12 ml.

2.2 Membranes Fabrication.
The tubular ceramic support used was 10 inch Doulton OBE Ceramic with 32 mm, 48 mm internal and outer diameter while the support length of 200 mm. The ceramic support was dipped into the prepared silica solutions, and dried at room temperature for 24 hours. After that, the samples were calcined at 773 K with holding time of three hours. The dipping, drying and calcine process was repeated for three times.

2.3 Membranes Characterization
The morphology of the membrane surface and support were determined using the field emission scanning electron microscope (FESEM) (JSM 6700F, JEOL). X-ray diffraction (XRD), was used to determine the element of the fabricated membrane. The Fourier transform infrared spectroscopy (FTIR) (Perkin Elmer System spectrum 100) was employed in determining the functional group of the membrane.

3. Results and Discussion

Figure 1 shows the FESEM image of the fabricated silica ceramic membrane for 8, 10 and 12 ml acid. From the FESESM images, morphology of the fabricated silica ceramic membrane was observed, where for FESEM image for 8 ml acid shows than the surface of the silica quite clear. However, with increasing on amount of acids, silica surface becomes rough as the agglomeration happens.

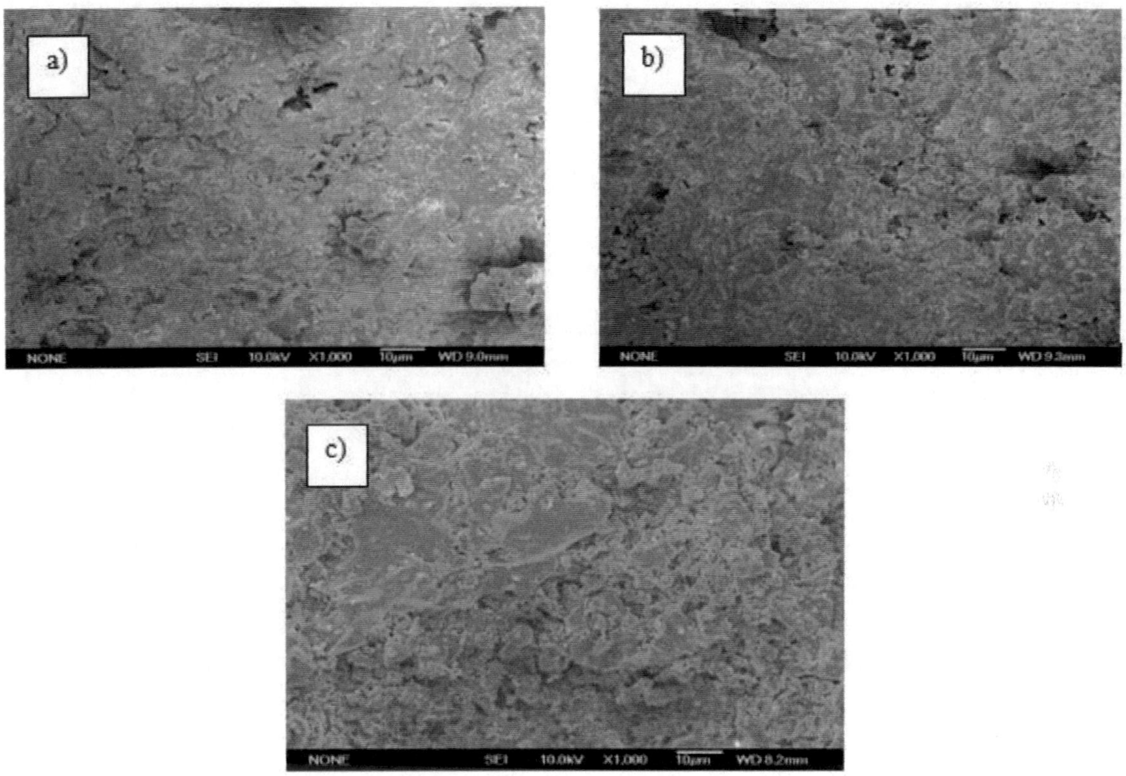

Figure 1. FESEM images of the fabricated silica ceramic membrane via sol-gel dip-coating at different acid amount; a) 8 ml, b)10 ml and c)12 ml.

Figure 2 shows XRD patterns fabricated silica ceramic membrane. The aim of this study is to fabricate silica ceramic membrane by using sol-gel dip-coating methods. From the XRD results, it is verified that silica was deposited on the substrate, as the detected reflections demonstrates the existence of silicate hydrates.

Figure 3 represents FTIR adsorption spectra of fabricated silica ceramic membrane by using different amount of nitric acid. It was discovered that the major peak in FTIR adsorption spectra of the silica, with different amount of acids as the catalyst, is around 1080 to 1090 cm^{-1}. The broad absorption band in the region is correlated to the Si-O-Si bond of silica, which then prove the presence of silica as the end results. However, different amount of acid used affect the intensity of the FTIR peak as shown, where increasing the amount of nitric acid used to reduce the peak intensity of FTIR.

Fabrication of the silica ceramics through sol-gel dip-coating technique is expected to be resulting in uniform pore size distribution. By varying the amount of nitric acid, the physical and chemical characteristic of the fabricated silica ceramic membranes is affected. The nitric acid amount are capable to control the hydrolysis and condensation process during fabrication. The 8 ml of nitric acid have shown the better physical and chemical characteristic of silica ceramic membrane compared to 10 ml and 12 ml of nitric acid.

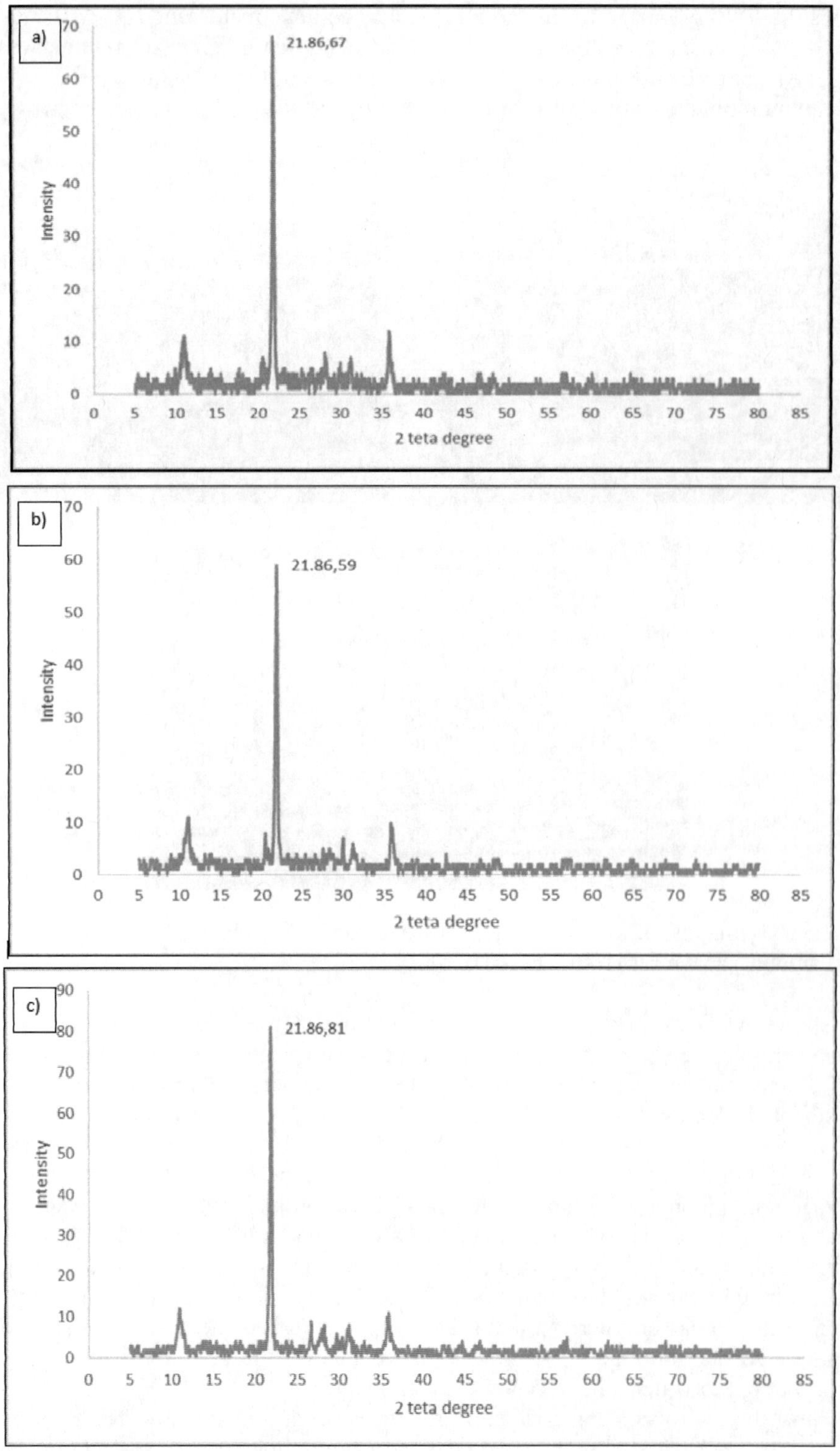

Figure 2. XRD pattern of the fabricated silica silica ceramic membrane via sol-gel dip-coating at different acid amount at; a) 8 ml , b)10 ml and c)12 ml.

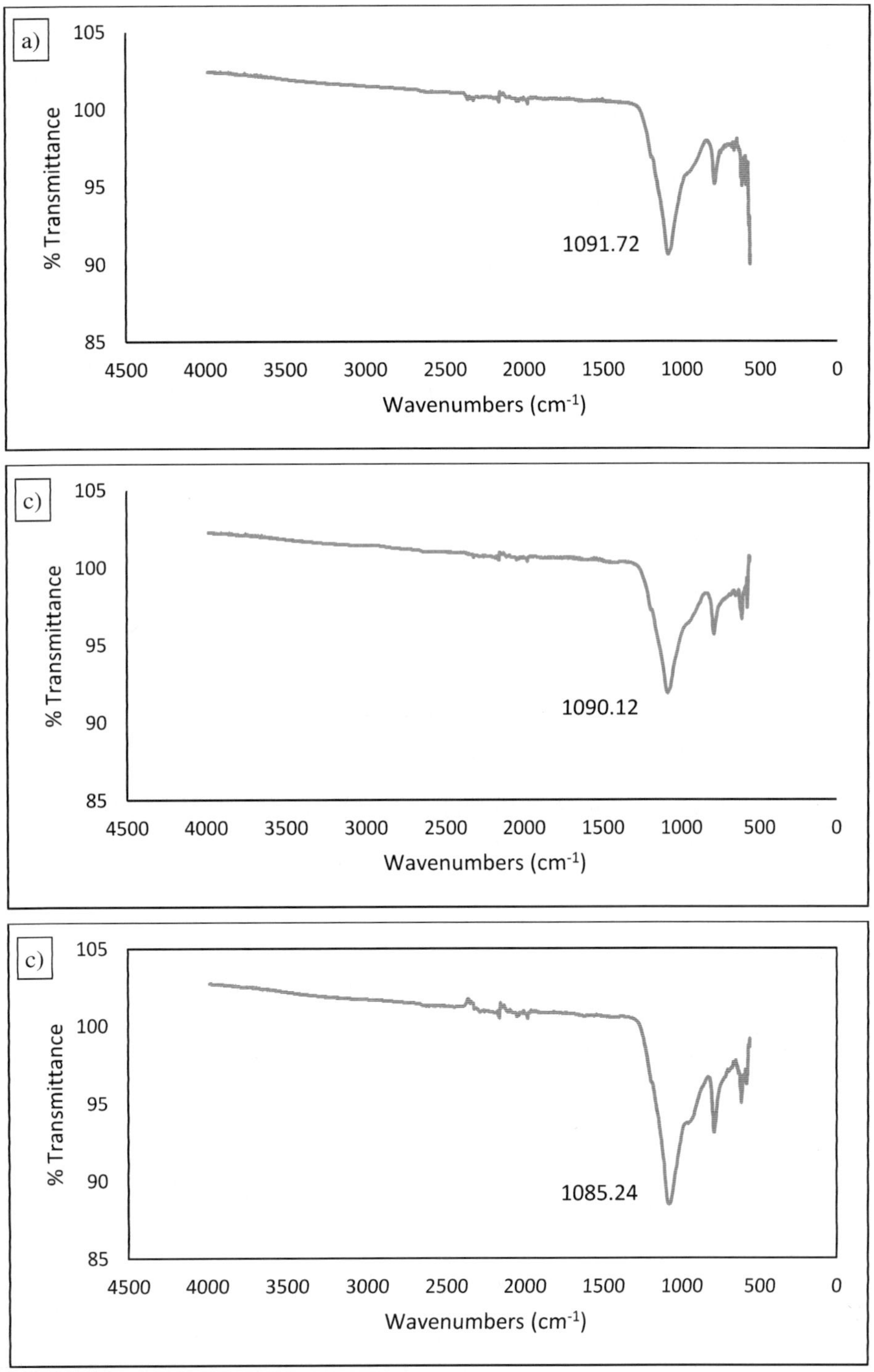

Figure 3. FTIR result of the fabricated silica ceramic membrane via sol-gel dip-coating at different acid amount for a) 8 ml, b)10 ml and c)12 ml.

IOP Conf. Series: Materials Science and Engineering **290** (2017) 012042 doi:10.1088/1757-899X/290/1/012042

4. Conclusion

In this study, silica ceramic membrane was fabricated by using the sol-gel dip-coating methods. The differences in characteristics of membranes were investigated as they were fabricated by using different amount of nitric acid. The differences in the amount of acid used as a catalyst, also control the concentration of the sol, which affect the properties of fabricated silica ceramic membrane. Increasing the amount of acid resulting in agglomeration on the silica surface. In FTIR analysis, the intensity decreased by increasing amount of acids. The fabrication of silica by sol-gel dip-coating method is considered as successful as the XRD pattern shows the existence of silicate hydrate in the end products. Therefore, the 8 ml of acid have presents the suitable amount of catalyst in producing good physical and chemical characteristic of silica ceramic membrane.

Acknowledgement

The authors would like to acknowledge for the financial supports given from RIGS (RIGS15-1490149) grant from IIUM.

References

[1] Buckley A M and Greenblatt M 1994 *J. Chem. Educ* **71** 599
[2] Milea C A, Bogatu C and Duta A 2011 *Bulletin of The Transilvania University of Brasov* **4** 53
[3] Schmidt H K, Geiter E, Mennig M, Krug H, Becker C and Winkler R P 1998 *Journal of sol-gel science and technology* **13** 397-404
[4] Musgo J, Echeverría J C, Estella J, Laguna M and Garrido J J 2009 *Microporous and Mesoporous Materials* **118** 280-287
[5] Fardad M A 2000 *Journal of Materials science* **35** 1835-1841
[6] Brinker C J and Scherer G W 1990 *Academic Press Inc*
[7] Sakka S 1994 *Journal of Sol-Gel Science and Technology* **2** 451-455.
[8] Coltrain B K, Melpolder S M., Salva J M, Uhlmann D R and Ulrich D R 1992 *Ultrastructure Processing of Advanced Materials Uhlmann, DR Ulrich, DR (Eds.)* pp 69-76
[9] Iler R K 1973 *The chemistry of silica* (New York: Wiley)
[10] Chiang C L, Ma C C M, Wu D L and Kuan H C 2003 *Journal of Polymer Science Part A: Polymer Chemistry* **41** 905-913
[11] De Lange R S A, Hekkink J H A, Keizer K and Burggraaf A J 1995 *Journal of Non-Crystalline Solids* **191** 1-16
[12] Vives S and Meunier C 2008 *Ceramics International* **34** 37-44
[13] Coradin T, Boissière M and Livage J 2006 *Current medicinal chemistry* **13** 99-108
[14] Topuz B and Çiftçioğlu M 2010 *Journal of Membrane Science* **350** 42-52
[15] Tüzün F N and Arçevik E 2010 *In Macromolecular symposia* **287** 135-142

IOP Conf. Series: Materials Science and Engineering **290** (2017) 012043 doi:10.1088/1757-899X/290/1/012043

Effect of compaction pressure of green body and heating current on photoluminescence property of ZnO crystal grown by electric current heating method

A G E Sutjipto[1], M H Mazwir[1], H L Yee[1], S R Misskon[2], A G M Shaitir[1], M A Jusoh[1] and R Othman[2]

[1]Material Technology Program, Faculty of Industrial Sciences & Technology, Universiti Malaysia Pahang (UMP), Gambang, 26300 Pahang, Malaysia
[2]Department of Science in Engineering, Faculty of Engineering, International Islamic University Malaysia (IIUM), Jalan Gombak, 53300 Kuala Lumpur, Malaysia

Corresponding author: agusgeter@ump.edu.my

Abstract. In this study, we reported the effect of applied compaction pressure on green body and electric current heating on ceramic bar on the ZnO crystal growth and its photoluminescence characteristic. Crystals grown on ZnO bar sintered by 1100 °C were mostly on (1 0 1) orientation. Sample with 3.0 ton and 3.0 A for applied pressure and current, respectively revealed the shortest photoluminescence (PL) wavelength of 409.5 nm with highest emission energy of 3.03 eV.

1. Introduction

ZnO has gained substantial interest in the research community because of its large exciton binding energy (60 meV) and wide direct band gap (Eg ~ 3.37 eV) at room temperature (R.T.). This large exciton binding energy paves the way for an intense near-band-edge excitonic emission such as lasing action at R.T. and higher temperatures, because this value is 2.4 times larger than that of the room temperature thermal energy (kBT = 25 meV). In addition, the interest in ZnO is fueled and fanned by its prospects in electronics and photonics applications, viz. transparent electrodes in solar cells, flat display devices and novel solid-state ultraviolet (UV) lasers owing to its direct wide band gap [1–7].

Many works have been done on the fabrication of ZnO micro/nanostructures with high aspect ratios using metalorganic chemical vapor deposition [8], thermal evaporation and thermal decomposition [9]. But these techniques appear to be involved process with many complex steps, require sophisticated equipment and rigorous experimental conditions. However, some of them have drawbacks like long reaction time, toxic templets and exotic metal catalysts, and low purity or poor crystallite quality of products, which may influence the quality and applications of ZnO. So there is still the need for developing a method that can produce the ZnO micro/nanostructures in laboratory environment with high quality, high repeatability and low cost process for a wide range of applications. Hence, a novel technique called ECH for the preparation of ZnO micro/nanocrystals [10-12].

When a certain direct current flowed through a sample of ZnO ceramic bar, the sample was Joule-heated, and the crystal growth occurred. ZnO is a key technological material. The lack of a centre of symmetry in wurtzite, combined with large electromechanical coupling, results in strong piezoelectric and pyroelectric properties and the consequent use of ZnO in mechanical actuators and piezoelectric sensors. In addition, ZnO is a wide band-gap (3.37 eV) compound semiconductor that is suitable for short wavelength optoelectronic applications. The high exciton binding energy (60 meV) in ZnO crystal can ensure efficient excitonic emission at room temperature and room temperature ultraviolet (UV) luminescence has been reported in disordered nanoparticles and thin films. The objective of this research was to investigate the effect of compaction pressure and current on the crystal growth and its PL characteristics.

2. Experimental procedures

The amount of 10 grams of ZnO powder with 99.00% purity was measured using digital weighing machine. The weighted ZnO powder was grinded manually with mortar and pestle for 6 hours at room temperature to achieve find particles. Ceramic bar of ZnO with average dimension of 13.00 mm x 2.50 mm x 1.00 mm each were formed by adopting powder metallurgy methods. It contained three basic steps which were powder grinding, die compaction and sintering. Firstly, half of the amount of 30ml ethanol was added into ZnO powder gently. The mixture was grinded on hotplate manually and heated with temperature around 70.0°C. Another half of the amount ethanol was added slowly into the mixture during grinding process. Grinding process was end once the mixture was dry. The amount of 1.5 gram ZnO powder was prepared to make green body using compaction die with dimension of 14.95 mm x 30 mm x 40 mm. Different green body was made by different in applying compaction pressure of 2 and 3 tons, respectively. The green body of different applied compaction pressure was then sintered under 1100 °C with rate of 16.5 °C/min for 3 hrs. The sintered ceramic was then cut into 13.00 mm x 2.50 mm x 1.00 mm each prior to ECH. The current applied into the ECH was vary from 2.5 to 3.5 A. Figure 1 shows the ECH process until the crystal grew on the surface of ceramic bar.

Figure 1. Electric current heating applied into ceramic bar for ZnO crystal growth. It is shown on the 3rd figure the crystal grew on the surface of ceramic bar.

Characterization and observation were carried out for XRD RIGAKU Miniflex II, UV-vis, scanning electron microscope Carl Zeiss EVO 50, and PL spectroscopy NIR 300/2.

3. Results and Discussion

Sample of SZ2, SZ3, SZ4, SZ12, SZ13, SZ14 and SZ15 were chosen to conduct ECH process. The treatment of each sample is listed in table 1.

Table 1. The percentage of relative density of ZnO bars after sintered used in this research.

Pressure Applied (Ton)	Current (A)	Sample	Volume (cm^3)	Density (g/cm^3)	Relative density (%)
2.0	2.50	SZ3	0.0249	5.1807	92.41
	3.00	SZ4	0.0248	5.2823	94.23
	3.50	SZ2	0.0263	5.3612	95.63
3.0	2.50	SZ15	0.0308	4.8377	86.29
	3.00	SZ12	0.0289	5.0519	90.12
	3.50	SZ13	0.0318	5.1887	92.56
	3.50	SZ14	0.0357	4.6779	83.44

Figure 2 shows the comparison of XRD of all samples where red colour of planes correspond to Cu powder and black colour of planes were contributed by ZnO crystal. The XRD results showed most of the crystals grew on the plane (1 0 1). The crystal was in rock like shape. This was proved by the plane (1 0 1) had the highest intensity in all the samples except SZ13 and SZ14. The highest intensity of crystal grew in SZ13 was plane (1 0 0). The highest intensity of crystal grew in SZ14 was contributed by Cu since large amount of Cu powder was placed on top of ZnO bar. This had affect the growth of ZnO crystal.

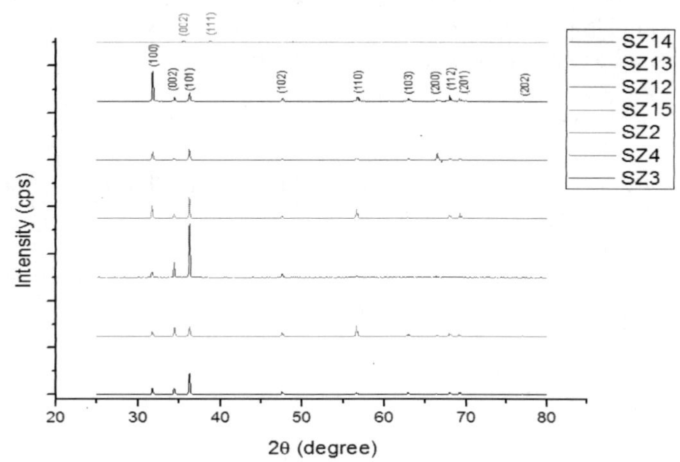

Figure 2. XRD of all samples where red colour of planes correspond to Cu powder and black colour of planes were contributed by ZnO crystal.

Figure 3 shows the comparison of the crystals grown and PL spectra between samples. Crystal like structures were grown on the surface of ceramic bars such as flower like (SZ12), grass like (SZ13),

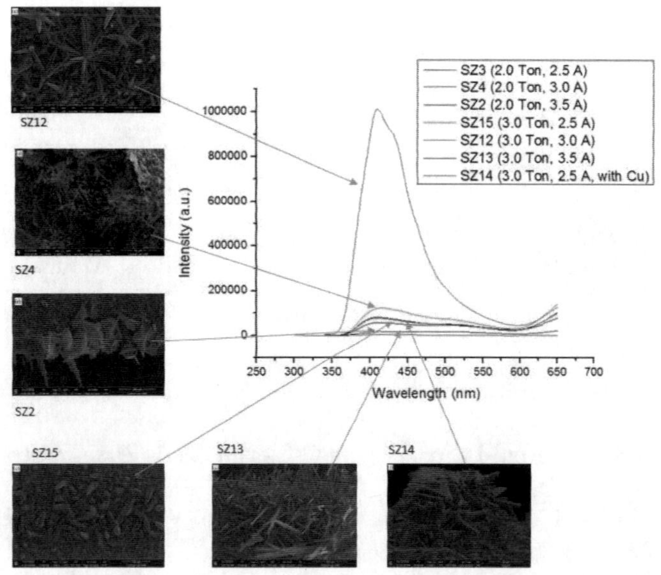

Figure 3. Comparison of SEM images and PL spectra between samples.

needle shape like (SZ4), KLCC tower like (SZ15), fish sting like (SZ2), and hexagonal shape like (SZ14). Under the PL spectrometer, the wavelength of samples SZ3, SZ4, SZ2, SZ15, SZ12, SZ13, and SZ14 were respectively 415, 411, 412, 413, 409, 412, and 415 nm. The shortest wavelength belonged to SZ12 has the highest intensity of about 1 million a.u. and 10 to 15 times higher than the other samples. The sample SZ12 which was under 3 tons and 3 A has the highest emission energy of 3.03 eV. This result was consistent with the result of UV-Vis spectrometer (see figure 4). The wavelength (in nm) and the excitonic energy (in eV) of SZ3, SZ4, SZ2, SZ15, SZ12, SZ13, and SZ14 are respectively 372 (3.34), 387 (3.21), 371 (3.34), 376 (3.30), 387 (3.20), 376 (3.30), and 376 (3.30). Sample 12 had emitted the longest wavelength and thus own the lowest excitonic energy. This showed that electron in SZ12 can be excited most easy among all samples.

4. Conclusion

In summary, it can be concluded that the sintered ZnO ceramic bar (SZ12) fabricated under 3 tons compaction pressure of green body with dimension of 13.0 mm x 2.5 mm x 1.0 mm and under ECH with input current of 3 A has the optimum value among the investigated samples. SZ12 sample revealed that it had the lowest exciton energy of 3.20 eV, the highest emission energy of 3.02 eV, the most significant crystal growth of flower like crystal.

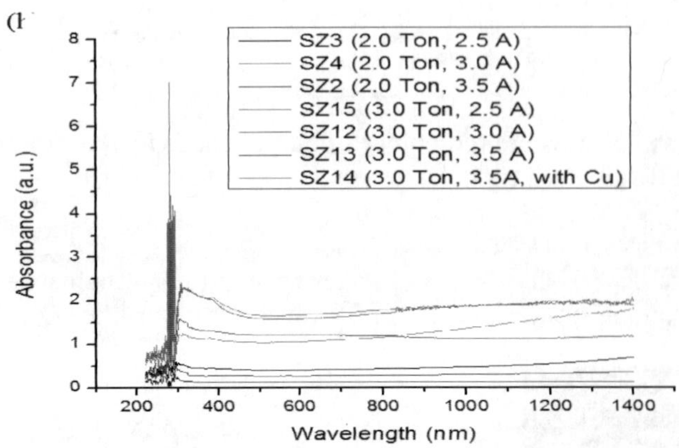

Figure 4. The result of UV-Vis spectrometer between samples.

Acknowledgement

This work was financially supported by Fundamental Research Grant Scheme 2016 (FRGS/1/2016/STG02/UMP/02/1), Ministry of Higher Education Malaysia and UMP Grant RDU1703194.

References

[1] R Sivakumar, T Tsunoda, Y Kuroki, T Okamot and M Takata 2012 *Mat. Chem and Phys* **134** 345– 349

[2] J Zhan, H Dong, S Sun, X Ren, J Liu, Z Chen, C Lienau and L Zhang 2016 *Adv. Optical Mater.* **4** 126–134

[3] D D Wang, G Z Xing, J H Yang, L L Yang, M Gao, J Cao, Y J Zhang and B Yao 2010 *J. Alloys Compd.* **504** 22

[4] R Sivakumar, K Punitha, C Sanjeeviraja, R Gopalakrishnan, 2014 *Materials Letters* **121** 141–144

[5] G H Nam, S H Baek, C H Cho and I K Park 2014 *Nanoscale*

[6] M Willander, L L Yang, A Wadesa, S U Ali, M H Asif, Q X Zhao and O Nur 2009 *J. Mater.*

Chem. **19** 1008
[7] A G E Sutjipto and L A Gafar 2012 *Contemporary Metallic Materials*, IIUM Press 141-146
[8] G Zhang, A Nakamura, T Aoki, J Temmyo and Y Matsui 2006 *Appl. Phys. Lett.* **89** 113112
[9] R Yousefi and B Kamaluddin 2009 *J. Alloys Compd.* **479** 11
[10] D Nezaki, S Takano, Y Kuroki, Y Kurihara, T Okamoto and M Takata 2000 *Trans. Mater. Res. Soc. Jpn.* **25** 205
[11] D Nezaki, T Okamoto and M Takata 2002 *Key Eng. Mater.* **228–229** 241
[12] H Yamasaki, K Minato, D Nezaki, T Okamoto, A Kawamoto and M Takata 2004 *Solid State Ionics* **172** 349

Optimizing the Synthesis of Alumina Inserts Using Hot Isostatic Pressing (HIP)

T F Ariff[1], A Z Azhar[1], M N Sariff [1], S N Rasid[1], S Z Zahari[1], R Bahar[1], M Karim[1] and AKM Nurul Amin[1]

[1]Department of Manufacturing and Materials Engineering, International Islamic University Malaysia (IIUM), Jalan Gombak, 53100 Kuala Lumpur, Malaysia

Email: tasnim@iium.edu.my

Abstract.
Alumina or Aluminium Oxide (Al_2O_3) is well known for its high strength and hardness. Its low heat retention and low specific heat characteristics make it attractive to be used widely as a cutting tool for grinding, milling and turning processes. Various synthesis methods have been used for the purpose of enhancing the properties of the alumina inserts. However, the optimization process using Hot Isostatic Pressing (HIP) has not been performed. This research aims in finding the optimum parameters in synthesizing the alumina inserts ($98Al_2O_3$ $1.6ZrO_2$ $0.4MgO$, $93Al_2O_3$ $6.4ZrO_2$ $0.6MgO$ and $85Al_2O_3$ $14.5ZrO_2$ $0.5MgO$) using HIP at different temperatures (1200, 1250 and 1300°C) and sintering time (10, 30 and 60 minutes). Hardness, density, shrinkage and microstructure using SEM were analysed. The optimum sintering condition for the alumina insert was found in $98Al_2O_3$ $1.6ZrO_2$ $0.4MgO$ sintered at 1300°C for 60 minutes for it exhibited the highest values of hardness (1917HV), density (3.95g/cm^3), shrinkage (9.6%).

1. Introduction

Various types of cutting inserts are available in machining industry and used widely in milling, turning, grinding and threading. A good cutting insert is used to achieve good performance. The performance of machined components can be seen through dimensional accuracy and surface roughness. The quality of the product is influenced by the cutting insert itself. Ceramic inserts exhibit excellent hardness, toughness and thermal conductivity. The synthesis of ceramic inserts are typically prepared by powder metallurgy technique and can be manufactured further to near net finish [1], [2].

One of the most potential cutting inserts is made from Al_2O_3 based materials which are popularly known to be chemically inert, possess high hardness, abundant and relatively cheap. Alumina is used as raw material for a broad range of advanced ceramic product and as an active agent in chemical processing [3], [4]. Zirconium dioxide (ZrO_2) is added to enhance the strength and toughness of the alumina, while magnesium oxide (MgO) is added to enhance the hardness of alumina [5], [6], [7].

Hot Isostatic Pressing (HIP) is a process commonly used for densification enhancement in powder metallurgy. The uniform heat and pressure applied to all surfaces enhances the mechanical properties of the part. The open pores in Al_2O_3 decreases linearly with increasing bulk density [8]. Various methods and compositions are used in the development of Al_2O_3 tool inserts. This paper aims in producing an optimum composition with optimum conditions in the synthesis of Al_2O_3 tool inserts using HIP.

2. Experimental procedure

Three different compositions of Al_2O_3 tool inserts were prepared (98wt%Al_2O_3 1.6wt%ZrO_2 0.4wt%MgO, 93wt%Al_2O_3 6.4wt%ZrO_2 0.6wt%MgO and 85wt%Al_2O_3 14.5wt%ZrO_2 0.5wt%MgO). These powders were weighed (Table 1) using a weighing scale (CP 224 S Sartorius Scale) and mixed accordingly in a ball mill (FRITSCH-pulverisette 5) using 23 steel balls with diameter of 8.89 mm. The powders were compacted using the VT manual hand press (MP-15T). A pressure of 200 kg/cm^2 was applied with a holding time of 5 minutes. 6 samples from each composition were produced. The samples were placed in the HIP machine (American-HP 630) at a temperature of 1300°C for 60 minutes. This part was first performed to identify the optimum composition.

Then, once the best composition was selected, another 6 samples were prepared. Optimum sintering temperature was selected by placing the samples into the HIP for 60 minutes at 1200, 1250 and 1300°C. Once the optimum temperature was selected, then the optimum sintering time had to be found. Experiment was repeated by preparing another 6 samples and placing them into the HIP at 1300C for 10, 30 and 60 minutes.

For every part of the experiment, the hardness (Micro-Vickers 401 MVA), density (GK-300) and the shrinkage were determined. Readings were repeated 3 times for each test. Results were plotted into graphs and charts. SEM images were obtained to analyze the microstructure and porosity (JEOL JSM-5600).

Table 1. Mass of powder used for each composition1ol inserts

Compositions	Al_2O_3 (g)	ZrO_2 (g)	MgO (g)
$98Al_2O_3$ $1.6ZrO_2$ 0.4MgO	4.508	0.0736	0.0184
$93Al_2O_3$ $6.4ZrO_2$ 0.6MgO	4.278	0.2944	0.0276
$85Al_2O_3$ $14.5ZrO_2$ 0.5MgO	3.91	0.667	0.023

3. Results and Discussion

3.1 Determining the optimum composition

Figure 1 shows the result of Vickers hardness. The hardest sample is $98Al_2O_3$ $1.6ZrO_2$ 0.4MgO (1917HV), followed by $93Al_2O_3$ $6.4ZrO_2$ 0.6MgO (1825 HV) and $85Al_2O_3$ $14.5ZrO_2$ 0.5MgO (1686 HV). Higher percentage of Al_2O_3 has exhibited better hardness. Increasing the percentage of ZrO_2 did not show any improvements in the overall hardness of the alumina insert.

The density and shrinkage values of the alumina inserts are shown in Figure 2. Density for $98Al_2O_3$, $93Al_2O_3$ and $85Al_2O_3$ are 3.89, 3.87 and 3.86 g/cm^3 respectively. The densities were found to be quite similar to each other for the 3 different compositions despite exhibiting different hardness values. The reduction in the densities are very insignificant with the reduction in the percentage of alumina. Nevertheless, the percentage of shrinkage has shown to increase 0.74% and 3.12% with the reduction in the densities for the 93 and 85wt% of Al_2O_3 respectively (Figure 2). Larger shrinkage values relate to larger porosities inside the compact.

IOP Conf. Series: Materials Science and Engineering **290** (2017) 012044 doi:10.1088/1757-899X/290/1/012044

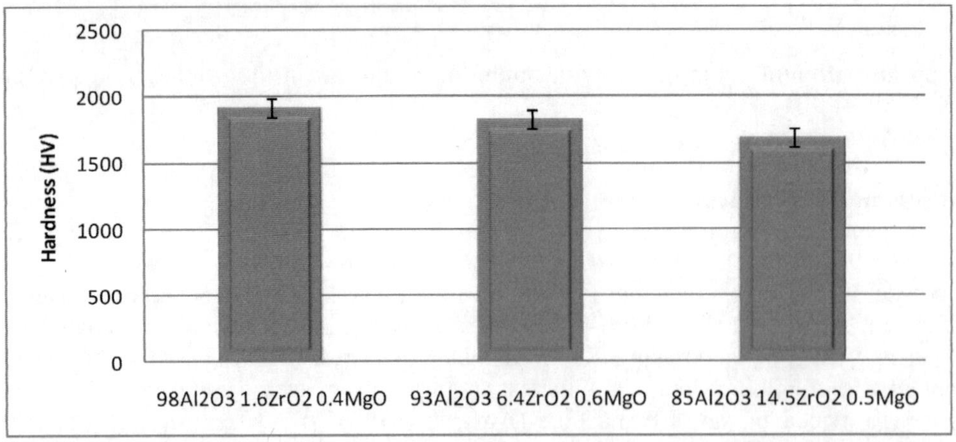

Figure 1. Hardness values for the 3 different compositions of alumina inserts

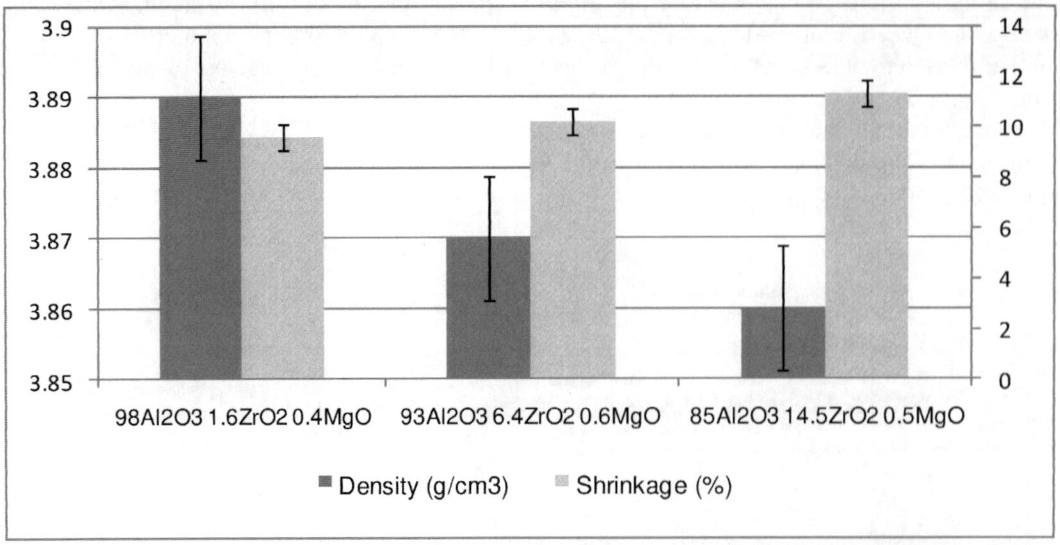

Figure 2. Density and shrinkage values for the 3 different compositions of alumina inserts

SEM images of the alumina inserts after HIP are shown in Figure 3. It can be seen that the surface of samples for $98Al_2O_3$ Figure 3(a) has less porosity when compared to $93Al_2O_3$ and $85Al_2O_3$ Figure 3 (b) and (c) respectively. This relates with the hardness and density values which decreases as the weight percentage of Al_2O_3 reduces from 98% to 85%. Full diffusion and densification can be seen in $98Al_2O_3$. Similar observation is also noticed in the commercialized insert (Sandvik Coromant) in Figure 3(d). $85Al_2O_3$ in Figure 3(c) indicates that less percentage of alumina resulted in larger porosity and lesser densification.

Since $98Al_2O_3$ has exhibited highest values of hardness, density and the lowest shrinkage, for the rest of the experiments, $98Al_2O_3$ $1.6ZrO_2$ $0.4MgO$ composition was selected. This appears to be the optimum composition to work on.

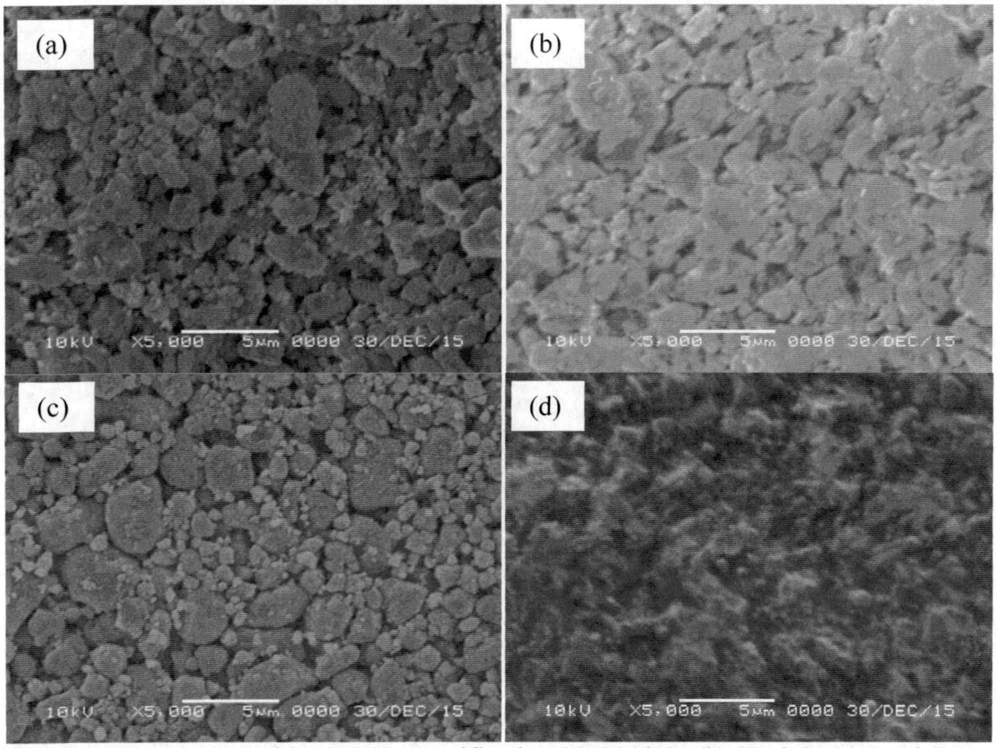

Figure 3. SEM image with 5000 X magnification (a) $98Al_2O_3$ (b) $93Al_2O_3$ (c) $85Al_2O_3$ (d) Sandvik Coromant (commercialized alumina insert)

3.2 Determining the optimum sintering temperature

$98Al_2O_3$ $1.6ZrO_2$ 0.4MgO is used in determining the optimum sintering temperature in HIP. Figure 4 shows graph of the hardness values obtained for the three different sintering temperatures (1200, 1250 and 1300°C). The highest hardness value (1917 HV) obtained is at the highest temperature 1300°C. There has been an increase in hardness by 57% when the sintering temperature increased from 1200°C to 1300°C. This is a significant increment. However, it only increased by 8% when sintered from 1250°C to 1300°C. Higher sintering temperature densifies the structure further, diffuses more completely and reduces porosity.

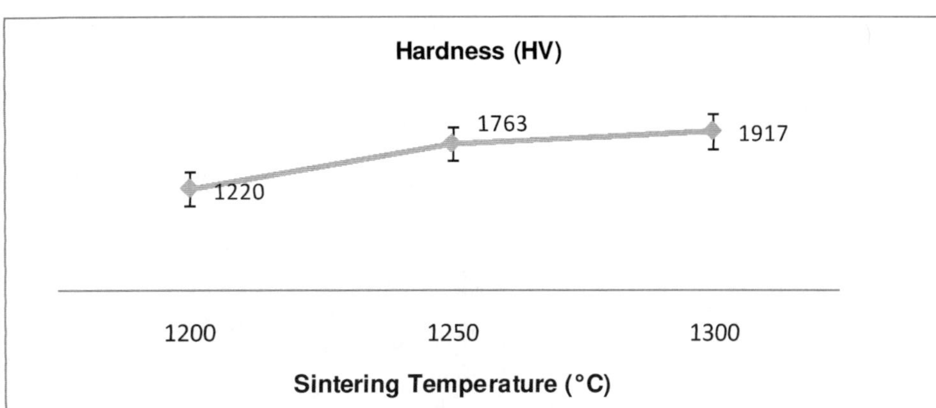

Figure 4. Hardness values for the $98Al_2O_3$ $1.6ZrO_2$ 0.4MgO Alumina Inserts sintered for 60 minutes

Figure 5 shows the density and shrinkage values for three different sintering temperatures obtained from the $98Al_2O_3$ $1.6ZrO_2$ 0.4MgO Alumina Inserts when sintered for 60 minutes. The values show that the highest density (3.89 g/cm^3) is achieved when the alumina insert is sintered at the highest temperature (1300°C). This is in line with the reduction in porosity where the highest shrinkage value (8.25%) obtained is also at this temperature.

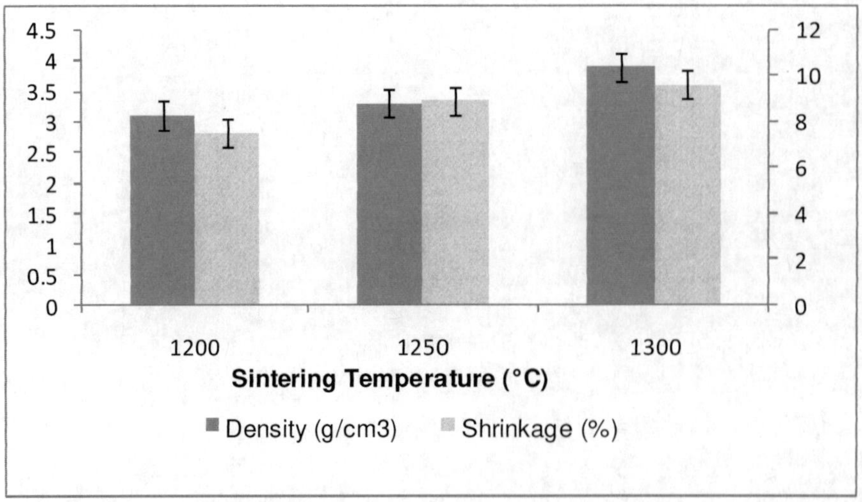

Figure 5. Density and Shrinkage for $98Al_2O_3$ $1.6ZrO_2$ 0.4MgO Alumina Inserts sintered for 60 minutes

3.3 Determining the optimum sintering time
98Al2O3 1.6ZrO2 0.4MgO sintered at 1300°C in HIP is used in determining the optimum sintering time in HIP. Figure 6 shows the graph of the hardness values obtained for the three different sintering time used in HIP (10, 30 and 60 mins). The highest hardness value (1917 HV) obtained is at the longest time (60 mins). There has been an increase in hardness by 37% when the sintering time increased from 10 to 30 minutes. This is a significant increment. Nevertheless, it has increased by 97% when sintered from 30 to 60 minutes. Longer sintering time has resulted in a more densified structure, enhanced diffusion and reduced porosity.

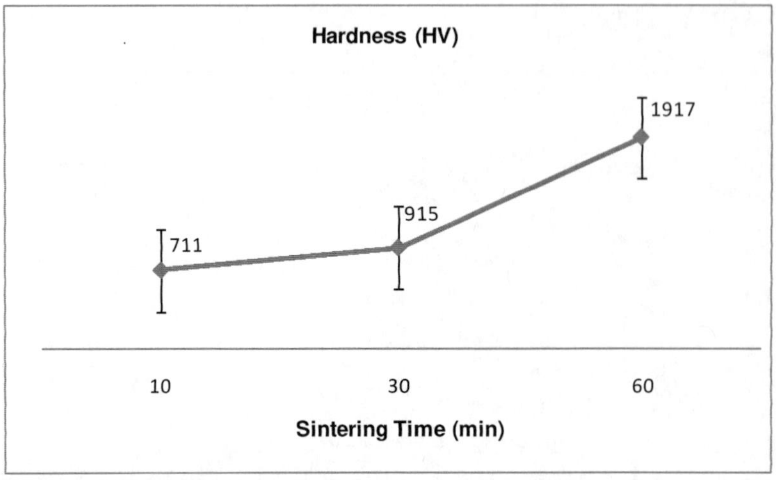

Figure 6. Hardness values for the 98Al2O3 1.6ZrO2 0.4MgO Alumina Inserts sintered at 1300°C

Figure 7 shows the density and shrinkage values for three different sintering time obtained from the $98Al_2O_3$ $1.6ZrO_2$ 0.4MgO Alumina Inserts when sintered at 1300°C. The values show that the highest density (3.95 g/cm^3) is achieved when the alumina insert is sintered for the longest time (60 mins). This is in line with the reduction in porosity where the highest shrinkage value (9.6%) obtained is also at this time.

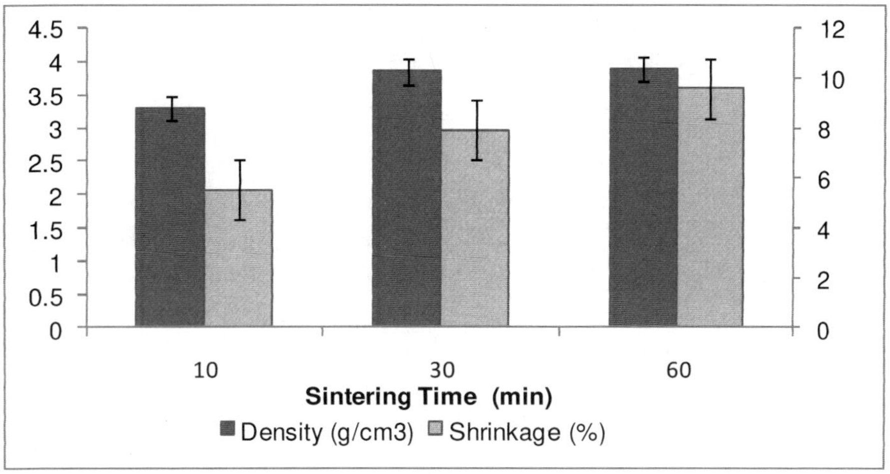

Figure 7. Density and Shrinkage for $98Al_2O_3$ $1.6ZrO_2$ 0.4MgO Alumina Inserts sintered at 1300°C

4. Conclusion

The optimum sintering composition for the alumina insert using HIP was found in $98Al_2O_3$ $1.6ZrO_2$ 0.4MgO sintered at 1300°C for 60 minutes for it exhibited the highest values of hardness (1917HV), density (3.95g/cm^3) and shrinkage (9.6%). The hardness of the synthesized alumina inserts using HIP may not be as high as the commercialised inserts (2400-2600 HV). However, further experiments in machining and tool wear analysis would enhance the findings, since hardness alone is not enough to justify longer tool life .

Acknowledgement

This work was financially supported by the Fundamental Research Grant Scheme, awarded by the Ministry of Higher Education Malaysia (FRGS 14-157-0398).

References

[1] Gutman E 1994 *Surf. Coat. Technol.* **67** 133-136
[2] Marinescu I 2004 Preface *Tribology of Abrasive Machining Processes* **VII-X**
[3] Liu J 2014 *Proceedings of the Institution of Mechanical Engineers, Part B: J. Eng Manuf* **45** 28–35
[4] Dimitrijevic M *et al* 2012 *Sci. Sint.* **44** 25-33
[5] Senthil Kumar A, Raja Durai A and Sornakumar T 2004 *Int. J. Refrac. Met. Hard Mater.* **22** 17-20
[6] Ahmad M A and Ebrahim M I 2014 *World J. Nano. Sci Eng.* **02** 50-57
[7] Ali A, Azhar A Z, Ratnam M M and Ahmad Z A 2015 *Adv. Mater. Res.* **1087** 101-105
[8] Dadbakhsh S and Hao L 2012 *J. Mater. Process. Technol.* **212 (11)** 2474-2483

The viability of MCM-41 as separator in secondary alkaline cells

S R Meskon, R Othman, M H Ani

Faculty of Engineering, International Islamic University Malaysia (IIUM), P.O. Box 10, 50728 Kuala Lumpur, Malaysia

Email: raihan@iium.edu.my

Abstract. The viability of MCM-41 membrane as a separator material in secondary alkaline cell is investigated. The inorganic membrane was employed in an alkaline nickel-zinc system. MCM-41 mesoporous material consists of arrays of hexagonal nano-pore channels. The membrane was synthesized using sol-gel route from parent solution comprising of quarternary ammonium surfactant, cethyltrimethylammonium bromide $C_{16}H_{33}(CH_3)_3NBr$ (CTAB), hydrochloric acid (HCl), deionized water (H_2O), ethanol (C_2H_5OH), and tetraethylortosilicate (TEOS). Both the anodic zinc/zinc oxide and cathodic nickel hydroxide electrodeposited film were coated with MCM-41 membrane. The Ni/MCM-41/Zn alkaline cell was then subjected to 100-cycle durability test and the structural stability of MCM-41 separator throughout the progression of the charge-discharge cycles is studied. X-ray diffraction (XRD) analysis on the dismantled cell shows that MCM-41 began to transform to lamellar MCM-50 on the 5th cycle and transformed almost completely on the 25th cycle. The phase transformation of MCM-41 hexagonal structure into gel-like MCM-50 prevents the mesoporous cell separator from diminished in the caustic alkaline surround. This work has hence demonstrated MCM-41 membrane is viable to be employed in secondary alkaline cells.

1. Introduction

The primary function of a separator is to provide physical barrier between the anode and cathode in order to prevent electrical short circuit. It also serves as electrolyte reservoir for the transport of ions during charge and discharge cycling of a battery [1]. Although a separator does not involve in chemical reactions of an electrochemical system, it has a direct influence on charge efficiency and discharge capacity of the cell. The wrong choice of separator may lead to premature cell failure [2].

The requirements of the separator often depend on the type of battery chemistry; for each chemistry has its typical cell behavior and concerns which need to be considered. In the case of lithium ion batteries for example, the separator should have very small pores and good mechanical strength to inhibit dendritic lithium penetration and conductive particles from the additives or active materials, in order to prevent short circuit and capacity loss [2]. The mechanical integrity of the separator should also remain preserved when the temperature rise during operation, i.e. its thermal shrinkage should be as low as possible [2]. As for Ni-Zn batteries, the separator should have the capability to lessen the solubility of zinc in the electrolyte to reduce capacity loss and the capability to protect the electrode from zinc dendrite penetration [1]. A separator for Ni-Zn battery typically consists of two different types and forms of materials i.e. wettable membrane (polypropylene and polyethylene) and nonwoven (polyamide) and is often constructed as multilayers [1-2].

Charging of Ni-Zn cell generates oxygen at the cathode. The oxygen needs to pass through the separator to reach the anode and recombine during discharge [2, 3]. Thus, it is essential that the separator to have a sufficient porosity for the mechanism to occur [2]. The gas evolution causes pressure buildup in the cell. Thus, the separator's mechanical integrity needs to be unimpaired. The separator also needs to supply sufficient electrolyte for the transport of ions so that the cell reaction can occur, since water is produced during charging and is consumed during discharging [2]. Thus, its water retention capability should be high enough.

MCM-41 mesoporous material stands out due to characteristics of narrow pore size distribution, tunable pore sizes, extremely high surface area, high pore volume and adjustable hydrophobicity [4-7]. The structure consists of hexagonally ordered arrays of pore channels with uniform diameter. Its high water-adsorption affinity is due to the presence of silanol (Si-O-H) group at the pore walls [8]. Saputra et al. [9] first reported the use of MCM-41 as a separator in zinc-air cells. The cell performance was comparable with that of the commercial cell. They further reported the transformation of MCM-41 silica into gel-like MCM-50 after 24 hours of controlled exposure in 6-M KOH [10]. However, the studies were not conducted in a complete cell i.e. only the reaction between MCM-41 and KOH.

Herein, we study the structural stability of MCM-41 silica as separator in alkaline Ni-Zn secondary cell as the cell was subjected to 100 cycles of charge-discharge cycling. The cell was then dismantled after the progression of several cycles and the structural changes of the membrane analyzed using XRD.

2. Experimental Procedures

Cell Description and Process Flow. The experimental Ni-Zn secondary cell consisted of a zinc anode, nickel cathode, cell electrolyte and separator. The zinc anode comprised of a Zn foil coated with zinc oxide (ZnO) while the nickel cathode comprised of electrodeposited nickel hydroxide ($Ni(OH)_2$) thin film on a copper substrate (of a blank printed circuit board (PCB)). A 6-M potassium hydroxide (KOH) solution was used as the electrolyte and MCM-41 silica thin film and powder were used as the separator. Three identical cells were prepared for the characterizations, in which each cell corresponded to different cycle number of charge-discharge cycling after which the separator material's condition would be studied.

Synthesis of MCM-41 Silica Thin Film and Powder. Two forms of MCM-41 were used in the cell fabrication, i.e. thin film and powder. Sol-gel method was employed for the synthesis. The MCM-41 thin film was formed from the precursor solution. In a typical synthesis, the precursor solution was prepared by mixing a quaternary ammonium surfactant, cetyltrimethylammonium bromide (CTAB, $CH_3(CH_2)_{15}N(Br)(CH_3)_3$; Merck, Germany) as an organic template, hydrochloric acid (HCl; J.T. Baker, Thailand), ethanol (C_2H_5OH; Systerm) as a solvent, distilled water (H_2O) and tetraethyl orthosilicate (TEOS, $Si(OC_2H_5)_4$; Merck, Germany) as the silica source, with a molar ratio of 0.05 CTAB: 1.0 TEOS: 0.5 HCl: 25 C_2H_5OH: 75 H_2O and stirring at 200 rpm, at a room temperature of 30 °C for about one hour [9]. On the other hand, the MCM-41 powder was synthesized by mixing CTAB, sodium hydroxide (NaOH, Sigma-Aldrich, Sweden), distilled water and TEOS, with a molar ratio of 0.05 CTAB: 1.0 TEOS: 0.5 NaOH: 100 H_2O, and stirring at 200 rpm, heating the mixture on a hotplate at around 50 °C for 15 hours. The white precipitate of the MCM-41 powder was separated by filtration, rinsed with distilled water and dried at 100 °C.

Preparation of Zinc Anode. The Zn foil was prepared by succesively grinding the surface of a Zn foil (2 cm x 1 cm) using P600 silicon carbide abrasive paper to obtain a new rough surface (free from oxide or grease), immersing (in a short time) in acetone and drying in ambient conditions. Subsequently, the Zn foil was dipped in a ZnO slurry (ZnO powder in ethanol). The ZnO-coated Zn was then let to dry at room temperature.

IOP Conf. Series: Materials Science and Engineering **290** (2017) 012045 doi:10.1088/1757-899X/290/1/012045

Preparation of Nickel Cathode. The nickel hydroxide ($Ni(OH)_2$) thin film was prepared by cathodic reduction (galvanostatically at current density of 8.5 mA/cm^2) of an electrolytic bath comprising 1.8-M nickel nitrate hexahydrate ($Ni(NO_3)_2{\cdot}6H_2O$; Merck) (49.4 vol. %), 0.075-M sodium nitrate ($NaNO_3$; HmbG) (16.6 vol. %) and ethanol (34 vol. %). A nickel foil was used as the working electrode and copper substrate as the counter electrode. The dimensions of the effective area of the copper film were 2 cm x 1 cm and the distance between the electrodes was 2.3 cm. The electrodeposition was carried out for 75 minutes using an Autolab PGSTAT 302N (Eco Chemie B.V., Netherlands) potentiostat/galvanostat. The thin film formed on the surface of Cu was lightly rinsed with distilled water and then dried at room temperature.

Assembly of Cell Components.

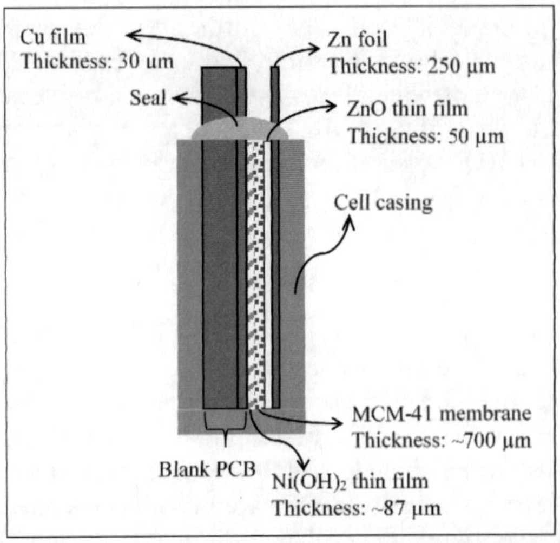

Figure 1 Schematic diagram of the cell construction layout.

The MCM-41 thin film was formed by drop-coating the as-synthesized MCM-41 precursor solution onto the surface of $Ni(OH)_2$ thin film using a pipette dropper. The solution would then self-spread over the entire surface of the thin film and left to dry at ambient conditions to form a layer of MCM-41 thin film. A total of 6 layers were prepared so as to obtain a firmly bonded $Ni(OH)_2$ thin film (to the surface of Cu substrate). The $Ni(OH)_2$ cathode and the Zn anode were then assembled into the cell casing (figure 1) and the gap in between the electrodes was filled with MCM-41 powder (0.35 g). The cell was activated by inserting KOH electrolyte (6 M, 0.5 ml) and finally sealed with acrylic resin.

Electrochemical Characterization. The cell was charged at a constant current of 2.0 mA for 2 hours, left at an open circuit potential for 5 minutes and then discharged at 0.2 mA until the cell potential reached 0.2 V. The steps were repeated for as many times until the cell was considered failed. A Neware battery testing system, model BTS-5V100mA (NEWARE Technology Ltd., China) was used.

Structural Stability Characterization. The main tool for the stability study was X-ray powder diffraction (XRD) and supported by surface morphology observation. A Bruker D8 ADVANCE X-ray diffractometer with a CuK_α radiation (λ = 1.5406 Å), operating at 40 kV, 40 mA and a scan rate of 2.4 deg. 2θ/min was used for XRD measurement and a field emission scanning electron microscope (FE-SEM) model JEOL JSM-6700F was used for surface morphology observation. Prior to measurements, all samples were dried in a vacuum oven at 50 °C for 3-4 hours to ensure they were sufficiently dry.

3. Results and Discussion

XRD of MCM-41 and $Ni(OH)_2$ Materials. The presence of MCM-41 silica can be noted from the four low-angle peaks at 2.16°, 3.78°, 4.32° and 5.8° of 2θ, which can be indexed as (100), (110), (200) and (210) Bragg peaks, respectively, corresponding to the hexagonal ordered array. Figure 2 shows the XRD pattern of the as-prepared MCM-41 silica powder; the corresponding *d*-spacing values are also displayed in the figure. The XRD pattern confirms the typical mesoporous structure of MCM-41. From the XRD data, the repeat distance, a_0 between two pore centres in the MCM-41 is 4.72 nm ($a_0 = (2/\sqrt{3})d_{100}$), i.e. using the interplanar spacing d_{100} value of 4.09 nm. The crystallite size of the MCM-41 powder based on the (100) plane, calculated from Scherrer's equation is 20.9 nm.

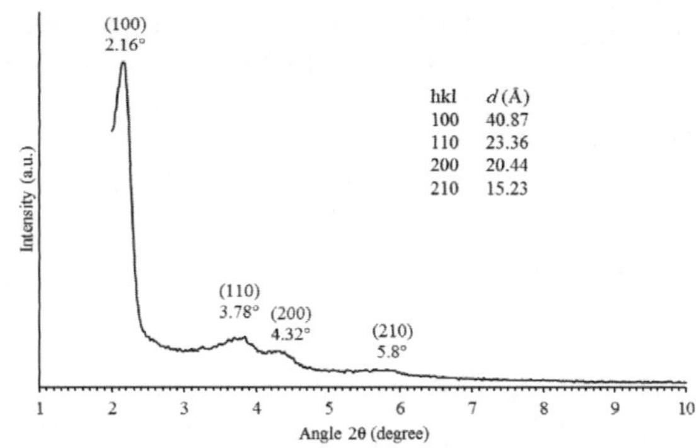

Figure 2 XRD pattern of the as-prepared MCM-41 silica powder.

Electrochemical Study of Ni-Zn Cell. Figure 3 shows the charge-discharge cycling result of the Ni-Zn cell, i.e. discharge capacity profile as a function of cycle number. The result describes the electrochemical behaviour of the cell throughout the course. The first 13 cycles show the cell's development or conditioning phase in which the capacity fluctuates prior to stabilizing i.e. up to the 13th cycle. Next, the cell underwent the mature phase (from the 14th to the 47th cycles) in which the cell capacity's becomes stable as marked by the plateau region. Subsequently, from cycle-48 until cycle-68 the capacity started to decline monotonously. Finally in the failure phase, the capacity ceases towards zero, i.e. from the 69th cycle onwards. Based on the profile observed, the structural observation of the separator material was thus studied for the 5th, 25th and 80th cycles of the development, mature and failure phases, respectively.

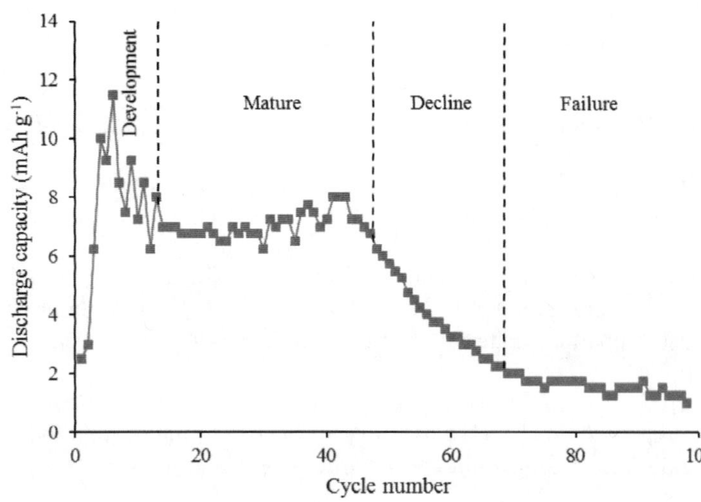

Figure 3 Discharge capacity profile of the Ni-Zn cell.

XRD of the MCM-41 Separator. Figure 4 shows the XRD patterns of the MCM-41 separator material for the 5th, 25th and 80th cycles (denoted as M5, M25 and M80, respectively). These XRD patterns denote the MCM-41's transformation under the influence of KOH electrolyte and the cell's charge-discharge conditions. From the results, the structural integrity of the membrane throughout the process can be analysed.

The peaks at 2θ of ca. 2.72° and 5.58° can be indexed to MCM-50's (100) and (200) planes, respectively [11, 12]. All XRD patterns show the presence of MCM-50 peaks, suggesting that the lamellar phase is predominant in the separator materials. However, the XRD pattern of MCM-41 are also present in M5, i.e. the (100) peak, at 2.18° of 2θ, indicating that MCM-41 coexisted with MCM-50. The shift of the (100) peak of MCM-41 from 2.16° in the as-prepared sample to 2.18° in M5 implies the lattice contraction of the mesoporous silica and thus decreased pores size. Similar trend was observed in [10] but the authors did not make any remarks on the shift of (100) peak position of MCM-41 under KOH (10–30 wt.%) loading effect. We attribute the lattice contraction to the formation of lamellar gel in the mix phase of MCM-50/MCM-41.

The (100) peak of MCM-41 is almost diminished in M25 and only a tiny peak appears at a substantially higher 2θ, i.e. 2.34° i.e. at this stage MCM-41 had almost entirely transformed to the lamellar MCM-50 phase. Whereas in M80, the predominant phase is MCM-50 and no trace of MCM-41 is observed.

Note that there are other peaks which are unindexed in M5 i.e. at 2θ of 3.06°, 3.38°, ~5.36°, ~6.1° and ~6.8°. Probably a transitional phase coexists with the diminishing two-dimensional (2D) hexagonal (*p6mm*) phase and the forming lamellar phase. These peaks are less distinct or absent in M25 and M80 samples.

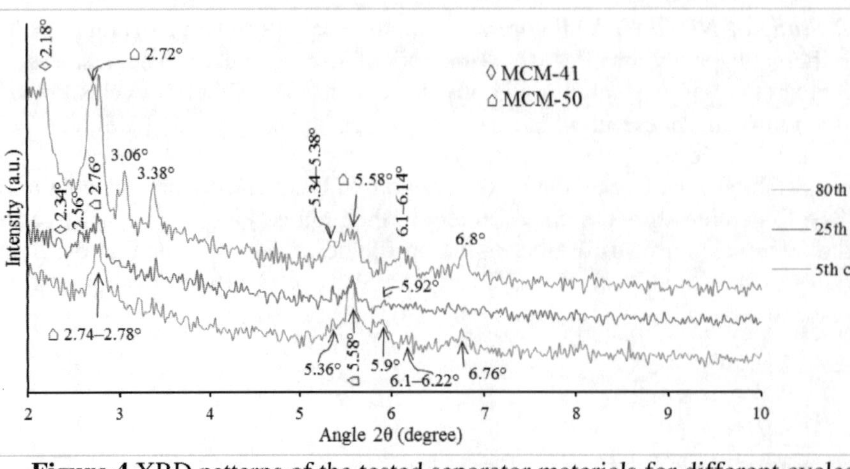

Figure 4 XRD patterns of the tested separator materials for different cycles.

Based on the XRD patterns of figure 4, it has been demonstrated that the MCM-41 membrane in the cell had partially underwent a phase transformation to MCM-50 as early as the 5th cycle and the material remained in the lamellar structure throughout the charge-discharge process. The phase transformation occurred as a result of the change of surfactant/silica molar ratio as the concentrated KOH reacts with the silica hexagonal matrix of MCM-41 [13]. This phenomenon prevented the collapse of the silica based separator material and hence the cell failure.

Surface Morphology of the Gelled MCM-41 Separator. The resulting formation of gelled MCM-41 (i.e. MCM-50) upon exposure to KOH was further inspected under the FE-SEM. The surface morphology of M5, figure 5(a), clearly indicates a particulate mixture morphology thus supporting the existence of MCM-41 phase. After 25 cycles (M25), figure 5(b), the particulate morphology turned into homogeneous agglomerates which most likely was the gel formation. Finally in figure 5(c), after 80 cycles (M80) the surface smoothened. At this stage, the gel structure might have shrunk due to water loss. Water loss is as a result of electrolysis during charging and heat induced evaporation. These observations are in agreement with the interpretations of XRD results; M5 comprised the mixed MCM-41/MCM-50 phase, and, M25 and M80 comprised of lamellar MCM-50 phase.

Therefore, the capacity fading of Ni/MCM-41/Zn secondary cell is not attributed to the use of silica based separator that underwent transformation in caustic alkaline surround. In fact, the formation of lamellar gel of MCM-50 prevented the disintegration of separator material. The collapse of the separator would have short-circuited the cell which did not happen throughout the progression of the 100-cycle charge-discharge test.

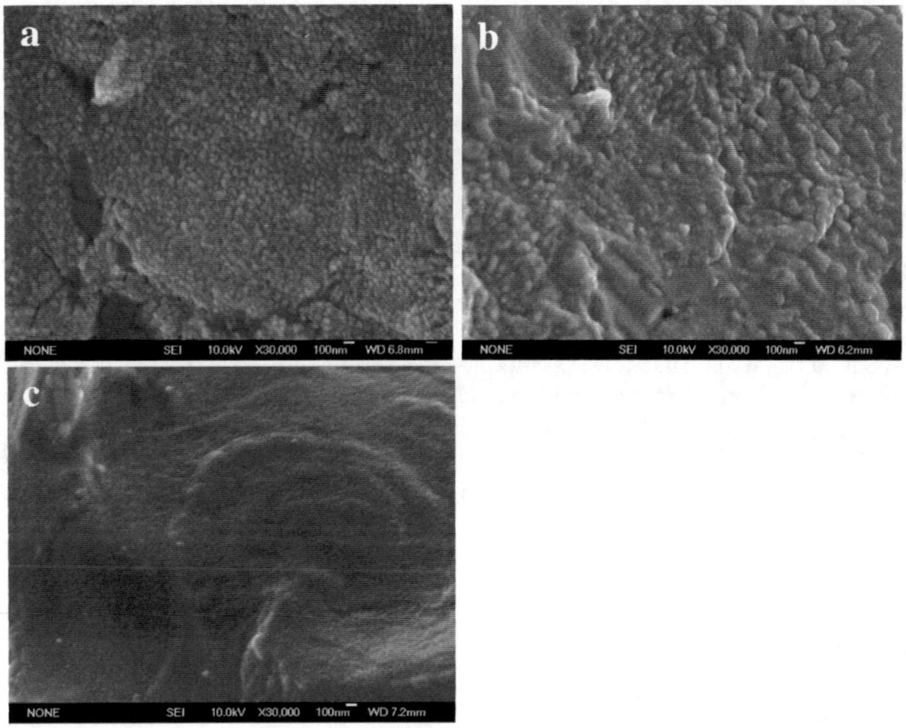

Figure 5 FE-SEM images of the separator materials which had gone through (a) 5, (b) 25 and (c) 80 cycles of charge-discharge cycling test.

4. Conclusions

The viability of MCM-41 as separator in secondary alkaline cells has been demonstrated in Ni-Zn electrochemical system. Investigations on dismantled Ni-Zn cell at various stages of charge-discharge cycle substantiated the transformation of hexagonal silica network of MCM-41 into lamellar gel structure of MCM-50. This phenomenon prevented the collapse of silica separator material and it remained in the lamellar gel structure throughout the 100 cycles of deep charge-discharge durability test. The ultimate Ni-Zn cell failure or capacity loss was not attributed to the use of MCM-41 membrane separator but rather most likely due to the electrolyte drying out and the well-known problem associated with zinc electrode i.e. mobility of zinc discharge products.

References

[1] H. Lee, M. Yanilmaz, O. Toprakci, K. Fu and X. Zhang, "A Review and Recent Developments in Membrane Separators for Rechargeable Lithium-Ion Batteries," *Energy Environ. Sci.*, 2014.

[2] P. Kritzer and J. A. Cook, "Nonwovens as Separators for Alkaline Batteries: An Overview," *J. Electrochem. Soc.,* vol. 154, no. 5, pp. A481–A494, 2007.

[3] M. Fetcenko, J. Koch and M. Zelinsky, "Nickel-Metal Hydride and Nickel-Zinc Batteries for Hybrid Electric Vehicles and Battery Electric Vehicles". In B. Scrosati, J. Garche and W. Tillmetz (Eds.), *Advances in Battery Technologies for Electric Vehicles* (pp.103-106), Cambridge: Woodhead Publishing, 2015.

[4] S. Beck, J. C. Vartuli, W. J. Roth, M. E. Leonowicz, C. T. Kresge, K. D.Schmitt, C. T.-W. Chu, D. H. Olson, E. W. Sheppard, S. B. McCullen, J. B. Higgins, and J. L. Schlenker, "A New Family of Mesoporous Molecular Sieves Prepared with Liquid Crystal Templates," *J. Am. Chem. Soc.*, vol. 114, pp. 10834–10843, 1992.

[5] N. Nishiyama, S. Tanaka, Y. Egashira, Y. Oku, and K. Ueyama, "Vapor-Phase Synthesis of Mesoporous Silica Thin Films," *Chem. Mater.,* vol. 15, pp. 1006–1011, 2003.

[6] F. Hoffmann, M. Cornelius, J. Morell, and M. Fröba, "Silica-based Mesoporous Organic–Inorganic Hybrid Materials," *Angew. Chem. Int. Ed.*, vol. 45, pp. 3216–3251, 2006.

[7] M. Schoeffel, N. Brodie–Linder, F. Audonnetx, and C. Alba–Simionesco, "Wall Thickness Determination of Hydrophobically Functionalized MCM-41 Materials," *J. Mater. Chem.*, vol. 22, pp. 557–567, 2012.

[8] J. C Vartuli, W. J. Roth, J. S. Beck, S. B. McCullen, and C. T. Kresge, "The Synthesis and Properties of M41S and Related Mesoporous Materials". In H. G. Karge and J. Weitkamp (Eds.), *Synthesis (Molecular sieves: Science and technology)* (pp. 101 – 109), Berlin, Germany: Springer-Verlag Berlin Heidelberg, 1998.

[9] H. Saputra, R. Othman., A. G. E. Sutjipto, and R. Muhida, "MCM-41 as a New Separator Material for Electrochemical Cell: Application in Zinc-Air System," *J. Membr. Sci.*, vol. 367, no. 1–2, pp. 152–157, 2011.

[10] H. Saputra, R. Othman, A. G. E. Sutjipto, R. Muhida and M. H. Ani, "Gel-Like Properties of MCM-41 Material and Its Transformation to MCM-50 in a Caustic Alkaline Surround", *Mater. Res. Bull.*, vol. 7, pp. 732–736, 2012.

[11] A. Monnier, F. Schuth, Q. Huo, D. Kumar, D. Margolese, R. S. Maxwell, G. D. Stucky, M. Krishnamurthy, P. Petroff, A. Firouzi, M. Janicke, and B. F. Chmelka, "Cooperative Formation of Inorganic-Organic Interfaces in the Synthesis of Silicate Mesostructures," *Science*, vol. 261 no. 5126, pp. 1299–1303,1993.

[12] F. Fajula, "Engineering of Mesostructured Silicas by Pseudomorphism," *Dalton Trans.*, pp. 291–294, 2007.

[13] J. C. Vartuli, K. D. Schmitt, C. T. Kresge, W. J. Roth, M. E. Leonowicz, S. B. McCullen, S. D. Hellring, J. S. Beck, J. L. Schlenker, D. H. Olson, and E. W. Sheppard, "Effect of Surfactant/Silica Molar Ratios on the Formation of Mesoporous Molecular Sieves: Inorganic Mimicry of Surfactant Liquid-Crystal Phases and Mechanistic Implications," *Chem. Mater.*, vol. 6, pp. 2317–2326, 1994.

IOP Conf. Series: Materials Science and Engineering **290** (2017) 012046 doi:10.1088/1757-899X/290/1/012046

XRD investigation of the Effect of MgO Additives on ZTA-TiO_2 Ceramic Composites

Ahmad Zahirani Ahmad Azhar[1], Hanisah Manshor[2], Afifah Mohd Ali[1]

[1]Department of Manufacturing and Materials Engineering, International Islamic University Malaysia (IIUM), Jalan Gombak, 53100 Kuala Lumpur, Malaysia
[2]Department of Science in Engineering, International Islamic University Malaysia (IIUM), Jalan Gombak, 53100 Kuala Lumpur, Malaysia

Email: zahirani@iium.edu.my

Abstract. Alumina (Al_2O_3) based ceramics possess good mechanical properties and suitable for the application of cutting inserts. However, this monolithic ceramics suffer from lack of toughness. Hence, there are some modification were made such as the addition of yttria stabilized zirconia (YSZ) to the Al_2O_3 helps in increasing the toughness of the Al_2O_3 ceramics. Some additives such as MgO and TiO_2 were used to further improve the mechanical properties of ZTA. In this study, high purity raw materials which consist of ZTA-TiO_2 were mixed with different amount of MgO (0.0 – 1.0 wt %). The mixture of materials was going through wet mixing, compaction and pressureless sintering at 1600°C for one hour. The samples were characterized for phase analysis, microstructure, shrinkage rate, bulk density, Vickers hardness and fracture toughness. Based on the XRD analysis results, the secondary phase ($MgAl_2O_4$) was detected in the sample with 0.5 wt% of MgO onwards which leads to grains refinement, thus improve the density and hardness of ZTA-TiO_2-MgO ceramics composites.

1. Introduction

Ceramics, with its excellence physical and mechanical properties, such as low density, strength, hardness and its inertness at high temperature, is widely known as a potential candidate for structural materials with very wide-ranging applications. The ever challenge in dealing with ceramic materials in any engineering application is their low brittleness. It is also having been a massive motivation in the field of ceramics research. In the motivation to enhanced the toughness of ceramic materials, one of the widely known ceramics Al_2O_3 based materials was often used the benchmark due its abundancy, relatively cost efficient, and outstanding mechanical properties.

Secondary component such as yttria-stabilized zirconia can be used to increase toughness of Al_2O_3 based materials. Work done by [1] has explored the effect of Al_2O_3/YSZ ratio to its mechanical and wear performance. His work summarized that 80:20 ratios would give the best wear performance in the turning process. Few authors has also experimented addition of solitary additives into zirconia toughened alumina to increase the toughness; the additives were MgO [2]–[5], Cr_2O_3 [6]–[9], $CaCO_3$ [10], CeO_2 [11] and TiO_2 [12]–[16]. Addition of MgO has shown to decrease Al_2O_3 grain size, resulting increase hardness and decrease toughness. Cr_2O_3, $CaCO_3$ and CeO_2 were shown to produce platelike Al_2O_3 grain, increasing to increase toughness and decreasing the hardness. Addition of the mentioned additives also reported to minimize the presence of monoclinic phase in YSZ which could withheld the t-m phase transformation during the presence of a propagating crack. Due to the additives unique contribution to the Al_2O_3 microstructure, research done by Rejab et al., [4],[17],[18] has focused on using multiple additive into ZTA. His work confirmed the formation of secondary phases which are $MgAl_{11}CeO_{19}$ and $MgAl_2O_4$ resulted from the addition of MgO into ZTA-CeO_2 ceramic composite. These newly form secondary phase tend to force propagating crack to travel further with crack deflection and crack bridging, resulted in higher fracture toughness.

In this paper, an approach of using two types of additives (3.0 TiO_2 wt % and 0 – 1.0 MgO wt %) was pursued as a continuation of the previous work [14]. This work aims to evaluate the effect of MgO additive on the ZTA-TiO_2 ceramic composite properties with respect to phase analysis.

2. Experimental procedures

The starting raw materials used were Al_2O_3 (Martinswerk, 99% purity), YSZ (Goodfellow, 5.4 mol % Y_2O_3 as stabilizer, 96 % purity), Anatase-TiO_2 (Fluka, Z99%) and MgO (Strem Chemicals, 99% purity). The initial mixture was 80 wt % Al_2O_3 and 20 wt % YSZ and was mixed with different amounts of MgO (0 – 1.0 wt %), whilst maintaining the ratio of Al_2O_3 to YSZ at 4:1. The compositional details are as per table 1. The mixtures were wet mixed using ABB-Mixer Mill with ZrO_2 balls. The slurry was then dried for 24 h at 100 °C in a Memmert-CU9760 oven, following which the dried cake was crushed and passed through a 75 mm sieve. The powders were then hydraulically pressed at 250 MPa for 120 s into pellets of 13 mm diameter and 4 mm thickness without binder. Sintering at 1600 °C for 1h of the pellets were carried out using a Udian furnace model KHT-16000X.

The sintered samples were then analyzed by X-ray diffractometer with CuKα (λ = 1.54056 Á) radiation operating at 40 kV, 40 mA with the 2θ range 10 – 90° and the scanning rate operated at 0.034°/s. The Rietveld refinement method was used to calculate the quantitative amount of each phase.

Table 1. Compositional details of the experiment.

Al_2O_3	YSZ	TiO_2	MgO
	wt %		
77.60	19.40	3.00	0.00
77.52	19.38	3.00	0.10
77.36	19.34	3.00	0.30
77.20	19.30	3.00	0.50
77.04	19.26	3.00	0.70
76.88	19.22	3.00	0.90
76.80	19.20	3.00	1.00

3. Results and Discussion

XRD analysis pattern for ZTA-TiO_2-MgO ceramic composites with varying composition of MgO is shown in figure 1. Three phases were identified in the ZTA-TiO_2 with 0 wt % MgO which are α-Al_2O_3 (ICSD 98-004-0015), t-ZrO_2 (ICSD 98-010-8646), and $Zr_{0.35}TiO_{0.65}O_2$ (ICSD 98-010-4677). The formation of secondary phase of $Zr_{0.35}TiO_{0.65}O_2$ is believed due to the reaction between ZrO and TiO_2 which was no longer incorporated into YSZ grains.

XRD analysis pattern for ZTA-TiO_2-MgO with the addition starting from 0.5 wt% MgO showed the formation of one new phase which is $MgAl_2O_4$ (ICSD 98-006-1951). This result is parallel to the previous work by Rejab et al. (2014). MgO reacts with Al_2O_3 to form fine secondary phase particles of $MgAl_2O_4$ due to an excess of MgO above the solubility limit. According to Rejab *et al.* (2014) the

$MgAl_2O_4$ peaks were more obvious at the Bragg angle of 37.02°. In this study, the $MgAl_2O_4$ peaks were seen approximated to angle of 37°. However, the peak only clearly observed in figure 1 at the sample with 1.0 wt% of MgO. Sintering at 1600°C caused the intensity of the corundum phase (α-Al_2O_3) decreased and on the opposite, the spinel phase was found to be increased (Sarkar *et al.* ,1999). The $MgAl_2O_4$ phase distribution was observed to be increased from 1.5% to 12.6% of phase distribution as shown in table 3.

Phase composition of tetragonal Zirconium Oxide, t-ZrO_2 was decreased from 0.0wt% to 1.0 wt%. The reduction of t-ZrO_2 phase distribution was almost 70%. The reduction is mainly due to the formation of secondary phase which is $Zr_{0.35}TiO_{0.65}O_2$andthe transformations of t-$ZrO_2$to m-ZrO_2 phase. According to Chaari *et al.* (2012), the formation of secondary phase will retard the densification and mechanical propertie. Azhar *et al.* (2014) reported that formation of monoclinic ZrO_2 can be prevented by the addition of stabilizing agent which is TiO_2. Although the amount of TiO_2 was fixed at 3.0 wt%, the formation of monoclinic Zirconium Oxide, m-ZrO_2 (ICDS 98-002-8057) still can be detected when 0.3 wt% of MgO was introduced to the ZTA-TiO_2 ceramics.

Figure 2 shows the enlarged XRD analysis of ZTA-TiO_2-MgO. Peak shifted to a larger angle indicates smaller lattice while broader peak signifies smaller particle sizes.

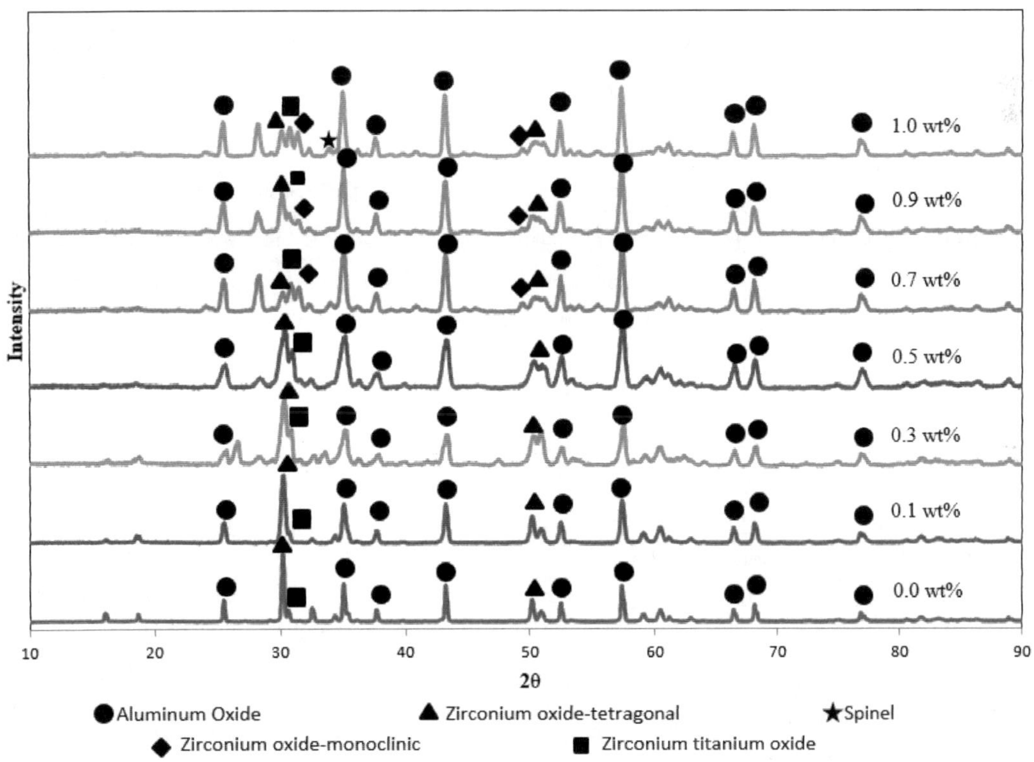

Figure 1. Normalized XRD analysis in ZTA-TiO_2 ceramic composites with different MgO compositions.

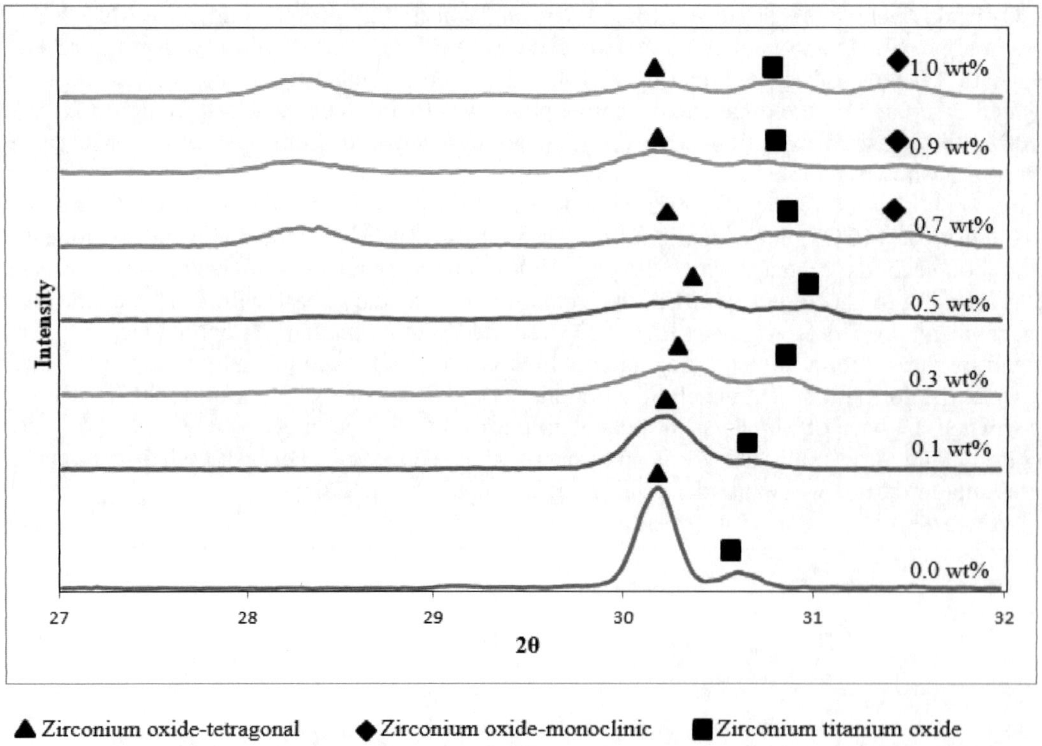

Figure 2. Enlarged XRD analysis in ZTA-TiO_2 ceramic composites with different MgO compositions.

Table 2. Lattice parameter of Al_2O_3 with different MgO composition.

MgO wt %	*a* [Å]	*c* [Å]
0	4.757786	12.992480
0.1	4.761736	13.003910
0.3	4.768100	13.023680
0.5	4.766066	13.016570
0.7	4.762318	12.999430
0.9	4.764477	13.006380
1.0	4.763065	13.001860

4. Conclusion

The effect of MgO addition on the phases of ZTA- TiO_2 ceramics composites was investigated. Addition of 0.5 wt % of MgO onwards only lead to formation of secondary phase, $MgAl_2O_4$. The formation of secondary phases is expected to change the physical properties of ZTA. The resulted phase are $MgAl_2O_4$. MgO reacts with Al_2O_3 to form fine secondary phase particles of $MgAl_2O_4$ due to an excess of MgO above the solubility limit.

Acknowledgement
This work was funded by the International Islamic University Malaysia (IIUM) under Grant RAGS13-021-0084 and Grant FRGS14-164-0405.

References

[1]Azhar A Z A, Ratnam M M and Ahmad Z A 2009 Effect of Al2O3/YSZ microstructures on wear and mechanical properties of cutting inserts *J. Alloys Compd.* **478** 608–14
[2]Wang D, Ismail N F and Badarulzaman N A 2012 Effect of MgO Additive on Microstructure of Al_2O_3 *Adv. Mater. Res.* **488–489** 335–9
[3]Azhar A Z A, Mohamad H, Ratnam M M and Ahmad Z A 2011 Effect of MgO particle size on the microstructure, mechanical properties and wear performance of ZTA–MgO ceramic cutting inserts *Int. J. Refract. Met. Hard Mater.* **29** 456–61
[4]Rejab N A, Azhar A Z A, Ratnam M M and Ahmad Z A 2013 The relationship between microstructure and fracture toughness of zirconia toughened alumina (ZTA) added with MgO and CeO2 *Int. J. Refract. Met. Hard Mater.* **41** 522–30
[5]Rittidech a., Portia L and Bongkarn T 2006 The relationship between microstructure and mechanical properties of Al2O3–MgO ceramics *Mater. Sci. Eng. A* **438–440** 395–8
[6]Azhar A Z A, Choong L C, Mohamed H, Ratnam M M and Ahmad Z A 2012 Effects of Cr2O3 addition on the mechanical properties, microstructure and wear performance of zirconia-toughened-alumina (ZTA) cutting inserts *J. Alloys Compd.* **513** 91–6
[7]Hirata T, Akiyama K and Yamamoto H 2000 Sintering behavior of Cr2O3 ± Al2O3 ceramics **20** 195–9
[8]Riu D, Kong Y and Kim H 2000 Effect of Cr2O3 addition on microstructural evolution and mechanical properties of Al2O3 **20** 1475–81
[9]Magnani G and Brillante A 2005 Effect of the composition and sintering process on mechanical properties and residual stresses in zirconia–alumina composites *J. Eur. Ceram. Soc.* **25** 3383–92
[10] Sktani Z D I, Azhar A Z A, Ratnam M M and Ahmad Z A 2014 The influence of in-situ formation of hibonite on the properties of zirconia toughened alumina (ZTA) composites *Ceram. Int.* **40** 6211–7
[11] Rejab N A, Azhar A Z A, Ratnam M M and Ahmad Z A 2012 The effects of CeO2 addition on the physical, microstructural and mechanical properties of yttria stabilized zirconia toughened alumina (ZTA) *Int. J. Refract. Met. Hard Mater.* **36** 162–6
[12] Wang C-J and Huang C-Y 2008 Effect of TiO2 addition on the sintering behavior, hardness and fracture toughness of an ultrafine alumina *Mater. Sci. Eng. A* **492** 306–10
[13] Maitra S, Das S and Sen a. 2007 The role of TiO2 in the densification of low cement Al2O3–MgO spinel castable *Ceram. Int.* **33** 239–43
[14] Manshor H, Md Aris S, Azhar A, Zahirani A, Abdullah E C and Ahmad Z A 2015 Effect of Titania (TiO2) Addition to Zirconia Toughened Alumina (ZTA) Phases, Hardness and Microstructure *Advanced Materials Research* vol 1087(Trans Tech Publ)pp 293–8
[15] BAGLEY R D, CUTLER I B and JOHNSON D L 1970 Effect of TiO2 on Initial Sintering of Al2O3 *J. Am. Ceram. Soc.* **53** 136–41
[16] Zu Y, Chen G, Fu X, Luo K, Wang C, Song S and Zhou W 2014 Effects of liquid phases on densification of TiO2-doped Al2O3–ZrO2 composite ceramics *Ceram. Int.* **40** 3989–93

[17] Rejab N A, Azhar A Z A, Kian K S, Ratnam M M and Ahmad Z A 2014 Effects of MgO addition on the phase, mechanical properties, and microstructure of zirconia-toughened alumina added with CeO2 (ZTA–CeO2) ceramic composite *Mater. Sci. Eng. A* **595** 18–24
[18] Rejab N A, Azhar A Z A, Ratnam M M and Ahmad Z A 2013 The effects of CeO2 addition on the physical, microstructural and mechanical properties of yttria stabilized zirconia toughened alumina (ZTA) *Int. J. Refract. Met. Hard Mater.* **36** 162–6

IOP Conf. Series: Materials Science and Engineering **290** (2017) 012047 doi:10.1088/1757-899X/290/1/012047

Vibration isolation analysis of new design OEM damper for malaysia vehicle suspension system featuring MR fluid

M H Unuh[1], P Muhamad[1], H M Y Norfazrina[2], M A Ismail[1] and Z Tanasta[1]

[1]Intelligent Dynamics & System Research Lab, Malaysia-Japan International Institute of Technology, Universiti Teknologi Malaysia, 54100 Kuala Lumpur, Malaysia
[2]Department of Mechanical Engineering, Kulliyyah of Engineering, International Islamic University Malaysia, 50728 Kuala Lumpur, Malaysia

Email: hishamunuh@gmail.com

Abstract. The applications of semi-active damper employing magnetorheological (MR) fluids keep increasing in fulfilling the demand to control undesired vibration effect. The aim of this study is to introduce the new design of damper for Malaysian vehicle model as well to evaluate its effectiveness in promoting comfort. The vibration isolation performance of the OEM damper featuring MR fluid was analysed physically under real road profile excitation experimentally. An experiment using quarter car rig suspension and LMS SCADAS Mobile was conducted to demonstrate the influence of current in controlling the characteristics of MR fluid in alter the damping behaviour under 5 cm bump impact. Subsequently, the displacement values were measured with respect to time. The new design OEM damper featuring MR fluid was validated by comparing the data with original equipment manufacturer (OEM) passive damper results under the same approach of testing. Comparison of numerical data of the new design OEM damper shown that it can reduce the excitation amplitude up to 40% compared to those obtained by OEM passive damper. Finally, the new design OEM damper featuring MR fluid has effectively isolated the disturbance from the road profile and control the output force.

1. Introduction

Since Karl Benz patent the creation of first car in the world in 1886, endeavour was completed so that the suspension system serves in a finest condition by enhancing the considerations of the suspension system. Rapid development and implementation of various vibration control system which are passive, active and semi-active have been introduced. Passive vibration control system provides design simplicity by typically consists of spring and conventional damper. This spring and damper have fixed characteristic and hence have no mechanism for feedback control. This is because the designed tuned frequency limits their performance and thus only yields nominal vibration control. Through applications of active forces, active vibration control systems have the ability to supply vibration isolation over large bandwidth disturbance. However, greater power sources is need to trigger the requisite active forces. If not accurately premeditated, it will disrupt the system. Contrariwise, by ensuring the consistency of passive yet sustaining the flexibility and compliance of fully active systems, semi-active control system had appeared between those two to utilize the superlative of both approaches [1]. Earliest findings point out that semi active control approaches hypothetically can succeed the majority of the performance of fully active systems because many active control systems work predominantly to adjust structural damping. Numerous approaches have been suggested to optimize semi active suspension. One of the approaches of semi-active suspensions uses controllable fluids. Magnetorheological (MR) fluids and Electrorheological (ER) fluids are viable challengers of controllable fluids [2]. ER fluid change its viscosity when an electric current is applied. Despite the development and research of ER fluid started early than MR fluid, there is very few practical applications using ER technology. This is due to ER fluid has limited shearing movement resistance. MR fluid on the other side able to produce ten times stronger shear strength than ER fluid [3].

MR fluid is the material upon an application of magnetic field. It exhibits quick, alterable and tuneable evolution from a continuous steady state to a highly viscous state. These fluids demonstrate dramatic changes in their rheological behaviours in reaction to a magnetic field [4]. Interest on the MR fluids keep increasing lately for their ability to deliver simple and immediate reaction interfaces between electronic control and mechanical devices and systems [5].

In order to attain high performance, one of the challenges of employing this knowledge is obtaining the accurate model to exploit the advantages of the excellent properties of MR fluids in real field [6]. Surprisingly, very few current published studies does focus on robustness of MR suspension control over real situation such as vehicle moving in uneven gravel road surface which is easily subjected to parameter of uncertainties and external disturbance in practice. The most recent study is done by Miao Yu and her co-worker using mix-mode monotube MR damper [7]. To the author's knowledge, there is no research done in Malaysia considering local technical standards and environment. For this purpose, a prototype of MR semi active damper will be prepared using standardized geometry dimension of original equipment manufacturer damper to evaluate the usefulness of MR technology in term of vibration isolation performance in Malaysian vehicle. Therefore, an experimental study will focus on vibration isolation of the quarter of the full vehicle suspension system on real field.

1.1. CAD Modelling of the Damper

Conceptual design (see figure 1) was developed using SolidWorks software. The objective is to establish full specification drawing before fabrication process. This approach is vital to prevent failure and argument of subsequent approaches. The design development with a precise and accurate dimension of OEM damper is important to be achieved initially before going to the next stage. Once the dimension is confirmed, a careful fabrication process is required before the experimental verification approach. The dimensions are summarized in table 1 and figure 1 shows the design of the damper.

Table 1. Geometry dimension of the OEM damper.

Part/ Assembly	Dimension (mm)
External housing outer diameter, $\emptyset_1$	39.00
External housing inner diameter, $\emptyset_2$	36.63
External housing height, h_1	342.29
Internal housing outer diameter, $\emptyset_3$	27.42
Internal housing inner diameter, $\emptyset_4$	24.82
Internal housing height, h_2	288.01
Piston head height, h_3	24.96
Stroke distance, h_4	299.52

The description of the conceptual design is serve to explain the principles of the present development. Figure 1 shows the view of the conceptual design in accordance with the principles of the present development. Referring to the figure, the design is a linearly-acting magnetorheological (MR) fluid damper, in particular a strut. In general, the strut is designed for operation as a load-bearing and shock-absorbing device within a vehicle suspension system. It is connected between the sprung body and unsprung masses (not shown).

The innovation consists of the followings:

- A magnetorheological fluid with serial number MR Fluid MRF-122 EG 0011369535.
- Electrical copper coil with standardized wire gauge system is 24 AWG
- Comprising two housing known as twin-tube. The internal housing holds the piston with respect of stationary exterior coil at the outside.
- Comprising a pair of rings to hold the electric coil.
- Comprising an external coil surrounding a portion of the internal housing, the external coil capable of generating a magnetic field across at least a portion of the flow gap.
- A piston assembly having disposed in the inner surface of internal housing with function to initiate a flow of the MR fluid between the external housing and internal housing.
- The bottom of the internal housing consists of foot valve as a flow gap permitting the flow of MR fluid.

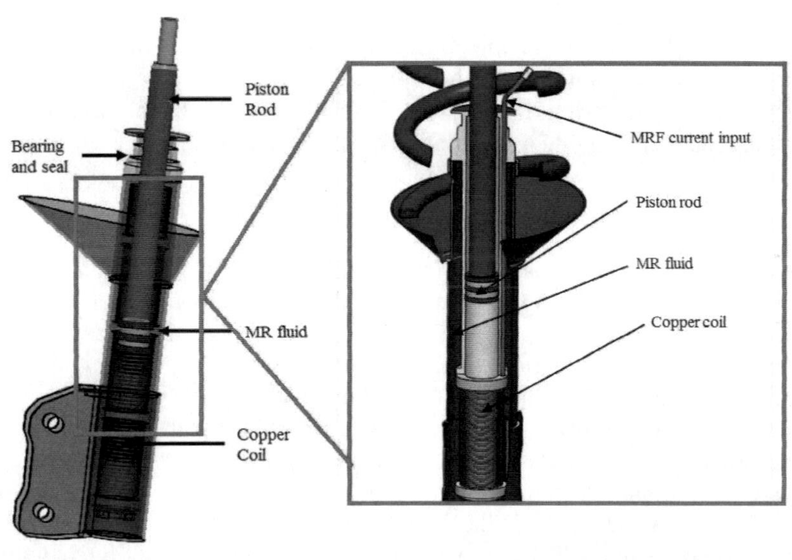

Figure 1. Conceptual design of OEM damper featuring MR fluid [8].

1.2. The Quarter Car Rig

The quarter car rig is designed (see figure 2) to provide specification of a real working system of a car. The characteristic of the quarter car rig are as followed [9]:

- Analogous dynamic behaviour as a car suspension. The dynamic behaviour which required being similar are nonlinearities, damping and resonance frequencies.
- Have the capability to sustain maximum excitation amplitudes within the frequency range of 25 Hz. For this experiment the excitation amplitudes is 5cm
- Have the ability to guide vertical movement of the chassis mass idyllically.

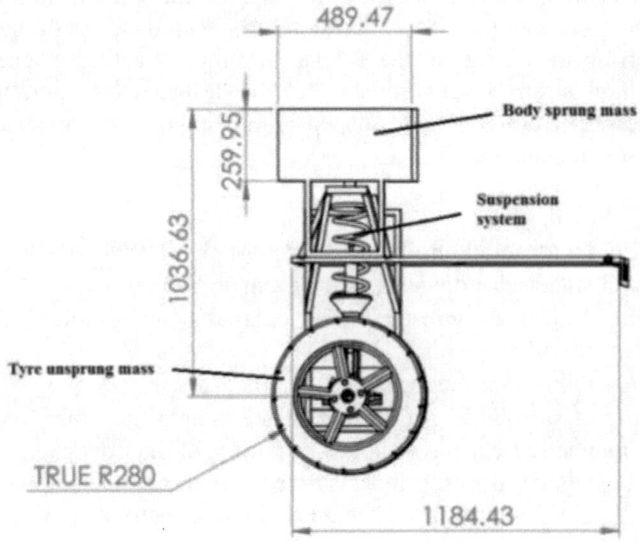

Figure 2. The Quarter Car Rig Drawing (Dimension in mm).

The mechanism of variable stiffness is the most vital embodiment in the quarter car design. The mechanism of various stiffness comprises the following:

- A linear spring encompassing damper with an upper end removable to a point of attachment on the sprung mass and having a lower arm that interlock to the tyre.
- Lower arm with ball joint as a receiving location attachable to an unsprung mass of tyre.

2. Experimental Set-Up

The experiment conducted on real road profile by pulling the complete quarter car rig behind a vehicle. The entire experiment set up is shown in figure 3. All the apparatus selected was connected to each other and kept ready to function respectively. Sprung mass frame was added to the quarter car rig frame to meet the requirement of quarter car mass which is around 250 kg. A bump was placed at appropriate distance (approximately 10m from the origin) to ensure the velocity of the quarter car rig was constant at 10 km/h when it hits the bump. The height of the bump is 5cm as shown in figure 4. All the wire connecting the accelerometer to LMS SCADAS Mobile and the electric coil to Agilent power supply were tie to their relevant position using tie rod to avoid loose of connection during experimentation. Other parameters of the experiment were stated in table 2. Initially the experimentation was done using OEM passive damper before being substituted with the fabricated new design OEM damper prototype.

Figure 3. Experimental set-up.

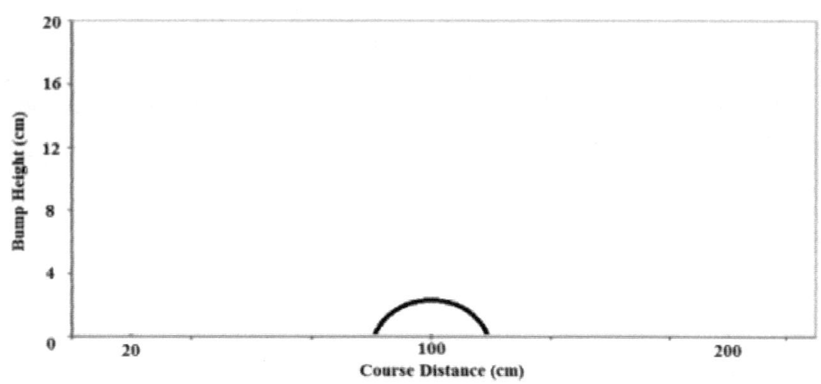

Figure 4. Equivalent suspension input graph showing suspension input versus course distance.

Table 2. Parameter of Malaysian quarter car.

Entry	Vehicle Parameter	Value	Description
1	m_s	250 kg	Mass of the body
2	m_u	25 kg	Mass of the tyre
3	k_s	17600 $N.m^{-1}$	Spring damping coefficient
4	k_t	15000 $N.m^{-1}$	Tyre damping coefficient
5	c	100 $N.s.m^{-1}$	Damping coefficient (only state for the OEM damper)

IOP Conf. Series: Materials Science and Engineering **290** (2017) 012047 doi:10.1088/1757-899X/290/1/012047

3. Results and Discussion

The discussion of the result begin with explaining the first set of experimental result by referring to figure 5. As observed the OEM passive damper achieved almost 20 mm maximum displacement response of sprung mass after impact. The subsequent figure 6 shows the displacement response results of the new design OEM damper prototype at off-state condition where 0A current is supplied. As shown its maximum displacement response is found below 8 mm which is approximately 48% from the displacement recorded by the OEM passive damper. The finding provides evidence that the presence MR fluid alone without the influence of magnetic field is able to offer great vibration isolation. Figure 7 is illustrated graphically to show the vibration isolation performance gap between those two dampers as explained previously. Figure 8 represents one of the results showing the effects of magnetic field on the damping force as function of displacement response. There is a slight increase of amplitude of the displacement response compared to the off state condition. It is believe the viscosity of MR fluid is increase due to the formation of fibrous structure by magnetic field, usually referred as magnetoviscous effect. Yet, the new design OEM damper prototype still isolates more vibration energy transmitted from the bump than OEM passive damper.

Figures 9 is illustrated for comparison and better understanding of the result between the OEM passive damper and the new design OEM damper featuring MR fluid prototype at its almost fully stiff condition. Take in account that the displacement recorded by the new design OEM damper prototype still small compared to OEM passive damper. As current applied increase from 0.8A to 1.0A, it is observed that the difference of the maximum displacement recorded between these two is equal to 0.581mm. The plausible explanation for these findings is that the MR fluids in the prototype was experiencing the maximum saturation. Hence, the suspension system was at the highest stiffness condition. The new design OEM damper prototype was no longer absorbing the excitation energy from the bump, the prototype damper itself almost played the same role as the spring. However, the displacement recorded from the new design OEM damper is still far more inferior to the OEM passive damper for applied current of 0.8A and above.

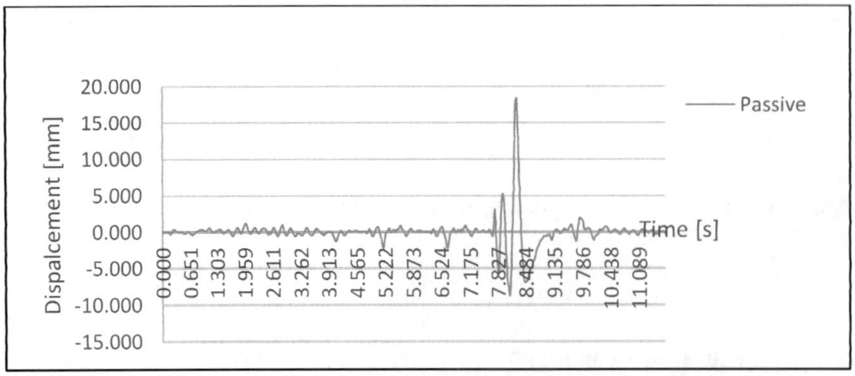

Figure 5. Displacement-time response of OEM passive damper for sprung mass.

IOP Conf. Series: Materials Science and Engineering **290** (2017) 012047 doi:10.1088/1757-899X/290/1/012047

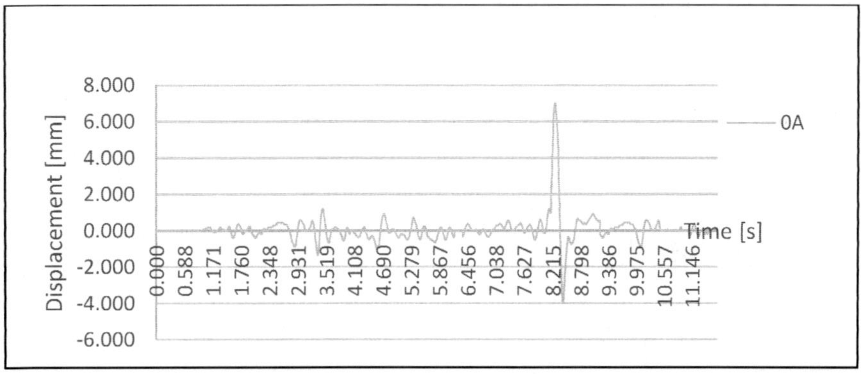

Figure 6. Displacement-time response of new design OEM damper featuring MR fluid at 0A.

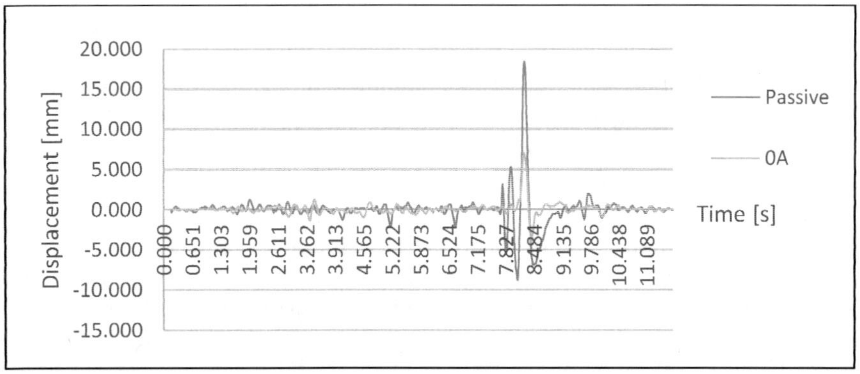

Figure 7. Displacement-time response of OEM passive damper and new design OEM damper featuring MR fluid at 0A .

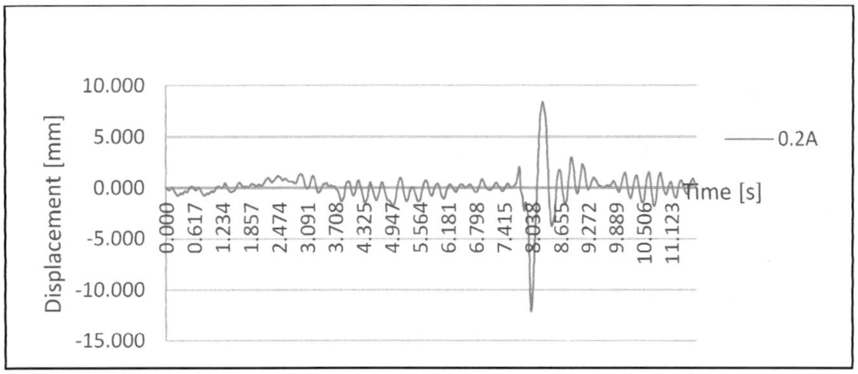

Figure 8. Displacement-time response of new design OEM damper featuring MR fluid at 0.2A.

IOP Conf. Series: Materials Science and Engineering **290** (2017) 012047 doi:10.1088/1757-899X/290/1/012047

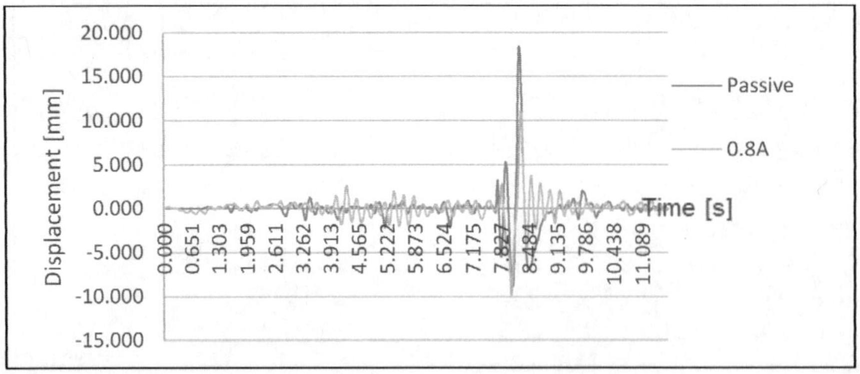

Figure 9. Displacement-time response of OEM passive damper and new design OEM damper featuring MR fluid at 0.8A.

Table 3 summarizes the results from all set of experiment. Notice that the new design OEM damper prototype has achieved a great level of performance. The results highlights that the new design OEM damper has reduced the excitation amplitude from the bump impact. In the condition of uneven road surface, the new design OEM damper prototype at lower supplied current is reliable such that it will caused the suspension system to become softer and has absorbed more vibration energy than the OEM passive damper. Low damping characteristic at low current supplied is needed to procure a good ride comfort. On the other hand, higher supplied current is needed to gain a good vehicle control on smooth road surface while speeding. This suggests that the new design OEM damper prototype can adapts damping characteristic to the profile of the road and the driver's gear shifting habits within just few milliseconds. The innovation had fully exploited the advantage of rheology properties of MR fluids.

Table 3. Displacement-time response result summarization.

	Max. Displacement (mm)	**Reduction (%)**
OEM Passive	18.423	-
0.0 A	7.008	61.96
0.2 A	8.399	54.41
0.4 A	9.400	48.98
0.6 A	10.282	44.19
0.8 A	11.014	40.22
1.0 A	11.595	37.06

The results from the experimental work indicates that the new design OEM damper prototype can control the input force and isolate the disturbance from the road profile and effectively control the output force to the sprung mass. MR fluid utilize in the development and the innovation in this work has micrometre-scale particles suspended in the main carrier fluid (oil) which align themselves and form a chains as soon magnetic field is applied. Our finding revealed that the act applying magnetic field in the foot valve of the prototype greatly increase the resistant for the fluid to flow from inner housing to outer housing and vice versa. This event will foster the damper to become stiff. Hence, it upsurge the handling control of the vehicle. The absence of magnetic field causes the damper became soft due to less viscosity of the MR fluid. A softer damper will offer mechanical grip as it will keep the tyres on the ground. Therefore, it will directly promote the increase to the satisfaction of riding comfort. The new design OEM damper prototype has directly proved its ability to solve problems in the constrained design of the OEM passive damper and can maintain its advantages over the active system. The prototype is able to alter the stiffness of the whole suspension system from soft to hard by utilizing the superior characteristics of MR fluid.

4. Conclusion

In this research, the development of passive original manufacturer (OEM) damper into semi-active based system is presented. The aim of the development is to use magnetorheological (MR) fluid to gain ultimate vibration isolation control. The new design OEM damper featuring MR fluid fabrication process was carried out and the prototype was tested physically in real environment rather than isolated environment. Vibration isolation performance is measured and the data is captured. The experimentation using the quarter car rig was successfully conducted. It is observed that there were significant reduction in displacement response recorded by the new design OEM damper prototype in compared with passive original equipment manufacturer results. At low supplied current, the prototype had greatly absorbed and isolated the vibration energy compared to OEM passive damper. The results demonstrates the new design OEM damper greatly isolate the excitation causes by the 5cm bump at applied current of 0.2A and 0.4A. The reduction computed were 54.41% and 48.98%. Optimum compromise between the level of driving comfort and vehicle control levels can be achieved by controlling the strength of magnetic field applied via the applied current to the prototype damper, tailored to the input or the state of the road travelled by the vehicle.

Acknowledgement

This work was financially supported by Universiti Teknologi Malaysia under Tier-1 Research University Grant (GUP) Project Vot. No. Q.K130000.2501.12H43.

References

[1] Guoliang H, Qianjie L, Ruqi D and Gang L 2017 *Journal of Advances in Mechanical Engineering* **9** 1-15
[2] Rashid M M, Rahim N A and Hussain M A 2011 *IEEE Transaction on Industry Applications* **47** 1051-1059
[3] Kulkarni A and Patil S 2013 *International Journal of Engineering Research and Application* **3** 1879-1882
[4] Premaltha S E, Chokkalingam R and Mahendran M 2012 *American Journal of Polymer Science* **2** 50-55
[5] Ashtiani M, Hashemabadi S H and Ghaffari A 2015 A*Journal of Magnetism and Magnetic Materials* **374** 716-730
[6] Miao Yu, Liao C R, Chen W M and Huang S L 2006 *Journal of Intelligent Material and Structures* **17** 801-806
[7] Jiang Z and RE Christenson 2012 *Journal of Smart Materials and Structure* **21** 65002-65013
[8] Ismail M A, Muhamad P, and Abu A 2014 *The International Conference on Kinematics* (Jakarta: Indonesia) **534** 111-116
[9] Lam A H F and Liao W H 2003 *International Journal of Vehicle Design* **33** 50–75

IOP Conf. Series: Materials Science and Engineering **290** (2017) 012048 doi:10.1088/1757-899X/290/1/012048

Effect of different reducing agents on phase formation and transformation behaviour of Ni-Ti shape memory alloy

H H Mohd Zaki[1], N A Bahrudin[1], J Abdullah[2], N Sarifuddin[1]

[1]Department of Manufacturing and Materials Engineering, International Islamic University Malaysia (IIUM), Jalan Gombak, 53100 Kuala Lumpur, Malaysia
[2]School of Mechanical Engineering, Universiti Sains Malaysia, Seri Ampangan, 14300 Nibong Tebal, Penang, Malaysia

Email: hafizahhanim@iium.edu.my

Abstract. In this study, NiTi synthesized via solid-state sintering from Ni-TiH_2 powders in reducing environments using (i) CaH_2 (ii) MgH_2 and (iii) CaH_2 and MgH_2 as in situ reducing agent were investigated. The phase formation was characterized by means of scanning electron microscope (SEM), energy-dispersive spectroscope (EDS) and powder X-ray diffraction (XRD), while transformation behavior was analyzed using differential scanning calorimetry (DSC) measurement. Among these three reducing agents, synthesis in reducing environment using CaH_2 as in situ reducing agent resulted in the formation of single phase NiTi with enthalpy change of 25-26 J/g, which is similar to melt-cast NiTi alloys. Specimens sintered in reducing environments using MgH_2 and MgH_2+CaH_2 as in situ reducing agents have lower enthalpy change, ~16-21 J/g compared to CaH_2. This work established the fact that, synthesis in different reducing environment appears to have profound effects on the phase formation and transformation behavior of NiTi.

1. Introduction

Near-equiatomic NiTi alloy has ranges of useful engineering properties besides their unique shape memory properties. While most current applications of NiTi are made of conventional alloys in solid forms such as wires, plates, and thin walled tubes [1], keen interests have been given to novel physical forms of NiTi. Such novel forms often require novel fabrication and processing techniques other than the conventional method of melting, casting, and metalworking. For example, thin films are produced by sputter deposition [2, 3], multilayer NiTi are produced by diffusion annealing of ultrafine laminates of Ni and Ti [4], ribbons are produced by melt-spinning method [5], porous NiTi [6, 7], and gradient composition NiTi [8] are obtained by solid state sintering of elemental powders. Active exploration of possible means to achieve porous single phase NiTi has included various solid state sintering such as conventional furnace sintering [9], hot-isostatic pressing (HIP) [10], self-propagating high temperature synthesis (SHS) [11], spark plasma sintering [12], and microwave sintering [13], though little success has been achieved in terms of satisfactory shape memory behavior as evidenced by the absence of visible applications of these materials.

The formidable challenge of obtaining NiTi via solid state with good shape memory behaviour, which can resemble the properties of melt-cast NiTi is yet to be achieved. Various sintering parameters and sintering techniques have been employed which did not lead to formation of single phase NiTi. Therefore, the question arises whether single phase NiTi can be formed via solid state synthesis from elemental powder. In addition, the study on oxidation by Shearwood, Fu [14] contributed towards an idea that the oxidation may retard the formation of single phase NiTi, but no further or repeatable works in the same nature have been done. Therefore, this research is conducted to investigate whether the formation of single phase NiTi via solid state synthesis from elemental powder is feasible. In this case, the sintering was conducted in protected environment i.e. reducing environment to protect the specimen from oxidation. The effect of using different reducing agents such as CaH_2, MgH_2 and a mixture of CaH_2 and MgH_2 on NiTi phase formation is discussed. The transformation behaviour of Ni-Ti synthesized in these reducing environments is also investigated.

2. Experimental procedures

2.1 Specimen Preparation

In this work, high purity Ni (average particle size < 30 μm) and TiH_2 (mean particle size = 44 μm) were used to form near equiatomic mixture. These powders were mixed at equiatomic composition by means of low energy ball milling for 24 hours to obtain homogenous mixing. After mixing, the powder mixture was cold compacted into cylindrical shape of ϕ12 x 4 mm to 3 tonnes of force. Then, the specimens were sintered at 930 °C for 3 hours in reducing environment. This reducing environment was achieved using CaH_2 and MgH_2 powder as in-situ reducing agent to prevent oxidation. Approximately 200 to 300 mg of CaH_2 or/and MgH_2 powder were poured on the exposed surface of the specimen.

For solid state diffusion synthesis, the sintering must be conducted at below 1310 °C temperature to prevent melting. There is also an eutectic point on the Ti side at 942 °C. Based on these considerations, the specimens were sintered below the eutectic point at 942 °C, to avoid partial melting and to ensure solid state reaction during sintering.

2.2 Specimen Characterization and Analysis

For sampling, a thin slice was cut off from the middle of the cylindrical sample for X-ray diffraction (XRD) analysis and scanning electron microscopy (SEM) characterization. Microstructural analyses were examined using backscattered electron imaging (Hitachi S-3400N) equipped with energy dispersive spectroscopy (EDS). After XRD analysis, a small piece was cut off from the middle of the XRD sample with approximately 1x2 mm for differential scanning calorimetric (DSC). Phase transformation behavior was studied by DSC measurement using TA Q10 DSC machine at a heating/cooling rate of 10 ^{0}C/min.

3. Results and Discussion

Figure 1 shows a collection of SEM micrographs of the Ni-TiH_2 specimens sintered at 930 °C for 3 hours in three different reducing environments. For the specimen sintered using MgH_2 as a reducing agent, there are many tiny spots, which correspond to Ti-rich phase with composition of Ti-57.3 at%. and Ni-42.7 at% in NiTi matrix. According to XRD result as indicated in figure 2, small peaks of Ti_2Ni were observed along with high intensity B2-NiTi peaks. Therefore, this Ti-rich phase is referred to as Ti_2Ni phase.

For the specimen sintered using CaH_2 as reducing agent, only single phase NiTi was observed with no Ti_2Ni phase. This observation is consistent with the XRD result where only peaks of B2-NiTi were formed. Meanwhile, specimen sintered in reducing environment of MgH_2 mixed with CaH_2 shows small patches of Ti_2Ni with composition of Ti-64.9 at%, Ni-35.1 at%. The formation of Ti_2Ni phase in NiTi matrix is similar to specimen sintered in MgH_2 reducing environment, except the composition of Ti is higher.

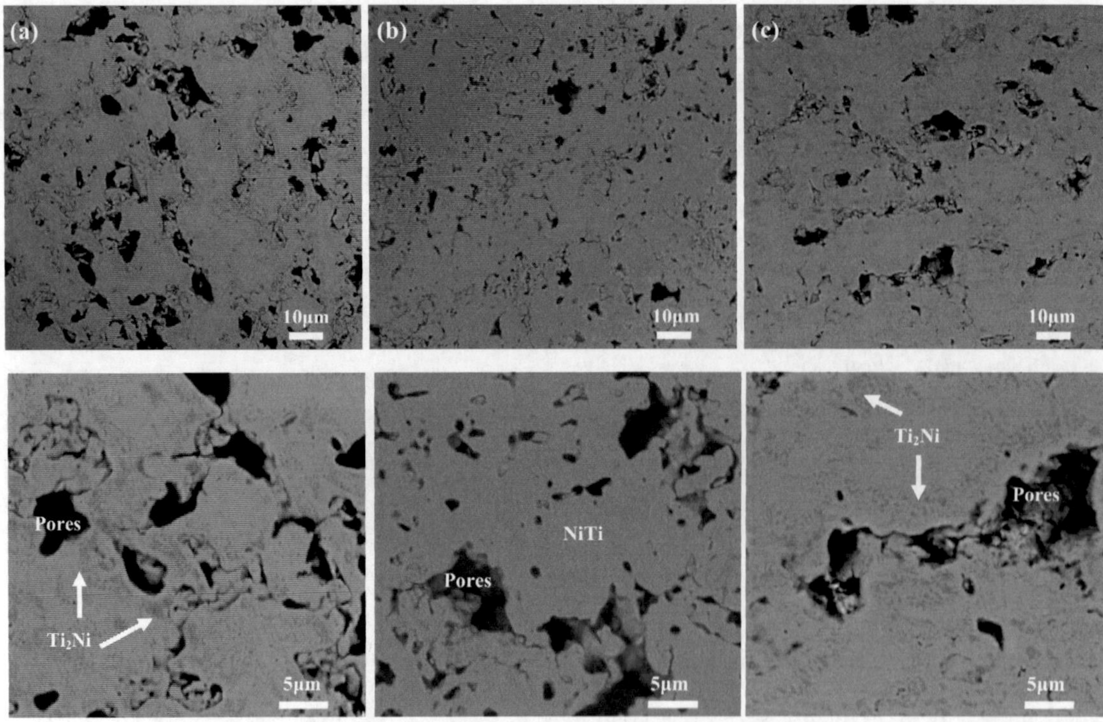

Figure 1. SEM micrograph showing single phase NiTi formed using different reducing agent of (a) MgH_2 (b) CaH_2 and (c) mixture of MgH_2 and CaH_2. The micrographs shown on the top are of lower magnifications and the micrographs shown on the bottom are of high magnifications

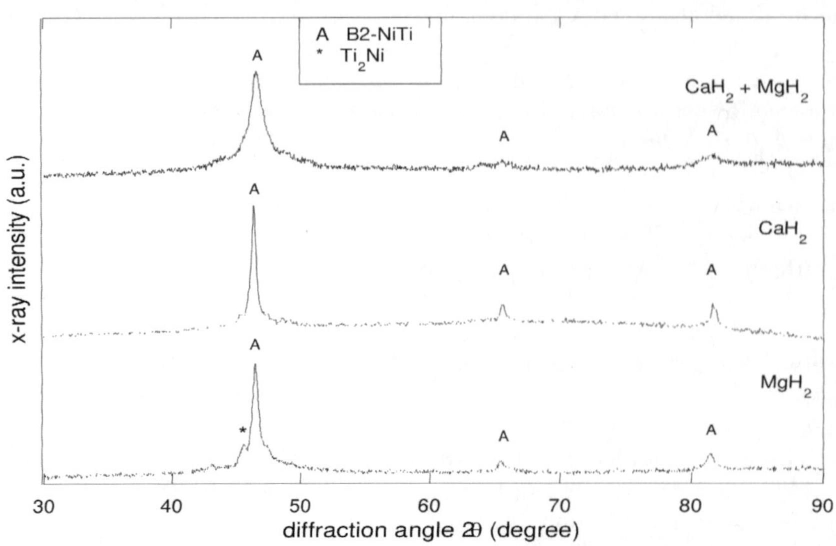

Figure 2. XRD spectrum of sintered specimen at 930 °C for 3 hours using different reducing agents

Figure 3 shows DSC curves for the specimen sintered at 930 °C for 3 hours in three different reducing environments. Referring to DSC curves, all specimens exhibited single peak A↔M martensitic transformation both upon cooling and heating, except for the specimen sintered in a CaH_2 reducing environment where the double-peaks referred to as A_{p1} and A_{p2} were observed upon heating.

These peaks are not associated with R-phase, instead this phenomenon is due to inhomogenous distribution of Ni composition exist within the individual grains as the transformation temperatures are very sensitive to composition.

For the specimen sintered using MgH_2 as a reducing agent, transformation peaks were detected both on cooling and heating with enthalpy change of $\Delta H_{A \rightarrow M}$ = 18.7 J/g and $\Delta H_{M \rightarrow A}$ = 21.6 J/g, respectively. For the specimens sintered in CaH_2 and MgH_2 reducing environment, intermediate value of enthalpy change was measured with 16.2 J/g and 19.6 J/g both upon cooling and heating, respectively. However, this value is still lower than the value measured for the specimens sintered in the CaH_2 reducing environment. This attributed to the formation of Ti_2Ni phase in NiTi matrix (refer figure 1) which lowered the enthalpy value as this phase does not involved in the martensitic transformation. However, this specimen still exhibited good transformation behaviour if compared to synthesis in non-reducing environment as reported by [9] .

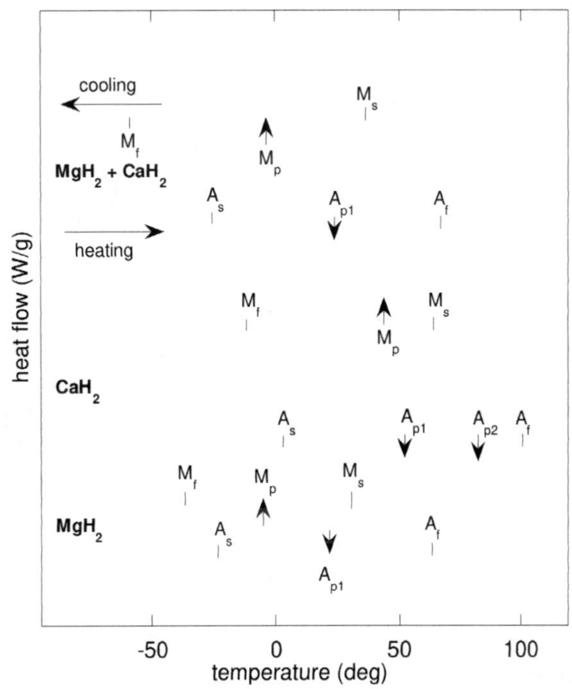

Figure 3. DSC curves of Ni-TiH_2 specimens sintered at 930 °C for 3 hours using different reducing agents

Figure 4 shows the EDS plot of Ni composition in the NiTi matrix. The overall Ni content in the NiTi matrix for the specimen sintered in a CaH_2 is below the martensitic transformation (MT) limit, therefore all the B2-NiTi phase is participating in the transformation, thus gives high value of enthalpy change. For the specimen sintered in the MgH_2 reducing environment, the Ni content is still below the MT limit. In addition, the Ni content is almost consistent and did not fluctuate too much as observed for the specimen sintered in the CaH_2 reducing environment, implying homogenous distribution of Ni content in NiTi matrix, therefore resulting in single peak transformation both on cooling and heating. As discussed previously, the only reason it has a lower enthalpy change value than CaH_2 is due to the formation of Ti_2Ni phase.

Meanwhile, for the specimen sintered in both MgH_2 and CaH_2 reducing environment, part of Ni content is exceeding MT limit, therefore this particular B2-NiTi did not participated in the transformation. This explains the lower value of enthalpy change obtained by this specimen compared to the specimen sintered in an MgH_2 reducing environment.

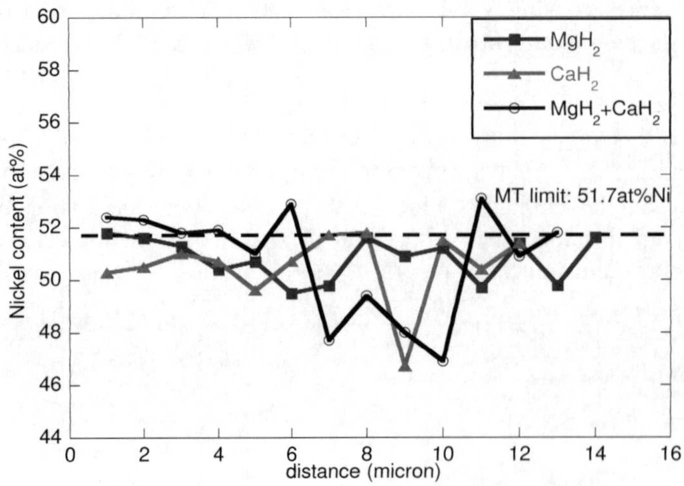

Figure 4. The EDS plot of Ni content in NiTi matrix of Ni-50TiH_2 specimen sintered at 930 °C for 3 hours in different reducing environments

4. Conclusion

In summary, synthesis of Ni-TiH_2 in reducing environment especially CaH_2 as an in situ reducing agent resulted in the formation of single phase NiTi with good enthalpy change (~25-26 J/g) similar to the melt-cast NiTi alloys (~24-28 J/g). Although specimens synthesized in MgH_2 and MgH_2+CaH_2 did not produced single phase NiTi, both specimen exhibited good transformation peak both on cooling and heating. This shows that synthesis in different reducing environment appears to have profound effects on formation of single phase NiTi and transformation behavior of NiTi synthesized in solid state.

Acknowledgement

The authors acknowledge the research funding through the Research Initiative Grant Scheme (RIGS); Project ID: RIGS16-080-0244, of the International Islamic University Malaysia (IIUM).

References

[1] Machado L G, Savi M A 2003 *Brazilian Journal of Medical and Biological Research* **36** 683-91
[2] Lehnert T, Grimmer H, Böni P, Horisberger M, Gotthardt R 2000 *Acta Materialia* **48** 4065-4071
[3] Cho H, Kim H Y, Miyazaki S 2006 *Materials Science and Engineering A* **438-440** 699-702
[4] Tomus D, Tsuchiya K, Inuzuka M, Sasaki M, et al. 2003 *Scripta Materialia* **48** 489-494
[5] Santamarta R, Cesari E, Pons J, Goryczka T 2004 *Metall. and Mater. Transactions A* **35** 761-770
[6] Chu C L, Chung C Y, Lin P H, Wang S D 2004 *Materials Science and Engineering A* **366** 114-119
[7] Li B Y, Rong L J, Li Y Y 1998 *Journal of Materials Research* **13** 2847-2851
[8] Bogdanski D, Koller M, Muller D, Muhr G, et al. 2002 *Biomaterials* **23** 4549-4555
[9]Abdullah J, Mohd Zaki H H 2013 *Springer Book Series: Developments in Sustainable Chemical and Bioprocess Technology* 389-394
[10] McNeese M D, Lagoudas D C, Pollock T C 2000 *Materials Science and Engineering A* **280** 334-48.
[11] Tosun G, Orhan N, Ã-zler L 2011 *Materials Letters* **66** 138-140

[12] Liu B, Huang S, Chen L, Van Humbeeck J, Vleugels J 2017 *Materials Letters* **191** 89-92
[13] Xu J L, Jin X F, Luo J M, Zhong Z C 2014 *Materials Letters* **124** 110-112
[14] Shearwood C, Fu Y Q, Yu L, Khor K A 2005 *Scripta Materialia* **52** 455-460

IOP Conf. Series: Materials Science and Engineering **290** (2017) 012049 doi:10.1088/1757-899X/290/1/012049

Altering Height Data by Using Natural Logarithm as 3D Modelling Function for Reverse Engineering Application

Nur Ilham Aminullah Abdulqawi[1] and Mohd Salman Abu Mansor[1]

[1]School of Mechanical Engineering, Engineering Campus, Universiti Sains Malaysia, 14300 Nibong Tebal, Pulau Pinang, Malaysia

Email: mesalman@usm.my

Abstract. The raw data extracted from reverse engineering based on vision mostly do not resemble the actual geometrical representation yet. Even though the higher object surface reflected the most visible light towards the camera and yield higher number of value based on Lambertian illumination model, this does not mean the curvature profile are always accurate. After all, there are many mathematical models to shape curvature profiles into the correct representation. However, one of the most appropriate models found is the natural logarithm function. The function itself has alteration properties towards the raw data generated from reverse engineering based on vision.

1. Introduction

Digitization in reverse engineering process can be categorized into 3 main types: contact method [1], scan method [2] and also vision method [3]. The vision method is less common as it uses vision information to generate digitized 3D data, also known as 3D point cloud. The 3D data generated is based on Lambertian illumination reflection as the light source used is visible light. Several techniques has been used for targeting on shading and reconstructing surface from vision information and deemed to be successful [4]. However, most techniques use half tone or greyscale vision information which has limited input data. Furthermore, the accuracy range is still unable to achieve reverse engineering acceptable standard since most algorithms are developed for computer graphics rendering purpose. To suit for reverse engineering purpose, the vision information as input data has to be as compact as possible and that is why RGB information data is used for this research. Furthermore converting the input as height distribution needs a more uniform representation of the RGB information and hence equivalent corresponding color information HSV is necessary for the correction of height data distribution before inversing the function of light structure for the vision data.

2. Related Works

Vision information contain integers as data type [5]. When the integers are inputted into a certain program, they can be manipulated and be transformed into other sets of data. Previous research shows how integers can be harvested from image data for decision making and selective storage use [6]. Besides, integers value can also be transformed into useful outputs such as graphical models [7]. Somehow, integers vision information can be manipulated into 3D point cloud data using numerical conversion method for reverse engineering application[8]. However, the output data were found distorted at the peak of the object surfaces. To overcome this issue, there must be a function that able

IOP Conf. Series: Materials Science and Engineering **290** (2017) 012049 doi:10.1088/1757-899X/290/1/012049

correct the distorted height data while retaining the acceptable region of the 3D point cloud data that resemble the actual objects.

3. Methodology

The reverse engineering based on vision used setup as illustrated in figure 1. The camera function is to capture the object surface details as vision information and the captured vision information are transferable to a desktop computer via USB cable.

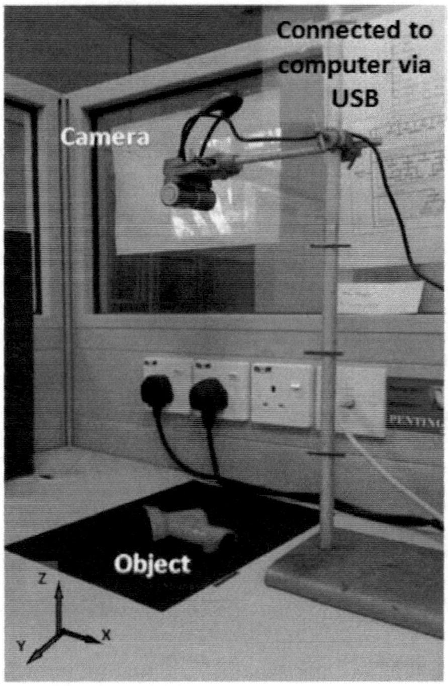

Figure 1. The setup of reverse engineering data extraction using camera to obtain vision information

The captured vision information from object surfaces are readily available as RGB format as default. Meanwhile, the RGB data are readily transferable to the computer memory as computer graphics. In computer graphics, RGB is usually being described as a 3D cube that has three axes of directions that each direction has a maximum value of 255, it is because RGB model is having three main components of color which is the red, green and blue with black is having value of 0 in each component and white is having value of 255 of each component.

Another type of color model that derived from RGB model is the HSV model, the HSV model is having three components namely the hue, saturation and value. HSV model is described as a hex-cone model because the colors are defined as angle in a 360 degree revolution, with the saturation increase by moving away from the centre axis of the hex-cone. The centre axis itself also contain value which the top of hex-cone has value of 1 while the bottom of the hex-cone has value of 0, the value direction is hence it means the increasing of value moves towards lighter colors of hue. The actual graphical description of both RGB model and HSV model is shown in figure 2.

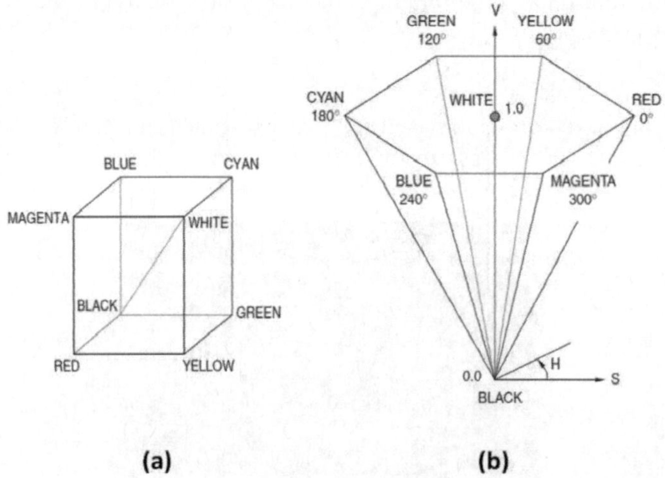

Figure 2. (a) RGB cube color model and (b) HSV hex-cone color model. Source from Elsevier [9]

Since HSV model data is the corresponding representation of the RGB model, while HSV model has the ability to interpret color as intensity distribution, it is used as the intermediate input for correcting the light structure of an object's surface captured as vision information.

The vision information structure of RGB and HSV is same accept of the composition of the components as shown

$$[RGB_{(i,j)}] \supseteq \{[Red_{(i,j)}] \cup [Green_{(i,j)}] \cup [Blue_{(i,j)}]\} \quad (1)$$

$$[HSV_{(i,j)}] \supseteq \{[Hue_{(i,j)}] \cup [Saturation_{(i,j)}] \cup [Value_{(i,j)}]\} \quad (2)$$

HSV model representation with subsequence process into height variation can be represented by the arrangement of X and Y and the equation can be represented by

$$[Grid_{X,Y}] = [pixel_{1,1}, ..., pixel_{m,1}, pixel_{1,2}, \ldots, pixel_{m,2}, ..., pixel_{1,n}, ..., pixel_{m,n}]^T \quad (3)$$

Meanwhile, the generated 3D data with the height variation of each X and Y is represented by

$$3D\ Data = [X, Y, Height_{X,Y}] \quad (4)$$

At this point the 3D data with height data as coordinates generated are actually the raw data. To verify on how the raw data was distorted, a small cylindrical machining workpiece was used as reference. Figure 3 shows data distribution of raw data generated.

IOP Conf. Series: Materials Science and Engineering **290** (2017) 012049 doi:10.1088/1757-899X/290/1/012049

(a)

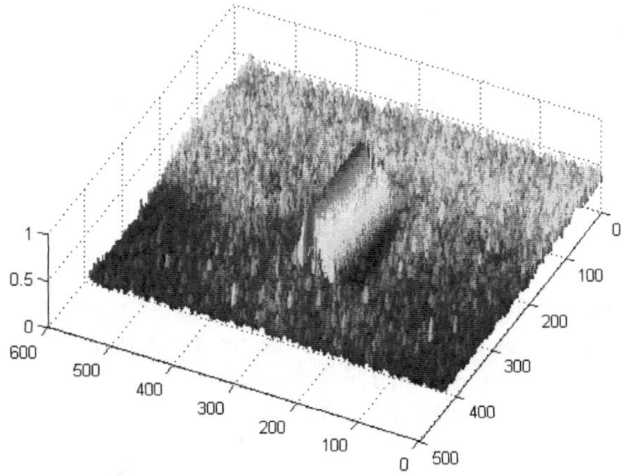

(b)

Figure 3. (a) The actual cylindrical profile used and (b) the raw data generated

By slicing the raw data into one side of the symmetrical profile, it shows that the generated raw data actually has distribution of exponential distribution. Figure 4 shows the detailed of the raw data with visible exponentially distributed pattern.

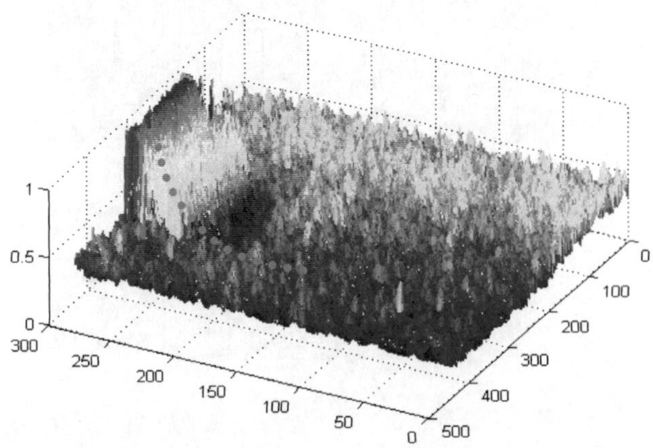

Figure 4. The sliced out data distribution from one side of the symmetrical profile

Meanwhile, by referring to the basic exponential equation of $y=e^x$, the starting y-value of the original exponential function for y is at 1, the equation was altered for interception at 0 to mimic the distributed pattern of HSV conversion model as

$$y = 0.582\,(e^x - 1) \tag{5}$$

Since the raw data has more of an exponential heights representation in surface profile, a counteracting function towards this exponentially rising pattern has to be implemented to the 3D data in order to correctly represent the shape of the actual object. By treating the 3D data as 2D profile, height data distribution is visualized as 2D plot using Desmos graph plotting tool in Figure 5.

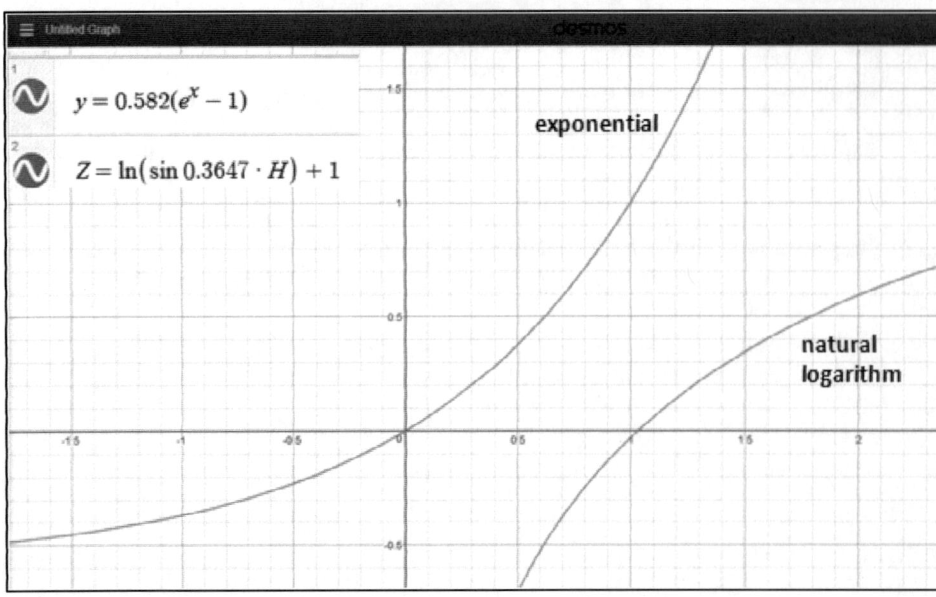

Figure 5. Exponential function of the distributed point cloud and the counteracting function is the natural logarithm function

To inverse the exponentially distributed height, the natural logarithm function used for this inverse effect is derived from the assumed point cloud's 2D profile distribution is

$$3D\ Data_{Altered} = [\ X,\ Y,\ ln\ (sin\ 0.3647 \times Height_{X,Y}) + 1] \quad (6)$$

At this point the 3D data should resemble the actual object surface profile. As to verify the equation, results and implementation was done and will be explained in the following section.

4. Result and Implementation

4.1. Intermediate result from test object

Using the cylindrical machining workpiece as test geometry, applying the mentioned natural logarithm function will enable the height data to be altered into the correct representation. Figure 6 shows the after applying natural logarithm function effect.

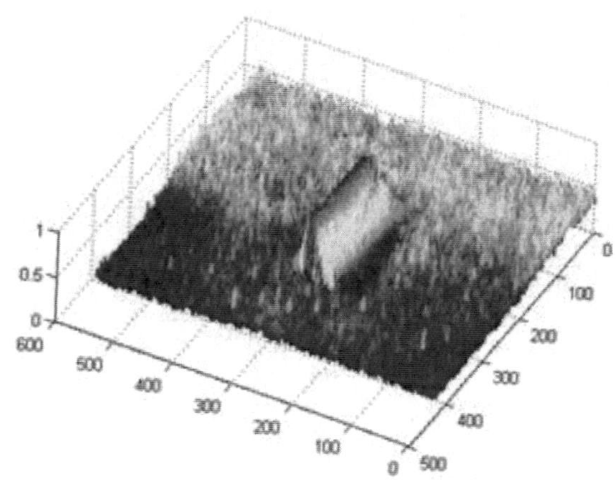

(a)

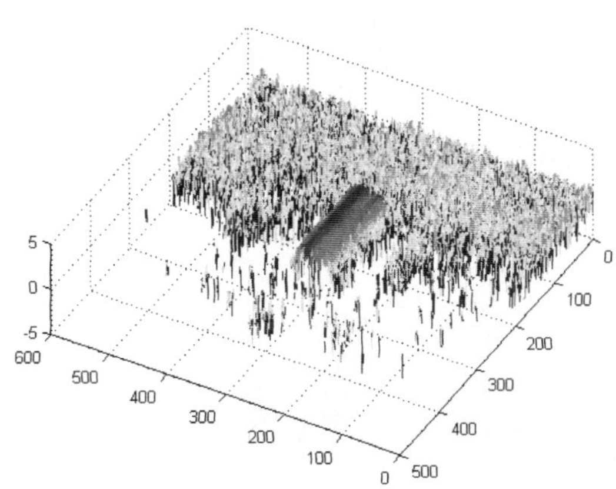

(b)

IOP Conf. Series: Materials Science and Engineering **290** (2017) 012049 doi:10.1088/1757-899X/290/1/012049

Figure 6. The results of (a) before and (b) after applying the natural logarithm function

Besides of observing the exponential profile being altered into natural logarithm profile, the background which initially were at 0 value altered into non-value as any logarithm of zero will resulted an indefinable output as log(0) will be infinity.

4.2. Implementation into actual object

Meanwhile for actual implementation, it was done on a bush cutter's cover and the so that the inversing effect can be visualised. The bush cutter cover is shown as in figure 7 and the object is captured from the top view before further going through additional processing into 3D point cloud.

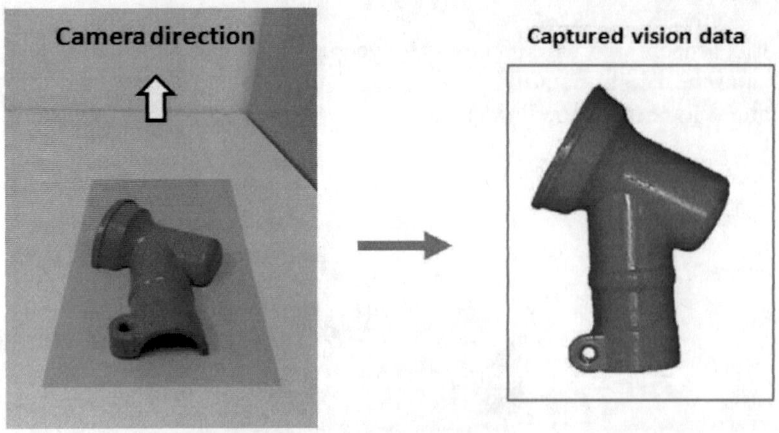

Figure 7. The specimen of bush cutter's cover placed below the camera to captured as vision data for further processing

The vision data that contains color and light intensity information as integers will be going through RGB to HSV conversion process and height extracting process in order to finally produced a set of 3D data or 3D point cloud in X,Y and Z coordinates which they resemble the surface detail of the object. The produce point cloud is readable by most computer aided design software. For this case, the software used is CATIA V5. Figure 8 shows the end result of the raw data point cloud right after processing of vision data.

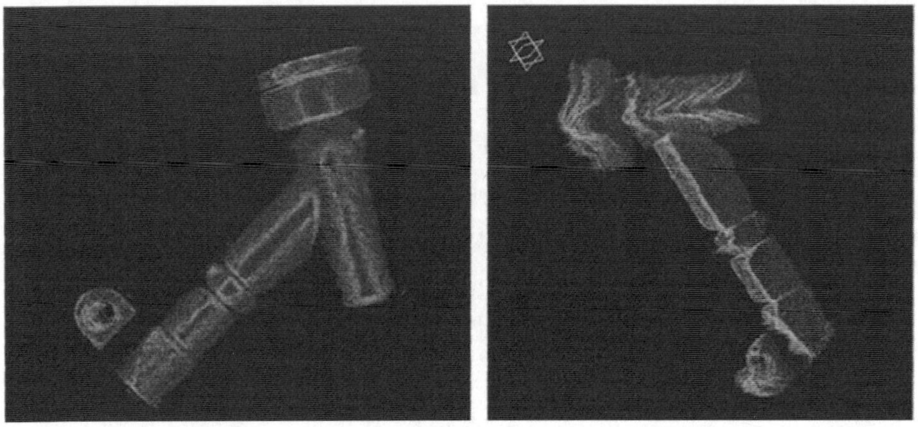

(a) (b)

Figure 8. The point cloud generated right after height distribution is process from vision information with (a) first view angle and (b) second view angle

The point cloud at this point shows incorrect distribution of height which has exponential characteristic where the middle area had sharp increase in height data, while the lower height data are at the bottom and hence removed. In order to correct the height data distribution on the point cloud, the inverse function mentioned in above section is applied to enhance on the final output so that it resemble the actual surface profile of the object. Figure 9 shows the after inverse effect by applying the mentioned natural logarithm to counteract on the exponentially distributed data.

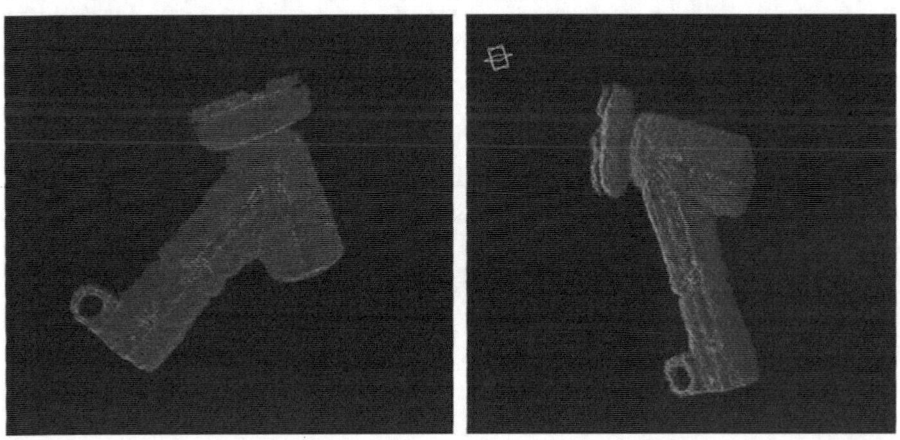

(a) (b)

Figure 9. The point cloud representation after inverse function of natural logarithm is implemented with (a) first view angle and (b) second view angle

After seeing the altered height data of the 3D point cloud to have better resemblance distribution, they can be used to further generate mesh and the height distribution can be visualised with 3D plot of 3D mesh.

Figure 10 shows the 3D mesh with different color according to the individual heights of each data point.

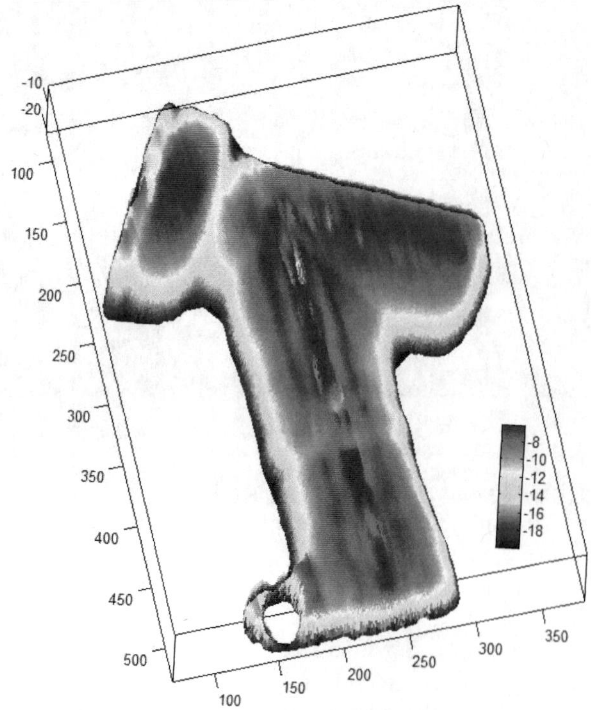

Figure 10. 3D mesh generated with different colors indicating the height data distribution.

5. Conclusion

Reverse engineering based on vision still faces problems in implementation. One of the problem is the generated height data distributions are inaccurate. Remodelling the height data distribution can correct the inaccuracy so that the following processes are errorless. Reverse engineering based on vision usually produced raw point clouds with exponentially distributed heights after applying HSV height extraction process. The proposed natural logarithm function is capable in solving the raw point clouds problem of having exponentially distributed heights. Furthermore, the proposed function improves the overall shape representation of the point clouds, and hence improves the overall accuracy for the outputs of reverse engineering that based on vision.

Acknowledgement

This research is supported by Universiti Sains Malaysia under the Research University Grant (No: 814247). The first author also would like to thank the tuition fee sponsorship of MyMaster by the Ministry of Higher Education Malaysia.

References

[1] Pedone P, Audrito E and Balsamo A 2014 Compensation of CMM geometrical errors by the GEMIL technique: Experimental results *CIRP Ann. - Manuf. Technol.* **63** 489–92

[2] Paulic M, Irgolic T, Balic J, Cus F, Cupar A, Brajlih T and Drstvensek I 2014 Reverse Engineering of Parts with Optical Scanning and Additive Manufacturing *Procedia Eng.* **69**

795–803
[3] Peng Q and Loftus M 2001 Using image processing based on neural networks in reverse engineering *Int. J. Mach. Tools Manuf.* **41** 625–40
[4] Chow C K and Yuen S Y 2009 Recovering shape by shading and stereo under lambertian shading model *Int. J. Comput. Vis.* **85** 58–100
[5] Chernov V, Alander J and Bochko V 2015 Integer-based accurate conversion between RGB and HSV color spaces *Comput. Electr. Eng.* **46** 328–37
[6] Verma M, Raman B and Murala S 2015 Local extrema co-occurrence pattern for color and texture image retrieval *Neurocomputing* **165** 255–69
[7] Piatkowski N, Lee S and Morik K 2016 Integer undirected graphical models for resource-constrained systems *Neurocomputing* **173** 9–23
[8] Abdulqawi N I A and Abu Mansor M S 2016 A Computer Method for Generating 3D Point Cloud from 2D Digital Image *J. Image Graph.* **4** 89–92
[9] Jack K 2008 *Digital Video and DSP* (Elsevier)

IOP Conf. Series: Materials Science and Engineering **290** (2017) 012050 doi:10.1088/1757-899X/290/1/012050

Mechanical performance of porous concrete pavement containing nano black rice husk ash

M Y Mohd Ibrahim[1], P J Ramadhansyah[1], H Mohd Rosli[1] and M H Wan Ibrahim[2]

[1]Faculty of Civil Engineering, Department of Geotechnics and Transportation, Universiti Teknologi Malaysia, 81310 UTM, Johor Bahru, Malaysia
[2]Faculty of Civil and Environmental Engineering, Department of Material and Structure Engineering, Universiti Tun Hussein Onn Malaysia, 86400 Batu Pahat, Johor, Malaysia

Email:ramadhansyah@utm.my

Abstract. This paper presents an experimental research on the performance of nano black rice husk ash on the porous concrete pavement properties. The performance of the porous concrete pavement mixtures was investigated based on their compressive strength, flexural strength, and splitting tensile strength. The results indicated that using nano material from black rice husk ash improved the mechanical properties of porous concrete pavement. In addition, the result of compressive, flexural, and splitting tensile strength was increased with increasing in curing age. Finally, porous concrete pavement with 10% replacement levels exhibited an excellent performance with good strength compared to others.

1. Introduction

Porous concrete pavement also known as a pervious or permeable concrete pavement and it has been used to control storm water runoff in road pavement and others applications. According to the previous study, the high content of voids in porous concrete lowers the ability to get the high strength [1], [2]. It also found in previous researches that the increasing in void percentages will decrease the strength of hardened concrete [3]-[5]. Mohd Ibrahim et al. [6] found in their literature that, Nano silica is one of the options in order to improve the strength loss in the porous concrete. By incorporating Nano silica in the mixture, it can increase the compressive strength of hardened cement paste and bond strengths of paste-aggregate interface [7]. As the major constituent of rice husk ash is silica, the Nano silica produced from black rice husk ash has been used in this study. In order to study the performance of using nano black rice husk ash, the objective has been set to investigate and to determine the optimum percentage of nano silica from black rice husk ash as a cement replacement in porous concrete specimens in term of mechanical engineering properties. Therefore, the compressive strength, tensile splitting strength, and flexural strength of the porous concrete specimens were examined.

2. Materials

2.1 Cement

Type I ordinary Portland cement (OPC) was used as the major binder material in this study. The cement used was supplied by Tasek Corporation Sdn. Bhd. in one batch for the entire experimental works. The chemical composition of the OPC was given in Table 1. The mean particle size of the OPC is 1353 nm.

2.2 Coarse aggregates

In this study, crushed granites were used as the aggregates source to include in the specimens mixtures. The aggregates were supplied by Hanson Quarry Product Sdn. Bhd. As the porous concrete specimen has little or no fine aggregates, only coarse aggregate used in this study. The aggregates were graded in the range of 12.5 mm to 4.75 mm nominal size. It has a specific gravity of 2.71, water absorption of 0.83%, and aggregates impact value was about 17%.

2.3 Nano black rice husk ash

Nano material was produced from the black rice husk ash. The black rice husk ash was ground using laboratory ball mill grinder. The mean particle size of nano used in this study is 66 nm. Table 1 shows the chemical composition of the black rice husk ash.

Table 1. The chemical composition of the OPC and nano silica.

Oxides	SiO_2	Al_2O_3	Fe_2O_3	CaO	MgO	Na_2O	K_2O	SO_3
OPC	22.68	4.72	3.5	62.27	1.89	-	0.31	4.29
BRHA	91.33	0.07	0.07	0.45	0.28	0.01	2.64	0.05

3. Experimental Procedures

3.1 Sample preparation

A control mix was prepared using OPC. Nano silica replacement levels of 10%, 20%, 30% and 40% by weight of cement were applied. Table 2 shows the mix proportion used in this study. In order to mix the porous concrete mixture, all the materials were placed into drum mixer in the following order: coarse aggregate and binder (OPC or blended cement). The materials were first mixed for 1 minute under dry condition. The water was then added into drum mixer and the mixing was performed for another 3 minutes. Then the mixture was casted in the steel mould. The specimen for compressive strength test, splitting tensile strength test and flexural strength test were casted in the cube (100 mm x 100 mm x 100 mm) steel mould, cylinder (100 mm diameter x 200 mm length) steel mould and prism (100 mm height x 100 mm width x 500 mm length) steel mould respectively. The specimens were casted in two layers and compacted using a bearing plate and Proctor hammer. Immediately after casting process, the specimens were kept and cover by wet hessian to avoid moisture loss.

Table 2. Mix proportions of porous concrete pavement containing nano black RHA

Nano replacement (%)	w/c	Water (kg/m^3)	Cement (kg/m^3)	Nano (kg/m^3)	Coarse Aggregate (kg/m^3)
0 (OPC)	0.34	153	450	0	1115
10	0.34	153	405	45	1115

20	0.34	153	360	90	1115
30	0.34	153	315	135	1115
40	0.34	153	270	180	1115

3.2 Curing condition

According to Mohd Ibrahim et al. [8], the curing condition followed BS1881-113:2011 [9] is the best option for porous concrete specimens. In this study, the curing condition as per BS 1881 – 113:2011 [9] has been selected. After demoulding the specimens, it was then immersed in the water. The specimens were drained and immediately placed in the polyethylene bag. Lastly, the polyethylene bag was seal and store until the testing date.

3.3 Compressive strength test

Compressive strength can be considered as the main property of concrete. In this study, the compressive strength test was conducted on 100 x 100 x 100 mm cube specimens. The specimens were compressed using MATEST compression strength test machine. The specimens were compressed with a loading rate of 3.5kN/s. The test was conducted according to the British standard test method BS EN 12390-3:2002 [10].

3.4 Tensile splitting strength test

Splitting tensile strength test on the concrete cylinder is a method to determine the tensile strength of the concrete. The test was carried out followed the British standard test method BS EN 12390-6:2009 [11], using cylindrical specimens of 200 mm length x 100 mm diameter. The specimens were compressed with a loading rate of 2.5kN/s.

3.5 Flexural strength test

The flexural strength test was conducted to evaluate the resistance of concrete to bending deflection. This test was conducted according to British standard BS EN 12390-5:2009 [12]. The loading rate of 0.4kN/s was applied during the test. The prism porous concrete specimens (100 mm height x 100 mm width x 500 mm length) were used for this testing.

4. Results and discussions

4.1 Compressive strength

The compressive strength of the concrete is one of the most important factors and it's usually used to judge the concrete quality [13]. Figure 1 shows the result of compressive strength for the various nano replacements. Generally, the compressive strength for all specimens increases with increasing in curing age. From the figure, it can be seen that the highest compressive strength at all ages is 10% replacement specimen which is 20.88 MPa, 24.45 MPa, 27.80 MPa, 34.20 MPa and 35.23 MPa for 7, 14, 28, 56 and 90 days respectively. The 40% replacement of NS is the lowest compressive strength which is 14.64 MPa, 16.28 MPa, 22.92 MPa, 27.53 MPa and 29.30 MPa for 7, 14, 28, 56 and 90 days respectively. As shown in the figure, the only specimen with 10% of nano replacement has the strength more than control specimens. Compared to strength with control specimens, the strength of specimens with 10% nano replacement increased 5.14%, 4.01%, 9.08%, 5.10% and 3.94% for 7 days, 14 days, 28 days, 56 days and 90 days respectively. The compressive strength for the each day of testing shows the same trend where the compressive strength starts to increase from 0 % replacement to 10 % replacement of nano. The compressive strength starts to decrease from 10 % to 20% replacement of nano. Same goes to 20 % to 30 % and 30 % to 40 % replacement. This finding similar to Siddharth et al. [14] where the compressive strength start to increase with 10% replacement and it is starting to decrease when the replacement beyond than 10%. This phenomenon happened probably due to workability and the lack of required water in the mixture. From this figure, the 10 % replacement of nano has the best property for the compressive strength evaluation.

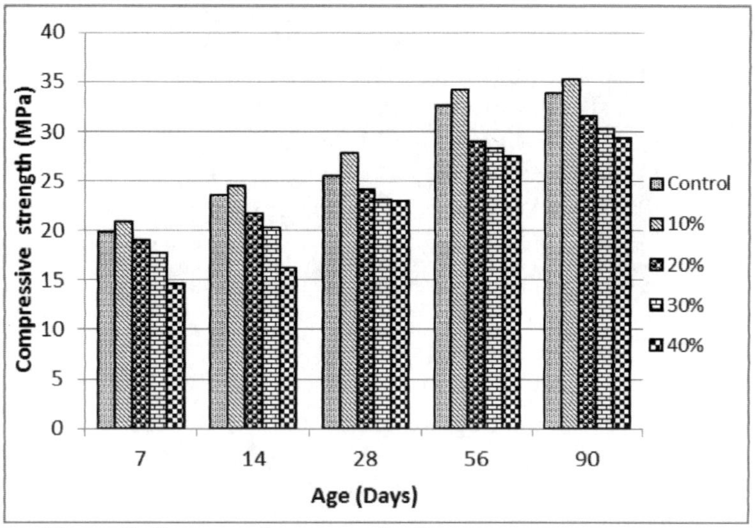

Figure 1. Compressive strength of porous concrete pavement

4.2 Tensile splitting strength

Figure 2 shows the results of tensile splitting strength for porous concrete specimens with different nano percentage replacement. The highest strength at 7, 14, 28, 56 and 90 days are 10% nano replacement which is 3.68 MPa, 3.75 MPa, 4.28 MPa, 4.57 MPa and 4.79 MPa respectively. However, the lowest strength at all age is at 40% nano replacement. The strength for 40% nano replacement is 1.67 MPa, 1.95 MPa, 2.49 MPa, 2.81 MPa and 2.93 MPa respectively. Generally, the strength starts to increase from 0% nano replacement to 10% nano replacement. Then, the strength starts to decrease at 20% nano. Same goes to 30% and 40% nano replacement, the strength continues to decrease. The result also shows that only specimen with 10% nano replacement has the strength more than control specimen. The strength of the specimens increased 11.98%, 10.63%, 8.26%, 10.56% and 11.14% at all ages respectively. The strength for specimens with 20%, 30% and 40% nano has strength below than control specimen. From the results obtained, the specimen with 10% nano replacement has the best property for the tensile splitting strength evaluation. This finding similar to Akeke et al. [15] where 10% is the optimum percentage of replacement.

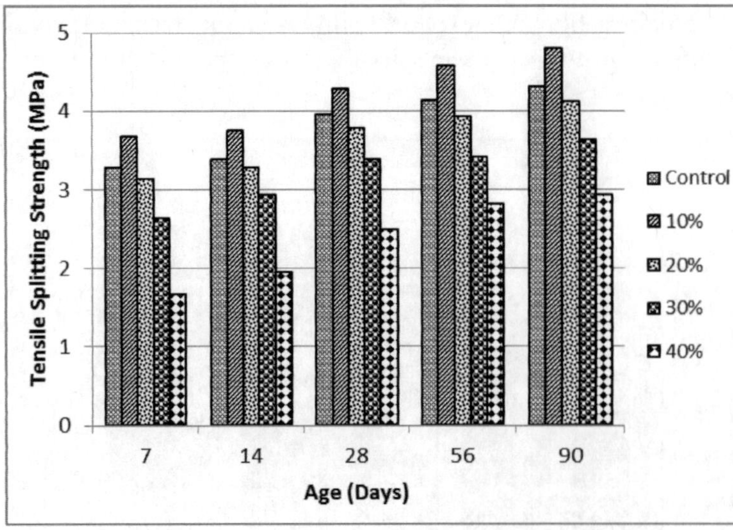

Figure 2. Tensile splitting strength of porous concrete pavement.

4.3 Relationship between compressive strength and tensile splitting strength

An overview of the relationship between compressive strength versus the tensile splitting strength of various percentage of nano replacement is presented in Figure 3, while Table 3 shows the R^2 value for each relationship. As reflected in Figure 3, the tensile splitting strength increased with the increase of compressive strength. As shown in Table 3, R^2 value for 7 days, 14 days, 28 days, 56 days and 90 days are 0.9948, 0.9739, 0.6732, 0.8142 and 0.8936 respectively. Among others, the R^2 value for 28 days specimens shows the lowest which is 0.6732. However, as presented in Figure 2, the graph still shows that the tensile splitting strength increased with the increase of compressive strength.

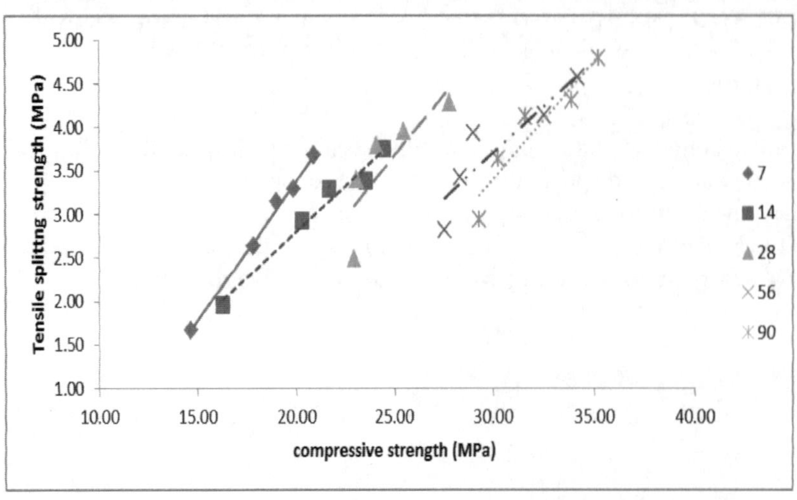

Figure 3. Relationship between compressive and tensile splitting strength.

Table 3. Coefficient of relationship between compressive and tensile splitting strength.

Age (days)	7	14	28	56	90

R^2	0.9948	0.9739	0.6732	0.8142	0.8936

4.4 Flexural strength

The result of flexural strength with different percentage of nano replacement was shown in Figure 4. It can be seen that the highest strength for 7 days, 28 days and 90 days age are specimens with 10% nano replacement which is 3.70 MPa, 4.43 MPa and 4.85 MPa respectively. Compared to control specimens, the strength for 10% nano replacement specimens increased 12.45%, 10.01% and 9.32% for 7 days, 28 days and 90 days respectively. The lowest strength for 7 days, 28 days and 90 days age are specimens with 30% nano replacement which is 2.92 MPa, 3.98 MPa and 4.37 MPa respectively. As shown in the figure, the strength starts to increase from 0% nano replacement to 10% nano replacement. The strength starts to decrease from 10% to 20% nano replacement. Same goes to 20% to 30% nano replacement, the strength continues to decrease. For all percentage nano replacement, the strength with 10% and 20% nano replacement has the strength higher than control specimen strength. It is different from compressive strength and tensile splitting strength that only 10% nano replacement specimens have strength higher than control specimens. Wang et al. [16] mentioned that NS had smaller particles and it can fill between cement particles, therefore the density of the cement paste increased. The presence of NS gives an improvement of the bond between the cement matrix and aggregate, resulting in an improvement in flexural strength. Although both specimens with 10% and 20% nano replacement has the strength higher than control specimen, the best property for flexural strength is specimens with 10% nano replacement. It is because, specimens with 10% nano replacement has the highest flexural strength compared to others.

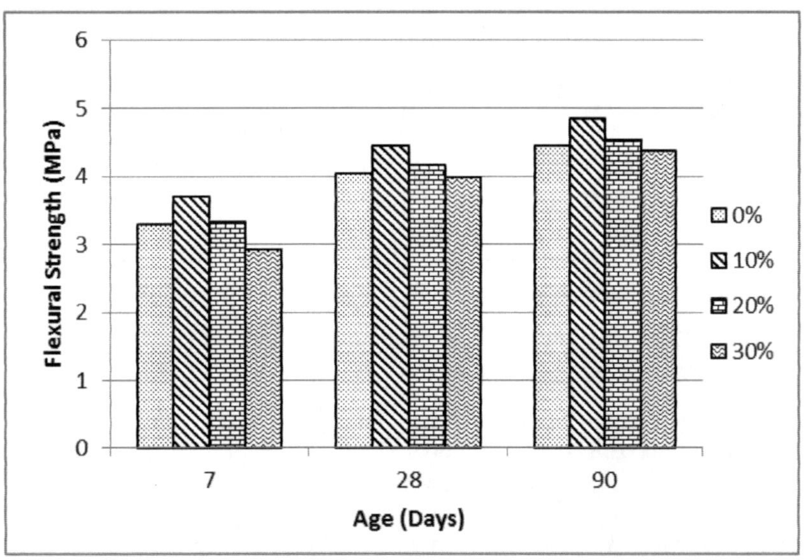

Figure 4 Flexural strength of porous concrete pavement

Figure 5 illustrated the relationship between compressive strength and flexural strength. From this graph, it can be seen that not all flexural strength increase proportionally with the increase in compressive strength. The graph shows that, the compressive strength of 20 % nano replacement lower than the control specimen. However, for the flexural strength, the specimen with 20 % nano replacement shows higher strength than control specimen. As reported by Ramadhansyah et al. [17], the advantages of using rice husk ash include increased in flexural strength. These phenomena happened at all age. It is obviously seen at 90 days age of specimen's relationship graph. Table 4 shows the R^2 value for the relationship between compressive strength and flexural strength. The R^2 value for 7, 28 and 90 days age is 0.9038, 0.7113 and 0.5826 respectively. The R^2 value for 90 days age is low due to phenomena as mentioned before.

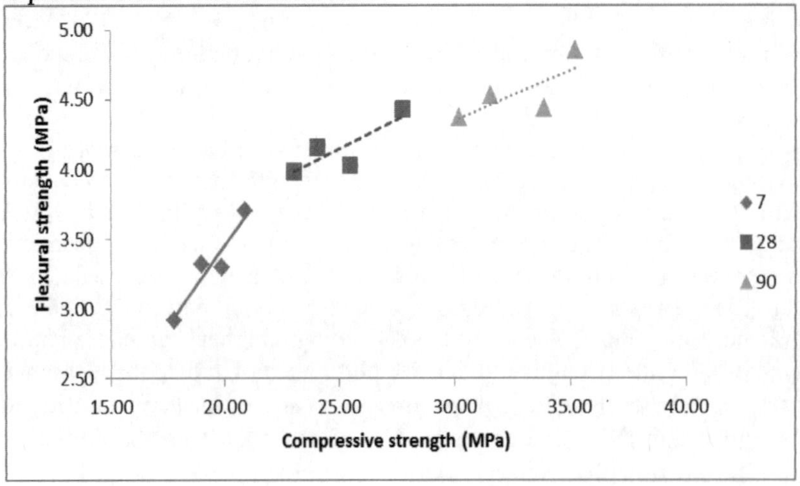

Figure 5 Relationship between compressive and flexural strength

Table 4 Coefficient of relationship between compressive and flexural strength

Age (days)	7	28	90
R^2	0.9038	0.7113	0.5826

5. Conclusion

The performances of nano from black rice husk ash as cement replacement on the mechanical properties of porous concrete pavement were investigated in this study. Results showed that the use of nano black rice husk ash in porous concrete pavements resulted in good compressive strength, tensile splitting strength, and flexural strength compared with the controlled specimen. Furthermore, 10% replacement of nano black rice husk ash to be found as an optimum replacement and indicated better performance than other mixtures.

Acknowledgement

The support provided by Malaysian Ministry of Higher Education and Universiti Teknologi Malaysia in the form of a research grant vote no. Q.J130000.2522.18H05 for this study is highly appreciated.

References

[1] Ramadhansyah P J, Mohd Ibrahim M Y, Mohd Rosli H and Wan Ibrahim M H 2014 *Appl Mech Mater* **554** 37-41

[2] Mohd Ibrahim M Y, Ramadhansyah P J, Mohd Rosli H, Wan Ibrahim M H 2014*Advanced Review on Scientific Research* **1(1)** 34-42

[3] Aamer Rafique Bhutta M, Tsuruta K, Mirza J 2012 *Constr Build Mater* **31** 67-73

[4] Kim H K and Lee H K 2010 *Appl. Acoust.* **71(7)** 607-615

[5] Pindado M A, Aguado A, Josa A 1999 *Cem. Concr.* **29(7)** 1077-1083

[6] Mohd Ibrahim M Y, Ramadhansyah P J, Mohd Rosli H, Wan Ibrahim M H 2016 *ARPN Journal of Engineering and Applied Sciences* **11(20)**11832-11835

[7] Ye Q, Zhang Z, Kong D and Chen R 2007 *Constr Build Mater* **21** 539-545

[8] Mohd Ibrahim M Y, Ramadhansyah P J, Mohd Rosli H and Wan Ibrahim M H 2015 *Jurnal Teknologi* **76(14)** 99-103
[9] British Standards Institution. BS1881-113: Testing concrete: Method for making and curing no-fines test cubes. British Standards. London: BSi 2011
[10] British Standards Institution.. BS EN 12390-3: Testing hardened concrete: Compressive strength of test specimens. British Standards. London: BSi 2002
[11] British Standards Institution. BS EN 12390-6: Testing hardened concrete: Tensile splitting strength of test specimens. British Standards. London: BSi 2009
[12] British Standards Institution. BS EN 12390-5: Testing hardened concrete: Flexural strength of test specimens. British Standards. London: BSi 2009
[13] Ramadhansyah P J, Badorul Hisham A B, Megat Azmi M J and Mohd Haziman W I *Cent Eur J Eng* **1(1)** 103-112
[14] Siddharth T, Jayeshkumar P. and Chetna M V 2015 *Int. J. Adv. Engg. Res.* **4(2)** 296-299
[15] Akeke G A, Ephraim M E, Akobo I Z S and Ukpata J O 2013 *Int. J. Eng.* **3(3)** 57-62
[16] Wang L, Zheng D, Zhang S, Cui H and Li D 2016 *Nanomaterials* **6(12)** 241
[17] Ramadhansyah P J, Mohd A A M N, Zainal A A and Zakaria M A 2013 *Adv Mat Res* **620** 87-93

IOP Conf. Series: Materials Science and Engineering **290** (2017) 012051 doi:10.1088/1757-899X/290/1/012051

Air ionizer application for electrostatic discharge (ESD) dust removal in automotive painting industry

M H Yosri[1], P Muhamad[1], M A Ismail[1] and N H M Yatim[2]

[1]Intelligent Dynamics & System Research Lab, Malaysia-Japan International Institute of Technology, Universiti Teknologi Malaysia, 54100 Kuala Lumpur, Malaysia
[2]Department of Mechanical Engineering, Kulliyyah of Engineering, International Islamic University Malaysia, 50728 Kuala Lumpur, Malaysia

Email: hafizanyosri@yahoo.com

Abstract. Dust and fiber have been identified among the highest contributor for the defect in automotive painting line with range from 40% to 50% of total defect breakdown. Eventually, those defects will effect on both visual appearance and also the performance of the parts. In addition, the significance of controlling dust in an assembly line is crucial in order to maintain the quality of the product, part performance yield and effect on workers' health [1]. By considering the principle and technology applied in electronic clean room technology, the ionizer have been introduce to control dust contamination in automotive painting line. The first auto maker industry whom found the effectiveness of the clean room application to reduce the defect and production line downtime was Chrysler [2]. By doing so, it's allowed the transmission plant to offer 50 000 mile guarantee on the transmission systems. The main objective of this research is to verify the effectiveness of ionizer device in order to reduce the rejection contribute by dust and fiber particle in the automotive painting line. Towards the main objective, a few sub areas will be explored, as a supporting factor to ensure the result gain from this study is solid and constructive. The experiment start by verifying the electrostatic value of the raw material (substrate) before and after the ionizer treatment. From here the correlation of the electrostatic value generated by the raw material that effect to production pass rate can be explored. At the meantime, the performance of the production pass rate after the ionizer treatment which related to the painted surface area can be determined.

1. Introduction

Recently, the top and highest rejection for almost all painting line in automotive industry is contributed from dust and fibre. The rejection contributed by dust and fibre range from 40% ~ 50% (see, figure 1 and 2) [9]. For coating industries, dust and fibre have effects on appearance issue and also contribute to the cost of poor quality. The rejection due to this defect will directly increase the operational cost to reworked and repair the defective parts which indirectly impact on the company revenue. By taking example and practice from electronic, the implementation of ionizer application in clean room environment to control dust and fibre can be adapted into painting line in automotive industry. A few considerations should be taken into account to ensure the method and technology is suitable for automotive application.

IOP Conf. Series: Materials Science and Engineering **290** (2017) 012051 doi:10.1088/1757-899X/290/1/012051

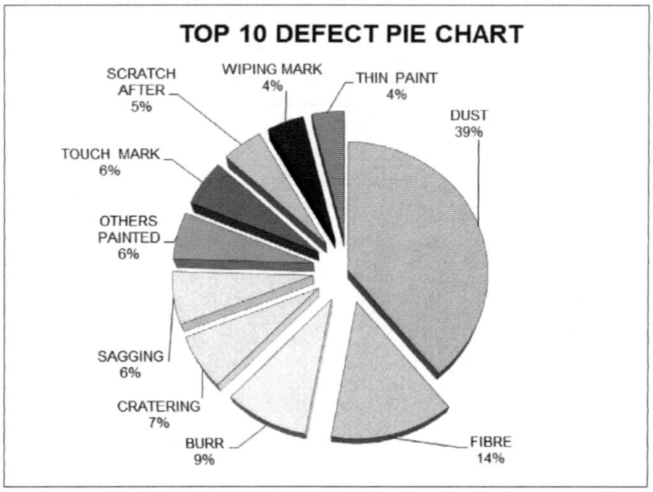

Figure 1. Pie chart of defect distribution for 2015 (Source from Combat Coating (M) Sdn Bhd 2015).

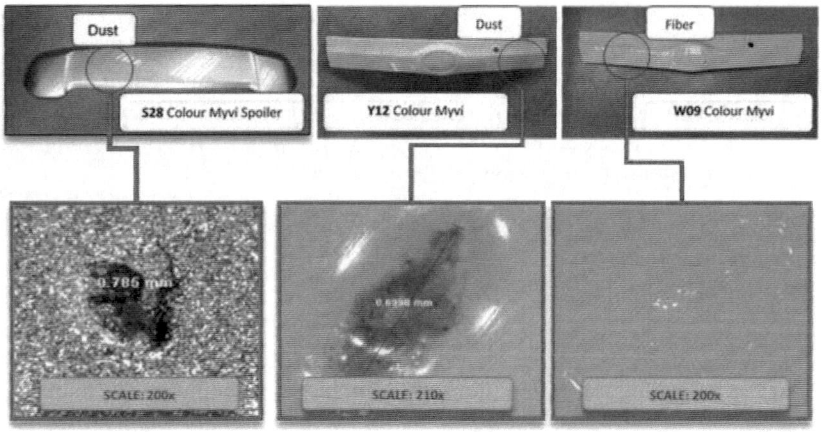

Figure 2. Sample contamination of dust and fibre in painting line (Source from Combat Coating (M) Sdn Bhd).

S. Arnold [4] suggests the best practice to reduce and control the ESD (electrostatic discharge) and contamination is by applying air ionization in clean room as shown in figure 3. Finding shows there is no laminar air flows able to reduce particle deposition in the clean room itself. The ESD and particle attraction are recognized as the major problem caused by the static charge in semiconductor manufacturing. The air ionization can reduce the contamination by creating a balance quantities of both negative (-) and positive (+) ions. The clean room and laminar air flows are installed with overheads ceiling or bar type ionizers to control the static charges to get better and effective results. Ionizers can make particle produced during certain production processes easier to be removed. Ionizers are used to control static electricity, which usually situated in the roof that can allow the static charge to dissipate. Bapat [5] presented the application of electrostatic precipitators (ESP) for controlling cement dust while Hindy [6] concluded that the application of ESP is very effective to control the cement dust.

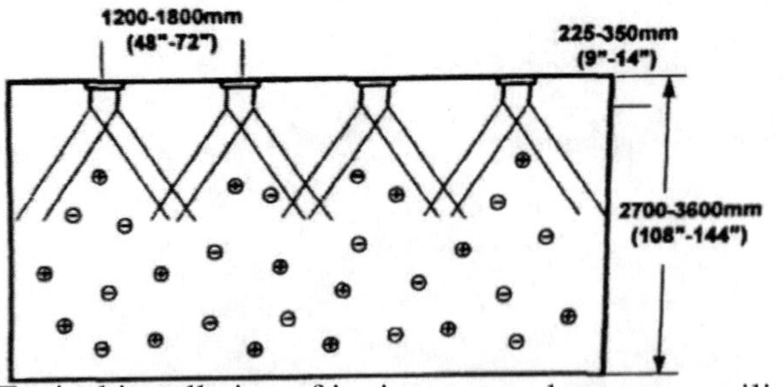

Figure 3. Typical installation of ionizers on a clean room ceiling.

G.W. Penney et al. [7] found the charged particles suspended in the air room create a space charge potential. The resulting voltage gradient drives the charged particle towards the wall. The chain effect from this condition will contribute to high dust particle where the effectiveness of the air cleaning device is not realized. This situation can be resolved by introducing the high efficiency cleaning and neutralizing space charge device such as the air ionizer [8].

Electrostatics phenomenon occurs when the particular object contacts with other surface and escalates the electrostatics charge of ion on the surface of objects. Even though electrostatics occur during the surfaces contact and separate, the effects of electrostatic flows charge exchange are usually happen when one of the object surfaces has a high resistance to electrical flow. The reason of this phenomenon is due to the charges that are embedded and trapped in the object surface for a certain period of time for the effects to be observed. These electrostatics charges are kept and preserved on the surface object until they either flow off to the ground or neutralized by a discharger [3]. An ion exposed to an electric field E will move with an average drift velocity v proportional to E, that is,

$$v = kE, \tag{1.1}$$

where k is the mobility of the ion.

Theoretically, when the air ion exposed to an electric field with a strength of 1 V/cm, its will moves at a velocity of about 1 cm/sec in the range of 1.0–2.0 cm2/V•s (centimeter^2 per volt-second). In addition, the movement of the negative ions is approximately 15% higher compared to positive ions. In summary, the air concentration n of positive ions with the mobility k and charge e, an electric field E will make the electric current to flow in the direction of E with the density j.

$$j = enkE = \lambda E \tag{1.2}$$

The constant λ that represent to enk, is the positive conductivity of the air or in another words the polar conductivity that contribute by the positive ions. In another hands, negative ions will flow in the opposite direction of the field. Nevertheless, the current density from negative ions can be calculate using equation (1.2) when e is taken as the numerical value of the ion charge.

Theoretically, an electric field is generate surrounding the object depends on the given charge, either positive or negative polarity. For example, whenever the objects is surrounded with air that contains both polarities, the opposite polarity air ion will moves and flows towards the object. However the same polarity will flow against and opposite of its. Even though the field will different from every position and area in space, it is always proportional to q. The movement of charge is an electric current. The neutralization current which is equivalent to the charge and to the relevant opposite conductivity of the surrounding air is moving toward to the object while carried by ions of polarity opposite to q.

The charge neutralization relative rate is constant in the event of the air conductivity does not change, while the charge will reduce proportionally with a time constant τ that rely on the air conductivity. Given an initial charge q_0, the charge remaining at a later time is given by

$$q = q_0^{-t/\tau,} \tag{1.3}$$

where the time constant τ is equal to the permittivity of the air ε0 divided by the air conductivity, λ.

$$\tau = \varepsilon_0/\lambda \tag{1.4}$$

thus, referring to equation (1.3),

$$q = q_0^{-t(enk/\varepsilon 0} \tag{1.5}$$

resulting the rate of charge neutralization proportional to the ion concentration.

2. Experimental procedures

In total there were 550 pieces of sample from 9 type of parts. From there, the entire sample can be grouped into 3 sub-groups which are small, medium and big parts. The purpose of dividing the parts into 3 categories is to monitor the factor that will potentially affect the quantity of dust particle to fall down into the painted parts. Among the factors are the painted surface area and geometrical design of the parts. The monitoring period had started from January to August 2016 when the installation was completed on earlier January 2016. From there, the machine were continuously used during normal production time.

Figure 4. Part position during ionizing process.

The ionizer machine was installed after the cleaning process in preparation area. The main objective was to monitor the effectiveness of ionizer machine to neutralize the electrostatic charge at the raw material before the painting process. The device was installed on a static condition with 45-degree angle to communize with all plastic parts. Consideration has been made based on the coverage of the ionizer air distribution with the painting jig position. With 45-degree position, the painting surface coverage with the ionizer treatment reaches almost 98% where the rest is hidden due to the parts' geometry and non-critical area for painting process. The ionizer device was installed after the air blow/ tag rag process. This was the last process before the parts enter the production line for the first primer coating. The experiment was conducted in actual painting line environment with fixed and constant parameter using exiting setting for current production. The air flow was supplied from Air Supply Unit (ASU) with the speed of 0.25 m/s while the temperature and humidity was set at 27.1 Celsius and 85% RH (relative humidity).

In this experiment, the Keyence Static Eliminator model SJ H108A was installed after the wiping process and before the first layer of primer coating. The SJ Series has adopted the pulse AC method that applies alternating high voltage to the electrode probe, producing ions of both polarities.

IOP Conf. Series: Materials Science and Engineering **290** (2017) 012051 doi:10.1088/1757-899X/290/1/012051

3. Results and Discussion

The initial data for the electrostatic value for the raw parts was collected to compare the values before and after the raw parts gone through the ionizer treatment. Most of the raw parts show zero value, which reflect the effectiveness of the ionizer to neutralize the electrostatic value on the raw parts. From table 1, it is seen that the production pass rate shows a slight reduction from 91.4% on 2015 to 90.8% on 2016 (data collected until August 2016) while the rejection trend for dust increases from 3.2% to 4.6%. However, the fibre defect shows some reduction from 1.3% to 0.9% in 2016. The lowest pass rate and highest rejection for dust and fibre is recorded on June 2016 exactly after the replacement of the ceiling filter. The data shows the production pass rate is reduced to 87.5% while the dust and fibre increased to 6.8% and 1.4%. However, the production pass rate recovers back on the next following month until August 2016 (Production pass rate: 93.3%, Dust: 3.2%, Fibre: 0.8%).

Table 1. Production Pass Rate.

Month	Jan-16	Feb-16	Mar-16	Apr-16	May-16	Jun-16	Jul-16	Aug-16	Sep-16	Oct-16	Nov-16	Average 2016	Average 2015
Pass Rate	91.10	91.56	91.45	91.17	88.76	87.50	91.10	93.30	93.95	92.20	87.20	91.21	91.36
Dust	4.33	3.93	4.32	3.94	5.86	6.80	4.06	3.23	2.30	2.55	4.55	4.13	3.21
Fiber	0.75	0.76	0.84	0.88	1.11	1.40	0.95	0.77	0.70	0.90	1.45	0.91	1.25

Table 2. Summary of correlation result of electrostatic value, material type and production pass rate.

No	Parts	Before (kV)		Material Type	PASS RATE (%)	Average		
		(+)	(-)			Pass Rate	(+) kV	(-) kV
1	Bumper	1.5	-2.0	Polypropylene (PP)	49.3	68.9	1.1	-1.1
3	Tail Gate	0.5	-0.4	Polypropylene (PP)	68.1			
4	Rocker Panel	0.5	-0.4	Polypropylene (PP)	93.1			
5	Side Sill	1.7	-1.6	Polypropylene (PP)	65.1			
6	Door Handle	1.4	-0.2	Polybutylene terephthalate (PBT)	95.0	95.0	1.4	-0.2
2	Spoiler	0.1	-0.1	Acrylonitrile butadiene styrene (ABS)	68.2	82.5	2.6	-2.6
7	Mirror Hood	0.7	-0.3	Acrylonitrile butadiene styrene (ABS)	87.2			
8	Scull Cap	3.1	-4.4	Acrylonitrile butadiene styrene (ABS)	86.1			
9	Back Door	6.4	-5.6	Acrylonitrile butadiene styrene (ABS)	88.4			

Acrylonitrile butadiene styrene (ABS) material shows the average pass rate is 82.5% even with high electrostatic value range from -2.6 kV to 2.6 kV. Polybutylene terephthalate (PBT) material shows the highest pass rate of 95.0% even with the lowest electrostatic value range from -0.2 kV to 1.4 kV while Polypropylene (PP) material shows the lowest pass rate of 68.9% with average electrostatic values ranging from -1.1 kV to 1.1kV. Ionizer machine shows good result to neutralize the electrostatic value of the parts despite the highest value shows at ABS material. From the experimental result, the material has no big influence to affect the outcome of the production pass rate with the application of the ionizer when the static value is neutralized to zero value after the treatment. However, other factors such as the part geometry and the painted surface provide higher potential for the dust particle to fall down into the surface. It aligns with the experimental result when the PP material parts especially bumper that has bigger painted surface provides less production pass rate due to this matter. In opposite of that, Back Door Garnish that is manufactured from ABS material with high electrostatic value still manage to get a good production pass rate (88.4%).

Among external factors that might influence the efficiency of the ionizer device is due to the static charge carrier during wiping process. As normal process flow, the wiping process was conducted before the ionizing process with a distance less than 2 meter between both stations and 60 seconds of time cycle between the carriers in the conveyor line. Hence, there was a possibility that the wiping process will generate electrostatic charge and reduce the efficiency of the device [11]. Furthermore, the distance and gap for the conveyor to travel after the ionizer treatment to the first painting booth (primer booth) will provide a chance for the particle to fall down into the parts. In other words, even though the parts were properly neutralized using the ionizer device, there was no guarantee that the particle will attach back to the part due the above mention factor.

However, the main cause that contributes to the downtrend of the production pass rate still cannot be determined. The only suggested hypothesis at this moment is due to the mechanical movement of the dust and fibre particle that has previously settled down in idle condition and is exposed to the painting line environment during the changing process of the filter itself.

4. Conclusion

The efficiency of ionizer device to reduce the rejection contributes by dust and fiber particle in the automotive painting line have been tested experimentally in actual environment. The outcome from this experiment is determined from the production pass rate after the ionizer treatment. Referring to the results from the actual manufacturing process, it shows a minor reduction of the fibre particle after the implementation of the ionizer machine. However, the rejection trend for dust defect increases around 2%. Nevertheless, the result gain from this period was doubtful due to uncertainty especially during the replacement of the ceiling filter on July 2016, which shows the highest rejection for dust particle. Further monitoring should be conducted in order to get more solid and certain data for the effectiveness of the ionizer application to control the dust and fibre particle. Other factors that should be considered are the parts' geometry and painting jig design which correlate to the painted surface area that is exposed to the environment which carries out the dust and fiber particle. In addition, the air flow, turbulence and particle concentration in paining line after the ionizing process shall be determined using a Computational Fluid Dynamic (CFD) [10] in order to improve the mechanical design of the painting jig, working layout and parameter setting of the painting line.

References

[1] Rosmaini Ahmad, S.K., Zahid A. Khan, Mohzani Mokthar, Indra Putra Almanar 2006 *Management of Environmental Quality, an International Journal* **7** pp 390-408
[2] Holbrook, D., 2009 *H.a. Technology Editor* pp 173-191
[3] Arnold Steinman 2006 Compliance Engineering, Annual Reference Guide, Copyright 2006 Canon Communications LLC
[4] Arnold., S.,1995 *EOS/ESD Symposium* pp 95-245
[5] Bapat, J.D., 2001 Journal of Hazardous Materials, **B81** pp 285-308
[6] Hindy, K.T. *Utilization of controlled cement dust as a concrete material in the United Arab Emirates*. in *Environmental Management and Health*. 1997
[7] Hewitt., G.W.P.G.W.,1949 *AIEE Transactions*
[8] Inculet, I. I., & Topping, D. R. 1971 *IEEE Transactions on Industry and General Applications* **2** pp 314-317.
[9] Combat Coating (M) Sdn Bhd, *Outgoing Quality Inspection Passrate Report*, 2015~2016
[10] Thongsri, J., & Pimsarn, M. 2015 *International Journal of Precision Engineering and Manufacturing* ***16*** pp 509-515
[11] Noh, K.C., Kim, H. S., and Oh, M. D. , 2010 *Building and Environment* **45** pp 825-831.

IOP Conf. Series: Materials Science and Engineering **290** (2017) 012052 doi:10.1088/1757-899X/290/1/012052

Modeling and simulation of graphene/palladium catalyst reformer for hydrogen generation from waste of IC engine

A Rahman[1], K M Aung[1]

[1]Department of Mechanical Engineering, International Islamic University Malaysia (IIUM), Jalan Gombak, 53100 Kuala Lumpur, Malaysia
International Islamic University Malaysia (IIUM), Jalan Gombak, 53100 Kuala Lumpur, Malaysia

Email:ataur7237@gmail.com

Abstract. A small amount of hydrogen made by on-board reformer is added to the normal intake air and gasoline mixture in the vehicle's engine could improves overall combustion quality by allowing nearly twice as much air for a given amount of fuel introduced into the combustion chamber. This can be justified based on the calorific value of Hydrogen (H_2) 141.9 MJ/kg while the gasoline ($C_{6.4}H_{11.8}$) is 47MJ/kg. Different weight % of Pd and GO uses for the reformer model and has conducted simulation by COMSOL software. The best result found for the composition of catalyst (palladium 30% and graphene 70%). The study shows that reformer yield hydrogen 23% for the exhaust temperature of 600-900^{0}C and 20% for 80-90^{0}C. Pumping hydrogen may boost the fuel atomization and vaporization at engine idle condition, which could enhances the fuel combustion efficiency. Thus, this innovative technology would be able to save fuel about 12% and reduce the emission about 35%.

1. Introduction

Hydrogen is a clean fuel and contributes as a combustible fuel of the IC engine due to wide range of flammability, prompt ignition, and high flame speed at stoichiometric ratios, high diffusivity, higher calorific value, and lighter and lower density. One of the systems for on-board hydrogen production is steam-reforming process. Steam reforming process is an endothermic process, which can convert unburned hydrocarbon of engine into hydrogen and carbon monoxide reaction with steam in presence of catalyst. The catalyst promotes the oxidation of the carbon to produce carbon dioxide.

Wyszynski et al. [1] stated that exhaust gas of fuel reforming with high temperature in the range of 800-900 °C produces high hydrogen yield, which is over 30% in the reformed gas. At exhaust temperature in the range of 500 – 700 °C, the hydrogen production can be obtained about 20% in the reformed gas [2,3]. The authors stated that a portion of hydrogen can be directly fed back to fuel input to increase the temperature of the 'hydrogen enriched mixture' for better fuel efficiency.

The hydrogen can be obtained from the reformating the exhaust gas by using catalyst reformer. The exaust gas purification apparatus and method has patented [4,5]. The inventor has claimed that the composition of partial oxidation of exhaust over nickel catalyst hydrogen can be generated. The system results for partial oxidation of gasoline obtained 78.5% thermal efficiency with hydrogen and carbon monoxide yields of Vol.21.6% and Vol.23.6% respectively. In his experimental result, the reformed fuel having hydrogen and carbon monoxide levels of 32.2% and 19.6% has observed at a reactor gas inlet with temperature setting of 1223 K and gas hourly space velocities (GHSV) of 1000 h^{-1}.

Catalysis is the hidden gem in environmental and chemical industries. While, the use of catalysts is most commonly known in automotive catalytic converters, they are routinely used in other applications including gasoline, automobile lubrication and its additives and the synthesis of fine chemicals. In addition to their economic impact, catalysts play a critical role in developing sustainable and environmentally friendly processes. The fifty highest-produced chemicals by volume in the US correlate with a substantial carbon footprint via the emission of carbon dioxide and other harmful gases into the earth's atmosphere. Currently thirty of these processes have employed catalysts to reduce the emissions released and waste generated [6]. Despite catalysis emerging in the forefront of developing sustainable and efficient processes, industries still struggle to successfully develop such processes. The E-factor, one of the most straight-forward and useful green chemistry metrics, is defined as kilograms of waste produced per kilograms of product [7]. While catalysts have proven to assist in product selectivity, future advancements in process efficiency by developing immobilized catalysts [8,9,10]. Over half of all chemicals produced are still made using traditional batch reactions, yet continuous reactions offer unique advantages including intensified mixing, smaller volume to heating ratio, and safer operating conditions [11,12,13]. The highly reproducible nature and therefore the robust application of continuous processes in the production of hydrogen from the exhaust of internal combustion engine have led to an increased interest in developing the proposed technology.

Hydrogen engines have demonstrated efficiencies (BTE) in excess of today's gasoline engines, NO_x is the only regulated tailpipe pollutant resulting from hydrogen combustion, and carbon dioxide is a non-issue, at least in terms of the driving (Tank-to-Wheels) component of the lifecycle [14]. Although current efforts by Ford and BMW reflect early stage development, both BMW and Sandia outlined technology approaches for subsequent generations of H2ICE systems that could deliver significant improvements in fuel economy and emissions reduction, while delivering additional power (Green car Symposium).

The aim of this study to develop a green technology to generate hydrogen from the waste of IC engine which would be able to reduce the fuel consumption and reduce emission.

2. Experimental procedures

Laminar flow physics is added to the heating pipes of the reformer for the steam flow. The wall velocity is set as zero and inlet velocity for heating tubes is set as 1 m/s. The flow properties are defined [5] by the following equation (1):

$$L_{entr}\overline{V}_t\left[-pl+\mu\left(\overline{V}_t u+\left(\overline{V}_t u\right)^T-\frac{2}{3}\mu l\left(\overline{V}.u\right)\right)\right] \\ =-np_{entr} \tag{1}$$

Then the outlet flow properties are defined [5] by the following equation (2):

$$\begin{aligned} &\eta^T n\left[-pl+\mu\left(\left[\overline{V}u+\left(\overline{V}u\right)^T\right]\right)-\frac{2}{3}\mu l(\overline{V}u\right] \\ &=-p_o \\ &with\ \ p_0\le p_0.u.t=0 \\ &According\ to\ Darcy's\ Law,\ \ \overline{V}.\left(\rho u\right)=Q_m \\ &u=-\frac{k}{\mu}\overline{V}p_i \end{aligned} \tag{2}$$

The mass flow rate of air, hydrogen and exhaust gas can be estimated by using the equation (3):

$$\dot{m}_{i,a} = \frac{C_{d,a}A_{v,a}P_a}{\sqrt{R_aT_a}}\left(\frac{P_m}{P_a}\right)^{\frac{1}{\gamma a}}\left(\frac{2\gamma_a}{\gamma_a-1}\left[1-\left(\frac{P_m}{P_a}\right)^{\frac{\gamma_a-1}{\gamma_a}}\right]\right)^{\frac{1}{2}}$$

$$\dot{m}_{i,H} = \frac{C_{d,h}A_{v,h}P_h}{\sqrt{R_hT_h}}\left(\frac{P_m}{P_h}\right)^{\frac{1}{\gamma a}}\left(\frac{2\gamma_h}{\gamma_h-1}\left[1-\left(\frac{P_m}{P_h}\right)^{\frac{\gamma_h-1}{\gamma_h}}\right]\right)^{\frac{1}{2}} \quad (3)$$

$$\dot{m}_{i,f} = \frac{C_{d,f}A_{v,f}P_f}{\sqrt{R_fT_f}}\left(\frac{P_m}{P_f}\right)^{\frac{1}{\gamma f}}\left(\frac{2\gamma_f}{\gamma_f-1}\left[1-\left(\frac{P_m}{P_f}\right)^{\frac{\gamma_f-1}{\gamma_f}}\right]\right)^{\frac{1}{2}}$$

where, $C_{d,a}$, $C_{d,h}$, and $C_{d,f}$ are coefficient of discharge for air, hydrogen and gasoline, respectively. $A_{v,a}$, $A_{v,h}$, and $A_{v,e}$ are restrictions flow areas for air, hydrogen and exhaust gas, respectively. P_a , P_h , and P_e the air, hydrogen, and exhaust pressures, and γ_a , γ_h , and γ_e the air, hydrogen, and exhaust specific heat ratios, respectively.

The mixture of exhaust gases and steam pass to the catalyst cells and form the chemical reaction as shown in figure 1 and generate the hydrogen (H_2), carbon dioxide (CO_2) and water (H_2O). The hydrogen pumps to the engine cylinder, CO_2 uses to heat the product of reformer H_2O and recycled to the reformer. Therefore, the rate of mass change inside this capacity can be given [15] by the continuity equation (4):

$$\frac{dm_m}{dt} = \dot{m}_i - \dot{m} \quad (4)$$

with $\dot{m}_o = \dot{m}_{co_2} + \dot{m}_{H_2} + \dot{m}_{NOx} + \dot{m}_{UHC} + \dot{m}_{steam(H_2O)}$ and

$$\dot{m}_i = \dot{m}_a + x\%.\dot{m}_{H_2(f)} + (1-x\%)\dot{m}_{C(f)}$$

where, $\dot{m}_m$ is the intake manifold mass content, $\dot{m}_i$ is the input mass flow to the intake manifold, and $\dot{m}_o$ is the outflow from engine cylinders, m_{UHC} is the outflow of unburn hydrocarbon, x% is the percentage of H_2 pumps to the engine cylinder, $m_{c(f)}$ is the mass rate of conventional fuel (gasoline) . The mass-balance equations (5) for the model are the Maxwell-Stefan diffusion and convection equations at steady state:

$$\bar{V}.\begin{pmatrix} \rho w_i u - \rho w_i \sum_{j=1} D_{ij}\left(\bar{V}x_j + (x_j - w_j)\frac{\bar{V}_p}{\rho}\right) - \\ D_i^T\frac{\bar{V}T}{T} \end{pmatrix} = R_i \quad (5)$$

where, ρ is the density, kg/m^3, ω_i the mass fraction of species *i*, x_j the molar fraction of species *j*, D_{ij} is the *ij* component of the multicomponent Fick diffusivity, m^2/s). D_i^T the thermal diffusion coefficient, kg/(m·s)), *T* is the temperature, K, and R_i the reaction rate kg/(m^3·s). The flow of heating gas in the tubes is described by the Navier-Stokes equation (6) at steady-state:

$$\rho u.\bar{V}u = \bar{V}.[-pI + \eta(\bar{V}u + (\bar{V}u)^T) - (2\eta/3)(\bar{V}.u)I\,] \quad (6)$$

where ρ is the density kg/m^3, u the velocity, m/s, viscosity, kg/(m· s), $\overline{V}$ the flow rate, kg/m^3 and p the pressure of the mixture in the tubes, Pa. At the outlet, viscous stresses are ignored and the pressure is set to 1 atmosphere.

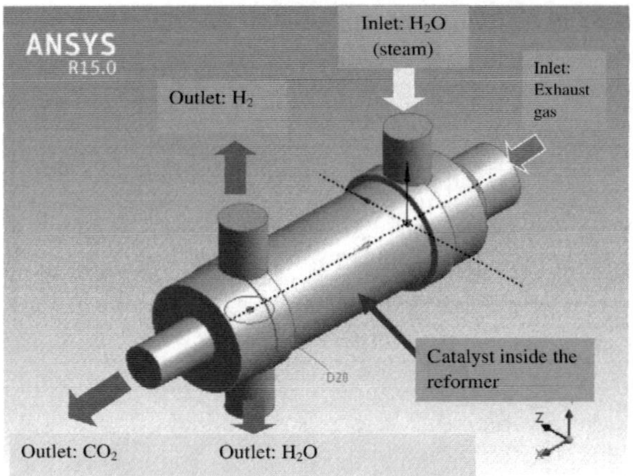

Figure 1. 3D exhaust reformer [5].

2.1 Estimation Of Exhaust Gas Temperature

Rahman et al. [15] carried out the effect of engine speed (rpm) to the heat transfer in the cylinder wall and volumetric efficiency by using GT-Suite software as shown in figure 2(a) and 2(b). At 6000 rpm, there will be less time for fuel intake and combustion, which caused the temperature of the engine higher and increase in unburn hydrocarbon. Reitz (2012) found that about 30% of the fuel energy is used by transferring heat energy to the cylinder wall. Therefore, the exhaust temperature becomes higher in a range of 300-900 °C as shown in figure 2(b) and it will affect the volumetric efficiency of the engine as the heat transfer at the cylinder wall of the engine becomes less at high speed. The result of the simulation literature for exhaust gas temperaute by GT-Suite is the pre-requirement for modelling of steam reformer as the exhaust temperature is crucial input in the simulation of exhaust reforming. For this project, average exhaust temperature of 900 K is used for the inlet of the steam reformer.

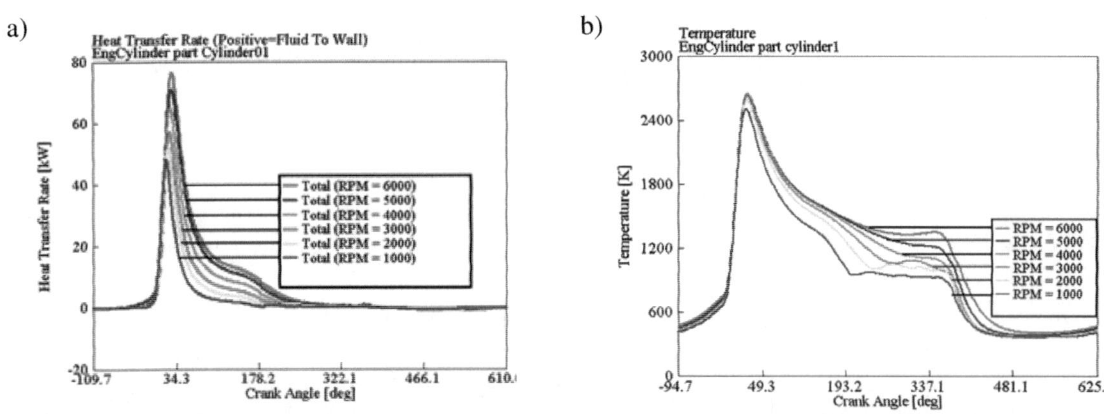

Figure 2. Affect of exhaust gas into engine cylinder at different crank angle [15]; heat transfer (a) and temperature (b).

2.2 Hydrogen Generation Process

The exhaust gas from the engine is passed to the turbocharger for boosting the flow rate of the exhaust gas. Figure 3 shows the hydrogen generation process from the exhaust gas and steam combination. The exhaust controller adjusts the amount of exhaust gas which distributes between the water tank and the reformer. About 5~7% of high temperature exhaust gas is considered to be utilized to heat the water of the tank and make steams which feeds into the reformer. The temperature of the hydrogen gas produced would be around 250~300 °C. In order to reduce the high temperature of hydrogen, the heat exchanger is placed immediately after the reformer to cool the hydrogen. When the desired temperature of 70~90 °C is achieved, the produced hydrogen is stored in the tank for usage in the IC engine as an alternative fuel source to save fuel consumption. The amount of hydrogen supplied to the engine about 7% of total fuel supply. The amount of hydrogen supplied to the engine is limited by the interior design of the engine. Using a crankshaft rpm sensor, the hydrogen supplied to the engine is controlled. Theoretically, the hydrogen production by the reformer increases for the engine operating rpm of 4000 rpm and above. The temperature of supplied hydrogen to the engine is expected to be 90 °C. The main reason for this temperature is to make the fuel atomized and vaporized in the engine cylinder, which helps the engine to achieve 95% fuel combustion. In the conventional engine, the amount of fuel combustion in the engine is about 80-90%. A heat exchanger is used to maintain the temperature of hydrogen at 90 °C [10]. The amount of hydrogen supplied to the engine is limited by the interior design of the engine. It is recommended that the hydrogen supply to the engine should not be more than the 7% of fuel.

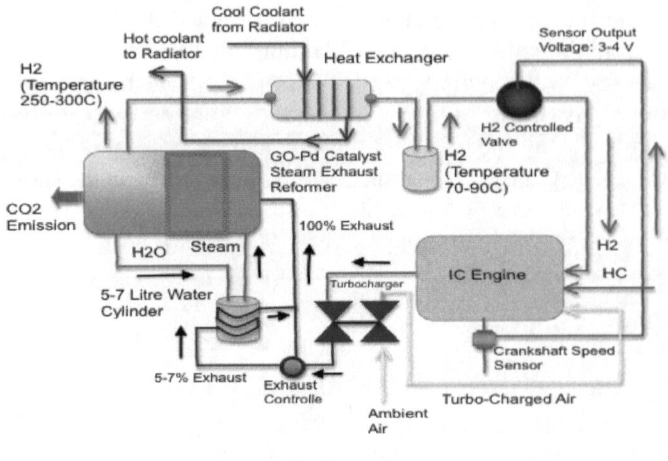

Figure 3. Hydrogen generation process.

2.3 Microwave Assisted Pd/GO Synthesis

Microwave assisted synthesis of GO/Pd is proposed based on the study [12,13]. Graphite oxide (1 to 8 mg) has been dissolved in 400 ml of water until a homogenous suspension is obtained. Palladium nitrate (210 μL) has been added to the homogenous mixture and agitated using a magnetic stir plate. The homogenous mixture then placed in an Emerson conventional kitchen microwave where 100 μL of hydrazine hydrate solution (25% in H_2O) has been added. The solution then immediately microwaved on full power (1000 W) in 30-second cycles (on for 10 s, off and stirring for 20 s) for a total reaction time of 60 seconds. The resulting black solution has been centrifuged at 5000 rpm for 20 minutes and dried under vacuum until black flakes obtained. The chemical reaction that incurs into

IOP Conf. Series: Materials Science and Engineering **290** (2017) 012052 doi:10.1088/1757-899X/290/1/012052

the reformer can be presented as:

$$\begin{aligned} &Exhaust(0.5\% H_2, 5\% CO, 11\% CO_2, NO_x, \\ &21\% HC\, and\, H_2O) + Steam(H_2O) \\ &\qquad \Big\downarrow \; Pd/GO\,Catalyst \\ &23\% H_2 + 10\% CO_2 + 5\% N_2 + H_2O \end{aligned} \tag{7}$$

3. Results and Discussion

The reformer has been developed with two zones; mixing zone and reforming zone as shown in figure 1. At the inlet of the reformer, exhaust and steam mix together to form the mixing zone. Immediate to the mixing zone is the reforming zone where the exhaust and steam reform throughout the graphene-palladium honeycomb composite catalyst to generate hydrogen. At the outlet, water will pass downwards hydrogen will be led upwards and remaining exhaust plus carbon dioxide will go straight through the respective ducts. The length of the catalyst is 150 mm with 60 mm diameter and 5mm holes are cut through the entire length equally to make honeycomb shape. Steam, water and hydrogen ducts are 19mm in diameter and 25 mm in length.

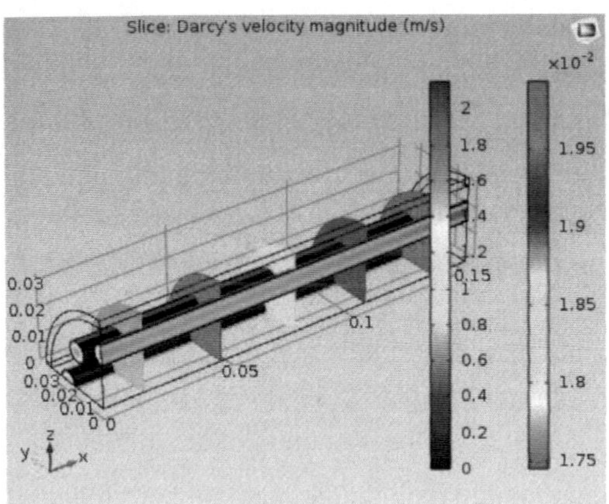

Figure 4. Darcy's velocity magnitude.

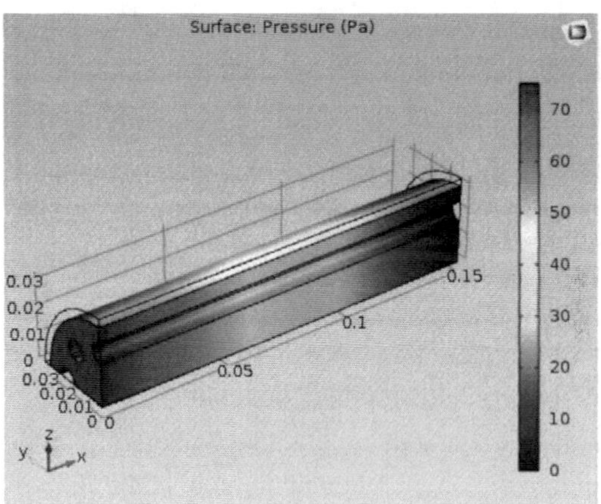

Figure 5. Pressure along the catalyst bed.

The flow of gases in the catalyst bed is simulated by the Darcy's velocity magnitude in figure 4. For the flow of gases in the catalyst bed, the flow started as 1m/s as defined by the input and steadily increased along the reactor. At the reformer exit, the velocity of the gases is about 2 m/s. For the steam flow in heating tubes, the initial velocity is 2 x 10^{-2} m/s at the very beginning of the tubes and decreased to 1.4 x 10^{-2} m/s throughout the length of the heating tubes. The pressure at the inlet of the reformer is 75 Pa as defined by the input and it is found that the pressure dropped uniformly throughout the length of the catalyst bed. At 0.05m, the input pressure dropped to about 50 Pa as shown in figure 5. At 0.1m of catalyst bed, the pressure again dropped to about 30 Pa. However, the outlet pressure of the catalyst bed is close to 0.

The surface temperature of the heating tubes begins with 700 K at the inlet as defined by the input. From there it decreased slightly along the heating tubes and increased to about 740 K just before the outlet of the heating tubes as shown in figure 6(a). The surface temperature of catalyst bed at the inlet is 900 K as defined by the parameter and that of insulated foam is about 800 K. However, the temperature dropped drastically to almost 620 K at 0.05m of catalyst bed length as shown in figure 6(b). This happened because the hydrogen formation reaction absorbed heat. However, due to the heating from the heating tubes, the temperature climbs back to 660 K at the outlet of the reformer.

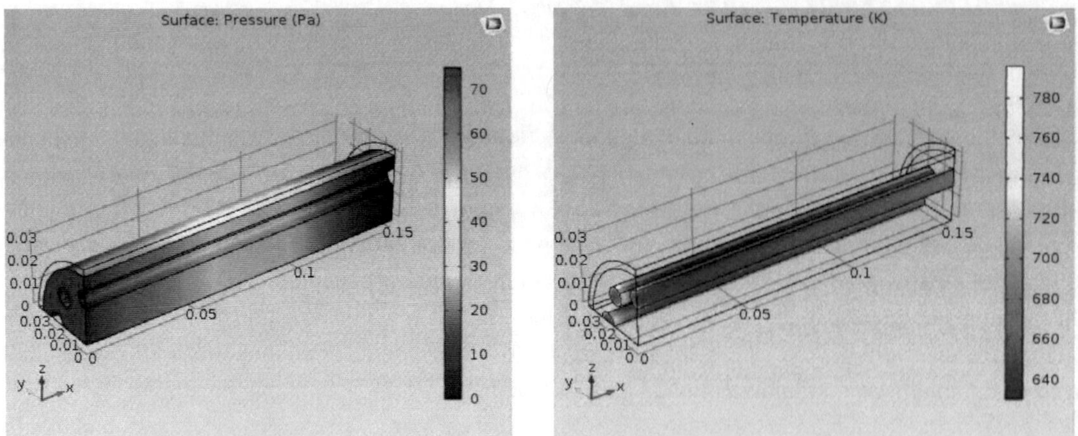

(a) (b)

Figure 6. Pressure (a) and temperature (b) of heating tubes.

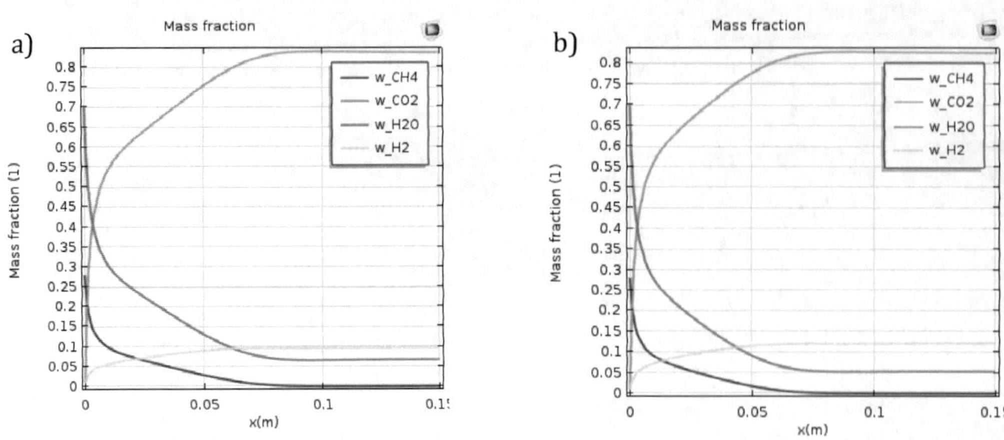

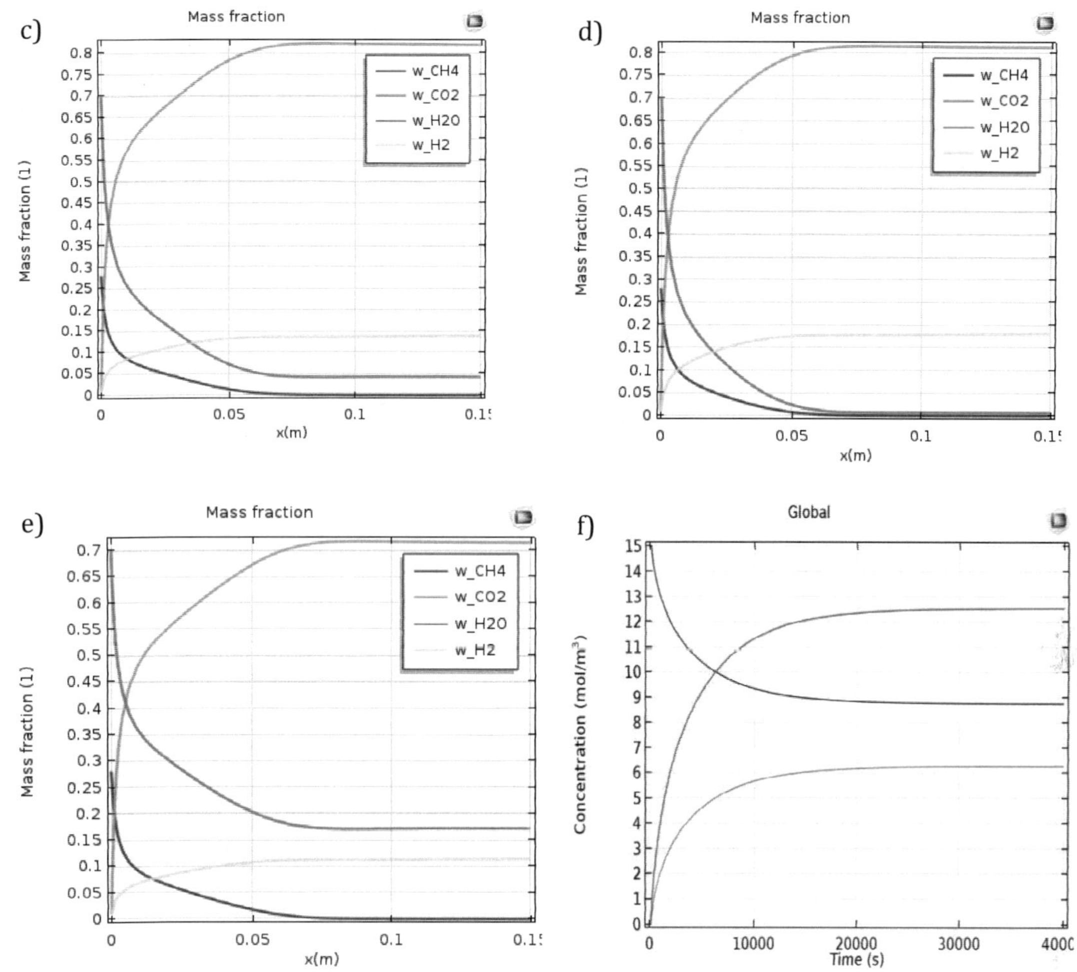

Figure 7. Mass fraction of gas species Pd 100% (a), Pd 70% Graphene 30% (b), Pd 50% Graphene 50% (c), 30% Pd and 70% Graphene (d).

At the inlet of the catalyst bed, the surface density of reacting gases 0.27kg/m^3. As the chemical reaction occurred, the density increased to 0.29 kg/m^3 at 0.05m. After that point, the density decreased gradually throughout the reformer length until 0.19 kg/m^3. The mass fraction of gas species is shown in figure 7(a) – 7(f) for the different composition of palladium and graphene. By considering only Palladium as catalyst, mass fraction of hydrogen is 0.1 and that of carbon dioxide is about 0.85 at the outlet as shown in Fig.7a. Although the mass fraction of methane is 0 at the outlet which means that all methane gases are consumed in this reaction, the mass fraction of water is 0.05. Therefore, there will be some steam left in the outlet products. Considering the Pd 70% and Graphene 30% as catalyst, the mass fraction of H_2 increased to almost 0.13 as show in in Fig.7b. This might be the effect of graphene in the catalyst. Also in this case, all the methan are consumed throughout the length of the catalyst bed. Temperature distribution by the energy exchange between heating pipes and reformed catalyst is investigated at 900 K and 1100 K. It is studied that the steam enters the heating pipes at 900 K and exit at 632 K roughly. However, the exhaust gas enters at 900 K reaching minimum temperature first and eventually exits at average temperature of 680 K. For the 1100K, it can be seen in figure 7(f),

IOP Conf. Series: Materials Science and Engineering **290** (2017) 012052 doi:10.1088/1757-899X/290/1/012052

the mass fraction of hydrogen increases with increasing the reformer heating pipes length, while the other decreases with increasing the length of the reformer.

Figure 8 shows the conversion of each species along the entire length of reformed catalyst. As the exhaust gas travels further along the catalyst bed, the deposition of hydrogen is increased because of faster reaction rate. In the plot for mass fractions, it can be seen that the entire reactor length is active in converting methane. Hydrogen concentration increased as the species goes into the entire reactor. However, the deposited mass fraction of hydrogen is less compared to CO_2 and H_2O due to the molecular mass of each species. The amount of hydrogen production for different catalyst concentration has presented in table 1.

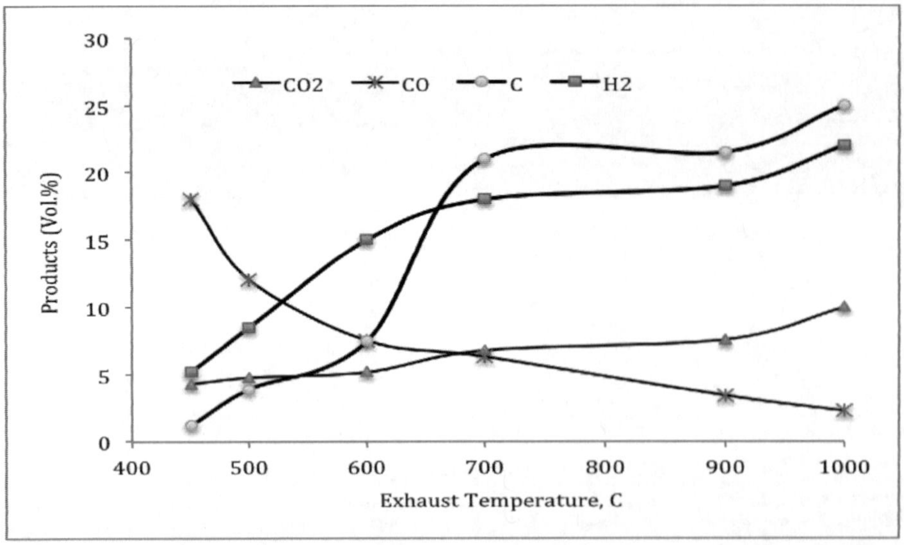

Figure 8. GO /Pd catalyst based steam reformer over the conventional reformer.

Table 1. Amount of H_2 produced and their energy.

Catalyst Concentration	H_2 Production (g) (T=900K)	Energy (Btu)
Pd 100%	8	908.304
Pd 70%/GO30%	10	1135.38
Pd 50%/GO50%	11.2	1271.625
Pd30%/GO70%	14.4	1634.95
GO 100%	8.8	999.13

Table 2. Performance comparison with reformer and without reformer.

Fuel Type		Performance Mileage (km/litre)	Performance Energy Content (Btu)
(i) without reformer			
CNG/kg		22	44589.6
Diesel/litre		20	34214
Petrol (without reformer)/litre		16	30118.89
(ii) with reformer			
	Pd 100%	20.8	39201.93
	Pd 70%/GO30%	22.1	414726.9
Petrol	Pd 50%/GO50%	22.7	42835.14
(with reformer) / litre	Pd30%/GO70%	24.7	46468.39
	GO 100%	21.3	40110.19

3.1 Model validation

The proposed GO/Pd catalyst based steam reformer simulation performance has been compared with the car of compact natural gas (CNG), diesel engine and petrol engine with and without reformer in terms of mileage, and energy contents. The comparison results has been presented in table 2. By utilizing the reformer with appropriate catalyst composition, the mileage of petrol-type vehicle can be improved by 8.685 km, which is 54.28% of the normal mileage per litre of petrol. By selecting the Pd 30%/GO70% catalyst composition, the amount of hydrogen obtained is 14.4 g and the energy content is 1634.95 Btu per mol of CH_4.

3.2 Influence of Pd/GO reformer on engine performance

Typically hydrogen engines are designed to use about twice as much air as theoretically required for complete combustion. At this air/fuel ratio, the formation of NOx is reduced to near zero. Unfortunately, this also reduces the power output to about half that of a similarly sized gasoline engine. To make up for the power loss, hydrogen engines are usually larger than gasoline engines, and/or are equipped with turbochargers or superchargers. The turbocharger has been proposed for the engine with GO/Pd Catalyst based steam reformer (figure 2). This study focuses to turn the waste energy into wealth by converting exhaust into hydrogen. Without interrupting the engine size the hydrogen is pumped to the engine about 7% of total gasoline injected into the engine. It is expected that the 7% of hydrogen, which not only reduces the fuel demand it also improve the combustion of air-fuel mixture by making the fuel atomization and vaporization into the combustion chamber. It is reported that the engine performance can be enhance about 7-10% by atomizing and vaporizing the fuel (Rahman et al., 2015 and Biplah, 2013). The combustion reaction of air/fuel mixture if the hydrogen is supplied 7% of C_8H_{18} (gasoline), the A/F (including 93% of C_8H_{18} and 7% of H_2):

$$0.93C_8H_{15} + 1.6H_2 + 11.75O_2 + 11.75(3.76)N_2$$
$$\Rightarrow 8CO_2 + 7.5H_2O + 11.75(3.76)N2$$

where, x is the percentage of reduction of fuel with adding x percentage of H_2. The calorific value of gasoline (C_8H_{18}) is 47 MJ/kg while hydrogen (H_2) is 141.9 MJ/kg. The 7.9 g of gasoline can be replaced with 3.2 g of hydrogen (1.6 moles of H_2). The air fuel ratio can be estimated:

$$0.93x8\,moles\,of\,C\times 12g/\,moles + 0.93x7.5\,moles\,of\,H_2\times 2g\,/\,moles = 103.23g$$
$$1.6\ moles\ of\ H_2 = 1.6\,moles\,of\,hydrogen\times 2\,g\,/\,moles = 3.2\,g$$
$$Mass\,of\,fuel = 103.23 + 3.2 = 106.43g$$
$$Mass\ of\ air = (1+3.76)(11.75)(29) = 1621.97g$$
$$The\ air\,/\,fuel\,(gasoline + hydrogen) = \frac{Mass\,of\,air}{Mass\ of\ fuel} = \frac{1621.97}{106.43} = 15.24$$

The 1.6 mole of hydrogen pumping to the engine increases the air intake into the engin, which makes the stoichiometric A/F ratio 15.27:1 while for 100% gasoline it is 14.67:1. This indicates that the hydrogen pumping can make the engine energy efficient by saving the engine fuel pumping losses and enhance the vehicle mobility with improving emission [17].

4. Conclusions

The proposed GO/Pd catalyst steam reformer can promote the production of clean fuel hydrogen from reforming of raw exhaust gas of automobiles and steam mixture. The principle of the exhaust reforming process can be utilized the hydrogen generation from waste of engine which can be used as clean fuel into IC engines, thus improving the fuel consumption, exhaust emissions and efficiency of the engine.

- Pd/GO catalyst steam exhaust reformer be able to generate hydrogen 23-33% for the exhaust temperature in the range of 300-1100 °C.
- The temperature distribution of the catalyst bed is in the range of reasonable working temperature 680-1100 K .
- Pumping H_2 about 7% of total fuel at temperature about 80-90 °C may boosted the fuel atomization and vaporization at idle condition could enhance the fuel combustible efficiency and engine efficiency in turns to increase the engine fuel consumption and reduces emission.
- The emission of NOx could be formed into the engine due to complete combustion. The temperature of the reformer will drop due to the presence of steam and the oxidation in presence of GO catalyst the NOx will be turn into N_2 and $3O_x$.
- The catalyst Pd and GO both are expensive for the reformer. However, it is very much reasonable to consider them to make the reformer efficient and compact.

References

[1] Wyszynski M L, Megaritis T and Lehrle R S 2017 *Hydrogen from Exhaust Gas Fuel Reforming: Greener, Leaner and Smoother Engines* (University of Birmingham)
[2] Abu Jrai A, Tsolakis A, Megaritis A and Glounski S E 2008 Chem. Eng. J. **141** 290-297
[3] Kalamaras C M and Efstathiou A M 2013 *Conference Papers in Energy* vol. 2013 p 9
[4] Takayanagi K 2015 *U.S. Patent No. US 20150231569 A1* (Washington, DC: U.S. Patent and Trademark Office)
[5] Rahman A, Kaw M A and Rafia A 2017 *Int. J. Eng. Technol. IJET-IJENS* **17** 13-24
[6] Hoffman A 1999 *Acad. Manage. J.* **42** 351-371
[7] Sheldon R A 2007 *Green Chemistry* **9** 1273-1283
[8] Lee M, Kim B, Lee Y, Kim B and Park J B 2014 *Bull. Korean Chem. Soc.* **35** (South Korea)
[9] Vaccaro A L, Lanari D, Marrocchi A,and Strappaveccia G 2014 *Green Chemistry* 16 3680-3704.
[10] Jing Q, Liu W, Pan Y, Silberschmidt V V, Li L and Dong Z L 2015 *Mater.Des.* **85** 808–814 (Singapore)
[11] Brinkley K W 2015 *The Synthesis of Solid Supported Palladium Nanoparticles: Effective Catalysts for Batch and Continuous Cross Coupling Reactions* (Unpublished doctoral

dissertation)
[12] Zhu Y, Murali S, Cai W, Li X, Suk J W, Potts J R and Ruoff R S 2010 J. Adv. Mater. 22 3906-3924
[13] Brinkley K W 2015 *The Synthesis of Solid Supported Palladium Nanoparticles: Effective Catalysts for Batch and Continuous Cross Coupling Reactions* (Unpublished doctoral dissertation)
[14] Rahman A, Razzak F A, Hawlader M N A and Rashid M 2013 *J. Renew. Sustainable Energy* **5** 1-13
[15] Rahman A., Fadhilah R., Rafia A, Mohiuddin A K M and Hawlader M N A 2015 *J. Renewable Sustainable Energy Rev.* **51** 382–395
[16] Rahman, A, Yahya A and Mohiuddin A K M 2007 *P. I. Mech. Eng. D-J. Aut.* **221** 789-800

The effects of excess calcium on the handling and mechanical properties of hydrothermal derived calcium phosphate bone cement

N N Razali[1], M A Sukardi[1], I Sopyan[1], M Mel[2], H M Salleh[2] and M M Rahman[3]

[1]Department of Manufacturing and Materials Engineering, Faculty of Engineering, International Islamic University Malaysia, 50728 Kuala Lumpur, Malaysia.
[2]Department of Biotechnology Engineering, Faculty of Engineering, International Islamic University Malaysia, 50728 Kuala Lumpur, Malaysia.
[3]Department of Mechanical Engineering, Faculty of Engineering, Universiti Tenaga Nasional, 43000 Kajang, Selangor.

sopyan@iium.edu.my

Abstract. The objective of this study is to determine the effects of excess calcium on the handling and mechanical properties of hydrothermal derived calcium phosphate cement (CPC) for bone filling applications. Hydroxyapatite powder was synthesized via hydrothermal method using calcium oxide, CaO and ammonium dihydrogen phosphate, $NH_4H_2PO_4$ as the calcium and phosphorus precursors respectively. The effects of calcium excess were evaluated by varying the CaO content at 0, 5 and 15 mole %. The precursors were then refluxed in distilled water at 90-100°C and dried overnight until the calcium phosphate powder was formed. CPC was then produced by mixing the synthesized powder with distilled water at the powder-to-liquid (P/L) ratio of 1.5. The result from the morphological properties of CPC shows the increase in agglomeration and particles size with 5 mole % of calcium excess but decreased with 15 mole % of calcium excess in CPC. This result was in agreement with the compressive strength result where the CPC increased its strength with 5 mole % of calcium excess but reduced with 15 mole % of calcium excess. The excess in calcium precursor also significantly improved the setting time but reduced the injectability of CPC.

1. Introduction

Calcium phosphate has been studied as bone repair materials over the decades. The use of calcium phosphate biomaterials, especially hydroxyapatite (HA) has gained popularity because attributed to the chemical similarity with the mineral components of natural bone and teeth [1]. Besides that, it also shows excellent bioactivity, biocompatibility and osteoconductivity for bone implant substitutes [2].

There are wide range of powder-processing techniques to synthesis calcium phosphate biomaterials with the aim to produce distinct particles morphology of HA [3]. Numerous synthesis techniques of HA powder have been developed such as solid state reaction, hydrothermal, hydrolysis, sol-gel and wet-chemical precipitation [4]-[6]. These preparative routes could affect the level of crystallinity, stoichiometry and morphology of synthesis HA. Among these methods, the hydrothermal method is one of the preferable technique due to the high yield of cement production, controlled morphology and stoichiometric Ca/P ratio [7].

For bone filling applications, calcium phosphate cement (CPC) is more attractive compared to HA powder because the ability to mould and shape to fill intricate bone cavities or narrow dental defect sites [8]. CPC element composed of ceramic compounds of calcium and phosphate that may be hardened upon contact with water or aqeous solution. There are two types of CPC: apatite transforming cement like HA, carbonated HA or calcium deficient HA (CDHA), and brushite tranforming cement like dicalcium phosphate dihydrate (DCPD) [9]. The apatite cement shows better mechanical properties but lower degradabality than the brushite cement [10].

The main advantage of CPC is the handling properties which can be injected and shaped perfectly in situ according to the defect dimension. Investigation on the effects of excess calcium precursor on the injectability, setting time, microstructure and compressive strength of calcium phosphate cement by employing low temperature hydrothermal technique is presented in this work.

2. Experimental procedures

2.1 Synthesis of powder and cement preparation.

The calcium phosphate cement (CPC) was produced by mixing the hydrothermal derived hydroxyapatite (HA) powder with distilled water. To synthesize HA powder, calcium oxide granules, CaO (Bendosen Laboratory Chemicals) and ammonium dihydrogen phosphate, $NH_4H_2PO_4$ (Friendmann Schmidt Chemical) were used as the precursors and distilled water was used as the solvent. The synthesis route of HA powder via hydrothermal method was reported elsewhere [11]. The chemical reaction to synthesize HA powder is as follows:

$$5CaO + 3NH_4H_2PO_4 \rightarrow Ca_5(PO_4)_3OH + 3NH_4OH + H_2O \quad (1)$$

Based on equation (1), the stoichiometric composition of hydroxyapatite (HA) produced is referred to 1.67 Ca/P ratio. Three synthesized powder were prepared by varying the CaO content: 0, 5 and 15 mole %. The synthesized HA was then dried overnight at 85°C in an oven until the dry powder was produced and then crushed. CPC was produced by mixing the powder and liquid phases homogenously for 3 min at the (P/L) ratio of 1.5. The as-synthesized powder was used as the powder phase, while distilled water was used as the liquid phase. Three different types of cements were prepared based on the variation of calcium excess labelled as CPC-0, CPC-5 and CPC-15. The samples were then fabricated by filling the cement into a teflon (PTFE) mould of 10 mm diameter and 15 mm height. After hardening, the cements were demoulded and left dried in room temperature overnight before characterization. Figure 1 shows the flowchart of the method to produce CPC with calcium excess.

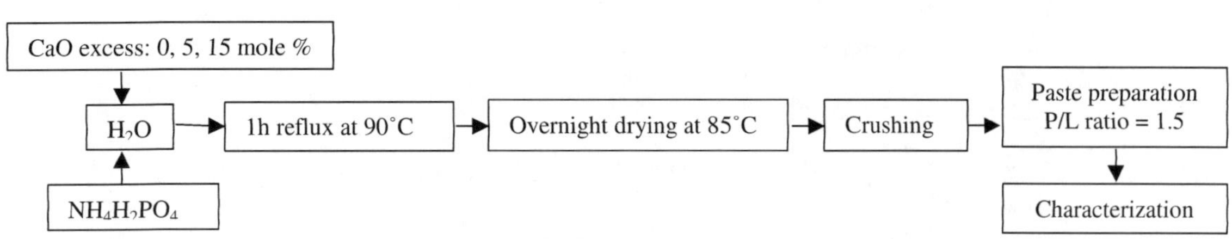

Figure 1. Flowchart of the methods to produce CPC-0, CPC-5 and CPC-15.

2.2 Physical and mechanical characterization

The microstructure analysis of the CPC samples with variation of calcium excess was performed by FESEM (JEOL, JSM 6700F) with a deposition current of 60mA and sputter coating using gold for

50s. Different acceleration condition and voltage were applied to the sample to observe the microstructure of samples.

2.3 Injectability test

The injectability test was done through extrusion of 5 ml non-needle plastic syringe with 2.68 mm inner diameter of canula filled with paste as the working sample. The injectability results of CPC with 0, 5 and 15 mole % of calcium excess were evaluated by observing the extrusion paste of the samples. The syringe was filled with mixture of as-synthesized powder and distilled water at P/L ratio of 1.5 before extruded using a Lloyd Universal Testing Machine, LR 10 K+ model. The crosshead speed and maximum load was constant for all samples at 50mm/min and 300 N, respectively. The injectability results was measured by recording the evolution of the extrusion force and extruding time of the syringes.

2.4 Setting time test

To identify the setting time, Gillmore needles method was used to measure the CPC with variation of calcium excess of 0, 5 and 15 mole %. The Gillmore method was done by following the ASTM C266-08 designation method [12]. After homogenously mixed, the CPC paste was moulded into a PTFE (teflon) mould. The initial and final setting times were determined by holding the light and heavy needles in vertical position and applied to the cement surface until no visible indentation formed.

2.5 Mechanical test

The compression test was done using the Lloyd LR 10 K+ Universal Testing Machine on 10 mm diameter x 15 mm length specimens at 1 mm/min crosshead rate for all samples of CPC.

3. Results and Discussion

The variation of calcium excess has shown some effects on the handling and mechanical properties of hydrothermal derived CPC. Figure 2 shows the extruded paste after injectability test was done on three samples (CPC-0, CPC-5 and CPC-15) with 0, 5 and 15 mole % of calcium excess in CPC, respectively. The P/L ratio and mixing time of all samples were constant at 1.5 and 3 min. CPC-0 and CPC-5 from Figure 2(a) and 2(b) exhibited excellent injectability because the whole paste extruded from the syringe. However, extrusion paste from CPC-0 is continuous compared to CPC-5 due to the reduction of water content of CPC-5 after more calcium content appeared in the paste. The CPC-5 paste started to increase its viscosity because of lack in water content which gives discontinuous form of paste. Figure 2(c) shows sample from CPC-15 where some paste has hardened and cannot be extruded out of the syringe after injected. Overall, increase in calcium excess of CPC can reduce the injectability performance of cement.

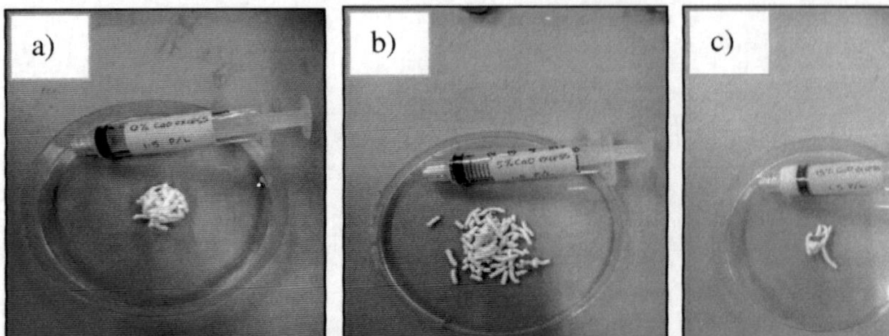

Figure 2. Extruded paste of samples after injected from the syringe (a) CPC-0, (b) CPC-5 and (c) CPC-15.

Figure 3 presents the initial and final setting times of CPC with the effects of calcium excess. The calcium excess has significantly shortened both the initial and final setting times of CPC. The initial setting time has reduced to 40, 30 and 15 min with 0, 5 and 15 mole % of calcium excess, respectively while the final setting time was 450, 360 and 210 min for CPC-0, CPC-5 and CPC-15, respectively. It can be concluded that the higher calcium excess can increase rate of self-hardening of CPC. This result is in agreement with injectability result where increase in calcium excess can harden the cement faster and reduced injectability.

The results of microstructure analysis using FESEM was shown in figure 4. The particles were in spherical and flakes shapes which represent HA powder. The powder also formed agglomerated particles of a needle like morphology. The particles size increased as the amount of calcium excess increased to 5 mole % but decreased with 15 mole % of calcium excess. Between three samples, CPC-5 shows the highest amount of agglomerated particles where all particles interconnected each other. This is in agreement with compressive strength results shown in figure 5. The highest compressive strength achieved by CPC-5 with 1.682 MPa compared to the compressive strength of CPC-0 and CPC-15 with 1.638 MPa and 1.508 MPa, respectively. This results indicated that, the increase in agglomerated particles may increase the strength of the cement. The agglomeration of particles also the main reason of poor injectability performance of CPC [13]. The mixture with higher calcium excess can be sticky on each other due to the inter-particles attractive force which may harden the cement faster [14].

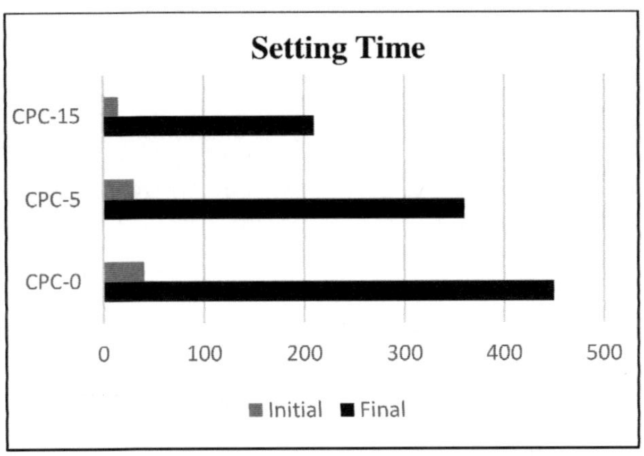

Figure 3. Effect of calcium excess on the setting time of CPC.

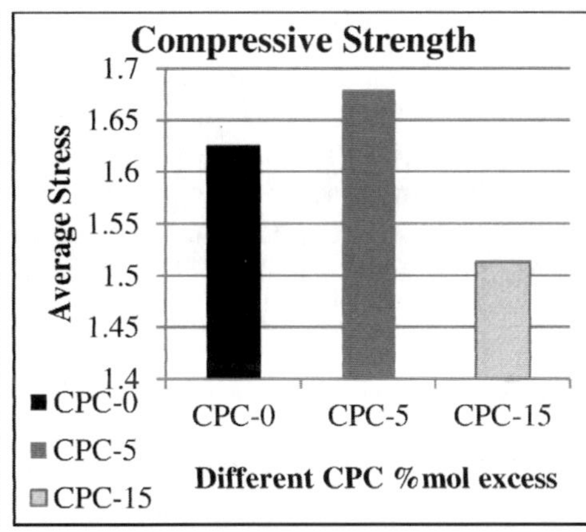

Figure 4. Effect of calcium excess on the compressive strength of CPC.

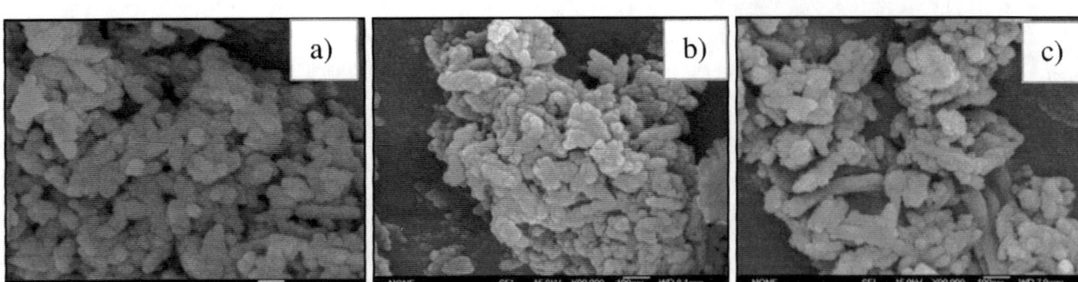

Figure 5. Microstructure of CPC with different amount of calcium excess (a) CPC-0 (b) CPC-5 and (c) CPC-15.

4. Conclusion

The effects of excess calcium on the injectability, setting time, morphology and compressive strength of hydrothermal derived CPC was successfully studied. The individual particles size and agglomeration increased with 5 mole % of calcium excess but decreased after 15 mole% of calcium excess in CPC. Increased the calcium content up to 5 mole % also increased the compressive strength in range of 1.638 to 1.682 MPa but decreased to 1.508 MPa with 15 mole % of calcium excess. The results also indicated that, both the initial and final setting times of all samples were improved as the calcium content increased. The initial setting time was 40, 30 and 15 min while the final setting time was 450, 360 and 210 min, respectively for CPC-0, CPC-5 and CPC-15. However, increase in calcium content of CPC could reduce the injectability performance of CPC due to lack of water content of CPC and increase in the rate of self-setting of cement. Overall, injectability is proportional to the hardenability of cement in term of setting.

Acknowledgement
This work was financially supported by Ministry of Higher Education (MOHE) through Fundamental Research Grant Scheme (FRGS) with the project ID of FRGS15-246-0487.

References

[1] Kokosha MS, Friedman CD, Castellano RD and Constantino PD 2004 *Arch Facial Plast Surg* **6** pp 290
[2] Sariibrahimoglu K, Leeuwenburgh SCG, Wolke JGC and Yubao L 2012 *J Biomed Mater Res A* pp 712
[3] Ramesh S, Tan CY, Bhaduri SB, Teng WD and Sopyan I 2008 *J Mater Process Technol* **206** pp 221
[4] Pramanik S, Agarwal A K, Rai K N and Garg A 2007 *Ceram Int* **33** pp 419
[5] Zhang C, Yang J, Quan Z, Yang P, Li C, Hou Z and Lin J 2009 *Cryst Growth Des* **9** pp 2725
[6] Bezzi G, Celotti G, Landi E, La Toretta TMG, Sopyan I and Tampieri A 2003 *Mater Chem And Phys* **78** pp 816
[7] Liu J, Ye X, Wang H, Zhu M, Wang B and Yan H 2003 *Ceramics International* **29** pp 629
[8] Manoj K and Varma KH 2003 *Bull Mater Sci* **4** pp 415
[9] Driessens FCM, Boltong MG, Bermudez O, Planell JA, Ginebra MP and Fernandez E 1993 *J Mater Sci : Mat in Med* **3** pp 164
[10] Bohner M and Gbureck U 2007 *J Biomed Mater Res B* **84B** pp375
[11] Alqap ASF and Sopyan I 2009 *Indian J Chem* **48A** pp 1492
[12] Zhang J, Liu W, Schnitzler V, Tancret F and Bouler J M 2013 *Acta Biomaterialia* **10** pp 1035
[13] Bohner M 2007 *J Mater Chem* **17** pp 3980
[14] Gbureck U, Barralet JE, Hofmann M and Thull R 2004 *J American Cer Soc* **87** pp 311

Energy Harvesting from Aerodynamic Instabilities: Current prospect and Future Trends

M Bashir[1], P Rajendran[1], S A Khan[2]

[1]School of Aerospace Engineering, Universiti Sains Malaysia, 14300 Pulau Pinang, Malaysia
[2]Department of Mechanical Engineering, International Islamic University Malaysia (IIUM), Jalan Gombak, 53100 Kuala Lumpur, Malaysia

Email:aeparvathy@usm.my

Abstract. This paper evaluates the layout and advancement of energy harvesting based on aerodynamic instabilities of an aircraft. Vibration and thermoelectric energy harvesters are substantiated as most suitable alternative low-power sources for aerospace applications. Furthermore, the facility associated with the aircraft applications in harvesting the mechanical vibrations and converting it to electric energy has fascinated the researchers. These devices are designed as an alternative to a battery-based solution especially for small aircrafts, wireless structural health monitoring for aircraft systems, and harvester plates employed in UAVs to enhance the endurance and operational flight missions. We will emphasize on various sources of energy harvesting that are designed to come from aerodynamic flow-induced vibrations, specific attention is then given at those technologies that may offer, today or in the near future, a potential benefit to reduce both the cost and emissions of the aviation industry. The advancements achieved in the energy harvesting based on aerodynamic instabilities show very good scope for many piezoelectric harvesters in the field of aerospace, specifically green aviation technology in the future.

1. Introduction

Energy harvesting (EH) deals with the energy conversion from external driving sources (like solar power, thermal energy, vibrational energy, mechanical deformation, etc.) into electrical energy which can then be captured and stored for small autonomous devices like those used in wearable electronics and wireless sensor networks [1]. The basic concept of EH technology is to obtain the electrical energy by energy conversion from a wasted heat, vibration, mechanical deformation, potential energy, etc. The general mechanism of this EH is depicted in figure 1. The energy converted derived from these ambient sources is either directly consumed, or stored in a rechargeable battery for future use.

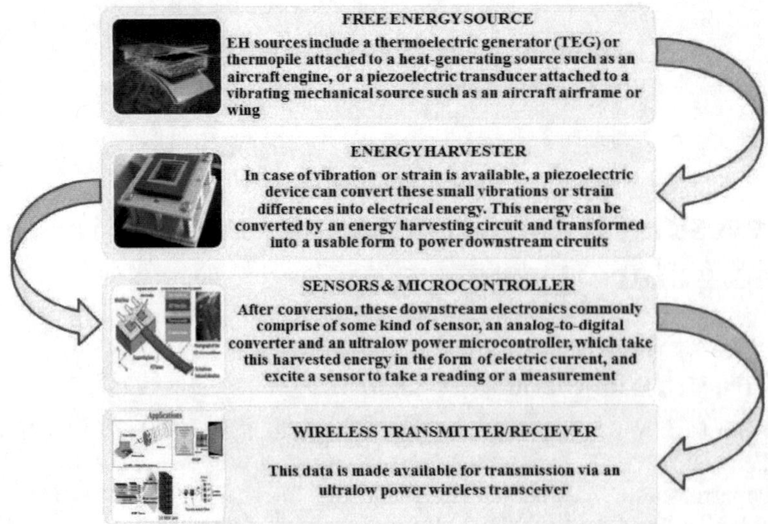

Figure 1. Schematic of general mechanism of energy harvesting principle.

One study has outlined three energy architecture networks as harvest-consume, harvest-store-consume, and harvest-store/consume to explain the final utilization of the harvested energy [2]. Each of the energy conversion methods used in EH has its own performance characteristics such as power density described by the power-to-weight, power-to-area or power-to-volume ratio. Therefore, due to its substantial applications, many investigators have been studying EH technology in various fields such as material science, electrical engineering, mechanical engineering, biomedical science, etc.

Moreover, in the past decade, many research investigations have been concerned with harvesting energy from base vibrations [3]. Hence, being such a fascination for research in EH from aeroelastic vibrations [3], the contemporary research focus of aeroelastic EH need to be reviewed. Recent developments in aeroelastic EH have been comprehensively penned by many authors, including the development and applications of piezoelectric (PZT) harvesters in aerospace applications. The significant progresses of EH capability of aeroelastic phenomena have been attracting engineers and researchers from different fields [3,4].

Aerospace EH application is one of these interested fields because billions of dollars are lost each year as maintenance cost, fuel consumption due to additional weight, and emissions from different systems. The high-efficient energy harvesters may convert vibrations, acoustic and dissipated heat into electrical energy. The potential for energy savings is enormous. The impact of energy efficiency improvement of aircraft systems on the environment has been highlighted by Lee [5].

As the conventional sources of energy need to be replaced by the renewable sources due to limited stock and high emission of pollutants, Boeing has started testing solar and thermal EH to power electronic windows, as a way to reduce wiring, weight, fuel use and carbon emissions. The research of EH has enhanced the possibility of green energy. One promising way for EH is vibrations, and hence our study will enumerate various EH techniques from different aerodynamic instabilities, that can employ different transduction methods to transfer these instability phenomena into electrical energy.

2. Aerodynamic instabilities and energy harvesting

Structures subjected to flow loads may exhibit different responses which include bifurcations, limit-cycle oscillations, internal resonances and chaotic motions [4,6]. Aerodynamic phenomena like vortex-induced vibrations, flutter, buffeting, and galloping, may cause unwanted and excessive vibrations which are to be avoided in large scale applications such as bridges, aircrafts, pipelines, buildings etc. Huge emphasis laid on research to design structures which are capable of sustaining damages.

When aeroelastic instabilities occur, the phase difference between the motions of the structure and the aerodynamic forces becomes such that the structure absorbs energy from the air leading to structural vibrations with increasing amplitude for which failure is the catastrophe. The aeroelastic phenomena were classified according to their fundamental causes and major characteristics as shown in fig. 2, and such a classification is also proposed in the general book by Naudascher and Rockwell dealing with flow-induced vibrations [7].

The three types of excitations as defined in figure 2 are 1) extraneously induced excitation (EIE); independent of any fluid instability due to the structure or its motion, e.g. turbulence in the oncoming flow-field or influence of the wake of an obstacle located upstream, 2) instability induced excitation (IIE); flow instability due to the presence of the structure, e.g. alternate vortex shedding (in the sense of Von Karman), or buffeting loads, and 3) motion induced excitation (MIE); fluctuating forces due to the motion of the structure. It leads to self-excited phenomena where energy transfers from the flow-field to the structure, e.g. flutter.

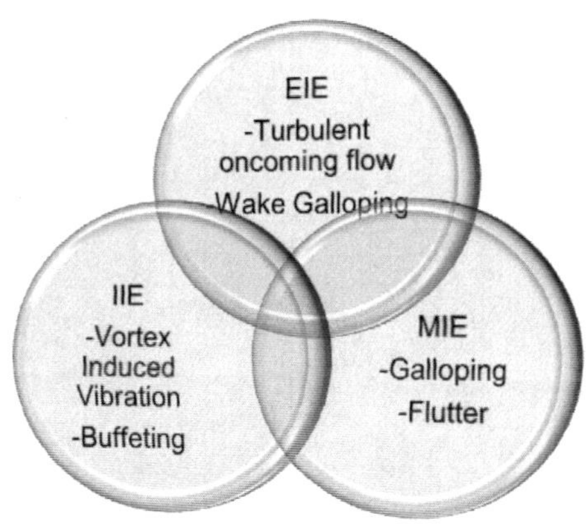

Figure 2. Classification of the aeroelastic instabilities of bluff-bodies.

In the literature of aero-elasticity, PZT materials (as actuators, converse effect) and other actuators have been used as active and semi-passive controllers to modify the aero-elastic behaviour of wings along with the effects of passive controllers have also been investigated by several researchers to increase the linear flutter speed. Researchers have also used PZT materials as actuators for morphing in aircrafts and macro-fibre-composite actuators have been used to suppress buffeting oscillations on the vertical fins of an F-18. On a flip side, these aerodynamic phenomena associated with structural nonlinearities have been proposed as a new source of power generation in small scale systems which are classified as distributed and concentrated based on the region of their existence.

Distributed nonlinearities arise generally from deformations of the entire structure whereas concentrated nonlinearities arise from loose or worn hinges of control surfaces. To harvest energy, the harvester is placed in a flow field and excited to undergo large limit-cycle oscillations that can be converted to electrical energy using PZT and/or electromagnetic transducers. Thus, findings of this research in chronological order; flutter in airfoil sections, vortex-induced vibrations in circular cylinders, galloping in prismatic structures, and wake galloping in parallel cylinders. The following subsections will enlighten about the contemporary investigations in the field of EH from aerodynamic instabilities and discuss possible designs of galloping (also wake galloping), flutter and vortex induced vibrations (VIV).

2.1 Energy Harvesting by Galloping

Galloping is a typical phenomenon of aeroelastic instability involving lightweight and lightly damped slender structures, usually with non-circular, bluff cross sections. It can be characterized as a velocity-dependent, damping-controlled instability problem with low-frequency, large-amplitude oscillations, mainly in the crosswind direction. Isolated structural elements, mono-tubular towers and cables subjected to icing conditions, lighting poles with heavy equipment at the top, tall buildings and high-rise structures are typical examples of structures susceptible to galloping when the mean wind velocity exceeds certain critical values.

Xu et al [8] investigated the EH by the aeroelastic galloping of a cantilever with attached prism, and found that the electric power output depends primarily on the length of the cantilever. As shown in figure 3, a PZT wake-galloping FEH consists of a mechanical oscillator coupled to an EH circuit through a PZT element. The oscillator is placed in the downstream of a bluff body or an obstacle. Laminar fluid flows once past the obstacle, undergoes symmetry breaking in the form of a Von Karman vortex sheet with periodic shedding vortices from the trailing edge thereby inducing alternating pressure forces which, in turn, generate a periodic lift on the mechanical oscillator.

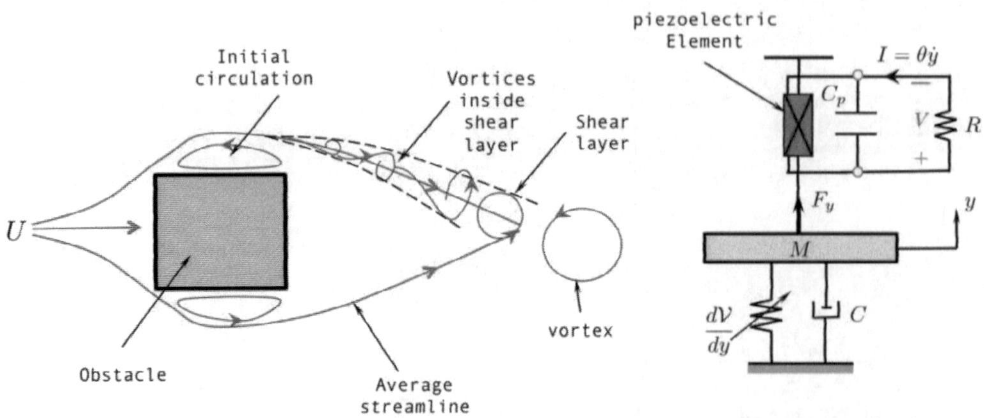

Figure 3. A schematic of a lumped parameters model of a nonlinear wake galloping energy harvester [9].

Alhadidi et al. [9] communicated that the lock-in region of a wake-galloping flow energy harvester can be improved by exploiting a bi-stable restoring force as shown in fig. 3. The results depicted the comparison with the linear design; bi-stability can be used to improve the steady-state bandwidth considerably. Raj et al [10] presented a vertical cantilever PZT energy harvester that exploits wind energy for vibration. It was concluded that the voltage, power output and onset of galloping wind force are significantly influenced by the attached load resistance.

Abdelmaula et al [11] investigated the potential of the electrical capacitance and inductance on the performance of galloping-based PZT systems for EH and control purposes. The results indicated that these experiments were not fruitful for energy harvesting. Javed et al [12] introduced an enhanced stability characterization for aero-elastic energy harvesters by using both the normal form of the Hopf bifurcation and shooting method. It is shown that this tool is strong in terms of designing reliable aero-elastic energy harvesters. The results show that this technique can accurately predict the harvester's response only near bifurcation, however, cannot predict the stable solutions of the harvester when subcritical Hopf bifurcation takes place. Then shooting method is employed which in turn proved to be beneficial in determining the stable and un-stable solutions of the system and associated turning points by covering draw backs.

Dai et al [13] presented an investigation on the suppression of vibration amplitudes of an elastically-mounted square prism subjected to galloping oscillations by using a non-linear energy sink. The results shows that this approach can be efficiently implemented to significantly reduce the galloping amplitude of the square prism. Wang et al [14] investigated the EH from both vortex-induced vibration and galloping phenomenon from self-excited vibration of a square cylinder under different wind speeds. For VIV, the maximum harvested power reached $1.14X10^{-5}$ watt at 1 $X10^{6}$ ohm resistance. For galloping, the harvested power is 1.4 watt at 2 $X10^{5}$ ohm resistance. It can be concluded that synchronization phenomenon can be utilized to harvest power when the velocity of wind is low.

2.2 Energy harvesting by Flutter

Flutter is an aeroelastic instability involving 2 degrees of freedom, one flap wise and one torsional. It is a self-feeding and potentially destructive vibration where aerodynamic forces on an object couple with a structure's natural mode of vibration to produce rapid periodic motion. Flutter can occur in any object within a strong fluid flow, under the conditions that a positive feedback occurs between the structure's natural vibration and the aerodynamic forces. That implies the vibrational movement of the object increases an aerodynamic loads which in turn drives the object to move further.

If the energy during the period of aerodynamic excitation is larger than the natural damping of the system, the level of vibration will increase. The vibration levels can thus build up and are only limited when the aerodynamic or mechanical damping of the object match the energy input; this often results in large amplitudes and can lead to rapid failure. Thus, structures exposed to aerodynamic forces including wings, aero-foils, but also chimneys and bridges are designed carefully within known parameters to avoid flutter.

Piñeirua et al [15] in particular investigated the energy transfer between a flow and a fluttering PZT plate as shown in figure 4, and benefits of the use of a synchronized switch harvesting on inductor (SSHI) circuit. The results presented show that a significant improvement of the harvested energy can be obtained using SSHI circuits compared to basic resistive circuits. Wang et al [16] proposed a novel structure for pre-rolled flexible PZT cantilevers that use wind energy to power a sub-munition electrical device. The PZT generator is composed of a PZT cantilever and a power management circuit with wind speed set to 40 m/s to ensure that the PZT cantilever would flutter at large amplitude, thereby ensuring that it could generate electrical energy.

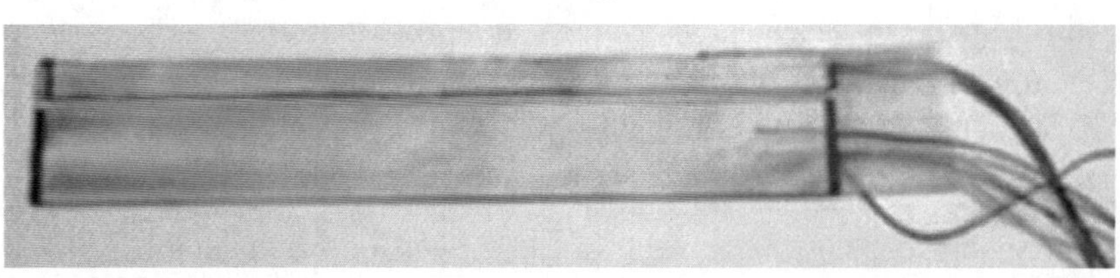

Figure 4. Schematic of the multilayer composition of the piezoelectric flag [15].

IOP Conf. Series: Materials Science and Engineering **290** (2017) 012054 doi:10.1088/1757-899X/290/1/012054

Akbar et al [17] presented an investigation on the EH exerted by the dynamic bending responses of a PZT embedded wing-box. A simulation for civil jet aircraft wing-box with PZT skin layer has been presented. Flutter is unlikely to be encountered during normal flight of civil jet aircraft and the structure constructed from more complicated configuration, i.e. skins, ribs and spars. Based on simulation results, the voltage and power responses could attain a promising level, in the order of 101–102 volts and 102–104 watts.

Hobeck and Inman [18] presented an analytical and experimental investigation of a flow-induced vibration phenomenon referred to as dual cantilever flutter (DCF). DCF occurs when two similar adjacent beams are placed normal cross-flow at a particular velocity and separation distance. Experimental results show that during DCF both beams become phase-locked 180 degrees out-of-phase while they undergo persistent large-amplitude vibration at their first bending mode frequency.
Bruni et al [19] presented an investigation to identify the best configuration of PZT elements for a typical condition of wing aeroelastic instability with attention to flutter behaviour of the structure. Best results are achieved with optimisation of frequency ratio, stiffness ratio, and load resistance value for flutter postponement and energy harvesting.

McCarthy et al [20] investigated the response of a fluttering PZT energy harvester in smooth flow and in aspects of replicated atmospheric boundary layer turbulence (12.7% intensity, 310-mm longitudinal integral length scale). The harvester was yawed and pitched, and the effects on the power output were examined. Key findings were the following: (1) off-axis flow conditions rapidly degraded the mean output power of the harvester; (2) turbulence, for this specific harvester, acted similarly to a dynamic damping mechanism.

2.3 Energy Harvesting by Vortex Induced Vibrations
Vortex induced-vibration is probably the most classical type of wind flow-induced vibration that takes place in flow velocities specified by the Strouhal number. In this flow, vortices are created at the back of the body and detach periodically from either side. Vortex-induced vibration is unlikely to be a major vibration problem for cable-stayed bridges. By adding a small amount of damping, vortex excitation can be suppressed effectively. When an elastic bluff body encounters uniform fluid flow, for high enough Reynolds numbers (RN) (RN > 100), the flow separates from the body surface and generates an unsteady broad wake. This wake is recognized by two vortices in each side of the body which are shed into the current, periodically. This periodic vortex shedding generates asymmetric pressure distribution around the body that provides periodic forces which consequently lead to a limited amplitude vibration in the body.

Zhang and Wang [21] investigated a rigid circular cylinder with two PZT beams attached on through vortex-induced vibrations (VIV) and wake-induced vibrations (WIV) by installing a big cylinder fixed upstream as seen in fig. 5, in order to study the influence of the different flow induced vibrations types. The VIV test shows that the output voltage increases with the increases of load resistance; an optimal load resistance exists for the maximum output power. The WIV test shows that the vibration of the small cylinder is controlled by the vortex frequency of the large one.

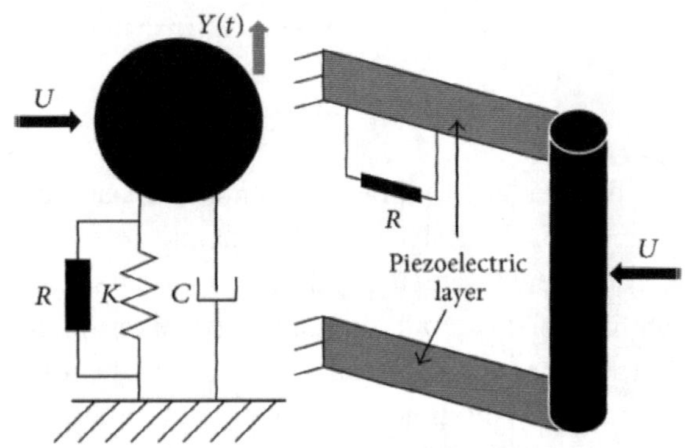

Figure 5. Schematic of a piezoelectric energy harvester [21].

Dai et al [22] investigated the characteristics and performances of four distinct VIVs PZT energy harvesters. Experiments show that the synchronization regions of the bottom, top, and horizontal configurations are almost the same at low wind speeds (around 1.5 m/s). The vertical configuration has the highest wind speed for synchronization (around 3.5 m/s) with the largest harvested power, which is explained by its highest natural frequency and the smallest coupled damping. Dai et al also [23] studied the mitigation of VIVs of a circular cylinder by utilizing a passive nonlinear targeted energy transfer. The chaotic vibrations of a VIV energy converter enhanced by a bistable stiffness element.

Wen et al [24] proposed a miniaturized VIV fluid energy harvester. A fluid-structure interaction simulation was carried out to optimize the structure of the energy harvester. As a result, the efficiency is greatly increased, which is experimentally proved. The demonstrations are working in the airflow velocity of 2 m/s, and reaches the maximum efficiency at 3.6 m/s. Jiang et al [25] investigated the VIV phenomena related to self-excited energy harvesters consisting of square cylinders by using the BGK incompressible lattice Boltzmann method. In this study, such a harvester is placed inside a channel flow and is allowed to oscillate without a structural restoring force in a direction normal to the flow. Arionfard and Nishi [26] investigated the VIV of a pivoted cylinder as a potential source of energy harvesting. The maximum efficiency of 31.4% has been observed for downstream placement of the pivot point by taking advantage of drag force and 2% for the upstream placement. Although the range of RN with high efficiency is wider when the pivot is located at the downstream, the performance of the system doesn't necessarily improve with increasing the RN in both cases.

Bhattacharya et al [27] investigated the angular oscillations of an elliptical cylinder attached to a torsional spring using computational methods, with axis placed perpendicular to a uniform flow, at RN of 100 and 200. The equilibrium angle and stiffness of the torsional spring is chosen such that the ellipse reaches stable equilibrium at an angle of roughly 45° with respect to the incoming flow. They found that the frequency spectra of fluid torque have only one peak up till density ratio of 3, and that a secondary peak emerges at higher density ratios. Power output is maximum when density ratios ranging from 3 and 10, and increases with RN. Peak efficiency of the generator is 1.7% at RN of 200. Xu et al [8] presented a theoretical study of the coupling between a VIV cylindrical resonator and its associated linear electromagnetic generator. It has been found that the proposed configuration has a maximum hydrodynamic to mechanical to electrical conversion efficiency (based on the VIV resonator oscillation amplitude) of 8%. For a cylindrical resonator of 10 cm long with a 2-cm diameter, this translates into an output power of 20 to 160mW for water stream velocities in the range from 0.5 to 1 m s^{-1}.

3. Future challenges in the research

A retrospect based on several flow-induced vibration phenomena has been presented. Different EH techniques with a particular emphasis on EH from aerodynamic instabilities and the exploration of using the harvested energy to contribute to green aviation were discussed. Many researchers addressed EH from aerodynamic instabilities for aircraft or unmanned aircraft applications [3, 4]. In these references, concepts and experimental prototypes have been introduced to harvest energy from base excitation, gust and other sources of vibrational energy by equipping the wing with a number of transducers. Several challenges that hinder the future development and application of aerodynamic instabilities-based EH in the aerospace have also been enlighten [4].

Nevertheless, many promising results have been obtained from EH research studies, but it is noticed that this technology has not been achieved in real aerospace applications. One of the challenges is that the harvested vibrational energy is low compared to the energy needed to operate sensors and electronics installed on-board aerial platforms. Also, there are many performance parameters that should be given keen consideration, like operating environment, cut-in speed and operating bandwidth of transducers to enhance output density/conversion efficiency, improve robustness and increase the life time.

4. Conclusion

EH technologies based aerodynamic instabilities of an aircraft have been presented in this paper. The retrospect on the EH literature has depicted that these types of EH techniques have achieved significant improvements in a very small period of time. In our study, most of the critical milestones in EH research has been presented on the applications of aircraft systems and other aerospace vehicles. The applications of these aircraft systems have employed ambient vibration EH and other forms of energy to reduce the dependency on conventional sources of energy like chemical batteries etc. In addition, the feasibility of EH principle allows for a safer environment and reliable energy cost management for the aviation industry. We have concluded our study by stating the future trends in the research of flow-induced EH techniques. It is demonstrated that harvesting vibration energy from induced aerodynamic instability phenomena such as galloping, flutter, and vortex induced vibration is one of the trending research areas among the researchers of aerospace.

ACKNOWLEDGMENT

This work was supported by the Universiti Sains Malaysia Short Term Grant No. 304/PAERO/6315002.

REFERENCES

[1] Kang M, Jung W, Kang C and Yoon S 2016 *Actuators* **5** 5
[2] Guruacharya S and Hossain E 2017 Self-sustainability of energy harvesting systems: concept, analysis, and design *Preprint* arXiv:1702.01648
[3] Abdelkefi A 2016 *Int. J. Eng. Sci.* **100** 112-135
[4] Rostami A B and Armandei M 2017 *J. Renew. Sustainable Energy Rev.* **70** 193-214
[5] Lee J and Mo J 2011 *Int. J. Environ. Res. Public Health* **8** 3777-3795
[6] Gilliatt H C, Strganac T W and Kurdila A J 2003 *Nonlinear Dynamics* **31** 1-22
[7] Naudascher E and Rockwell D 2012 *Flow-induced vibrations: an engineering* guide (Courier Corporation)

[8] Xu-Xu J, Vicente-Ludlam D and Barrero-Gil A 2016 *Eur. J. Mech. B Fluids* **60** 189-195
[9] Alhadidi A and Daqaq M 2016 *Appl. Phys. Lett.* **109** 033904
[10] Raj A, Garg A and Dwivedy S. 2016 *Procedia Eng.* **144** 936-944
[11] Abdelmoula H and Abdelkefi A 2016 *J. Sound Vib.* **370** 191-208
[12] Javed U, Abdelkefi A and Akhtar I 2016 *Communications in Nonlinear Science and Numerical Simulation* **36** 252-265
[13] Dai H, Abdelkefi A and Wang L 2016 *Int. J. Non-Linear Mechanics* **81** 83-94
[14] Wang J, Wen S, Zhao X, Zhang M and Ran J 2016 *J. Sensors* **2016** 12
[15] Piñeirua M, Michelin S, Vasic D and Doaré O 2016 *Smart Mater. Struct.* **25** 085004
[16] Wang P, Sui L, Shi G and Liu G 2016 *AIP Adv.* **6** 055002
[17] Akbar M and Curiel-Sosa J 2016 *Compos. Struct.* **153** 193-203
[18] Hobeck J and Inman D 2016 *J. Fluids Struct.* **61** 324-338
[19] Bruni C, Cestino E and Frulla G 2016 *Aircr. Eng. Aerosp. Technol. J.* **88** 382-388
[20] McCarthy J, Watkins S, Deivasigamani A and John S 2016 *J. Sound Vib.* **361** 355-377
[21] Zhang M and Wang J 2016 *J. Sensors* **2016** 7
[22] Dai H L, Abdelkefi A, Yang Y and Wang L 2016 *Appl. Phys. Lett.* **108** 053902
[23] Dai H L, Abdelkefi A and Wang L 2017 *Communications in Nonlinear Science and Numerical Simulation* **42** 22-36
[24] Wen Q, Schulze R, Billep D, Otto T and Geßner T 2016 *Semiconductor Technology Int. Conf.* (China) p 1-3
[25] Jiang X, Andreopoulos Y, Lee T and Wang Z 2016 *Comput. Fluids* **124** 270-277
[26] Arionfard H and Nishi Y 2017 *J. Fluids Struct.* **68** 48-57
[27] Bhattacharya A and Shahajhan S S S 2016 *J. Fluids Struct.* **63** 140-154

IOP Conf. Series: Materials Science and Engineering **290** (2017) 012055 doi:10.1088/1757-899X/290/1/012055

Transient flow in a compressor blade row for a periodic vibration motion

Moumen Idres [1], Mohamed Labanie [1] and Mohamed Okasha [1]

[1]Department of Mechanical Engineering, International Islamic University Malaysia (IIUM), Jalan Gombak, 53100 Kuala Lumpur, Malaysia

Email: midres@iium.edu.my

Abstract. The goal of this work is to conduct a transient compressor blade row flow simulation as part of blade flutter modeling. An integral step of blade flutter modeling is the calculation of the aerodynamic damping factor as a function of the possible vibration mode shapes. Using Fourier method, the number of blade passages required for transient flow analysis is kept to a minimum of two for all vibration modes. In this work, a compressor rotor blade row is considered. The vibration modes are obtained using ANSYS mechanical, then, unsteady flow is obtained for vibrating blades with a harmonic motion. Work of the flow on the blade is calculated and hence the aerodynamic damping is obtained.

1. Introduction

Aircraft gas turbine engine designers were faced with compressor blade flutter in the middle of 50s while developing the second generation of jet engines. At the present day, a huge theoretical and practical experience has been accumulated. Typically, regions of the various flutter types are plotted on the compressor operating map [1-6] as shown in Figure 1. This map clearly shows that possible flutter regions are mainly located near surge line or significantly lower than the compressor operating line. When operating in these regions, the air flow in blade passages is unsteady with a vortices and shocks. The only exceptions are regions bounded by lines 3, 4, and 6. They lead to the blade flutter near operating line for unstalled flow.

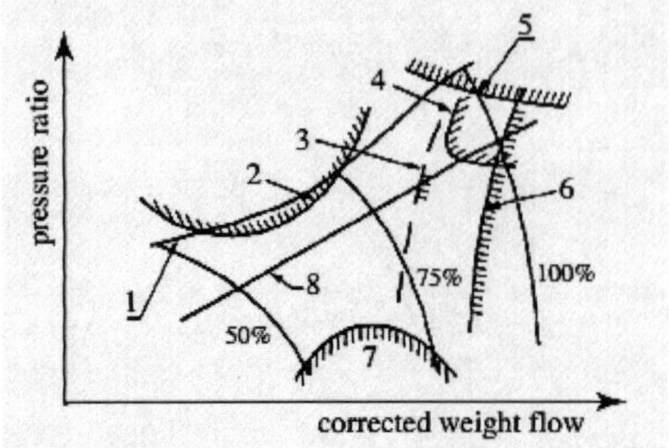

Figure 1. Compressor operating map [2]: 1- surge line; 2-subsonic stalled flutter; 3- bending-torsional flutter line (expected); 4- supersonic unstalled flutter induced by shocks;5-supersonic stalled flutter; 6-supersonic unstalled flutter; 7 – choking flutter; 8-operating line.

The problem of numerical flutter prediction for compressor or fan blade wheels is associated with coupled aeroelastic problem, which needs a solution for unsteady flow in blade passages. Modern computational codes, such as ANSYS CFX, Star-CCM, Fluent, FlowVision, etc. provide reliable airflow parameters in blade passages near the operating line, where the flow is unstalled. For surge line and for stalled flow, it is very difficult (and usually impossible) to determine reliable flow parameters numerically. However, for the design phase of modern gas turbine engine compressors and fans the most important is to suppress flutter around the operating line with required safety margins.

In this paper, a rotor blade row of an axial compressor is considered. The blades are vibrating with mode shape and frequency corresponding to one of the vibration modes. Using Fourier Transformation model, transient flow requires the usage of a double passage. Simulation is performed using ANSYS CFX 16.2 employing Transient Blade Row Model. The Fourier Transformation method is less expensive and is reliable when it comes to the point of flutter and aerodynamic damping.

2. Methodology

For a metal blade vibrating in air, the influence of the flow on vibration modes is negligible. This assumption is valid for the case of sufficiently stiff blades, when flow disturbances excited by small blade vibration have no significant effect on blade eigenmodes. Therefore, the airflow can result only in small additional damping (for stability case), or additional energy inflow (for flutter case) without change of natural modes and frequencies.

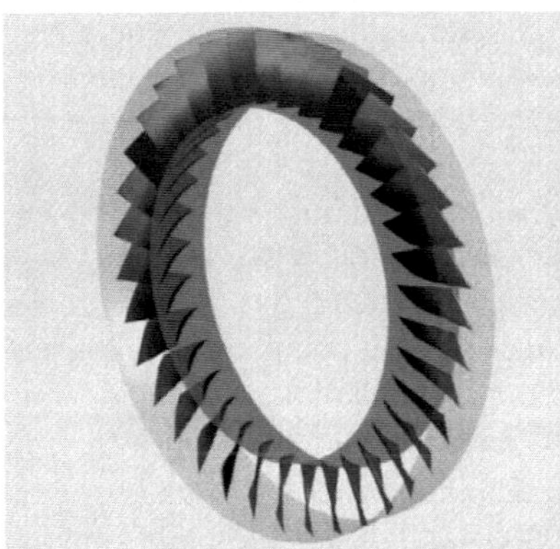

Figure 2. Geometry of rotor stage in consideration.

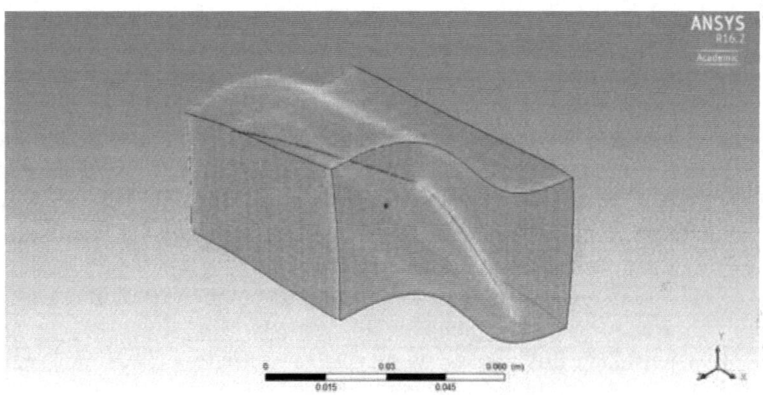

Figure 3. Single airflow passage.

Neglecting structural damping and viscous forces of the flow, we assume that the only force is the pressure distributed along the blade surface. Then change of kinetic energy over the cycle of oscillation is:

$$\Delta E = W = \int_{t_0}^{t_0+T} \int_S p(x,y,z,t) \cdot \vec{n}(x,y,z,t) \cdot \vec{v}(x,y,z,t)\, ds\, dt \qquad (2.1)$$

Where $T=1/f$ is the blade oscillation period (f is the natural frequency), S is the blade surface, p is the flow pressure, n is the blade surface normal, v is velocity of the blade. The key goal is then estimating the work per cycle performed by the fluid on the blades. Positive work per cycle of the fluid on the blades results in negative aerodynamic damping. Unless sufficient mechanical damping is available in the system to cancel out the negative aerodynamic damping, flutter would occur.

An accurate prediction of the unsteady pressure distribution consisting of its amplitude and phase with respect to the blade displacement provides a mean of validation of the CFD method when compared to experimental results. Under the assumption of having the blades of the disk vibrating at the same frequency, mode shape and amplitude with a phase lag on time when they reach their maximum amplitude, it is said that the blades are vibrating in a travelling wave mode (TWM). The phase lag between the vibratory motions of two adjacent blades is known as Interblade Phase Angle (IBPA) and it depends on the number of blades (N) and the nodal diameter (ND). The nodal diameters are lines of zero displacement during the vibration of the disk. They go across the disk passing through its center. For each nodal diameter, a forward and a backward travelling wave mode exists. Figure 4 demonstrates an example of the nodal diameter pattern and the resulting instantaneous wave mode.

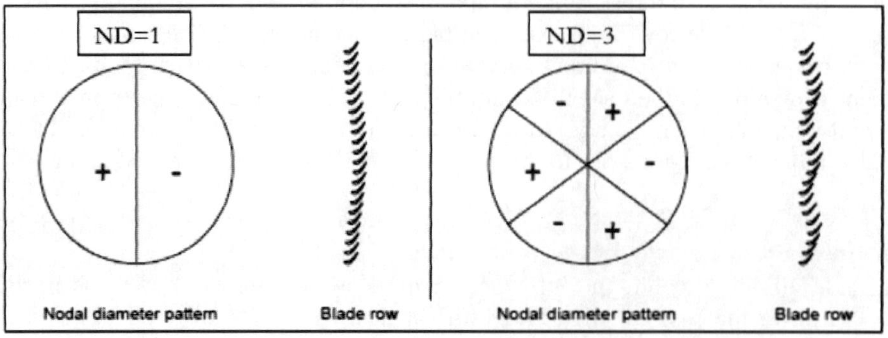

Figure 4. Nodal diameter modes and wave modes for a blade row.

For modeling forward (or backward) traveling wave, which is typical for cascade flutter phase lag ($-\alpha$) and lead ($+\alpha$) with respect to the middle blade are specified for neighboring blades, where the phase shift (α) $=2\pi m/N$ (m is the number of nodal diameters). The work done by steady component of pressure during one harmonic oscillation is zero. The work associated with the steady pressure can be non-zero if there is a phase shift between different blade points (i.e., eigenmodes are complex), as explained in [7]. However, we assume that each blade oscillates in form of standing wave, such that there is no phase shift between points of each blade (though the phase shift between different blades is non-zero), hence the eigenmodes are real. In this case, the work done by the steady pressure is zero. Therefore, we will assume that the pressure (p) is the unsteady pressure due to blade oscillation. Flutter analysis consists of four stages:

1. Modal analysis of elastic blades. Interpolation of mode shapes by Lagrange polynomials.
2. Steady-state flow analysis.
3. Unsteady flow analysis with blades oscillating (i.e. fluid mesh moving) in a specified vibration.
4. Calculation of work done by pressure for the middle blade and check of flutter criterion.

Steps 3 and 4 are executed for each vibration mode potentially sensitive to flutter. Work is calculated for the last of several simulated cycles of oscillations, such that the flow response to the blade oscillations is pure harmonic. Three periods is typically enough to have harmonic response. Structural modal analysis is performed using ANSYS Mechanical FE software. For fluid flow analysis, we use ANSYS CFX 16.2.

3. Results and Discussion

In this work, we will model a nodal diameter (ND) of four using the Fourier transformation approach with two passages. The machine is rotating at 1800 [rad/s]. The inlet boundary condition is modeled as total pressure and total temperature in the stationary frame, with a specified flow direction in the cylindrical components. The outlet boundary condition is set to an average static pressure of 138 [kPa], varying in the radial direction only. The blade vibration is modeled as forced periodic motion at a fixed frequency with a specified interblade phase angle. The frequency and displacement profile (mode shape) are obtained from cyclic symmetry calculations in ANSYS Mechanical using a single blade model, and exported to a file that is been used later as input for *ANSYS CFX 16.2*. For this case the vibration frequency is 1152.13 [Hz], and the maximum displacement for the mode shape is 0.00129 [m]. The surface of revolution mesh motion boundary condition is used at the shroud to model the sliding of the mesh along the surface. Figure 5 shows the two passages needed for Fourier method. The interface between the two passages is used to store Fourier coefficients. The mesh and boundaries are shown in figure 7. Phase-shifted periodic boundary conditions are used at the interface between passages.

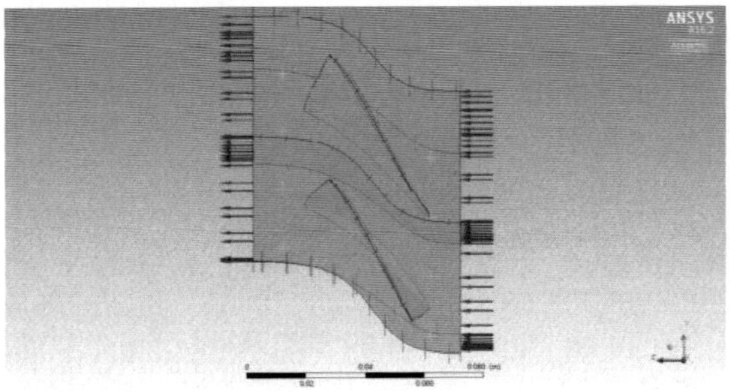

Figure 5. Double passages for Fourier Transformation Method.

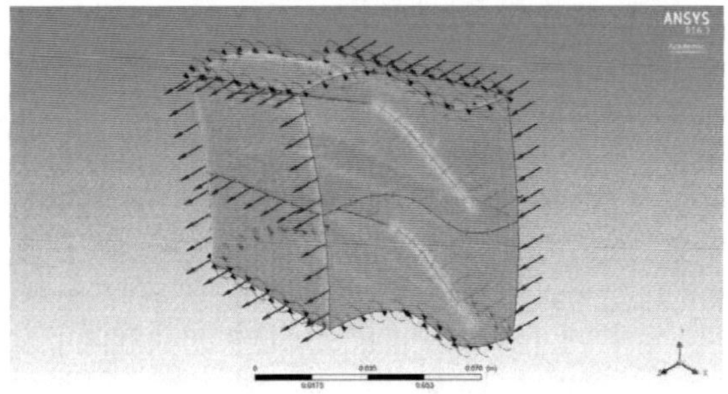

Figure 6. Mesh and boundary condition conditions.

The main settings used in the simulations are provided in Table 1. It should be noted that double precision in the calculations is required when using the FT method to avoid truncation errors when storing the temporal Fourier series data on the sampling interface. According to recommendations from the ANSYS CFX user's manual, the number of time steps per period when using the FT method should be dependent on the nodal diameter. Convergence criteria for steady and transient simulations are different. For steady simulations, the solution is considered converged when the residuals of mass, momentum, energy and turbulence equations reach 10^{-6}. The solution of transient simulations is considered converged when fluid properties as pressure, velocity, temperature and integrated flow quantities as force and work on the blade reach a periodic behavior. The transient simulations included in this study converged to less than 0.01% variation in work per cycle between periods.

Table 1. CFX-Solver settings

Property	Value/Option
Analysis Type	3D RANS - Transient Blade Row (TBR)
TBR Model	Fourier Transformation
Flow Type	Subsonic (axial inlet velocity = 1 m/s)
Turbulence	K-ε (5% intensity)
Inlet Pressure	101 kPa
Inlet Temperature	288 K
Outlet Pressure	138 kPa
Vibration Mode	1152.13 Hz, max. amplitude of the mode shape 0.00129 m, ND=4
Periods	10
Cycles Per Period	72
Mesh Stiffness	Finite Volume Dependent
Number of Nodes	116190
Number of Elements	105568

The results showed in figure 7 represent the cycle and number of runs per cycle that was needed till the aerodynamics damping coefficients reached to the convergence results. The run was set to have 10 cycles, with a 72 steps per cycle. Eventually, the simulation needed less than the specified number of cycles to reach convergence values. Curves in figure 7 represent the aerodynamic damping for the two blades considered using one cycle time integration or full time integration. All curves converge to final value of the aerodynamic damping (0.0053). In figure 8, a contour plot of instantaneous work on the is shown. The whole geometry of the cascade is assumed to be perfectly tuned in terms of mass, geometry and material stiffness. Higher work near to the tip of the blade is observed leading to

IOP Conf. Series: Materials Science and Engineering **290** (2017) 012055 doi:10.1088/1757-899X/290/1/012055

instability effect. Work distribution along the blade surface might be used to suggest geometry modifications to reduce flutter instability.

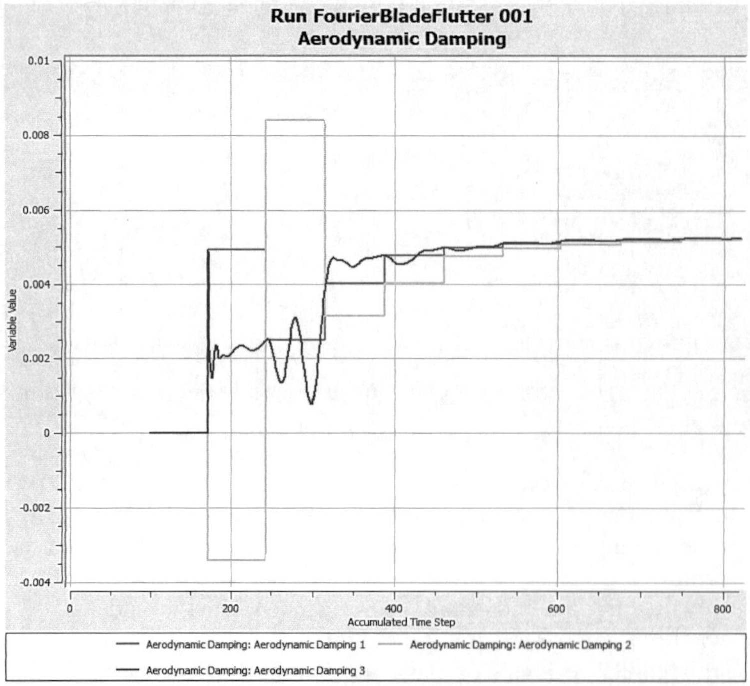

Figure 7. Aerodynamic damping convergence.

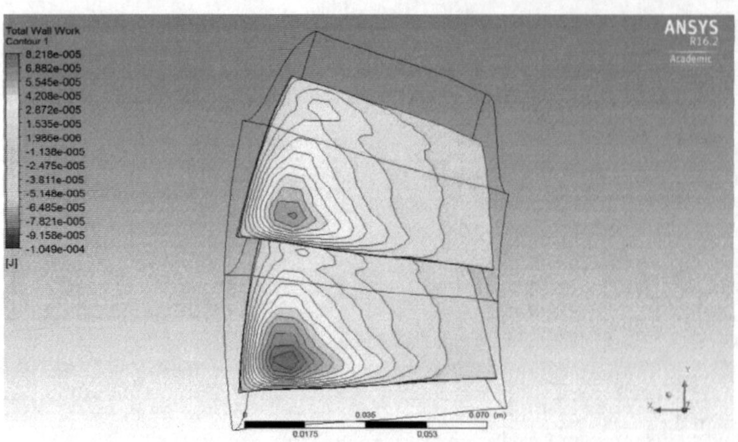

Figure 8. Work distribution along the rotor blades.

4. Conclusion

Unsteady flow simulation for a vibrating compressor rotor blade is conducted using Fourier transformation method in ANSYS CFX. The method allows the use of two passages for all vibration modes. Blades are forced to vibrate at one of the natural modes, then, the unsteady flow forces are calculated. Solution was converged after seven oscillation cycles. The aerodynamic damping is calculated which allows the evaluation of flutter in later aeroelastic analysis.

IOP Conf. Series: Materials Science and Engineering **290** (2017) 012055 doi:10.1088/1757-899X/290/1/012055

Acknowledgement
The authors gratefully acknowledge the financial support from kulliyah of engineering, International Islamic University Malaysia (IIUM).

References

[1] Shrinivasan A V 1997 *Journal of Engineering for Gas Turbines and Power* **119** 742-775
[2] Bendiksen O 1988 *Journal of Propulsion and Power* **4** 164-172
[3] Petrie-Repar P, McGhee A, Jacobs P and Gollan R 2006 *ASME Turbo Expo* (Barcelona: Spain)
[4] Marshall J G and Imregun M 1996 *Journal of Fluids and Structures* **10** 237-267
[5] Bölcs A and Suter P 1986 *Transsonische Turbomaschinen* Wissenschaft und Technik Taschenausgabe, Karlsruhe, G. Braun.
[6] Khorikov A A, Danilkin S Y 2011 *Herald of Samara State Aerospace University* **2** 57-63
[7] Kielb R., Barter J, Chernysheva O and Fransson T 2006 *Unsteady aerodynamics, aeroacoustics and aeroelasticity of tubomachines* Springer.

IOP Conf. Series: Materials Science and Engineering **290** (2017) 012056 doi:10.1088/1757-899X/290/1/012056

Experimental investigation on frequency shifting of imperfect adhesively bonded pipe joints

F N Haiyam[1], I Hilmy[1], ESulaeman[2],T Firdaus[1], E Y T Adesta[1]

[1]Department of Manufacturing and Materials Engineering, International Islamic University Malaysia (IIUM), Jalan Gombak, 53100 Kuala Lumpur, Malaysia
[2]Department of Mechanical Engineering, International Islamic University Malaysia (IIUM), Jalan Gombak, 53100 Kuala Lumpur, Malaysia

Email:ihilmy@iium.edu.my

Abstract. Inspection tests for any manufactured structure are compulsory in order to detect the existence of damage.It is to ensure the product integrity, reliability and to avoid further catastrophic failure. In this research, modal analysis was utilizedto detect structural damage as one of the Non Destructive Testing (NDT) methods. Comparingthe vibration signal of a healthy structure with a non-healthy signalwas performed. A modal analysis of an adhesively bonded pipe joint was investigated with a healthy joint as a reference. The damage joint was engineered by inserting a nylon fiber, which act as an impurity at adhesive region. The impact test using hammer was utilized in this research. Identification of shifting frequency of a free supported and clamped pipe joint was performed.It was found that shiftingfrequency occurred to the lower side by 5%.

1.Introduction

There are many non-destructive methods/testing that are currently used in the engineering structure such as electromagnetic, radiographic, magnetic particle, ultrasonic, vibration andvisual testing. Vibration Analysis (VA) refers to the process of monitoring the vibration signatures of the structure. With VA, a decision whether urgent action is necessary for example to replace the critical component or whether it is still able to withstand for duration of time. Vibration analysis benefits the maintenance technician in which it enables the identification of machine faults, provides information on root causes, localizes the affected components and allows early planning of maintenance measures. Many evidences show that the recovery cost that needed to replace a failure system due to no predictive maintenance is much higher than the cost of regular predictive maintenance using any kind of NDT.Mobley [1] stated that VA is a useful tool for predictive maintenance techniques and diagnostics that isused to monitor, analyze and evaluate critical

IOP Conf. Series: Materials Science and Engineering **290** (2017) 012056 doi:10.1088/1757-899X/290/1/012056

structures, machines and systems in a typical plant. VA can be displayed in time domain. Nevertheless, to have an easier interpretation, presenting VA in frequency domain is preferred as shown in figure 1 (a) and (b). In frequency domain, contribution of vibration signal from particular component can be identified. Each component in a multi body structure has a unique vibration signal. This is the basis of the analysis in this research.

Previous findings show that damage of an engineering structure reduced the stiffness of the damaged region [2], [3], [4]. It was a practical application to detect damage-induced loss of stiffness by measuring frequency shifting. It will be usefulas long as the damage part contribute significantly in its dynamic behavior as stated by Huston [5]. Some also introduced the impurity or the auxiliary mass instead of the existence of void as proposed by Wang et. al[6]. Shifting frequency was preferable than mode shape analysis because the measurement of frequency gives better accuracy as identified by Zhang[7]. Shifting frequency wasutilized to identify the damage evolution of an engineering structure. Piping failures due to vibration was one of the major problems occur in industrial plant over the past years as stated by Wachelet. al[8].

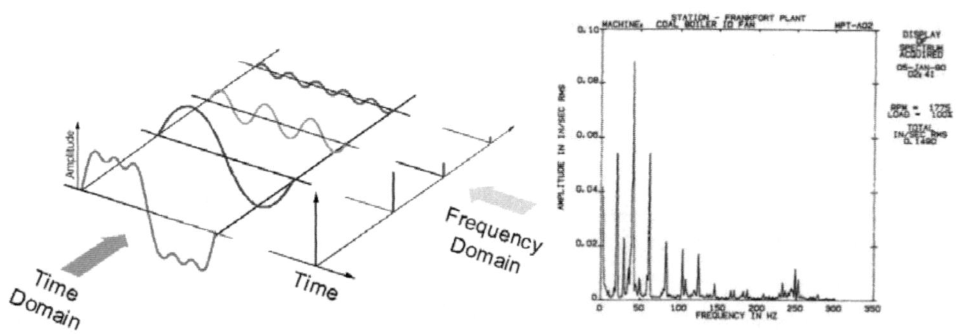

(a) (b)

Figure 1. (a) Time vs Frequency domain, (b) Vibration measurement in Frequency domain.

Structures like pipe joint are widely used in all sorts of areas such as plumbing, constructions, and mechanicals. However, undesirable damage like void or micro cracks cannot be seen in the structures and it will develop gradually through time.It will lead to a catastrophic failure. Because the damage of the pipe joint was not known, the lifetime of the structure could not be predicted too. This will be hard for the maintenance technician to get prepare for any potential problems in the field. Due to the unnoticeable damage in the pipe joint, any maintenance could not be performed before it been officially manufactured. Thus, every time the pipe joint is damage the company need to replace a new one because the damage cannot be repair anymore. One damage pipe joint also can affect the failure of the whole plumbing system and this will cost the owner more for the maintenance and replacement.

The objectives of this research were: (1) Design and fabricating the pipe joint with the "perfect condition" and "damage condition", (2)design the set up experiment of the impact test, (3) Detect the shifting phenomenon of natural frequency.The outcome of this research will give positively impact on the community especially the design engineers and industries. This research will be able to identify the natural frequencies of a structure and analyze the responses when forces excite the structure. From this, it will helps the engineers to design a better structure and also be able to resolve the potential maximum of the structures before it will destruction.

2.Methodology

Since many type of pipe joints are available, thisresearch focused on the adhesively bondedbutt joint of pipe. The consideration was the practicality to introduce or fabricate damage in the adhesive region. Damage defined as the presence of the void, micro crack or impurity. The damage was defined by introducing the impurity. To achieve the damage condition, a set of nylon strings were inserted in the adhesive region. The vibration characteristics in term of modal analysis between the damage pipe joint and undamaged pipe jointwere investigated. Flowchart of the research is shown in figure 2(a).

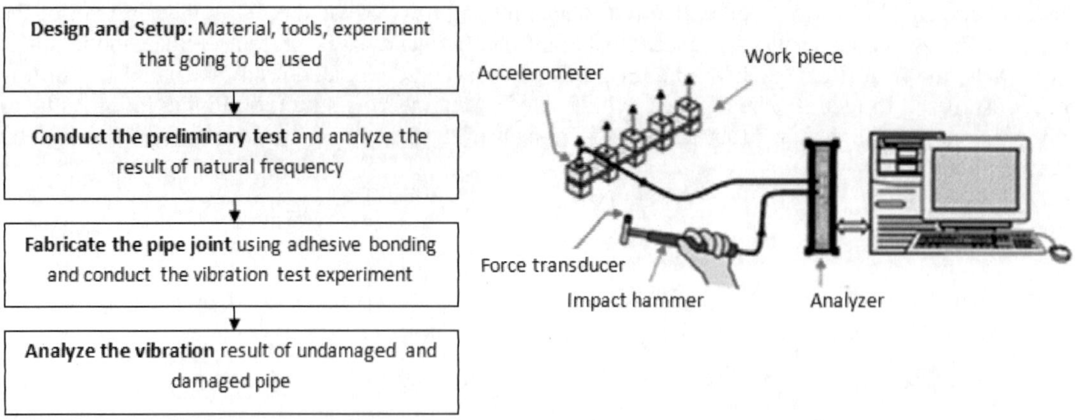

(a) (b)

Figure 2. (a) Methodology, (b) Setup configuration.

Configuration of the experimental setup is shown in figure 2(b). Hammer DYNAPULSE with model 5800B2 was usedas the exciter. This stainless steel hammer has a range of 222.4 N with maximum force of 4448 N. The attached load cell measure the force generated during excitation. To read the response signal, the DYTRAN 3035B1G was utilized torecord the accelerationlevel. LMS Scadasdata acquisition system with the installed LMS Test Lab software wasutilized to analyze the response.

Four hollow identical cylinders were cut equally from the same aluminum pipe. The dimension of each pipe was measured thoroughly using digital vernier caliper as well as the weight. Before the dimensions were taken, post processing were conducted due to the surface of the pipes were rough and unsymmetrical. Thus, facing operationswere performed on both surfaces of each pipe. The constant gap lengthbetween the pipes need to be maintained.

Epoxy resin wasusedas the adhesive bonding substrate due to the popularity in many industrial applications. The adhesive consists of an epoxy resin and a hardener. Bothwere mixed together in equal portion. The resin and hardener reacted to produce a tough, rigid, high strength bond towards variety of materials for gap bonding, surface repairs and others. Fail to maintain the equal portion will cause the adhesive either too flexible or too brittle.figure 3 (a) shows the attachment of epoxy resin mix. figure 3 (b) shows the insertion of the nylon thread. To fabricate "damage condition" of pipe joining, nylons were inserted before applying epoxy. This treatment was to represent the impurity condition of the join. Gap gauge was inserted to determine the distance or gap between pipe was attached as shown in figure 3 (c).

The upper part of pipe was hanged to maintain the position during the curing.

(a) (b) (c)

Figure 3. (a) Applying epoxy, (b) Inserting nylon thread and (c) curing while maintaining gap

For the fabricated damage pipe joint, the similar procedure was performed refer to figure 3 (b). An additional step is inserting nylon threads at every 30° angle locations as shown in figure 3 (b).

Figure 4 (a) shows the test on the free-free condition. To simulate this condition, the specimen was supported with a soft platform as shown in figure 4 (b). Figure 4 (c) shows the point of hammering.

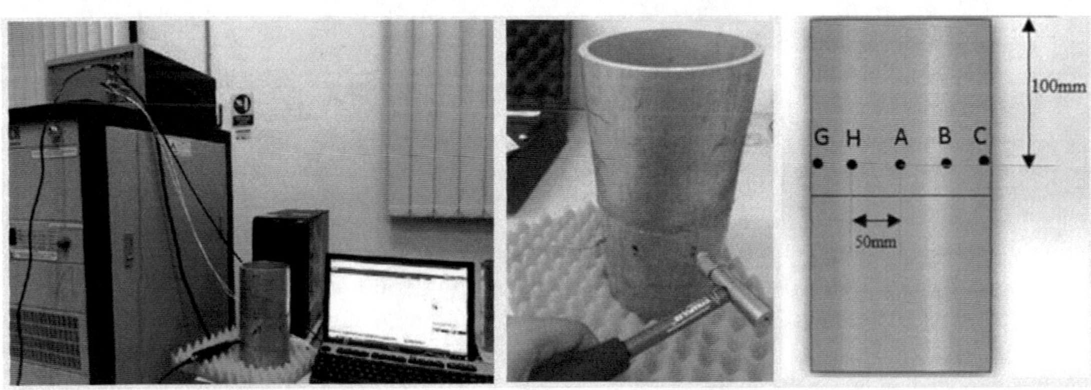

(a) (b) (c)
Figure 4. (a) Setup of FRF Testing (b) Free-free configuration, (c) points of the hammering

Figure 5 (a) shows the test on the fixed-free condition. To simulate this condition, the specimen was clamped at one end that acts as a cantilever. Figure 5 (b) shows the points of hammering.

IOP Conf. Series: Materials Science and Engineering **290** (2017) 012056 doi:10.1088/1757-899X/290/1/012056

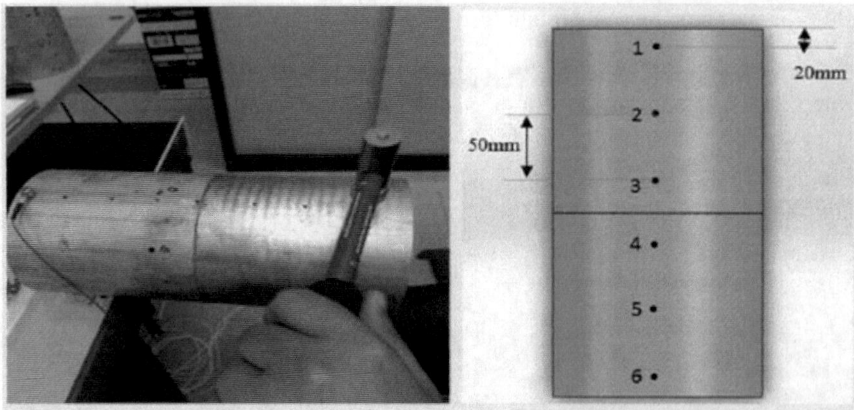

(a) (b)

Figure 5.(a)FRF analysis of Fixed-free condition and (b) points of hammering.

3. Results and Discussion

Modal analysis tests were conducted with one accelerometer sensor attached at one end and excitation at several point of interests as shown in figure 4 (b) for free-free condition and figure 5 (a) for fixed end condition respectively. Details of the results werelisted in Table1.

Table 1. Frequency Result of impact hammer test of Free-Free condition (in Hz)

	Undamaged			Damage		
Point	F_1	F_2	F_3	F_1	F_2	F_3
A	4.61	32.07	978.32	3.98	29.05	966.57
B	3.98	31.19	976.38	3.98	29.05	964.68
C	2.97	32.95	968.75	2.85	30.05	965.72
D	4.11	28.92	977.87	3.85	29.93	966.57
E	3.98	28.04	968.75	3.98	29.05	966.57
F	3.98	28.16	968.75	3.98	28.04	965.39
G	3.85	28.04	975.43	4.11	29.17	970.12
H	3.98	30.18	976.38	4.11	30.18	968.35

From Table 1, the first two columns for both Undamage and Damage (F1 and F2) are ignored since the result came from the noise. The third frequency for both Undamage and Damage are considered. The first column with the title Point refers to figure 4 (c). From this Table, the chart comparing F3 for both cases are generated and displayed in figure 6. The drop of frequency or shifting to the lower value occurred consistently in all point of hammering.

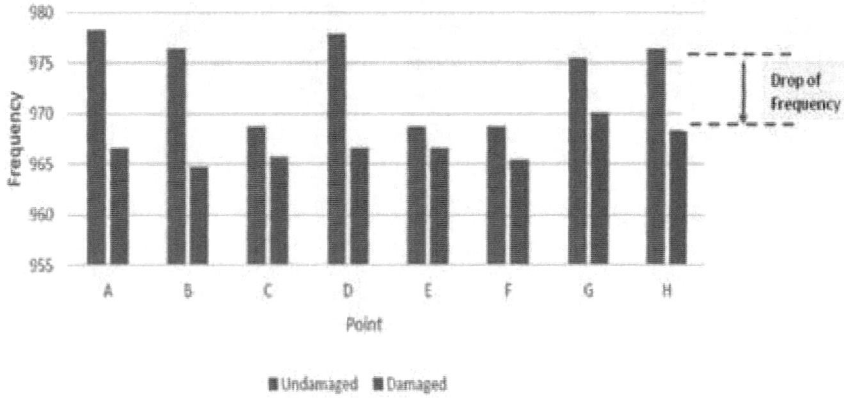

Figure 6. Frequency comparison of free-free configuration (in Hz)

Table 2 shows the FRF result for fixed-free condition. The first columns of F1 for both cases were ignored as well due to the low value. The analysis will be done to the second and third column (F_2 and F_3 for both cases). In this analysis, second mode of frequency which is the F_3 from both conditions was evaluated. The first column with the title Point refers to figure 4 (c). From this Table, the chart comparing F_3 for both cases were generated and displayed in figure 7. The drop of frequency or shifting to the lower value occurred consistently in all point of hammering except in point 4. This was due to measurement error. If the second column was evaluated (column F_2), consistency of the drop of frequency were also occurred.

Table 2. Frequency Result of impact hammer test of Fixed-Free condition

	Undamaged			Damage		
Point	F_1	F_2	F_3	F_1	F_2	F_3
1	23.88	911.34	1245.74	22.12	901.23	1232.61
2	24.01	922.22	1242.54	21.89	905.89	1237.18
3	24.01	912.56	1242.54	22.87	894.78	1232.61
4	24.01	911.34	1226.87	21.87	907.65	1232.61
5	24.13	921.11	1236.76	23.13	902.55	1228.96
6	24.01	910.17	1242.54	23.00	905.89	1233.73

IOP Conf. Series: Materials Science and Engineering **290** (2017) 012056 doi:10.1088/1757-899X/290/1/012056

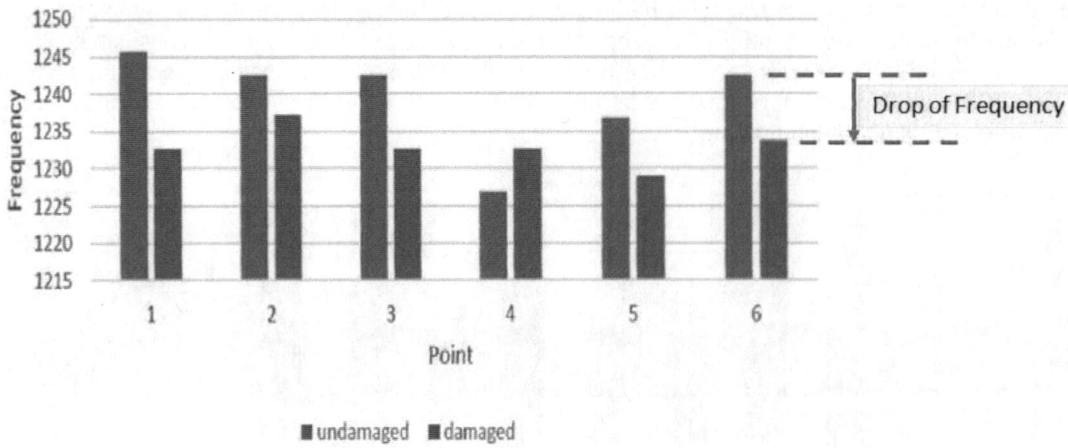

Figure 7. Frequency comparison of Fixed-free condition (in Hz)

4. Conclusion

Different location of excitation or hammering will make the level of amplitude different. In the other hand, the resonance or natural frequency will be remaining the same because it is a characteristic or identity of the structure. Both cases that were obtained clearly showthe drop of the natural frequencies between theundamaged pipe and damaged pipe.Damaged of the pipe joint contain impurities and air bubbles that will slightly disturb the adhesive bonding between the pipes.The stiffness of the pipe joint were affected and the frequencies decrease.Therefore, the changes of the natural frequencies can become an indicator that there is a degradation of the integrity of a structure.

References

[1] Keith Mobley, 1999, *Vibration fundalmentals,* Butterworth–Heinenman, United States of America
[2] IHilmy, MA Wahab, EYT Adesta, TFirdaus, 2015, Journal of Physics Conference Series *628(1)* 42
[3] MA Wahab et al., 2013, *On the Use of Low and High Cycle Fatigue Damage Models*, Key Engineering Materials
[4] M M Abdel Wahab et al., 2011, Journal of Adhesion Science and Technology
[5] D Huston, 2011, *Structural Sensing Health Monitoring and performance Evaluation,* Taylor and Francis Group LLC
[6] LWang, ST Lie, YZhang, 2016, *Mechanical Systems and Signal Processing*, 66-67 pp 298–313
[7] Y Zhang, LWang, ST Lie, Z Xiang, 2013, Journal of Sound and Vibration, 332 pp 6665–6684
[8] JC Wachel, SJ Morton and KE Atkins, 1990, *19th Turbomachinery Symposium,* Texas A&M University, September

Determining system boundaries on commercial broiler chicken production system using ISO 14040/14044 guideline: A case Study

'A A Sidek[1], S A Suffian[1], M H F Al-Hazza[1], H M Yusof[1]

[1]Department of Manufacturing and Materials Engineering, International Islamic University Malaysia (IIUM), Jalan Gombak, 53100 Kuala Lumpur, Malaysia

Email:atiahas@iium.edu.my

Abstract. The demand of poultry product in Malaysia market shows an escalation throughout the year and expected to increase in the future. The expansion of poultry production has led to environmental concern in relation to their operational impact to environmentAt present, assessment of waste management of poultry production in Malaysia is lacking. A case study research was conducted in a commercial broiler farm to identify and assess the system boundaries in the lifecycle supply chain of broiler chicken production using ISO 14040/44 guidelines. ISO 14040/44 standard includes Life Cycle Assessment (LCA) framework guidelines to evaluate environmental influence associated with a product/process throughout its life span. All attributes associated with broiler operation is defined and the system boundaries is determined to identify possible inputs and outputs in the case study. This paper discuss the initial stage in the LCA process, which set the context of the research and prepare for the stage of Life Cycle Inventory.

1. Introduction

Chicken meat has become one of the most extensively consumed food products in the world [1]. Apart from its low price, chicken is popular as it is an important source of proteins, vitamin B and minerals and is low in saturated fats [2],[3]. Broiler chicken is the type of chicken that are reared for their meat. Commercial broiler chickens are bred to be very fast growing in order to gain weight quickly. The life of chickens destined for meat production consists of two distinct phases. They are born in a hatchery and moved to a growing farm at 1 day old. They remain here until their weight is optimum; normally up to 3kg and sent to the slaughterhouse or chicken supplier.

Figure 1. Broiler chicken in the case study.

ISO 14040 provide guidelines of Life Cycle Assessment (LCA) to analyse and evaluate the environmental performance of a product/process/activity at a various points in its life cycle span [4],[5]. In order to address the environmental aspects of the operation, the system boundaries of the broiler production are determined. The performance characteristics of the systems is specified where it will provide as a reference which the input output data are normalized [6]. This research will be analyzing the system boundaries of broiler farm from cradle to gate where it analyzed the process from the day one old broiler chick until to when they are matured and ready to be transported to slaughterhouse/supplier.

This research intended to define the goal and scope definition of broiler chicken production system by identify their system boundary. In order to achieve the objective of study, process flow of the chicken production must be identified.

2. Research Methodology

Case study used in the research is a commercial broiler chicken farm in Kalumpang, Selangor. The farm used a closed house system where all broilers placed in two large double-storey chicken coops. Capacity of all the chicken coops is about 106,000 chicks. This commercial broiler farm was established in 2016 and has been rearing the broilers for the fifth batch. Their records on mortality rate is rather impressive which is below than 10% and the lowest is 3% of the total broilers. The broiler farm use semi-automated technology in handling the feed, water and temperature of the broilers.

Operation lifecycle in a commercial broiler farm is identified and quantitative data for insertion in the inventory are collected for each unit process specified within the system boundary.Functional unit in the system boundary is defined to be 3kg of broiler chicken meat at the point of sale.The unit is chosenas it is the amount of raw chicken that contains recommended daily protein supply for an average family (four people). In addition, it is the average weight of a broiler chicken (excluding packaging) available for consumption in Malaysian markets.

3. Results and Discussion

Main function of the system under study is to identify the system boundaries and to study the chicken production flow throughout the life cycle of broiler chicken production system through a cradle-to-gate perspective.

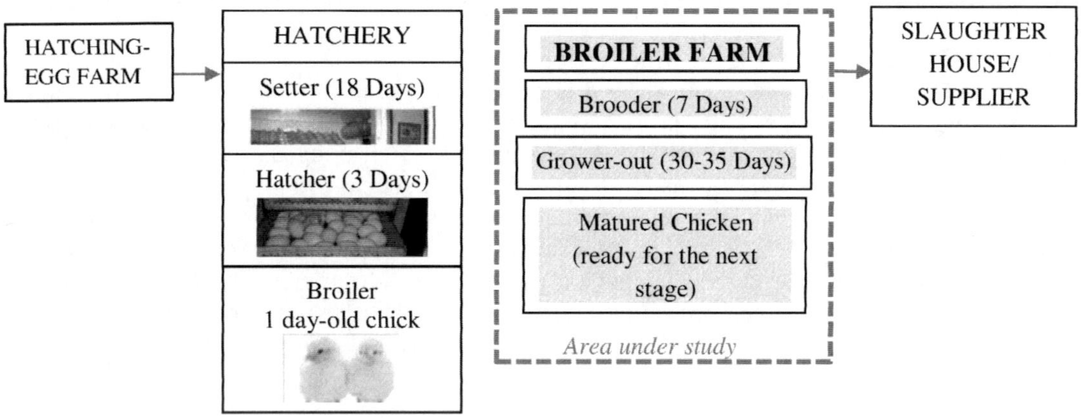

Figure 2. Process flow of chicken production system in Malaysia.

According to figure 2, process flow of the chicken production system in Malaysia has been identified. Starting from hatching-egg farm, the chicken will transported to the hatchery and then to the broiler farm for fattening period. Area under study which starts from cradle-to-gate will cover from the day the chicks enter broiler farm until they are matured enough to leave the broiler farm.

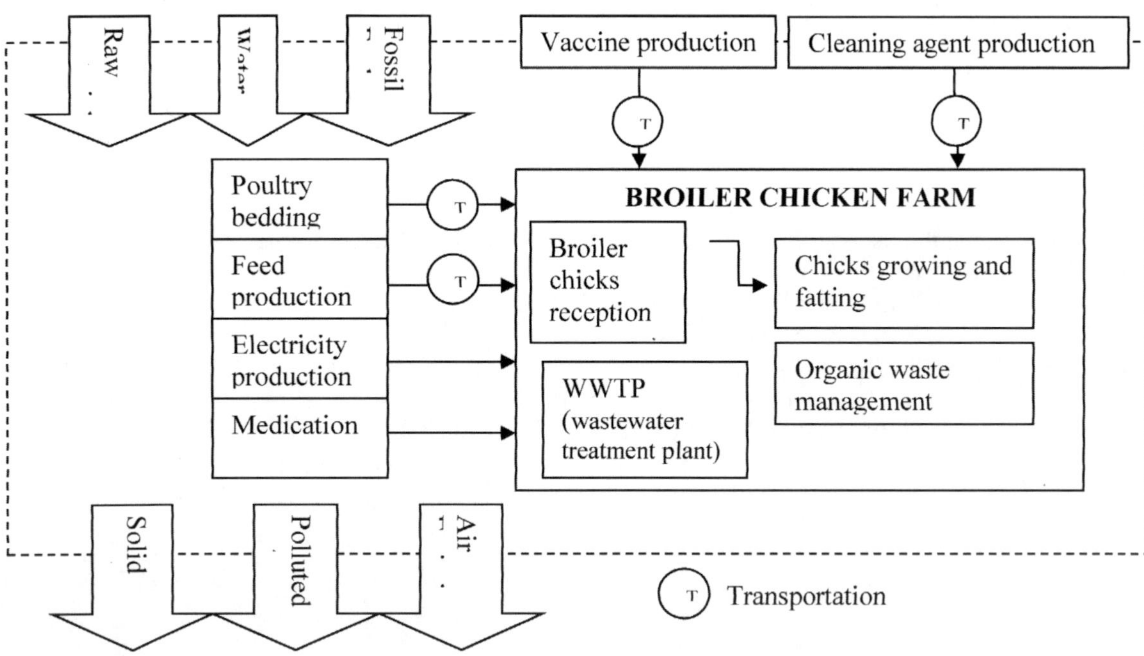

Figure 3. System boundaries of the broiler chicken production system.

Production system under study encompassed all activities related to the chicken farm according to cradle-to-gate analysis (Figure 3). In this way, the product (broiler chicken) could be monitored until it was ready for distribution to wholesales. Activities carried out directly in this subsystem starts with the reception of chicks at the farm gate. Chicks are transferred to a production pen for a fattening period of 35 days.

Broilers are housed in poultry beds, which are commonly composed of a mixture of materials such as sawdust, bark, wood shavings or chips. Feed is mainly composed of maize, wheat, soybean cake, soybean oil, monocalcium phosphate, protein concentrate and fats. Throughout the growth period, vaccine and medication are given to the broilers. Activities of cleaning the chicken farm not only involve water but also the cleaning agents (detergents and disinfectants).

Wastewater generated from cleaning activities and farm-related uses is treated in a small wastewater treatment plant (WWTP) which is located in the broiler chicken farm. Manure and wasted poultry beds produced in the chicken farm will cause emission towards environment. Manure is an important source of nutrient which can be processed to be organic fertilizer for agriculture product. Possible air emission that involved in the broiler chicken farm are NH_3, N_2O, CH_4, NO^-_3 and PO_4^{-3}.

4. Conclusion

In summary, it can be concluded that a complete system boundary of broiler chicken production system especially in the case study was developed. Throughout the system boundary that is from cradle-to-gate, several inputs parameter were defined such as water supply, poultry bedding, feed production, electricity production, vaccine and medication and also cleaning agent for cleaning process. Output parameters of the system include solid waste (manure, dead chicken, bedding waste), polluted water and air emission (NH_3, N_2O, CH_4, NO^-_3 and PO_4^{-3}).System boundaries was conducted to identify possible input and output used in the production system.

Acknowledgement

Ministry of Higher Education (MOHE) through Fundamental Research Grant Scheme (FRGS) financially supported this work, to which the author are grateful. Special thanks also goes to respected broiler chicken farm, all interviewed broiler chicken workers and other participants who helped to carry out every activity.

References

[1] Magdelaine P, Spiess M P and Valceschini E 2008 *World's Poultry Science J* **64** 53-64.
[2] Windhorst H W 2006 *World's Poulty Science J* **62** 585-602
[3] Silvestre J D, Brito J, and Pinheiro M D 2013 *Proceedings of the Institution of Civil Engineers-Engineering Sustainability* **167** 53-63
[4] ISO 14040:2006, ISO 14040:2006 2017-*Environmental management--Life cycle assessment: Principle and Framework*
[5] ISO 14044:2006, ISO 14040:2006 2017-*Environmental management--Life cycle assessment: Requirement and guidelines*
[6] Protosh R 2009 *A Review of Life Cycle Assessment (LCA) on some food products.*

IOP Conf. Series: Materials Science and Engineering **290** (2017) 012058 doi:10.1088/1757-899X/290/1/012058

Inventory Data on Commercial Broiler Chicken Production System using Life Cycle Assessment Approach: A Case Study

S A Suffian[1], A A Sidek[1], H M Yusof[1], M H F Al-Hazza[1]

[1]Department of Manufacturing and Materials Engineering, International Islamic University Malaysia (IIUM), Jalan Gombak, 53100 Kuala Lumpur, Malaysia.

Email:syakiraafiqah93@gmail.com

Abstract.An inventory analysis of the life cycle of broiler chicken production from cradle-to-gate perspective was carried out with the aim to identify possible input and output parameters involved in the system. To do so, broiler chicken production in Myra Chicken Farm and Services was investigated in detail. Result shows the inventory data on feed consumption, transportation, physical performance parameter and other utilities that affect the product which is broilers. Broilers production in fact shows escalation year by year because of high demand from consumer. A cradle-to-gate assessment was conducted based on ISO 14040/14044 guidelines. Inventory data was gathered from farmers and available literature. Improving all the input and output system will increase the level of productivity and the cost of the production. Thus, at the end of the research, it will able to make industry player to understand and take into consideration the solutions in order to promote a green broiler chicken production.

1. Introduction

Broiler meat is a primary protein source for the majority of Malaysian populations [1]. Chicken production systems can be classified into two groups according to the farming scheme [2]: free range chicken, which is reared outdoors with a purely vegetable diet and broiler (or conventional) chicken, which is reared on farms with industrial feed as the base diet [3].The overall production of broiler has expanded steadily, in line with the growth in local demand and could be exported to some countries [1]. The development of broiler industry in Malaysia is supported by good production technology.

One of the most widely accepted international methods of quantifying inventory analysis of food products is Life Cycle Assessment (LCA) [4]. Life Cycle Assessment (LCA) is a standardized and hollistic methodology that allows the identification of the environmental consequences of the life cycle of the product [5]. By using LCA framework, life cycle inventories must be defined in order to quantify the environmental impact of the food product.

In LCA studies, it is important to consider real data to obtain representative and relevant results. Life cycle inventory (LCI) or inventory analysis is a data collection stage. At this stage, input and output data will be collected. Inputs are anything that enter the system and help to produce the product during their processing stages such as any energy, water and raw materials. On the other hand, outputs are anything that release from the system or processing stages such as waste. This phase summarize the processes, input and output of the product.

2. Research Methodology

The broiler chicken farm was assessed and inventoried in detail to gather representative data. In this study, life cycle inventory (LCI) data for the subsystem was collected through interviews and visit to the broiler chicken farm. All data recorded are valid for all four chicken coops that consist of 106,000 broiler chicks. In order to get the data accurately, information was gathered from all workers to see if there is any uncertainty in data collection. Functional unit, 3kg of broiler chicken meat ready to be distributed to the point of sale, was chosen because a) it is the amount of raw chicken that contains recommended daily protein supply for an average family (four people). b) it is the average weight of a broiler chicken (excluding packaging) available for consumption in Malaysian markets.

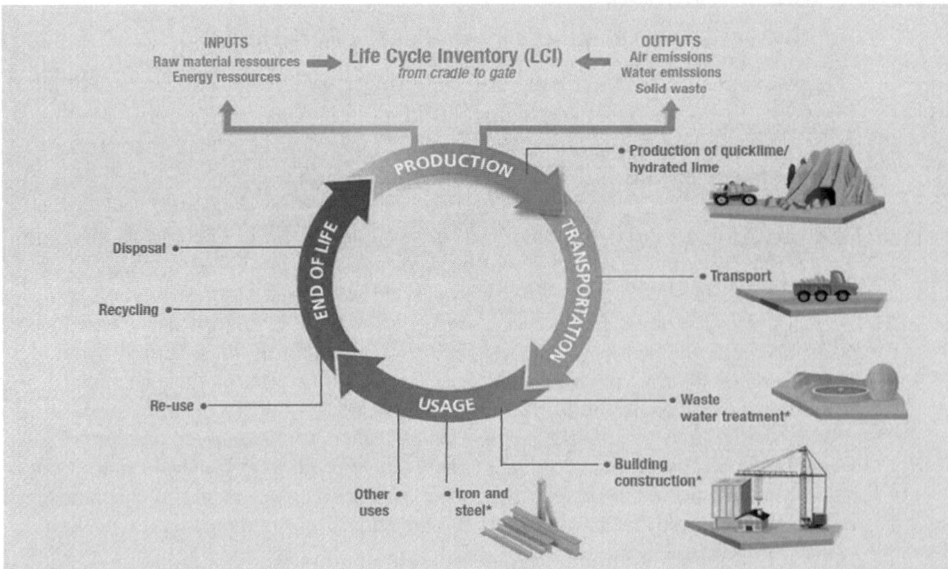

Figure 1. Steps in collecting data on inventory analysis of cradle-to-gate for broiler chicken production system [7].

Figure 1 shows steps in collecting data on inventory analysis of cradle-to-gate for broiler chicken production system. Inventory analysis consists of detail measurement of input and output parameters that enters and exits the system boundaries.

3. Results and Discussion

In LCA studies, it is important to consider real data in order to obtain representative and relevant environmental results. In this study, life cycle inventory (LCI) data was collected. The broiler chicken farm was assessed and inventoried in detail to gather theinformation. Information concerning all feed intake, transportation related, physical performance of the chicken and utilities were provided by Myra Chicken Farm and Services.

Table 1. Inventory data of the chicken farm per functional unit.

Type of Input	Measurement/Unit	Types of Output	Measurement/unit
Feed Intake:		**Solid waste:**	
Pre-starter	325g	Municipal solid waste	1.10g
Starter	807g		
Grower	2.4kg		
Transportation related:		**Air Emission:**	
Distance	20588km	NH_3	21.5g
Amount of inputs	106 000 chicks	CO	15.7g
Type of transport	lorry	CO_2 fossils	92.6mg
Physical Performance parameter:		**Water Emission:**	
Number of birds	106 000 chicks	NO_3^-	5.31g
Start weight	100g	PO_4	0.12g
Finish weight	2.5-3kg		
Feed intake per bird	3.5kg/bird		
Feed conversion ratio	1.78		
Mortality rate	3.6		
Utilities:			
Electricity use	0.02kWh		
Amount of water for drinking	0.1m^3		
Amount of bedding used	0.3kg		

Table 1 shows the inventory data for all broilers in the case study farm. In order to obtain the inventory data per functional unit, all data must be divided by the amount of broilers in the broiler farm. Output data was obtained from available literature from previous researcher.

3.1 Feed Intake

Throughout the growth period, chicken are fed for 35 days with a ratio of 85g of feed per day per chicken. The feed is mainly composed of maize, wheat, soybean cake, soybean oil, monocalcium phosphate, protein concentrate and fats.

Table2. Total weight of feed consumed by the broilers throughout one cycle.

Type of feed	No. of bags	Total weight (1000kg)
Starter	689	34.45 kg
Starter crumble	1711	85.55 kg
Grower pellet	5023	251.15kg
	Total	371.15 kg

3.2 Transportation

Transport activities required to provide inputs to the production line were also taken into account. From the data obtained, carbon footprint of fuel can be calculated.

Table3. Total distance and number of trips for transportation involved in the system.

Vehicle	No. of trip	Distance	Total distance (km)
Lorry (feed)	13	Kalumpang - Port Klang (106 km)	1378
Lorry (waste)	6	Kalumpang - Cameron Highland (140km)	840
Lorry (broiler in)	10	Kalumpang- Desaru, Johor (479 km)	4790
Lorry (broiler out)	97	Kalumpang - Cameron Highland (140 km)	13580
Total			20588

Distance travel per unit of fuel used in km per liter (km/l) = 9.67 km/l (2)
Total liter usage = 20588 liter x 9.67 km/liter = 199085.96 liter

3.3 Electricity use

Electricity is one of the major input that affect the performance of the broiler chicken and contribute the most to the carbon footprint. Equipment involved are lamp, pump and fan.

Table 4. Inventory data of the chicken farm per all broilers.

Equipment	No. of equipment	Consumption per day	Day per batch	Total consumption per equipment
Lamp	124	0.207kWh	33	847.044 kWh
Pump (water)	2	7.2kWh	33	475.2 kWh
Pump (feed)	3	0.27kWh	33	26.73 kWh
Fan (ventilation)	-	-	-	805 kWh
Total				2153.97 kWh

The total consumption for equipment considering the number of equipment used for each farm. For lamp, there are 124 bulbs, 2 water pump and 3 feeding pump. The total amount of electricity consumption for one batch of broiler production is:

$$2153.97 \text{ kWh} * 4 = 8{,}615{,}160 \approx 8.61 \text{ Wh}$$

4. Conclusion

The result of the study shows inventory data for broiler chicken production system in Myra Chicken farm and Services. From the analysis, input and output parameters are identified successfully. Thus, further research on environmental impact assessment can be done.

Acknowledgement

Ministry of Higher Education (MOHE) through Fundamental Research Grant Scheme (FRGS) financially supported this work, to which the author are grateful. Special thanks also goes to respected broiler chicken farm, all interviewed broiler chicken workers and other participants who helped to carry out every activity.

References

[1] Syauqi M et al 2015 *Broiler Industry in Malaysia*
[2] Bengtsson J and Seddon J 2013 *Journal of Cleaner Production* **41** 291-300
[3] Castellini C et al 2006 *J Agriculture Ecosystem Environment* **114** 343-350
[4] Prorosh R 2009 A Review of *Life Cycle Assessment (LCA) on some food products*
[5] ISO 14040:2006, ISO 14040:2006 2017-*Environmental management--Life cycle assessment: Principle and Framework*
[6] Sara G et al 2014 *Jornal of Cleaner Production* **74** 125-134
[7] EuLa. *Environmental Impact Assessment (LCA) Life Cycle Analysis*

An examination on the influence of small and medium enterprise (SME) stakeholder on green supply chain management practices

M Z Shahlan, A A Sidek, S A Suffian, M H F A Hazza, M R C Daud

Department of Manufacturing and Materials Engineering, International Islamic University Malaysia (IIUM), Jalan Gombak, 53100 Kuala Lumpur, Malaysia

Email:atiahas@iium.edu.my

Abstract. In this paper, climate change and global warming are the biggest current issues in the industrial sectors. The green supply chain managements (GSCM) is one of the crucial input to these issues. Effective GSCM can potentially secure the organization's competitive advantage and improve the environmental performance of the network activities. In this study, the aim is to investigate and examine how a small and medium enterprises (SMEs) stakeholder pressure and top management influence green supply chain management practices. The study is further advance green supply chain management research in Malaysia focusing on SMEs manufacturing sector using structural equation modelling. Structural equation modelling is a multivariate statistical analysis technique used to examine structural relationship. It is the combination of factor analysis and multi regression analysis and used to analyse structural relationship between measure variable and latent factor. This research found that top management support and stakeholder pressure is the major influence for SMEs to adopt green supply chain management. The research also found that top management is fully mediate with the relationship between stakeholder pressure and monitoring supplier environmental performance.

1. Introduction

The fastest growing of manufacturing industries worldwide has led to the various environmental problem such as global warming, acid rain, release of greenhouse and toxic gases such as CO2, SO2, NO2 and other pollutants [1]. To overcome these problems, industrial awareness towards environment is very important. Green supply chain management (GSCM) is one of the concept that can be implemented in the industry to initiate the environmental awareness and innovation. in this approach, the sustainability of the environment is consider from combination of the intra-firm management, inter-firm management of the upstream and downstream supply chain [2] Green supply chain management (GSCM) is very crucial in organization's operation management whether in it marketing strategy, operations, logistics or distributions. In recent years, the interaction between supplier and end users are important to achieve the objective of green supply management. The big and well-known companies usually have sustainable strategies such as Panasonic Malaysia, Toyota Malaysia and others which had lunched the green supplied chain program to promote and give education to the consumer [3]. Green supply chain managements is the involvement of various parties in the organization towards environmental sustainability managements which include suppliers, materials, process, organization managements, consumers and product life cycle. The influence of stakeholder in implementing the GSCM in the market is very significant. It is because; they have the ability to push the company to produce green products in order to become their supplier. Large companies such as Intel and Apple have the ability and power to ask the vendor to adopt GSCM practices in the firm's [4]. Furthermore, many non-governmental organization (NGO) given environment education to the society in how the functionality of environment and how human can manage the environment and maintain the sustainability. When more and more people concern about the environment and the eco-friendly products, it will push the company to take one-step forward towards environmental sustainability due to consumer pressure. This project emphasize on exploring the relationship between stakeholders pressure and company green supply management and to study the either the stakeholder pressure will gives positive influence towards company green supply managements. Successful implementation of GSCM will be the key to improve products environmental performance and management systems toward facilitate a better environmental future.

2. Experimental procedures

The hypothesis was carefully choose by intensive literature review and research to answer all the objectives of this study. In this paper, the hypothesis will cover all the stakeholder pressure and top management supports in influencing SMEs to adopt green supply chain management. Stakeholder has greater influence in changing the mind of top managements because of internal and external pressure given. Stakeholder comprehend of employees, clients/ customers, governments, shareholders, nongovernmental organization and the community [5]. This groups of people and organization have the biggest influence on company, the research understand that stakeholder can give different strategies in influencing SMEs in making judgments [6]. Top management is the key players in implementing the GSCM on the firms. The influence and pressure of stakeholder to the top managements is higher so the stakeholders is the key players in push the top managements to implementing GSCM in SMEs. There are many example of research and company that success by implementing this green supply managements practices such as Apple will change the material use in they are products and becoming more prior compare to his rivals Dell [7]. By adopting this environmental managements company can improve the managements and boost company capital. Thus, the highest managements should consider deeply on the stakeholder pressures. Hence, we believed that:

H1: The greater the stakeholder pressure to implementing green supply chain management, the greater top management support for green supply chain management initiatives.

SMEs believe that by following the internal and external pressure can make them more competitive in the markets. Employee is the internal stakeholders, they have the power to make move and give the pressure to SMEs in making more environmental friendly managements [8]. The company must give input and education to the employee to make them talented and have concern about the environments in supply chain, company also need to create proactive work place and promote new innovation and ideas in the company.

This research believe that without the support and point of views of the top managements toward the implementing green supply chain managements it is difficult to make it success in the firm's or organizations. Thus, we believe that:

H2: The greater the top management support of green supply chain management initiatives, the greater the firm's monitoring supplier's environmental performance.

Target population (*n*) in this study involved only the Manufacturer sector of the Small and Medium Enterprises (SMEs) in Malaysia. Manufacturer sector definition for SMEs has a turnover 1 not acceding RM50 million or full-time employee not acceding 200 people [9]. The positive aspects of Manufacturing SMEs, there are plenty of room for improvements, innovation and growth to compete with larger manufacturing industries. This organisation can make major improvements without spending a lot of money compare to larger industries in term of innovation and improvements because of normally SMEs focus only in one productions and small target market. The SMEs industry will make a breakthrough in the market in term of innovation, sustainability and health economical developments [10]. This study focus on manufacturing sector because its play an important role in Malaysia economy, the production of raw material and finish product is crucial. Eventhough the effect of SMEs on the environment is still doubtful but we believe it will be accurately measure in the future research. However, because it high in number the environment impact is still huge. Thus, this research believe that it is a stepping stone for manufacturing sector in SMEs to implement GSCM in the company to remain competitive.

3. Result and Analysis

Evaluating validity in construct is an important aspect in this project. Validity defined as the ability of the instrument to measure what its intended to measure [11]. The validity test is convergent validity, discriminant validity and common method bias. For convergent validity it measure the items of each construct should agree with each other and for discriminant validity each construct should disagree with measure of other construct [12]. After removing the item with loading factor less than 0.60,Cronbach's alpha reliability test, convergent validity and discriminant validity conducted. Convergent validity happen when group of items measure one common factor. Convergent validity shows by statistical significant at alpha (eg. $p=0.05$). The reliabilities of this model is range from 0.760 to 0.842 (see table 1). The reliability result of composite reliability (CR) is above the cut-off point of 0.7. The average variance of the data is 0.619 and above. Cronbach's alpha values of all the construct are well over 0.7, as shown in table 1 with all the value for convergent validity. Discriminant validity was tested by assessing whether the intercorrelation within the construct is less than 0.7 which suggest the construct have less than half of their variance in common. All of this construct meet the threshold. Furthermore for discriminant validity it been tested by comparing the average variance extracted (AVE) for each construct with all possible pairs of construct. In this analysis, the AVE greater than square of all the correlation between all possible pairs off construct (table 1**).** Overall, the results shows positive discriminant validity within the construct.

Table 1. Convergent validity and reliability.

	CR	AVE	MSV	MaxR(H)	stake	supp	top
stake	0.842	0.728	0.143	0.843	**0.853**		
supp	0.760	0.619	0.420	0.913	0.378	**0.787**	
top	0.841	0.726	0.420	0.942	0.334	0.648	**0.852**

Common method bias also known as common method variance or mono-method bias frequently occurs when same method used to measure correlation between construct. For example, in ordinary survey an individual responds to items at one particular survey at one point in time, this will expose to common method bias. In this research, we analyse common method bias by using Harmon's single factor test for survey data using confirmatory approach to assess the degree of common method bias. The factor is fully introduce for the analysis. Table 2 shows that single factor explain 38.744% of the total variance, which is below the 50% cutoff value. Thus, the following factor does not effected by common method bias.

Table 2. Harmen's single factor test.

	Initial Eigenvalues			Extraction Sums of Squared Loadings		
Factor	Total	% of Variance	Cumulative %	Total	% of Variance	Cumulative %
1	2.941	49.021	49.021	2.325	38.744	38.744
2	1.299	21.643	70.664			
3	0.840	14.008	84.672			
4	0.379	6.309	90.982			
5	0.290	4.829	95.810			
6	0.251	4.190	100.000			

Hypothesis Testing: Then, we examine our hypothesis using structural equation modeling. Structural equation modelling is a multivariate statistical analysis technique used to examine structural relationship. It is the combination of factor analysis and multi regression analysis and used to analyze structural relationship between measure variable and latent factor. The measurement model of fit indices in this project meet the cutoff point given.

All the hypothesis in this project are proven to be positive. As expected, shows in table 3 stakeholder pressure are positively related (β=0.299, p=0.006) to a firm's top managements support for implementing green supply chain management initiative (H1). Secondly, top management support for implementing green supply chain management initiative have positive significant impact (β=0.499, p=0.000) on monitoring supplier environmental achievement.

Table 3. Model fit measure.

Measure	**Estimate**	**Threshold**	**Interpretation**
CMIN	4.719	-	-
DF	7.000	-	-
CMIN/DF	0.674	Between 1 to 3	Excellent
CFI	1.000	›0.95	Excellent
SRMR	0.030	‹0.08	Excellent
RMSEA	0.000	‹0.06	Excellent
Pclose	0.797	›0.05	Excellent

Next, in this model the top management support on green supply chain management initiative fully mediates the relationship between stakeholder pressure on environmental initiative and monitoring supplier environmental initiative. To test mediation analysis we conducted two more statistical analysis using two different model M1 and M2 by constructing two structural equation-modelling models. Firstly, M1 we concentrate on the relationship of dependent variable monitoring supplier green supply chain initiative towards independent variable stakeholder pressure. Secondly, M2 we include the second independent variable of top management support. According to the result in Table 4, M1 the independent variable stakeholder pressure have positive significant to the dependent variable monitoring supplier to implement green supply chain initiative. After that, we control the mediator by adding top management support as independent variable mediator; the result is stakeholder pressure to monitoring supplier environment initiative is not significant. Thus, the results is top management is fully mediate.

Table 4. Mediation test.

PATH	M1	M2
Stakeholder pressure to monitoring supplier	0.287***	0.115
Stakeholder pressure to top management support		0.265***
Top management support to monitoring supplier		0.469***

****significant at 0.001 level*

4. Conclusion

Overall results of this study show that stakeholder pressure influence the implementation of green supply chain management initiative in SMEs by the important of top management support of environmental management initiative. Top management support plays an important role to secure crucial resources and to give leadership in risky environment. To achieve and maintain effective response to the environmental initiative and stakeholder pressure top management support is required. In addition, the research finding shows that the top management supports is fully mediate with stakeholder pressure and monitoring supplier environmental performance. We find that stakeholder pressure have significant influence for SMEs company to adopt green supply chain management same as top management support. We also find that top management support influenced company to monitor suppliers' environmental performance.

Acknowledgement

This research is supported by International Islamic University Malaysia (IIUM), SME Corporation Malaysia and funded by Fundamental Research Grant Scheme (FRGS16-019-0518), Ministry of Higher Education Malaysia to which the authors are grateful.

References

[1] Ni M, Leung D Y, Leung M K and Sumathy K 2006 *Fuel Processing Technology* **87** pp 461-472
[2] Walker H, and Jones N 2012 *Supply Chain Management: An International Journal* **17** pp 15-28
[3] Malaysia Production Corporation (MPC), 2010. *Sustainable Development Initiatives in Malaysia*
[4] Mollenkopf M, Stolze H, Tate W L and Ueltschy M, 2010 *Green, lean, and global supply chain* **40** pp 14-41
[5] Sarkis J 2010 *Journal of operation Management* pp163-176
[6] Frooman 1999 *Academy of management review* pp 191-205
[7] Srivasta 2007 *International Journal of Management Reviews* pp 53-80
[8] Cantor 2012 *Journal of Supply Chain Management* pp 31-51
[9] Malaysia SME Corp 2013 retrieved on 05 July 2016 from: http://www.smecorp.gov.my/.
[10] Han, Dong Ren 2004 *Study of the growth of small and medium-sized enterprises of science and technology type*
[11] Ali, Rosmah 2016 *Rasch measurement analysis for validation instrument to evaluate students technical readiness for embedded system*
[12] Li, Suhong 2005 *Journal of Operations Management* pp 618-641

IOP Conf. Series: Materials Science and Engineering **290** (2017) 012060 doi:10.1088/1757-899X/290/1/012060

Environmental Management Competitive Pressure Effect on SME Environmental Innovation Activities: A Green Supply Chain Perspective

A A Rashid[1], A A Sidek[1], S A Suffian[1] and M R C Daud[1]

[1]Department of Manufacturing and Materials Engineering, International Islamic University Malaysia (IIUM), Jalan Gombak, 53100 Kuala Lumpur, Malaysia

Email:atiahas@iium.edu.my

Abstract. The idea of assimilating green supply chain is to integrate and establish environmental management into the supply chain practices. The study aims to explore how environmental management competitive pressure influences a SME company in Malaysia to incorporate green supply chain integration, which is an efficient platform to develop environmental innovation. This study further advances green supply chain management research in Malaysia by using the method of quantitative analysis to analyze the model developed which data will be collected based on a sample of SMEs in Malaysia in manufacturing sector. The model developed in this study illustrates how environmental management competitive pressure from main competitors affects three fundamental dimensions of green supply chain integration. The research findings suggest that environmental management competitive pressure is a vital driving force for a SME company to incorporate internal and external collaboration in developing green product innovation. From the analysis conducted, the study strongly demonstrated that the best way for a company to counteract competitor's environmental management success is to first implement strong internal green product development process then move to incorporate external environmental management innovation between their suppliers and customers. The findings also show that internal integration of green product innovation fully mediates the relationship of environmental management competitive pressure and the external integration of green product innovation.

1.Introduction

Supply chain is a cumulative network of multiple business companies which involved in upstream and downstream effort to produce and deliver products or services to the customer. Environmental innovations and development in these supply chain are crucial to maximize performance and develop sustainable competitive advantage. Development of environmental management is indispensable for an organization in order to accomplish a particular end goal, which is to gain competitive advantage in marketplace. Small and medium sized enterprises (SMEs) in Malaysia is among the major contributor to the economic growth and play and important role the Malaysia's overall production network [1]. Increasingly, SMEs present key roles in supply chain management and provide significant impact on the supply chain processes.

This project aims to examine how environmental management competitive pressure from main competitors affects and influences an SME to produce new environmental innovations and implement green supply chain through its environmental management. This project applies Schumpeterian perspective of competition as theoretical frameworks and research model. The Schumpeterian perspective is an out and out theoretical view that is used to provide vital information on why firms engage in competitive behaviors [2]. By using the Schumpeterian perspective of competition as guidelines, the study investigates how rivalry with main competitors influences an SME organization to engage in environmental innovation activities.

2.Experimental procedures

The hypotheses will cover the internal and external integration of green product innovation and also green supply chain integration. It is vital for a company to always observe its' competitor's practices and strategies in the market. A company will gain competitive advantage based on Schumpeterian view on competition by taking action as counteract against its' main competitor. Developing green product innovation is one of the main act of countermove in respond to main competitor's green success. The green product development in this case will involve information sharing process in real-time within the company itself. The integration inside the company will involve product design, procurement, production, sales and distribution. Thus hypothesis number one is:

H1: Environmental management competitive pressure from main competitors positively influence a company internal integration of green product innovation.

H2a: A company internal integration of green product innovation will have positive influence on the company's customer integration of green product innovation.

H2b: A company internal integration of green product innovation will have positive influence on the company's supplier integration of green product innovation.

A company that has implemented internal integration of green innovation will have the ability to create strong external integration of green innovation. The strong level of coordination in internal integration will also be demonstrated in external integration. According to Hillebrand & Biemans [4] internal integration is vital and essential part in external integration. In order to form strong external collaboration, efficient internal functions is needed [5].

Target population: SMEs in Malaysia are mainly divided into 5 major sectors. They are services, manufacturing, construction, agriculture, mining & quarrying. However, this research will only focus on SMEs in the manufacturing sector. The research concentrates on SMEs with 20 to 200 employees with turnover from RM5 million to RM 200 million which satisfies the definition of SME provided by SME Corporation Malaysia [1]. Less than that number will be considered insufficient to produce desirable result for the study; and anything larger than that is considered as large enterprise and does not fall into SME definition in Malaysia.

Convergent validity exists if a group of measures that are theoretically related to each other are proven to be related or they measure one common factor. Convergent validity is assessed by ensuring that the statistical significance of the loadings. For the minimum level of item loadings, value of 0.7 is used as suggested by [6]. For the Composite reliability (CR), all items are well above 0.70 and the average variance extracted (AVE) for all construct is also above 0.50. Hence, convergent validity is proven for the measurement model. Discriminant validity for the constructs was tested assessing the inter-correlation between the constructs. The inter-correlation between the constructs is less than 0.70 which means the constructs have less than their variance in common [7]. The discriminant validity was also tested by comparing the average variance extracted (AVE) for each construct with the maximum square of correlation among all possible pair of constructs [8]. table shows that AVE for all cases is greater than maximum square of the correlation between all possible pairs of constructs.

Table 1. Convergent Validity and Realibility.

	CR	AVE	MSV	1	2	3	4
1. Environmental managemet competitive pressure from main competitors	0.786	0.647	0.635	**0.805**			
2. Internal integration of green product innovation.	0.815	0.688	0.635	0.797	0.830		
3. Supplier Integration of green product innovation	0.841	0.727	0.375	0.612	0.541	0.852	
4. Customer integration of green product innovation	0.740	0.59 0	0.370	0.60 8	0.568	0.568	0.768

Common method bias is demonstrated when the variations in the survey responses is causes by the measurement instrument rather than the actual disposition of the respondents the instrument trying to measure. To test potential effects of the common method bias, Harman's single factor test was aplied [9]. Using exploratory factor analysis, all factors are loaded onto a single factor. The factors are also constrained to make sure there is no rotation. This factor is introduced solely for the analysis. The table shows that the single factor explains 45.031% of the total variance which is below 50% cutoff value [10] where common method bias exist. Hence, common method bias is mitigated based on this test.

Table 2. Harman's Single Factor test.

Factor	Initial Eigenvalues			Extraction Sums of Squared Loadings		
	Total	% of Variance	Cumulative %	Total	% of Variance	Cumulative %
1	4.146	51.825	51.825	3.603	45.031	45.031
2	.983	12.289	64.114			
3	.926	11.569	75.683			
4	.615	7.681	83.365			
5	.464	5.806	89.171			
6	.365	4.563	93.733			
7	.256	3.195	96.928			
8	.246	3.072	100.000			

3.Result and Analysis

Global and local test are two type of classification of the hypotheses tested. The first step to test the hypotheses is to conduct the global test of model fit. The fit indices of the research model meet the cut-off values suggested by Hooper et.al [11]. This demonstrates that the model has good fit. Next, the researcher tests the local test of regression squared and significant p-values.

Table 3: Model Fit Measures.

Measure	Estimate	Threshold	Interpretation
CMIN	19.313	--	--
DF	17	--	--
CMIN/DF	1.136	Between 1 and 3	Excellent
CFI	0.992	>0.95	Excellent

SRMR	0.052	<0.08	Excellent
RMSEA	0.040	<0.06	Excellent
PClose	0.534	>0.05	Excellent

Environmental management competitive pressure from main competitors positively influence (β=0.861, $p<0.001$) a company internal integration of green product innovation (H1). A company internal integration of green product innovation positively influences (β=0.651, $p<0.001$) the company's customer integration of green product innovation (H2a). Finally, a company internal integration of green product innovation also positively related (β=0.737, $p<0.001$) with the company's supplier integration of green product innovation (H2b).

Next, the researcher assessed the mediating effects of internal integration of green product innovation on the relationship between environmental management competitive pressure from main competitors on customer integration of green product innovation and also on the relationship between environmental management competitive pressure from main competitors and supplier integration of green product innovation. To examine mediation within this model, the researcher incorporated the approach suggested by [11]. Two SEM models is incorporated in order to conduct this analysis. The first model (Model 1examines the direct relationship between the independent variable (environmental management competitive pressure from main competitors) and the dependent variables (customer and supplier integration of green product innovation). The second model (Model 2 introduces mediator variable, internal integration of green product innovation.

The results listed in table 4 shows that in Model 1, the direct effect of independent variable (environmental management competitive pressure from main competitors) on dependent variables (customer and supplier integration of green product innovation) is positively significant. In Model 2, independent variable (environmental management competitive pressure from main competitors) significantly influence the mediator variable (internal integration of green product innovation). The results also show that the mediator also has a positive and significant effect on dependent variables (customer and supplier integration of green product innovation). To analyze the type of mediation, the researcher examines the direct effect of the independent variable and dependent variables in Model 2. The results show that after the effects of mediator is controlled, the influence of environmental management competitive pressure from main competitors on customer and supplier integration of green product innovation is not statistically significant. This advocate that internal integration of green product innovation has full mediation effect on the dependent variables (customer and supplier integration of green product innovation).

Table 4: Mediation Analysis.

Paths	**Model 1**	**Model 2**
Competitive pressure to customer integration	0.705***	0.428
Competitive pressure to supplier integration	0.732***	0.429
Competitive pressure to internal integration		0.809***
Internal integration to customer integration		0.224***
Internal integration to supplier integration		0.317***

4.Conclusion

IOP Conf. Series: Materials Science and Engineering **290** (2017) 012060 doi:10.1088/1757-899X/290/1/012060

The model developed in this study illustrates how environmental management competitive pressure from main competitors affects three fundamental dimensions of green supply chain integration. This research suggests that environmental management competitive pressure is a vital driving force for a SME company to incorporate internal and external collaboration in developing green product innovation. Furthermore, the research findings demonstrated that internal integration of green product innovation fully mediates the relationship of environmental management competitive pressure and the external integration of green product innovation. The findings can also provide guidelines for SME companies managers how to react to their competitor's environmental management competitive pressure success.

Acknowledgement

This research is supported by International Islamic University Malaysia (IIUM), SME Corporation Malaysia and funded by Fundamental Research Grant Scheme (FRGS16-019-0518), Ministry of Higher Education Malaysia to which the authors are grateful.

References

[1] SME Corp Malaysia. 2016. *SME Profile and Importance to the Economy*. Retrieved from SME Corp Malaysia
[2] C M Grimm, H Lee and K G Smith 2006 *Strategy as action: Competitive dynamics and competitive advantage*
[3] B Hillebrand and W G Biemans 2004 *Journal Production Innovation Management* **vol 21** no 2 pp. 110–122
[4] W G Biemans 1991 *Technovation* **vol. 11** no. 3 pp 163–182
[5] D Westen and R Rosenthal 2003 *J. Pers. Soc. Psychol.* **vol. 84** no. 3 pp. 608–618
[6] D Alarcón, J A Sánchez and U P De Olavide 2015 *Assessing convergent and discriminant validity in the ADHD-R IV rating scale : User-written commands for Average Variance Extracted (AVE*), Composite Reliability (CR), and Heterotrait- Monotrait ratio of correlations (HTMT).," pp. 1–39
[7] J Henseler, C M Ringle and M Sarstedt 2014 *J. Acad. Mark. Sci.* **vol. 43** no. 1 pp. 115–135
[8] J F Hair, W C Black, B J Babin and R E Anderson 2010 *Multivariate Data Analysis, Vectors*. p. 816
[9] B R Eichhorn 2014 *Midwest SAS Users Gr.* pp. 1–11
[10] P M Podsakoff, S B MacKenzie and N P Podsakoff 2012 *Annu.* Rev. Psychol. **vol. 63** pp. 539–69
[11] D Hooper J Coughlan and M R Mullen 2008 *Structural Equation Modelling : Guidelines for Determining Model Fit* **vol. 6** no. 1 pp. 53–60
[11] D P MacKinnon, A J Fairchild and M S Fritz 2007 *Mediation analysis.*, Annu. Rev. Psychol. **vol. 58** pp. 593–614

IOP Conf. Series: Materials Science and Engineering **290** (2017) 012061 doi:10.1088/1757-899X/290/1/012061

Mechanical Properties of Gracilaria Lichenoides Reinforced Bioplastic Film

M Othman [1], N Hasmida [1], Z Halim[1], N A Jamal [1], K Khalid [2], N Zakaria[3], S A M Al-Bat'hi [1]

[1]Department of Manufacturing and Materials Engineering, International Islamic University Malaysia (IIUM), Jalan Gombak, 53100 Kuala Lumpur, Malaysia
[2]Laboratory Management Program, Technical Service and Technology, Malaysian Agricultural Research and Development Institute (Mardi), 42400 Serdang
[3] Faculty of Biocomposite Technology, UniversitiTeknologi MARA, 40450 Shah Alam, Selangor, Malaysia

Email:maizatulnisa@iium.edu.my

Abstract. In this study, the mechanical properties of gracilaria lichenoides with additional of plasticizer and filler were evaluated. For samples with the addition of 5.5% of plasticizer, produced low tensile strength and this results is vice versa with elongation at break results. The tensile strength of the bioplastic continuously decreases from 14.8 to 2.7MPa as the plasticizer increases up from 1.5% to 5.5%. This phenomenon was analyses under scanning electron microscope (SEM), it shows that, the formation of pores and crystal agglomeration at sample with 5.5% glycerin. To alter these flaws, squid bone is introduce as filler to the bioplastic. Based on the analysis, additional of 6% filler content did alter the tensile strength up to 8 MPa with 3% of the elongation at break.

1. Introduction

Almost every commercial type of plastic currently used in industry is based on petrochemical which is chemical products derived from petroleum [1]. However, petroleum is a finite resource because it formed over millions of years and is being used faster than it is being formed. They will eventually run out. Besides, as the petrochemical based polymers produced, consumed and discarded into the environment, they ending up as non-degradable waste [1,2]. Recently, concerns focus on the potential impact of artificial substances on human health, waste management and environmental damage. While recycling and incineration represent as the alternatives to landfill disposal of plastic products, they understandably also have their own weaknesses [3]. Recycling plastics will deteriorate the quality of the plastic itself and at the same time the costs can also be high. Therefore, intensive research and industrial efforts have focused on the movement in the plastics industry towards sustainability by production of biopolymers as a green alternative to synthetic polymers. Basically, there are four main types of biopolymer based respectively on which are starch, sugar, cellulose and synthetic materials. The main strategies may be followed in synthesizing a polymer which is to build up the polymer structure from a monomer through chemical polymerization. The alternative is to use a naturally occurring polymer and chemically modify it to give it the desired properties. The current and future development in plastics industry is mainly concerned with using these biopolymers to form more viable materials for mass plastic production by increasing its availability and reducing prices. Algae based bioplastics have been a recent trend in the era of bioplastics compared to traditional methods of utilizing feedstock of corn and potatoes as plastics [4]. While algae-based plastics are in their commencement, once they are into commercialization they are likely to find applications in a wide range of industries. Algae are an interesting natural resource because they proliferate quickly and not impinging on food production [5]. They need nothing but sunlight and a bit of waste water to grow on. Hence, Algae are not harmful to marine's life, instead considered unsightly and customers want to control or eliminate it. They serve as an excellent feedstock for plastic production owing to its many advantages such as high yield and the ability to grow in a range of environments.

2. Experimental Procedures

The gracilaria lichenoides powder used for preparation of biopolymer plastic and plasticizer which was glycerin (99.0-101.0%, HmbG Chemicals) were purchased from Progressive Scientific Sdn Bhd. The acetic acid used as additive was purchased from local market. Squid bone used as filler was obtained from Pasar Besar Gombak, Selangor Malaysia. The gracilaria lichenoides powder, water, acetic acid and glycerin were mixed together in a single beaker. After that, the mixture was heated on the hot plate, set temperature of 90°C and magnetic stirrer velocity of 3 rpm. Different formulations were prepared by varying glycerin content in each formulation. A series of agar blends were processed into films by casting method. Pure gracilaria lichenoides films were prepared as a control. The film-forming solution was cast into a square mold. After drying in an oven at temperature of 50°C for 10h, the film was obtained. Meanwhile, with the addition of squid bone filler, the squid bone was washed and dried in the lab dryer for temperature of 40°C within 30 min. Then, it was ground by using blender, model Panasonic MX-798S. Then, it followed the same process took place was same as making algae based matrix biopolymer.

Tensile test was applied according to the technical standard method ASTM D882 using the tensile machine (SHIMADZU, Autograph AGS-X). Measurements were carried out at 25°C and crosshead rate of 5 mm/min with the load of 5 kN. For each sample, 7 specimens were tested. The sample preparation first been done by using Tensile Cutter machine (Standard Toggle Press) to form dumbbell cut. Morphological analysis for the thin film of algae based bioplastic samples were carried out by using Scanning Electron Microscope (JEOL JSM 5600). Before the electron microscopy observation, the surfaces were placed on the aluminum stubs by using two-sided adhesive tape and sputter coated with an alloy of 60% gold (Au) and 40% palladium (Pd) using (JEOL JFC-1600). SEM imaging was performed at an acceleration voltage of 7 kV.

3. Results and Discussion

The tensile strength and elongation at break of the algae sample without filler and different percentage of glycerin content are shown in figure 1, the tensile strength of bioplastic sample decreases with incerasing amount of plasticizer used. As the content of glycerin increase from 1.5% to 5.5%, the tensile strength (TS) values of the bioplastic decreases from 14.8 to 2.7 MPa. Polyol, such as glycerol, has two plasticizing effects; first as a result of its presence in the film and second because its intensively hygroscopic character tends to draw additional water into the matrix [6]. Therefore, as the plasticizer amount increased, the numbers of plasticizer molecules that reside between polymer chains were also increased. Thus, the intermolecular spacing between the polymer chains networks also increased and allowing the water molecules to go through the films, resulting in lowered the tensile strength. In addition, absorption of water vapor from the surrounding due to hydrophilic properties of plasticizer can cause swelling of the polymer film. This can also be the possibility of reduction in interaction between the polymer chains. The results obtained was in agreement with the previous research conducted by Amman Ullah et al.,[2] whereby the weak tensile properties of albumin glycerol plastics could have been explained through disordered conformations as the relatively large chemical structure of glycerol prevented any long range plasticization to occur and vice versa [7].

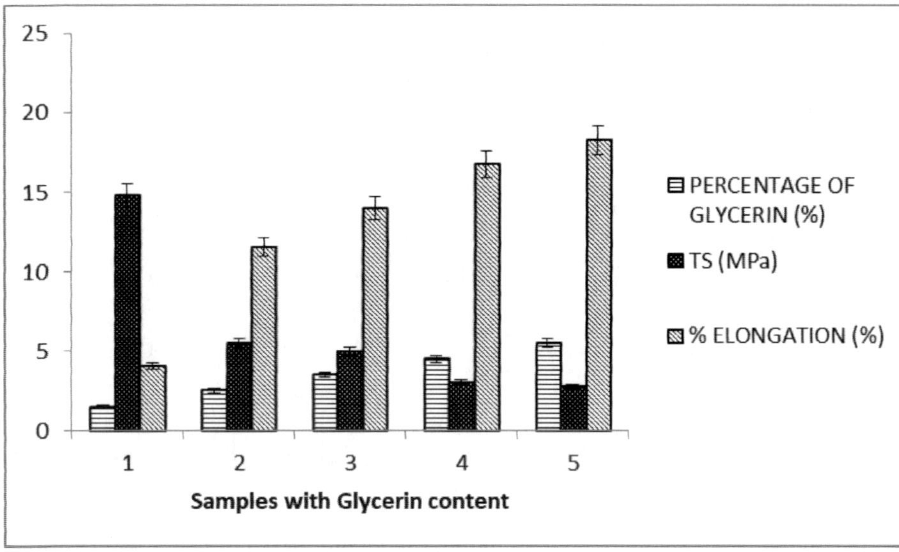

Figure 1. Samples with glycerin content on the tensile strength and elongation at break.

In terms of the extension that occurred before breaking, the sample with higher amount of glycerin used as plasticizer produced higher amount of extension which elongate nearly 18.3% on average before a ductile break. Meanwhile, the other sample with lower amount of plasticizer only extended around 4.1% before breaking. This clearly showed that, the plasticizer molecules affect the strength of the interactions between adjacent molecules, increasing the distance and thus reducing the internal force which gives more flexibility to the films. The possible reason that can be suggested was the bonding that occurred within the structure as plasticization occurred, which had been found in previous research. The study suggested that the tensile test results show incorporating the red algae bioplastics with glycerol did not improved their ductility and the use of glycerol even reduced their tensile strength and energy absorption [8].

Due to higher tensile strength poses by the sample with 1.5% of glycerin, further experimental research had been conducted on it with the addition of squid bone filler to enhance its properties. Based on figure 2, reinforcement of sample by squid bone filler led to increase in tensile strength up to concentration of 3%. At that concentration, squid bone filler seem to form skeleton with interconnected grain in bioplastic that bear mechanical load. However, at concentration of 5%, the mechanical strength was lower. This shows that, maximum amount of filler which gave the highest value of tensile strength was up to 3% only and the value obtained was two times smaller than pure bioplastic matrix. This result in accordance to Sundaram et al. [4] and Zhang et al.[7] that mention the trend of increasing tensile strength of bio-composite does not follow a straight line relationship with filler concentration [9,10]. The filler content affects the interparticle distance and the stress state of matrix polymer surrounding the voids [11]. At low filler concentration, the particle distance is long so the interparticle matrix ligament lies in plane strain state which is hard to yield so that mechanical strength cannot be improved. But if the filler concentration is increased so that the interparticle distances reach the suitable range, the interparticle matrix ligament lies in plane press state which can be plastic yield easily and the mechanical strength can be improved [12].

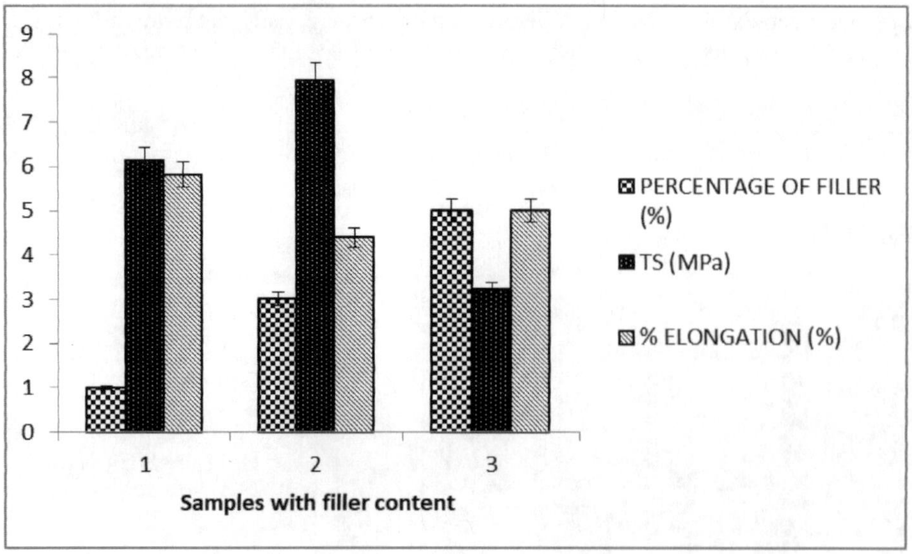

Figure 2: Sample with filler content on the tensile strength and elongation at break.

It was clear that as shown in figure 2 that increasing content of squid bone filler improved its tensile strength. However, this strength achieved maximum value for sample with 3% of squid bone filler, at 5% the tensile strength was going down due to the limited binder bonding the particle.

When all bioplastic were analyzed through scanning electron microscopy, the images illustrated that samples with 1.5% glycerin content gave more fairly homogenous mixture of algae and the plasticizer microstructure compared to samples with 5.5% glycerin content. The structure of thin film with lower amount of plasticizer is considered as homogenous and dense due to the continuous cross section without defects up to the highest magnification used which was 1000x as in figure 2.

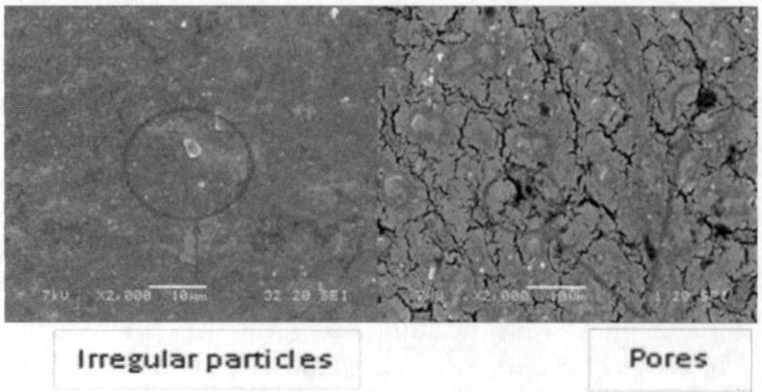

Figure 3: SEM micrographs of Algae thin films at magnifications of 1000x.

When higher amount of plasticizer being added in to the biofilm, up to 5.5%, a few irregular particles were visible on the surface of the biofilm. The microstructure observed have pores and cracks which clearly can be seen started at the magnification of 500x and above. From the analysis, it was found that, the important to use the right amount of plasticizer in order to alter the structure and morphology surface of the biofilm. This finding was supported by the fact that glycerol is known to enter polysaccharides chain interior, disrupted intermolecular and intramolecular hydrogen bonds and made the polymer plastic forming a continuous phase of plasticized film [13]. Furthermore, at higher percentage of plasticizer, the surface becomes uneven and heterogeneous as pores and cracks structure appeared especially at the highest magnification as shown in the figure 3. This phenomenon can be explained with the reference to previous study conducted by Jones et.al.,[12] regarding the moisture and glycerol effect occurred during the SEM been operated [9-12]. According to them, glycerol may leach from the plastic on a much smaller scale with the moisture being removed from the sample when SEM chamber was sealed under high vacuum. As the glycerol slowly released into the environment, it left pores inside the plastic and the pores would be able to absorb water from the surrounding environment. Furthermore, the crack formation resulted when the SEM chamber was sealed and moisture was vacuumed from the testing chamber. This induced strong and rough dehydration on the sample which creates cracks visible in the observation. Based on the results from this morphological analysis, it can be concluded that the properties of plastic thin films may altered as they were able to absorb moisture from the atmosphere and influenced the homogeneity of structure.

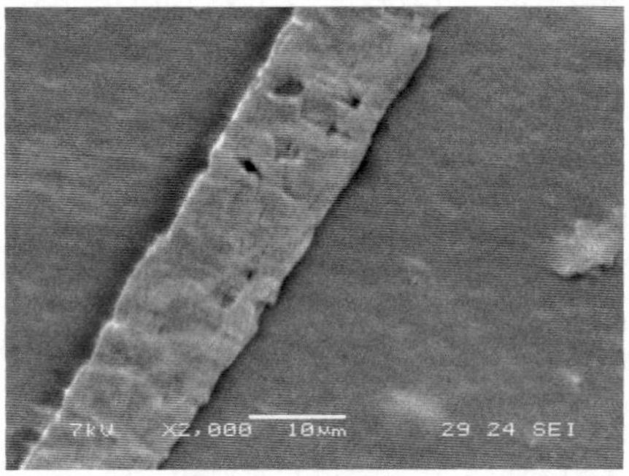

Figure 4: SEM micrographs of biofilms reinforced at magnifications of 2000x.

The morphology of squid bone filler and Algae matrix thin films were investigated by SEM. figure 4 shows the surface morphology of Algae bioplastic blends with squid bone filler contents of 1g, 3g and 5g. With the addition of 1g and 3g of fillers, it was possible to see a compatible microphase in the algae matrix biopolymer even though some squid bone flakes appeared on its structure. Generally, these indicate that the two materials are compatible to each other. A quite different morphology was observed for 5g content of filler while there were some small holes may have been caused by the exit of filler and air during the processing procedure. Besides, there were some small particles could be found. This change in behavior means that the adhesion between two components has become poorer as the filler content increases.

4. Conclusion

From the study, it can be concluded that roles of glycerin have a considerable influence on the mechanical properties of algae based bioplastic. Based on the mechanical properties of the bioplastic, it was revealed that the amount of glycerol used as plasticizer at 1.5%, the properties observed of the fabricated thin films were the best by having the highest tensile strength but in reverse effect on the percentage of elongation. In addition, the squid bone as filler for the algae matrix bioplastic gave highest tensile strength when the maximum amount of filler achieved at 3%. The difference in amount of glycerol and filler used did play a significant role affected the mechanical properties of the bioplastic. Evaluation work on morphological study also been carried out under SEM observation which pointed out that the change in surface morphology and increase in amorphous content on the addition of plasticizer and filler.

Acknowledgement

This work was financially supported by Research Iniatiative Grant Scheme (RIGS 16-176-0340), Universiti Teknologi MARA (UiTM), Malaysian Agricultural Research and Development Institute (MARDI) and International Islamic University Malaysia (IIUM).

IOP Conf. Series: Materials Science and Engineering **290** (2017) 012061 doi:10.1088/1757-899X/290/1/012061

References

[1] Coupland, IN., Shaw, N.B., Monahan, FJ., a'Riordan, D. & O'Sullivan, M., 2000 *Journal of Food Engineering* **43** pp 25-30
[2] Aman Ullah T.V., Bressler D., Elias A.L.,& Jianping W., 2011 *Biomacromolecules* **12** pp 3826–3832
[3] M. Nizar Machmud, Reza Fahmi,Rohana Abdullah &Coco Kokarkin,. 2013 *J. Sci. Eng*. pp 81-88
[4] Sundaram, J., Durance, T.D., and Wang, R., 1997 Acta Biomaterialia **4** pp 932-942
[5] Zhang, Q.; Liu, L.; Ren, L. & Wang, F. *2008 Journal of Applied Polymer Science* **64** pp 2057–2263
[6] Bartezak, Z., Argon, A.S., Cohen, R.E., and Weinberg.M., 1999 *Polymer* **40** pp 2347- 2365
[7] Zhang, S., Cao, X.Y., Ma, Y.M., Ke, Y.C., Zhang, J.K., and Wang, F.S., 2011 *eXPRESS Polymer Letter* pp 581-590
[8] Cides L.C.S., Araújo A.A.S., Filho M.,& Matos., 2006 **84** pp 441
[9] Castelló M., Dweck J., &Aranda. 2009 *J Therm Anal Calorim* **97** pp 627–630
[10] Sudip Mondal, Biswanath Mondal, Apurba Dey,and Sudit S. Mukhopadhyay 2012 *Journal of Minerals & Materials Characterization & Engineering* **11** pp 55-67
[11] Norazowa I, Wan M.Z.W.Y., Maizatulnisa O., Khalina 2011 *Journal of Polymer Reinforced Plastics and Composites* **30** pp 381-388
[12] Jones, A., Zeller, M.A., & Suraj S., 2013 *Progress in Biomaterials* **2** pp 12
[13] Cadman. J., Chen, Y., Zhou, S., Li, Q., 2012 *Mater. Sci. Eng. C-Biomimetic Supramol.*

IOP Conf. Series: Materials Science and Engineering **290** (2017) 012062 doi:10.1088/1757-899X/290/1/012062

Extraction of caustic potash from spent tea for biodiesel Production

Sarina Sulaiman[1], Ahmad Faiz Che Fisol[1], Atikah Mohamed Sharikh[1], Dzun Noraini Jimat[1], Parveen Jamal[1]

[1]Department of Biotechnology Engineering, International Islamic University Malaysia (IIUM), Jalan Gombak, 53100 Kuala Lumpur, Malaysia

Email:sarina@iium.edu.my

Abstract. Biodiesel is an alternative to non-renewable fossil fuels due to its low gas emission and economical value. This study aims to extract caustic potash (KOH) from spent tea and to optimize the transesterfication process based on parameters such as amount of catalyst, reaction temperature and methanol to oil ratio. The spent tea was first dried at 60°C prior to calcination at 600°C for two hours. Caustic Potash were extracted from the calcined spent tea. The transesterification process was done based on Design of Experiments (DOE) to study the effects of amount of catalyst ranging from 0.5 wt % to 2.5 wt %, reaction temperature from 55°C to 65°C and methanol to oil ratio from 6:1 to 12:1 at a constant agitation rate of 300 rpm for three hours. The calcined spent tea produced was recorded the highest at 54.3 wt % and the extracted catalyst was 2.4 wt %. The optimized biodiesel yield recorded was 56.95% at the optimal conditions of 2.5 wt % amount of catalyst, 65°C reaction temperature and 9:1 methanol to oil ratio.

1. Introduction

The utilization of diesel fuel in the world is tremendous, ranging from agriculture and transport towards commercial and domestic industries in generating power and mechanical energy. However, diesel fuel is a non-renewable source of energy that is running low on natural sources. This could be a major issue if no new alternatives can replace diesel fuel. In this research, biodiesel fuel made from vegetable oils or animal fats are studied for its promising features that attract huge attention all over the world for its availability, eco-friendly, non-toxicity and biodegradability [1].

The process of converting waste cooking oil (WCO) to biodiesel is called transesterification, a process that requires catalyst in order to react. By using WCO for biodiesel production, it can reduce the cost of raw materials for biodiesel as well as the problem of environmental pollution. Phan & Phan, 2008 study shows that the amount of usage of the waste cooking oil affects the amount of catalyst needed in the production process [2]. In their study, the WCO acid values ranges from 0.67 to 3.64 mg KOH/g when the sample WCOs are used once or a few times. Low free fatty acid content results in lower concentration of potassium hydroxide (KOH) needed to be added. On the contrary, the higher free fatty acid requires an addition of KOH to compensate for this acidity. Current biodiesel production uses conventional catalysts such as strong acid-based or basic catalysts. These catalysts are highly corrosive to the equipment, relatively expensive due to the production and downstream separation required, and has serious environmental effect.

Another occurring problem is the usage of homogenous catalyst in the process of transesterification. Homogenous catalysts cannot be regenerated or reused, as it is consumed during the reaction. The separation of catalyst will also be much harder due to the homogeneity, which uses more equipment thus increasing the production cost.

Spent tea was chosen as the restaurant waste for low cost catalyst in transesterification because tea is one of the most abundant wastes in Malaysia. Annually, Malaysia produces less than five million kg of tea while the per capita consumption of tea in Malaysia is about half kg and has the annual demand of nine million kg. Such biowaste can be reused as precursor in catalyst production can bring significant change to the environment. Kumar [3] found that materials containing alkali metals such as potassium, calcium and sodium may produce potassium carbonate upon heating [3]. When burnt and oxidized in the presence of air, metal oxides are formed. Carbon dioxide produced through burning will react with potassium oxide to produce potassium carbonate while in the presence of water vapour will produce potassium bicarbonate and potassium hydroxide. Thus, by calcination of tea spent, KOH can be produced to be later used in the transesterification process.

In this study, spent tea, which is a waste from restaurant, was chosen as catalyst. Caustic potash was extracted from spent tea in order to render the transesterification process achievable and able to attain high yield of biodiesel. This paper studies the extraction of caustic potash from spent tea as low cost catalyst and the optimization of methanol to oil ratio, amount of catalyst and temperature in biodiesel production from WCO.

2. Experimental procedures

For extraction of catalyst, used teabags were collected from cafeterias around International Islamic University Malaysia (IIUM), the spent tea was first washed with tap water and then distilled water a few times. Then, it was dried in an oven at 60°C for five hours. After drying, calcination of the spent tea was conducted as the final step of extracting caustic potash from the waste. The dried tea was put in an aluminum foil and into the furnace at 600°C for two hours. After that, it was left to cool inside the furnace for five hours.

WCO was also collected from the cafeterias and methanol was bought from Sigma Aldrich Sdn Bhd. Transesterification was done by mixing WCO with the calcined spent tea and methanol with the amount based on the DOE. The catalyst caustic potash KOH is stirred in methanol and dissolved in a 250 ml beaker. 20-30g of WCO is then transferred into the biodiesel incubator shaker along with the KOH/methanol mixture. The incubator was set to be running for 3 hours at 300 rpm with varying parameters set based on the DOE designed by the Design Expert 10 software as shown below.

Table 1. Reaction parameters as per generated by DOE

Run	Factor 1: Catalyst amount, A (wt %)	Factor 2: Temperature, B (°C)	Factor 3: Methanol:oil	Response: Yield (%)
1	1.5	60	6	
2	2.5	55	12	
3	1.5	60	9	
4	0.5	60	9	
5	2.5	55	6	
6	2.5	65	12	
7	0.5	55	6	
8	1.5	55	9	
9	1.5	60	9	
10	1.5	60	9	
11	1.5	60	9	
12	2.5	60	9	
13	0.5	55	12	

14	2.5	65	6
15	1.5	60	12
16	1.5	60	9
17	1.5	60	9
18	0.5	65	12
19	1.5	65	9
20	0.5	65	6

Next, the samples were transferred into a 50ml disposal tube for centrifugation and then heated in the oven overnight at the temperature of 68°C. It is then left overnight to cool down and separated by gravity in a separating funnel. The layer formed on the top is the methyl ester while the bottom layer is the by-product glycerol. The bottom layer is removed and put in a separate container while the upper layer was heated to remove excess methanol.

Based on Hosseini et al. [4], to analyze the biodiesel produced gas chromatography (GC) was conducted to find the value of biodiesel produced yield content and the value of conversion from triglycerides to methyl esters. According to Cheng et al. [5], using hexane as the solvent, the chromatography was done under the following conditions: Helium as carrier gas, oven temperature of 150 °C initially for 1 minute and then increased to 250°C at the rate of 15 °C /min. The FAME content percentage is calculated using the formula below:

$$FAME\ Content\ \% = \frac{\text{weight of FAME contained in crude biodiesel}}{\text{weight of biodiesel}} \times 100\% \quad (1.0)$$

3. Results and Discussion

3.1. Extraction of Caustic Potash from Spent Tea

Based on the titration done, the calcination of catalyst from spent tea produced low amount of KOH, only 3.43 wt %. This lead to low conversion of biodiesel during the transesterification process as the amount of catalyst used in the reaction was weighted based on the weight of calcined spent tea, not the weight of KOH.

3.2. Optimization of the Transesterification Process

Based on the ANOVA study, the relationship of the three parameters was observed. In figure 1, while keeping the methanol to oil ratio at 9:1, the three dimensional response surface showed that the biodiesel yield increases with the amount of catalyst and temperature of reaction. From 55°C to 65°C, the yield of biodiesel showed a significant increase to 56.945% when the amount of catalyst is at 2.5 wt %. According to Vyas et al. [6], high amount of KOH leads to saponification and decreases the conversion yield [6]. However, in this case the amount of KOH in each gram of catalyst is only 0.034g thus the biodiesel yield is the highest when 2.5wt % of catalyst was used.

Figure 2 shows that at 0.5 wt % to 1.0 wt % catalyst amount the biodiesel yield did not deviate much from the value of 42.3% when the reaction temperature was kept at 60°C. However, when the amount of catalyst was at 2.5 wt %, the highest biodiesel yield was shown to be at 49.61% with methanol to oil ratio at 9:1. The yield value dropped when the methanol to oil ratio deviates further from the 9:1 ratio despite the amount of catalyst is at 2.5 wt %. Excess methanol is required to shift the reaction equilibrium forward. However when high amount of methanol was mixed the basic catalyst, it produces water that will cause saponification to occur [7]. Another reason to the decrement of biodiesel yield is the excess methanol interfering with the methyl ester and glycerol separation by increasing glycerol solubility [8].

IOP Conf. Series: Materials Science and Engineering **290** (2017) 012062 doi:10.1088/1757-899X/290/1/012062

In figure 3, when the amount of catalyst was kept constant at 1.5 wt %, the increment in conversion yield occurred when the methanol to oil ratio increased from 6:1 to 9:1. The yield reached its peak at 9:1 but decreased when the ratio was further increased to 12:1. Relative to the reaction temperature, the biodiesel conversion yield increased as the temperature was increased to 65°C. Generally, at high temperature speeds up the transesterification reaction and increases the yield [2]. Low yield at lower temperatures may due to incomplete reaction. Low yield at high temperature was probably caused by acceleration of the saponification reaction [9].

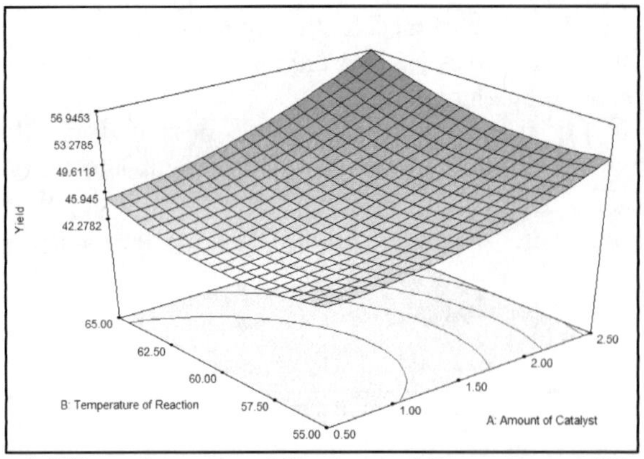

Figure 1. Effect of Amount of Catalyst and Temperature of Reaction

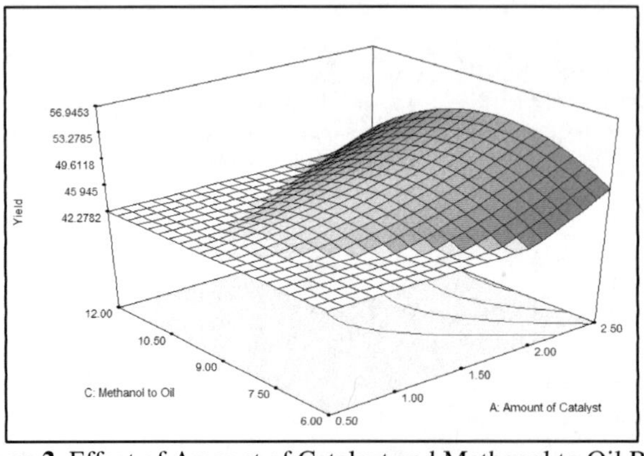

Figure 2. Effect of Amount of Catalyst and Methanol to Oil Ratio

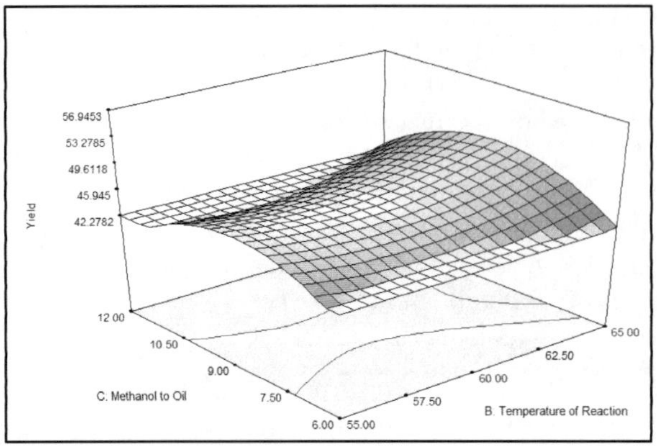

Figure 3. Effect of methanol to oil ratio and temperature of reaction

4. Conclusion

This research was conducted to extract caustic potash from spent tea as a low cost catalyst and to optimize the process of transesterification for biodiesel production by varying the methanol to oil ratio, amount of catalyst and temperature of the reaction based on the DOE. The amount of catalyst extracted from the spent tea averaged about 53.32% in weight. However, titration was done on the basic catalyst and it was found only 0.34 KOH was found in 10g of extracted catalyst. This low value of KOH resulted in low conversion of WCO to biodiesel. For the transesterification process, the optimum condition for the reaction was proposed by ANOVA to be 2.31 wt % of catalyst, 10.32 methanol to oil ratio at 53°C for three hours reaction time.

Acknowledgement

This research is funded by Ministry of Higher Education-FRGS 2013 grant (FRGS 13-079-0320) and TWAS-COMSTECH Research Grant_REF:15-333 RG/REN/AS_C – FR3240288948.

References

[1] Demirbas A 2009 *Energy Conversion and Management.* **50** 923-927
[2] Phan A NandPhan T M 2008 *Fuel.* **87** 3490-3496
[3] Kumar A 2013 *International Journal of Chemical Sciences.* **11**
[4] Hosseini M, Nikbakht A MandTabatabaei M 2012 *Modern Applied Science.* **6** 40
[5] Cheng J, Huang R, Yu T, Li T, Zhou JandCen K 2014 *Bioresource technology.* **151** 415-418
[6] Vyas A P, Verma J LandSubrahmanyam N 2011 *Advances in Chemical Engineering and Science.* **1** 45
[7] Myint L LandEl-Halwagi M M 2009 *Clean Technologies and Environmental Policy.* **11** 263-276
[8] Musa I A 2016 *Egyptian Journal of Petroleum.* **25** 21-31
[9] Agarwal A 2007 *Prog Energ Combust.* **33** 233 - 271

IOP Conf. Series: Materials Science and Engineering **290** (2017) 012063 doi:10.1088/1757-899X/290/1/012063

Improvement of MRR and surface roughness during electrical discharge machining (EDM) using aluminum oxide powder mixed dielectric fluid

A A Khan[1], A K M Mohiuddin[2], M A A Latif[1]

[1]Department of Manufacturing and Materials Engineering, International Islamic University Malaysia (IIUM), Jalan Gombak, 53100 Kuala Lumpur, Malaysia
[2] Department of Mechanical Engineering, International Islamic University Malaysia (IIUM), Jalan Gombak, 53100 Kuala Lumpur, Malaysia

Email: aakhan@iium.edu.my

Abstract. This paper discusses the effect of aluminium oxide (Al_20_3) addition to dielectric fluid during electrical discharge machining (EDM). Aluminium oxide was added to the dielectric used in the EDM process to improve its performance when machining the stainless steel AISI 304, while copper was used as the electrode. Effect of the concentration of Al_20_3 (0.3 mg/L) in dielectric fluid was compared with EDM without any addition of Al_20_3. Surface quality of stainless steel and the material removal rate were investigated. Design of the experiment (DOE) was used for the experimental plan. Statistical analysis was done using ANOVA and then appropriate model was designated. The experimental results show that with dispersing of aluminium oxide in dielectric fluid surface roughness was improved while the material removal rate (MRR) was increased to some extent. These indicate the improvement of EDM performance using aluminium oxide in dielectric fluid. It was also found that with increase in pulse on time both MRR and surface roughness increase sharply.

1. Introduction

Electrical discharge machining (EDM) is one of the most extensively used non- conventional material removal processes. EDM works by using the principle of erosion of metals by using spark discharges [1]. The surface finish of the workpiece produced depends on many factors. The basic die sinking EDM system consists of an electrode and workpiece connected to a direct current (DC) power supply and placed in dielectric fluid. Usually the tool is connected to the negative terminal and the workpiece is connected to the positive terminal of the power source.

The dielectric breaks down and a transient spark discharges through the fluid when the potential difference between the tool and the workpiece is sufficiently high. This will remove a very small amount of metal from the workpiece surface [2]. Since the surface of the workpiece at the location of the previous discharge is now separated from the tool by a greater distance, this location is less likely to be the site of another spark until the surrounding regions have been reduced to the same level or below. The individual discharges occur hundreds or thousands of time per second to ensure that constant erosion of the entire surface occurs in the area of gap [3]. The discharge occurs at the location where the two surfaces are closest.

Electrical Discharge Machining (EDM) is one of the most reliable and precise manufacturing processes applicable for creating complex geometries [4]. The superior performance of EDM has been validated in applications involving materials with high strength, high hardness, or complicated shapes. EDM erodes material with the high heat of electrical discharges generating sparks between an electrode tool and the workpiece. EDM manufacturing is economical and relevant when high accuracy is required. Nevertheless, EDM process has some drawback such as formation of white layer on the workpiece surface due to deposition of materials melted during electrical discharges. As a result, the quality of surface diminishes. Thus, the aim of this study is to develop an EDM process for improved work surface quality and increase MRR. A comparatively new approach of powder mixed electrical discharge machining (PMEDM) has been developed for this purpose. In the PMEDM, a suitable fine abrasive powder material is mixed with the dielectric fluid.

Aluminium oxide is a chemical compound of aluminium and oxygen with the chemical formula (Al_2O_3). It specifically identifies as aluminium (III) oxide and commonly called alumina. PMEDM is used to produce a surface coated with hard layer of Al_2O_3.

2. Experimental procedures

This project was aimed to study the effect of powder added dielectric fluid. The workpiece and the electrode material for this study were stainless steel and copper. Dimension of the workpiece was 13mm x 14mm x 44mm. DOE version Design Expert 6.08 was used in order to reduce the number of experiments. Experiments were conducted on an EDM machine shown in figure 1. Dielectric fluid used was kerosene. During the experiments current and pulse off time were kept constant at 6.5 A and 9.0 μs respectively. Variable parameters were powder concentration of Al_2O_3 and pulse off time (t_{off}). The size of the powders was 35-40μ. The experimental plan is shown in table 1. Square holes of 6 mm x 6 mm were machined on the workpiece to a depth of 1 mm. The weight of the workpiec was measured before and after each of the experiments. The responses were material removal rate (MRR) and work surface finish (R_a). Ra was measured using a surface roughness-measuring machine shown in figure 2.

Figure 1. EDM Die-Sinking Machine

IOP Conf. Series: Materials Science and Engineering **290** (2017) 012063 doi:10.1088/1757-899X/290/1/012063

Table 1. Experimental plan

Std	Run	Block	Factor 1 A : Powder Concentration (mg/L)	Factor 2 B : Pulse on time (μs)
3	1	Block 1	0.0	7.5
6	2	Block 1	0.3	10.0
1	3	Block 1	0.0	6.0
5	4	Block 1	0.0	10.0
2	5	Block 1	0.3	6.0
4	6	Block 1	0.3	7.5

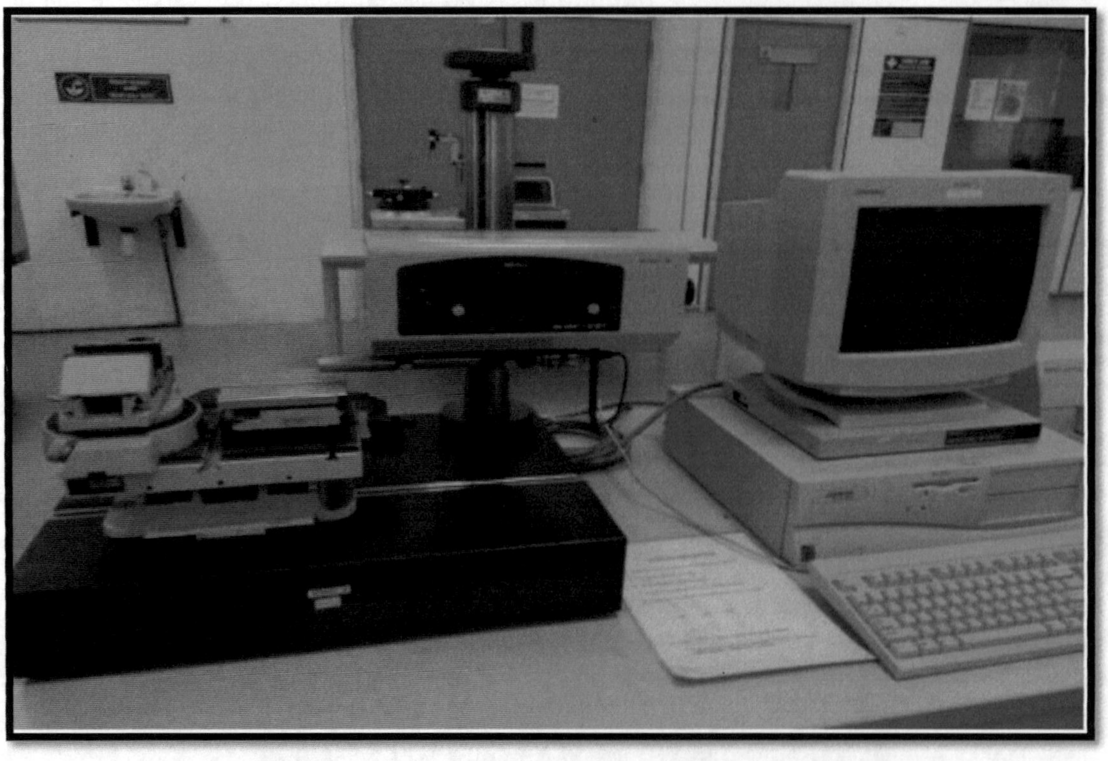

Figure 2. Mitutoyo Surface Test Equipmen

3. Results and Discussion
Measurements of Ra and MRR are shown in table 2.

Table 2. Experimental results

No	T_{on}	Powder Concentration (mg/L)	Time	Surface roughness Ra (µm)	Weight of workpiece Before (g)	Weight of workpiece After (g)	MRR (g/min)
1	6	No powder	23min 5s	8.63µm	192.0451	191.1802	0.0375
2	7.5		5min 1s	15.45µm	191.1802	190.4516	0.1452
3	10		3min 1s	25.41µm	190.4516	189.2405	0.4015
4	6	0.3	6min 29s	8.7µm	189.2405	189.0251	0.0332
5	7.5		6min 18s	13.11µm	189.0251	187.7185	0.2074
6	10		1min 47s	20.53µm	187.7185	186.9711	0.4191

Material Removal Rate (MRR). The contour plot in figure 3 shows the relationship between powder concentration and pulse on time on MRR. It can be seen from the figure that MRR increases with increase in pulse on time. The maximum MRR is 0.4191 g/min with the highest pulse on time at 10µs and powder concentration of 0.3mg/L. Due to the presence of powders the electrical discharge is scattered and electrons passes more uniformly in the discharge area. As a result, MRR increases [5, 6, 7]. It can also be observed that with increase of T_{on}, MRR increases more sharply. As T_{on} increases, the workpiece gets more time to absorb heat energy that melts and vaporizes the work material [8]. 3D plot for influence of powder concentration and Ton on MRR is illustrated in figure 4.

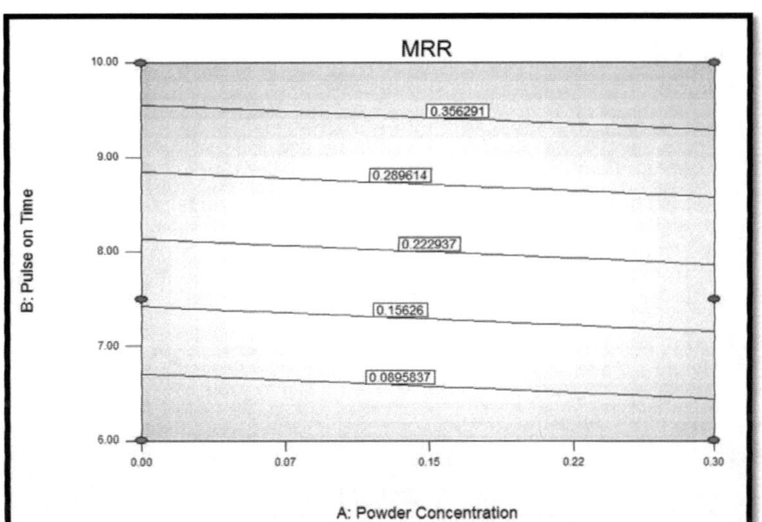

Figure 3. 2D plot for powder concentration and T_{on} on MRR

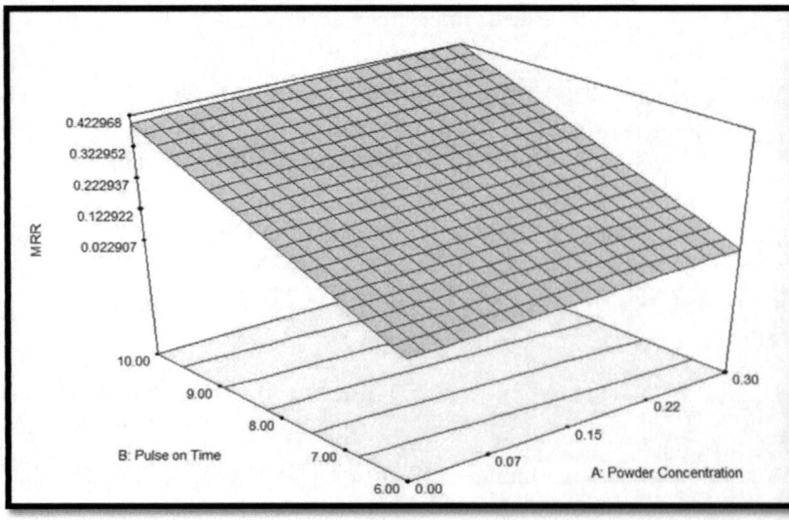

Figure 4. 3D surface plot of MRR versus powder concentration and T_{on}

The contour plot in figure 5 shows the relationship between powder concentration and pulse on time on surface roughness. From the graph, it was observed that minimum surface roughness at Ton 6μs. It can be observed that with increase in powder concentration surface roughness decreases. The electrical discharge is broken into many small discharges due to the presence of powders [9, 10]. As a result many small cavities are formed with smaller depths on the work surface. This in turn produces a smoother surface. and without any powder concentration. Figure 6 shows the 3D surface plot of surface roughness versus powder concentration and peak current. It can be observed that surface roughness improves with increase in Ton. As Ton is increased, heat energy is absorbed during a longer time making the surface less stressed that produces a better surface [11].

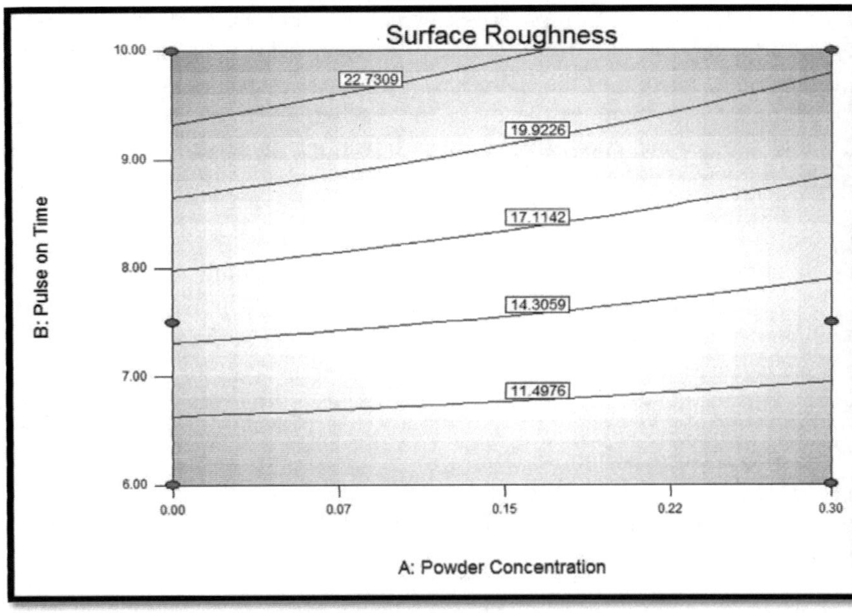

Figure 5. 2D plot for Ra depending on powder concentration and R_a

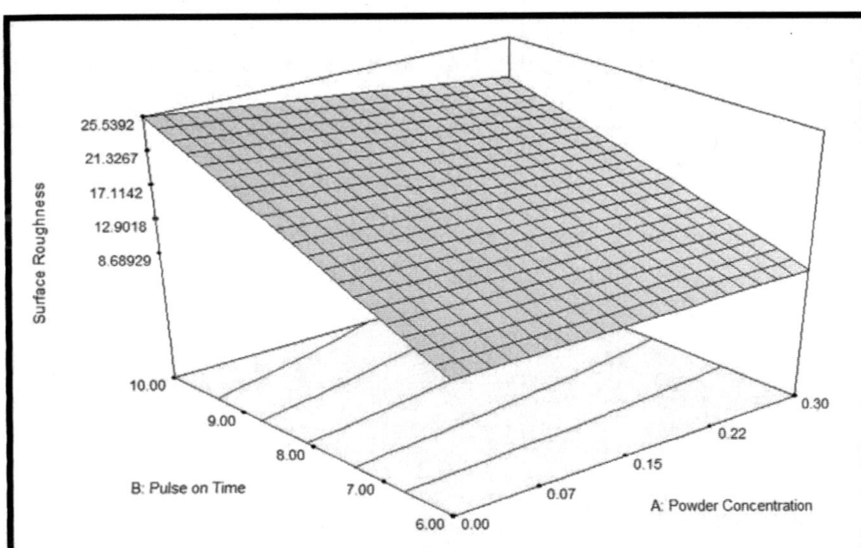

Figure 6. 3D surface plot of surface roughness versus powder concentration and T_{on}

4. Conclusion

In this study, the effect of EDM parameters like addition of aluminium oxide in dielectric fluid and pulse on time on material removal rate (MRR) and surface roughness, Ra have been investigated. The workpiece material and electrode materials were stainless steel AISI 304 and copper respectively. Based on the experimental results, by adding 0.3mg/L of Al_20_3 in dielectric fluid, the surface roughness was improved and MRR was increased. The electrical discharge is broken into many small discharges due to the presence of powders. As a result, many small cavities are formed with smaller depths on the work surface. This in turn produces a smoother surface. As Ton is increased, heat energy is absorbed during a longer time making the surface less stressed that produces a better surface. MRR increases with increase in pulse on time. Due to the presence of powders, the electrical discharge is scattered and electrons passes more uniformly in the discharge area. As a result, MRR increases. As T_{on} increases, the workpiece gets more time to absorb heat energy that melts and vaporizes the work material. This results increase in MRR.

Acknowledgement
The work was supported by RIGS project, RMC, International Islamic University Malaysia (IIUM). The authors acknowledge the support and help from all the staff of the Department of Manufacturing and Materials Engineering, IIUM.

References

[1] Sabur A, Yeakub MY, Maleque MA and Khan AA 2013 *Procedia Engineering* **56** 696 – 701
[2] Kalpakjian S and Steven Schmid S 2017 *Manuf. Eng. Technol.* 754-758
[3] Groover M P 2007 *Fundamentals of Modern Manufacturing* 100-125
[4] Simao J, Lee H G, Aspinwall D K, Dewes R C and Aspinwal EM 2003 *Int. J. Mach. Tools Manuf.* **43** 121-128
[5] Ho K H, Newman S T 2003 *Int. J. Mach. Tools Manuf.* **43** 1287-1300
[6] El-Hofy H 2015 *Advanced Machining Processes Nontraditional and Hybrid Machining Processes* (McGraw-Hill)
[7] Amoljit S Gil, Kumar S 2015 *Materials Today: Proceedings* **2** 1723-1730
[8] Marashi H, Jafarlou D M, Sarhan A A D and Hamdi M 2016 Precision Engineering **46** 11-33
[9] Sanjeev Kumar A and Batra U 2012 *J. of Manuf. Processes* **14** 35-40
[10] Wong Y S, Lim L C and Tee W M 1998 *J. of Materials Process. Technol.* **79** 30-40
[11] Peças P and Henriques E 2008 *J. of Materials Process. Technol.* **79** 30-40 250-258

IOP Conf. Series: Materials Science and Engineering **290** (2017) 012064 doi:10.1088/1757-899X/290/1/012064

Analysis of design tool attributes with regards to sustainability benefits

S Zain[1,a], A F Ismail[2,b,*], Z Ahmad[1,c] and E Y T Adesta[1,d]

[1]Department of Manufacturing and Materials Engineering, International Islamic University Malaysia (IIUM), Jalan Gombak, 53100 Kuala Lumpur, Malaysia
[2] Department of Mechanical Engineering, International Islamic University Malaysia (IIUM), Jalan Gombak, 53100 Kuala Lumpur, Malaysia

Email : [a] sz1396@gmail.com, [b] faris@iium.edu.my, [c] zuraidaa@iium.edu.my, [d] eadesta@iium.edu.my

* Correspondence

Abstract. The trend of global manufacturing competitiveness has shown a significant shift from profit and customer driven business to a more harmonious sustainability paradigm. This new direction, which emphasises the interests of three pillars of sustainability, i.e., social, economic and environment dimensions, has changed the ways products are designed. As a result, the roles of design tools in the product development stage of manufacturing in adapting to the new strategy are vital and increasingly challenging. The aim of this paper is to review the literature on the attributes of design tools with regards to the sustainability perspective. Four well-established design tools are selected, namely Quality Function Deployment (QFD), Failure Mode and Element Analysis (FMEA), Design for Six Sigma (DFSS) and Design for Environment (DfE). By analysing previous studies, the main attributes of each design tool and its benefits with respect to each sustainability dimension throughout four stages of product lifecycle are discussed. From this study, it is learnt that each of the design tools contributes to the three pillars of sustainability either directly or indirectly, but they are unbalanced and not holistic. Therefore, the prospective of improving and optimising the design tools is projected, and the possibility of collaboration between the different tools is discussed.

1. Introduction

The traditional practice of manufacturing is obviously biased towards the economic dimension [1]. The stimulus behind product development and production is how to retain market share and become more competitive. As a result, products are normally designed based on the perspective of the users and with a strong consideration on the return on investment. In doing so, the focus of product design is to improve the product quality, shorten the manufacturing time and reduce the operational cost. To achieve these objectives, numerous design tools have been successfully developed and are widely practiced in the industry. Since the introduction of sustainable development and sustainable manufacturing, another criterion that has been taken into account in product design is environmental performance. Hence, numerous design tools are also created to facilitate the design tasks.

IOP Conf. Series: Materials Science and Engineering **290** (2017) 012064 doi:10.1088/1757-899X/290/1/012064

In this study, four well established design tools in industry are analysed, namely Quality Function Deployment (QFD), Failure Mode and Element Analysis (FMEA), Design For Six Sigma (DFSS) and Design for Environment (DfE). The justification behind the selection is due to their unique attributes with regards to their powerful customer-driven and market oriented method [2], effective analysis and solution of potential failure modes tool [3], comprehensive tool created under the Six Sigma umbrella to design products and processes to achieve greater variations [4] and systematic approach to improve the environmental performance of products [5], respectively.

Firstly, the background and approach of each design tool are studied. Next, by studying their attributes, their contribution to the three dimensions of sustainability, namely, social, economic and environment throughout the product lifecycle are analysed. There are four phases of the product lifecycle, starting with engineering, where the product is designed, followed by production, use and finally end of life. Since sustainability is a vast framework, the focus of this research is limited to the attributes of the design tools related to the customers and manufacturers only.

2. Attributes of Design Tools in response to the sustainability call

Over half a century ago, various design tools were developed and successfully applied in the manufacturing industry. In general, the function of the design tools is to support the design team in finding solutions to address specific design problems with regards to quality (e.g. performance, safety), cost (e.g. materials, waste) and time (e.g. time to market). However, according to Gmelin [6], the previous work on product development is not well-covered with respect to all dimensions of sustainability. In this section, analysis of each of the selected design tools is conducted to identify its main attributes and benefits with respect to each sustainability dimension throughout the product lifecycle. The criteria that will be taken into account in this study for each phase of product life cycle and sustainability dimension are shown in Table 1.

2.1 Quality Function Deployment (QFD)

Originally developed in Japan in the 1970s [7], QFD is a customer-driven quality tool for new product development that has the unique capability of systematically bridging and communicating the demands of customers with the engineering characteristic of a product. From a larger perspective, QFD effectively allows the communication of three entities, namely, the Voice of Customer (VoC) as the dominant component, Voice of Business (VoB), and Voice of Engineer (VoE) [2], using a series of matrices called House of Quality (HoQ), throughout the product development and production stages. The matrices deal with the requirements and constraints from customers (e.g. quality, functionality, budget), organisation (e.g. resources, profit) and technical components (e.g. technology, manufacturing facilities) [2]. After being successfully implemented in Japanese companies, in 1983, QFD then was introduced in the USA and Europe [7] and since then has become increasingly popular in various sectors globally [2].

Previous studies show that QFD dominantly contributes to both social and economic dimensions, especially in engineering, production and usage phases. Throughout its four stages, QFD promotes active communication [8] between management and cross-functional units in the firm, as well as customers, which consequently leading the firm to achieve the expected results. The party that enjoys the social benefits the most is understandably the customers. After providing inputs regarding their demands and feedback on the product in the design stage, the requirements are subsequently translated into engineering characteristics. QFD also provides significant economic benefits. Although the QFD process involves many parties and phases, it is capable of shortening the design cycles with fewer and earlier design changes, reducing lead time and start-up costs [9] as well as improving documentation and operational procedures [10]. Bicknell & Bicknell [10] even claim that if QFD is utilised properly, the company can shorten the design cycles by 30-50%, achieve a reduction in design changes by 30-50% and enjoy 20-60% lower start-up costs.

Table 1. (a) Criteria of Sustainability Performance for each phase of the Product Life Cycle - Phase 1 & 2;
(b) Criteria of Sustainability Performance for each phase of the Product Life Cycle - Phase 3 & 4.

(a)

Phase 1: Engineering			Phase 2: Production		
Social	**Economic**	**Environment**	**Social**	**Economic**	**Environment**
Manufacturers: ▪ Development of experts (e.g., design team) ▪ Communication, integration and collaboration between inter functional units Users: ▪ Communication with manufacturer	Manufacturers: ▪ Product Development cost ▪ Utilisation of facilities and technology ▪ Design cycle ▪ Time to market ▪ Documentation ▪ Design optimisation (functionality, materials etc.)	▪ Selection of Materials (e.g., environmentally friendly/renewable materials) ▪ Waste reduction (functionality, materials, etc.)	Manufacturers: ▪ Development of experts and skilled workers ▪ Safety & Heath (employees) ▪ Comfort and satisfaction (employees) ▪ Communication, integration and collaboration between inter functional units	Manufacturers: ▪ Production cost ▪ Effectiveness and efficiency of manufacturing processes, assembly and testing ▪ Operation cost (electricity, water, etc.) ▪ Utilisation of facilities and technology ▪ Defect reduction ▪ Waste reduction (materials, defects, etc.)	▪ Consumption of raw materials ▪ Efficiency of operation (electricity, machines, water, etc.) ▪ Minimising wastages and pollution (air, water, sound) ▪ Utilisation of green energy and technology

(b)

Phase 3: Use			Phase 4: End of Life		
Social	**Economic**	**Environment**	**Social**	**Economic**	**Environment**
Manufacturers: ▪ Competent staff for service after sales (e.g., repair) Users: ▪ Safety & Health ▪ Ergonomic & Comfort ▪ Aesthetic & Perceived Quality ▪ Satisfaction	Manufacturers: ▪ Warranty ▪ Maintenance (spare parts, service, etc.) Users: ▪ Value for money ▪ Operation cost ▪ Maintenance cost ▪ Protection of product (from theft, corrosion, dirt, flood, etc.) ▪ Resale value (if applicable)	▪ Efficiency of operation (electricity, water, fuel, etc.) ▪ Minimising wastages (e.g., spoiled parts) ▪ Minimising pollution (air, water, sound) ▪ Utilisation of green energy and technology	Manufacturers: ▪ Develop expertise in disposal science (e.g., adverse effect of plastic materials on animals, plants) Users: ▪ Awareness/training to reuse products/parts or send for recycling remanufacture	Manufacturers: ▪ Remanufacture & reuse products/parts Users: ▪ Conserve by reusing products/parts	▪ Reusable, recyclable & remanufacturable products/parts ▪ Efficient waste management ▪ Minimising pollution (air, water, sound) ▪ Efficient and environment friendly disposal

As a result, these advantages as a whole, are translated into the reduction in both development and production costs and finally result in improved products with more competitive prices [9], [10]. Manufacturers can benefit from this financial gain due to fewer warranty claims [10], while end users benefit from the value for money. In addition to the aforesaid benefits, by implementing QFD properly, manufacturers may significantly improve their competitiveness in the market [8], which can be translated into an increase in loyal and new customers, and most importantly, increase in revenue and market share. As for the environment, the attributes of QFD to minimise environmental impact throughout the entire product lifecycle is hardly found in the literature.

2.2 Failure Mode and Element Analysis (FMEA)

Initially introduced in the early 1950s in the US [11], FMEA is a design tool for systematically addressing failure issues while simultaneously improving the reliability of product through design. The initial approach in the FMEA mechanism is to identify and prioritise potential failure modes in the critical parts, functions and components of products or processes. Subsequently, a decision is made on the appropriate mitigating actions to overcome the issues in order to avoid production loss and prevent failures from reaching the end users [25,26]. By doing so, this will significantly improve the safety, quality and reliability characteristics of the product [14]. Although initially used in the aerospace and automobile industry [3], nowadays, FMEA is widely applied in various industries such as the military, nuclear, healthcare, electronics and electro-technical industry [13], [14].

Similar to QFD, previous studies show that the benefits of FMEA are mostly associated with social and economic dimensions in engineering, production and usage phases. In engineering and production phases, the application of FMEA may involve active communication and teamwork between the management, design team and production team as well as the customers [3], [14], [15] in order to identify, prioritise and finally overcome the potential risks. When the products reach the customers, the end users may enjoy the benefits mostly due to the fact that the probabilities of catastrophic failure of the product have been mitigated earlier, and as a result, the risks of death or injuries may be reduced as much as possible. On top of the safety enhancement, the improvement in quality and reliability characteristics also will increase the customers' satisfaction [3] and ultimately, their loyalty. It seems that the economical dimension enjoys most of the benefits of FMEA, both for manufacturers and customers. The comprehensive technique of the FMEA increases the efficiency of product development and production in terms of implementation time and cost [3]. Besides that, Huang, Shi and Mak [15] claim that industrial users of FMEA managed to improve product quality by around 15-45% as well as time to market. The introduction of FMEA application in most recent computer aided software such as Computer Aided Design (CAD) and Computer Aided Process Planning (CAPP) increases the effectiveness and efficiency of the risk assessment tremendously [13]. Furthermore, when the product reaches the end users, due to increased quality and reliability attributes [15], a lower number of failures is expected, hence the repair cost can be reduced. In addition, FMEA can be applied to optimise the maintenance by suggesting preventive maintenance [3], [14], as a result, this will improve the lifespan of the product. We can conclude here that FMEA benefits end users due to higher reliability of product, meanwhile manufacturers enjoy lower warranty cost and fewer complaints, hence increasing their competitiveness. Besides that, FMEA can also be applied to address the potential risks associated with the negative impact on the environment [3]. Nevertheless, the benefits of FMEA application in the disposal phase are scarcely found in the literature.

2.3 Design for Six Sigma (DFSS)

Design for Six Sigma (DFSS) is the complementary tool to Six Sigma. It is a proactive design tool that leads to the design or redesign of the products, services or processes to minimise process variability and unpredictability [16] without compromising the product's quality [4]. In contrast to Six Sigma, which focuses on improving the consistency of the processes, the DFSS approach is focused more on improving the effectiveness of the processes[17]. The commonly used approach in DFSS is DMADV, which is the process of Define, Measure, Analyse, Design and Verify. The first three steps are associated with process characterisation (existing process), while the Design and Verify steps are associated with process optimisation (new process) [18]. Since its establishment at Motorola in the 1980s [19], Six Sigma has been applied widely in various sectors, nevertheless, DFSS application is still growing, especially in automotive, manufacturing, energy, medical and aerospace industries [4].

Most design tools work individually to achieve a specific objective, however, DFSS employs other widely used tools such as QFD, FMEA, Design of Experiments (DOE), Robust Design etc. [4,30,32]. This, as a result, makes DFSS more comprehensive and hence understandably offers more benefits. In term of sustainability benefits, economic performance is the winner by far, followed by social impact. In the design stage, both the product development cost [16], [18] and development time [4,30] could be reduced as a result of the application of DFSS. The production phase is where DFSS greatly contributes economically. This is because the preventive action performed by DFSS at the design stage is substituted with the application of Six Sigma in this manufacturing phase. Consequently, the quality, reliability and robustness of the products can be increased [4], time to market can be shortened [19] and overall production costs can be reduced greatly as a result of reduction in defects [18] and waste [19]. Furthermore, when the products reach end users, the number of claims and costs can be diminished significantly [17], [19].

In term of social benefits to the manufacturers, the execution of DFSS encourages involvement and extensive teamwork from cross-functional units [4] as well as substantial support from the management [18]. Moreover, most of the personnel especially those from the management and technical team needs to undergo extensive training such as Black Belt and Green Belt [4], hence this will offer added value to their career development. Customers, meanwhile, are also involved and contribute to voice their demands and feedbacks during the design stage. As a result, they may enjoy the benefits when they use the products thanks to the great work done by DFSS and Six Sigma to produce quality and robust products that can be translated into great value for money and eventually customer satisfaction [4]. These social and economic benefits consequently spur an increase in productivity and sales [18], hence leading to improved market share [16]. These claims have been proved by Samsung when they implemented DFSS in 2000, they managed to increase Samsung's R&D projects, reaching 80 percent of commercialization in 2014 as compared to 61 percent in 2002. Meanwhile the total sales of the Samsung Group and its 63 affiliates reached $122 billion in 2004, up from $102 billion in 2003, which led Samsung to be a global leader in the electronic business [20]. In the end of life phase, however, the attributes of DFSS with regards to social and economic benefits are less discussed in the literature. The discussion of the specific attributes of DFSS that contribute to the environmental domain in the literature, however, is very limited.

2.4 Design for Environment (DfE)

DfE was introduced in the early 1990s [21] as an initiative in the manufacturing industry in response to the global concern for the environment, as a result of environmental neglect for decades [22]. DfE is a systematic guideline that encompasses any design activity which aims to enhance the environmental performance of a product [5] and to minimise the impact of manufacturing on the environment across the entire life cycle of the product [23]. DfE covers several related environmental issues in design such as human health and safety, hazardous material management, recycling, disassembly and disposal [24]. There is no exclusive approach in DfE. A variety of methods have been developed which range from general to specific to support the design team in making the optimal decision [5], [24] such as Guidelines and Checklist Document, Product Design Matrix, Environmental Effect Analysis (EEA), etc. DfE has been applied in various industries such as automotive, medical and electronics [5] and the interest is growing globally.

Through the implementation of DfE, environment is arguably the dominant dimension that benefits throughout the four phases of the product's life cycle. In the product development, production and usage stages, DfE contributes significantly to the environmental sustainability through the reduction of the consumption of material extraction and processing, conservation of input materials, selection of materials and processes that are less harmful, increase in energy and resource efficiency, reduction of waste and toxin containment and enhancement of product durability and maintainability [22], [24]. The product's end of life phase, meanwhile, is where DfE prominently offers a range of solutions to address environmental problems neglected by most design tools as discussed previously. At the disposal stage, the main issue is the treatment of huge volume of waste generated because of our consumption. In order to minimise the end of life impacts, DfE increases the environmental performance of products in terms of their remanufacturability, reusability and recyclability [5]. This, as a result, may prolong the life of the products/parts and reducing the volume of waste and harmful materials going to landfills.

The aforementioned environmental advantage can be translated into social and economic benefits. DfE is an interdisciplinary approach that encourages interaction and collaboration of multi-disciplinary teams [23]. In the production stage, employees can benefit from better and safer working conditions [22], whilst end users may enjoy greener and eco-efficient products that can lead to an increase in their levels of satisfaction. The efficiency of materials management and resource operation as well as the improvement of the eco-performance of products can be translated into a major reduction in operational cost and a foreseeable increase in sales and profits. After the end of the products' life, the ability of the used products/parts to be remanufactured, reused and recycled will turn them into new products [5]. This, in turn, can generate new opportunities for business and employment in recycling, green technology, waste management and related industries. Overall, DfE is a very effective tool in reducing pollution and the long-term impact on the environment [22] caused by the manufacturing industry, and hence contributes greatly to the sustainability of the planet.

3. Discussion

From the above analysis, it is discovered that all of the selected design tools contribute to each of the sustainability dimensions throughout the entire product lifecycle, either directly or indirectly, as summarised as in Table 2. The three pillars of sustainability, although they are disparate, they are closely interrelated and interplayed. In this regard, the analysis shows that economy, noticeably, is the sustainability dimension that benefits the most. This is expected because economic profitability is the main reason behind the establishment of the manufacturing players. Moreover, the economy is the aspect that is often given attention by both the government and customers.

Table 2. Summary of the contribution of design tools with regards to sustainability benefits.

No.	Design Tools	Engineering			Production			Use			End of Life		
		S	E	Env.	S	E	Env.	S	E	Env.	S	E	Env.
1	QFD	M:X U:X	M:X	O	M:X	M:X	O	M:O U:X	M:X U:X	O	M:O U:O	M:O U:O	O
2	FMEA	M:X U:X	M:X	O	M:X	M:X	O	M:O U:X	M:X U:X	O	M:O U:O	M:O U:O	O
3	DFSS	M:X U:X	M:X	O	M:X	M:X	O	M:O U:X	M:X U:X	O	M:O U:O	M:O U:O	O
4	DfE	M:X U:O	M:X	X	M:X	M:X	X	M:O U:X	M:X U:X	X	M:X U:X	M:X U:X	X

Note: S - Social, E - Economic, Env. - Environment, M - Manufacturers, U - Users,
X - Contribute directly, O - Contribute indirectly.

The above study also shows that each design tool has its own advantages and disadvantages with regard to its contribution to the sustainability demand. By employing QFD, products that accurately fulfil customer requirements in term of functionalities, performance, features etc. can be produced, as a result, the customers' satisfaction and their loyalty to the products and firms would be increased. Moreover, huge savings can be made across the product lifecycle as most of the unnecessary requirements are designed out as early as during the design stage. Otherwise, the needless characteristics would be useless and eventually turned into waste in various forms and contribute substantially to pollution. FMEA, meanwhile, has a special capability in improving the reliability of a product which is crucial when it is utilised by the end users. This would be translated into mutual benefits, socially (e.g. high safety, decent performance) and economically (e.g. fewer defects and maintenance) to the users. The advantage of DFSS, however, is its ability to reduce the process variability and unpredictability in the production phase through design, which is crucial in order to achieve the Six Sigma quality level. As a result, DFSS collectively is not only able to fulfil the customer demand but is also capable of improving the accuracy and precision of the manufactured products, which contribute extensive benefits to both manufacturers and customers.

Nevertheless, QFD, FMEA and DFSS share similar disadvantages when it comes into environmental performance, where none of the tools possesses specific attributes that directly contribute to the environmental sustainability. Furthermore, in the end of life phase, product recovery efforts seem to be obviously ignored by all of the design tools. Consequently, the contribution of these three design tools to the three pillars of sustainability at this phase is expectedly minor. These weaknesses, however, could be addressed by employing DfE. Even though these characteristics are not necessarily what the customers require and do not promise better quality and a cheaper product, this would contribute greatly to the social and economic sustainability in the long run, whether directly and indirectly. For instance, if the degree of pollution can be reduced comprehensively, this would provide a better atmosphere for social life. Furthermore, taxes associated with environmental recovery can be reduced, and new business opportunities with regards to recycling, remanufacturing and waste management activities, etc., can be created.

4. Conclusion

Considering the limits of resources to accommodate an increasing global population and development activities, sustainability has become an urgent action to ensure the survival of existing communities and the future generation. In the manufacturing sector, design tools have been employed widely in order to systematically increase the competitive merits of the products as well as the manufacturers. Therefore, it is crucial to study the attributes of these design tools and their contribution to sustainability. In this study, four design tools, namely QFD, FMEA, DFSS and DfE and their unique attributes have been analysed in terms of their contribution to the sustainability benefits. From the analysis, it is discovered that each design tool contributes directly and indirectly to certain dimensions of sustainability, but they are unbalanced and not comprehensive. Therefore, it is necessary to enhance the capability, flexibility and attributes of the existing design tools. Besides that, the effort required to improve the collaboration among the design tools must be explored and integrated to the product development process. In addition, development of a new design tool to specifically address the sustainability imbalance can be discovered as a potential alternative.

Acknowledgement

The authors would like to thank the Ministry of Higher Education (MOHE) for sponsorships of Myra incentive Research Grants (MiRGS 13-02-002-0001) and the Research Management Center (RMC) of the International Islamic University Malaysia (IIUM) for assistance in the success of this study.

References

[1] S Walker 2013 *Des Issues* vol. **29** pp 89–107
[2] J R Sharma, A M Rawani and M Barahate 2008 *Int J Data Anal Tech Strateg.* **1** pp 78–103
[3] M. Ben-Daya 2009 *Handbook of Maintenance Management and Engineering* pp 75–90
[4] I Gremyr and J B Fouquet 2012 *Int. J. Lean Six Sigma* **3** pp 45–58

[5] M Z Hauschild, J Jeswiet and L Alting 2004 *Ann. - Manuf. Technol.* **53** pp 1–4
[6] H Gmelin and S Seuring 2014 *J. Clean. Prod.* **69** pp 1–9
[7] Y. Akao and G. Mazur 2003 *Int. J. Qual. Reliab. Manag.* **20** pp 20–35
[8] L K Chan and M L Wu 2002 *Qual. Eng.* **15** pp 23–35
[9] J R Hauser, A Griffin, and . Robert, "Quality function deployment (QFD)," in *Wiley International Encyclopedia of Marketing*, 2010.
[10] B. A. Bicknell and K. D. Bicknell, *The Road Map to Repeatable Success Using QFD to Implement Changes*. CRC Press: Boca Raton, FL, 1994.
[11] B. S. Dhillon, *Maintainability, Maintenance, and Reliability for Engineers*. Taylor & Francis Group, LLC, 2006.
[12] V Ebrahimipour, K Rezaie, and S Shokravi 2010 *Expert Syst. Appl.* **37** pp 671–677
[13] L Y Zheng, Q Kiu, and C A McMahon 2010 *Proc. 6th CIRP-Sponsored Int. Conf. Digit. Enterp. Technol.* **66** pp 1673–1686
[14] H Arabian-Hoseynabadi, H Oraee, and P J Tavner 2010 *Int. J. Electr. Power Energy Syst.* **32** pp 817–824
[15] G Q Huang, J Shi, and K L Mak 2000 *Int. J. Adv. Manuf. Technol.* **16** pp 603–608
[16] T. T. Allen, "Design for Six Sigma," in *Introduction to Engineering Statistics and Lean Sigma : Statistical Quality Control and Design of Experiments and Systems*, Springer London, 2010, pp. 511–518.
[17] R Bañuelas and J Antony 2003 *TQM Mag.* **15** pp 334–344
[18] S Taghizadegan, "Design for Six Sigma : Roadmap for Successful Corporate Goals," in *Essentials of Lean Six Sigma*, Academic Press 2006 pp 49–58.
[19] J Antony 2002 *Int. J. Product. Perform. Manag.* **51** pp 6–8
[20] S Park and Y Gil 2006 *Resarch Technol. Manag.* **49** pp 24–29
[21] US Environmental Protection Agency, "History of Safer Choice and Design for the Environment," 2016. [Online]. Available: https://www.epa.gov/saferchoice/history-safer-choice-and-design-environment. [Accessed: 03-Dec-2016].
[22] M DeMendonça and T E Baxter 2001 *Environ. Manag. Heal.* **12** pp 51–56
[23] K Soylu and J C Dumville 2011*Marit. Econ. Logist.* **13** pp 29–43
[24] D. P. Fitzgerald, J. W. Herrmann, P. A. Sandborn, L. C. Schmidt, and T. H. Gogoll, "Design For Environment (DfE): Strategies, Practices, Guidelines, Methods, and Tools," in *Environmentally Conscious Mechanical Design*, M. Kutz, Ed. Wiley & Sons, 2007, pp. 2–21.

IOP Conf. Series: Materials Science and Engineering **290** (2017) 012065 doi:10.1088/1757-899X/290/1/012065

Preparation and characterization of jackfruit seed starch/poly (vinyl alcohol) (PVA) blend film

N Sarifuddin[1], N A Shahrim[1], N N S A Rani [1], H H M Zaki[1], A Z A Azhar[1]

[1]Department of Manufacturing and Materials Engineering, International Islamic University Malaysia (IIUM), Jalan Gombak, 53100 Kuala Lumpur, Malaysia

Email: norshahida@iium.edu.my

Abstract. From the environmental point of view, biodegradable materials have been rapidly developed in the past years. PVA is one of the biodegradable synthetic polymers commonly used, but its degradation rate is slow. As an alternative to reduce plastic waste and accelerate the degradation process, PVA frequently blended with other natural polymers to improve its biodegradability. The natural polymer such as starch has high potential in enhancing PVA biodegradability by blending both components. The usage of starch extracted from agriculture wastes such as jackfruit seed is quite promising. In this study, jackfruit seed starch (JFSS)/poly (vinyl alcohol) (PVA) blend films were prepared using the solution casting method. The effect of starch content on the mechanical (tensile strength and elongation to break %) and physical properties of the tested films were investigated. The optimum tensile strength was obtained at 10.45 MPa when 4 wt. % of starch added to the blend. But, decreasing trend of tensile strength was found upon increasing the amount of starch beyond 4 wt. % in starch/PVA blend films. Nevertheless, elongation at break decreases with the increase in starch content. The mechanical properties of the blend films are supported by the Field Emission Scanning Electron Microscopy (FESEM), in which the native JFSS granules are wetted by PVA continuous phase with good dispersion and less agglomeration. The incorporation of JFSS in PVA has also resulted in the appearance of hydrogen bond peak, which evidenced by Fourier Transform Infrared (FTIR). Additionally, the biodegradation rate of JFSS/PVA was evaluated through soil burial test.

1. Introduction

Synthetic polymer materials are necessaries, especially in the packaging industry, but they have caused major problems to the environment after being disposed as solid wastes. This is because the degradation of these materials can take up to a thousand years. Therefore, a high inclination towards the effort in developing a substitute biodegradable polymer materials have been increased interest by the researchers to overcome this problem. Having said that, numerous attempts have been made to develop biodegradable materials by adding natural polymers such as starch to synthetic polymers like poly (vinyl alcohol) (PVA) because it is well known as a synthetic biodegradable polymer and possesses excellent mechanical and physical properties [1]-[2].

Among natural polymer resources, starch is the most appealing candidates because of its low cost, pervasive availability and renewable [3]. Little information existed on modification of jackfruit (*Artocarpus heterophyllus*) seed starch by blending with PVA. It has been investigated that, the seeds mainly contain protein and starch, due to that, these seeds are considered as a promising source of starch [4]. Furthermore, PVA is petroleum-based polymer obtained from hydrolysis of polyvinyl acetate that is non-toxic, flexible, soluble in water, chemical resistance, good film forming and barrier properties towards oxygen [2]. Ample interest lies in blending starch with PVA because starch have demonstrated excellent compatibility in the blend [1].

This study aims to prepare biodegradable blends based on starch and PVA by solution casting method and to investigate the physical properties, tensile strength (TS) and elongation at fracture

(E%). Biodegradability studies through soil burial test were performed to investigate the effect of starch incorporation in the film's degradation.

2. Experimental procedures

The starch/PVA blend was prepared using solution casting method according to Ismail and Zaaba [5] with slight modification. The experimental procedures were carried out on six samples coded as SP0, SP2, SP4, SP6, SP8 and SP10 which containing starch from 0 to 10 wt. %, as shown in figure 1. PVA content was fixed at 3 wt. % and the amount of starch was varied from 0 to 10 wt. % with total mass of the system is 100 g.

A mixture of 3 wt. % PVA and distilled water were heated to 95°C. JFSS was then added to the mixture and stirred continuously to 70°C to obtain good homogeneity for the mixture. Then, the film solution was casted on an acrylic plate and dried at 35°C for 24 hours. The film was peeled off from the molds and stored in the desiccator at 23°C and 30% relative humidity until further testing.

Table 1. The ratio of jackfruit seed starch to PVA in blend film

Sample	Compositions of blends (weight percentage, wt. %)		
	Starch	PVA	Distilled water
SP0	-	3	97
SP2	2	3	95
SP4	4	3	93
SP6	6	3	91
SP8	8	3	89
SP10	10	3	87

The mechanical properties (tensile strength and percent elongation at break) of the JFSS/PVA blend films were verified in accordance to ASTM standard D882-02 by using Universal Tensile Tester (Shimadzu; Material Testing System) with 20 mm/min of speed and 5 kN of load. The samples of cast films were cut into strips with dimension of 70 mm × 10 mm based on standard ISO–527. Next, the morphology of JFSS/PVA blend films was revealed using Field Emission Scanning Electron Microscope (FESEM; JEOL JSM-6700F). The samples were coated with a layer of gold using a coating machine (Auto Fine Coater; JEOL JFC-1600) and observed under FESEM at a voltage of 10 kV. Furthermore, the infrared (IR) spectra of the film was recorded on FTIR spectrometer (Perkin Elmer; Spectrum 100). The samples were measured at a resolution of 4 cm^{-1} between 4000 cm^{-1} to 600 cm^{-1} in the average of 32 scans per sample. Finally, the SP4 film was buried in the composted soil and exposed to the natural weather. The weight loss was measured according to Azahari et al. [1] with slight modifications. The samples were cut into smaller pieces, weighted and buried in the soil. The degradation rate of the samples was studied at regular time intervals (5 days) for 20 days by weighting the mass of each sample. The weight loss of the samples is calculated using equation (1).

$$Weight\ loss\ (\%) = \frac{w_i - w_d}{w_i} \times 100 \quad (1)$$

where w_i is the initial weight of the sample and w_d is the weight of the sample after 5 days.

3. Results and Discussion

The mechanical properties in figure 1 shows the variation of the tensile strength (TS) and percent elongation at break (E%) with different blend ratios of JFSS/PVA blend films. It is expected that pure PVA film (SP0) shows the highest TS at 19.45 MPa and E% at 57.44%, while addition of starch considerably decreased both TS and E%. The reduction of these mechanical properties probably due to the amorphous nature of starch and poor filler-matrix interaction when starch is in excess [1]. The TS increased initially for the blend films containing 4 wt. % starch. This is because higher starch amount promotes filler-filler interactions to occur rather than filler-matrix interactions [1]. After the initial increase of TS again start decreasing with increasing starch content more than 4 wt. %. Thus, considering the starch/PVA blend based on the aim of study, when JFSS was introduced, the SP4 sample gave the highest result of TS which about 10.45 MPa and optimum E% at 2.28% compared to other JFSS/PVA films. The trends obtained are in agreement with the study of rice starch/PVA blend film [6].

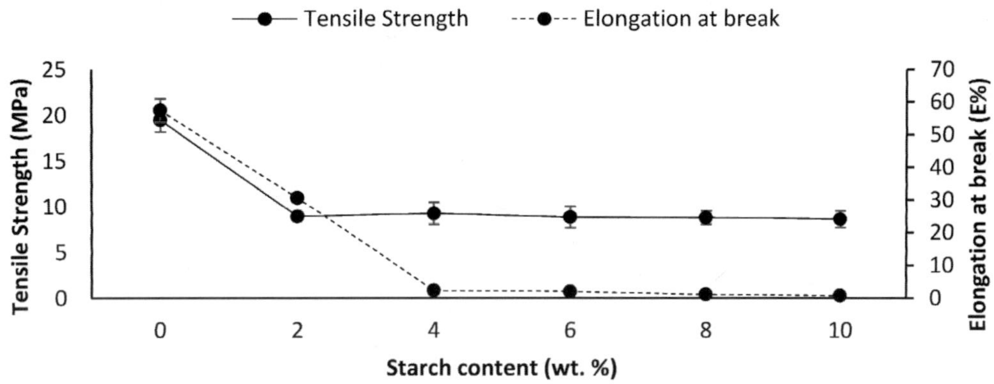

Figure 1. Effect of increasing starch content on the mechanical properties of JFSS/PVA blend films

The morphological structure of JFSS/PVA blend films (magnification 100×) was presented in figure 2. Figure 2 (a) revealed the unfilled PVA film has smooth and transparent surfaces. When 2 wt. % of JFSS is added to PVA, the starch granules are seemed to be wetted by PVA continuous phases as shown in figure 2 (b). However, it can be postulated that due to insufficient amount of starch, it results in poor interaction between starch and PVA. Meanwhile, addition of 4 wt. % JFSS demonstrated the optimum interactions between starch and PVA compared to other blend ratio as shown in figure 2 (c). It can be observed that the starch granules are well dispersed, less agglomerated and no formation of voids took place. These interactions are responsible for high TS of the film. However, further addition of starch content more than 4 wt. % resulted in the formation of agglomerates and voids as shown in figure 2 (d) – 2 (f). The formation of agglomerate occurred because starch and PVA are not soluble when mixed together due to excess of starch in the blend [7].

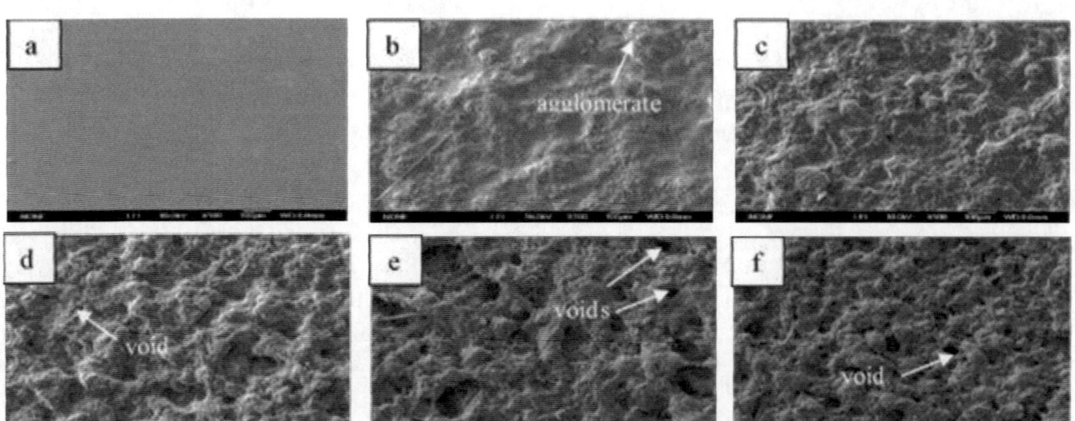

Figure 2. Surface morphology of (a) 0 wt. %, (b) 2 wt. %, (c) 4 wt. %, (d) 6 wt. %, (e) 8 wt. % and (f) 10 wt. % starch content in PVA blend film

The FTIR spectra of PVA and starch/PVA blend films are shown in figure 3. The wave numbers related to vibration peaks of the functional groups in PVA for each sample are depicted in table 2. All spectra exhibit the characteristic absorption bands of pristine PVA which are 3274, 2919, 1732, 1374-1243 and 1087 cm^{-1}. The vibrational peaks are assigned to O–H stretching, C–H stretching, C=O stretching, C-H wagging and C-O stretching in C-O-C group of PVA, respectively. Consequently, starch/PVA blends also exhibit similar characteristic peaks in the FTIR spectra, indicating the success of blending of PVA with starch. The stretching vibration of the hydrogen bonding -OH group of PVA and starch/PVA blends was shifted from 3269 to 3274 cm^{-1}. These indicate the -OH group of starches was involved in the hydrogen bond formation. The characteristic peaks at 1732 and 1716 cm^{-1} in PVA are associated to the residual acetate groups due to the manufacture of PVA from hydrolysis of polyvinyl acetate [2].

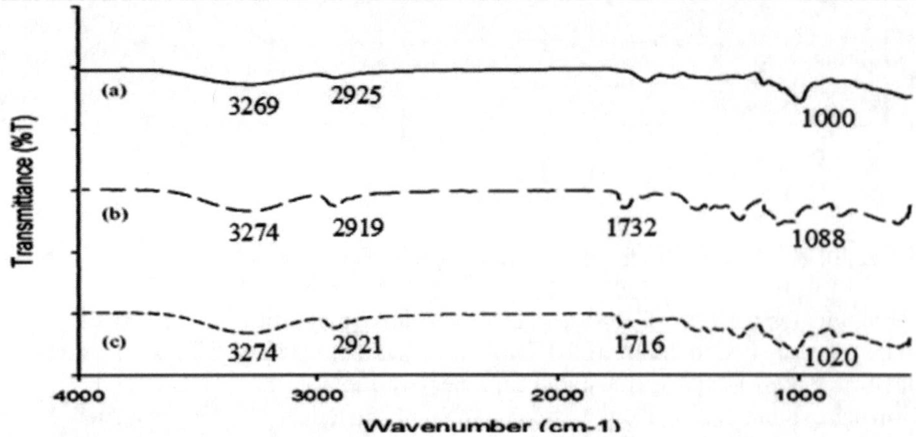

Figure 3. FTIR spectra for (a) JFSS powder, (b) PVA (SP0) and (c) 4 wt% starch in PVA (SP4) films

Table 2. Peak assignment in the FTIR spectra of JFSS, SP0 and SP4 samples

Peak assignment	Samples' wave number (cm^{-1})		
	JFSS	**SP0**	**SP4**
C-O stretching in C-O-C	1000	1088	1020
C-H wagging	1367-1234	1374-1243	1373-1246
C=O stretching characteristics carbonyl vibration in the residual acetate in PVA	-	1732	1716
C-H stretching	2925	2919	2921
-OH stretching	3269	3274	3274

The biodegradation rate of 4 wt. % starch in PVA blend film (SP4) is determined based on the respective weight loss at regular time interval (5 days) for 20 days through soil burial method. As seen in figure 4, the buried sample was suffered of weight loss, which the weight loss increased as the burial time increased. Azahari et al. [1] reported that pure PVA has the lowest weight loss about 77.19%% after being buried in soil for 56 days. Whereas, the weight loss of the SP4 was found to be 27.42% after being buried in soil for 20 days. This finding was attributed to the JFSS content in the films which is believed more biodegradable than pure PVA. The high weight loss ascribed to starch/PVA blends are due to both starch and PVA content. Starch is inherently biodegradable and PVA is soluble in water. At the same time, weight loss also indicates the initial degradation process due to the metabolic activity of microorganisms in which they consume the constituents of the degraded material as source of nutrients [2]. In fact, the white spots observed on the surface of the film samples (figure 5) was believed due to microorganisms' attacks. After 20 days of soil burial, it seems that the sample size become smaller, which is indicating the degdaration and weight loss of the samples. Thus, it is noted that the weight loss of the sample might be continuous over time and the degradation rate of the starch/PVA is improved compared to the pure PVA.

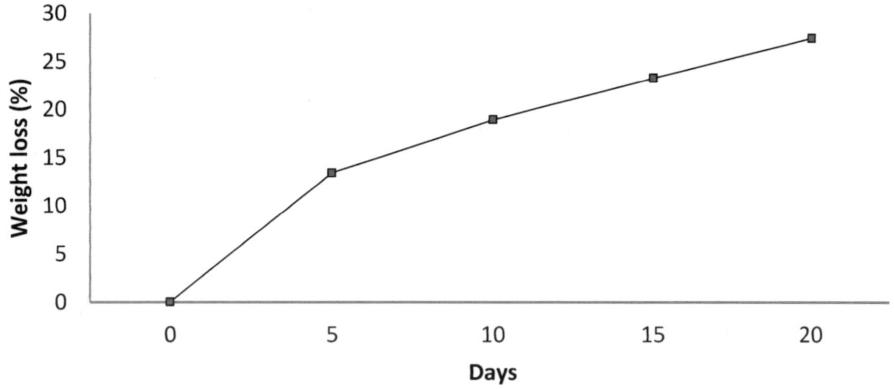

Figure 4. Weight loss of starch/PVA blend films after being buried in soil for 20 days

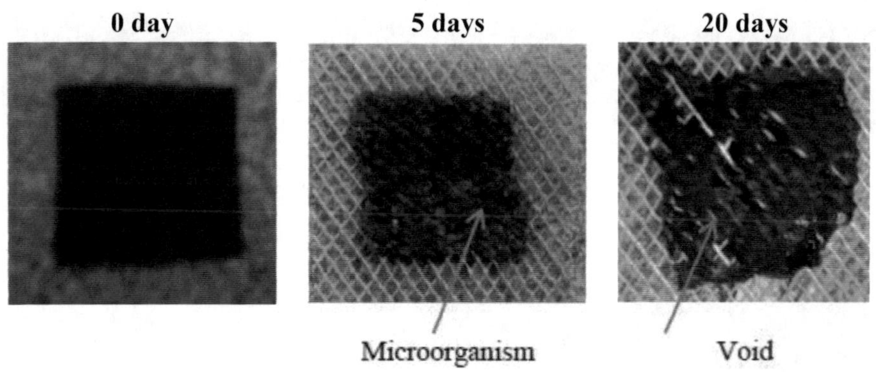

Figure 5. Macroscopic appearance of starch/PVA films after being buried for 20 days

4. Conclusion

The PVA blend containing jackfruit seed starch from 2 wt. % to 10 wt. % were prepared via solution casting method and their effect on the mechanical and physical properties were investigated. This study has shown that starch effectively disperse in the PVA blend and good compatibility. The optimum tensile strength 10.45 MPa and elongation at break 2.28% and form hydrogen bond interactions in PVA/starch. In addition, the results proposed that the addition of jackfruit seed starch to the PVA improved degradation behavior under natural weather compared to pure PVA.

Acknowledgement

The authors would like to acknowledge the Research Management Centre, International Islamic University Malaysia for the research grant RIGS16-072-0236.

References

[1] Azahari N A, Othman N and Ismail H 2011 *J. Physical Science* **22** 15-31
[2] Râpă M, Grosu E, Stoica P, Andreica M and Hetvary M 2014 *J. Environ. Prot.* **11** 34-42
[3] Mao L J, Imam S, Gordon S, Cinelli P and Chiellini E 2000 *J. Polym. Environ.* **8** 205-211
[4] Noor F, Rahman M J, Mahomud M S, Akter M S, Talukder M A I and Ahmed M 2014 *Int. J. Food Sci. Nutr.* **3** 347-354
[5] Ismail H and Zaaba H F 2014 *J. Elastom. Plast***46** 96-109
[6] Parvin F, Rahman M A, Islam J M M, Khan M A and Saadat A H M 2010 *Adv. Mater. Res.* **123** 351-354
[7] Razzaq W and Javaid A 2015 *J. Fac. Eng. Tech.* **22** 1-12

Investigation of tool engagement and cutting performance in machining a pocket

E Y T Adesta[1], R Hamidon[2], M Riza[1], R F F A Alrashidi[1], A F F S Alazemi[1]

[1] Department of Manufacturing and Materials Engineering, International Islamic University Malaysia (IIUM), Jalan Gombak, 53100 Kuala Lumpur, Malaysia
[2]PPK Pembuatan, University Malaysia Perlis, Kampus Pauh Putra, 02600 Arau, Perlis, Malaysia

Email:roshaliza@unimap.edu.my

Abstract. This study investigates the variation of tool engagement for different profile of cutting. In addition, behavior of cutting force and cutting temperature for different tool engagements for machining a pocket also been explored. Initially, simple tool engagement models were developed for peripheral and slot cutting for different types of corner. Based on these models, the tool engagements for contour and zig zag tool path strategies for a rectangular shape pocket with dimension 80 mm x 60 mm were analyzed. Experiments were conducted to investigate the effect of tool engagements on cutting force and cutting temperature for the machining of a pocket of AISI H13 material. The cutting parameters used were 150m/min cutting speed, 0.05mm/tooth feed, and 0.1mm depth of cut. Based on the results obtained, the changes of cutting force and cutting temperature performance there exist a relationship between cutting force, cutting temperature and tool engagement. A higher cutting force and cutting temperature is obtained when the cutting tool goes through up milling and when the cutting tool makes a full engagement with the workpiece.

1. Introduction

Tool engagement can be defined as the portion of the tool that engages with the workpiece during machining process. In pocket machining process, the engagement angle is an important cutting parameter that affects the cutting forces and cutting temperatures. The engagement is constant during straight cutting. However, when the cutting tool turns along a corner profile of the pocket, variation in engagement occurs. As a consequence, it will affect the surface finish of the machined surface. Many researches have looked into the problem of fluctuation of the cutting force caused by variation of engagement. One of early study of tool engagement was done by Kramer [1]. He developed an algorithm to detect the minimal engagement for pocket with island using zig zag tool path. As improvement to the Kramer technique, the tool path modification technique was introduced by later researchers to maintain the engagement angle during machining especially at the corners. Most of the studies focused on modification of tool path strategy to achieve a constant tool engagement and hence indirectly maintaining the cutting force. One of the earliest studies in tool path modification was by Stori et al. [2]. They introduced an offset tool path to maintain engagement during convex profiling of pocket. Practically, this method is unreliable because in a pocketing operation, the tool travels in various profiles to make a pocket such as convex, triangle etc. In addition, the approach is suitable for rough cutting only as excessive left and additional cutting is required.. However, this approach is distinctive and as initial step for other researchers to make improvement in pocketing operation. Due to the drawbacks of the Stori approach, Ibaraki et al. [3] presented a new algorithm for tool path modification known as backward and forward tool path generation. The application of forward tool path with the same value feed rate was compared with the cutting force. Another improvement regarding tool path modification to maintain tool engagement had been made by Uddin et al. [4] whom expended the Ibaraki approach by presenting an algorithm for semi-finishing cutting. From the results, cutting forces can be maintained throughout the cutting process. Apart from that, Dumitrache [5] had improved the Ibaraki backward algorithm by setting the reference value of tool engagement angle to 20°. As an improvement to previous methods, by applying this algorithm, machining time can be reduced. However, this method is suitable for only rough machining process.

Cutting temperature is closely related to cutting forces. Primarily, the rise of cutting temperature would affect cutting tool and lead to bad surface finish. For pocketing operations, the cutting tool needs to travel to make both straight and cornering cuttings. At the corner, the spindle speed needs to slow down to change direction of the tool. The changes in spindle speed and feedrate during the corner cutting operation thermally affects the work material and cutting tool Held et al. [6]. The investigation of tool engagement and cutting temperature not much explore by researcher. The recent study conducted by Riza [7]. The study conducted is specifically for contour tool path strategy only.

Hence, this study intended to investigate the behavior of cutting force and cutting temperature for different tool engagements in end milling process. Tool engagements model were developed for different tool path strategies. The relationship between tool engagement and cutting forces and cutting temperature was analyzed.

2. Experimental procedures

Experiments were conducted to investigate the effect of different tool engagements on cutting force and cutting temperature for the machining of a rectangular pocket with dimension 110 mm x 80 mm. Two types of tool path strategies were used in this study; contour and zig zag tool path strategy. The cutting parameters used for this experiment is shown in table 1. The machining operation was done by using CNC Vertical Machining Centre (VMC) by Mazak Nexus VCN 410-II. A Kristler 9257B dynamometer was attached to the workpiece and that supported by a charge amplifier and signal conditioning to measure the cutting force. The cutting temperature was measured using infrared thermal camera ThermoPro TP8. The IrAnalyzer software was used to record the temperature as well as to analyse the thermal image. To perform the machining operation, CoroMill 490 cutting tool with coated carbide insert was used. The tool diameter was 20 mm with two indexable insert positions. The two types of tool path strategies used for the machining the pocket was the contour and zig-zag tool path strategies.

Table 1. Cutting Parameters

Tool Path Strategy	Cutting Speed	Feed Per Tooth	Depth of Cut
Contour tool path strategy	150m/min	0.05mm/tooth	0.1mm
Inclination tool path strategy	150m/min	0.05mm/tooth	0.1mm

3. Results and Discussion

3.1 Tool engagement modeling

In modeling the tool engagement, two types of cutting were considered which were slotting and peripheral cuttings. Slotting is when the radial depth of cut or step over of the path is the same as the cutting tool diameter. During slotting, half of the cutting tool circumference is engaged with the machining workpiece. In this case, the cutting tool makes a full engagement with the workpiece. Entry angle for slotting is zero degree and exit angle 180°. Peripheral cutting is when the radial depth of cut is less than cutting tool diameter. Compared to slotting, smaller portion of the cutting tool circumference is engaged with the workpiece. The amount of tool engagement depends on radial depth of cut as shown in Figure 1. The engagement angle for straight cutting can be obtained as discussed by Tsui et al. [8].

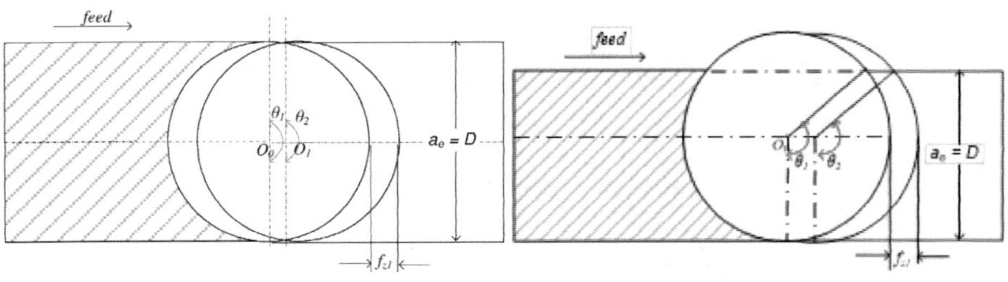

(a) (b)

Figure 1. Tool engagement for straight cutting

In machining a pocket, the tool maintains full engagement along a straight path. However, when it turns 90° direction, the angle is reduced. It depends on the amount of feed per tooth and radial depth of cut of the path. However, the value of tool engagement angle remains constant for the case of radial depth of cut less than diameter of cutting tool. The value of engagement angle when the tool turns 90° corner is given by the following relationship;

$$\theta = 90 + \cos^{-1}\left(\frac{r - f_{z1}}{r}\right) \tag{1}$$

Then, the engagement increases until it reaches 180° again where the value of $(r-(f_{z1}+f_{z2}+\ldots+f_{zn})=0)$ as in Figure 2. The increasing of engagement angle depends on the position of cutting tool in which the value of feed per tooth is;

$$\theta = 90 + \cos^{-1}\left(\frac{r - (f_{z1} + f_{z2} + \cdots \ldots \ldots \ldots . + f_{zn})}{r}\right) \tag{2}$$

$$f_{z1} + f_{z2} + \cdots \ldots \ldots \ldots . + f_{zn} < 10 \text{ (for D=20 mm)}$$

$$f_{z1} = f_{z2} = \cdots \ldots \ldots \ldots . f_{zn} = f_z$$

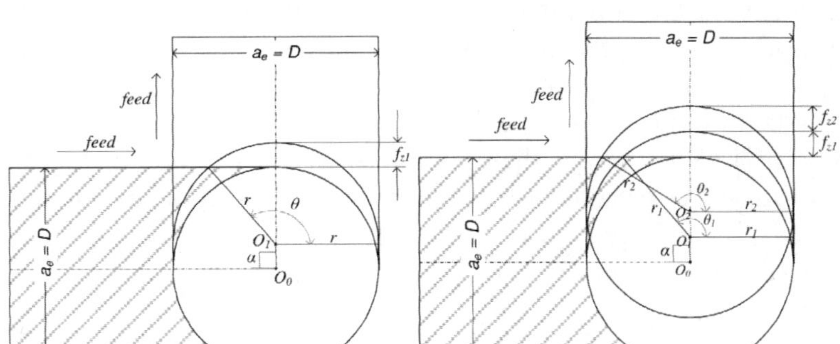

Figure 2. Tool engagement after 90° turn

For the case of less than 90° corner, the engagement depends on the angle of turning corner (α) as shown in Figure 3. The radial depth of cut decreased after the tool changes direction. Tool engagement starts to vary at the on-set of cutting and then increases until the radial depth of cut is equal to the diameter of cutting tool. Radial depth of cut for can be determined by:

$$a_e = f_z \tan \alpha \quad (3)$$

Then, tool engagement can be calculated as:

$$\text{if } a_e \leq r$$

$$\theta = \cos^{-1}(1 - (a_e/r) \quad (4)$$

$$\text{if } a_e > r$$

$$\theta = 90 + \sin^{-1}(a_e/r - 1) \quad (5)$$

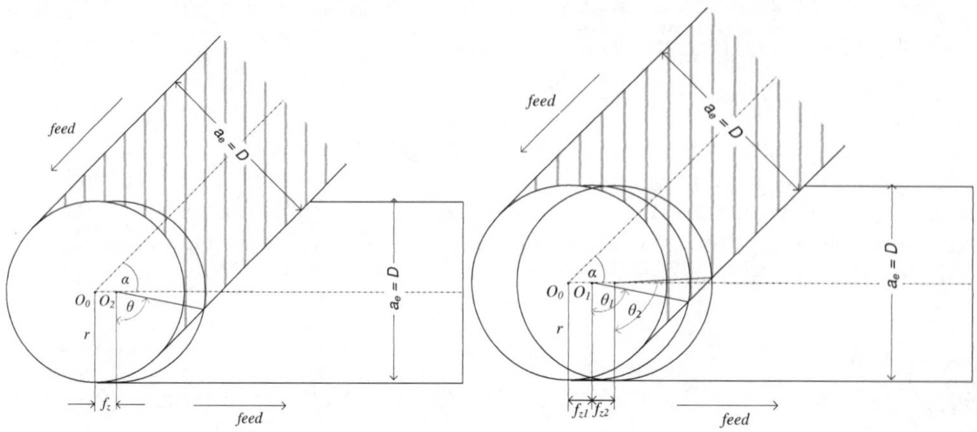

Figure 3. Tool engagement after less 90° turn

The calculated amounts of tool engagement during tool path strategies were plotted in the graph. Tool engagement angle was calculated by assuming that the first tooth of the cutting tool engages at the start of the circumference of radial depth of cut. The calculation of tool engagement angle is based on cutting tool with 20 mm diameter and 0.05 mm/tooth of feed. figure 4 and figure 5 shows the changes of engagement during contour and zig-zag tool path strategy respectively. For contour tool path strategy, the tool engagement remains constant during straight cutting. The changes occurred when the tool changes it direction.

For contour tool path strategy, there are sharp changes of tool engagement angle when the cutting tool turns 45° corner and when the tool enters to the slotting operation. For zig zag tool path strategy, the engagement angle changes alternately when the cutting tool changes its direction from full (100%) engagement to 75% engagement.

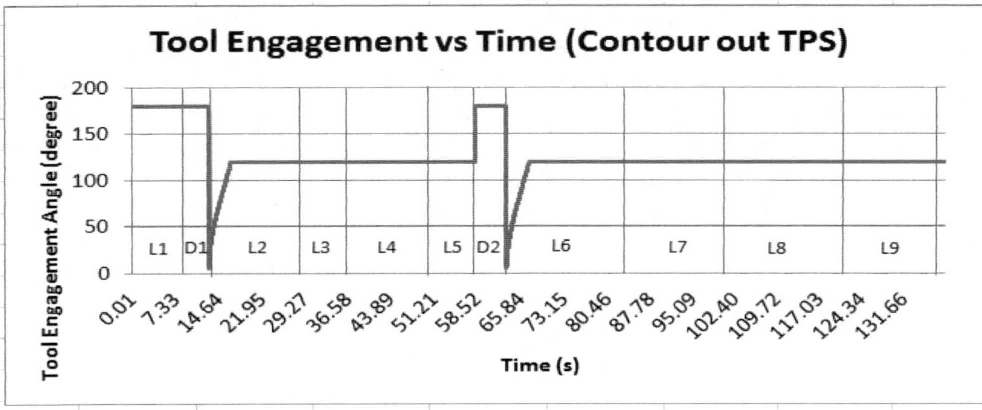

Figure 4. Tool engagement angle for contour tool path strategy

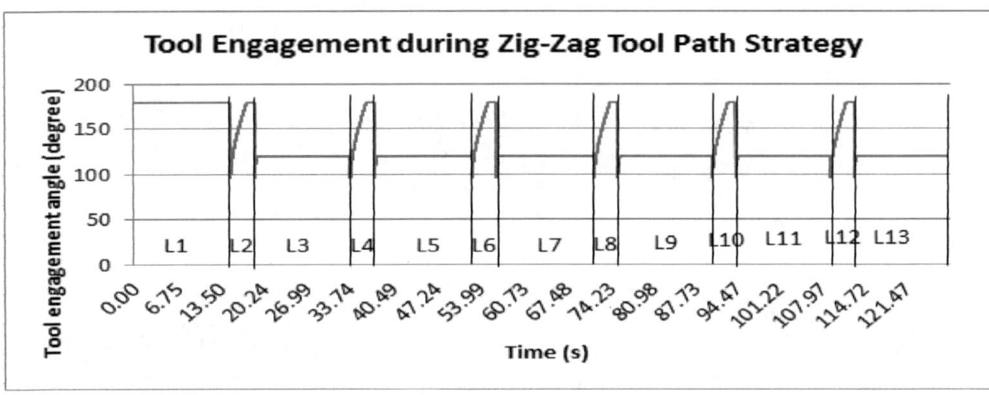

Figure 5. Tool engagement angle for zig zag tool path strategy

3.2 Cutting force and cutting temperature

The graph in Figure 6 shows the cutting force and cutting temperature for the contour tool path strategy. The cutting force and cutting temperature is the highest along L1 compared to the rest of the cutting process. This is due to the full engagement (180°) between the cutting tool and workpiece. In milling operation, when tool makes a full engagement with the workpiece, the cutting tool experiences both up milling and down milling. Chip thickness is zero at the starting and at the end of the process. Maximum chip thickness occurs in the middle of the process. The changes of tool engagements occur every time the tool changes its direction. The engagement angles for L2 to L4 and L6 to L9 is the same which is 120°. However, the cutting tool experiences down milling during L2, L3, L6 and L7. When entering L4, L5, L8 and L9, up milling occurs. From the graph, the cutting force is higher when the cutting tool transverses along the up milling path compared to down milling path. The same trend can also be seen for cutting temperature. Theoretically, the cutting forces in up-milling are higher compared to down- milling. For up milling, chip thickness is zero at the start of the process whereas for down milling chip thickness is zero at the end of the tool rotation. In up-milling, the cutting tool rotation is in the opposite direction of the feed direction which results in a higher cutting force. However, in down-milling, the cutting rotation is in the same direction as the feed direction which results in a lower magnitude of the cutting forces.

In the zig-zag tool path strategy, the tool engagement changes alternately from 180° in the horizontal direction to 120° for vertical direction. Based on the graph in Figure 7, there are significant changes to the cutting force when the zig-zag tool path strategy was used. The changes occur when the

IOP Conf. Series: Materials Science and Engineering **290** (2017) 012066 doi:10.1088/1757-899X/290/1/012066

tool turns its direction. As in the contour tool path strategy, the tool experiences both up milling and down milling throughout the path. For cutting temperature, it is noted that a higher cutting temperature occurs along L4 and L8 where there is full engagement between the cutting tool and workpiece. Before entering these paths, the cutting tool experiences up milling. This generates a large amount of heat at the cutting tool flank and workpiece resulting in higher cutting temperature [9]. At the end of the process, the cutting force is more stable. This is because after a certain period of time, the cutting parameters as well as other factors affecting the machining process become more significant compared to tool engagement.

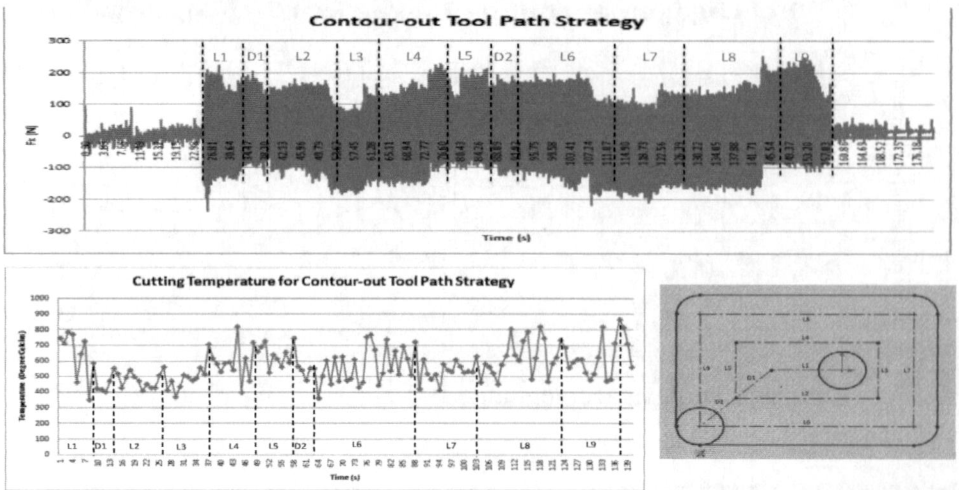

Figure 6. Cutting force and cutting temperature for contour tool path strategy

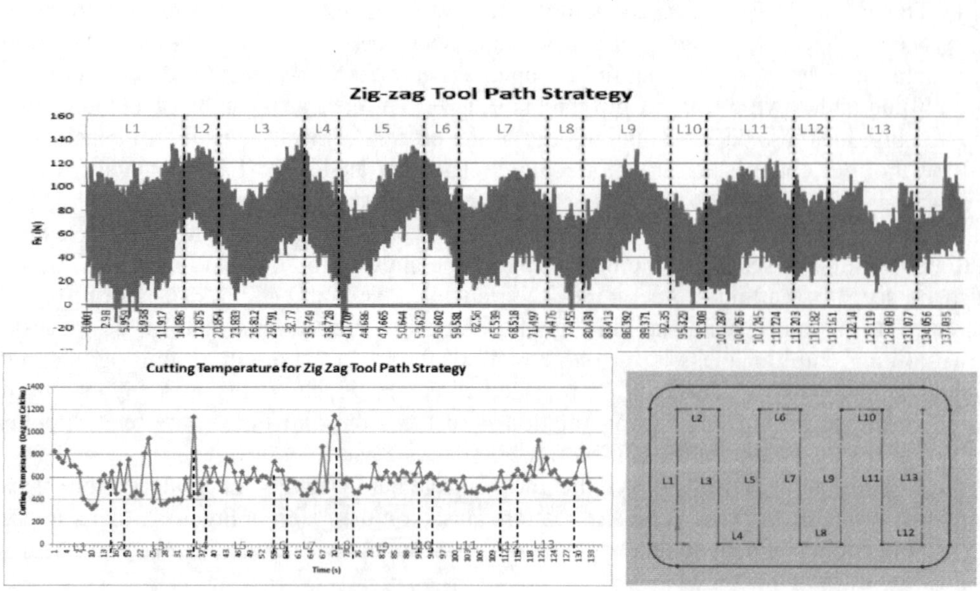

Figure 7. Cutting force and cutting temperature for zig zag tool path strategy

4. Conclusion

The investigation of tool engagement in machining a pocket is very important. This paper reports on the cutting force and cutting temperature performances for different tool engagements. The findings from this study are as follows:

i. The variations of tool engagement occur every time the cutting tool changes it directions. These variations affect the performance of cutting force and cutting temperature.

ii. When applying the tool path strategies, the tool experiences up milling and down milling when the chip thickness is varied. A higher cutting force and cutting temperature is obtained when the cutting tool goes through up milling and when the cutting tool makes a full engagement with the workpiece.

This study can be improved by performing a detailed analysis of tool engagement and chip thickness for the different tool path strategies.

References

[1] Kramer T R 1991 *J.Manuf. Syst.* **11** 114-123
[2] Stori J and Wright P 2000 *J. Manuf. Syst.* **19** 172–184
[3] Ibaraki S, Ikeda D, Yamaji I, Matsubara A and Kakino Y. 2004 *Japan-USA Symposium on Flexible Automation* pp. 6–9
[4] Uddin M S, Ibaraki S, Matsubara A, Nishida S and Kakino Y 2006 *JSME Int. J. Series C* **49** 43–49
[5] Dumitrache A, Borangiu T and Dogar A 2010 *IFAC Proc. (IFAC-PapersOnline)* vol 10 pp. 252-257
[6] Held M and Spielberger C 2009 *Computer-Aided Design* **41** 539–550
[7] Riza M, Adesta E Y T and Suprianto M Y 2015 Adv. Mater. Res. **1115** 47-50
[8] Tsui K S and Chan K W 2005 *Int.Conf. on Computational Intelligence and Multimedia Applications* pp. 132–137.
[9] Abukhshim N A, Mativenga P T, Sheikh M A 2006 *Int. J. Mach. Tool Manuf.* **46** 782–800

IOP Conf. Series: Materials Science and Engineering **290** (2017) 012067 doi:10.1088/1757-899X/290/1/012067

Experimental study of electrical discharge drilling of stainless steel UNS S30400

E A H Hanash and M Y Ali

Department of Manufacturing and Materials Engineering
Faculty of Engineering, International Islamic University Malaysia
P.O.Box 10, 50728 Kuala Lumpur, Malaysia

E-mail: mmyali@iium.edu.my

Abstract. In this study, overcut and taper angle were investigated in machining of stainless steel UNS S30400 against three different electrical discharge machining parameters which are electric current (I_p), pulse on-time (Ton) and pulse off-time (Toff). The electrode used was of 1 mm diameter with aspect ratio of 10. Dimensional accuracy was measured by evaluating overcut and taper angle. Those two measurements were performed using optical microscope model (Olympus BX41M, Japan). The experimentation planning, evaluation, analysis and optimization have been carried out using DOE software version 10.0.3 RSM based method with total number of twenty experiments. The research reveals that, discharge current was found to have the most significant effect on overcut and taper angle followed by pulse on-time and pulse off-time. As the discharge current and pulse on-time increase, overcut and taper angle are increased. However, when pulse off-time increases, overcut and taper angle decrease. The outcome result of this study will be very useful in the manufacturing industry to select the appropriate parameters for the selected work material. The model has shown a great accuracy with percentage error of less than 5%.

1. Introduction

The current demand of miniaturized parts with high quality output is growing rapidly due to the fast development of technology in all fields such as aerospace, automotive, medical, biomedical and electronic industry. Moreover, Machining of hard materials such as stainless steel requires, high friction force between the cutting tool and the workpiece; provided that the cutting tool is stronger and harder than the work material. Therefore, the traditional machining operations, machining of hard materials with high accuracy and in the micro-level is seemed to be impossible. However, with the development of the non-traditional machining operations, it is made possible with considerable precision and accuracy. Among the non-conventional machining processes to achieve a good accuracy of the drilling of small holes is electrical discharge machining (ED-Drilling). Therefore, in this study, presents general overview of the main categories of the machining operations and the main study focuses on the electrical discharge machining (EDM) specifically ED-Drilling. Therefore, non-conventional processes such ultrasonic machining, ion beam machining, water jet machining, laser machining, electrical discharge machining; just to name few among others, are processes which have the ability to machine complex geometrical shapes on difficult-to-cut materials such as tungsten carbide, diamond and steel alloys [1].

Researchers have investigated the machinability of the stainless steel and different machining performances such as MRR, EWR, surface roughness and other EDM machining characteristics [2, 3]. The study of machining performance, the researchers study the MRR, EWR and surface tribology and quality. Rajmohan, Prabhu [4] have investigated the effect of pulse on-time (T_{on}), pulse off-time (T_{off}), voltage (V) and the current (I) on the MRR in 304 stainless steel. They used Taguchi method and

IOP Conf. Series: Materials Science and Engineering **290** (2017) 012067 doi:10.1088/1757-899X/290/1/012067

ANOVA to analyse the results and it was found that the pulse of time and current have the most significant effect on MRR. However, dimensional accuracy has not been investigated.

Dimensional accuracy is considered of great importance in machining the different parts for different applications. It is more sensitive in the applications such as nozzles or on the cooling fans of turbines whereby high accuracy is required. The dimensional accuracy can be measured by the amount of overcut and taper angle [5]. Many researchers have done analysis of the quality, performance and dimensional analysis of the EDM-Drilling holes of different work materials with different hole diameters [2, 5-14]. As previous researchers focused on the measurement of machining performance, this study investigates the dimensional accuracy of the drilled hole made by electrical discharge drilling so that the quality of machined parts is maintained. Therefore, the effect of different electrical discharge parameters are investigated.

1.1 Hole Overcut

Hole overcut is defined as the difference of the average diameter of the entrance and exit hole with respects to the electrode diameter. The diameter of the hole will always be larger than the diameter of the electrode. This phenomenon is due to the spark gap and due to the secondary spark formation, which is generated by the debris that are removed by the pressurized liquid. Overcut can be calculated from figure 1. From the figure, the formula for overcut can be easily deduced. It is the difference between the hole diameter and the electrode diameter and divide by two, as expressed by Equation 1.

$$\mathrm{OC}=\left(\frac{da-de}{2}\right) \tag{1}$$

1.2 Hole Taper Angle

Hole taper is also one of the characteristics to measure the dimensional accuracy of the small drilled hole. It is occurred due to the wear of the electrode and debris issue that made the entrance diameter larger than the exit diameter. Therefore, taper is the angular measurement between the opening and the end of the drilled hole. There is a strong relationship between taper angle of the hole and the corner wear of the electrode. Taper can be calculated using equation 2 where θ is the taper angle; D_h, D_e, and D_{ex} are hole diameter at entry, electrode diameter, hole diameter at the exit respectively and t is the thickness of the workpiece. Figure 2 shows the schematic diagram representing taper angle occurrence.

$$\theta = \tan - 1\left(\frac{\mathrm{dh}-\mathrm{dex}}{2\mathrm{t}}\right) \tag{2}$$

2. Experimental Setup

The experiment is planned using the response surface method to design, analyse and optimise the results with total numbers of 20 experimental runs. These runs have been executed in the ED-Drilling machine model OCT3525NA (Taiwan) to investigate the effect of the input machining parameters such as electric current, pulse on-time, and pulse off-time on the dimensional accuracy of the drilled hole measures by the hole overcut and taper angle. The electrode that have been used in this experiment is made of a brass tube material with 1mm outer diameter and inner diameter of 0.3 mm. prior to the measurement of the machining characteristics such as hole overcut and taper angle, the set of holes in one line in the 304-stainless steel were cut exactly from the middle of the holes by the aid of the wire electric discharge machining, so that more accurate results is obtained. The measurement of the hole overcut and taper angle were done using the optical microscope (Olympus BX41M, Japan). The electro discharge condition are shown in table 1.

Table 1. Electro discharge drilling condition

Fixed Parameters	Value
Workpiece Material	stainless steel UNS 30400
Electrode tool material	Brass
Electrode diameter (mm)	1
Capacitance (F)	0
Polarity	-ve
Dielectric flushing pressure (Mpa)	6
Hole depth (mm)	9.65
Voltage gap (v)	22
Feed rate (mm/s)	2
Spark power (VDC)	105

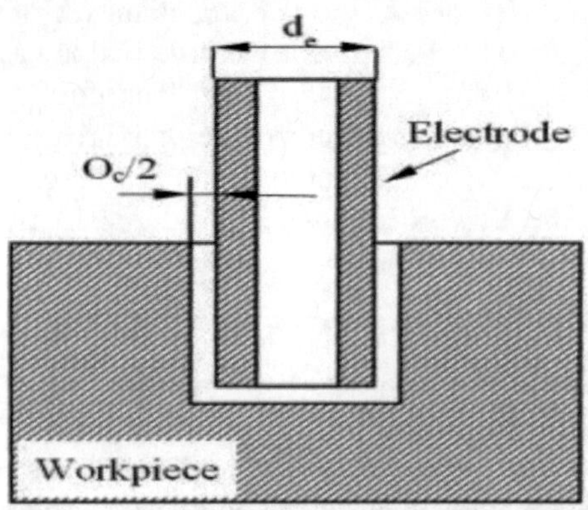

Figure 1. Hole overcut schematic diagram [5]

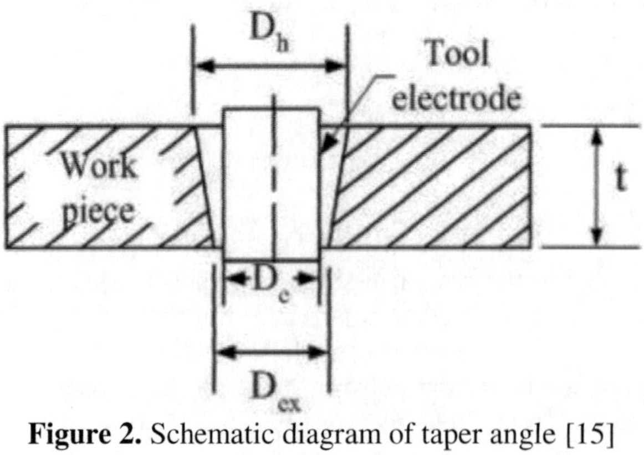

Figure 2. Schematic diagram of taper angle [15]

2.1 Work Materials

The work material is selected based on the application of the material itself in the industry. After studying different literature studies, different materials have been investigated such as D2 steel, SS316, SS 304, Inconel 718, and various types of ceramic based materials. The investigation was mainly to measure the machining performance such as surface roughness, MRR, EWR and dimensional accuracy such as overcut, taper angle, and circularity, however, the examining of the dimensional accuracy of 1mm holes have not been done before on UNS 30400.

Table 2. Process parameters and level using RSM

Variable Parameters	Lower Level	Upper Level	Response
Pulse Current (I_p)	5	21	Overcut Taper Angle
Pulse on-time (T_{on})	15	75	
Pulse off-time (T_{off})	5	45	

Figure 3. Electrode made of brass tube [8]

Therefore, in this study, UNS 30400 has been selected to investigate the dimensional accuracy of the drilled holes. UNS 30400 enters into various applications in aerospace, automotive, biomedical, biotechnology, and home appliance manufacturing industry. It is mostly used in valves, refrigeration equipment, skyscraper trim, turbine blades, evaporators, sinks, splash back, a basic machining method for the manufacturing industries of aerospace, automotive, nuclear, medical and die-molding industry.

2.2 Electrode Materials

Brass electrode was used to perform drilling holes on 304 stainless steel due to its good physical properties and it is relatively cheap. Figure 3 shows the electrode which made of brass. It is a hollow electrode which allow the pressurized water to go through the internal diameter of the electrode to remove the debris from the hole while drilling.

2.3 Experimental Procedures

The experimental procedures were implemented on stainless steel grade 304 using different machining parameters such as electric current, pulse on-time and pulse off-time with 1mm electrode diameter made of a brass tube. The experimental design, evaluation and analysis were done using the response surface method (RSM) with total number of 20 experiments with three factors and 5 levels. Table 2 depicts the process parameters and their levels.

3. Results and Discussion

The effect of the different parameters such as electric current, pulse on-time and pulse off-time on the dimensional accuracy of the drilled hole were studied. Different parameters have different effect on overcut and taper angle. As it can be seen from figure 4 and figure 5 that electric current has the significant effect on overcut and taper angle because it has the largest factor which is a small change in electric current can directly affect the change in overcut and taper angle as can be seen in the models equations 3 and 4 where A, B, C are electric current, pulse on-time and pulse off-time respectively.

$$OC = 0.028 + 0.008219A + 0.005056B - 0.002156C - 0.001149A2 - 0.001374B2 \quad (3)$$

$$\theta = 0.40 + 0.085A + 0.042B - 0.024C \quad (4)$$

This is because a small change in the electric current will cause higher change in these two criteria. In the case of overcut and taper angle, the result is also confirmed with previous researchers' results [2, 5, 15]. This is due to the higher spark gap, which melts and vaporizes more material from surface and this also could be explained due to the material properties of the SS304 work material. However, this is an opposite to some of the researchers which states that with the increases of the discharge current overcut and taper angle decrease [16]. The effect of pulse on-time on overcut and taper angle is that with the increasing of the pulse on-time, taper angle increases in a way less than what electric current does. This is because the pulse on-time and pulse off-time affect the formation of the secondary spark during removal of the debris by pressurized liquid. The pulse off-time was with the least effect on both responses so that as the pulse off-time

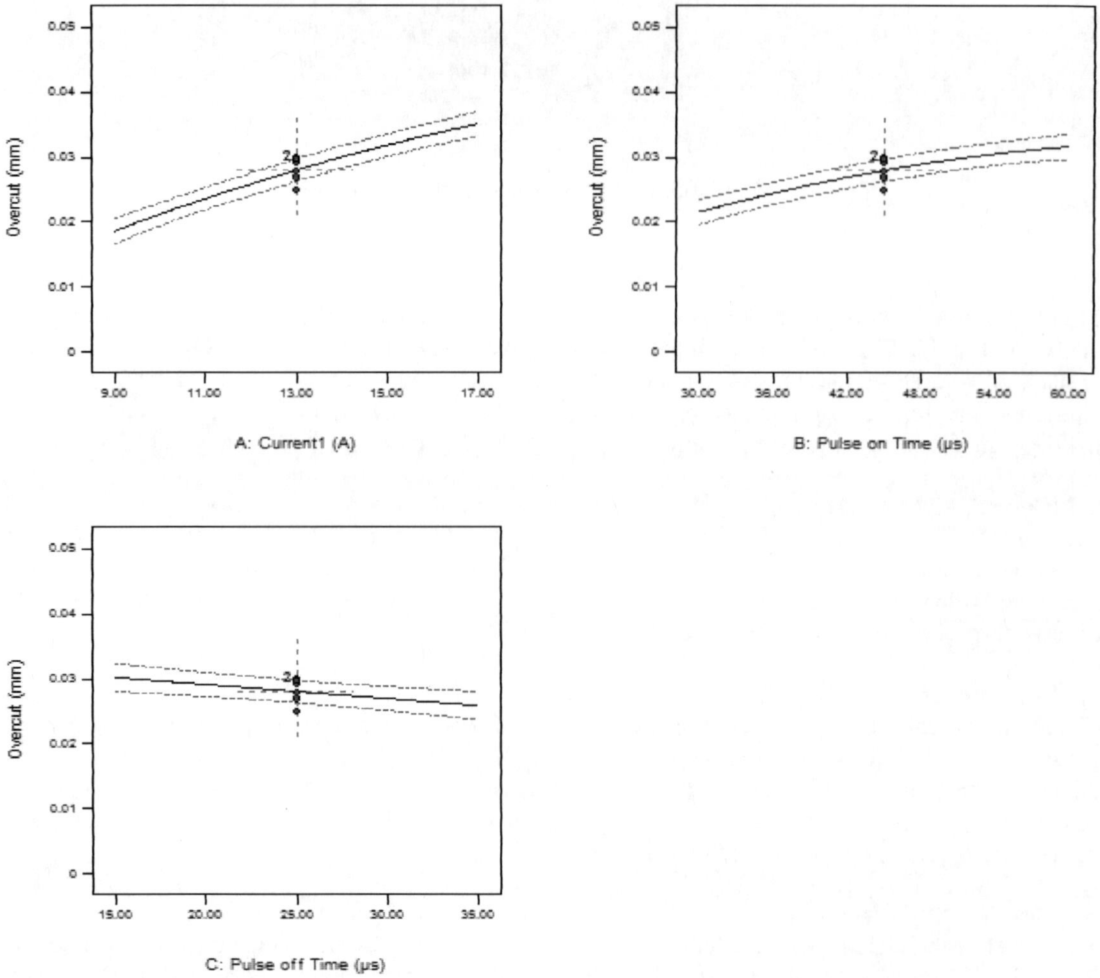

Figure 4. The effect of the three different EDM parameters namely electric current, pulse on-time and pulse off-time on overcut

IOP Conf. Series: Materials Science and Engineering **290** (2017) 012067 doi:10.1088/1757-899X/290/1/012067

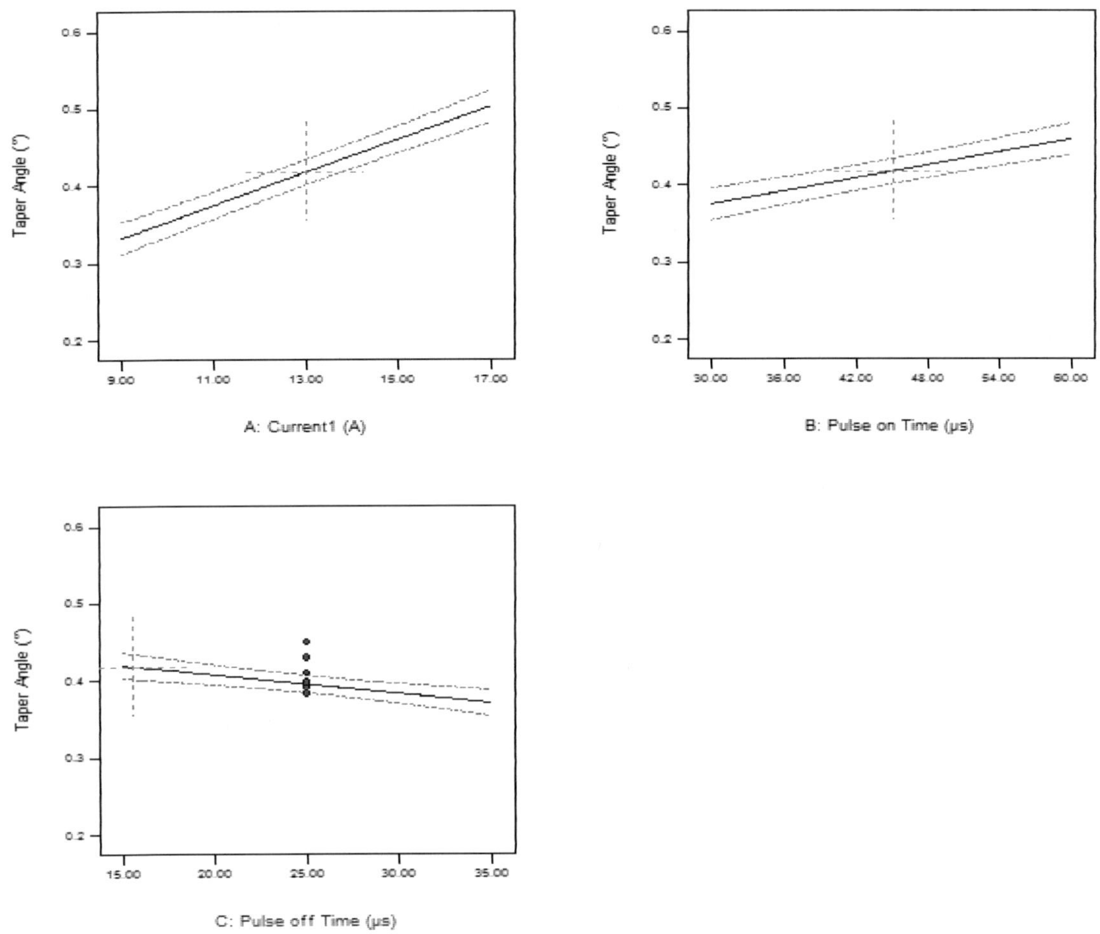

Figure 5. The effect of the three different EDM parameters namely electric current, pulse on-time and pulse off-time on taper angle

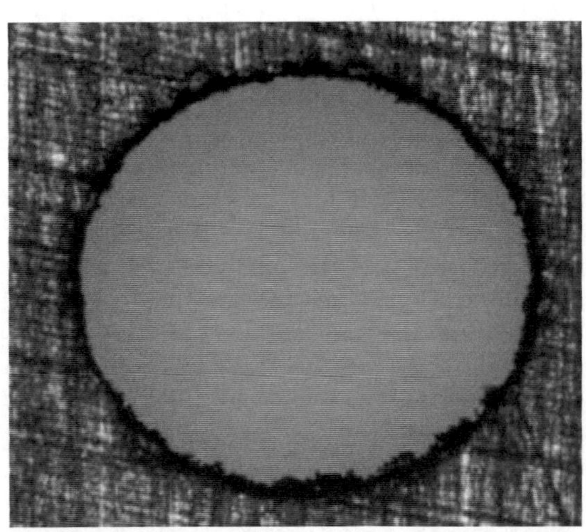

Figure 6. Optical microscope image (current = 9A, pulse on time = 30 μs and pulse off time = 35 μs)

3.1 Study of Hole Morphology

Figure 6 depicts that when the machine tool is running at current 30 A, pulse on time 30 µs, pulse off time 35 µs, it achieves lower overcut and reduced the taper angle. It can be observed also some sharp edges appeared in the hole and this could be because of the flushing pressure need to be studied and adjusted.

3.2 Model Validation

Two sets of experimental runs were performed several times and the average data were taken and compared against the predicted data obtained from the model equations presented previously. Table 3 presents the data used to confirm the validity of the model. Two set runs are performed. The first run is 7 A, 25 µs, and 30 µs for Current, Ton and Toff respectively and the second run are 12 A, 40 µs and 20 µs. The percentage error were calculated and was found to be less than 5% which is promising result.

Table 3. Conformation runs for overcut and taper angle models with percentage errors

No.	Overcut (mm)		Taper Angle°		% Error Overcut	% Error Taper
	Predicted	Actual	Predicted	Actual		
1	0.0028	0.0029	0.199	0.203	3.929	2.061
2	0.0251	0.0258	0.372	0.374	2.789	0.403

4. Conclusion

In this study, overcut and taper angle are investigated in machining of stainless steel 304, against three different parameters such as electric current (I_p), pulse on-time (T_{on}) and pulse off-time (T_{off}). The electrode is made of a brass tube with diameter of 1mm. Dimensional accuracy was measured by evaluating overcut and taper angle. Those two measurements were performed using optical microscope. The experimentation planning, evaluation and analysis had been carried out by the RSM with total number of 20 runs. According to the experimental outcomes, followings are the specific conclusions that have been drawn.

1. The research reveals that, discharge current was found the most significant factor affecting overcut and taper angle as well followed by pulse on-time and pulse off-time. This is due to the higher spark gap, which melts and vaporizes more material from surface and this also could be explained due to the material properties of the SS304 work material
2. As the discharge current and pulse on-time increase, overcut and taper angle are increased. In the case of the pulse off-time, as it increases, overcut and taper angle decreases.
3. The optimized parameters that give the minimum overcut and taper angle response are 9 A current, 30 µs Pulse on-time and 35 µs Pulse off-time which gives 0.010 mm overcut and 0.244° taper angle.
4. The validation of the model has been performed and it was found that the percentage errors are about 3.36% for overcut and 1.23% for taper angle.

Acknowledgements

The authors would like to thank the Ministry of Science, Technology and Innovation (MOSTI), Malaysia for financial support under Science Fund Research Project SF 15-016-0066.

IOP Conf. Series: Materials Science and Engineering **290** (2017) 012067 doi:10.1088/1757-899X/290/1/012067

References

[1] Venkatesh, V., N. Swain, G. Srinivas, P. Kumar, and H.C. Barshilia. 2016. *Materials and Manufacturing Processes*
[2] Natarajan, N. and P. Suresh. 2015 *The International Journal of Advanced Manufacturing Technology*, **77**(9-12), 1741-1750
[3] D'Urso, G., G. Maccarini, and C. Ravasio 2015 *The International Journal of Advanced Manufacturing Technology*, 1-13
[4] Rajmohan, T., R. Prabhu, G.S. Rao, and K. Palanikumar 2012 *Procedia Engineering*, **38**, 1030-1036
[5] Eyercioglu, O., M.V. CAKIR, and G. Kursad 2013 *Proceedings of the Institution of Mechanical Engineers, Part B: Journal of Engineering Manufacture*, 0954405413498730
[6] Sultan, T., A. Kumar, and R.D. Gupta 2014 *International Journal of Manufacturing Engineering*
[7] Parmar, K.V. and R. Patel. 2014. Experimental Investigation of Process Parameters of Micro EDM Drilling on Carbide (K20) material
[8] Janmanee, P. and A. Muttamara 2011 *International Transaction Journal of Engineering, Management, & Applied Sciences & Technologies*, **2**(4), 471-481
[9] Tamang, S., N. Natarajan, and M. Chandrasekaran 2016 *Journal of the Brazilian Society of Mechanical Sciences and Engineering*, 1-11
[10] Bozdana, A.T., O. Yilmaz, M.A. Okka, and I.H. Filiz 2010 Paper presented at the Responsive Manufacturing-Green Manufacturing (ICRM 2010), 5th International Conference on
[11] Singh, B., P. Singh, G. Tejpal, and G. Singh 2012 *Int. J. of Adv. Engg. Tech*, **3**(4), 130-133
[12] Singh, H. and E. Singh. 2013. Effect of Pulse on/Pulse Off on Machining of Steel Using Cryogenic Treated Copper Electrode
[13] Li, Y., W. Hou, J. Xu, and H. Yu. 2016 IEEE
[14] Bhaumik, M., K. Maity, and K.D. Mohapatra. 2016 Paper presented at the Applied Mechanics and Materials
[15] Jahan, M.P., Y. San Wong, and M. Rahman. 2010 *The International Journal of Advanced Manufacturing Technology*, **46**(9-12), 1145-1160
[16] Siva, M., M. Parivallal, and M.P. Kumar. 2014 *Procedia Materials Science*, **5**, 1829-1836

IOP Conf. Series: Materials Science and Engineering **290** (2017) 012068 doi:10.1088/1757-899X/290/1/012068

The optimization study on the tool wear of carbide cutting tool during milling Carbon Fibre Reinforced (CFRP) using Response Surface Methodology (RSM)

M K Nor Khairusshima[1], B Muhammad Hafiz Zakwan[1], M Suhaily[1], I S S Sharifah[1], N M Shaffiar [1], M A N Rashid [1]

[1]Department of Manufacturing and Materials Engineering, International Islamic University Malaysia (IIUM), Jalan Gombak, 53100 Kuala Lumpur, Malaysia

Email:norkhairusshima@iium.edu.my

Abstract. Carbon Fibre Reinforced Plastic (CFRP) composite has become one of famous materials in industry, such as automotive, aeronautics, aerospace and aircraft. CFRP is attractive due to its properties, which promising better strength and high specification of mechanical properties other than its high resistance to corrosion. Other than being abrasive material due to the carbon nature, CFRP is an anisotropic material, which the knowledge of machining metal and steel cannot be applied during machining CFRP. The improper technique and parameters used to machine CFRP may result in high tool wear. This paper is to study the tool wear of 8 mm diameter carbide cutting tool during milling CFRP. To predict the suitable cutting parameters within range of 3500-6220 (rev/min), 200-245 (mm/min), and 0.4-1.8 (mm) for cutting speed, speed, feed rate and depth of cut respectively, which produce optimized result (less tool wear), Response Surface Methodology (RSM) has been used. Based on the developed mathematical model, feed rate was identified as the primary significant item that influenced tool wear. The optimized cutting parameters are cutting speed, feed and depth of cut of 3500 rev/min, 200 mm/min and 0.5 mm, respectively, with tool wear of 0.0267 mm. It is also can be observed that as the cutting speed and feed rate increased the tool wear is increasing.

1. Introduction

CFRP is an expensive Fibre Reinforced Plastic (FRP) which is very favourable material especially in aircraft industrial due to their mechanical and physical properties. The combination of the matrix and fiber reinforcement at high strength levels in engineering components provides high fracture toughness, high specific stiffness, excellent in creep, and corrosion and thermal resistance. It is also stronger than steel and stiffer than titanium while retaining its lighter weight compared to conventional single-phase materials. As the applications for CFRP composites expand in various fields, the parameters for machining have also diversified. The necessity for machining arises as CFRP parts are produced near-net-shape, a process that eliminates rough machining operations. Thus, machining operations such as milling is still necessary to give the CFRP parts their final shape [1,2,3].

In spite of growing demand and usage, problems have arisen with CFRP in terms of machining. Knowledge and experience acquired for conventional materials cannot be applied to these newer materials. It has been reported that machining CFRP is not only difficult to handle because of it inhomogeneity and anisotropic structure, but also because of its abrasive nature which comes from the carbon, and that aspects differ from metal [4,5]. In general, the challenges face during machining CFRP can be classified into two which are excessive tool wear and surface quality [1]. Unlike metal, it has been reported that machining CFRP not only involved the tool edge chipping but also the excessive abrasive wear due to the high strength carbon fibers [3, 5]. Those machining difficulties make machining CFRP time-consuming and expensive. To achieve high productivity and associated cost reduction, many approaches have been attempted such as finding the optimization of machining CFRP via modeling as it is necessary to understand the mechanics of the process and to improve the machinability of CFRP [7].

2. Experimental Materials and Methodology

The milling process was performed on the unidirectional CFRP with dimensions of 200 x 200 x 3 mm. CFRP panel, which was made using vacuum bagging method was illustrated in Figure 1. The material of panel includes the mixing 58% of carbon fibres and 42% of resin. The detailed of physical properties was shown in Table 1.

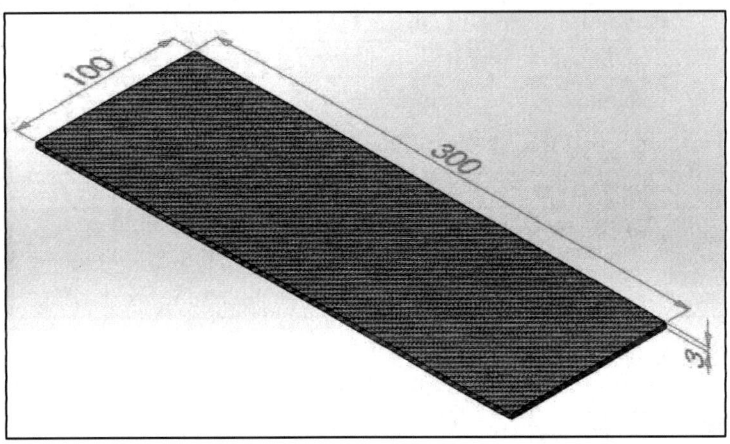

Figure 1. Illustration of CFRP used

Table 1. Physical Properties of Resin

RESIN		EPOLAM 2017 RESIN		
HARDENERS			**EPOLAM 2017 HARDENER**	**EPOLAM 2018 HARDENER**
Mix Ratio by Weight		100	30	30
Aspect		Liquid	Liquid	Light Amber
Viscosity at 25^{0}C (MPa,s)	BROOKFIELD LVT	2,850	25	20
Density 25^{0}C		Liquid	Liquid	Light Amber
Pot life at 25^{0}C	ISO 1675:1985	1.17	0.96	0.96
Viscocity at 25^{0}C	BROOKFIELD LVT	2MPa.s	550	450

IOP Conf. Series: Materials Science and Engineering **290** (2017) 012068 doi:10.1088/1757-899X/290/1/012068

Two flute solid uncoated carbide end mills (S2FE-080) was used to machine CFRP composite. The cutting tool has a diameter of 8.0 mm, a 20–24° relief angle, and a 30° helix angle. The actual photo and geometrical properties of the solid uncoated carbide cutting tool are shown in Figure 2 and listed in Table 2 respectively. The chemical composition and physical properties of the cutting tool are shown in Table 3 and Table 4.

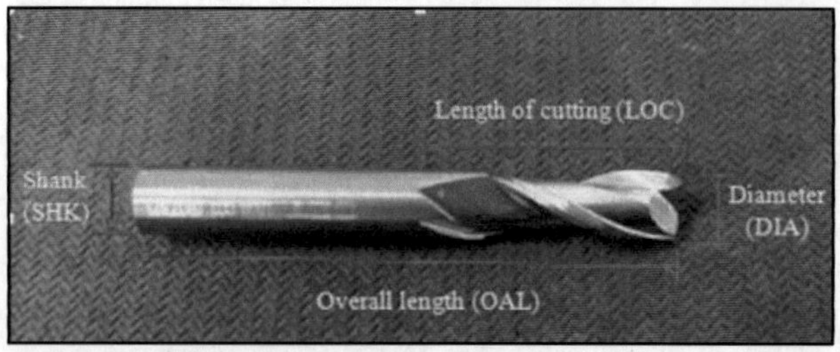

Figure 2 Carbide Cutting Tool

Table 2. Geometry of uncoated carbide cutting tool

DIA (mm)	SHK (mm)	OAL (mm)	LOC (mm)
8.0	8.0	60.0	20.0

Table 3. Chemical composition of uncoated carbide cutting tool

Element	Weight %
Tungsten Carbide, WC	88.4–90.0
Cobalt, Co	9.5–10.5
VC+Cr3C2	0.5–1.1

Table 4. Physical properties of uncoated carbide cutting tool

Density, g/cm3	Hardness, HRA
14.35 ± 0. 1	9.18 ± 0.5

A Computer Numerical Control (CNC) milling machine (DECKEL MAHO DMG DMU 35M) was used to machine the CFRP panel in the experiment. The tool wear of the cutting tool were measured by using a tool maker Nikon Microscope MM-40. During machining process, at every distance travelled of 200 mm, tool wear are recorded. The picture of the tool wear is captured for every distance travelled after machining process. The proper clamping method is the one of the important step to ensure that CFRP is not bending during machining process. The setup of the experiment is shown in Figure 3.

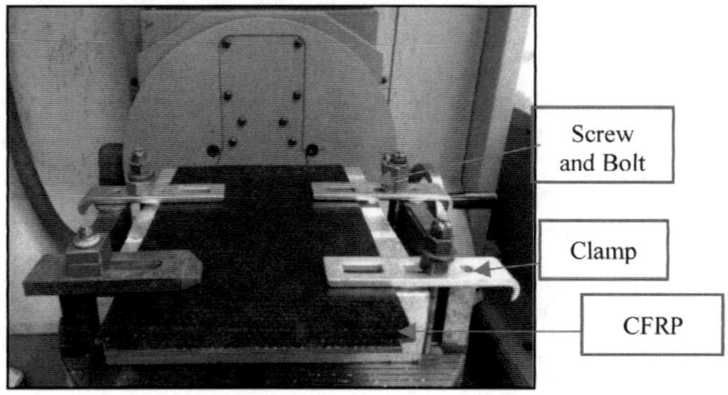

Figure 3. Machining Setup

RSM was utilized to conduct the experimental runs in this study. To reduce the number of runs, small Central Composite Design (CCD) was selected. Table 5 shows the five input variable levels which were used.

Table 5. Range of Cutting Parameters

Coded /Actual	Input Variable				
	Lowest $-\sqrt{2}$	**Low** **–1**	**Centre** **0**	**High** **1**	**Higher** $\sqrt{2}$
A:Cutting speed (rpm)	2361	3500	6250	9000	10139
B: Feed rate (mm/min)	190	225	200	250	260
C: Depth of cut (mm)	0.2	0.5	1.25	2.0	2.3

3. Results and Discussion

3.1. Effect of cutting speed, feed rate and depth of cut on tool wear

In industry, tool wear is the most important factor affecting the profitability and productivity of cutting process and cutting parameters such as cutting speed, feed rate and depth of cut, which have been identified as one of the factors that influence tool wear. Hence, it is important to find the suitable cutting parameters that can produce the optimize result (low tool wear) during machining. In this paper, analysis of tool wear was analyzed by using the analysis of variance (ANOVA). **Error! Reference source not found.** displays the ANOVA model to determine the tool wear. According to this model, the model F-value of 15.68 implies that the model is significant with value of "Prob > F" less than 0.05. There is only 0.0003% that a "Model F-Value" this large could exist due to noise. In this model, the significant model terms are main effect of cutting speed (A), feed rate (B) and depth of cut (C). The lowest "Prob>F" values indicates that feed rate (B) which is (0.0005) gives more influences in machining process compare to other main effects. Lack of Fit for this model is not significant with 15.89% relative to the pure error, so that, non-significant lack of fit model is good. The R^2 is 0.81 which is the value high and close to 1. The R^2 predicted is 0.59 which is reasonable agreement with the Adj. R^2 which is 0.75. A ration greater than 4 is desirable. Ratio in this model is 13.066 indicates an adequate signal. Figure 4 displays the normal probabilities plot of standardized residuals. It is shown that the plot is reasonable because of the point close to the straight line which presents that the error was normal distribute

Table 6. ANOVA Model for the Tool Wear

Response 1 Tool Wear ANOVA for Response Surface Linear Model Analysis of variance table [Partial sum of squares]						
Sources	Sum of	D	Mean	F	Prob>F	
Model	0.00050123	3	0.0001670	15.685	0.0003	significant
A	1.62855E-	1	1.62855E-	1.5289	0.242	
B	0.00024625	1	0.0002462	23.118	0.0005	
C	0.00023869	1	0.0002386	22.409	0.0006	
Residua	0.00011716	1	1.06517E-			
Lack of	9.79687E-	7	1.39955E-	2.9157	0.1589	not
Pure	0.0000192	4	0.0000048			
Cor	0.0006184	1				
R^2= 0.812	Adj. R^2 = 0.758		Pred. R^2=0.596		Adeq. Precession= 13.065	

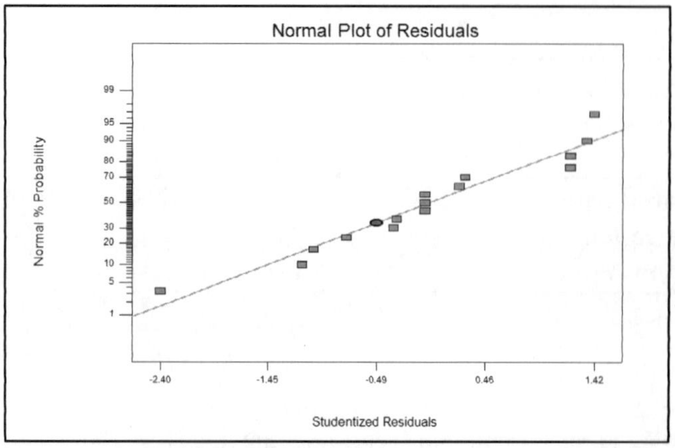

Figure 4. Normal probabilities of residuals for tool wear

Meanwhile Figure 5 shows the three-dimensional graph of the response surface for tool wear. It can be observed that as the cutting speed and feed rate increase, the value of tool wear is also increased which indicate that the tool life of the cutting tool is decreasing. According to Nor Khairusshima [8] at higher cutting speed, the cutting temperature is high, which worn out the cutting tool easily. Meanwhile, at higher feed rate, high rapid traverse of cutting tool is the primary reason of high tool wear during machining CFRP.

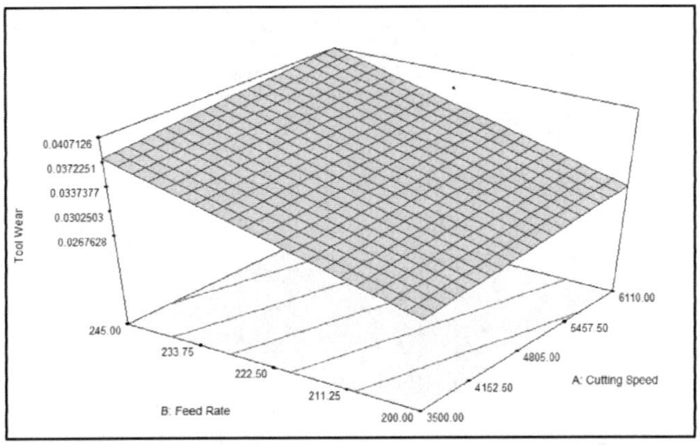

Figure 5 Three-dimensional graph of tool wear

Clearer view of tool wear can be seen in Figure 6. It can be seen that the shinning area which has been identify by Nor Khairusshima [8] as abrasive wear is getting bigger as the feed rate increased. This Figure 6 is supporting the graph in Figure 5 which is developed by RSM.

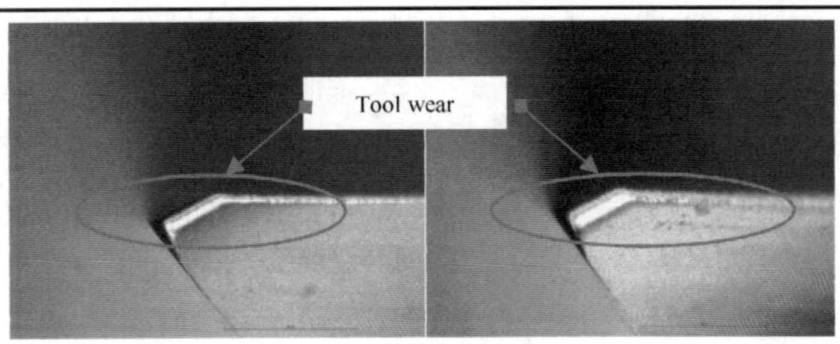

Figure 6. Growth of Tool Wear as Feed Rate Increased from (a) 191 mm/min and (b) 254 mm/min) as cutting speed and depth of cut are constant

The equation that describes linear surface Figure 5 is shown in equation 3.1

$$Tool\ wear = 0.039 + 1.427x10^{-03} * A + 5.584x10^{-03}B + 5.426x10^{-03} * C \quad (3.1)$$

Where A is cutting speed, B is feed rate and C is depth of cut

3.2. Optimization
In machining operation, the determination of the most suitable and optimum cutting parameter is very important. By having optimization, the machining process can be conducted with guided cutting parameters that produce high quality results. There are 8 solutions of the optimization according to Design Expert software and it is tabulated in Table 7. According to the Table 7, highest desirability index obtained is 0.916. The best result (low tool wear in the range of cutting parameters) can be obtained at cutting speed of 3500.0 rev/min, feed rate 200 mm/min and depth cut 0.50 mm. These conditions yield optimum value of tool wear is 0.0267 mm respectively.

Table 7. Optimized solution of tool wear for machining CFRP

	Optimized Cutting Parameter			Optimized		
No.	**Cutting Speed**	**Feed Rate**	**Depth of Cut**	**Tool Wear (mm)**	**Desirability**	
1	3500	200	0.50	0.0267	0.916	(Selected)
2	3515	200	0.50	0.0267	0.915	
3	3558	200	0.50	0.0268	0.913	
4	3999	200	0.50	0.0273	0.891	
5	4116	200	0.50	0.0274	0.883	
6	3500	203	0.51	0.0275	0.876	
7	4936	200	0.50	0.0283	0.841	
8	5733	200	0.50	0.0292	0.799	

4. Conclusion

The following conclusions are drawn from the results of this experiment:

1. Based to the ANOVA analysis, all the main cutting parameters (cutting speed, feed rate and depth of cut) are the significant factors to the tool wear.
2. Feed rate has been identified as the most pronounced factor which affect tool wear during milling CFRP compared to the cutting speed and depth of cut.
3. From the 3-D graph, it can be observed that as the cutting speed and feed rate increase, the value of tool wear is also increased which indicate that the tool life of the cutting tool is decreasing.
4. The shinning area on the cutting tool which has been identified as abrasive wear is getting bigger as the feed rate increase especially which indicate the tool wear is increased.
5. From the model developed, best optimization for tool wear (0.0267 mm) can be achieved by applying cutting speed 3510 rev/min, feed rate 200 mm/min and depth of cut 0.5 mm

References

[1] Brinksmeier E, Fangmann S, and Rentsch, R 2011. *CIRP Annals* Manuf Tech **60(1)** 57-60
[2] Hosokawa A, Hirose N, Ueda T and Furumoto, T 2014 *CIRP Annals – Manuf Tech*, **63(1)** 89-92
[3] Karpat Y, Bahtiyar O, and Değer B, 2012 *Procedia CIRP* **1** 460-465
[4] Dandekar C R, and Shin Y C 2012 *Int. J. Machine T Manuf* **57** 102-121
[5] Hintze W, Hartmann D and Schütte C. 2011 *Comp Sci & Tech* **71(15)** 1719-1726.
[6] Park K H, Beal A, Kim D, Kwon P, and Lantrip J, 2011. *W* **271(11–12)** 2826-2835.
[7] Schulze V, Becke C, Pabst R 2011 *CIRP Annals Manuf Tech* **60(1)** 69–72.
[8] Nor Khairusshima M K, Che Hassan C H, Jaharah A G, Amin A K M, and Md Idriss A N 2013 *W* **302(1–2)** 1113-1123.

Assembly meshing of abrasive waterjet nozzle erosion simulation

N H Kamarudin[1], A Mebrahitom[1] and A Azhari [1]

[1]Department of Manufacturing, Universiti Malaysia Pahang (UMP), 26600 Pekan, Pahang Malaysia

Email:naqibhbk@gmail.com

Abstract. Computational Fluid Dynamics (CFD) softwares have been prevalent in Abrasive Waterjet (AWJ) Modelling for optimization and prediction. However, there are many different methods in approaching a single problem especially in predicting the erosion rate of nozzle which is critical in influencing kerf quality of AWJ cutting. In this paper, three main methods of assembly meshing for an abrasive waterjet erosion were simulated which is Quadrilateral, Cutcell and Tetrahedrons and each processing time, quality of convergence and accuracy of results are discussed. Results shows that Quadrilateral mesh prevails in the mentioned category followed by Tetrahedrons and Cutcell.

1. Introduction

Computational Fluid Dynamics (CFD) is a technique to solve and analyze problems that involves fluid flows using numerical methods and algorithms. Using computers to perform modeling, simulations and analysis of fluid flow enables CFD practitioners to have insights of flow patterns that have been difficult to be performed using traditional experimental techniques. Currently, CFD applications can be found in the field of aerospace, turbo machinery, automotive and maritime. The capabilities of CFD also made it way in meteorology, oceanography, astrophysics, oil recovery and architecture. A comprehensive explanation of CFD principles and applications have been done by Blazek [1], Versteeg and Malalasekera [2] and Anderson and Wendt [3]. Currently, the development of CFD software has been very promising. ANSYS Fluent software is well known and widely used CFD engineering software for modeling fluid flow. Abrasive Waterjet (AWJ) Machine have been known for its versatility and widely used in the industry for multiple fabrication process and cleaning [4-6]. An AWJ related study was conducted by Junkar, Jurisevic, Fajdiga and Grah [7] to study the effect of erosion impact of a single particle abrasive with the target workpiece which is stainless steel (AISI 304). The research was done using LS Dyna where the crater spherity in the simulation was used to compare with the observation of experimental shapes of craters on the workpiece material. The study is emphasized only in the measurements of the sphericity of the impact crater and limited parameters of abrasive and workpiece material.

Maniadaki, Kestis, Bilalis and Antoniadis [8] proposed a finite element-based model for pure waterjet process simulation and the main objective was to investigate and analyse in detail the material behaviour of a work piece under waterjet impingement. The model developed using LS-DYNA 3D code to simulates the erosion of the target material caused by the high-pressure waterjet flow.

Maniadaki, Kestis, Bilalis and Antoniadis [8] proposed a finite element-based model for pure waterjet process simulation and the main objective was to investigate and analyse in detail the material behaviour of a work piece under waterjet impingement. The model developed using LS-DYNA 3D code to simulates the erosion of the target material caused by the high-pressure waterjet flow.

Deepak, Anjaiah, Karanth and Sharma [9] had also used ANSYS software to examine the effect of inlet pressure on skin friction coefficient and jet exit kinetic energy. Per the results, the effect of increasing the inlet pressure causes significant increase to the skin coefficient friction and increased in the jet kinetic energy correspondingly. Further analysis then suggested that by increasing the volume fraction of abrasives causes a significant decrease on both skin coefficient friction and jet kinetic energy.

Baisheng, Hui, Lei, Jufeng, Hua, Zhen, Longkang and Hailong [10] have performed numerical simulation of the flow field of abrasive waterjet nozzle under submerging conditions based on FLUENT software. The research used RNG κ-ε turbulent model and simple algorithm for simulation to simulate flow field from generated from an abrasive waterjet nozzle. The research indicated that there are three zones which are free jet zone, shock zone and wall jet zone. Shock zone suggest the distance for the best cutting distance which is shown to be within the scope of 2 – 7 times the nozzle exit diameter.

Deepak, Anjaiah, Karanth and Sharma [9] had also applied CFD in AWJ to examine the effect of inlet pressure on skin friction coefficient and jet exit kinetic energy of the AWJ nozzle. Per the results, the effect of increasing the inlet pressure causes significant increase to the skin coefficient friction and increased in the jet kinetic energy correspondingly. Further analysis then suggested that by increasing the volume fraction of abrasives causes a significant decrease on both skin coefficient friction and jet kinetic energy. It is noted that the research does not undertake erosion studies of the nozzle.

On the other hand, studies using CFD in investigating erosion in AWJ nozzle are very limited. Mostofa, Kil and Hwan [11] used the ANSYS CFX software to simulate multiphase flow of water, air and abrasives in the mixing tube to monitor erosion rate at the mixing tube wall and predict the influence of abrasive particle size with different parameters of tube lengths. The erosion model is based on Finnie's erosion model of ductile material. The result showed that abrasive shape and jet velocity have effect on erosion rate. The erosion rate has similar linear results as obtained by Nanduri, Taggart and Kim [12]. Further CFD based for AWJ studies was conducted by Kamarudin, Rao and Azhari [13] by using Discrete Phase Method (DPM) to predict the erosion of AWJ nozzle. Present study shows how different the results of the erosion rate of the nozzle affected by manipulating the assembly of the mesh. The information for the processing time and convergence are also added to provide alternative options for the best method of AWJ simulation.

Assumptions and Theoretical Formulation

The flow is assumed to be multiphase and the fluid is treated as continuum and incompressible. The particle's abrasive velocity is assumed to be the same as those of flowing fluid. The Langrangian-Eulerian model was used to solve the coupling of the fluid flow and abrasives, also called discrete particle modeling, which solves the equations of motion individually for each particle, whereby the continuous phase is modeled using a Eulerian framework and the trajectories of the particles are simulated within a Lagrangian framework. In a two-way coupling, each particle exchanges mass, momentum and energy with the fluid phase valid for the current work [9], [14]-[17]. The theoretical waterjet velocity was used by Mostofa, Kil and Hwan [11] to validate their simulation model. Similarly, the same theory was used to validate the simulation model for the present work. The equation for the water jet velocity at the orifice can be obtained using the following theoretical formulation.

Theoretical waterjet velocity, V_{th}:

$$V_{th} = \sqrt{\frac{2p}{\rho}} \quad (1)$$

Compressibility of water:

$$\frac{\rho}{\rho_\circ} = \left(1 + \frac{P}{L}\right)^n \quad (2)$$

Where p is the operating pressure, ρ is water density (kg/m3), L = 345 Mpa and n = 0.162 at 25[0] C. Note that value of L depends on the pumping pressure of the AWJ machine for the pump plunger to pumped before any water begins to exit the check valve.
The resulting equation for the waterjet velocity:

$$V_j = \sqrt{\sqrt{\frac{2L}{(1-n)\rho_0}\left[\left(1+\frac{P}{L}\right)^{1-n} - 1\right]}} \quad (3)$$

Compressibility factor:

$$\varphi = \sqrt{\frac{V_j}{V_{th}} = \sqrt{\frac{L}{P\,(1-n)}\left[\left(1+\frac{P}{L}\right)^{1-n} - 1\right]}} \quad (4)$$

The waterjet velocity can then be expressed in the following equation:

$$V_J = C_d \varphi V_{th} \quad (5)$$

Where discharge coefficients, $C_d = 0.85$. The value is based on the function of the jet size from past research [11].
Using the particle erosion and accretion method to enables erosion rates to be monitored at wall boundaries. The erosion rate equation is defined as below:

$$R_{erosion} = \sum_{p=1}^{n} \frac{\dot{m}_p C(d_p) f(a) v^{b(v)}}{A_{face}} \quad (6)$$

Where $C(d_p)$ is a function of particle diameter,a is the impact angle of the particle path with the wall face, $f(a)$is a function of impact angle, v is the relative particle velocity, $b(v)$ is a function of relative particle velocity, and A_{face} is the area of the cell face at the wall. Default values are $C = 1.8 \times 10^9$, $f = 1$, and $b = 0$. The erosion rate density is presented in the unit of kg/s/m^2, noted that this value just represent the qualitative value and not the physical value which reflect the actual materials being used.
K-ε turbulence model is used to simulate turbulence for mixed flow of water, air and abrasive of an AWJ process. The turbulence quantity is specified using the Percentage Intensity, I, which is defined as the root-mean-square of the velocity fluctuations, u', to the mean velocity $u_{average}$. The value of the turbulence intensity specified at water pressure inlet boundary conditions are defined as below:

$$I = \frac{u'}{u_{average}} = 0.16(Re_{Dh})^{-1/8} \quad (7)$$

The Reynold's number, Re, for given size of the hydraulic diameter, D_h, of the AWJ waterjet circular tube were obtained using the following equation:

$$Re_{Dh} = \frac{\rho u D_h}{\mu} \quad (8)$$

Where, Re is Reynolds number (non-dimensional), u is velocity based on the actual cross section area of the duct or pipe (m/s), μ is the dynamic viscosity (Ns/m^2) and L is the characteristic length (m) and ν is the kinematic viscosity (m^2/s).

2. Geometry and Parameters

The numerical model is based on the 3-D model of commercial AWJ cutting head from past research [11], [12]. The geometrical and parameters are defined as shown in table 1. The material properties for the nozzle and the abrasives are shown in table 2 and 3 respectively. High pressured waterjet and abrasive is released from their corresponding inlet and mixed into the mixing chamber where entrainment occurs and focused to the nozzle tube and eventually released. The model was then meshed using the standard meshing in FLUENT 15.0. The wall of the nozzle is meshed using inflation by selecting the surface selection option to simulate wall interaction with the fluid flow.

(a)

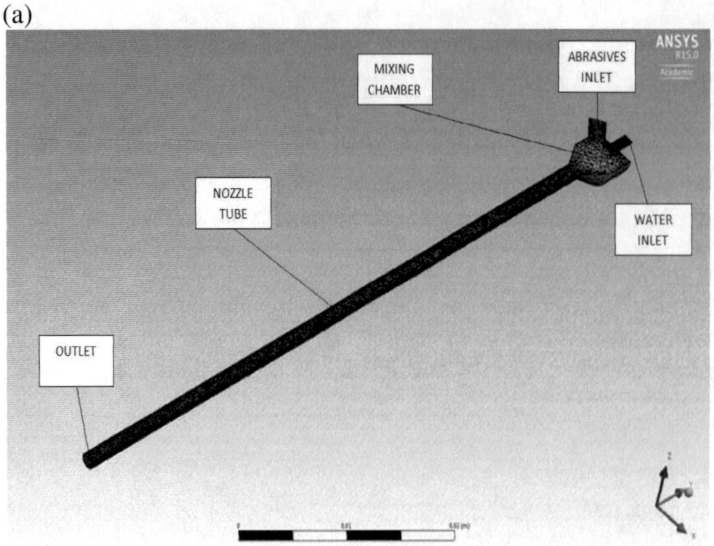

(b)

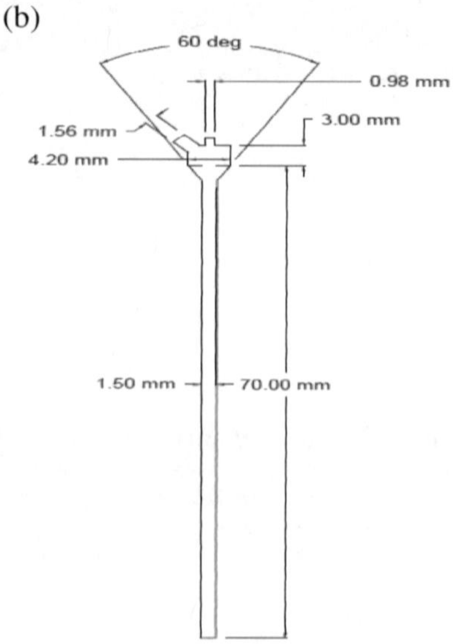

Figure 1. Schematics of the Cutting head (a) Geometry parameters designations (b) Geometry values.

Table 1. Boundary condition and parameter values.

Parameters	Typical Values
Nozzle length (mm)	50.8
Nozzle inlet angle (^{0})	60^0
Nozzle diameter (mm)	1.14
Orifice diameter (mm)	0.38
Water pressure (Mpa)	310
Abrasive flow rate (g/s)	3.8
Mesh size (#)	80# (0.177 mm)

Table 2. Nozzle material properties.

Material	Density (kg/m^3)	Specific Heat, Cp (J/kg-k)	Thermal Conductivity
Tungsten Carbide	15680	39.8	110

Table 3. Abrasives material properties.

Material	Density (kg/m^3)	Specific Heat, Cp (J/kg-k)	Thermal Conductivity
Aluminium Oxide (al2o3)	3950	880	-

Table 2 shows the parameters for boundary conditions. The shape factor is a parameter for the abrasive particles irregularity. A shape factor of 1.0 signifies a particle with a circular shape and decrease of value signifies increase of irregularity. The typical value for particle mesh size chosen was 80 or approximately 0.1 mm in diameter and is assumed to be uniform for simplification of the solution.

3. Results and Discussion

The geometry of the nozzle were set and mesh methods were generated. In figure 2 (a) shows the assembly methods for Quadrilateral mesh which mainly partition the parts to four sides cell shape. Figure 2 (b) shows the mesh generated by the Cutcell assembly methods which the areas zone were partitioned to almost exact cubic shaped. It simplified when dealing with tubular zone. Figure 2 (c) shows the mesh generated by tetrahedrons methods which are observed to be very fine in details. Tetrahedrons partitioned the zones to four vertices, six edges and bound to four triangular faces.

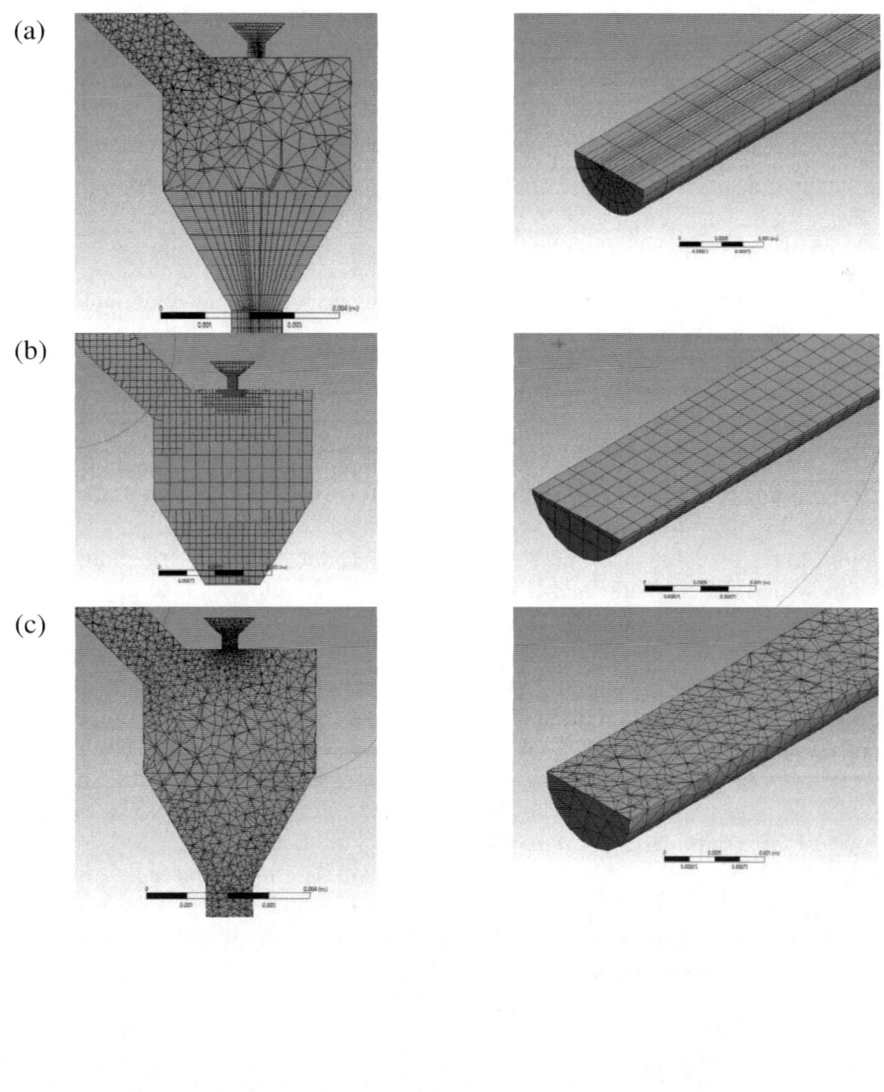

IOP Conf. Series: Materials Science and Engineering **290** (2017) 012069 doi:10.1088/1757-899X/290/1/012069

Figure 2. Meshing structure of Quadrilateral mesh (a), Cutcell mesh (b) and Tetrahedrons mesh (c).

Table 4 shows the comparison of the assembly methods in term of the cell size, nodes, number of elements, processing time, convergence quality and the adjusted under relaxation factor (URF). The cell size was adjusted to be equal so that the comparison will be relevant. However, there were some limitations to ensure the same cell size as the software automatically adjusted the value to prevent ratio error. Therefore, the cell size has been to be as close as possible which also includes the number of nodes and elements. The simulation were done by comparing different nozzle size and compared with results of past studies [12]. In term of processing time, the Quadrilateral mesh methods is the fastest followed by Cutcell and Tetrahedrons methods, respectively. This is mainly due to cell size which is less fine compared to tetrahedrons which took around seven days for a simulation to complete [18]. The simulations tend to diverged more for both Cutcell and Tetrahedrons which shows that the meshing methods proved to be difficult for complex structure and multiphase flow. The URF were also adjusted to negate the divergence problem.

Table 4. Comparison of different Assembly Methods in term of Processing time, Convergence Quality and Under Relaxation Factor (URF).

Assembly Methods	Quadrilateral	Cut cell	Tetrahedrons
Min Size	3.03E-05	4.69E-05	4.69E-05
Max Face Size	3.03E-03	6.07E-03	6.07E-03
Nodes	22205	26037	20416
Elements	30482	23682	95680
Processing time	3 Days	3 – 4 Days	7 Days
Convergence Divergence	Converged Well and rarely diverged	Divergence is Prevalent	Divergence is Prevalent
Under Relaxation Factor	None	Minimal	High

Figure 3 shows the effect of nozzle length with the erosion wear rate. It is revealed that the length of the nozzle has a direct influence on the nozzle wear rate. However, the simulated nozzle wear shows varied results as the length of the nozzle increases. In table 5, the information of the nozzle rate presented in (kg/m^2-s) and the accuracy between the simulated and the experimental erosion rate in percentage were revealed. The Quadrilateral meshing shows the closest accuracy out of the other assembly methods. This is followed by Tetrahedral with a minor accuracy difference between Quadrilateral mesh and Cut Cell assembly which have major differences in accuracy. It also presented that Quadrilateral meshing have a consistent trend with the actual erosion rate in which the erosion rate declines as the length of the nozzle increases. The Tetrahedral assembly have a considerable accuracy at the beginning nozzle value of 32.5 mm. The erosion wear however had increased it erosion rate value when nozzle was increased to 50.8 mm. The erosion rate value reduced again as the length increases. Whereas for Cut Cell mesh assembly, a contrasting value were shown when the nozzle length is at value of 32.5 mm. The erosion rate begins to be decrease in trend but were slightly inconsistent as the length increases. Given the parameters of length is concerned, the Quadrilateral assembly shows considerable accuracy and consistent trend of erosion rate value and trend. General pattern indicates that the erosion of nozzle decreases as the nozzle length increases which is similar to results acquired in past experimental results [12],[19],[20].

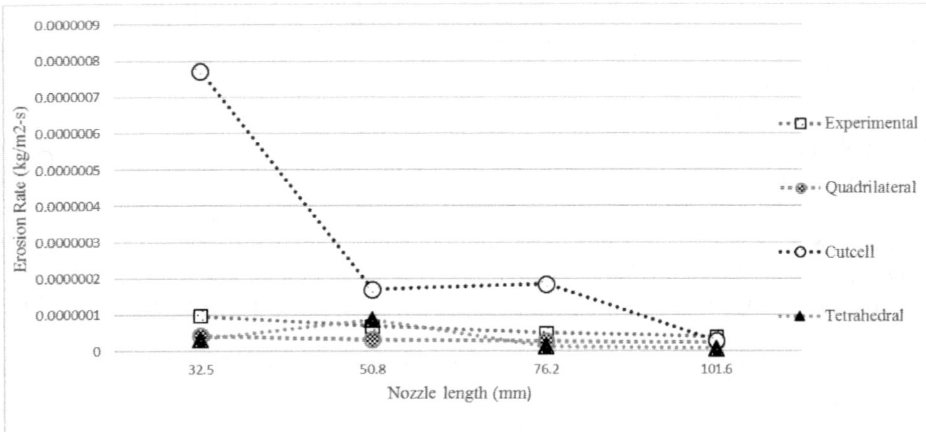

Figure 3. Effect of Nozzle length on erosion wear.

Table 5. Comparison of Assembly methods in term of nozzle length.

Experimental model		**Quahedral**		**Cut cell**		**Tetrahedral**	
Nozzle length (mm)	**Erosion Rate (kg/m2-s)**	**Erosion rate (kg/m2-s)**	**Accuracy (%)**	**Erosion rate (kg/m2-s)**	**Accuracy (%)**	**Erosion rate (kg/m2-s)**	**Accuracy (%)**
32.5	9.723E-08	4.078E-08	-58%	7.722E-07	694%	3.261E-08	-66%
50.8	6.802E-08	3.180E-08	-53%	1.701E-07	150%	8.670E-08	27%
76.2	4.918E-08	2.621E-08	-47%	1.854E-07	277%	1.290E-08	-74%
101.6	3.907E-08	2.315E-08	-41%	2.676E-08	-31%	6.588E-09	-83%

4. Conclusion

In summary, different assembly method of meshing was generated and simulated to solve the AWJ nozzle erosion. Out of the three methods, the Quadrilateral mesh showed promising results in term of processing time, convergence quality and erosion rate accuracy followed by Cutcell and Tetrahedrons mesh, respectively. The Cutcell and Tetrahedrons methods requires significant processing time and are prevalent in encountering divergence which eventually leads to repetitions of simulation. It also required to adjust the URF which will affect the accuracy of the calculation. Therefore, it is suggested that the simulation to be proceed using the Quadrilateral mesh method.

Acknowledgement

The authors would like to gratefully acknowledge the financial support from the Universiti Malaysia Pahang (UMP) Research Grant (RDU150347).

References

[1] Blazek J 2005 *Computational Fluid Dynamics: Principles and Applications:(Book with accompanying CD)*: Elsevier

[2] Versteeg H K and Malalasekera W 2007 *An introduction to computational fluid dynamics: the finite volume method*: Pearson Education

[3] Anderson J D and Wendt J 1995 *Computational fluid dynamics* vol 206: Springer
[4] Azhari A, Schindler C, Kerscher E and Grad P 2012 *The International Journal of Advanced Manufacturing Technology* **63** 1035-46
[5] Folkes J 2009 *Journal of Materials Processing Technology* **209** 6181-9
[6] Momber A W and Kovacevic R 2012 Springer Science & Business Media)
[7] Junkar M, Jurisevic B, Fajdiga M and Grah M 2006 *International Journal of Impact Engineering* **32** 1095-112
[8] Maniadaki K, Kestis T, Bilalis N and Antoniadis A 2007 *The International Journal of Advanced Manufacturing Technology* **31** 933-40
[9] Deepak D, Anjaiah D, Karanth K V and Sharma N Y 2012 *Advances in Mechanical Engineering* **4** 186430
[10] Baisheng N I E, Hui W, Lei L I, Jufeng Z, Hua Y, Zhen L I U, Longkang W and Hailong L 2011 *Procedia Engineering* **26** 48-55
[11] Mostofa M G, Kil K Y and Hwan A J 2010 *Journal of mechanical science and technology* **24** 249-52
[12] Nanduri M, Taggart D G and Kim T J 2002 *International Journal of Machine Tools and Manufacture* **42** 615-23
[13] Kamarudin N H, Rao A P and Azhari A 2016 In: *IOP Conference Series: Materials Science and Engineering*: IOP Publishing) p 012016
[14] Hu G, Zhu W, Yu T and Yuan J 2008 In: *Industrial Informatics, 2008. INDIN 2008. 6th IEEE International Conference on*: IEEE) pp 1700-5
[15] Meakhail T A and Teaima I R 2013 *ratio* **2**
[16] Ng E Y-K and Guannan D 2014 *The International Journal of Advanced Manufacturing Technology* **78** 939-46
[17] Zhang H, Han B, Yu X and Ju D 2013 *Shock and Vibration* **20** 895-905
[18] Hosain M L and Fdhila R B 2015 *Energy Procedia* **75** 3307-14
[19] Pi V N and Tuan N Q 2009 In: *Advanced Materials Research*: Trans Tech Publ) pp 345-50
[20] Hashish M 1994 *Journal of tribology* **116** 439-44

Undercut feature recognition for core and cavity generation

Mursyidah Md Yusof[1], Mohd Salman Abu Mansor[1]

[1]School of Mechanical Engineering, Engineering Campus, Universiti Sains Malaysia, 14300 Nibong Tebal, Pulau Pinang, Malaysia

Email: mesalman@usm.my

Abstract. Core and cavity is one of the important components in injection mould where the quality of the final product is mostly dependent on it. In the industry, with years of experience and skill, mould designers commonly use commercial CAD software to design the core and cavity which is time consuming. This paper proposes an algorithm that detect possible undercut features and generate the core and cavity. Two approaches are presented; edge convexity and face connectivity approach. The edge convexity approach is used to recognize undercut features while face connectivity is used to divide the faces into top and bottom region.

Keywords: Computer-aided design, undercut, two-plate mould

1. Introduction

In the current plastic injection mould industry, the mould design works commonly utilize commercial CAD software. In order to operate the software, design engineers are required to have experience, knowledge and skills. This however could lead to inconsistency in the design works due to the variation of work from one designer to another. Moreover, time consumption is a known factor that increases operational cost and the design work beginning from product design towards manufacturing design is significantly time consuming. Mould design is a very critical and essential phase as it defines the quality of the final product and also acts as a link which takes information from the product design team, evaluating and processing the information and transferring the manufacturing data to the manufacturing process team.

Many research have been made to develop and automate the mould design process which would support the mould design team and help provide consistent quality, increase productivity and reduce designing time. In the plastic industry, injection moulding is one of the manufacturing process that is widely used and mould design is one of the stages is injection moulding process. In mould design, undercut feature recognition is the initial step to start the designing process. Undercut feature recognition such as through holes and pockets were previously studied by researchers and among the approaches that were used to recognize them are graph-based recognition [1], polyhedron face adjacency graph [2], surface visibility [3], feature-based [4,5], volume-based [6] and plane projection [7].

This paper aims to propose an algorithm that generates core and cavity by (1) recognizing the undercut features using edge convexity approach and (2) dividing the faces into top and bottom regions using face connectivity approach.

2. Methodology

Figure 1 shows the proposed algorithm which is implemented and simulated using a computer programming software which uses C++ and ACIS language, and solid modeller software. The 3D CAD design of any part is input into the system and the topological and geometrical information of the part is

IOP Conf. Series: Materials Science and Engineering **290** (2017) 012070 doi:10.1088/1757-899X/290/1/012070

extracted and stored in the system. All edges in the body is then analysed for its convexity. Two types of convexity is categorized, concave and convex. Next the undercut features are recognized based on the edge convexity. The number of undercuts are then determined based on the parting direction inserted by the user. All faces are then analysed using face connectivity to be divided into top and bottom regions. If there are undercut present the system will exit, while if there are no undercut existing the system will proceed. Shut off surfaces are created for through holes and the core and cavity is finally generated.

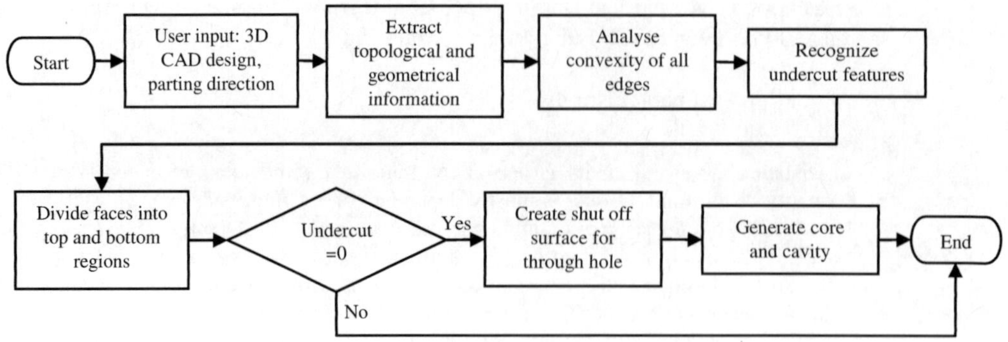

Figure 1. Flowchart of proposed algorithm

Step 1
The user input an existing 3D CAD design of a part into the system and the vector of parting direction, Pd.

Step 2
The system extracts the topological and geometrical information from the 3D CAD part and stores all entity such as vertices, edges and faces into an array of storage data.

Step 3
All edges stored are analysed for its convexity. The convexity of an edge is analysed by intersecting two extruded bodies created from the faces sharing the edge. The two faces that share the edge is extruded along the face normal, creating an extruded body for each face. The extruded bodies are then intersected, if the intersection results a solid intersection, the edge is a concave edge as shown in figure 2 (a), however if the intersection results a NULL, the edge is a convex edge. This test is proven consistent with the theory of a concave edge where the angle between the faces sharing the edge must be less than 180° as shown in figure 2 (b).

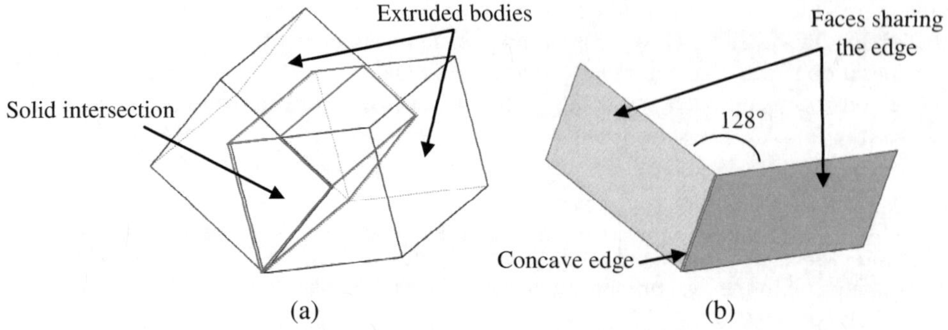

Figure 2. Concave edge (a) intersection of two extruded bodies (b) faces sharing the edge is less than 180°

Step 4
Faces that share a concave edge is grouped together creating a face region. This face region is a possible undercut region. The face region is then extruded along both negative and positive parting direction to infinity, if the extruded body intersects other faces at both direction, the face region is recognized as an undercut feature. However if the extruded body only intersect one of the directions, for example it intersects only along positive parting direction, the face region is considered to belong to the bottom region and vice versa.

Step 5
All remaining faces that are not included in any regions are then grouped into top or bottom region using the face connectivity approach. A face may be grouped into the top region if it abide these rules: (1) the face when extracted along positive parting direction must not intersect with any other faces and (2) the face is sharing an edge with any faces in the top region. These rules are similar for bottom region with negative parting direction. The second rule however is exempted for the initial face in each top and bottom region.

Step 6
The number of undercut must not be more than zero for the system to proceed. If not, the system will exit.

Step 7
Shut off surface is created for any existing through hole.

Step 8
Finally the core and cavity are generated [8].

3. Results and Discussion

The system is executed using two 3D CAD inputs, Part A and Part B as shown in figure 3 (a) and (b) with both parting direction along z-axis. The dimension of Part A is length 56mm, width 23mm and height 32mm, and Part B is length 160mm, width 140mm and height 10mm. Part A has a total of 17 faces while Part B has 29 faces.

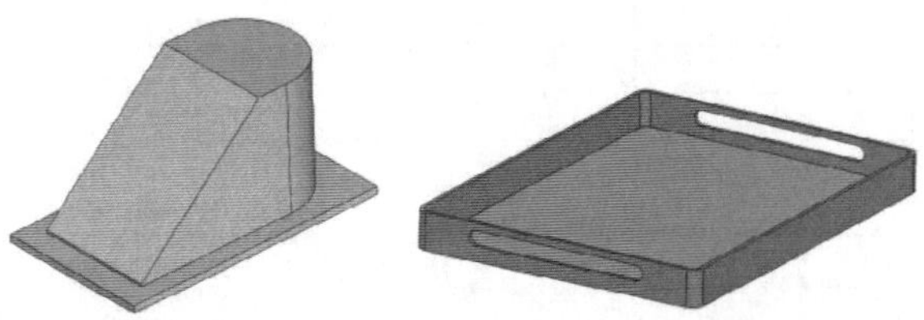

(a) (b)

Figure 3. 3D CAD inputs (a) Part A (b) Part B

Table 1 shows the face region and the edge convexity for Part A and Part B. Random edges and faces were selected in the table to show the analysed faces and edges. Each convexity and region is assigned a value where, top: 1, bottom: 0, none: 2, concave: 3, convex: 4. The remarks for Part A shows that no undercut feature were recognized while Part B had two region of undercut features. Figure 4 and figure 5 shows the top, bottom and undercut regions that were grouped based on the remarks in table 1.

This is proven by the output generated which shows that for Part A, the top and bottom regions were grouped and the core and cavity were able to be generated as shown in figure 6. However due to detection of undercuts in Part B, the core and cavity were not able to be generated and only the top and bottom regions were grouped as shown in figure 7. Part A had 6 faces in the top region and 5 faces in the bottom region while Part B had 9 faces in the top region and 12 faces in the bottom region. Part B also had undercut features with 4 faces at each undercut region as shown in figure 8.

The computational time taken from start to finish for Part A is 4.21 seconds while Part B is 5.13 seconds. The time difference is due to the number of faces that the system had to process where Part B had more faces than Part A thus more computational time taken.

Table 1. Face connectivity and edge convexity of Part A and Part B

Part A				Part B			
Face	**Shared Edge**	**Convexity/ Region**	**Remark**	**Face**	**Shared Edge**	**Convexity/ Region**	**Remark**
3		1	Top region	0		2	Undercut region
	9	4			73	3	
4		1		26		2	
	11	4			75	3	
5		1		25		2	
	12	4			77	3	
6		1		28		2	
...	...	...		...	...	...	...
8		0	Bottom region	1		2	Undercut region
	15	4			72	3	
9		0		22		2	
	16	4			70	3	
10		0		21		2	
	17	4			71	3	
11		0		24		2	
...	...	...	...	...	...	...	...
14		0	Bottom region	5		0	Bottom region
	39	3			6	4	
15		0		8		0	
	38	3			11	4	
16		0		6		0	
	37	3			14	4	
17		0		11		0	

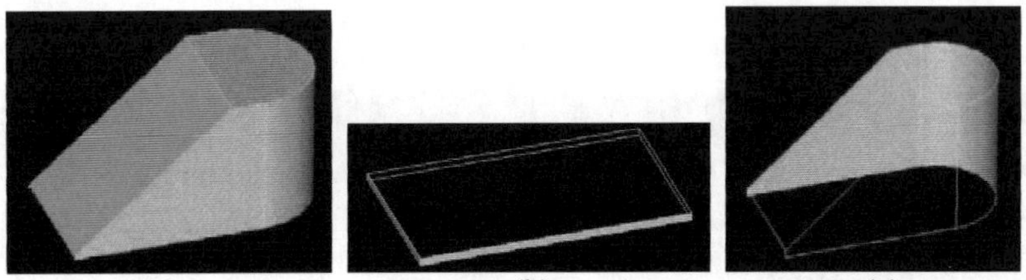

(a) (b) (c)

Figure 4. Part A face regions (a) top region (b) bottom region (c) bottom region

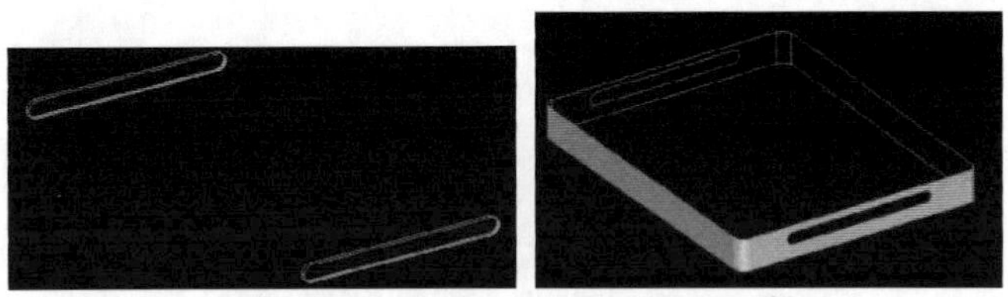

(a) (b)

Figure 5. Part B face regions (a) undercut region (b) bottom region

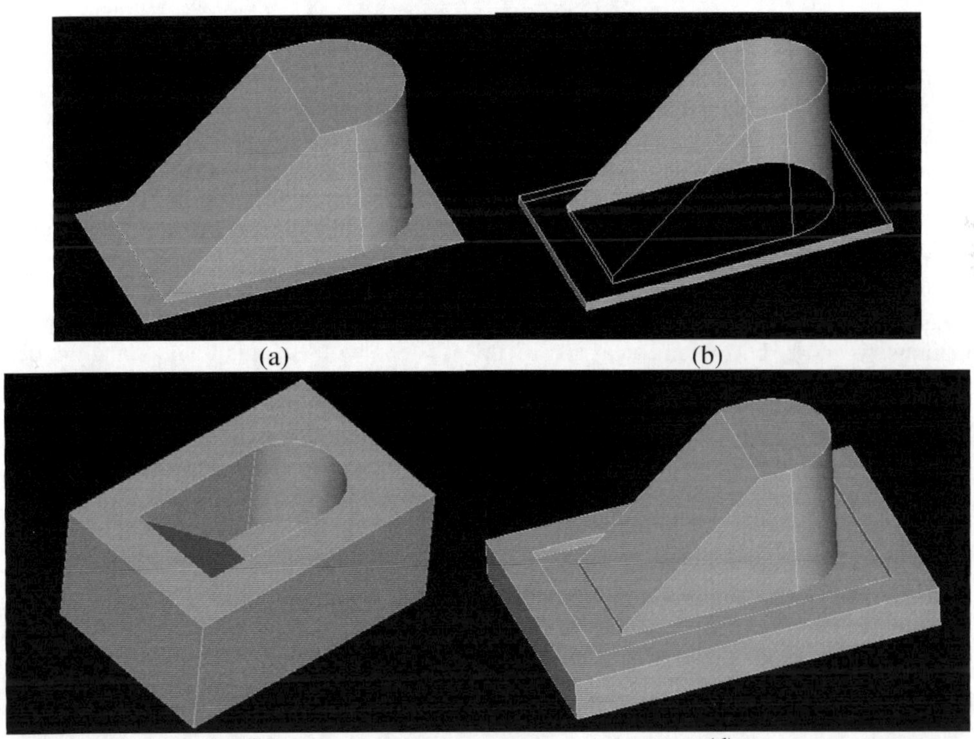

(a) (b)

(c) (d)

Figure 6. Part A output (a) top region (b) bottom region (c) cavity (d) core

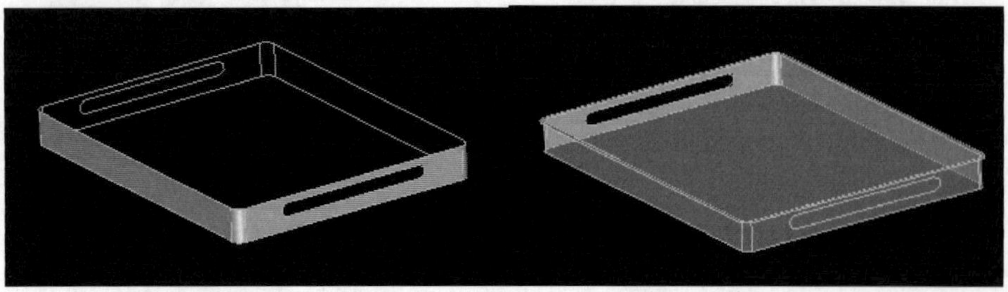

(a) (b)

Figure 7. Part B output (a) bottom region (a) top region

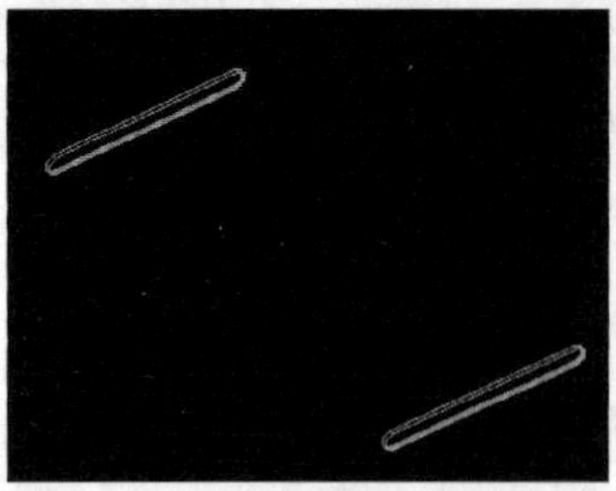

Figure 8. Undercut features of Part B

4. Conclusion

The results and outputs have proven that the algorithm using edge convexity approach was able to recognize the undercut features and the connectivity approach was able to divide the faces into top and bottom regions automatically. The system was also able to generate core and cavity for parts without undercut and detect undercut features for parts with undercuts which makes it easier for the user to adjust the part design and remove the undercut. The computational time taken for both parts were satisfactory and obviously faster than using commercial CAD software. Even though this system is still not robust and has limitations for now, but extension work of this system will eventually help mould designers to increase productivity by reducing the designing time.

Acknowledgement

This research is supported by Universiti Sains Malaysia under the Research University Grant (No: 814247).

References

[1] Shao J, Shen G. Research on graph-based recognition of undercut features from molded part. 2nd Int Conf Inf Sci Eng ICISE2010 - Proc 2010:1468–71. doi:10.1109/ICISE.2010.5689693.

[2] Kumar N, Ranjan R, Tiwari MK. Recognition of undercut features and parting surface of moulded parts using polyhedron face adjacency graph. Int J Adv Manuf Technol 2006;34:47–55. doi:10.1007/s00170-006-0574-7.

[3] Fu MW, Nee AYC, Fuh JYH. The application of surface visibility and moldability to parting line generation. CAD Comput Aided Des 2002;34:469–80. doi:10.1016/S0010-4485(01)00117-8.

[4] Bok AY, Abu Mansor MS. Generative regular-freeform surface recognition for generating material

removal volume from stock model. Comput Ind Eng 2013;64:162–78. doi:10.1016/j.cie.2012.08.013.

[5] Zhou ZZZ, Gao SGS, Gu ZGZ, Shi JSJ. A feature-based approach to automatic injection mold generation. Proc Geom Model Process 2000 Theory Appl 2000. doi:10.1109/GMAP.2000.838238.

[6] Yin, Z., Ding, H. & Xiong, Y., 2001. Virtual prototyping of mold design : geometric mouldability analysis for near-net-shape manufactured parts by feature recognition and geometric reasoning. CAD Computer Aided Design, 33, pp.137–154

[7] Ran, J.Q. & Fu, M.W., 2010. Design of internal pins in injection mold CAD via the automatic recognition of undercut features. CAD Computer Aided Design, 42, pp.582–597

[8] Md Yusof M, Abu Mansor MS. Core and Cavity Generation Using Slicing and Boolean-Based Algorithm. Appl Mech Mater 2015;761:175–9. doi:10.4028/www.scientific.net/AMM.761.175.

IOP Conf. Series: Materials Science and Engineering **290** (2017) 012071 doi:10.1088/1757-899X/290/1/012071

3D documenatation of the petalaindera: digital heritage preservation methods using 3D laser scanner and photogrammetry.

Harlina Md Sharif[1], Hazman Hazumi[2] and Rafiq Hafizuddin Meli[1]

[1]Kulliyyah of Architecture and Environmental Design, International Islamic University Malaysia, Jalan Gombak, 53100 Kuala Lumpur, Malaysia
[2] Global Heritage Consultancy Sdn. Bhd., Starparc Point Commercial Centre, Off Jalan Genting Kelang, Setapak 53100 Kuala Lumpur, Malaysia

Email:msharlina@iium.edu.my

Abstract. 3D imaging technologies have undergone massive revolution in recent years. Despite this rapid development, documentation of 3D cultural assets in Malaysia is still very much reliant upon conventional techniques such as measured drawings and manual photogrammetry. There is very little progress towards exploring new methods or advanced technologies to convert 3D cultural assets into 3D visual representation and visualization models that are easily accessible for information sharing. In recent years, however, the advent of computer vision (CV) algorithms make it possible to reconstruct 3D geometry of objects by using image sequences from digital cameras, which are then processed by web services and freeware applications. This paper presents a completed stage of an exploratory study that investigates the potentials of using CV automated image-based open-source software and web services to reconstruct and replicate cultural assets. By selecting an intricate wooden boat, *Petalaindera,* this study attempts to evaluate the efficiency of CV systems and compare it with the application of 3D laser scanning, which is known for its accuracy, efficiency and high cost. The final aim of this study is to compare the visual accuracy of 3D models generated by CV system, and 3D models produced by 3D scanning and manual photogrammetry for an intricate subject such as the Petalaindera. The final objective is to explore cost-effective methods that could provide fundamental guidelines on the best practice approach for digital heritage in Malaysia.

1. Introduction

3D laser scanners use laser technology to deliver highly detailed 3D Documentation. 3D images of complex environments and geometries can be generated in mere minutes. It can be used for accident reconstruction, 3D modeling of process plants, architectural preservation tasks, digital factory layouts or deformation monitoring. Photogrammetry is the art and science of using overlapping photographs to reconstruct three dimensional scenes or objects. The practice has been around for over a century, but has become especially popular over the course of the past decade with the proliferation of digital cameras and free or relatively inexpensive, but accurate, processing software.

2. Equipment and Software

2.1. 3D Laser Scanner

IOP Conf. Series: Materials Science and Engineering **290** (2017) 012071 doi:10.1088/1757-899X/290/1/012071

The hardware and equipment used for the 3D laser scanner process are as follows:

i. FARO® Focus3D X130 Scanner
ii. FARO® Freestyle3D Scanner
iii. Microsoft Surface Pro 3 Tablet
iv. Checkerboard
v. Tripod

3D Laser Scanning data processing will mainly utilise two (2) major software, which are FARO SCENE and MeshLab. SCENE is a software specifically designed to process 3D point clouds collected by FARO® FocusS, Focus3D and Freestlye3D Laser Scanners. SCENE processes and manages scanned data easily and efficiently by using automatic object recognition as well as scan registration and positioning. SCENE can also generate high-quality colorized scans very quickly, while providing the tools for automated target less or target-based scan positioning.

MeshLab is an open source system for processing and editing 3D triangular meshes. It provides a set of tools for editing, cleaning, healing, inspecting, rendering, texturing and converting meshes. It offers features for processing raw data produced by 3D digitization tools/devices and for preparing models for 3D printing.

2.2. Photogrammetry

The hardware and equipment used for the 3D laser scanner process are as follows:

i. DSLR Camera
ii. Tripod
iii. Target

Photogrammetry data processing will mainly utilise two (2) major software, which are Agisoft PhotoScan and Autodesk MAYA. Agisoft PhotoScan is an advanced image-based 3D modelling solution aimed at creating professional quality 3D content from still images. Based on the latest multi-view 3D reconstruction technology, it operates with arbitrary images and is efficient in both controlled and uncontrolled conditions. Photos can be taken from any position, providing that the object to be reconstructed is visible on at least two photos. Both image alignment and 3D model reconstruction are fully automated.

Maya 3D animation software offers a comprehensive creative feature set for 3D computer animation, modelling, simulation, rendering, and compositing on a highly extensible production platform. Maya has next-generation display technology, accelerated modelling workflows, and tools for handling complex data.

3. Workflow – 3D Laser Scanner

For this project, two type of 3D laser scanner will be used to obtain the point cloud data. This is decided based on the type and the scanning area of artefact.

The workflow for FARO Focus3D Scanner involves two (2) major step:

i. Scan – Scanning process using checkerboard marker around the artefact.
ii. Data Processing – Post processing to merge the multiple scans.

The workflow for FARO Freestyle3D Scanner involves two (2) major step:

i. Scan – With an intuitive plug and play system, capture real-time point cloud visualisation data effortlessly of almost any surface type in a wide range of environments.
ii. Data Processing – Register and process point cloud data using FARO CAPTURE and SCENE.

IOP Conf. Series: Materials Science and Engineering **290** (2017) 012071 doi:10.1088/1757-899X/290/1/012071

3.1. Scanning – FARO Focus3D Scanner

Before starting the scan, set up the workplace placing checkerboard markers on the wall. The checkerboard placement is decided with the visibility to the scanner in mind in order for the software to be able to register the scan placement during data processing. The scanner is set up on a tripod and the settings of the scanner are determined on the quality vs. duration of scan.

Then, the scanning process can begin. During the scan, the scanner must remain in place and no object/person should be blocking the view between the scanner and the artefact and the checkerboards. The scan processes are repeated on several locations surrounding the artefact ensuring there are no blind spot on the scan data.

It is recommended to have 3 to 5 metres of buffer area around the artefact for comfortable work area. On several occasion the scanner will be place on the ground or the tripod is boosted on a stool to capture low-level area or high-level area respectively.

Figure 1. Checkerboard markers area.

Figure 2. Scanning process.

3.2. Data Processing – FARO Focus3D Scanner

The data processing for the FARO Focus3D Scanner is done in FARO's own software, SCENE. The multiple scan points that has been captured on site are merged together to form a point cloud model of the scanned object. The software used the checkerboard markers scanned together with the object as reference points to merge the multiple scan points into one. The software also offers tools to manipulate the point cloud such as deleting unwanted data and colorising the point cloud model.

Figure 3. Cloud point model of Petalaindera.

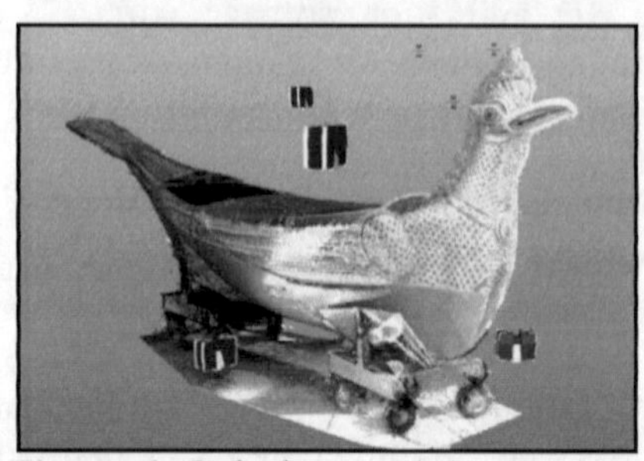

Figure 4. Colorize result in 3D view.

3.3. Scanning – FARO Freestyle3D Scanner

With the faro scanner freestyle, you can capture effortlessly almost any surface type simply by pointing the handheld device to the surface of object. Through the real time visualization on the tablet, the intuitive user interface and the lightweight (less than 1kg) of the scanner, the data acquisition process scan is fast and can easily be done even by semi-trained operators.

Figure 5. Scanning process of Freestyle3D.

Figure 6. Scanning process Freestyle3D.

The scanning is done by pointing the scanner toward the artefact and steadily and slowly moving to cover the whole scan area. The scanned data can be viewed in real-time within the tablet to ensure all data are collected. The data that have been fully captured will be displayed in green cloud.

It is recommended to finish the scanning process in only one scan as multiple scans may impose deviation in data when merging them together. The scanner can be calibrating for scanning distance between 0.5-3 metres depending on the artefact size, for a large sized artefact like the ***Petalaindera***, a 3-4 metres buffer zone surrounding the artefact is required for a comfortable workspace and optimal scanning process.

3.4. Data Processing – FARO Freestyle3D Scanner

Similar with Facos3D Scanner, Freestyle3D Scanner data are processed in the software SCENE. All of the unwanted scan data are removed and the cloud point are processed into a cloud point model.

Figure 7. Deleting unwanted point clouds.

Figure 8. Result in 3D view.

3.5. Combining Focus3D and Freestyle3D Scan Data

Once both of the scanned data are processed, the next step is to merge them together. This is also done inside SCENE, by marking several matching reference points on the both scan data the software will be able to identify and merge the point cloud models into one.

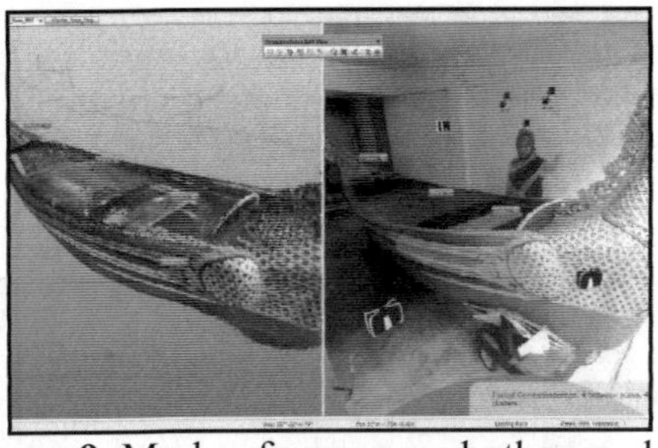

Figure 9. Mark references on both scan data.

Figure 10. Final result of the merge.

IOP Conf. Series: Materials Science and Engineering **290** (2017) 012071 doi:10.1088/1757-899X/290/1/012071

3.6. Building Mesh Using MeshLab

MeshLab is an open source, portable, and extensible system for the processing and editing of unconstructed 3D triangular meshes. The scan data from SCENE are imported into MeshLab for mesh construction. The software works by identifying the individual points in order for surface reconstruction in the form of triangular faces. It also provide multiple tools to clean and edit the mesh. The mesh can then be exported as various formats for further actions including STL (STereoLithography) file format for 3D printing.

Figure 11. Imported scan data in MeshLab.

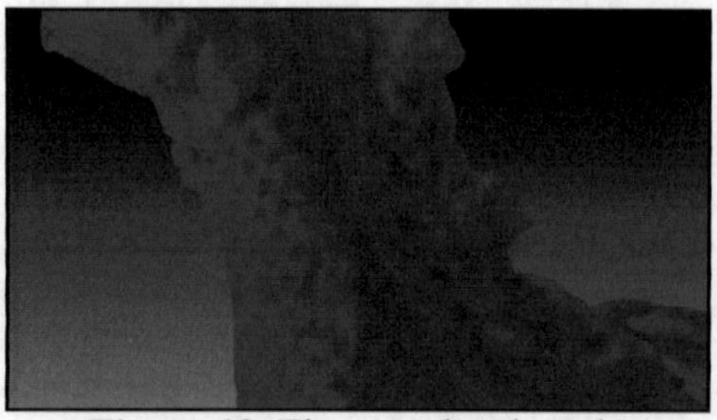

Figure 12. The completed mesh.

4. Workflow – Photogrammetry

The workflow for Photogrammetry involves two (2) major step:

i. Photography – Capturing multiple subsequence overlapping photographs of the artefact.
ii. Data Processing – Processing the photos into point cloud and then into mesh model.

4.1. Capture Overlapping Photographs

The following figure represent several typical capturing scenarios:

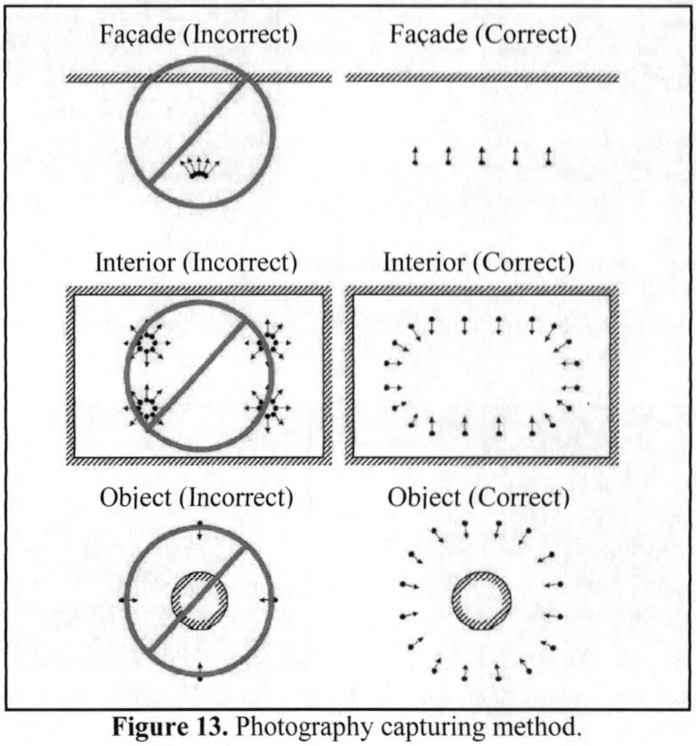

Figure 13. Photography capturing method.

Before start taking photos, set camera mode to manual mode and set a fixed exposure level to get a consistent sharp image with little or no noise. After that, shoot 100 - 300 photos of artefact depend on size. Take photos with different level such as eye level, below and above or many angles possible. It is quite possible to generate usable 3D data by taking two overlapping photographs. The more data will make better to take a sequence of images, according to the size and detail of the subject.

There are some ground rules when taking images for photogrammetry. The photos must overlap by no less than 60%, the camera must be kept at consistent distance for each shot, and must keep the camera perpendicular to the subject. As an example, a single footprint might take 3 shots with 60% overlap at the first pass with the camera in its normal orientation, back to the original and rotate the camera -90 degrees and take 3 more, rotate the camera 180 degrees and take the next sequence.

Figure 14. Setup place.

Figure 15. Taking photos of artefact.

4.2. Data processing – Photogrammetry

The software PhotoScan makes the order of operations easy to follow via its Workflow menu. Basic operations can be accomplished by stepping through the menu and performing each of the following tasks in turn.

i. Add Photos (or Add Folder containing all photos)
  - This first step loads all of raw images into the software's interface.

ii. Align Photos
  - The first processing step compares the pixels in the photos to find matches and estimate camera locations and 3D geometry from them

iii. Build Dense Cloud
  - Once satisfied with the alignment, the sparse point cloud (a mere fraction of the total data) is processed into a dense cloud in which each match able pixel will get its own X, Y, Z location in 3D space

iv. Build Mesh
  - This step connects each set of three adjacent points into a triangular face, which combine seamlessly to produce a continuous mesh over the surface of your model

v. Build Texture
  - In the final step, the original images are combined into a texture map and wrapped around the mesh, resulting in a photo-realistic model of original object.

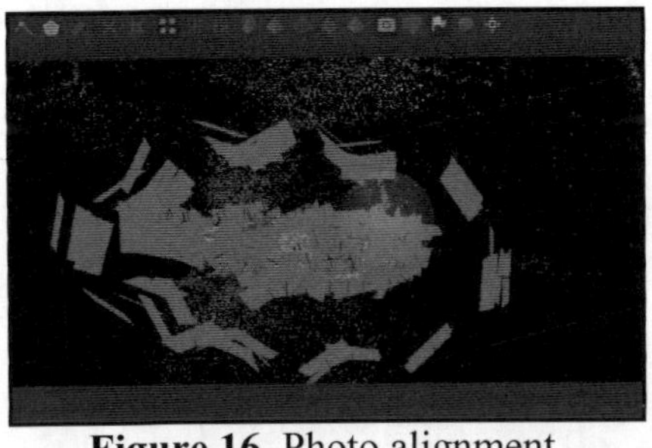

Figure 16. Photo alignment.

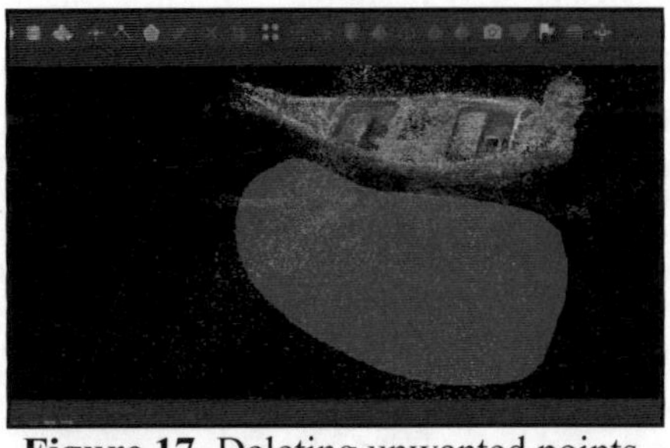

Figure 17. Deleting unwanted points.

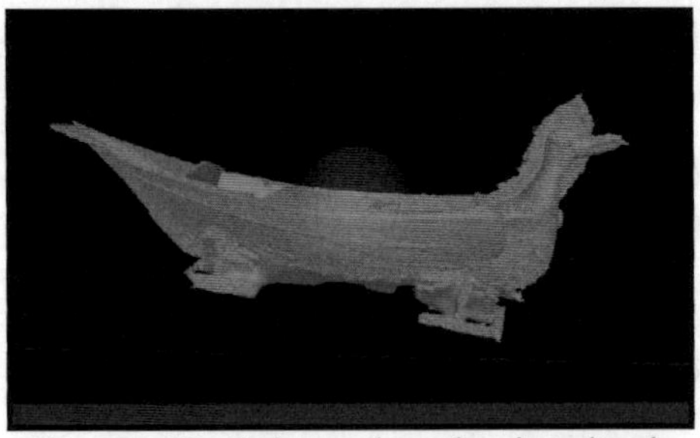

Figure 18. Color-stripped point cloud.

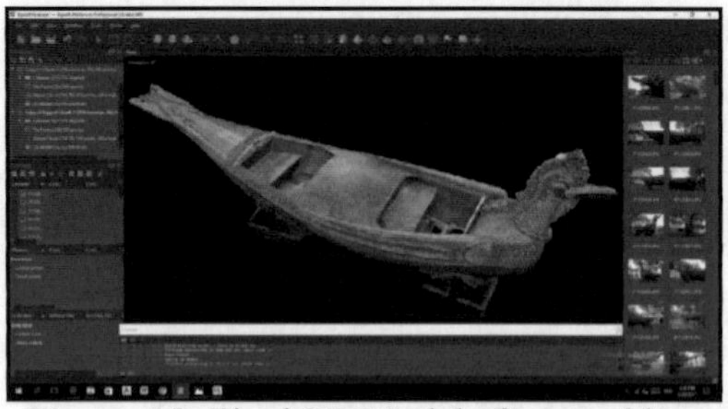

Figure 19. Final 3D Model after meshing.

The finished model can be exported in various formats for display or import into other 3D analysis suites, animation software or game engines. Common formats are:

1. 3D PDF: an interactive format that is widely accessible, as the model can be viewed, manipulated and even measured using the ubiquitous and free Adobe Reader.
2. Wavefront (.obj) and Collada (dae) are the most portable 3D mesh formats. If you are working with other animation platforms like Blender or game engines like Unity 3D, these formats can be easily added as assets to your project.
3. Pointcloud (LAS, .txt, etc.) these formats are probably the most future proof for long term storage, and also offer many options for secondary analysis in other tools like Geomagic and Meshlab.

The model can also be can be transferred to Autodesk Maya for 3D development and final adjustment. Apply lighting and texturing to artefact to enhance the 3D model. The 3D model can also be manipulated to add or subtract any of its elements using this software.

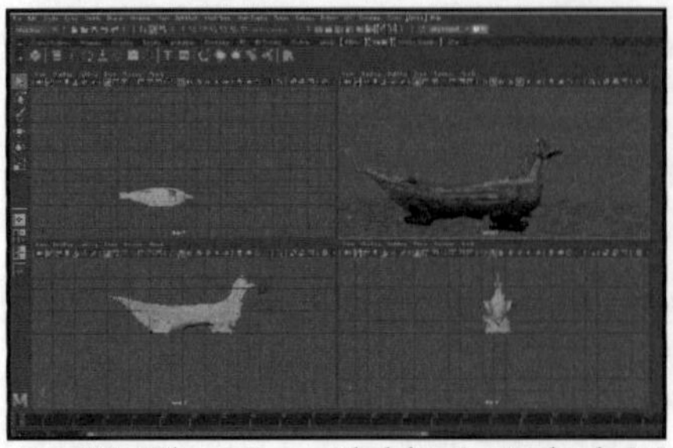

Figure 20. The 3D model in Autodesk Maya

5. Discussion and conclusion

Cultural heritage is very core of a nation's self-respect and identity and without it, the history can be influenced for reinterpretation or even rewritten. However, many of these heritage assets are left undocumented and unknown to the public. Thus, the awareness for the preservation of cultural heritage had brought exploration of new sustainable methods, one of them being digital heritage. There are many methodologies in digital documentation; 3D laser scanner and photogrammetry are two examples of the most current technology available in the industry.

In term of the availability and accessibility of these methods, each stand on the end of the spectrum respectively, where photogrammetry is on the cheaper side while 3D laser scanner in on the expensive side. This is because for hardware, photogrammetry utilises the standard everyday DSLR camera, while 3D laser scanner requires its own specialised expensive equipment. Software wise, there are numerous freeware available online for photogrammetry and some paid version for extra tools and features, while 3D laser scanner requires specialised software that are compatible with the data captured by the scanner. While both benefits from processing data in a high-end computer, processing 3D scanner data are a lot more demanding than photogrammetry.

IOP Conf. Series: Materials Science and Engineering **290** (2017) 012071 doi:10.1088/1757-899X/290/1/012071

Photogrammetry also have a low skill ceiling in comparison to 3D laser scanner as it can be done by semi-trained personnel and there are numerous how-to guide available online for those who want to experiment at home. 3D laser scanning however requires trained personnel to handle both the scanning and the data processing as the workflow are lot more complicated than photogrammetry.

However the output from 3D laser scanner does justify its high demand as it produce a lot more refine and accurate data compared to photogrammetry. The output from photogrammetry is more suitable for light usage, such as animation for display, 3D printing for souvenirs, etc. While the data produced by 3D laser scanners are more versatile in which they can be used for both research and managerial sector, as the data can produce measured drawing, used in simulation, digital to real restoration, etc. This is because the data obtain from the 3D laser scanner are measurable and to scale with around ±2mm accuracy with the scanner model used, while photogrammetry data requires manual scaling.

Acknowledgement
This work is financially supported by the Research Acculturation Grant Scheme (RAGS15-059-0122), Ministry of Higher Education Malaysia and International Islamic University Malaysia (IIUM). Application of 3D scanning technology and replication is supported through the cooperation with Global Heritage Consultancy Sdn. Bhd.

References
1. Agisoft Photoscan User Manual: Professional Edition, version 1.2 (2016). Retrieved from http://www.agisoft.com/pdf/photoscan-pro_1_2_en.pdf
2. Meshlab features and tutorials (2017). Retrieved from http://www.meshlab.net/#references
3. FARO products and features (2017). Retrieved from http://www.faro.com/en-sg/home
4. Autodesk Maya features and tutorials (2017). Retrieved from http://www.autodesk.com/products/maya/overview
5. Abby Crawford (2015,July 18), *How to set up a successful photogrammetry project*. Retrieved from https://blog.sketchfab.com/how-to-set-up-a-successful-photogrammetry-project/
6. Westoby, M.J., Brasington J., Glasser N.F., Hambrey M.J. and Reynolds J.M., *'Structure-from-Motion' photogrammetry: A low-cost, effective tool for geoscience applications* in Geomorphology 179 (2012) 300–314.
7. McCarthy, J., *Multi-image photogrammetry as a practical tool for cultural heritage survey and community engagement* in Journal of Archaeological Science 43 (2014) 175-185.
8. Remondino, F., *Heritage Recording and 3D Modeling with Photogrammetry and 3D Scanning* in Remote Sensing (2011), 3, 1104-1138.
9. Linguaa A, Piumatti P and Rinaudo F *The International Archives of the Photogrammetry, Remote Sensing and Spatial Information Sciences* **XXXIV** Part 5/W12.

The effect of cutting parameters on the performance of ZTA-MgO cutting tool

A M Ali[1], N E Hamidon[1], N K M Zaki[1], S Mokhtar[1], A Z A Azhar[1], R Bahar[1] and Z A Ahmad[2]

[1]Department of Manufacturing and Materials Engineering, Kulliyyah of Engineering, International Islamic University Malaysia, 53100 Gombak, Selangor, Malaysia
[2]Structural Materials Niche Area, School of Materials & Mineral Resources Engineering, Universiti Sains Malaysia, 14300 Nibong Tebal, Penang, Malaysia

Email: afifahali@iium.edu.my

Abstract. The effect of cutting parameters on the performances of ZTA-MgO ceramic cutting tool investigated. The aim of this project is to discover the effect of cutting speed and feedrate on the performance of the ZTA-MgO cutting tool via wear and surface roughness measurement. CNC turning machining performed using the cutting speed, *Vc* range from 354 to 471 m/min and the feed rate, *f* 0.1, 0.3 and 0.5 mm/rev while the depth of cut, *d* is kept constant at 0.2 mm. The flank wear, crater wear, and chipping were measured accordingly using optical microscope, Matlab programming and SEM. Surface roughness of machined stainless steel 316L surface were measured using the surface roughness tester (Mitutoyo MTR097-8. The result showing the increment trend of flank wear with increment of cutting speed and feed rate with the lowest value of flank wear, 0.061 mm achieved at *Vc* = 354 m/min and *f* = 0.1 mm/rev while the highest flank wear is 0.480 mm at *Vc* = 471 m/min and *f* = 0.5 mm/rev. The increasing pattern also observed in the crater wear results. The lowest area of crater wear is 2.2736 mm^2 at *Vc* = 354 m/min and *f* = 0.1 mm/rev while the highest value is 4.8524 mm^2 at *Vc* = 471 m/min and *f* = 0.5 mm/rev. As for the surface roughness, the higher the cutting speed, the lower the average roughness (*Ra*) value. Cutting speed, *Vc* = 471 m/min with *f* = 0.1 mm/rev has the lowest value of *Ra* which is 0.72μm.

1. Introduction

Zirconia toughened alumina (ZTA) is one of the ceramic cutting tool that has been commercialised and well accepted in the industry. Due to their great potential as cutting tool, the scholars have prioritized improvement on the properties of ZTA based ceramic cutting tool especially on the hardness and fracture toughness. One of the methods to improve the properties of ZTA is via introduction of additives into the ceramic compositions. Many additives discovered by the researchers that may improve the properties of the cutting tool. The example of the additives are chromium oxide, (Cr_2O_3)[1][2], ceria (CeO_2)[3]–[5], titanium carbide (TiC)[6]–[8] and many more. One of the additives is nano particle magnesium oxide (MgO) shows a promising result in improvement of properties. Studies by [9] [10] showing great improvement of the ZTA-MgO properties where the hardness hav been increased to 1740 HV and fracture toughness, 4.32 MPa√m compared to without the additives at 1650 HV and 3.40 MPa√m.

Although the improvement have been made on the ZTA cutting tool, the potential of the cutting tool has not been fully discovered, as the wear of the cutting tool has not been thoroughly studied especially in high speed machining. Previous study [11], [12] showing a good result of wear on ZTA-MgO cutting tools in machining process especially when the cutting speed is increased to high value. Due to that, the investigation on the tool wear of the ZTA-MgO cutting tool with enhancement of parameters selection at higher value will be conducted. Flank wear, crater wear, chipping condition of the ZTA-MgO cutting tool and surface roughness of the machined surface will be presented in this research paper.

2. Experimental procedures

In order to achieve the objectives, the experiments were conducted by using the fabricated ZTA-MgO cutting tool with hardness 1568.69 HV and fracture toughness 8.60 MPa.√m as shown in figure 1. Turning process conducted by using CNC lathe machine (ROM-Bridgeport) with ZTA- MgO cutting tool mounted on Sandvik Coromant (Coro turn CCLNR 164D-4) tool holder. The machining performed on stainless steel 316L at parameter presented in table 1. The wears on the cutting tool were measured accordingly. Flank wear is measured by using the optical microscope (Nikon measuring microscope MM400/L) while the crater wear is measured by capturing the image of the wear on the optical camera (OLYMPUS BX41M) before measured by using the Matlab programming. Chipping analysis performed on the scanning electron microscope (SEM). Surface roughness of machined surface measured by using the surface roughness tester (Mitutoyo - MTR097-8). Five readings were taken and the average value was calculated and analysed.

Figure 1. Fabricated ZTA-MgO ceramic cutting tools.

Table 1. Parameter and cutting condition used in the experiment.

Parameter	Condition / Value
Cutting Speed, V_c	354, 393, 432, 471 m/min
Feed rate, f	0.1, 0.3, 0.5 mm/rev
Depth of cut, d	0.2 mm
Cutting length, l	20 mm
Cutting Condition	Dry

3. Results and Discussion

3.1. Flank Wear

Flank wear is the most common wear that discovered on the cutting tool during the turning process. The flank wear occurs on the relief face of the cutting tool and it occurs as a result of friction on the tool flank between the workpiece and machined surface. From this experiment, the data on the flank wear were gathered and tabulated into graph in figure 2. From this graph, an increasing trend of flank wear with increment of cutting speed, (Vc) and feedrate, (f) can be observed. This result is expected as the higher the cutting speed and feedrate, thus higher friction between the flank surface and the workpiece. Among all of the causes, abrasion is the main cause to the flank wear due to rubbing at lower temperature on the flank surfaces. However, due to some conditions, the mechanism that cause crater wear which is increased of the V_C and cutting temperature also contribute to the flank wear development [13].

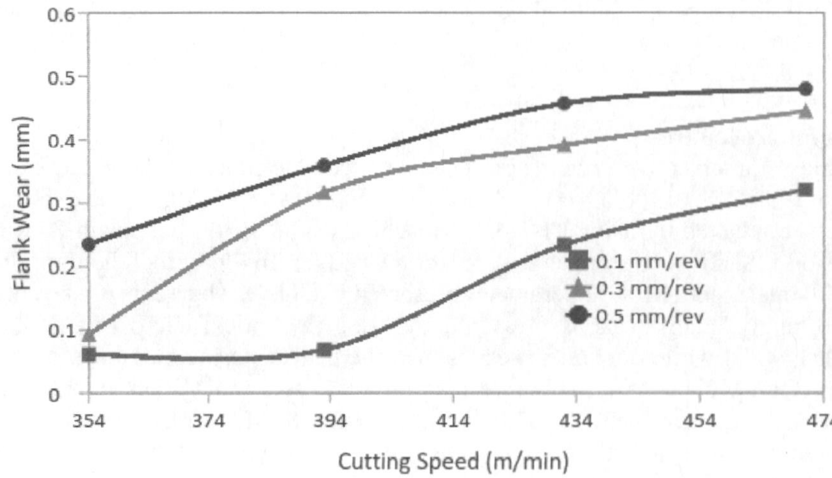

Figure 2. Flank wear at different cutting speed and federate.

The lowest flank wear measured at Vc= 354 m/min and feedrate 0.1 mm/rev at 0.06 mm value while the highest flank wear is at Vc = 471 m/min and feedrate 0.5 mm/rev at 0.48 mm. This value is close to the allowable V_B value, which is 0.5-0.6 mm. If the amount of flank wear is more than its critical value ($V_B > 0.5 \sim 0.6$ mm), then it will cause tool failure due to the excessive cutting force [13]. It can also be seen that the flank wear start at higher value when machining perform at f=0.5 mm/rev and Vc = 354 m/min with the value 0.36 mm compared to f= 0.1 and 0.3 mm/rev which are at 0.06 and 0.09 mm respectively. Therefore, it is recommended for machining to be performed at low feedrate for ZTA-MgO cutting tool.

3.2. Crater wear

Crater wear occurs on the rake face of the cutting tool and it consists of a concave section on the tool face caused by the action of the chip pass through the machined surface [14]. Result of crater wear were tabulated into graph presented in figure 3. Similar to the flank wear trend, the crater wear also showing an increment of wear area when the cutting speed and feedrate is being increased.

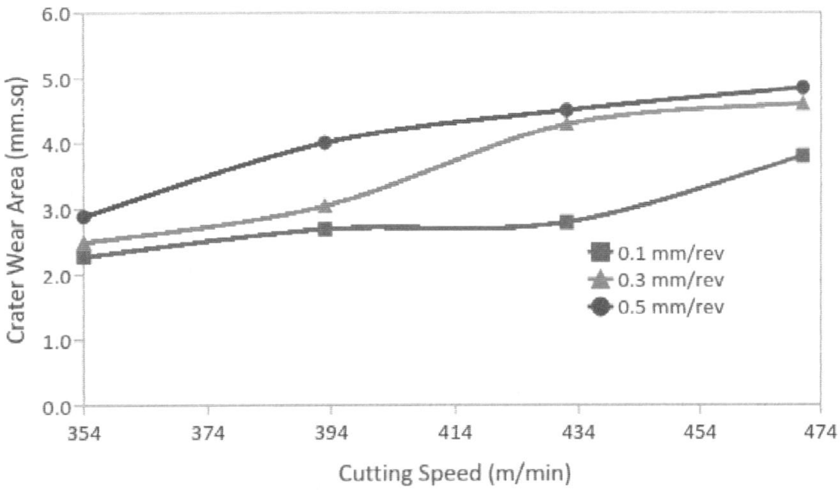

Figure 3. Crater wear at different cutting speed and federate.

The lowest reading of crater wear is at parameter Vc = 354 m/min and f= 0.1 mm/rev while the highest at Vc = 471 m/min and f= 0.5 mm/rev with the value 2.2736 and 4.8524 mm² respectively. Kumar et al. [15] stated that common cause of crater wear is due to the action of chip flow over the face at high temperature and short distance. Other than that, the factors that are influenced the flank wear may influence the crater wear too.

3.3. Chipping

Chipping always associated to brittle cutting tool and ceramic cutting tool can severely affected by chipping during the machining process. The analysis of chipping on ZTA-MgO ceramic cutting tool presented in figure 4 shows that there is no significant changes of chipping area when the *Vc* is being increased from 354 to 471 m/min. However, the chipping area found to be increased when the feedrate is increased from 0.1 to 0.5 mm/rev. This condition can be visualize from figure 4 where the increment of chipping can be seen as the feedrate is increased from 0.1 mm/rev as shown in figure 4(a) to 0.3 mm/rev in figure 4(b).

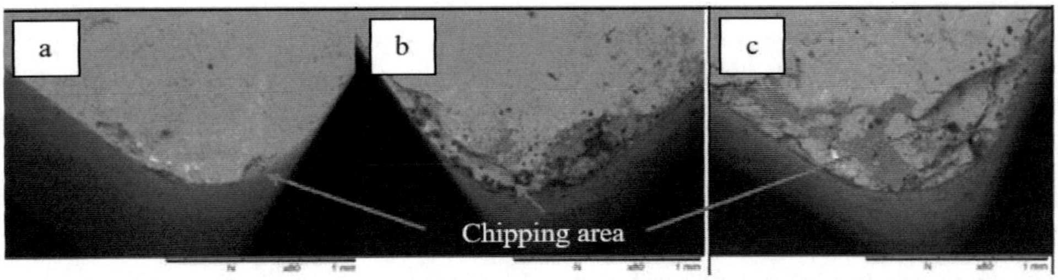

Figure 4. SEM micrograph of the chipping area at different feedrate a) 0.1 mm/rev, b) 0.3 mm/rev and c) 0.5 mm/rev at Vc = 471 m/min.

Further increment of the feedrate to 0.5 mm/rev increase the chipping area to the largest area as shown in figure 4(c). According to Deng et al. [16], chipping occur as a result of an overload of mechanical shock due to interruption during machining process such as very high depth of cut or feed rate and excessive wear on the insert. Chipping started with the crack initiation on defect surface and due to excessive load to the cutting tips and higher feed rate, the crack network spread wider until the chipping occurred. It also may cause by the high temperature that oxidize the wear, and the alternate and impact stress [17].

3.4. Surface roughness

Surface roughness is one of the methods to evaluate the quality and performance of machining. The result of surface roughness versus the cutting speed at three level of feedrate presented in figure 5.

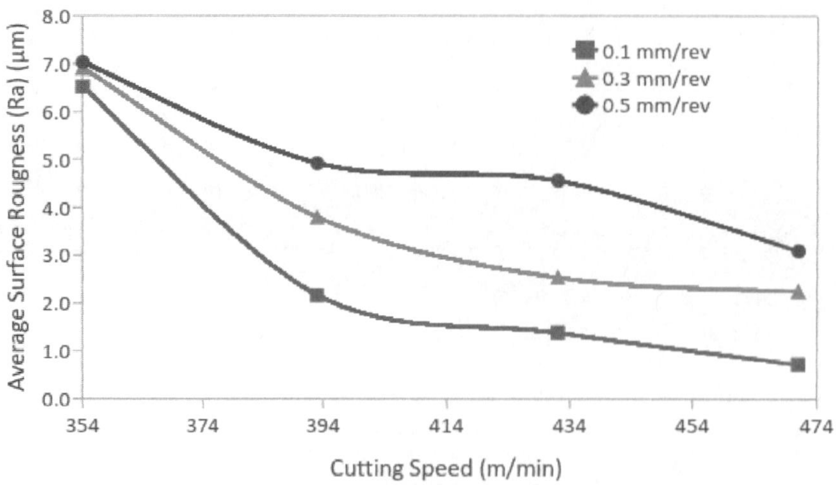

Figure 5. Surface roughness of Stainless Steel 316L at different parameters.

Although the wears are showing the increment pattern when the cutting speed is being increased, surface roughness (Ra) showing the inverse pattern where the Ra value decreased when the cutting speed is increased. The increment pattern of Ra occur when the feedrate is increased. The lowest Ra value is 0.72 µm at Vc=471 m/min and f =0.1 mm/rev while the highest value of Ra is 7.04 µm at Vc=354 m/min and f = 0.5 mm/rev. During high speed machining, high friction between the cutting tool and the workpiece is higher causing the temperature on the cutting tip of the cutting tool and the workpiece to be higher. At this moment, the surface of workpiece will become softer and making the chip formation process to be smoother. This result in better surface roughness but in the same time causing higher wear as the cutting tip also affected with the temperature increment. As the temperature rises, the abrasion and oxidation on the tool surface also raised causing higher flank and crater wear. The poor surface roughness at high feedrate related to the chipping on the cutting insert as excessive chipping give poor surface finish which streaked the machine surface cause by the fractured tool [16][18].

4. Conclusion

From this research, as conclusion both cutting speed and feedrate give influence to the wear of ZTA-MgO cutting tool where increment of both parameter increased the flank and crater wear on ZTA-MgO cutting tool. Chipping mainly caused by the excessive load on the cutting edge when the feedrate is increased. Higher feedrate will give higher chipping area. Surface roughness decreased with increment of cutting speed but increased with increment of feedrate. The best parameter combination when machining by using the ZTA-MgO cutting tool is at high cutting speed (471 m/min) and low feedrate (0.1 mm/rev).

References

[1] Lin H T et al. 2007 *Key Eng. Mater.* **351** 93–97

[2] Azhar A Z A et al. 2012 *J. Alloys Compd.* **513** 91–96
[3] Lin J D and Duh J G 2003 *Mater. Chem. Phys.* **78** (1) 253–261
[4] Kumar A K et al. 2007 *Int. J. Refract. Met. Hard Mater.* **25(3)** 214–219
[5] Rejab N A et al. 2013 *Int. J. Refract. Met. Hard Mater.* **36** 162–166
[6] Barry J. and Byrne G. 2001 *Wear* **247** 139–151
[7] Xing Y et al. 2014 *Int. J. Refract. Met. Hard Mater.* **43** 46–58
[8] Cheng Y et al. 2016 *Int. J. Refract. Met. Hard Mater.* **55** 39–46
[9] Azhar A Z A et al 2010 *Adv. Mater. Res.* **173** 29–34
[10] Azhar A Z A et al 2012 *Adv. Mater. Res.* **364** 450–454
[11] Ali A M et al. 2015 *Adv. Mater. Res.* **1087** 101–105
[12] Ali A M et al. 2016 *Mater. Sci. Forum* **840** 315–320
[13] Dutta A K et al. 2006 *Wear* **261** 885–895
[14] Marinov V 2010 *Manuf. Tech.* 2010 77–80
[15] Kumar A S et al 2006 *Tribol. Int.* **39** 191–197
[16] Deng Jet al. 2012 *Int. J. Refract. Met. Hard Mater* **35** 17–26
[17] Wang Y et al. 2012 *Int. J. Refract. Met. Hard Mater.* **30** 200–207
[18] Xing Y et al. 2014 *Wear* **318** 12–26

IOP Conf. Series: Materials Science and Engineering **290** (2017) 012073 doi:10.1088/1757-899X/290/1/012073

A secondary, coplanar design Ni/MCM-41/Zn microbattery

S R Meskon, R Othman, M H Ani

Faculty of Engineering, International Islamic University Malaysia (IIUM), P.O. Box 10, 50728 Kuala Lumpur, Malaysia

Email: raihan@iium.edu.my

Abstract. A secondary Ni/Zn microbattery (~200 μm thick) has been developed in a coplanar electrode configuration. The cell is essentially of a circular shape (~30 mm in diameter) consisting of a fine circular ring (cathode) and a circle (anode) split apart (~800 μm). Unlike the stacking cell architecture, coplanar configuration offers simple design, ease of fabrication and eventually cost saving. The use of MCM-41 mesoporous silica as the membrane separator cum electrolyte reservoir enables the successful implementation of coplanar configuration. The fabrication of Ni/Zn microbattery first begins with electrodeposition of zinc (Zn) and nickel hydroxide ($Ni(OH)_{2)}$ thin films onto patterned FR4 printed circuit board, followed by deposition of zinc oxide (ZnO) slurry onto the zinc active layer, and finally ends by multiple drop-coating procedures of MCM-41 from its precursor solution at ambient temperature. Once a potassium hydroxide (6 M KOH)/MCM-41 electrolyte-separator mixture is incorporated, the cell is sealed with an acrylic sheet and epoxy adhesive. The fabricated microbatteries were capable to sustain around 130 deep charge-discharge cycles. When rated at 0.1 mA, the energy density of the microbattery was around 3.82 Wh l^{-1} which is suitable for low rate applications and storage for micro energy harvesters such as piezoelectric generators.

1. Introduction

MCM-41 material stands out due to its highly unusual pore texture resulting from long-range ordering, extremely high surface areas in the order of 1000 m^2g^{-1} and very large pore volume i.e. 1 cm^3g^{-1} [1-4]. The mesoporous material also possesses the properties of adjustable hydrophobicity and very good thermal stability [5]. The MCM-41 structure on the nanometer scale is reported as a hexagonal stacking of porous tubes of uniform diameter. Its pore diameters have sizes from about 1.5 to 10 nm [4, 6, 7].

The zinc-air cells fabricated by Saputra et al. [8] employed MCM-41 silica membrane as separator-cum-electrolyte reservoir; the membrane was dip-coated on a Zn foil or Zn electrodeposit. The cells possessed a maximum power density of 32 mW cm^{-2} and an energy density of 360 Wh l^{-1}. The transformation of MCM-41 silica into gel-like MCM-50 after 24 hours in controlled exposure of 6-M KOH solution was reported in [9]. Soon after, Mohd Zawi et al. [10] employed the MCM-41 silica membrane as separator in Zn/MnO_2 Leclanché button cells. The cells were able to produce a maximum power of 0.7 mW and 1.3 mW for monopolar and bipolar configurations, respectively, rated at current densities of 0.44 and 0.76 mA cm^{-2}. However, the cells were quite thick i.e. ~2 mm. The viability of MCM-41 incorporated in the primary cells drive us to test the material in the Ni/Zn secondary battery system which is more challenging due to concerns of a pressure build-up from gas evolution and dendritic growth of Zn.

Humble et al. [11] fabricated Ni/Zn microbatteries for use in autonomous microsystem applications, employing a side-by-side electrode design. The cell had a footprint area of 0.02 cm^2 and could produce power in the range of 1 – 2 mW. The cell had a capacity of 0.389 mAh cm^{-2} at a current density of 2.5 mA cm^{-2}, which corresponds to 55 Wh l^{-1} of specific energy. Although the performance was remarkable, the cell runtime was only around 11 minutes, i.e. the cell was discharged with only 10% of depth of discharge (DOD). A Ni/Zn microbattery fabricated with a three dimensional (3D) architecture was reported by Chamran and co-workers [12]. The microbattery consisted of interdigitated electrode post arrays with a footprint area of 0.25 cm^2. Despite its high surface area, the cell could only sustain 6 cycles of charge and discharge due to the severe etching of the Zn electrode by the KOH. The microbattery was charged and discharged at 10 μA and the capacity on the sixth cycle was 0.0025 mAh cm^{-2}.

The commonly used microbattery substrates includes epoxy-coated silicon [11], polycrystalline alumina [13], silica (on silicon) [14] and glass [12, 15], while the current collectors used were usually electron-beam evaporated [11] or electroplated [12] zinc and nickel metals, or pyrolyzed carbon [14]. However, some of these processes prevent mass production of microbattery fabrication and involve applications at high temperatures, which thus increase fabrication cost.

In this paper, a relatively simple microbattery fabrication route comprising photolithography, electrodeposition and drop-coating is presented, using copper (Cu) printed circuit board (PCB) (an easily obtained material) as current collector cum microbattery substrate. The microbattery had a coplanar (side-by-side) electrode configuration which further simplified the fabrication steps. The secondary Ni/MCM-41/Zn electrochemical system has been selected for the test. MCM-41 serves as the membrane separator cum KOH electrolyte reservoir.

2. Experimental Procedures

Microbattery Fabrication.

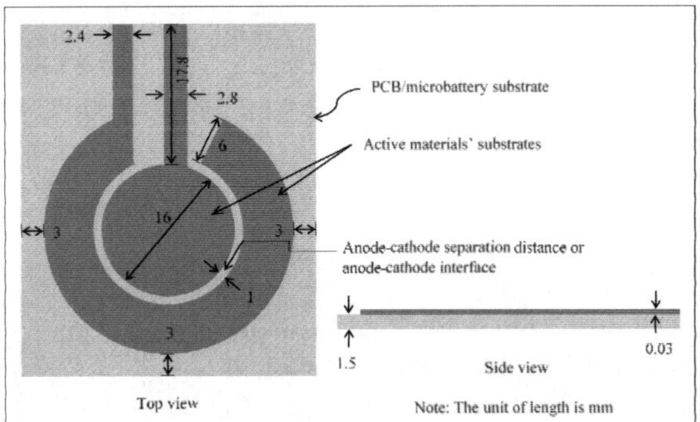

Figure 1 Schematic diagram of microbattery design.

The cell's current collector was patterned on a FR4 PCB substrate by photolithography processes (Figure 1). Next, Zn and $Ni(OH)_2$ thin films were electrodeposited successively to form the anode and cathode, respectively. The electrodeposited Zn was later coated with zinc oxide (ZnO) slurry (~50 μm). Once dried, the entire area of cell including the anode-cathode interface area was coated with MCM-41 thin film by successive drop-coating procedures (a total of 6 layers) of the precursor solution, intercepted by drying in ambient conditions. The KOH electrolyte was applied in the form of a KOH/MCM-41 solution/powder mixture (weight ratio of 70:30). Finally, the microbattery was capped using a Perspex® sheet and thereafter sealed with an epoxy adhesive; the depth of well of the cap was 200 μm.

Synthesis of MCM-41 Silica Thin Film and Powder. Two forms of MCM-41 were used in the microbattery fabrication, i.e. thin film and powder. Sol-gel method was employed for the synthesis. The MCM-41 thin film was formed from the precursor solution. In a typical synthesis, the precursor solution was prepared by mixing a quaternary ammonium surfactant, cetyltrimethylammonium bromide (CTAB, $CH_3(CH_2)_{15}N(Br)(CH_3)_3$; Merck, Germany) as an organic template, hydrochloric acid (HCl; J.T. Baker, Thailand), ethanol (C_2H_5OH, Systerm) as a solvent, distilled water (H_2O) and tetraethyl orthosilicate (TEOS, $Si(OC_2H_5)_4$; Merck, Germany) as a silica source, with a molar ratio of 0.05 CTAB: 1.0 TEOS: 0.5 HCl: 25 C_2H_5OH: 75 H_2O and stirring at 200 rpm, at a room temperature of 30 °C for about one hour [8]. On the other hand, MCM-41 powder was synthesized by mixing CTAB, sodium hydroxide (NaOH, Sigma-Aldrich, Sweden), distilled water and TEOS, with a molar ratio of 0.05 CTAB: 1.0 TEOS: 0.5 NaOH: 100 H_2O, and stirring at 200 rpm, heating the mixture on a hotplate at around 50 °C for 15 hours. The white precipitate of the MCM-41 powder was separated by filtration, rinsed with distilled water and dried at 100 °C.

Electrodeposition of Zn and $Ni(OH)_2$ Thin Films. Zn electrodeposits were prepared by cathodic reduction of an electrolytic bath comprising of 2 M zinc chloride ($ZnCl_2$) (50 vol. %) and 3 M ammonium chloride (NH_4Cl) (50 vol. %) galvanostatically (12.43 mA/cm^2) for 60 minutes, at room temperature.

$Ni(OH)_2$ electrodeposits were prepared from an electrolytic bath of of 1.8 M nickel nitrate hexahydrate ($Ni(NO_3)_2{\cdot}6H_2O$) (49.4 vol. %), 0.075 M sodium nitrate ($NaNO_3$) (16.6 vol. %) and absolute ethanol (34 vol. %), reduced cathodically (4.40 mA/cm^2) for 180 minutes at room temperature.

The electrodepositions were performed using an Autolab PGSTAT302N (Eco Chemie B.V., Netherlands) potentiostat/galvanostat and the working electrode-counter electrode separation distance was fixed at 2.45 cm.

Cell Testing. Galvanostatic charge-discharge cycling test was performed on the assembled microbattery cells using a Neware battery testing system, model BTS-5V100mA (NEWARE Technology Ltd., China) at a room temperature of 30 °C to determine the performance of the secondary Ni/Zn microbattery. The cells were charged at 1.0 mA for 2 hours, left at open circuit potential for 5 minutes before being discharged at 0.1, 0.2 and 0.3 mA, respectively, until the potential reached 0.49 V. The steps were repeated for as many cycles until the cell failed.

3. Results and Discussion

Cell Dimensions. Based on the FE-SEM cross sectional observations (not shown here), the apparent thickness (excluding the cap and substrate) of the coplanar-designed microbattery was ca. 195 μm, i.e. comprising the Cu layer (30 μm), Zn electrodeposits (65 μm) and ZnO layer (~50 μm) and KOH/MCM-41 layer (~50 μm); the thickness of $Ni(OH)_2$ electrodeposits (27 μm) is not included since the anode was much thicker than the cathode and the electrodes were arranged side-by-side. The footprint area of the microbattery was 6.41 cm^2 and thus its effective volume was 0.128 cm^3.

Electrochemical Performance of the Microbatteries.
The discharge capacity and efficiency profiles of the microbattery discharged at 0.1 mA (denoted as MB-0.1) are shown in Figure 2; the profiles correspond to the 1st cycle until the 134th cycle (complete failure) of the charge-discharge cycling. The first 16 cycles show that the capacity increases by 11.8 times the capacity of the 1st cycle. The increase in capacity for the first few cycles was related to the gradual transformation of $Ni(OH)_2$ to nickel oxyhydroxide (NiOOH) [12]; the 17th cycle could be the point in which the cell was fully charged. The capacities for cycles 17th to 43rd fluctuate considerably from 8.93 to 15.83 mAh g^{-1}. The trend for cycles 44th to 59th is stabilizing, showing good point-to-point continuity, with an average capacity of 14.82 mAh g^{-1}; this corresponds to a value of 0.07 mAh cm^{-2}, i.e. when normalized by the total area of the microbattery including the anode-cathode separation

area. Cycles from the 76th to the 80th show a descending trend before the capacity continues to fluctuate in the remaining cycles.

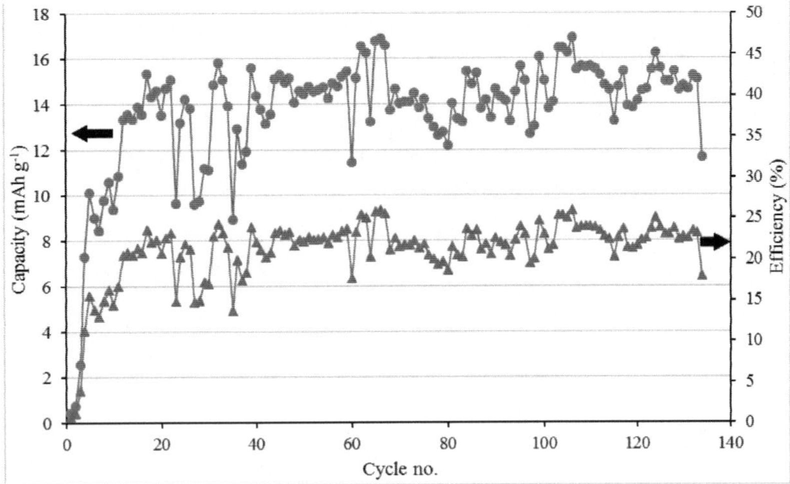

Figure 2 Discharge capacity and efficiency profiles of the microbattery discharge at 0.1 mA as a function of cycle number.

Figure 3 shows some of the discharge curves of MB-0.1; the number accompanying each curve denotes the corresponding cycle number. In the beginning the discharge curve shows the typical Ni-based battery discharge curves featuring distinct plateaus and slopes (refer to cycle number 8, 13, 17, 25 and 32). Then, the sagging of the plateaus took place i.e. the plateau diminishes and is substituted with a slope (refer to cycle number 53 and 66), implying the onset of polarization. It could be due to the resistance or concentration polarization as the electrolyte was in the "semi-dry" or gel form. As the ion mobility are greatly reduced when the electrolyte is gelled, the cell becomes sensitive to water loss resulting from electrolysis and heat generated during repeated charge cycles. Discharge curves from the 66th cycle onwards evolve substantially as "bulges" and grow in size successively before the cell failed completely. The cell energy and power output varies from 0.388 – 0.577 mWh and 0.091 – 0.129 mW, respectively.

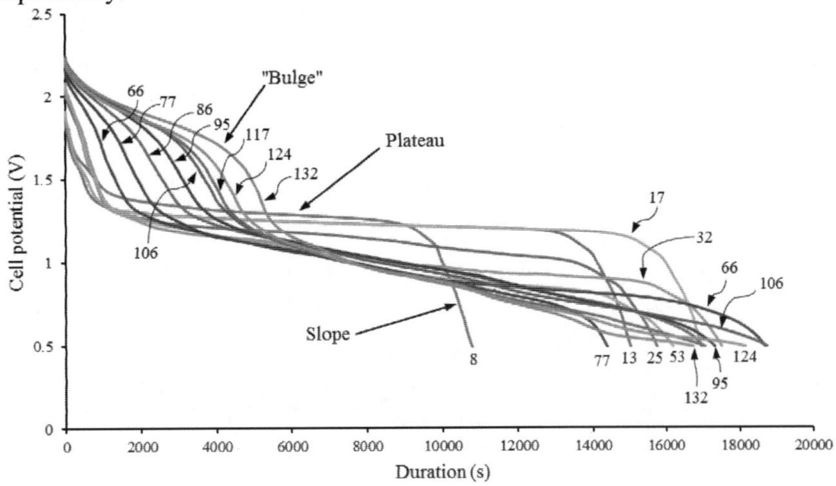

Figure 3 Discharge curves of the microbattery discharged at 0.1 mA for the selected cycle numbers.

The discharge capacity and efficiency profiles of the microbattery discharged at 0.2 mA (MB-0.2), as a function of cycle number are shown in Figure 4. The profile of MB-0.2 is better than MB-01 since it has more cycles with increasing capacity trend (in the early stage), i.e. 26 cycles as compared to only 17 cycles for MB-0.1, and the capacity of MB-0.2 does not fluctuate as much as that of MB-0.1. However, the increase in capacity up to the 16th cycle is only 4.4 times the capacity of the 1st cycle while for MB-0.1 the increment after 16 cycles is 11.8 times the capacity of the 1st cycle. The capacities of MB-0.2 between the 27th to the 49th cycles fluctuate between 11.07 to 14.00 mAh g^{-1} and later stabilizes with a mean value of 10.91 mAh g^{-1} (53rd-68th cycle). From then onwards, the cell capacity starts to decrease (cycle number 76 to 91), fluctuates considerably starting from the 112th cycle until the failure cycle (138th).

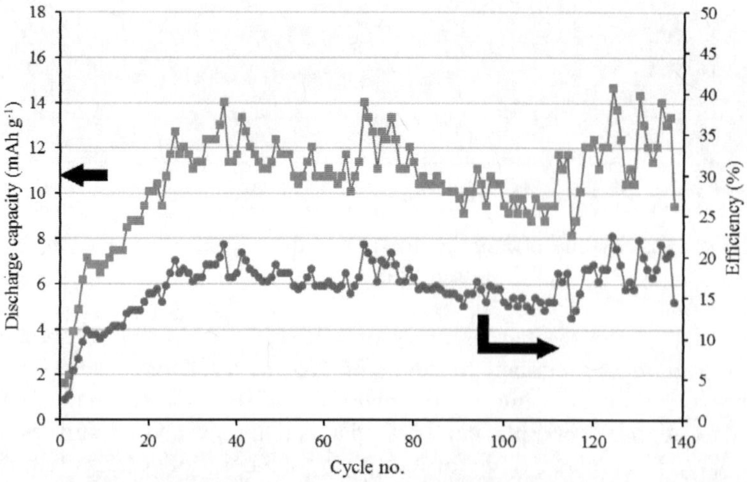

Figure 4 Discharge capacity and efficiency profiles of the microbattery discharge at 0.2 mA as a function of cycle number.

Figure 5 shows some of the discharge curves of MB-0.2 with similar features of MB-0.1 as described above. Overall, however, MB-0.2 possesses higher power output i.e. 0.157 – 0.252 mW, but lower energy output i.e. 0.284 – 0.389 mWh. The average capacity for cycles from the 53rd to the 86th is 11.3 mAh g^{-1} or 0.05 mAh cm^{-2}.

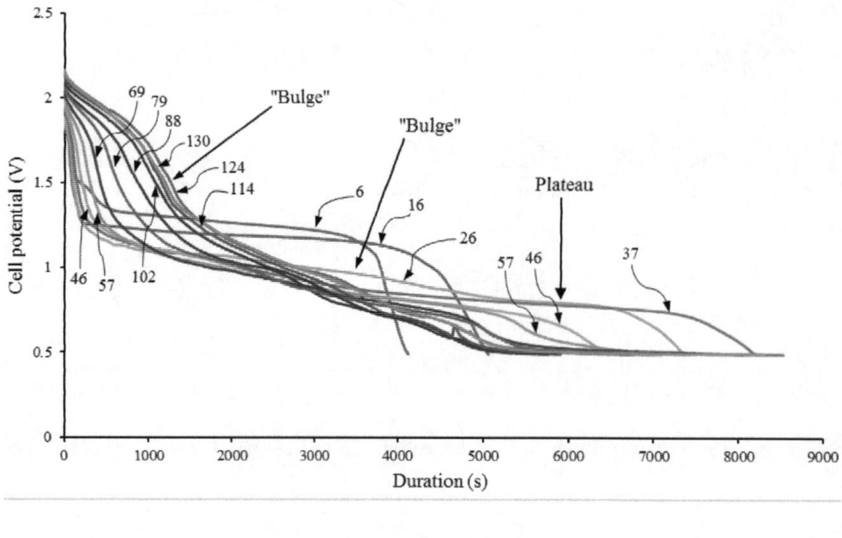

Figure 5 Discharge curves of the microbattery discharged at 0.2 mA for the selected cycle numbers.

When the discharge current was increased to 0.3 mA (MB-0.3), the cell could only sustained 11 cycles of charge-discharge cycling before failed completely (Figure 6). The discharge curves of figure 6 clearly illustrate the onset of polarization from the early cycles indicating 0.3 mA is the limiting discharge load of the cell.

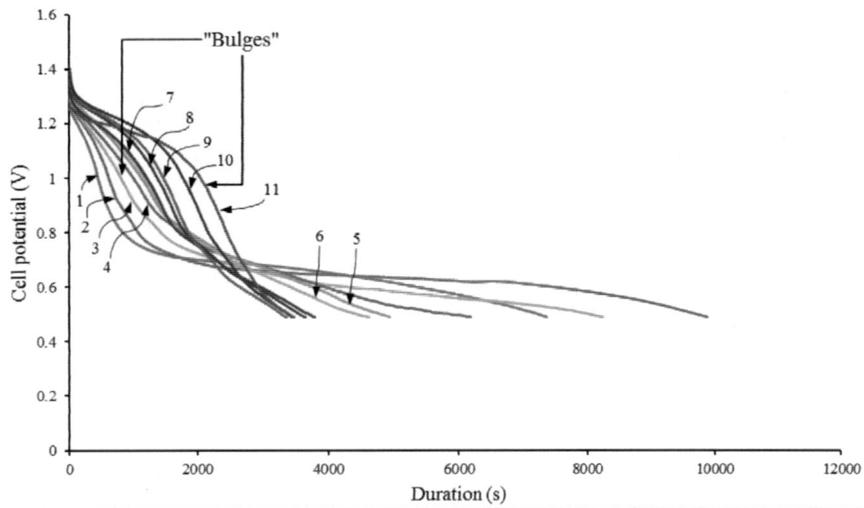

Figure 6 Discharge curves of the microbattery discharged at 0.3 mA for all cycle numbers.

The use of MCM-41 membrane separator enables the successful implementation of co-planar configuration Ni/Zn microbattery. The membrane could be prepared sufficiently thin simply by employing drop-coating procedures which make it compatible with existing microfabrication processes. Unlike polymeric membrane separator that requires additional layer as electrolyte reservoir, MCM-41 material essentially serves as a cell separator, ionic exchange membrane and electrolyte reservoir. Its transformation into lamellar MCM-50 gel has probably limited the cell's high discharge rate capability. Although the Ni/Zn microbattery could sustain more than 100 deep charge-discharge cycles, the discharge load could not exceed 0.3 mA. As such the microbattery of the present co-planar configuration is suitable for low discharge rate applications. Among the potential applications are as a power source for remote sensing devices and power storage for piezoelectric energy harvesters.

One cell design aspect to be implemented is to coat the bare copper substrate with an inert, conducting material such as carbon. This is due to the poor electrochemical stability of copper in the highly-concentrated KOH electrolyte [16, 17]. In fact, XRD measurements on dismantled cells show the evidence of dissolved copper and cupric oxide (CuO) (results are not presented here) due to the caustic alkaline surround.

4. Conclusions

A secondary Ni/Zn microbattery of co-planar design has been fabricated onto a FR4 printed circuit board. The use of inorganic MCM-41 membrane separator permits the cell to be designed sufficiently thin (ca. 200 μm). The preparation processes adopted for the cell fabrication are compatible with existing manufacturing processes involving printed circuit board, thus potentially cost saving. The microbattery demonstrated an acceptable cycling stability performance (> 130 cycles) and was tested under a high depth of discharge and long cycling period (exceeded one month).

References

[1] J. S. Beck, J. C. Vartuli, W. J. Roth, M. E. Leonowicz, C. T. Kresge, K. D.Schmitt, C. T.-W. Chu, D. H. Olson, E. W. Sheppard, S. B. McCullen, J. B. Higgins, and J. L. Schlenker, "A New

Family of Mesoporous Molecular Sieves Prepared with Liquid Crystal Templates," *J. Am. Chem. Soc.*, vol. 114, 10834–10843, 1992.

[2] C. T. Kresge, M. E. Leonowicz, W. J. Roth, J. C. Vartuli, and Beck, J. S. "Ordered Mesoporous Molecular Sieves Synthesized by a Liquid-Crystal Template Mechanism," *Nature,* vol. 359, pp. 710–712, 1992.

[3] N. Nishiyama, S. Tanaka, Y. Egashira, Y. Oku, and K. Ueyama, "Vapor-Phase Synthesis of Mesoporous Silica Thin Films," *Chem. Mater.,* vol. 15, pp. 1006–1011, 2003.

[4] M. Schoeffel, N. Brodie–Linder, F. Audonnetx, & C. Alba–Simionesco, "Wall Thickness Determination of Hydrophobically Functionalized MCM-41 Materials," *J. Mater. Chem.*, vol. 22, pp. 557–567, 2012.

[5] N. Nishiyama, H. Saputra, D.-H. Park, Y. Egashira, and K. Ueyama, "Zirconium-Containing Mesoporous Silica Zr-MCM-48 For Alkali Resistant Filtration Membranes," *J. Mem. Sci.*, vol. 218, pp. 165–171, 2003.

[6] T. Sano, T. Kasuno, S. Arazaki, and Y. Kawakami, "Sorption of Water Vapor on HZSM-5 Type Zeolites," *Stud. Surf. Sci. Catal.*, vol. 105, pp. 1771–1778, 1997.

[7] D.-H. Park, N. Nishiyama, Y. Egashira, & K. Ueyama, "Enhancement of Hydrothermal Stability and Hydrophobicity of a Silica MCM-48 Membrane by Silylation," *Ind. Eng. Chem. Res.*, vol. 40, no. 22, pp. 6105– 6110, 2001.

[8] H. Saputra, R. Othman., A. G. E. Sutjipto, & R. Muhida, "MCM-41 as a New Separator Material for Electrochemical Cell: Application in Zinc-Air System," *J. Membr. Sci.*, vol. 367, no. 1 – 2, pp. 152–157, 2011.

[9] H. Saputra, R. Othman, A. G. E. Sutjipto, R. Muhida & M. H. Ani, "Gel-Like Properties of MCM-41 Material and Its Transformation to MCM-50 in a Caustic Alkaline Surround", *Mater. Res. Bull.*, vol. 7, 732–736, 2012.

[10] N. L. Mohd Zawi, R. Othman and M. H. Ani, Zn/MCM-41/MnO_2 Leclanché Button Cell R2025 for Low Rate Applications," *Adv. Mater. Res.*, vol. 576, pp. 374 – 377, 2012.

[11] P. H. Humble, J. N. Harb, and R. LaFollette, "Microscopic Nickel-Zinc Batteries for Use in Autonomous Microsystems. *J. Electrochem. Soc.*," vol. 148, no. 12, pp. A1357–A1361, 2001.

[12] F. Chamran, Y. Yeh, H.-S. Min, B. Dunn, and C. J. Kim, "Fabrication of High Aspect-Ratio Electrode Arrays for Three-Dimensional Microbatteries," *J. Microelectromechanical Syst.*, vol. 16, pp. 844–852, 2007.

[13] N. J. Dudney, "Solid-State Thin-Film Rechargeable Batteries," *Materials Science and Engineering B*, vol. 116, pp. 245–249, 2005.

[14] H.-S. Min, B. Y. Park, L. Taherabadi, C. Wang, Y. Yeh, R. Zaouk, M. J. Madou, and B. Dunn, "Fabrication and Properties of a Carbon/Polypyrrole Three-Dimensional Microbattery," *J. Power Sources*, vol. 178, pp. 795–800, 2008.

[15] J. H. Pikul, H. G. Zhang, J. Cho, P. V. Braun, and W. P. King, "High-Power Lithium Ion Microbatteries From Interdigitated Three-Dimensional Bicontinuous Nanoporous Electrodes," *Nat. Commun.*, vol. 4, pp. 1732, 2013.

[16] M. R. Othman and Riyanto, "Electrochemical stability of Cu, Ni, Co, Pt and Ir metals sheet and their composite electrodes in potassium hydroxide solution," *Int. J. Electrochem. Sci.*, vol. 7, pp. 8408–8419, 2012.

[17] G. Garg and S. Basu, "Studies on Degradation of Copper Nano Particles in Cathode for CO_2 Electrolysis to Organic Compounds", *Electrochimica Acta*, vol. 177, pp. 359–365, 2015.

IOP Conf. Series: Materials Science and Engineering **290** (2017) 012074 doi:10.1088/1757-899X/290/1/012074

Effect of solvents on the morphology and performance of Polyethersulfone (PES) polymeric membranes material for CO_2/CH_4 separation

M S Ahmad[1], D F Mohshim[1], R Nasir[2], H A Mannan[3], H Mukhtar[3]

[1]Department of Petroleum Engineering, Universiti Teknologi PETRONAS, Bandar Seri Iskandar, 32610, Perak, Malaysia
[2]Department of Chemical Engineering, NFC Institute of Engineering and Fertilizer Research, Faisalabad, 38090, Pakistan
[3]Department of Chemical Engineering, Universiti Teknologi PETRONAS, Bandar Seri Iskandar, 32610, Perak, Malaysia

Email: dzetifarhah.mohshim@utp.edu.my

Abstract. Membrane technology has several advantages such as the ability to separate chemical species within compact plant footprints, low thermal energy requirements and simple process flow schemes. By optimizing the available materials and analysis, this work comes with the objective to synthesize the polymeric membrane, which has the best separation performance. In this work, three (03) membranes have been synthesized and a comparative analysis were conducted based on different types of solvent namely N, N-dimethylacetamide (DMAc), N, N-dimethylformamide (DMF), N-methyl-2-pyrrolidinone (NMP). In characterizing the synthesized membrane, Thermo Gravimetric Analysis (TGA), Field Emission Scanning Electron Microscopy (FESEM) and Fourier Transform Infrared Spectroscopy (FTIR) analysis were used. A comparative study was carried out to compare the effects of each solvent towards CO_2 separation performance.

1. Introduction

Gas separation membranes have gained great attention in different industries as compared to conventional separation technologies [1]. Consequently, many researchers have worked on the improvement of membrane performance with the effects of key performance parameters [2], [3]. One of the most important and effective parameter is solvent. The selection of solvent can affect the gas-permeation properties and the morphology of the membranes [4]. Various solvents, depending on their specific chemical and physical properties, interact differently with polymers, and these results in different morphologies [5]-[7]. The physical properties of the different solvents have been given in table 1.

Table 1. Physical properties of solvents [8, 9].

Solvent name	Molecular formula	Mole weight (g/mol)	Molar volume (cm^3/mol)	Boiling Point (°C)
N-methyl-2-pyrrolidinone (NMP)	C_5H_9NO	99.13	96.430	202-204
N, N-dimethylacetamide (DMAc)	$CH_3CON(CH_3)_2$	87.12	92.978	164.5-166
N,N-dimethylformamide (DMF)	$HCON(CH_3)_2$	73.09	77.426	152-154

The main objective of this work is to perform a comparative study on the morphology and gas permeation of dense PES membranes prepared from various solvents e.g. N, N-dimethylacetamide (DMAc), N,N-dimethylformamide (DMF), N-methyl-2-pyrrolidinone (NMP). The high performance glassy polymer PES was selected as the membrane polymer. The repeating unit in the structure of PES possesses a certain degree of rigidity such that it has a glass transition temperature of 225°C. Field emission scanning electron microscopy (FESEM) and Thermo gravimetric analysis (TGA) were performed to characterize the membrane morphologies and correlated them to their gas-permeation results.

2. Experimental Procedures

2.1. Materials

A commercial grade polyethersulfone (PES), ULTRASON E6020P®, with good thermal and mechanical properties, was purchased from BASF chemicals, Germany. Three (03) solvents e.g. N, N dimethylacetamide (DMAc), N, N-dimethylformamide (DMF), N-methyl-2-pyrrolidinone (NMP) were supplied by Merck (Germany) and used without further treatment. Test gases (CH_4 and CO_2 > 99.9 %) were purchased from Gas walkers Sdn. Bhd, Malaysia.

2.2. Synthesis of Membranes

Pre-dried and known percentage of PES was dissolve in different solvents e.g. DMAc, DMF and NMP to prepare casting solution in a closed container. The casting solution was stirred for 24 hour at room temperature to completely dissolve the polymers followed by degassing for 30 minutes to remove the air bubbles formed during the mixing process. The casting knife was adjusted at 180-μm opening and the casting solution was poured onto a clean glass plate to cast the membrane films. The casted membranes were isolated in the vacuum oven at 90° C for the 24 hours. After drying process, the membranes were secured. After number of experimental attempts, the optimized compositions of casting solution are given in table 2.

Table 2. Membrane compositions.

	Solvent (wt %)	Polyethersulfone (PES) (wt%)
NMP	83	17
DMF	82.5	17.5
DMAc	85	15

2.3. Characterization of Membranes

The morphological analysis of synthesized membranes was characterized by Field Emission Scanning Electron Microscope (VPFESEM, Zeiss Supra55 VP) and Thermogravimetric Analyzer (TGA) (Perkin Elmer Simultaneous Analyzer STA 6000) in order to investigate the thermal stability of membranes.

2.4. Gas Permeability Studies

The permeability of pure CO_2 and CH_4 gas across the membrane was recorded in the pressure range of 6-10 bar and room temperature. The CO_2 permeability was calculated according to the following formula:

$$P = \frac{22414}{A} \times \frac{l}{(p_2 - p_1)} \times \frac{p_1}{RT} \times \frac{dV}{dt} \quad (1)$$

where A representing membrane area (cm^2), P_1 and P_2 are the feed and permeate pressure, respectively, R is the universal gas constant (6236.56 cm^3cmHg/mol.K), T represents absolute temperature (K), dV/dt is the volumetric displacement rate (cm^3/s) and 22,414 is the number of (STP) of penetrant per mole [9-11]. The schematic diagram of membrane module is shown in figure 1.

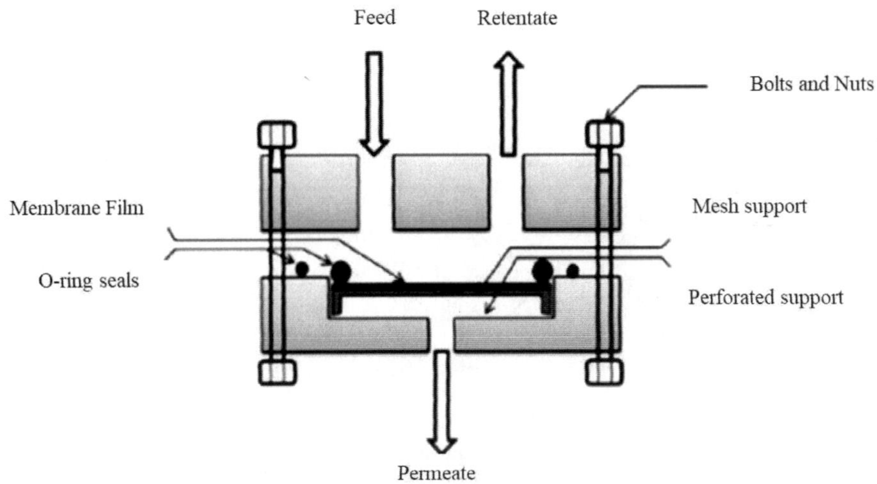

Figure 1. Schematic diagram of membrane module [12].

IOP Conf. Series: Materials Science and Engineering **290** (2017) 012074 doi:10.1088/1757-899X/290/1/012074

3. Results and Discussion

3.1. Morphological Analysis

The cross sectional views of the PES-NMP, PES–DMF and PES-DMAc polymeric membranes are shown in figure 2(a-c), respectively. As it can be seen in figure 2, the dense symmetric cross-section of all membrane was observed. figure 2a shows the dense, non-porous structure of PES membrane. No voids were present in this magnification. figure 2b and 2c also illustrated dense and nonporous structure. It shows the compatibility of polymer and solvent.

Based on the cross-sectional images of the membranes, it confirmed that all the membranes are dense in structure. This means that permeate can diffuse through whether by pressure, concentration potential gradient. In this context, the permeate which was CO_2 diffuse through pressure gradient. On the other hand, when the values for thickness of each membrane were compared to each other, the thickness for PES-DMAc membrane is quite higher than the other two (2) membranes. Table 3 shows the summary of membranes and the thickness.

Table 3. Membranes thickness.

Membranes	Thickness, μm
PES-NMP	44.67
PES-DMF	48.03
PES-DMAc	124.4

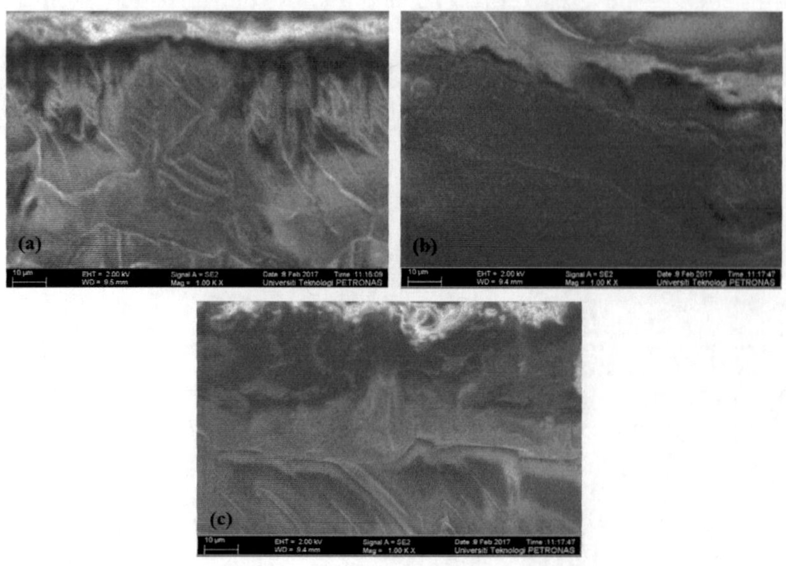

Figure 2. FESEM micrographs of (a) PES-NMP (b) PES-DMF and (c) PES-DMAc polymeric membranes.

The membranes synthesized are having different thickness because of their different rates of solvent evaporation during the heating process. For instance, PES-DMAc membrane has very large thickness due to the low rate of solvent evaporation during heating process. To conclude, all concentration of solvents managed to form homogeneous compatible solution with PES polymer matrix.

3.2. *Thermal Analysis*

Figure 3 shows the weight loss of PES-NMP, PES–DMF and PES-DMAc polymeric membranes as a change of temperature. It was observed that TGA curve has two weight loss curves. The first weight loss may be due to a small quantity of the trapped solvent. It was observed that the amount of residual solvent is in the order of NMP > DMF > DMAc. The amount of solvent residue depends on drying temperature of the membrane and increases with the increase in boiling point of solvent [12]. The second weight loss was found that the membranes are stable up to 450 °C. This is due to the decomposition temperature of PES which around this temperature. From the curves, the thermal stability behavior for each membrane can be determined. For instance, the major weight loss for each membrane samples can be obtained and analyzed. The major weight loss might be due to the decomposition of main chains of the polymer. Thus, TGA can be used to get the decomposition temperature for each membrane. The decomposition temperature will determine until what temperature the membrane can withstand before start to degrade and can no longer be used. It will be an indicator whether the synthesized membrane can be used in industry condition or not.

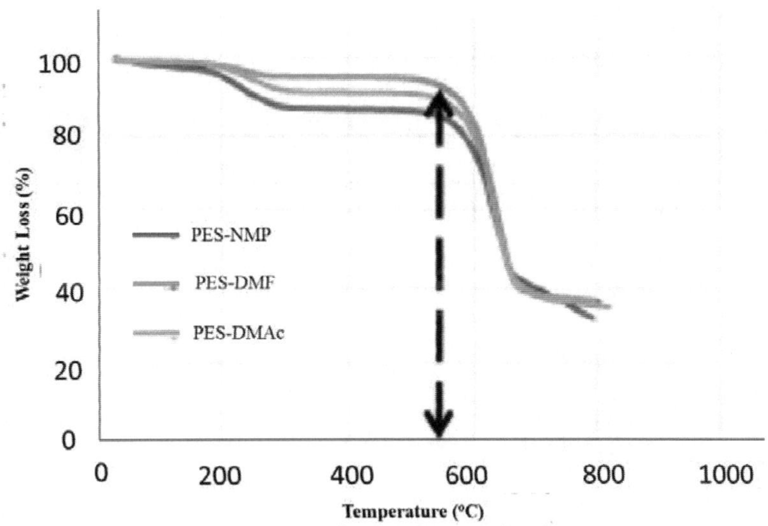

Figure 3. TGA spectra of PES-NMP, PES-DMF and PES-DMAc polymeric membranes.

3.3. *Gas Permeation Study*

Based on the result obtained from gas permeability test, the gas permeance for each type of membrane samples decreasing as the pressure increase. This reflects the normal behavior for PES membrane. However, at certain pressure, the value of permeance for PES-NMP and PES- DMF membranes start to increase back. This is might be resulted from plasticization of the membrane. Plasticization causing the changes in structure of the membranes which allowing more permeate to pass through the membrane. It was observed that all polymeric membranes have higher values of selectivity at 2 bar pressure. It was also observed that PES-DMF shown higher selectivity at 2 bar pressure as compared to PES-NMP and PES-DMAc. In summary, plasticization increases the amount of free volume for the membrane which causing increase in diffusivity. However, for PES-DMAc membrane, there was no effect of plasticization as the barrer continue to decrease when pressure increase. Summary for gas permeability results is shown in table 4.

Table 4. Summary for gas permeability test result.

Pressure (bar)	P_{CO2}, (barrer)	P_{CH4}, (barrer)	Ideal Selectivity, P_{CO2}/P_{CH4}	Membranes
2	1.91	0.80	2.40	PES-NMP
4	1.35	0.60	2.25	
6	0.99	0.48	2.06	
8	1.17	0.57	2.07	
2	45.72	17.89	2.56	PES-DMF
4	38.58	15.43	2.50	
6	29.40	13.72	2.14	
8	30.87	16.24	1.81	
2	52.42	24.60	2.13	PES-DMAc
4	38.10	19.99	1.95	
6	31.34	17.76	1.76	
8	25.79	13.78	1.87	

4. Conclusion

In summary, PES was added in three different solvents to develop the polymeric membrane with CO_2 and CH_4 permeation properties. FESEM analyses confirmed the dense and no porous structure. TGA analysis revealed that the developed polymeric membranes are thermally stable up to 450°C. The highest value of selectivity (2.56) was achieved by the PES-DMF membranes at operating condition of 02 bar and 25°C. The result from permeability test indicating that PES-DMF and PES-DMAc displayed higher gas permeance unit (GPU) when compared to PES-NMP membrane. However, in term of membrane selectivity, all membrane samples displayed the same trend, which was decrease in selectivity as the pressure increase.

Acknowledgement

The authors acknowledge the financial support provided by Universiti Teknologi PETRONAS. Malaysia.

References

[1] Bondar V, Freeman B and Pinnau I 2000 *Journal of Polymer Science Part B: Polymer Physics* **38** 2051-62
[2] Ghosal K, Chern R, Freeman B, Daly W and Negulescu I 1996 *Macromolecules* **29** 4360-9
[3] Fu Y-J, Hu C-C, Qui H-z, Lee K-R and Lai J-Y 2008 *Separation and Purification Technology* **62** 175-82
[4] Şener T, Okumuş E, Gürkan T and Yilmaz L 2010 *Desalination* **261** 181-5
[5] Shao L, Chung T-S, Wensley G, Goh S H and Pramoda K P 2004 *Journal of membrane science* **244** 77-87
[6] Wu L, Sun J and He C 2010 *Journal of applied polymer science* **116** 1566-73

[7] Isanejad M, Azizi N and Mohammadi T 2017 *Journal of Applied Polymer Science* **134**
[8] Moradihamedani P, Ibrahim N A, Yunus W M Z W and Yusof N A 2013 *Journal of Applied Polymer Science* **130** 1139-47
[9] Adewole J K, Ahmad A L, Ismail S, Leo C P and Sultan A S 2015 *Journal of Applied Polymer Science* **132** 42205
[10] Merkel T, Bondar V, Nagai K, Freeman B and Yampolskii Y P 1999 *Macromolecules* **32** 8427-40
[11] Ghadimi A, Sadrzadeh M, Shahidi K and Mohammadi T 2009 *Journal of Membrane Science* **344** 225-36
[12] Nasir R, Mukhtar H, Man Z, Shaharun M S and Abu Bakar M Z 2015 *RSC Advances* **5** 60814-22

IOP Conf. Series: Materials Science and Engineering **290** (2017) 012075 doi:10.1088/1757-899X/290/1/012075

Antiwashout behavior of calcium phosphate cement incorporated with Poly(ethylene glycol)

S Hablee[1], I Sopyan[1], M Mel[2], H M Salleh[2], M M Rahman[3]

[1]Department of Manufacturing and Materials Engineering, International Islamic University Malaysia, Jalan Gombak, 53100 Kuala Lumpur, Malaysia
[2]Department of Biotechnology Engineering, International Islamic University Malaysia, Jalan Gombak, 53100 Kuala Lumpur, Malaysia
[3]Department of Mechanical Engineering, College of Engineering, Universiti Tenaga Nasional, 43000 Kajang, Selangor Darul Ehsan, Malaysia

Email:sopyan@iium.edu.my

Abstract. The effect of powder-to-liquid ratio and addition of poly(ethylene glycol) on the antiwashout behavior of calcium phosphate cement has been investigated. Calcium hydroxide, $Ca(OH)_2$, and diammonium hydrogen phosphate, $(NH_4)_2HPO_4$, were used as precursors with distilled water as the solvent in the wet chemical precipitation synthesis of hydroxyapatite powder. Cement paste was prepared by mixing the as-synthesized powder with distilled water at certain ratios, varied at 1.0, 1.3, 1.5 and 1.6. Poly(ethylene glycol) was added into distilled water, varied at 1, 2, 3, 4 and 5 wt% using the powder-to-liquid ratio of 1.3. The antiwashout properties of the cement has been investigated by soaking in Ringer's solution for 3 and 7 days. The evolution of compressive strength of calcium phosphate cement before and after soaking have been determined. After 7 days soaking, the strength of the cement increased by 94.4%, 2.98%, 11.39% and 111.29% for powder-to-liquid ratios 1.0, 1.3, 1.5 and 1.6 respectively. The addition of poly(ethylene glycol) up to 3% shows an increase in strength after 7 days soaking, with 57.75%, 16.4% and 19.97% increase for 1, 2 and 3% poly(ethylene glycol) contents respectively. The calcium phosphate cement produced in this current study shows excellent antiwashout behavior since no cement dissolution happened and the compressive strength of the cement increased with soaking time throughout 7 days soaking in Ringer's solution.

1. Introduction

Calcium phosphate cement (CPC) is clinically accepted ceramic and has been widely used in biomedical application, predominantly in bone cement application. This is because CPC shows remarkable biological properties, injectability and potential to set *in vivo* [1]. CPC can be injected and molded to fill and take the shape of defect sites. Injectable CPC consisted of calcium phosphate powder of various phases and water mixed at certain ratios to set *in vivo* via dissolution-precipitation mechanism [2]. The mixing of phases such as tetracalcium phosphate (TTCP) and α–tricalcium phosphate (α-TCP) with water induced the formation of apatite cements of hydroxyapatite (HA) [3].

Biodegradability of CPC is significant to help the healing process through substitution of the hardened CPC by the newly forming bone [2]. Generally, the rate of cement resorption should be the same with the rate of new bone formation. Too fast resorption rate will cause collapse at the defect site [2]. However, disintegration of CPC tend to occur upon early contact with body fluids because of its

IOP Conf. Series: Materials Science and Engineering **290** (2017) 012075 doi:10.1088/1757-899X/290/1/012075

weak cohesion [1]. The solubility of CPC is affected by the surrounding pH value, and HA has the lowest solubility compared to that of dicalcium phosphate dihydrate (DCPD), octacalcium phosphate (OC) and β-tricalcium phosphate (β-TCP) [4].

The incorporation poly(ethylene glycol) (PEG) into CPC is one of the approach to adjust the dissolution of CPC, so that the dissolution rate is similar to the rate of new bone formation. PEG is remarkably accepted in biomedical fields due to its non-toxicity, water solubility, flexibility and anti-coagulant activity [5]. PEG also has been used as thickening agent in premixed CPC to keep the cement paste stable [5,6]. Addition of PEG into CPC is expected to improve setting time and anti-washout performance, but lower injectability [6].

The evaluation of dissolution behavior of CPC has been done *in vitro* using Ringer's solution. The composition of Ringer's solution is similar to the inorganic components of extracellular body fluid to mimic the biological environment for bone, which aid the formation of new bone tissues. In this study, HA powder was synthesized via wet chemical precipitation method. The purpose of this study is to determine the effect of powder-to-liquid (P/L) ratio and addition of PEG on the antiwashout behavior of CPC.

2. Experimental procedures

2.1. Synthesis of HA powder

HA powder has been synthesized via wet chemical precipitation method following the procedure reported elsewhere [7]. This route employed calcium hydroxide, $Ca(OH)_2$, and diammonium hydrogen phosphate, $(NH_4)_2HPO_4$, as the precursors and the solvent used was distilled water. The calcium-to-phosphorus (Ca/P) ratio was fixed at 1.67. Each precursor was dissolved in distilled water to prepare calcium and phosphorus solution. The two solutions were mixed together through drop wise addition of calcium solution into the phosphorus solution. 25% ammonia solution was added to adjust the pH of the mixture at pH 11. Then, the mixture was refluxed at 90°C. The precipitate formed underwent overnight (18 hours) aging at room temperature, washing with distilled water and filtering. The filtered precipitate was then dried overnight at 85°C, and finally crushed.

2.2. Preparation of CPC

Cement paste was prepared by mixing the synthesized HA powder with distilled water. Two types of cement were prepared, without and with PEG addition. Cement paste without PEG was prepared at different P/L ratios, varied at 1.0, 1.3, 1.5 and 1.6. Cement paste incorporated with PEG (MW300, Sigma) was prepared with the P/L ratio of 1.3. PEG was added into the liquid phase, varied at 1, 2, 3, 4 and 5 wt%. The cement paste was put into a Teflon mold and left for 48 hours to dry. The sample dimension used is 10 mm diameter x 15 mm length.

2.3. Antiwashout test

The antiwashout behavior of CPC was investigated by soaking the cement in a Ringer's solution for up to 7 days. The mechanical strength of CPC after the immersion was measured at 3 and 7 days. At each interval, samples were washed with distilled water and dried before compression test. The compression test of CPC was done by using Lloyd LR 10 K+ Universal Testing Machine on the 10 mm diameter x 15 mm length specimen at 1 mm/s crosshead rate.

2.4. Porosity measurement

The porosity of CPC was calculated using the following equation:

$$\% \text{ Porosity} = 100\% - \text{Relative density}$$

$$\text{Relative density} = \frac{\rho_{app}}{\rho_{th}} \times 100\% \quad (2.2)$$

Where, ρ_{app} is the apparent density measured by densitometer and ρ_{th} is the theoretical density of HA taken as 3.156 g/cm^3 [8],[9].

3. Results and Discussion

The dissolution behavior of CPC was investigated by soaking in Ringer's solution for 7 days. The effect of P/L ratio and PEG content on the dissolution properties of CPC was investigated. CPC specimens was prepared using the P/L ratio of 1.0, 1.3, 1.5 and 1.6. The addition of PEG into the liquid phase of CPC was varied at 0, 1, 2, 3, 4, and 5%. From the observation, all CPCs show excellent integration properties since no cement dissolution shown throughout 7 days soaking in Ringer's solution.

The evolution of compressive strength of CPC before and after soaking have been evaluated in this present work. Figure 1 shows the result of compressive strength of CPC with different P/L ratios up to 7 days soaking in Ringer's solution. All cements show increasing compressive strength with the increase in soaking period. The strength increased by 94.4%, 2.98%, 11.39% and 111.29% for CPC with the P/L ratios 1.0, 1.3, 1.5 and 1.6 respectively, with the range of 0.416 MPa to 1.344 MPa before soaking and increased to the range of 0.879 MPa to 1.384 MPa after 7 days soaking. This might be attributed to the formation of apatite via biomineralisation mechanism. As reported by Mohammadi et al. [10], CPC was immersed in human blood plasma and ringer's solution revealed that the compressive strength of CPC increased with increasing soaking time. The formation of apatite also enhanced throughout the soaking period. Study done by Verma and Sinha [11] reported the cohesiveness of CPC in distilled water for different P/L ratios, (1:1, 1:1.1 and 1:1.2 g/ml). Their study revealed that no structural degradation of CPC after 7 days immersion for CPC after 20 min incubation. However, structural disintegration was observed when the paste was directly injected into the distilled water and the decrease in P/L ratio increased the rupture time.

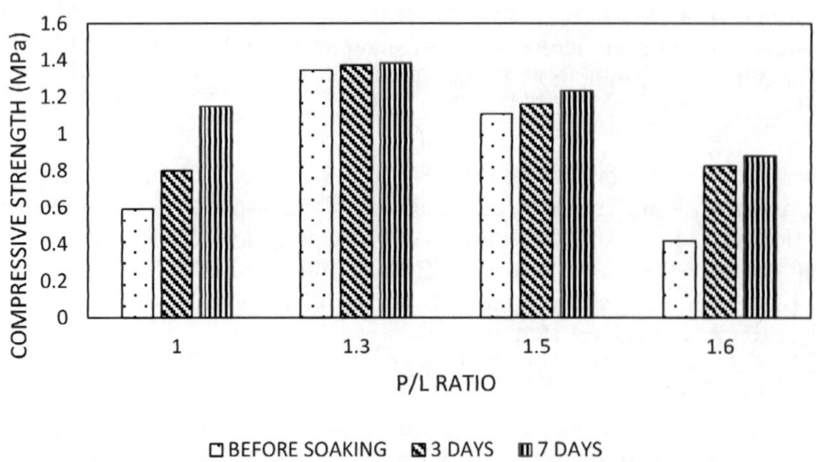

Figure 1. Effect of P/L ratio on the dissolution behavior of CPC after soaking in Ringer's solution.

The effect of PEG addition on the compressive strength of CPC after soaking in Ringer's solution is shown in figure 2. After 7 days immersion, the compressive strength of CPC increased with the increase in immersion time for CPC with 0, 1, 2 and 3% PEG contents. The strength increased by 2.98%, 57.75%, 16.4% and 19.97% for the CPC with 0, 1, 2 and 3% PEG content,s respectively, with the range of 1.167 MPa to 1.786 MPa before soaking and increased to the range of 1.384 MPa to 2.079 MPa after 7 days soaking period. However, CPC with 4 and 5% PEG contents show an increase in compressive strength up to 3 days soaking with 7.18% and 3.89% increase respectively, and then the

IOP Conf. Series: Materials Science and Engineering **290** (2017) 012075 doi:10.1088/1757-899X/290/1/012075

compressive strength decreased by 25.5% and 40% respectively after 7 days soaking. The strength decreased from 1.371 MPa to 1.1 MPa and from 1.285 MPa to 0.801 MPa for 4 and 5% PEG contents respectively. This might be because of hydrophilicity of PEG, which causes PEG to dissolve in the liquid medium leading to the increase in porosity and hence reducing the compressive strength of CPC. Roy et al. [12] has reported that addition of PEG slightly delayed the nucleation and formation of DCPD crystals since its compressive strength was high initially but dropped significantly after 3 days soaking in PBS and increased again after 7 days soaking.

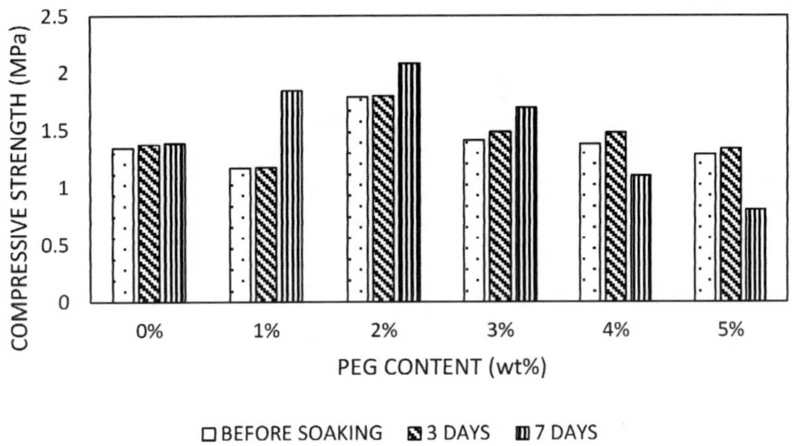

Figure 2. Effect of PEG addition on the dissolution behavior of CPC after soaking in Ringer's solution.

Porosity of CPC before soaking have been determined. The average porosity of CPC ranged from 39.2% to 47.1%. Their apparent density ranged between 1.67 to 1.92 g/cm^3. From table 1, the porosity of CPC decreases with the increase in P/L ratio. This is attributed to the increase in powder content in the cement paste. High porosity of CPC is presumably due to water entrapped between the crystals and porosity introduced during cement preparation [13]. When PEG is added into CPC, the porosity of is in the range of 39.6% to 51.4% and the apparent density is between 1.53 to 1.90 g/cm^3. The result shows that incorporation of PEG into CPC increases porosity of CPC as tabulated in table 2. The higher the PEG content, the lower its porosity because PEG particles filled up the pores in CPC. This is true only for 1% to 4% PEG content. The porosity started to increase again when 5% PEG was added. This indicates that an appropriate amount of PEG is needed to reduce porosity of CPC.

Table 1. Porosity of CPC with different P/L ratios.

P/L ratio	Porosity (%)
1.0	47.1
1.3	45.2
1.5	40.8
2.0	39.2

Table 2. Porosity of CPC with different PEG contents for the P/L ratio of 1.3.

PEG content (wt%)	Porosity (%)
0	45.2
1	51.4
2	48.5
3	43.8
4	39.6
5	45.7

4. Conclusion

This present work successfully investigated the effect of P/L ratio and addition of PEG on the antiwashout behavior of CPC. The prepared CPC shows remarkable antiwashout properties because no cement dissolution was observed throughout the 7 days soaking in Ringer's solution. In addition, the evolution of compressive strength of CPC after soaking shows strength improvement with the increase in soaking period. The strength of CPC before soaking ranged from 0.416 MPa to 1.344 MPa and increased to the range of 0.879 MPa to 1.384 MPa after soaked for 7 days. PEG addition up to 3% showed an increase in strength by 2.98%, 57.75%, 16.4% and 19.97% for CPC with 0, 1, 2 and 3% PEG contents, respectively, which ranged from 1.167 MPa to 1.786 MPa before soaking and increased to the range of 1.384 MPa to 2.079 MPa after 7 days soaking.

Acknowledgement

This work was financially supported by Ministry of Higher Education (MOHE) through Fundamental Research Grant Scheme (FRGS) with the project ID of FRGS15-246-0487.

References

[1] Zhang J, Liu W, Schnitzler V, Tancret F and Bouler J 2014 *Acta Biomaterialia* **10** 1035-1049
[2] Dorozhkin S V 2011 *International Journal of Materials and Chemistry* **1(1)** 1-48
[3] Alqap A S F, Sopyan I, Husni M and Athirah N 2012 *Applied Mechanics and Materials* **110-116** 8-12
[4] Sakai S, Anada T, Tsuchiya K, Yamazaki H, Margolis H C and Suzuki O 2016 *Dental Materials Journal* **35(2)** 216-224
[5] Chen F, Liu C, Wei J, Chen X, Zhao Z and Gao Y 2011 *Materials Chemistry and Physics* **125** 818-824
[6] Perez R A, Kim H W and Ginebra M P 2012 *Journal of Tissue Engineering* **3(1)** 1-20
[7] Hablee S, Sopyan I, Mel M, Salleh H M, Rahman M M and Singh R 2017 *IOP Conf. Series: Materials Science and Engineering* **205** 012012
[8] Unosson J E, Persson C and Engqvist H 2015 *J Biomed Mater Res Part B* **103B** 62-71
[9] Pazarlıoglu S and Salman S 2017 *J Aust Ceram Soc* 1-11
[10] Mohammadi M, Hesaraki S and Hafezi-Ardakani M 2014 *International Journal of Engineering Practical Research (IJEPR)* **3(1)** 1-7

[11] Verma N P and Sinha A 2013 *J Mater Sci: Mater Med* **24** 53-59
[12] Roy M, DeVoe K, Bandyopadhyay A and Bose S 2012 *Materials Science and Engineering C* **32** 2145-2152
[13] Öhman C, Unosson J, Carlsson E, Ginebra M P, Persson C and Engqvist H 2015 *J Mater Sci: Mater Med* **26** 210

IOP Conf. Series: Materials Science and Engineering **290** (2017) 012076 doi:10.1088/1757-899X/290/1/012076

Preparation of SS316L MIM feedstock with biopolymer as a binder

A A Abdullah[1], H Norita[1], H N Azlina[1], A B Sulong[2], N N Mas'ood

[1]Department of Manufacturing and Materials Engineering, International Islamic University Malaysia (IIUM), Jalan Gombak, 53100 Kuala Lumpur, Malaysia
[2]Faculty of Engineering and Built Environment, UniversitiKebangsaan Malaysia, 43600 UKM-Bangi, Selangor, Malaysia

Email: norita_hassan@iium.edu.my

Abstract. This paper focus on feedstock preparation for SS316L metal injection molding (MIM) part. The primary step of feedstock preparation, critical powder loading determined by two method; maximum filled volume calculation model and torque analysis. The critical powder loading determined by calculation was 70 vol% to 77 vol% while for experimental approaches shows the value of 75 vol%. The feedstock was prepared by mixing SS316L powder and polymer binder with ratio 70:30 at 175 °C with speed of 50 rpm. The feedstock was analyzed by thermogravimetric analysis (TGA) and Scanning electron microscope (SEM). The composition for the feedstock after preparation step was confirmed by TGA. It was found that the prepared feedstock component was compatible to each other and composition is maintain along the mixing step.

1. Introduction

Metal injection molding (MIM) is an economics and environmental friendly metal working process. The advantages of MIM such as high volume of production and able to produce intricate shape with minimum finishing process [1]. This process consist of four main step. The first step is feedstock preparation; the metal powder mixed with polymer binder. The second step is injection molding followed by debinding. The final step of MIM is sintering, in this step the debind part known as brown part is densified. Each step parameter and condition needs to consider in detail due to failure chaining from one-step to another.The first step of MIM is considered as crucial part. Numbers of studied done on the preparation of MIM feedstockdetermine the optimum binder composition[2], to improve the bonding between metal powder and binder[3-4], to increase the dispersion of powder and binder in the feedstock[5],and the best metal powder characteristics to ensure free defect part produced[6]. Various paper reported on the optimum composition of feedstock in MIM. Petroleum based polymer has been widely accepted as the binder for MIM feedstock and it is found that bio based polymer can reduced the debinding process of MIM part. Polyhydroxyalkanoates (PHA) is one of bio based polymer that recently been studied by researcher. By replacing petroleum based polymer with bio based polymer, the processing time of MIM part will reduced.

In this paper, MIM feedstock consist of SS316L and (PHA) were investigated concerning the critical powder loading determined by two method and the feedstock properties.

2. Experimental procedures

The SS 316L was supplied by CNPC Powder with density of 7.92g/cm^3. The shape and the particle size distribution of the powder were determined by field emission scanning electron microscope (FESEM JEOL JSM-7000F) analysis and laser particle size analyzer, respectively. The critical powder loading was calculated by the ratio of tap density to pycnometer density relation and measured by mix the powder with oleic acid gradually 1ml for every 3 minutes with speed of 50rpm. The feedstock was prepared with powder loading 5vol% less than critical powder loading determined and binder composition 55, 40, and 5 vol% of paraffin wax (PW), polyhydroxyalkanoates (PHA) and steric acid (SA) respectively. The feedstock was mixed in Barbender internal mixer with 50 rpm at 180°C. The feedstock was analyzed using TGA and FESEM. The composition of the binder in the prepared feedstock confirmed by TGA and supported by SEM analysis.

3. Results and Discussion

The theoretical value of critical powder loading for SS316L is determined by maximum filled volume calculation model. The tap density to the pycnometer density ratio show the maximum packing state of the powder [7]. It is widely accepted that the critical powder loading will be 10% to 20% more than the tap density: pycnometer density value [8]. Based on the pycnometer density and tap density 7.92 g/cm^3 and 5.1 g/cm^3 respectively, the critical powder loading calculated is in range of 70 vol% to 75 vol %. However, this method is not accurate as the result from calculation is commonly much lower compared to the experimental result [9]. Hence, torque analysis was done to support the theoretical finds. Mixing torque between SS316L and oleic acid is illustrated in Figure 1. The critical powder loading for SS316L was identified at the highest torque value, which is at 75 vol% of powder loading. At this point, the space between the powder particles was fully filled by the binder. As the volume of oleic acid keep increasing, the mixing torque decrease as the liquid be the major constituent and act as lubricant. This critical powder loading value is much higher for the similar average size of powder found by other researcher [10-11]. This is due to the difference in the particle size distribution as indicated in Figure 2. The SS316L powder has wide distribution for size less than D_{50} and the pore between the large particle filled results in high powder loading.

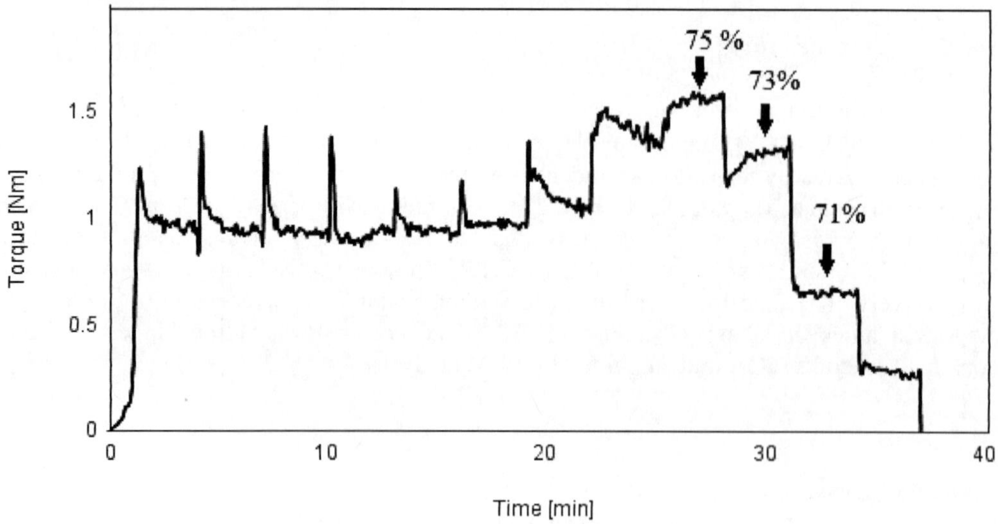

Figure 1 Mixing torque between SS316L and oleic acid

Table 1: Critical and possible optimum powder loading range for different determination methods

Method	Critical powder loading	Optimum powder loading
Maximum filled volume calculation	70 – 75vol%	65- 73vol%
Mixing torque analysis	75vol %	70-73vol%

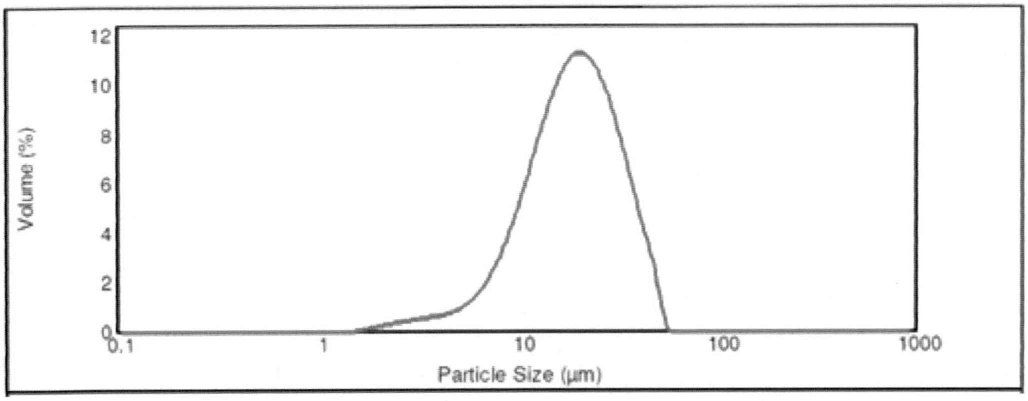

Figure 2: Particle size distribution of SS316L powder

Figure 3 shows the morphology of the feedstock before(a) and after(b) been analyzed under TGA at 450°C. From figure 3(a), shows that the feedstock dispersion between the powder and binder is homogenous after been prepared with the set parameter and condition. The binder has completely removed and some binding particle melted forming physical bonding between the large powder particles as shown in figure 3(b). This phenomenon provide mechanical support to the brown part produce after thermal debinding which may reduce the brittleness of the part. It is confirmed that this temperature is the lower point temperature range for the debinding temperature.

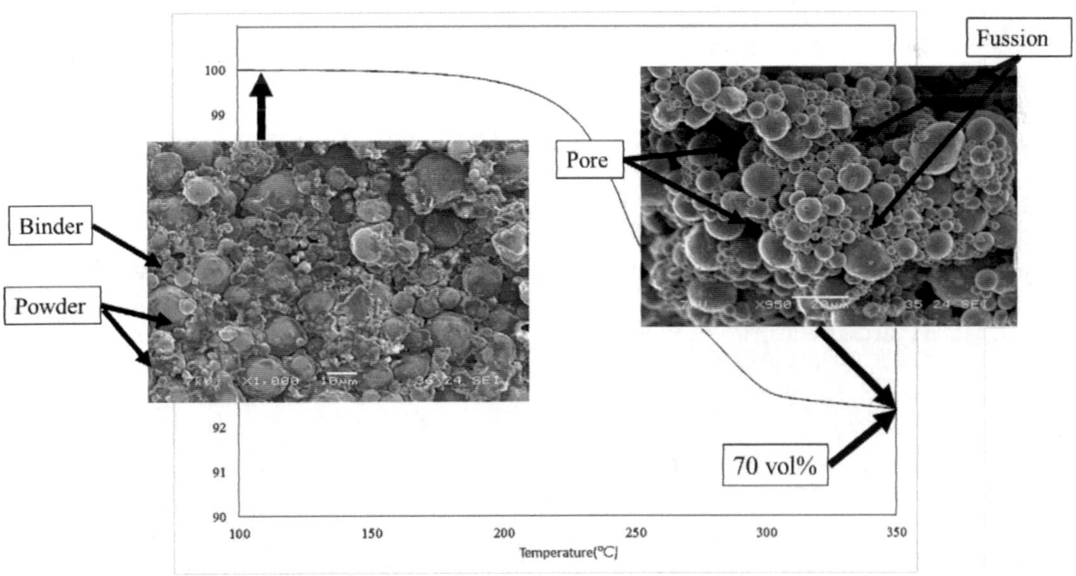

Figure 3 Morphology of MIM feedstock under SEM (a) dispersion of powder and binder are homogenous at 100 vol% (b)feedstock after TGA test done, binder has fully removed from the feedstock and small particle difused into the large particle increace the strength of the brown part.

4. Conclusion

The paper summarised the primary step in feedstock preparation of SS316L with critical powder loading of 75 vol% determined by mixing torque analysis and 70 vol% to 77 vol% by theoretical approach. The preparation process gives no effect to the composition of the feedstock. A maximum powder loading for MIM feedstock is important to reduce the shrinkage factor and defect caused by excessive amount of binder.

References

[1] German R M 1990 Powder Injection Moulding, Metal Powder IndustriesFederation *Princeton New Jersey* 161-170.

[2] Vyal' E Y & Laptev A M 2005 Use of Certain Binders to Increase The Strength of Green Compacts *Powder Metallurgy and Metal Ceramics 44* **11** 614-618.

[3] Xie Z P, Yang J L, & Huang Y 1997 The Effect of Silane Contents on Fluidity and Green Strength for Ceramic Injection Moulding. *Journal of materials science letters 16* 15 1286-1287.

[4] Fan Y L & Hwang K S 2007 Properties of Metal Injection Molded Products Using Titanate Containing Binders. *Materials transactions*, *48* **3**, 544-549

[5] Supati R, Loh N H, Khor K A & Tor S B 2000. Mixing and Characterization of Feedstock for Powder Injection Molding. *Materials Letters*, *46* **2**, 109-114.

[6] Heaney D F, Zauner R, Binet C, Cowan K &Piemme J 2004 Variability of Powder Characteristics and Their Effect on Dimensional Variability of Powder Injection Moulded Components. *Powder metallurgy*, *47* **2**, 144-149.

[7] Dihoru L V, Smith L N and German R M 2000 Experimental Analysis and Neural Network Modelling of The Rheological Behavior of Powder Injection MouldingFeedstocks Formed with Bimodal Powder Mixtures', *Powder Metall.*, *43*, **1**, 31–36.

[8] Contreras J M, Jimenez-Morales A, &Torralba J M 2010 Experimental and Theoretical Methods for Optimal Solids Loading Calculation in MIM Feedstocks Fabricated from Powders With Different Particle Characteristics. *Powder metallurgy*, *53* **1**, 34-40.

[9]Hausnerova B, Mukund B N &Sanetrnik D 2017 Rheological Properties of Gas and Water Atomized 17-4PH Stainless Steel MIM Feedstocks: Effect of Powder Shape and Size. *Powder Technology 312* 152-158.

[10] Kong X, Barriere T, &Gelin, J C 2012 Determination Of Critical and Optimal Powder Loadings for 316L Fine Stainless Steel Feedstocks for Micro-Powder Injection Molding. *Journal of Materials Processing Technology*, *212* **11**, 2173-2182.

[11]Warren J and German R M 'The Effect of Powder Characteristics on Binder Incorporation for Injection Molding Feedstock', in 'Compendium on metal injection molding II'; 1989, Princeton, NJ, Metal Powder Industries Federation.

IOP Conf. Series: Materials Science and Engineering **290** (2017) 012077 doi:10.1088/1757-899X/290/1/012077

Mechanical properties, morphology, and hydrolytic degradation behavior of polylactic acid / natural rubber blends

Y F Buys[1], A N A Aznan[1], H Anuar[1]

[1]Department of Manufacturing and Materials Engineering, International Islamic University Malaysia (IIUM), Jalan Gombak, 53100 Kuala Lumpur, Malaysia

Email: yose@iium.edu.my

Abstract. Due to its biodegradability and renewability, polylactic acid (PLA) has been receiving enormous attention as a potential candidate to replace petroleum based polymers. However, PLA has limitation due to its inherent brittleness. In order to overcome this limitation, blending PLA with elastomeric materials such as natural rubber (NR) are commonly reported. In previous, several researches on PLA/NR blend had been reported, with most of them evaluated the mechanical properties. On the other hand, study of degradation behavior is significance of importance, as controlling materials degradation is required in some applications. This research studied the effect of blend composition on mechanical properties, morphology development, and hydrolytic degradation behavior of PLA/NR blends. Various compositions of PLA/NR blends were prepared by melt blending technique. Tensile test and impact test of the blends were performed to evaluate the mechanical properties. Addition of NR improved the elongation at break and impact strength of the blends, but reduced the tensile strength and stiffness of the specimens. Dynamic Mechanical Analysis (DMA) measurements of the blends displayed two peaks at temperature -70°C which corresponded to T_g of NR and 65°C which corresponded to T_g of PLA. Field Emission Scanning Electron Microscopy (FE-SEM) micrograph of 70/30 PLA/NR specimen also showed two distinct phases, which lead to indication that PLA/NR blends are immiscible. Hydrolytic degradation behavior was evaluated by measuring the remaining weight of the samples immersed in sodium hydroxide solution for a predetermined times. It was shown that the degradation behavior of PLA/NR blends is affected by composition of the blends, with 100 PLA and 70/30 PLA/NR blend showed the fastest degradation rate and 100 NR displayed the slowest one.

1. Introduction

Nowadays, most of common type of plastic shopping bag uses polyethylene. It is made from petroleum-derived polymer and technically non-biodegradable products. Non-biodegradable materials do not break down easily. It took more than hundreds year to degrade and certain materials might not be destroyed over time. The harm caused by products that non-biodegradable will continue to pile up over time, demanding more and more land devoted to holding waste with increasing of demand on plastics production. Besides, it brings harmful effect toward the environment such as global warming and pollution. Therefore, it is very important to have a biodegradable plastic to replace the petroleum based plastics which caused the environmental issues.

Among the many types of disposable plastic packaging, Polylactic Acid (PLA) has been intensively attracting attention as it is one of the biodegradable thermoplastic polyester. It can be used to reduce the environmental impact on plastic production and waste since it is biodegradable. PLA is an ideal material for the food packaging and for other consumer products since it is biodegradable thermoplastic polyester which gives excellent physical and mechanical properties [1]. PLA is strong but it is very brittle [2]. In order to overcome these problems, blending PLA with flexible materials has been widely used.

Natural rubber (NR) is an elastic material and it has been considered as a good candidate to the blended with PLA to improve the brittleness of PLA. The addition of 10 % of NR into PLA improves the ductility and its crystallization [3]. Several studies had been reported on PLA/NR blend specifically on physical and mechanical properties [3]. However, there are no studies that related to degradation behavior of PLA/NR blend.

It is very significance to aware with rate of degradation since difference application required dissimilarity rate of degradation. Degradation behavior of polymer blend is affected by composition miscibility and morphology of the blends. Therefore, this research attempts to address this issue.

2. Experimental procedures

2.1 Materials

In this experiment, PLA 3051 D from NatureWorks LLC was used. The density of the PLA is 1.24 g/cm^3 and its melt flow index is 10-25 g/10 min. Meanwhile the NR used was the Standard Malaysian Rubber (SMR-L) supplied by Rubber Research Institute of Malaysia (RRIM), with 0.93 g/cm^3 density.

2.2 Sample Preparation

Prior of being used, PLA was dried in an oven at 80°C for one hour to remove the moisture. Received NR was extruded through a two-roll mill (Kodaira) at room temperature for three min to obtain the standard thickness for mixing procedure. PLA was melt blended with NR in various blend composition (vol %) by using internal mixer (Brabender Plastograph) with 50 rpm rotating speed at temperature 190°C for 10 min. The mixtures were then crushed by a crusher machine before compression moulded by using XH-406B Tablet Press at temperature 190°C under 15 MPa for 5 min.

2.3 Tensile Test

In order to determine the tensile strength, Young's modulus and elongation at break of the samples, tensile tests were conducted under ASTM D638 standard. The tests were performed by a Universal Tensile Machine LR 10K (Lloyd) with a 5 mm/min strain rate at room temperature. The values of tensile strength, Young's modulus, and elongation at break were obtained by Nexygen software support

2.4 Impact Test

Impact tests were performed to determine the amount of energy absorbed by the material during fracture. The tests were carried out according to ASTM D256 standard by using Advance Pendulum Impact (Dynisco Polymer), with 7.5 J hammer. Samples dimension was 60 mm x 13 mm x 3 mm. Prior to the tests, a V-notch was used to notch the rectangular specimens.

2.5 Dynamic Mechanical Analysis (DMA) Measurement

DMA measurements were performed to study viscoelastic properties as well as determine the miscibility of PLA/NR blends. The measurements were carried out by a Pyris Diamond DMA (Perkin Elmer Instruments) with single cantilever using 3 point bending mode at 1 Hz frequency. The dimension of the samples were 40 mm x 13 mm x 3 mm. Liquid nitrogen were used to cool the samples and nitrogen gas were purged during the measurement to avoid oxidation. The DMA tests were performed from -100°C to 100°C, with 5°C/min heating rate.

2.6 Scanning Electron Microscopy (SEM) Observation

In order to study the morphology of PLA/NR blends, SEM observation were performed by using a Field Emission SEM (FE-SEM) Jeol JSM-670 at room temperature. Observations were conducted on liquid nitrogen fractured surface of the specimens. Before inserted to the observation chamber, the samples were coated with Au in order to impart conductivity.

2.7 Hydrolytic Degradation Test

Alkaline hydrolytic degradation tests were performed in order to determine the degradation behavior of PLA/NR blends. Specimens sized 10 mm x 10 mm x 1 mm were placed in 25 ml of NaOH aqueous solution with pH 13 for a predetermined time at room temperature. After hydrolysis, the samples were washed in distilled water and dried in open air at room temperature for a few hours. Subsequently, the samples were dried in oven with temperature 80°C for 24 hours to remove remaining moisture and then the remaining weight of the samples were recorded.

3. Results and Discussion

3.1 Mechanical Properties

Figure 1 shows the Young's modulus, tensile strength, as well as elongation at break of PLA/NR blends obtained from tensile tests. The error bars represents the standard deviation. From figure 1 (a) and (b), it can be observed that as NR content increased, the Young's modulus and tensile strength of the blends decreased. Meanwhile, figure 1 (c) indicates that the elongation at break of the blends experienced slight increase when the NR content increase. These trends are understandable as the elastic and flexible nature of the NR reduce the stiffness of PLA. Similar trends had been also reported elsewhere [3-7]. However, only slight increase in the elongation at break also indicates that in current system, the interface between PLA and NR has not fully optimized. It is believed that existence of suitable compatibilizer may optimize the increase in elongation at break [7-9].

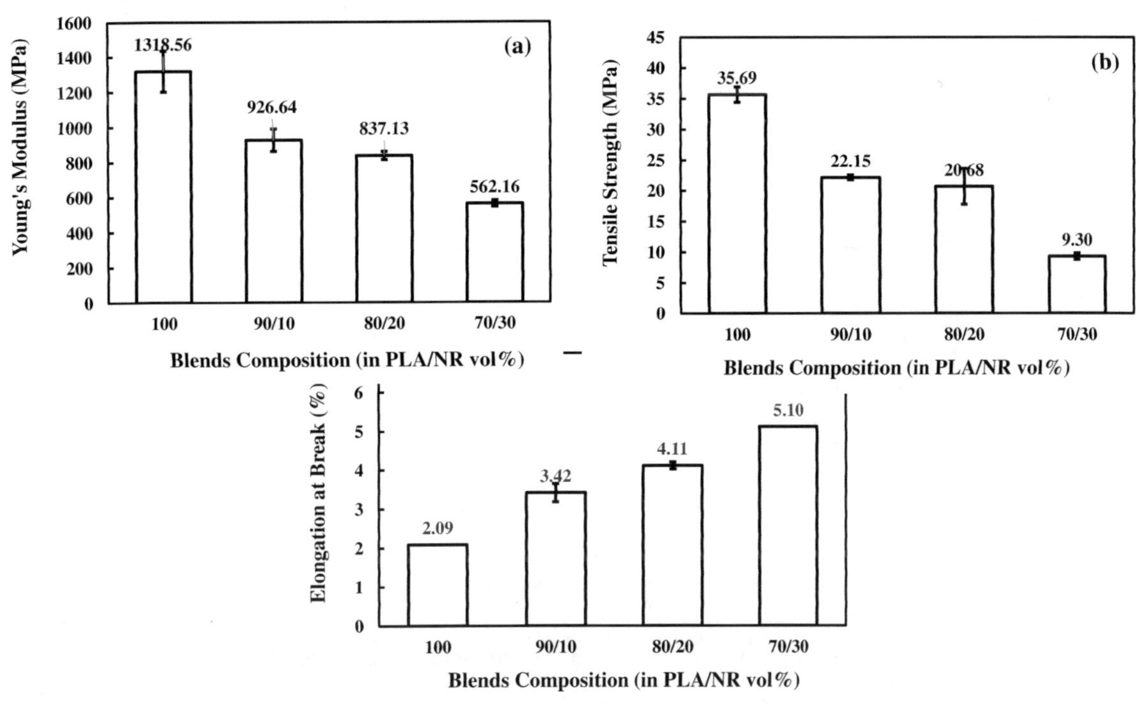

Figure 1. (a) Young's modulus, (b) tensile strength, and (c) elongation at break of PLA/NR blends.

The impact strength data of PLA/NR blends is shown in figure 2. The error bars represents the standard deviation. As NR content in the blends increased, increment trend in the impact strength was observed. The value of impact strength had dramatically increased from 4.14 kJ/m² of the neat PLA to 17.97 kJ/m² of 50/50 PLA/NR. From this figure, it is clear that the low toughness of PLA had been improved by NR blending. It is also interesting to note that 10% addition of NR showed higher value compared to 95/5 or 85/15 blends, which had also been highlighted by other research in previous [9]. Nevertheless, in general, it can be deduced that addition of NR into PLA results in increase in the impact strength, as also reported elsewhere [10].

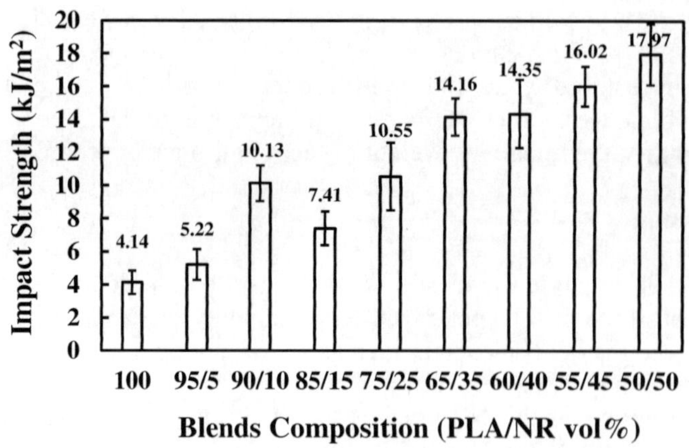

Figure 2. Impact strength of PLA/NR blends.

3.2 Miscibility and Morphology

Figure 3 illustrates the tangent delta versus temperature of PLA, NR and PLA/NR blends obtained from DMA measurement. The peak represented glass transition temperature (T_g). From figure 3, it is observed that pure PLA has a T_g around 65°C, and pure NR shows its T_g around -70°C. Meanwhile for composition of 65/45 PLA/NR, 60/40 PLA/NR and 55/45 PLA/NR, two tangent delta peaks were observed at every curve, i.e. at temperature of -70°C and 65°C. Therefore, it can be concluded that PLA/NR blends are immiscible in conjunction with the presence of two peaks.

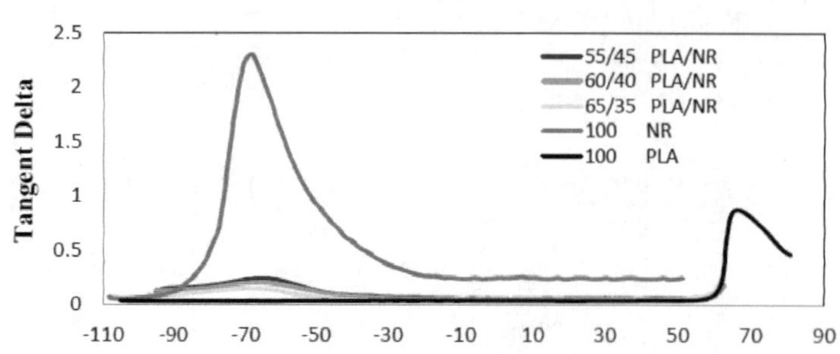

Figure 3. Tangent delta of PLA, NR and PLA/NR blends

Field Emission Scanning Electron Microscope (FE-SEM) was used to examine the fracture surface morphology between PLA/NR blends. Figure 4 represents the composition of 70/30 PLA/NR blends with magnification of 2000 times. It was observed that the fracture surface of the blends shows the sea-island structure in composition of 70/30 PLA/NR. Since the composition of NR is 30%, it is considered that NR is the minority phase (island phase) meanwhile PLA is majority (sea phase) as its composition is 70%. Since two distinct phases are observed, this figure also re-emphasize the deduction that PLA/NR is immiscible.

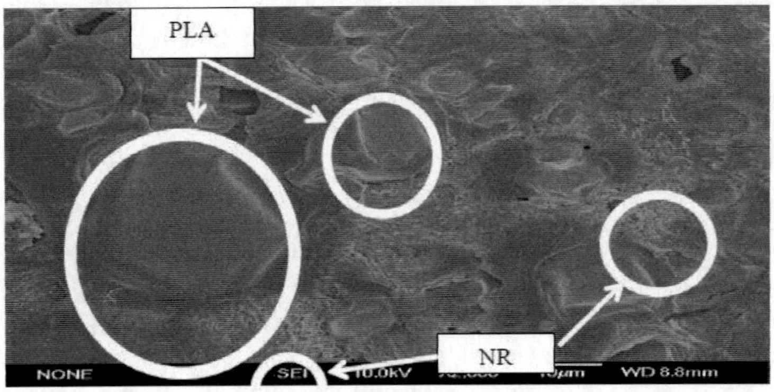

Figure 4. SEM images of 70/30 PLA/NR blend

3.3 Degradation Behavior

Figure 5 indicated the weight loss of the samples with the time consuming (days) for the samples to degrade. It took 35 days for PLA to completely degrade. Composition of 70/30 PLA/NR blends similar degradation behavior with that pure PLA. However, figure 5 of pure NR displayed that there was slight changed in the result. It showed that the rate of degradation at NR was very slow compared to the pure PLA and PLA/NR blends. The detail of physical change of PLA, NR and PLA/NR blends can be seen in figure 6.

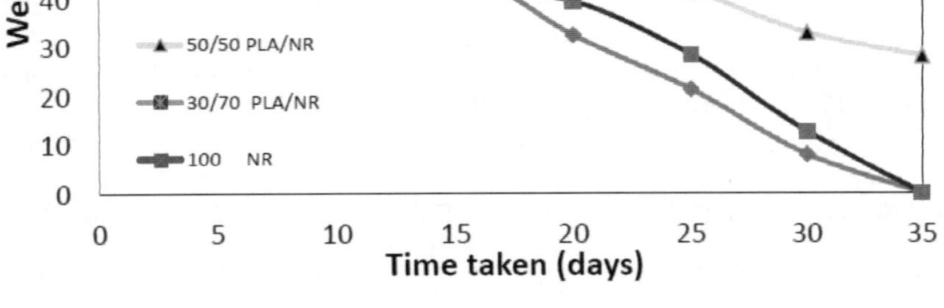

IOP Conf. Series: Materials Science and Engineering **290** (2017) 012077 doi:10.1088/1757-899X/290/1/012077

Figure 5. Weight loss of PLA,NR and PLA/NR blends

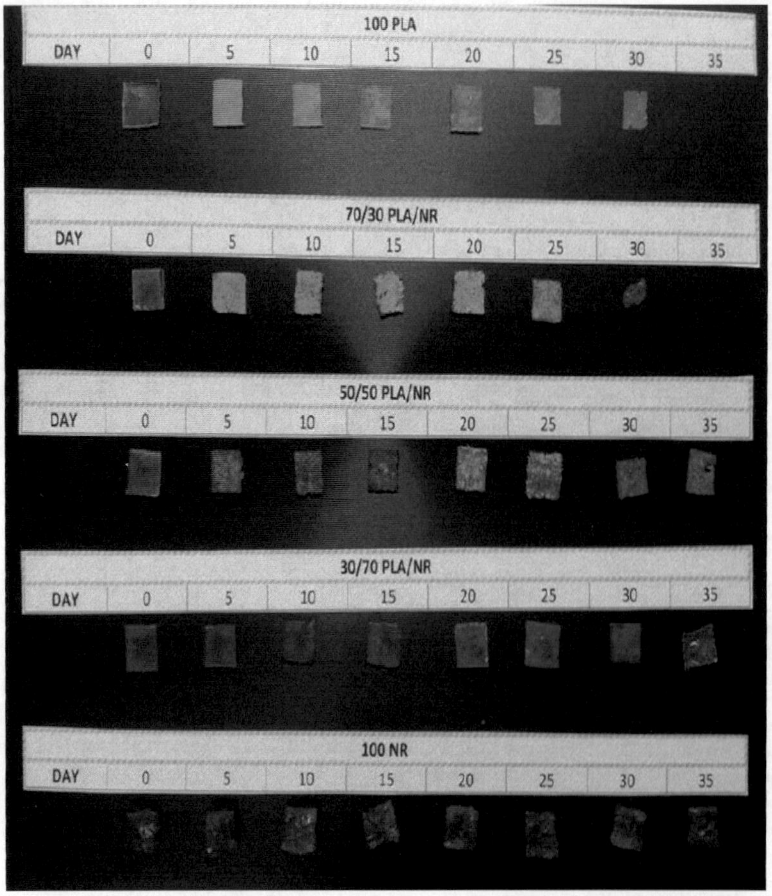

Figure 6. Physical change of PLA, NR and PLA/NR blends

The rate of degradation was increased when the amount of PLA increased in the compositions. When the amount of NR increased in the composition, the rate of degradation became slower.

4. Conclusion

Polylactic Acid (PLA) is a biodegradable matrix which is blend with the Natural Rubber (NR) in order to overcome the brittleness of the PLA. Through the tensile test, it proved that PLA is brittle. By addition of NR in PLA, it can reduce the brittleness of PLA. The result clearly can be seen in elongation of break in tensile test and impact test as it increased along the addition of NR. Meanwhile for DMA, there were two peaks visible in the tangent delta graph which represented the PLA and NR in the blend. From the storage modulus graph in DMA also provided information of Tg which constituent revealing the immiscibility of blends. The immiscibility of the blends also can be proved through the FE-SEM morphology. The imaged of blends produced the sea island structure at composition 70/30 PLA/NR blends and co-continuous structure at 50/50 PLA/NR. Degradation test result showed that PLA faster in degradation rate compared to NR. The addition of NR caused degradation rate decreased. Pure PLA and 70/30 PLA/NR were completely degraded after 35 days. Therefore, it can be concluded that the miscibility and morphology of blends strongly affected the degradation rate.

Acknowledgement
This work was supported by Research Initiative Grant Scheme RIGSl6-085-0249, International Islamic University Malaysia

References

[1] Conn R, Kolstad J, Borzelleca Dixler J F, File R Jr, La Du J R and Pariza M W, 1995. Safety assessment of polylactide (PLA) for use as a food-contact polymer. *Food and Chemical Toxicology,* **33**, 273-283.

[2] Gupta B, Revagade N, Hilbornb J, 2007. Poly(lactic acid) fiber: an overview. *Progress in Polymer Science*, **32**, 455-482.

[3] Bitinis N, Verdejo R, Cassagnau P and Lopez-Manchado M A, 2011. Structure and properties of polylactide/natural rubber blends. *Materials Chemistry and Physics*, **129 (3)**, 823-83.

[4] Jaratrotkamjorn R., Khaokong C and Tanrattanakul V, 2012. Toughness enhancement of poly (lactic acid) by melt blending with natural rubber. *Journal of Applied Polymer Science*, **124 (6)**, 5027-5036.

[5] Pongtanayuta K, Thongpina C, and Santawitee O, 2013. The effect of rubber on morphology, thermal properties and mechanical properties of PLA/NR and PLA/ENR blends. *Energy Procedia*, **34**, 888 – 897.

[6] Mohammad N N B, Arshad A, Rahmat A R and Abdullah Sani N S, 2015. Effects of maleated natural rubber on mechanical properties of polylactic acid/natural rubber blends. *Material Science Forum*, **819**, 284-289.

[7] Mohammad N N B, Arshad A, Rahmat A R, Abdullah Sani N S and Ali Mohsin M E, 2016. Influence of compatibilizer on the structure properties of polylactic acid/natural rubber blends. *Polymer Science, Series A*, **58 (2)**, 177–185.

[8] Chumeka W, Tanrattanakul V, Pilard J F, and Pasetto P, 2013. Effect of poly(vinyl acetate) on mechanical poperties and characteristics of poly(lactic acid)/natural rubber blends. *Journal of Polymers and the Environment*, **21**, 450-460.

[9] Juntuek P, Ruksakulpiwat C, Chumsamrong P and Ruksakulpiwat Y, 2012. Effect of glycidyl methacrylate - grafted natural rubber on physical properties of polylactic acid and natural rubber blends. *Journal of Applied Polymer Science*, **125 (1)**, 745-754.

[10] Zhang C, Man C, Pan Y, Wang W, Jiang L, and Dan Y, 2011. Toughening of polylactide with natural rubber grafted with poly(butyl acrylate). *Polymer International*, **60**, 1548–1555.

IOP Conf. Series: Materials Science and Engineering **290** (2017) 012078 doi:10.1088/1757-899X/290/1/012078

Hardness and adhesion performances of nanocoating on carbon steel

J N Hasnidawani[1], H N Azlina[1], H Norita[1] and N N Bonnia[2]

[1]Department of Manufacturing and Materials Engineering, International Islamic University Malaysia (IIUM), Jalan Gombak, 53100 Kuala Lumpur, Malaysia
[2]Material Technology Programme, Faculty of Applied Sciences, Universiti Teknologi Mara (UiTM), Shah Alam 40450 Selangor, Malaysia

Email:noorazlina_hassan@iium.edu.my

Abstract. Nanocoatings industry has been aggressive in searching for cost-effective alternatives and environmental friendly approaches to manufacture products. Nanocoatings represent an engineering solution to prevent corrosion of the structural parts of ships, insulation and pipelines industries. The adhesion and hardness properties of coating affect material properties. This paper reviews ZnO-SiO_2 as nanopowder in nano coating formulation as the agent for new and improved coating performances. Carbon steel on type S50C used as common substrate in nanocoating industry. 3wt% ZnO and 2wt% SiO_2 addition of nanoparticles into nanocoating showed the best formulation since hardness and adhesion of nanocoating was good on carbon steel substrate. Incorporation of nanoparticles into coating increased the performances of coating.

1. Introduction

Nanocoating offer great potential for various applications due to their superior characteristics that are not typically found in conventional coatings. Nanocoating gains the most practical application in industry that is produced by usage of some components at nanoscale to obtain desired properties. Through nanocoatings, a wide variety of performance characteristics can be achieve which includes self-cleaning, self-healing, anti-microbial, UV protection, corrosion resistance and waterproofing. Nanoparticles also reduced contact tension, moisture penetration and surface roughness. On the other hand, nanoparticles increased surface area, enhanced physical, chemical, mechanical, optical and magnetic properties. Nanoparticles also absorb more resins, reduced free surface. So, the transport path of corrosive species increased and enhanced the properties and performance of nanocoating. In conventional nanocoating, when there was a drop of water on the surface of nanocoating, the contact tension was normal and the moisture can penetrate through the surface surface rougnhess much more compared to nanocoating [1,2-4].

Nanocoating refers to covering nanoscale entity to produce structured materials and nanocomposite or it is related to covering materials with nano meter scale layer of 10 to 100 nm thickness. Adhesion test methods are used to establish whether the adhesion of a coating to a substrate is at a generally adequate level. The differences in adhesion ability of the coating surface can affect the results obtained with coatings having the same inherent adhesion. In these test, a lattice pattern with either six or eleven cuts in each direction is made in the film to the substrate, pressure-sensitive tape is applied over the lattice and then removed, and adhesion is evaluated by comparison with descriptions and illustrations. If a coating is to fulfill its function of protecting or decorating a substrate, it must adhere to it. The crosscut test is a simple and easily practicable method for evaluating the adhesion of single- or multi-coat systems. For hardness study, nanoindentor was used to study the mechanical properties of nanocoatings on medium carbon steel substrate grade S50C. The type of indentor tip used was diamond Berkovich. The thickness of the nanocoating was about 0.01mm-0.03mm or 10000nm-13000nm.

2. Experimental procedures

These test methods cover procedures for assessing the adhesion of coating films to metallic substrates by applying and removing pressure-sensitive tape over cuts made in the film according to ASTM D3359. Test Method is suitable for use in the laboratory. Sharp razor blade, scalpel, knife or other cutting device having a cutting edge angle between 15 and 30° that will make either a single cut or several cuts at once. The space of the cuts 2 mm apart and make six cuts. The tape placed at the center of the grid area. To ensure good contact with the film, the tape was firmly rub with the eraser on the end of a pencil. The color under the tape is a useful indication of when good contact has been made. Within 30s of application, the tape was removed rapidly. The parameter of the nanoindentation test was tabulated in table 1. More than 10 indentation tests were conducted on each sample for accuracy.

Table 1. Parameter of Nanoindentor test

Loading rate (mN/s)	0.5
Load applied (mN)	6.0
Depth (nm)	10000

3. Results and Discussion

For the addition of nanoparticles into epoxy coating within 1-3wt% as figure 1 (a,b,c and f,g,h) small flakes of the coating are detached along edges and at intersections of cuts. The area affected is 5 to 15 % of the lattice. However, when the addition of nanoparticles too much (4-5wt%) in figure 1 (d,e and i,j), the coating has flaked along the edges and on parts of the squares. The area affected is 15 to 35 % of the lattice. In order to make sure that the coating was satisfied, coatings must adhere to the substrates. Based on figure 1 (k,l,m and n), hybrid between ZnO and SiO_2 was good compared to single addition of nanoparicles into epoxy coating. 3wt% of hybrid showed the best result. The edge of the cuts was completely smooth and none of the squares of the lattice is detached. Only small flakes of the coating are detached at intersections. Less than 5% of the area is affected. The coating has flaked along the edges of cuts in large ribbons and whole squares have detached in epoxy only picture without any addition of nanoparticles as figure 1 (o). The area affected is 35 to 65 % of the lattice with flaking and detachment worse than coating without addition of nanoparticles.

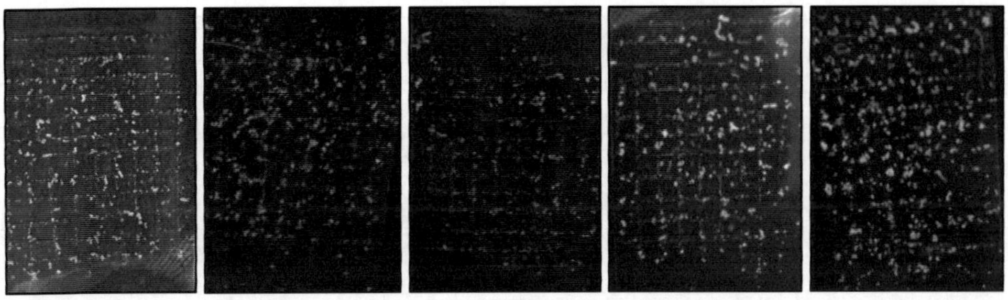

(a) (b) (c) (d) (e)

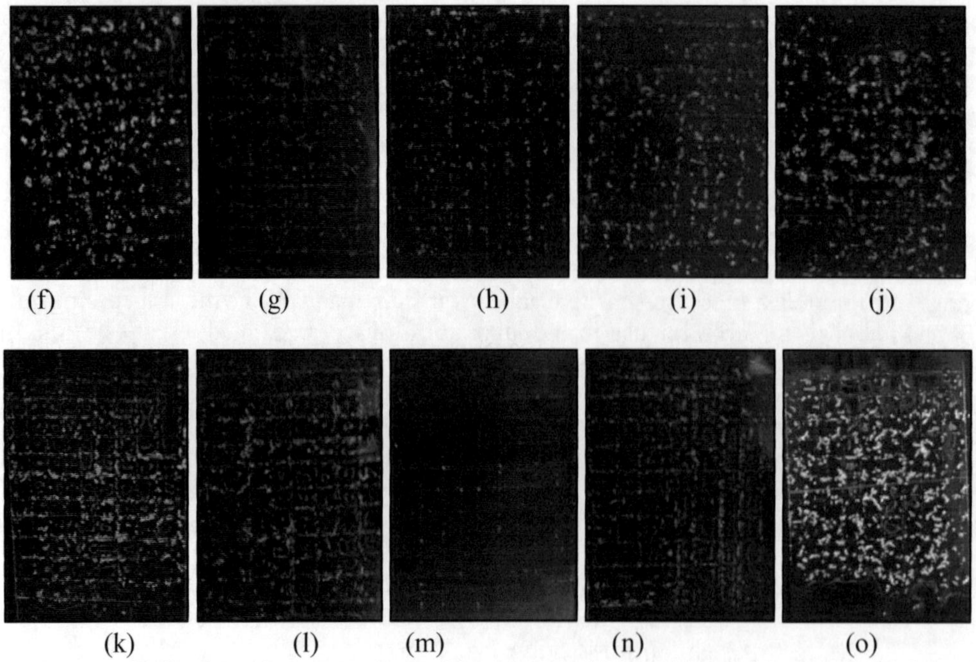

Figure1. Images of adhesion of nanocoatings. a)1wt% ZnO; b)2wt% ZnO c)3wt% ZnO; d)4wt% ZnO; e)5wt% ZnO; f)1wt% SiO_2; g)2wt% SiO_2; h)3wt% SiO_2; i)4wt% SiO_2; j)5wt% SiO_2; k)1wt% hybrid; l)2wt% hybrid; m)3wt% hybrid; n)4wt% hybrid; o)epoxy only.

Table 2. Result of hardness for different samples by Nanoindention test

Samples	Weight Percentage (%)	Hardness (GPa)
0	epoxy only	0.36461
1	$1ZnO4SiO_2$	0.39284
2	$2ZnO3SiO_2$	0.39194
3	$3ZnO2SiO_2$	0.39438
4	$4ZnO1SiO_2$	0.40767
5	ZnO only	0.39060
6	SiO_2 only	0.39006
7	Steel	7.2

Table 2 showed the result of hardness for different samples by nanoindention test. The average hardness of the samples of different loading of nanoparticles is respectively 0.39284 GPa, 0.39194 GPa, 0.39194 GPa and 0.47067 GPa, which are all giving better hardness than sample without nanoparticles. By comparing sample 0 with 1-6 that containing nanoparticles, it can be found that nanoparticles has higher hardness. The reason is that, the nanoparticles was distributed uniformly into matrix (epoxy) and having good bonding process. Samples without nanoparticles showed lower hardness because higher porosity as shown in figure 2. Higher porosity will result in lower hardness. All these results are agreeing with the conclusions in Ref [5] and [6]. Small size of nanoparticles makes it beneficial to increase the performances of nanocoating [7]. Increase the amount of nanoparticles increasing the hardness. However too much of nanoparticles also decrease the hardness because aggregation [8]. The hardness of steel type S50C showed in figure 3 below.

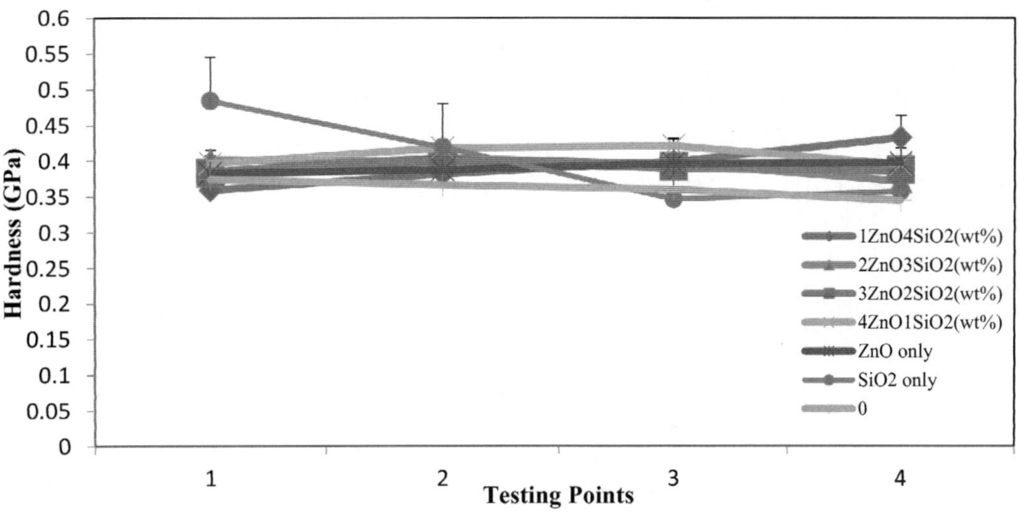

Figure 2. Hardness of nanocoatings by Nanoindentation test

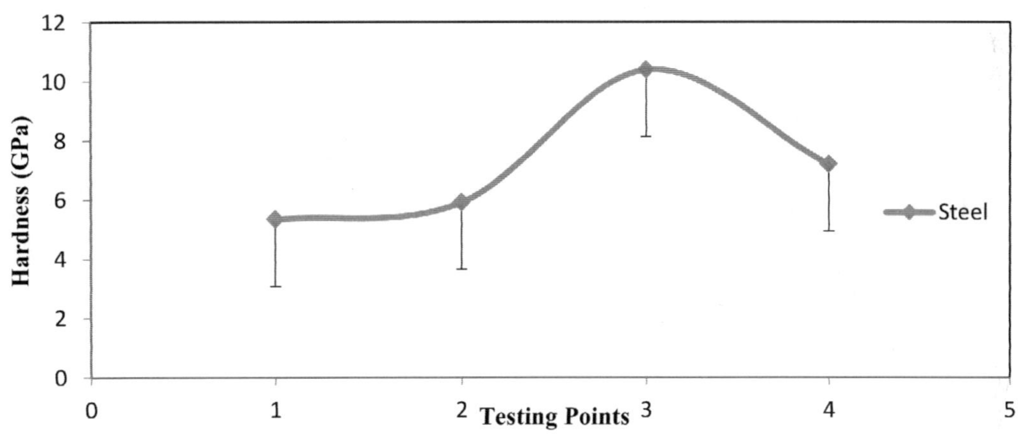

Figure 3. Hardness of Steel by Nanoindentation test

4. Conclusion

In summary, it can be demonstrate that 3wt% ZnO and 2wt% SiO_2 addition of nanoparticles into nanocoating was the best formulation since hardness and adhesion of nanocoating was good on carbon steel substrate. Incorporation of nanoparticles into coating increases the performances of coating.

Acknowledgement

The authors wish to thank the Ministry of Higher Education, Malaysia, for the grant FRGS13-077-125-0318, which made this research work possible and IIUM for their support of this research.

References

[1] Kumar S S, Venkateswarlu P, Rao V R, and Rao G N 2013 *International Nano Lett.* **3** 28 30

[2] Trueba Monica, Trassati Stefano P and Flamini Daniel O 2010 *Advance Material Research.* **138** 63-78

[3] Li-Ping Liu, Xue-Feng Yan, Zhi-Ming Zhang & Liang-Min Yu 2011 *Advance Material Research.* **337** 37-40

[4] Podrabinnik Pavel Anatolyevich and Shishkovsky Igor Vladimirovich 2015 *Materials Science Forum.* **834** 113-118

[5] Oliver W C and Pharra G M 2004 *J. Mater.* **19** 3-20

[6] Chung Cheng I and Andrea Hodge M 2012 *Advanced engineering materials.* **14** 219-226
[7] Hodge A M, Biener J, Hayes J R, Bythrow P M, Volkert C A, Hamza A V 2007 *Acta Materialia.* **55** 1343-1349
[8] Masataka Hakamada and Mamoru Mabuchi 2007 *Scripta Materialia.* **56** 1003-1006

IOP Conf. Series: Materials Science and Engineering **290** (2017) 012079 doi:10.1088/1757-899X/290/1/012079

Protective Behavior of Poly(m-aminophenol) and Polypyrrole Coatings on Mild Steel

Sabrina M Yahaya[1], M K Harun[1], R. Rosmamuhamadani[2], N N Bonnia[2], S Ratim[2]

[1]Department of Applied Chemistry, Faculty of Applied Sciences, UniversitiTeknologi MARA, 40450 Shah Alam, Selangor, Malaysia
[2]Department of Material Technology, Faculty of Applied Sciences, UniversitiTeknologi MARA, 40450 Shah Alam, Selangor, Malaysia

Email:sabrina@salam.uitm.edu.my

Abstract. Electrodeposition of polypyrrole (PPy) and poly (m-aminophenol) (PMAP) films on mild steel (MS) substrate was achieved in 0.3M oxalic acid solution and 0.3M NaOH, water:ethanol (70:30) solvent respectively using cyclic voltammetry technique. The morphology of the films constructed was determined by scanning electron microscope (SEM) while energy dispersive X-Ray analyzer (EDX) was used to establish the presence of organic PMAP and PPy film coating and its compositions. The corrosion performance of MS coated with both polymer films were investigated after 0.5 hours immersed in 0.5M NaCl aqueous solution by using polarization curves. It was found that PPy coating provides anodic protection while PMAP coating provides cathodic protection towards corrosion protection of mild steel substrate.

1. Introduction

Electropolymerization of thin polymeric film from various aromatic compounds have shown great potentials and applications in areas such as electrical, electronics, thermoelectric, electrochemical, electromagnetic, electromechanical, electro-luminescence, electro-rheological, chemical, membrane, sensors and corrosion control [1]. Over the past decade, there have been innumerable published reports on the application of polymer films based on phenol derivatives, intrinsically conducting polymers (ICPs) such as polyaniline (PANI), polypyrrole (PPy), poly(N-ethyl aniline), polythiophene, polyindole on corrosion protection of metals [2]−[7]. However, PPy remain dominant as corrosion control of metals. PPy has some advantageous with its ease of synthesis electrochemically on metal surface in a single step, this eliminates many time consuming and expensive processes [8]. Nevertheless, PPy also has high thermal stability (150°C in the air) and low monomer oxidation potential and that PPy film keeps its high conductivity in a wide pH range and the mechanical properties of this coating are very good. In contrast, aminophenols possess two functional groups namely (–NH_2 and –OH) which can be oxidized. Technically, they can be conductive, thus resemble those of anilines and phenols [9]. Phenol and its derivatives such as 2-allylphenol, 3-methylphenol, 2, 6- dimethylphenol, m-aminophenol and others were mainly used as thin layers of insulating polymers, intended for the protection of metals against corrosion. According to Hur et al. [9], PMAP protect the mild steel substrate by shifting the corrosion potential of mild steel substrate to more cathodic values. This indicates that PMAP films are protecting mild steel substrate by hindering the cathodic corrosion reaction. Thus, cathodic protection effect of PMAP arises from the barrier effect for diffusion of oxygen. Therefore, it can be understood that electrodeposited PMAP films are acting as barrier for oxygen diffusion, which in turn reduces the rate of cathodic reaction.

In the present work, PMAP and PPy coatings were electrodeposited on mild steel surface by using cyclic voltammetry technique and characterized by SEM EDX analyses. An attempt has been made to understand the behavior of corrosion protection provided by PMAP and PPy coatings on mild steel substrates.

2. Experimental procedures

The electrosynthesis of PPy film on bare mild steel surface was not practical as the dissolution of the active metal led to oxidation of monomer at the surface at the reaction potential [10]. Therefore, the electrosynthesis of polypyrrole (PPy) polymer was performed by using cyclic voltammetry technique in two steps. The method is based on the work of Tuken, in 2004 [7] and Tuken et al. in 2006 [11]. In first step, the mild steel surface was electrochemically passivated in 0.3 M oxalic acid solution containing 0.1 M pyrrole monomers. The potential range was –0.5 V to +0.3 V (Ag/ AgCl) with scan rate used was 4 mV/s. Low scan rate was used because it was considered that the slow nucleation of ferrous oxalate crystals could provide a better passivation for mild steel electrode [3] [4]. All mild steel samples were electrochemically passivated first before electropolymerization step. The second step is electrosynthesized of the polymer films. The synthesis was performed in a sequence of 10 cycles with a higher scan rate which was 50 mV/s. The electrosynthesis process of PPy film was started after passivation step without changing the polymerization solution in same potential range from -0.5 to 1.5V. To examine the adherence of synthesized coating on mild steel electrode, the standard test methods by tape test was applied according to ASTM D3359 method and very satisfactory results were obtained for all synthesized coating. Electrosynthesis of m-aminophenol was done with the mild steel electrode immersed in 0.3M NaOH, water:ethanol (70:30) solvent containing 0.1M m-aminophenol. A total number of ten cycles were conducted during the electrodeposition process between a potential range of -0.4V and +1.4V at a scan rate 20 mV/s. The method used was following Hue et al. [12] and Harun et al. [13]. After the electrodeposition, the coated samples were characterized by SEM EDX for surface morphology studies and polarization resistance for corrosion performance measurement. All experiments and measurements were performed at room temperature.

3. Results and Discussion

Cyclic voltammogram obtained at first, fifth and ten cycles of PMAP are shown in Figure 1. The results show a peak which was associated to the oxidation of m-aminophenol potential at about +1.0 V (Ag/AgCl) which is consistant with the value obtained by other workers in the same solvent [12]. The first cycle produced a peak current density of 47.89 mA/cm^2 followed by 35.71 mA/cm^2 and 29.98 mA/cm^2 on 5th and 10th cycle respectively. This indicated that the peak current density for the cyclic voltammetry of m-aminophenol decreases with increasing numbers of cycles. Thus as the number of cycles are increased the resistance of the insulative coating become higher. The results also indicated that the oxidation of m-aminophenol is a non-reversible reaction as no reduction peak was detected during the reverse cycle which were consistent with previous findings reported by other researchers [12]–[14]. The current suppresion phenomenon and a non reversible reaction is consistent with the formation of an insulative coating on the mild steel electrode. As the number of cycles increased electropolymerization process can also become more difficult because the active sites on the surface of the anode electrode decreased leading towards difficulty of electron transfer as indicated by the suppression of current on the cyclic voltammogram. A comparison scan of mild steel in the same solvent as a control (monomer free) is illustrates in Figure 2. The figure indicated no oxidation peak at the range of the potential. This evident the growth of insulating poly(m-aminophenol) on mild steel surface.

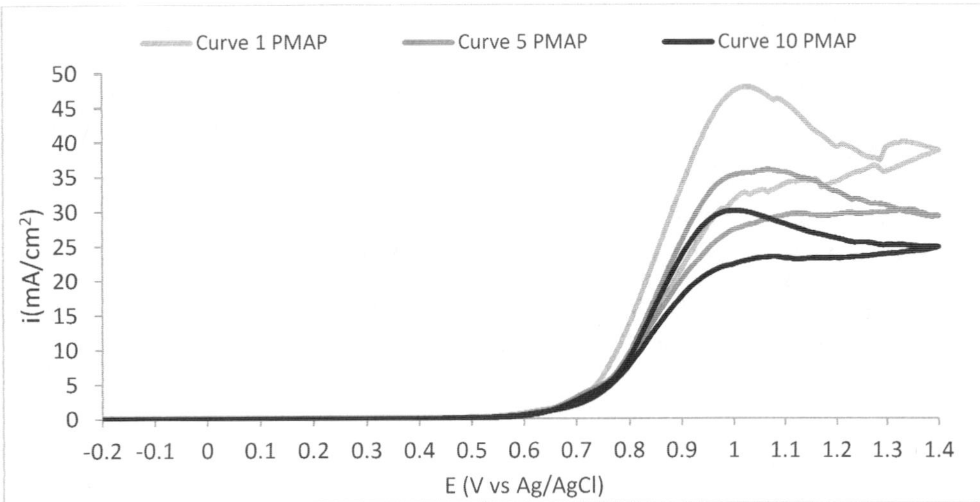

Figure 1: Cyclic voltammogram of PMAP on passivated MS in 0.3M NaOH + water:ethanol (70:30) containing 0.1M m-aminophenol solution. The scan rate was 20 mv/s.

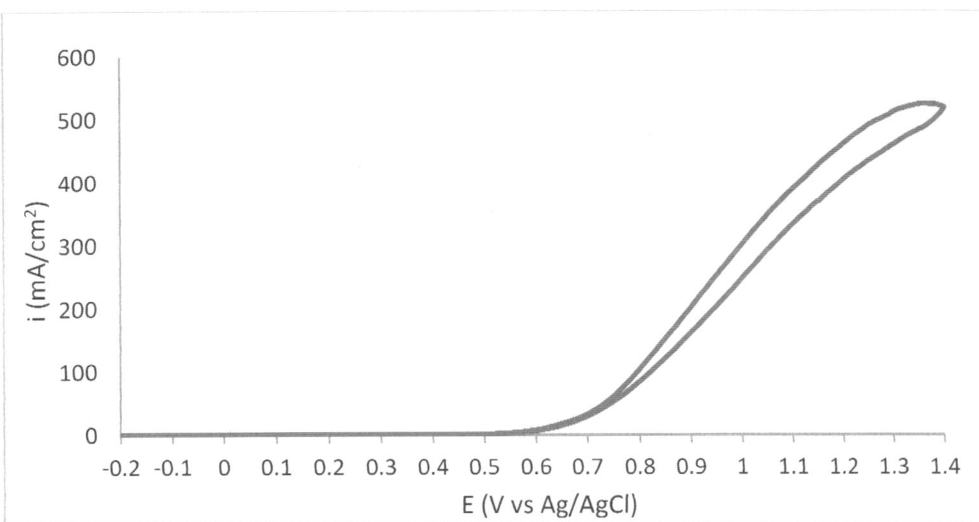

Figure 2: Cyclic voltammogram behavior of the PMAP film in monomer free solution. The scan rate was 20 mV/s.

Figure 3 shows the passivation of mild steel electrode in 0.3 M oxalic acid solution (without monomer). As can be seen in the curve, the active dissolution of mild steel started at around -0.5V followed by the appearance of passivity region at applied potential around -0.1V. The peak potential values (E_p) for characteristic oxidation–passivation process for mild steel was determined to be at −0.35V.

In 0.1M pyrrole monomer containing solution, the first, second, fifth and ten cycle voltammograms are illustrated in Figure 4. In first cycle, a well-defined anodic peak attributed to the oxidation and passivation of mild steel surface was observed as in monomer free solution. The current value then, remained almost constant at around zero up to +0.67 V. After that the current value started to increase

IOP Conf. Series: Materials Science and Engineering **290** (2017) 012079 doi:10.1088/1757-899X/290/1/012079

sharply due to the oxidation process of the monomer and a black colored polymer like film covered the surface rapidly, at the first cycle [11]. The re-passivation peak of mild steel was disappeared in reverse scan. This was evident that the freshly produced thin polypyrrole (PPy) film must have been compact and adherent enough that the mild steel surface was covered with polymer film [7], [11]. In second, fifth and ten cycles voltammograms the forward scan shows an increased current, which begins at an applied potential of around +0.65V, indicating the oxidation of pyrrole monomer. This oxidation process could not be observed as a separate peak in pyrrole containing solution and the current values corresponding to oxidation potential region did not decrease at the following cycles. A broad cathodic peak appeared at around −0.10V at the reverse scan, was believed to have been caused by de-doping of oxalate ion making the polymer to be in a reduced form [11]. The current values corresponding to polypyrrole polymer in a reduced form increased regularly with increasing numbers of cycle. This was attributable to growth of pyrrole polymer films.

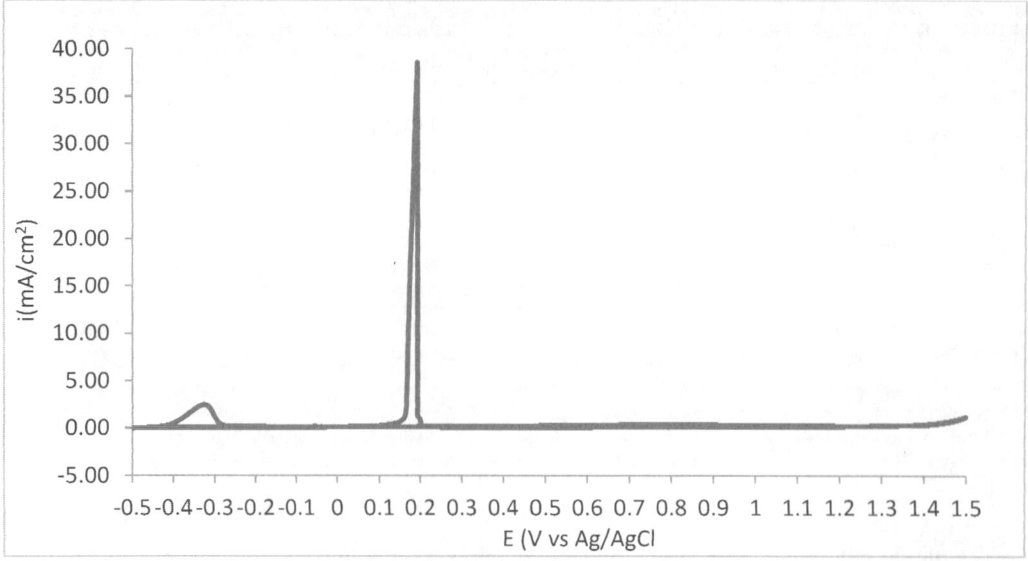

Figure 3: Cyclic voltammograms of mild steel in 0.3M oxalic acid solutions, scan rate: 4 mV/s

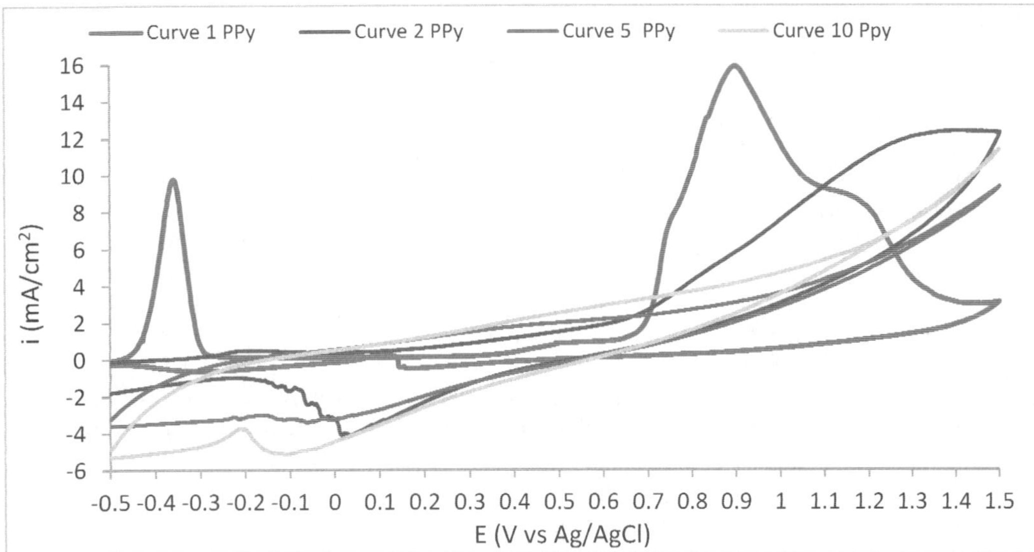

Figure 4: The electropolymerization of polypyrrole films on mild steel electrode in 0.3 M oxalic acid. The scan rate was 50 mV/s.

Scanning electron microscopy coupled with energy dispersive X-Ray analyzer (SEM-EDX) was used to establish the presence of organic PMAP and PPy film coating and its compositions. The examination of uncoated mild steel surface morphology (Figure 5(a)) reveals that a smooth surface, while the micrographs of the mild steel coated with PMAP in Figure 5(b) clearly indicate the formation of a continous organic film. Figure 5(c) illustrates the micrograph of polypyrrole films deposited on mild steel surface. The polypyrrole films present a globular texture as reported by Zu et al., 2015 with particle sizes ranging between 5 to 10μm in diameter. In EDX analysis of bare mild steel three atoms have been detected namely C, O and Fe with atomic percentage of each atoms are 19.12%, 5.48% and 75.40% respectively. Comparatively to coated mild steel, the presence of high compositions of carbonacious, oxygenacious, nitrogenous and the absence of Ferum atoms in EDX analysis further confirmed the formation of the a continous PMAP and PPy organic films formed on the mild steel surface. On the other hand, the absence of Ferum atom also indicates that the continous organic films formed covers the mild steel surface uniformly and homogeneously.

Tafel polarization curve is measured to investigate the protective performance of PMAP and PPy films. The polarization curve for uncoated mild steel substrate, PMAP and PPy coated mild steel electrodes after 0.5 h exposed in 0.5 M NaCl solution at room temperature were plotted in Figure 6. Information on current density (I_{corr}), corrosion potential (E_{corr}) and corrosion rate were collected from Tafel extrapolations method. Table1 summarized the corresponding results calculated by extrapolation based on polarization plots. These results indicate that PMAP coating reduce the corrosion rate of uncoated MS about 3 times. Concurrently, the corrosion potential (E_{corr}) value of PMAP coating was shifted to more negative (cathodic) potentials as compared to E_{corr} value of uncoated mild steel suggesting that PMAP coating protecting the mild steel electrode by retarding the cathodic corrosion reaction. The mechanism below represents the corrosion behavior of mild steel in NaCl solution. An anodic reaction starts with diffusion of Cl^- ions through coating and produce soluble irons compounds.

Anodic Reaction:

1) $Fe^{2+} + 2H_2O + 2Cl^- \rightarrow Fe(OH)_2 + 2HCl$ (1)

2) $FeOOH + H_2O \rightarrow Fe^{3+} + 3OH^-$ (2)

3) $FeOOH + Cl^- \rightarrow FeOCl + OH^-$ (3)

4) $FeOCl + H_2O \rightarrow Fe^{3+} + Cl^- + 2OH^-$ (4)

IOP Conf. Series: Materials Science and Engineering **290** (2017) 012079 doi:10.1088/1757-899X/290/1/012079

Cathodic Reaction:

1) $O_2 + 2H_2O + 4e- \rightarrow OH^{-}$ (5)

Therefore, from the above mechanisms it is suggested that PMAP coating act as oxygen diffusion barrier, which in turn reduces the rate of cathodic reaction which agrees with Hur et al. [12]. The above findings were also supported by SEM microphotograph of PMAP coating which revealed that the coating has a dense continuous structure thus could slow down diffusion of oxygen molecules into coating body. These results agreed with Bereket et al [14] and Hur et al [12]. They concluded that cathodic protection effect of PMAP coating was contributed from the barrier effect for diffusion of oxygen.

In polypyrrole coating, corrosion potential increased significantly to a more positive value, while gradually reduced current density. Lower corrosion current density (i_{corr}) and higher corrosion potential (E_{corr}) means better corrosion resistance. This behavior is typical to an anodic inhibition behavior. The corrosion rates of the polypyrrole coated mild steel decreased because of the reduction in current density. The inhibition efficiency provided by the polypyrrole film is 75.5%. These results indicate that a reduction in the corrosion rate occurred for polypyrrole coated mild steel. Besides, polarization resistance (R_p) is also an important factor to assess corrosion resistance of surfaces and a higher R_p value lead to better corrosion resistance performance. The polypyrrole films show a much higher R_p which is about 4 times that of the bare mild steel. It is evident that the polypyrrole films possesses a better corrosion resistance and exhibits an effective protection for the mild steel substrate. It is anticipated that the reduction in the corrosion rate of polypyrrole films may have been due to the inhibitory effect of oxalate anions. We suggest that oxalate anions may have diffused to the metal/polymer interface and interact with Fe^{2+} ions to form a passive layer of iron (II) oxalate dihydrate ($FeC_2O_4.2H_2O$) on the electrode surface. Oxalate anions were also probably released during the reduction of the polymer. It was reported that polypyrrole coatings have auto de-doping properties during immersion in corrosive medium [15]. This could be explained by auto-dedoping mechanism proposed by Beck et al. [16]. According to this mechanism when polypyrrole is interact with iron with the presence of aqueous electrolytes, iron as relatively non-noble metal, is able to reduce polymer, while oxalate anions are released according to;

$$[PPy^{y+}(C_2O_4^{2-})_y]_n + 2nye \leftrightarrow (PPy)_n + nyC_2O_4^{2-} \quad (6)$$

where y is the doping degree. Iron dissolves through the pores of polymer as Fe^{2+}.

$$Fe \rightarrow Fe^{2+} + 2e \quad (7)$$

This anodic reaction is responsible for the reduction of polypyrrole films. In the anodic protection, polypyrrole coating works as an oxidant to the mild steel electrode. The polypyrrole coating could oxidize the mild steel electrode by shifting the corrosion potential (E_{corr}) of mild steel substrate to the passive state thus the passive state of the substrates is kept under a state of low dissolution rate.

IOP Conf. Series: Materials Science and Engineering **290** (2017) 012079 doi:10.1088/1757-899X/290/1/012079

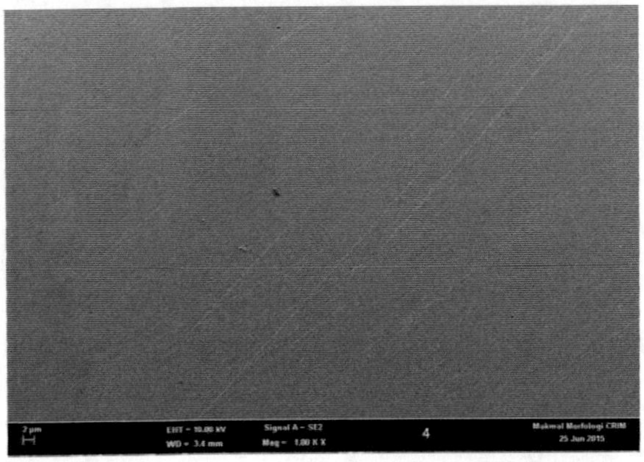

a)

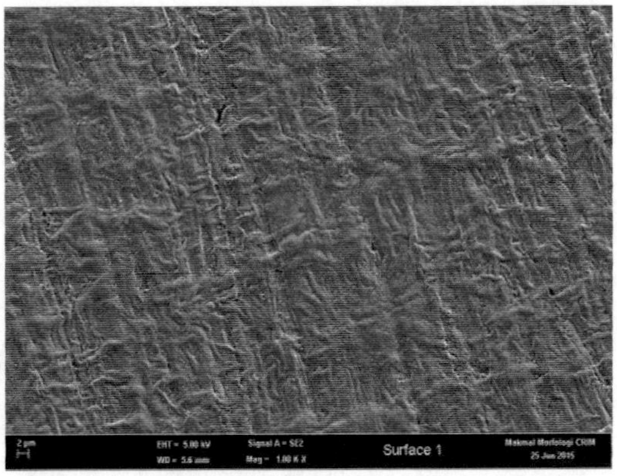

b)

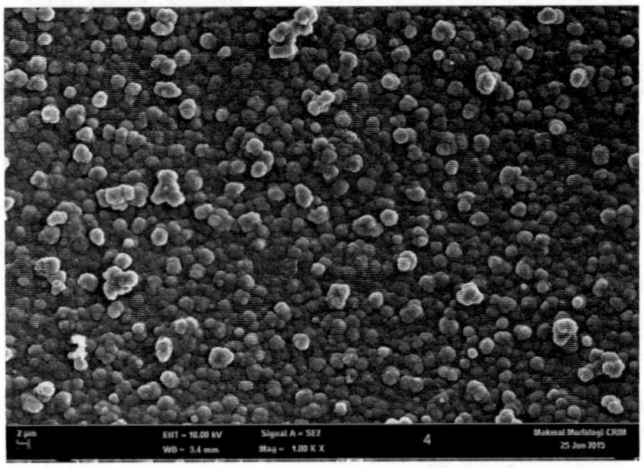

c)

Figure 5: SEM micrographs of (a) uncoated MS (b) PMAP coated MS and (c) PPy coated MS at x1.00K magnification

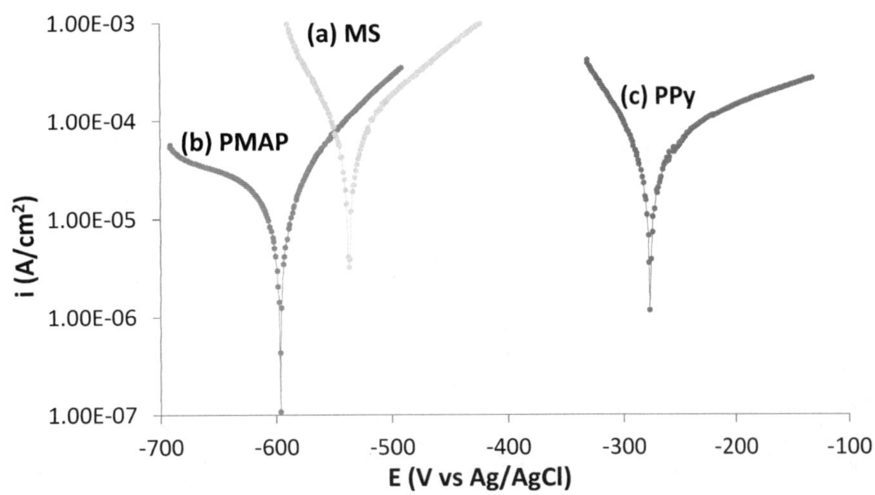

Figure 6: Potentiodynamic polarization curves of (a) uncoated MS substrate and (b) PMAP and (c) PPy films after exposed for 0.5 h in 0.5 M NaCl solution.

Table 1: Corrosion potential (E_{corr}), corrosion current density (I_{corr}) and corrosion rate for bare mild steel, PMAP and PPy coatings as evaluated from the polarization curve.

STATE	TIME (H)	E_{corr} (mV) (Ag/AgCl)	I_{corr} (A/cm^2)	Corrosion Rate (mpy)	β_a (V dec^{-1})	β_c (V dec^{-1})	R_p (ohm. cm^2)	PE (%)
MS	0.5	-538	161.1E-6	73.60	112.8E-3	57.30E-3	102.50	
PMAP	0.5	-597	64.84E-6	25.55	114.5E-3	1 E15	776.5	86.8
PPY	0.5	-277	60.70E-6	27.74	265.4E-3	74.90E-3	418.4	75.5

4. Conclusion

The results of this study have shown that two remarkable effects of PMAP and PPy films inhibited the dissolution of mild steel substrate. PMAP films act as oxygen diffusion barrier, which in turn reduces the rate of cathodic reaction, while PPy film shifts the electrode potential towards more positive potentials. In summarized, the positive shifting may lead to the passivation of mild steel and induce the formation and growth of a passive oxide layer underneath PPy film. On the other hand, the oxidative property of PPy films works as an oxidant to the mild steel substrate further allows the oxide layer to be stably maintained.

Acknowledgement

This work was financially supported by LESTARI Grant Scheme (600-IRMI-MyRA 5/3/LESTARI (046/2017), Universiti Teknologi MARA (UiTM) Shah Alam.

References

[1] Biallozor S and Kupniewska A 2005 *Synthetic Metals*. **155** 443 – 449
[2] Ohtsuka T 2012 *International Journal of Corrosion*. **2012**
[3] Deshpande P P, Jadhav N G, and Sazou D 2014 *Journal Coating Technology*. **11** 473-494
[4] Min Gao, Liying Lu, Xiaoguang Xu, Zhe Yao, and Yong Jiang 2017 *Electrochimica Acta*. **228** 552-527
[5] Hasanov R and Bolgic R 2009 *Progress in Organic Coating*. **64** 435-445.
[6] Qi K, Chen Z, Zhang G and Guo X 2011 *The Open Corrosion Journal*. **4** 18-26.
[7] Tuken T, Ozyilmaz A T, Yazici B and Erbil M 2004 *Applied Surface Science*. **236** 292-305.
[8] Nguyen T, Hubbard J B and Pommersheim J M 1996 *Journal of Coatings Technology*. **68** 45-56.
[9] Hur E, Bereket G, Duran B, Özdemir D and Şahin Y 2007 *Progress Organic Coating*. **60** 153-160.
[10] Narayanasamy B and Susai R 2010 *Progress in Organic Coatings*. **67** 246-254
[11] Tuken T 2006 *Surface and Coatings Technology*. **201** 2782-2790.
[12] Evrim Hür, Gözen Bereket, Berrin Duran, Derya Özdemir and Yücel Şahin 2007 *Progress in Organic Coatings*. **60** 153-160
[13] Harun M K, Lyon S B and Mars J 2005 *Progress in Organic Coating*. **52** 246-252.
[14] Bereket G 2009 *Progress in Organic Coatings*. **65**: 116–124
[15] Bernard M C, Garcia B and Deslouis 2004 *Synthetic Metal*. **140** 287-293
[16] Beck F Michaelis R Schloten F and Zinger B 1994 *Electrochimica Acta*. **39** 229-234

IOP Conf. Series: Materials Science and Engineering **290** (2017) 012080 doi:10.1088/1757-899X/290/1/012080

Effect of Different Concentration of Sodium Hydroxide [NaOH] on Kenaf Sandwich Structures

M Aziz[1], Z Halim[1] and M Othman[1]

[1]Department of Manufacturing and Materials Engineering, International Islamic University Malaysia [IIUM], Jalan Gombak, 53100 Kuala Lumpur, Malaysia

Email:zahureen@iium.edu.my

Abstract. Sandwich panels are structures that made of three layers, low-density core inserted in between thin skin layers. This structures allow the achievement of excellent mechanical performance with low weight, thus this characteristic fulfil requirement to be use in aircraft application. In recent time, sandwich structures have been studied due to it has multifunction properties and lightweight. The aim of this study is to fabricate a composite sandwich structures with biodegradable material for face sheet [skin] where the fibre being treat with different concentration of sodium hydroxide [NaOH] with 10 and 20 hours of soaking time. Kenaf fibre [treated] reinforced epoxy will be used as skins and Nomex honeycomb is chosen to perform as core for this sandwich composite structure. The mechanical properties that are evaluated such as flexural strength and impact energy of kenaf fibre-reinforced epoxy sandwich structures. For flexural test, the optimum flexural strength is 13.4 MPa and impact strength is 18.3 J.

1. Introduction

Composites are the most widely chosen materials because of their adaptability to different situation to combine with other materials to serve specific purposes and exhibit desirable properties. Recently, due to its' cost-effectiveness, composites are used in processing of components for heavy transport vehicles Basic requirement of good composite materials are good reproductively and resilience handling by semi-skilled workers. Composite can be very dynamic because of its strong and stiff yet very light in weight, so that ratio of strength to weight and stiffness to weight are better than steel of aluminium [1]

Nowadays the development of using natural fibre in high performance engineering products is increasing worldwide due to renewable and environmental issues. Among many type of natural fibres, recently kenaf fibres has been extensively exploited [2]. According to [3], natural fibre in sandwich composites received a great deal in research, development and industrial sector due to high demand in many applications such as automotive, construction and aerospace industries.

Due to it has low weight and good mechanical properties and also it is inexpensive, Kenaf is one of good properties of natural fibre [4]. Kenaf is a valuable fibre plant of the Hibiscus cannabinus family cultivated mainly in India. Kenaf is an industrial crop with high potential for cultivation in a tropical climate and it is a source of raw materials for fibre-based industries. Since 2000, research has been conducted to explore its adaptability and utilization in Malaysia.

Natural fibre composites are gaining more attention due to their eco-friendliness, biodegradability and cost effectiveness [5]. However, natural-fibre reinforced composites are lacking in terms of water

IOP Conf. Series: Materials Science and Engineering **290** (2017) 012080 doi:10.1088/1757-899X/290/1/012080

resistance properties and mechanical performance compared to the conventional glass-fibre reinforced composites [6].

For aviation industry, composite materials are important because they provide structural strength comparable to metallic alloys, but at a lighter weight. This situation will result on improving fuel efficiency and performance for an aircraft. The most common composite material is fiberglass, and consists of glass fibres reinforced resin matrix. Besides that, because of the ability of sandwich composite structure to absorb large amount of energy under impact loads that result in high structural crashworthiness make it widely used in aerospace industry.

2. Experimental procedures

This chemical treatment of kenaf fibre with sodium hydroxide [NaOH] result in increasing hydrogen bonding between the fibres and also remove impurities that can make the kenaf fibres behavior more efficient. Kenaf fibres are cut into pieces to make it easy for fabricating process. Then the kenaf was rinsed with distilled water to remove dirt. Then kenaf was dried in oven for 2 hours before immersed in sodium hydroxide for 2 hours to make the solution totally absorb into the fibres. After two hours, wash and rinsed again with distilled water then manually press and dry in room temperature for 2 days.

After treatment of kenaf done, kenaf fibres are ready to fabricate as the skin layer of sandwich structure. Mixture of epoxy and hardener is prepared. This ratio of this mixture is [2:1], epoxy is 100ml and hardener is 55ml. The mixtures were applied on the kenaf fibre using hand lay-up method to fabricate the skin layer. After applying all the mixture, roller should be applied on the top surface to avoid formation of air bubble trapped inside the skin layer. Then the skin layer is left for 24 hour to make sure the epoxy and hardener mixture is dried.

3. Results and Discussion

SEM result shows that microstructure of kenaf fibre for the treated kenaf is less impurities to be compare to untreated kenaf. The lignin and hemicelluloses of the fibre peel off during the treatment of kenaf fibre with alkaline solution. Figure 1 shows SEM result of treated kenaf.

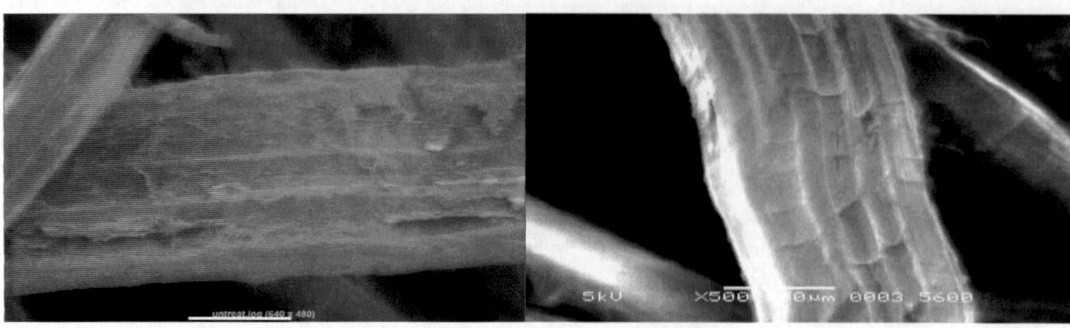

Figure 1. Microstructure of treated kenaf under SEM

To determine flexural strength of sandwich panels, flexural test were conducted. Through the maximum stress of the load, the flexural strength can be obtained. The experimental result of flexural test of each specimen between ranges 9.2 Mpa to 13.4 Mpa. The experimental data is as shown in figure 2.

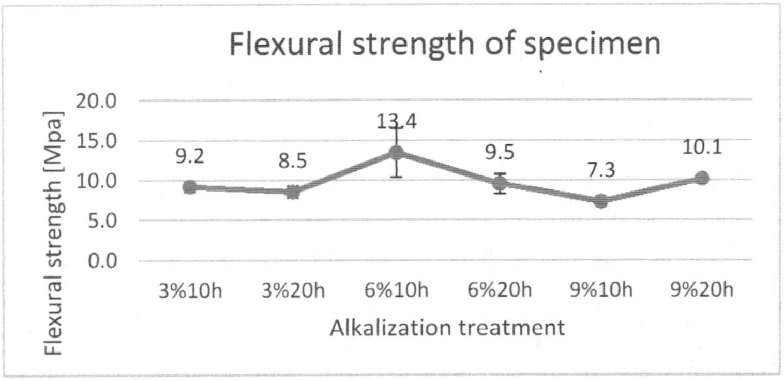

Figure 2. Flexural strength of different alkalized treatment kenaf fibre composite

Treatment of 6% NaOH for 10 hours soaking time sandwich structure displayed the highest flexural strength [13.4 Mpa], while the lowest alkalization treatment of 9% NaOH with 10 hours soaking time. The decrease of flexural strength when the concentration of NaOH was increased up to 9% might be because the high concentration of NaOH could be damage the fibres. From the data, the result of alkalization treatment for 6 % NaOH with soaking time of 10 and 20 hours both gives a good result of flexural strength. This shows that 6% of NaOH gives the best result for flexural properties. In term of soaking time, it also gives effect to the flexural strength. There was displayed that flexural strength increased with increased in concentration of NaOH but slightly decreased in flexural strength when soaking time increased. However, the flexural strength of 9% NaOH is slightly different. Flexural strength increased after soaking time increased from 10 hours to 20 hours.

Charpy Impact Test. can obtain the energy absorption capability of every different alkalization treatment of kenaf composite structure. This test is conducted to obtained impact energy, which is required to fracture a part subjected to shock loading. The result of energy absorption capability of different alkalization treatment is shown in figure 3.

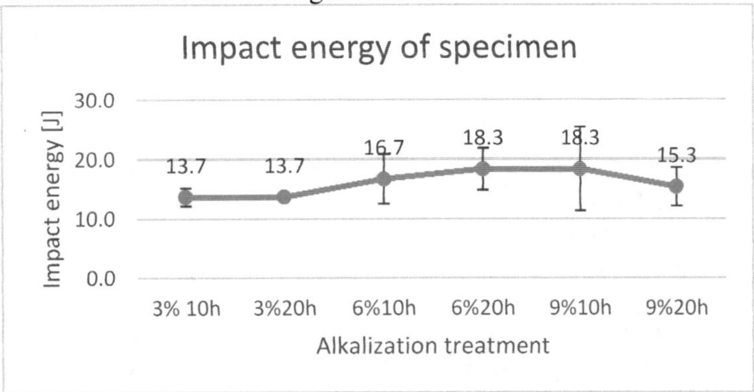

Figure 3. Impact strength of different alkalized treatment kenaf fibres reinforced epoxy sandwich structures

The sandwich structures tested displayed low strength for 3% of NaOH alkalization treatment for both 10 hours and 20 hours soaking time, which are the average of 13.7 for both treatments. The highest result were shows in the alkalization treatment of 6 % NaOH with 20 hours soaking time and 9% NaOH with 10 hours of soaking time which gives impact strength of average 18.3 J. Impact energy tested result showed the soaking time does not really gives effect. Regardless of 10 or 20 hours soaking time applied in the alkalization treatment, the result of impact energy does not give big changes. According to Edeerozey A M et al., [7] the impact energy was dissipated by debonding, fibre fracture and fibre pull out. Fibre fracture dissipates less energy than fibre pull out.

Scanning electron microscopy provides an excellent technique for examination of surface morphology and fracture surface of composite. The morphology changes that occur in every alkalization treatment for failure fracture of impact test were examined.

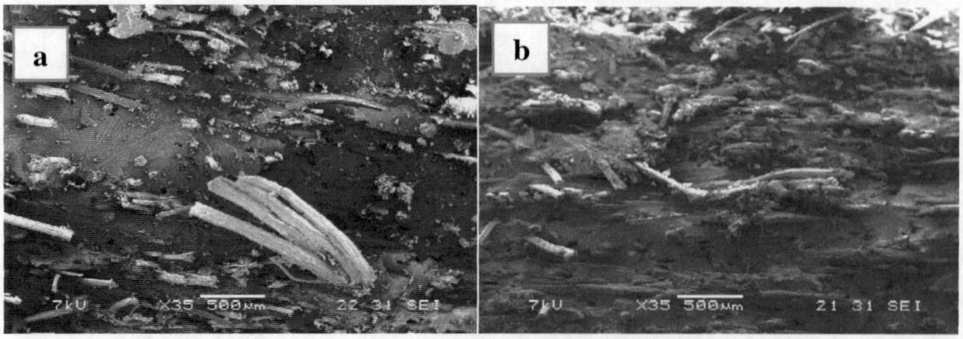

Figure 4. SEM micrograph for impact fracture of [a] 3% 10 hours alkalization treatment and [b] 3% 20 hours of alkalization treatment

Figure 4 shows the microstructure of SEM for impact fracture of alkalization treatment 3% NaOH with 10 and 20 hours soaking time. The microstructure for both treatments shows that natural compound such as hemicellulose, wax and lignin, which are still present on the fibre surface after treatment of 3 % solution of NaOH. The microstructure of fibre surface is clean after treated but the deposits contribute to poor matrix interfacial bonding by decreasing the interfacial area of contact between matrix and fibre and the failure is cause by fibre break-age might be because of improper layup technique [8].

4. Conclusion

The objective of this study to investigate the properties of treated kenaf with different concentration of sodium hydroxide [NaOH] and different soaking time by characterization using SEM is successfully achieved. Besides that, the fabrication of sandwich structure using the different treatment of kenaf also successfully achieved. The mechanical properties of different alkalization treatment of sandwich structure were analyzed using three different mechanical testing which are Flexural tests, Impact test, and indentation test. The overall result of mechanical testing shows the higher concentration of NaOH used, the better result of mechanical properties, but 9 % of NaOH sometimes shows slightly decreased of mechanical properties perhaps because of the higher concentration might burned the fibre. In term of soaking time, some mechanical properties shows effect of soaking time such as indentation test. 20 hours of soaking time gives higher result of indentation load of each alkalization treatment. However, since testing is on sandwich structure, the result might be influence by the core also.

The production of this sandwich structure by using treated kenaf reinforced epoxy may support the commercialization of using natural fibres widely in the industry, thus this will improve the production of biodegradable materials that is environmental friendly. The result obtained from this study perhaps will help other researcher in the same field and will contribute to find information about fabrication and mechanical properties of treated kenaf in sandwich structure.

References

[1] Akil H, Omar M F, Mazuki A A M, Safiee S Z A M, Ishak Z M and Bakar A A 2011 *Materials & Design* **32** 4107-4121
[2] Asumani O M L, Reid R G and Paskaramoorthy R 2012 *Composites Part A: Applied Science and Manufacturing* **43** 1431-1440
[3] Azwa Z N and Yousif B F 2013 TIn 3rd Malaysian postgraduate conference [MPC2013] 4-5

[4] Bajuri F, Mazlan N, Ishak M R and Imatomi J 2016 *Procedia Chemistry* **19** 955-960
[5] Banea M D and da Silva L F 2009 *Composites Part A: Applied Science and Manufacturing Journal of Materials Design and Applications* **223** 1-18
[6] Chawla K K 2012 Composite materials: science and engineering. Springer Science & Business Media.
[7] Chawla K K 1998 Fibrous materials. Cambridge University Press, United Kingdom.
[8] Edeerozey A M, Akil H M, Azhar A B and Ariffin M Z 2007 *Materials Letters* **6** 2023-2025

IOP Conf. Series: Materials Science and Engineering **290** (2017) 012081 doi:10.1088/1757-899X/290/1/012081

Effect of thermoplastic polyurethane (TPU) on the thermal and mechanical properties of polylactic acid (PLA)/ curcumin blends

I S S Sharifah[1], M D A Adnan[1], M K Nor Khairusshima[1], N M Shaffiar[1], Y F Buys[1]

[1]Department of Manufacturing and Materials Engineering, International Islamic University Malaysia (IIUM), Jalan Gombak, 53100 Kuala Lumpur, Malaysia

Email:shaimihezri@iium.edu.my

Abstract. Polylactic acid (PLA) is known to be brittle by nature and thus limits the flexibility of the polymer. A possible solution to enhance the flexibility of PLA is to add a flexible polymeric based material such as thermoplastic polyurethane (TPU). In this study, 30-50 wt% of TPU was added into PLA/curcumin blends to improve its flexibility. Thermal analysis using differential scanning calorimetry shows that further additions of TPU at the expense of PLA did not affect the glass transition temperature, crystallisation temperature and melting temperature of the blends. Fibers of PLA/curcumin/TPU were successfully drawn and Single Fiber Tensile Test (SFTT) showed vast improvement in elongation at break. The initial addition of 30 wt% of TPU to the brittle PLA/curcumin composition causes a significant increase in elongation at break by 39 times and further additions at 50 wt %, the elongation at break increases by 105 times. However, with the increase in elongation, a decrease in strength and Young's modulus was observed.

1. Introduction

Currently there is an increasing demand for polymeric based materials that are being made from renewable starting materials and are biocompatible as well as biodegradable. Amongst the most studied biocompatible polymers are polylactic acid (PLA) and thermoplastic polyurethane (TPU). Generally, PLA has a high strength but its brittleness with a very low impact resistance and flexibility, restricts its application as a general-purpose plastic [1]. Nevertheless, PLA has been proposed for biomedical application due to its ability to fulfill complex requirements on biocompatibility, complete biodegradability, non-toxic degradation products and thermoplastic processability [2] . Whilst TPU, an effective flexibilizer has attracted wide attention in many fields due to its excellent mechanical properties such as high elasticity, great flexibility and toughness, and high resistance to tear, oxidation, and humidity [3]. Thus the combination of PLA and TPU in various ratios would lead to tailored properties suitable for a wide variety of tissue engineering scaffold applications [4, 5]. Degradation tests in phosphate buffered saline of PLA/TPU scaffolds showed a balanced degradation behavior in-between the more easily degraded PLA and the more stable TPU as well as good biocompatibility [6].

Curcumin is a polyphenolic compound derived from Curcuma domestica (Zingiberaceae) that possesses diverse pharmacological effects including anti-inflammatory, antioxidant, antimicrobial, and anticarcinogenic activities [7]. In vivo and in vitro studies have uncovered many important bioactivities of curcumin such as inducing cell apoptosis, inhibiting cell proliferation, anti-cell adhesion and motility, anti-angiogenesis and anti-microbe properties [8]. The blend of curcumin and PLA have been produced in the form of nanofiber mat and have been observed to generate faster wound closure at treated animals [9, 10].

The present study investigates the effect of TPU on PLA/curcumin blends. Different weight percentages of TPU was added at the expense of PLA whilst curcumin amount was maintained at 5 wt%. The blends were thermally characterized prior to being processed via melt drawing process into fibers. The fibers were then mechanically tested to obtain their strength, modulus and elongation at break.

2. Experimental procedures

PLA (Natureworks IngeoTM Biopolymer 3052D grade) and chloroform (R&M Chemicals, United Kingdom) were of analytical grade hence they were used without further purification. The curcumin used in this study with Bioperine extract was manufactured by Pahang Pharmacy Sdn. Bhd., Malaysia. The TPU used in this study is TPU Desmopan 8785A from Bayer Material Science.

2.1 Sample preparation

A weight amount of PLA was dissolved in the following ratio of 1g PLA: 10 ml chloroform. Curcumin powder (5 wt%) was then mixed in PLA solution. The mixture was stirred for 2 days to achieve a homogenous solution using magnetic stirrer at room temperature. The mixture was then poured into a Teflon covered tray and was left to dry in the fume cupboard. The dried PLA/curcumin mixtures were then crushed prior to melt blending with TPU via the internal mixer (Thermo Sientific Haake Rheomix). The mixer was set at temperature of 170°C, speed of 50 rpm and duration of 15 minutes. The PLA/curcumin/TPU blend was then crushed prior to the melt spinning process into fibers.

The fibers were spun at the speed of 115 rpm and at a temperature range between 160-180°C to accommodate the change in viscosity as more TPU was added to the PLA.

2.2 Differential Scanning Calorimetry (DSC)

Differential scanning calorimetry (DSC) was performed employing a Sapphire DSC (Perkin Elmer Instruments, USA) within the temperature range 30 °C to 320 °C at a constant heating rate of 10°C/min under nitrogen gas flow 50mL/min.

2.3 Single Fiber Tensile Test

The single fiber tensile test (SFTT) was performed according to ASTM D 3882 (Standard Test Method for Tensile Properties of Single Textile Fibers). Essentially, 10 fibers randomly chosen from the fiber bundle were individually mounted on a custom-made sample holder made from paper. A slot with a gage length of 25 mm is cut out in the middle of the tab as shown in Figure 1. The ends of each fibers were bonded to the plastic tab with an epoxy adhesive, Devcon Flow-Mix (USA). The diameter of the fiber was measured individually via Celestron Handheld Digital Microscope PRO (USA). The specimen was then mounted on a Shimadzu Tensile Tester (Japan). Without disturbing the setup, both sides of the paper tab were cut at the middle of the gage length. SFTT was then performed at room temperature with a load capacity of 50 N and a chart speed of 10 mm/min.

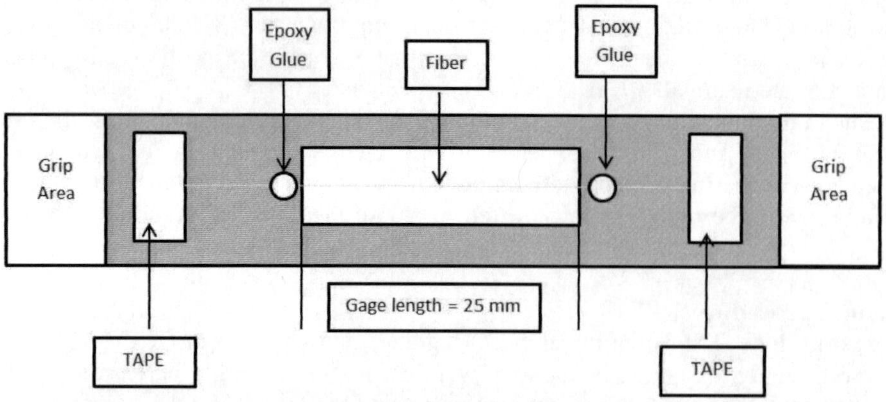

Figure 1. Single fibre specimen sample holder for tensile testing

3. Results and Discussion

The followings in table 1 are the abbreviations used for the composites/blends in this study. DSC results in table 1 shows that initial addition of 30 wt% of TPU in the PLA/curcumin blend did not affect the glass transition (T_g) of the blend. The range of glass transition temperatures (T_g) obtained were in the range of 57 – 60 °C. It is also observed that there is no difference between the crystallization temperatures (T_c) as TPU was incrementally added from 30 to 50 wt% into the mixture at the expense of PLA. The thermal stability of the PLA/Curcumin blends initially dropped as 30 wt% of TPU was added. However further additions of TPU did not affect the thermal stability of the PLA/Curcumin/TPU blends. Lack of improvement in thermal stability with increased amount of TPU suggests that there is a lack of chemical bonding between TPU and PLA. The addition of TPU into the blends created a second melting point at approximately 157°C which suggests the presence of two crystal phases in the blends (see Figure 2).

Table 1. Sample composition (wt%) and thermal properties via DSC.

Sample	PLA (wt %)	Curcumin (wt %)	TPU (wt %)	T_g (°C)	T_c (°C)	T_{m1} (°C)	T_{m2} (°C)	Thermal stability (°C)
PLA_Cur_0TPU	95	5	0	57.1	128.0	158.5	-	70.9
PLA_Cur_30TPU	65	5	30	58.3	117.5	151.1	156.8	59.2
PLA_Cur_40TPU	55	5	40	59.6	116.2	151.8	157.2	56.6
PLA_Cur_50TPU	45	5	50	60.3	116.5	152.1	156.6	56.2

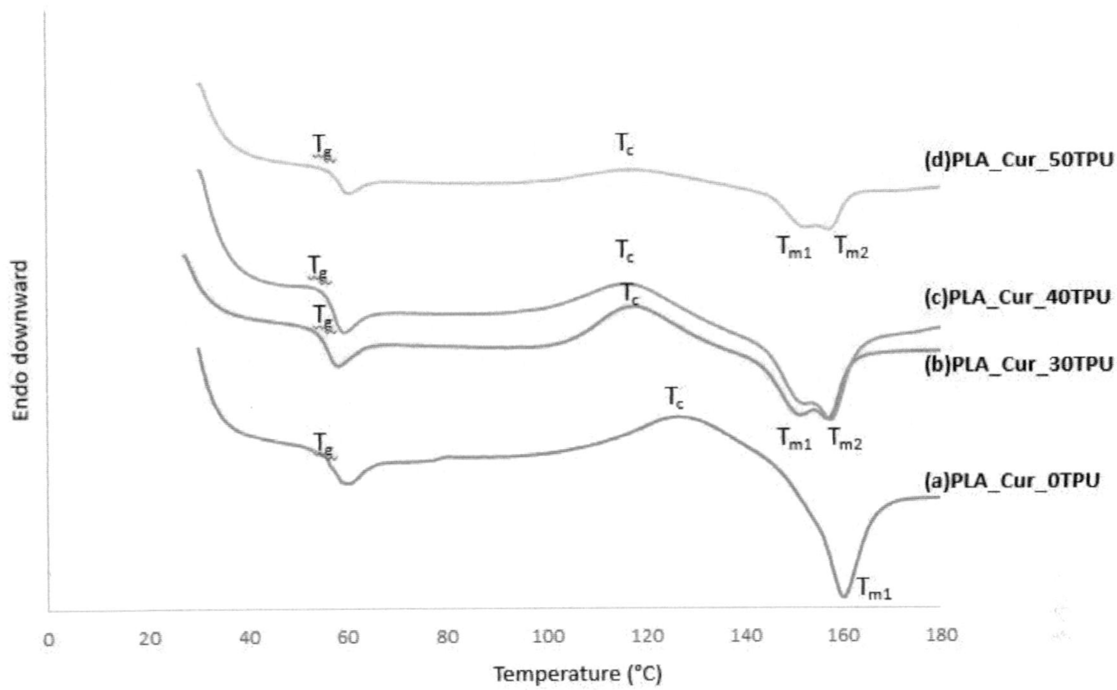

Figure 2. DSC thermograms of (a) PLA_Cur_30TPU (b) PLA_Cur_40TPU (c) PLA_Cur_50TPU

Fibers with diameters in the range between 100 to 200 μm as shown in Figure 3 were successfully melt drawn using an in house built fiber drawing tower. Initial addition of 30 wt % of TPU caused an obvious decrease in strength and modulus of more than 90% (see Figure 4 and Figure 5). Further addition of TPU only causes a slight drop in both strength and modulus of the fibers. However, initial addition of 30 wt% of TPU to the brittle PLA/curcumin composition causes a significant increase in elongation at break by 39 times and further addition at 50 wt %, the elongation at break increases by 105 times (see Figure 6).

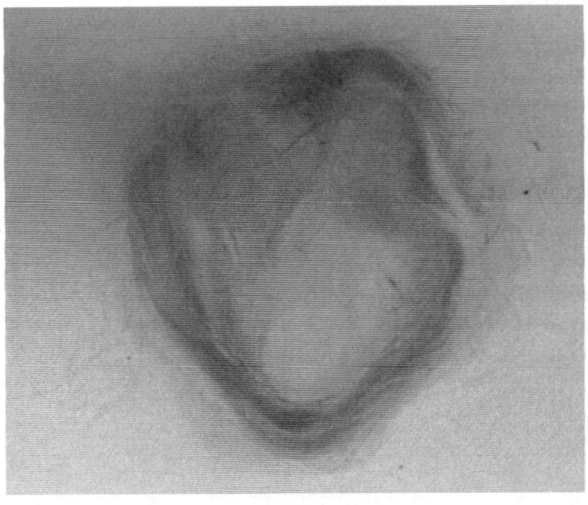

Figure 3. Samples of PLA/Cur/TPU fibers obtained via melt spinning.

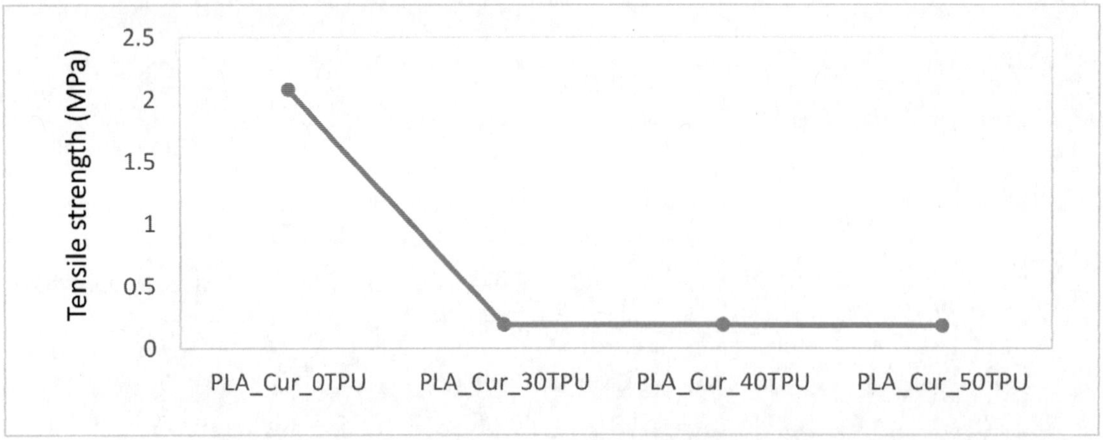

Figure 4. Tensile strength properties of PLA/Curcumin/TPU blends

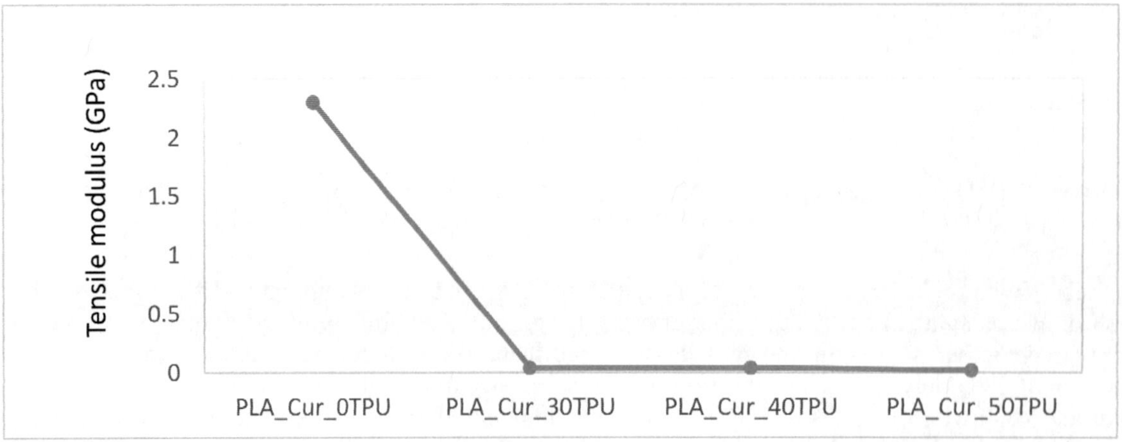

Figure 5. Tensile modulus properties of PLA/Curcumin/TPU blends

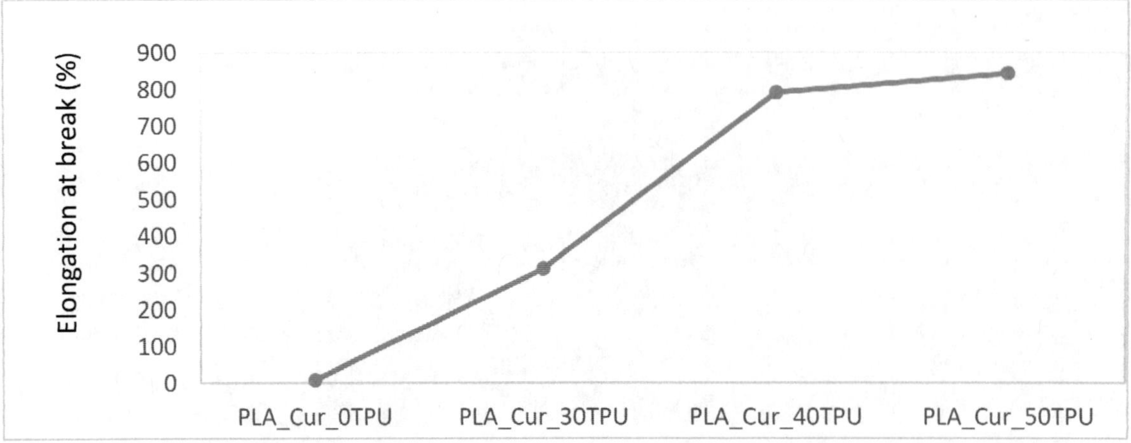

Figure 6. Elongation properties of PLA/Curcumin/TPU blends

TPU is a block co-polymer consisting of alternating sequences of hard and soft segments. The soft block, consists of a polyol and an isocyanate, gives the flexibility and elastomeric property of a TPU. Therefore, when TPU was added into PLA/curcumin blends, the flexibility of the composite blend increases and the brittle characteristics of PLA decreases. Feng and Ye [1] observed the transition of PLA from brittle fracture to ductile fracture with the addition of TPU into PLA and suggested a partially miscible system due to the hydrogen bonding between the molecules of TPU and PLA. The increase in flexibility seen in this study with the addition of TPU can also be achieved if a plasticizer is added instead. Similar effect as seen in a plasticized blend was also observed in this study in which some tensile strength would be lost, but higher elongation characteristics would be achieved [11]. In a study by Jia et al. [12] also found that the addition of TPU by 40% in blends of Polylactic acid (PLA)/organo-montmorillonite (OMMT) increased the elongation at break by 30 times. From the view of manufacturing process, elongation is important in manufacturing as it determines the maximum bending and shaping a material can withstand without breaking. Therefore there is high potential to tailor the strength, modulus and elongation of PLA/Curcumin/TPU blends to be used in biomedical application such internal wound suture or as filaments for rapid prototyping.

4. Conclusion

This study was carried out to study the effect of TPU on PLA/curcumin blend. DSC results showed that the thermal stability, T_g and T_m of the blends was not affected significantly with further addition of TPU. PLA/curcumin fiber produced in a previous study by the author was found to be brittle. The brittleness nature of these fibers was found to have improved with the addition of TPU. It was found that the elongation at break increased significantly at the expense of strength and modulus up to 105 times with 50 wt% of TPU.

References

[1] Fei F and Lin Y 2011 *J Appl Polym Sci* 2778–2783
[2] Wang K, Strandman S and Zhu X X 2017 *Front. Chem. Sci. Eng.* 1-12
[3] Yueyun Z, Lifa L, Wenyong L, Guangsheng Z and Yi C 2015 *Adv Mater Sci Eng* 1-9
[4] Mi HY, Salick MR, Jing X, Jacques BR, Crone WC, Peng X F and Lih-Sheng T 2013 *Mater Sci Eng C Mater Biol Appl* **33** 4767-4776
[5] Qiyi C, Mangadlao J D, Wallat J, Leon A D, Pokorski J K and Advincula R C 2017 *ACS Appl Mater Interfaces* **9** 4015–4023
[6] Jing X, Hao-Yang M, Salick M R, Cordie T, Crone W C, Xiang-Fang P and Lih-Sheng T 2014 *J Cell Plast* **50** 361-379
[7] Rachmawati H, Yanda Y L, Rahma A and Mase N 2016 *Sci Pharm.* **84** 191-202
[8] Fan X, Zhang C, Liu D B, Yan J and Liang H P. *Curr. Pharm. Des.* 2013 **19**(11) 2011-2031
[9] Ghosh C, Nguyen T T T, Park J S and Hwang S 2012 *FASEB J.* **26**
[10] Dhurai B, Saraswathy N, Maheswaran R, Sethupathi P, Vanitha P, Vigneshwaran S and Rameshbabu V 2013 *Front. Mater. Sci.* 7 350-361
[11] Kingsland C 2010 "PLA : A Critical Analysis," Italian Packaging Technology Award, Canada
[12] Shikui J, Jinping Q, Rongyuan C, Chengran W, Zan H, Shufeng Z, Weifeng L and Yanhong F *Polym Eng Sci* 2014 54 2292–2300

IOP Conf. Series: Materials Science and Engineering **290** (2017) 012082 doi:10.1088/1757-899X/290/1/012082

Experimental study of low-velocity impact on foam-filled Kraft paper honeycomb structure

N Abd Kadir, Y Aminanda, M S Ibrahim, H Mokhtar

[1] Mechanical Engineering Department, Faculty of Engineering, International Islamic University Malaysia, 53100,Kuala Lumpur, Malaysia

Email: yulfian@iium.edu.my

Abstract. Low-velocity impact tests of unfilled and foam-filled Kraft paper honeycomb are carried out to investigate the effect of foam, indenter size and location of indentation on maximum or peak force and energy absorption capability. In this study, three indenter sizes (10mm, 12mm, 15mm) and three different locations of indentation (vertical edge, double wall and single wall) were used and compared. The test results show that the foam is given a significant increment of peak force and specific energy absorption to the honeycomb structure subjected to indentation load. The peak force and energy absorption capability also effected by indenter size which due to the contact area of indentation. As for the location of indentation, vertical edge gives highest peak force and energy absorption by the fact that vertical edge is the intersection of three walls of honeycomb cell.

1. Introduction

Honeycomb core has been extensively used in energy absorbing application since they are relatively low weight and high energy absorbing capability. However, one of the main concern is the fact that the honeycomb structure are susceptible to indentation failure due to localized loading [1]. The load-carrying capability of the structure significantly reduced by the presence of the damage which can lead to a decrease of the structure strength by 50%. Hence, it is important to make a good understanding of the indentation response on the honeycomb structure. Castanie et al. [2] studied the core crush criterion to determine the residual strength of impacted sandwich structure using finite element analysis. Meanwhile Aminanda et al. [3] investigated the crushing mechanism of honeycomb structure and found that the cell wall buckled very quickly and the failure of the honeycomb due to the folding of the cell wall. In order to overcome this problem, a number of work has been carried out on the effect of foam as a filler to hollow core [4–9]. As results, they found that filling the low-density foam inside the cell has strengthened the cell wall and enhanced the energy absorption capability and damping properties of the structure. Besides, the mean crushing strength and energy absorption capability of foam filled structure are greater than the summation of the individual component [8,9]. However, most of the previous works focused on the uniform compression loading, and least research work on low-velocity impact or indentation test on foam-filled structure. Also, low-velocity impact is considered as the first step of study prior to the higher velocity impact.

In this work, low-velocity impact tests are conducted on foam-filled Kraft paper honeycomb to investigate the influence of foam as a filler in Kraft paper honeycomb. Then, the effect of indenter size and indentation location on the peak forces and energy absorption capability of foam-filled structure were studied. The deformation and failure mechanism of unfilled and foam-filled Kraft paper honeycomb are observed and described as well.

2. Experiment materials and methods

The Kraft paper honeycombs are fabricated from commercial Kraft by using handmade honeycomb maker. They are coated with vanish and cured at room temperature for 24 hours. Polyurethane foam with density of 52 kg/m^3 are used as filler in the Kraft paper honeycomb structure. It is made by mixing of polyol and isocyanate in liquid foam with ratio 110:100. The mixer was poured inside the honeycomb cells until its blow and left for 24 hours to harden. Then, the indentation tests are carried out on foam, unfilled honeycomb and foam-filled specimen. The bottom surface of specimen is fixed and the upper surface are subjected to indentation loading using spherical indenter with diameter, d = 10mm, 12mm, 15mmand with speed 2mm/min. Three different locations of indentation are identified; vertical edge, double wall and single wall as shown in figure 1. The force-displacements curves are plotted for further analysis.

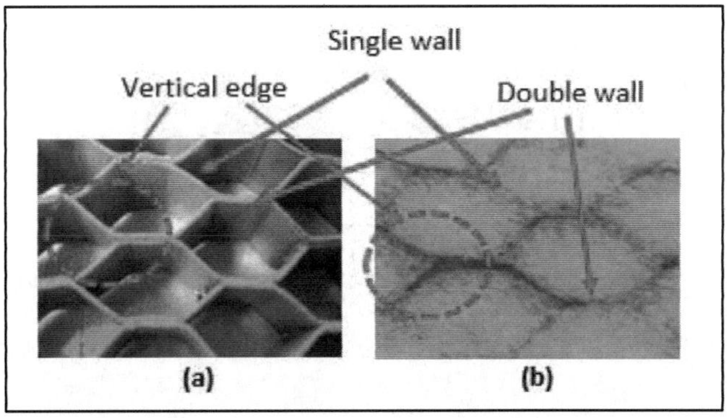

Figure 1. The location of vertical edge (VE), double wall (DB) and single wall (SW) for a) unfilled and b) foam filled Kraft paper honeycomb.

3. Results and Discussion

3.1 Unfilled Kraft paper honeycomb

A sample of force-displacement curves of unfilled Kraft paper honeycomb after indented with 10mm diameter size of indenter at three different location (vertical edge, double wall, single wall) are shown in figure 2. It can be observed that unfilled Kraft paper honeycomb exhibited nonlinear behavior for all locations and the curves demonstrate the same features where the force initially increases linearly until its peak force which corresponds to the maximum force needed to create the first folding of crushed cell wall. After initial folding of the cell wall, the force remains constant over the plateau which corresponds to the succession of fold. The plateau is importance for energy absorption capability calculation and the length of the plateau depend on the location of the indentation.

The peak force for vertical edge is the highest followed by double wall and single wall. The deformation of unfilled Kraft paper honeycomb at different locations can be observed in figure 3. It can be seen that the unfilled Kraft paper honeycomb collapse due to the folding of the cell wall. The single wall and double wall shows the same fold shape compared to vertical edge which indicates tearing or opening phenomenon. The tearing is due to the fact that vertical edge is formed by the intersection of three cell walls which lead to the increment of buckling strength and therefore the peak force. Vertical edge also behaves as a stiffener in Kraft paper honeycomb which has been discussed by Castanie et al. [3] and Aminanda et al. [10] in their works.

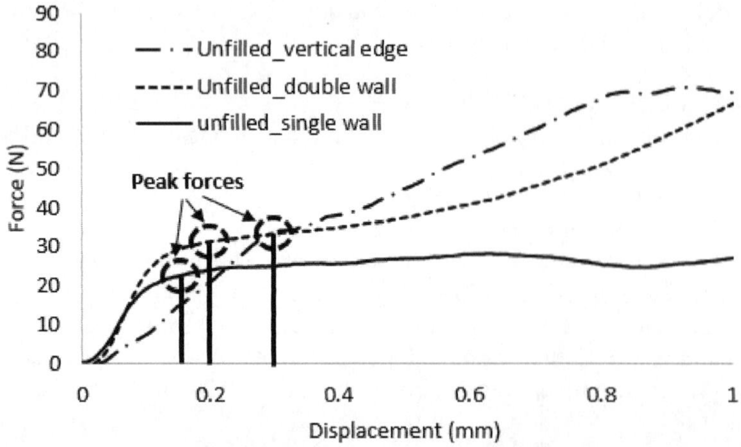

Figure 2. Force-displacement curves of unfilled Kraft paper honeycomb at different location subjected to indentation with diameter of indenter d= 10mm.

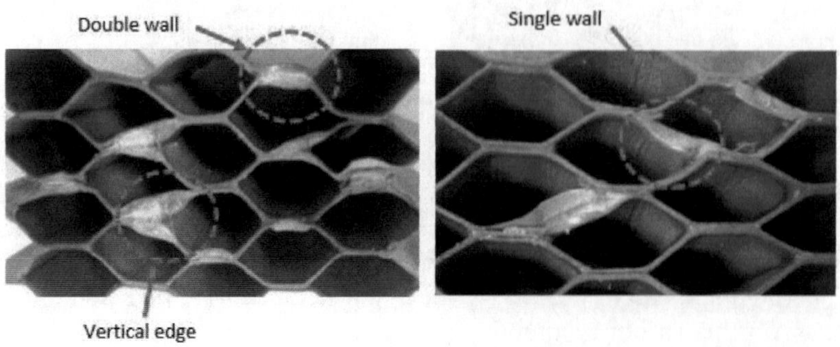

Figure 3. Condition of unfilled Kraft paper honeycomb after indentation.

3.2 Foam-filled Kraft paper honeycomb structure

Using the same parameters, the force in function of indentation displacement for unfilled, foam alone and foam-filled and using different indenter size and indented at different location are drawn in figure 4. From figure 4 (a), the curve shows a nonlinear behavior for all specimens. The peak force of foam-filled is higher than the additional of foam and unfilled one which shows the effect of foam in increasing significantly the peak force and energy absorption capability. Figure 4 (b) shows the effect of indenter sizes on the force of foam-filled Kraft paper honeycomb where 15mm indenter size exhibited higher peak force followed by 12mm and 10mm indenter size. Figure 4 (c) shows the force magnitude depending on the locations of indentation.

Table 1 represent the comparison of peak forces and energy absorption for different parameters mentioned previously. The energy absorption is calculated from the area under the force-displacement curve up to maximum indentation which id the radius of indenter. The table show that, the peak force and energy absorption of foam-filled honeycomb are always higher than unfilled and foam alone. Based on the failure mechanism of foam-filled specimen (figure 5), the foam filling up the folded wall which increases the buckling strength of walls. From table 1 and figure 6, it can be observed as well that higher indenter diameter resulting higher peak force and energy absorption. From damage area observation, higher diameter creates bigger indentation contact area, hence increases the peak force and energy absorption.

IOP Conf. Series: Materials Science and Engineering **290** (2017) 012082 doi:10.1088/1757-899X/290/1/012082

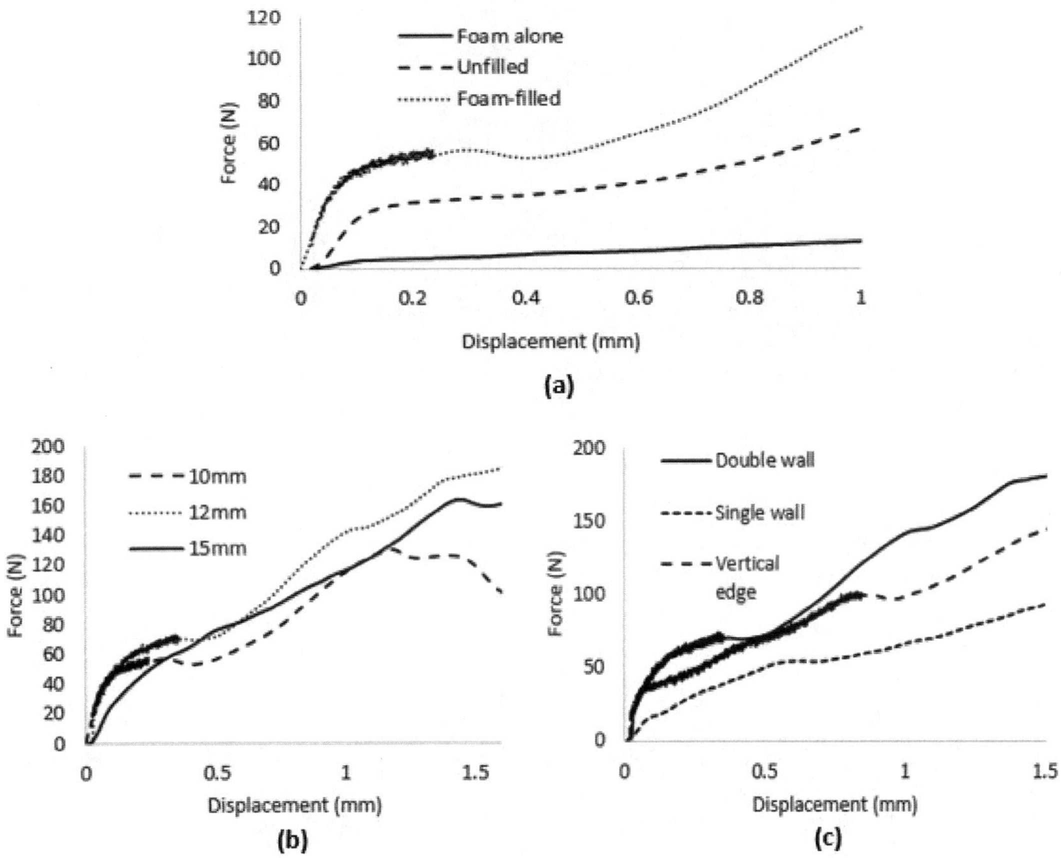

Figure 4. Force-displacement curves of a) foam-filled, unfilled and foam alone at different location b) foam-filled at different indenter size c) foam-filled at different location of indentation.

Table 1. Comparison of peak forces between foam filled paper honeycombs, unfilled and foam alone with three different indenter size and location.

	Peak Force (N)					Energy Absorption (J)				
	Foam	unfilled	Foam-filled			Foam	unfilled	Foam-filled		
			10mm	12mm	15mm			10mm	12mm	15mm
Vertical edge		33.6	85.7	181	244		43	230	720	792
Double wall	3.1	32.5	77.9	99.2	159	7.3	38	180	514	622
Single wall		24.8	54.6	55.7	89.5		24	167	383	603

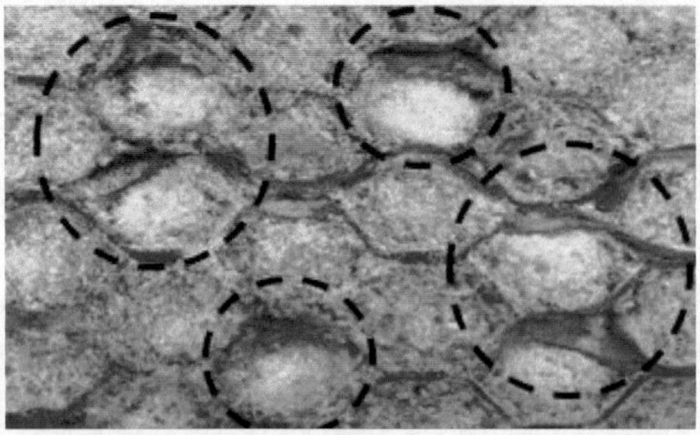

Figure 5. The deformation of foam-filled Kraft paper honeycomb after indentation test.

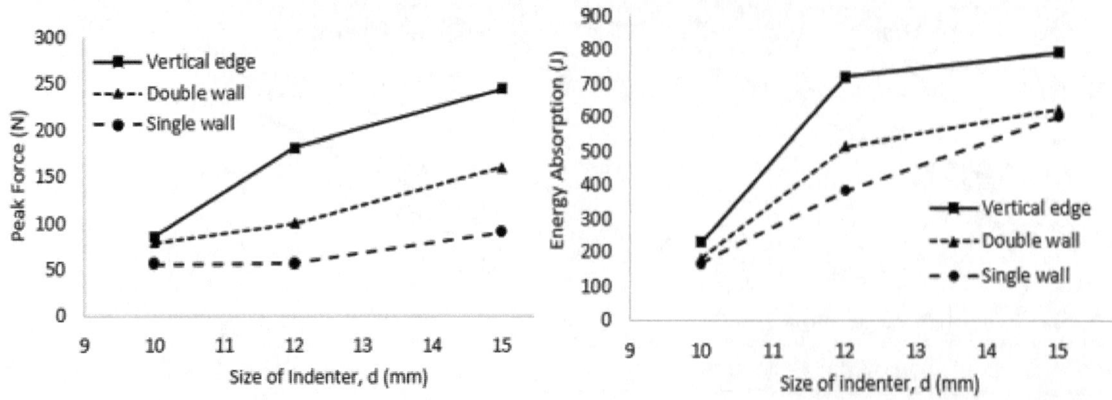

Figure 6. Peak forces and energy absorptions of foam foam-filled Kraft paper honeycomb for different indenter size and location of indentation.

4. Conclusion

Low-velocity impact tests are carried out to study the peak forces and energy absorption performances of Kraft paper honeycomb. The results showed that foam-filled Kraft paper honeycomb exhibited high peak force and energy absorption compared to unfilled and foam alone due to the strengthened of cell wall by foam as a filler. Then, the effect of indenter size and location of indentation were discussed. Bigger indenter size is prone to increasing the peak force and energy absorption capability due to large indentation area. Meanwhile, the indentation at vertical edge gives higher peak force value and energy absorption compared to double wall and single wall due to the indentation area of the Kraft paper honeycomb which increase the resistance to buckling.

Acknowledgements

This research was supported by Ministry of Higher Education Malaysia and Research Management Centre of the International Islamic University Malaysia (Grant no: FRGS14-155-0396)

References

[1] Li Z, Zheng Z, Yu J and Yang J 2014 *J. Reinf. Plast. Compos.* **33** 1671–81
[2] Castanié B, Aminanda Y, Bouvet C and Barrau J-J 2008 *Compos. Struct.* **86** 243–50
[3] Castanié B, Bouvet C, Aminanda Y, Barrau J-J and Thevenet P 2008 *Int. J. Impact Eng.* **35** 620–34
[4] Alavi Nia A and Sadeghi M Z 2010 *Mater. Des.* **31** 1216–30

[5] Burlayenko V N and Sadowski T 2010 *Compos. Struct.* **92** 2890–900
[6] Mahdi M, Niknejad A, Hossein G and Zamani M 2012 *Int. J. Mech. Sci.* **65** 134–46
[7] Vaidya U K, Ulven C, Pillay S and Ricks H 2003 *J. Compos. Mater.* **37** 611–26
[8] Wan Abdul Hamid W L H, Aminanda Y and Shaik Dawood M S I 2013 *Appl. Mech. Mater.* **393** 460–6
[9] Kadir N A, Aminanda Y, Ibrahim M S and Mokhtar H 2016 *IOP Conf. Ser. Mater. Sci. Eng.* **152** 12048
[10] Aminanda Y, Castanié B, Barrau J-J and Thevenet P 2005 *Appl. Compos. Mater.* **12** 213–27

IOP Conf. Series: Materials Science and Engineering **290** (2017) 012083 doi:10.1088/1757-899X/290/1/012083

A study of tensile test on open-cell aluminum foam sandwich

N A Ibrahim[1], M H F Al Hazza[1], E Y. T. Adesta[1], Atiah Bt. Abdullah Sidek[1] and N A Endut[1]

[1]Department of Manufacturing and Materials Engineering, International Islamic University Malaysia (IIUM), Jalan Gombak, 53100 Kuala Lumpur, Malaysia

Email: muataz@iium.edu.my

Abstract. Aluminum foam sandwich (AFS) panels are one of the growing materials in the various industries because of its lightweight behavior. AFS also known for having excellent stiffness to weight ratio and high-energy absorption. Due to their advantages, many researchers shows an interest in aluminum foam material for expanding the use of foam structure. However, there is still a gap need to be fill in order to develop reliable data on mechanical behavior of AFS with different parameters and analysis method approach. Least of researcher focusing on open-cell aluminum foam and statistical analysis. Thus, this research conducted by using open-cell aluminum foam core grade 6101 with aluminum sheets skin tested under tension. The data is analyzed using full factorial in JMP statistical analysis software (version 11). ANOVA result show a significant value of the model which less than 0.500. While scatter diagram and 3D plot surface profiler found that skins thickness gives a significant impact to stress/strain value compared to core thickness.

1. Introduction

Demand for the use of aluminum foam sandwich (AFS) is rapidly increasing in various applications because of their lightweight material, high specific stiffness and strength and also excellent energy absorption [1]. Sandwich panels with porous structure core material had been used in many application in industries such as lightweight construction, aerospace industry, ship building, railway industry, machine construction and biomedical industry [2]. Earlier researchers had found several new classes of material that can be used in diverse applications and one of them are metallic foams. There are two types of metallic foam structure which is open-cell and closed-cell aluminum foam as shown in the figure 1 below.

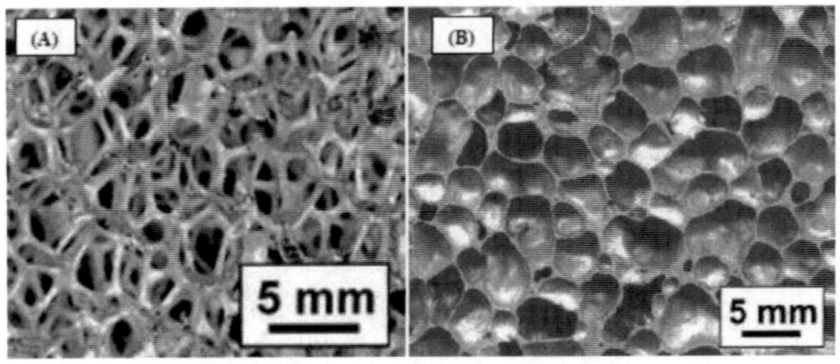

Figure 1. Open-cell metallic foam (a) and Closed-cell metallic foam (b) [3]

At the beginning of creation of foam structure, previous research had developed foam structure material using polymer. However, polymer foam has many limitations such as low heat resistance. Hence, in order to fix the limitation, previous researchers had developed metallic foam. Nevertheless, foam itself is weak, so sandwich structure consist of metallic foam as core and different type of metal as a faces was develop.

It is important to determine the mechanical behavior of new sample produce for identifying their suitable application. Therefore, many researchers had prepared the research on metal foam with different types of skin material and testing method. There are researches had been done for tensile test on steel foam only [4], tensile test on closed-cell metal foam [5], compression and tensile on closed-cell aluminum foam [6], impact test on aluminum foam [7], tensile test on aluminum foam only [8-9] and many more. Most of the previous researchers focusing on closed-cell metal foam compared to open-cell metal foam.

To increase the area of the metallic foam material application and for future use, there is a need for researchers to create many reliable data on the mechanical behavior of the material [4]. After many years, the development of data for aluminum foam sandwich properties are growth in variety parameters and application and most of them focusing on closed-cell aluminium foam. Least of them focusing on the tensile test of open-cell aluminum foam sandwich and also statistical analysis method. Therefore, the aim of this paper were to present analysis of the tensile test data on open-cell aluminium foam sandwich using full factorial in JMP statistical analysis software.

2. Experimental procedures

The paper discussed about the analysis of the tensile test data of aluminum foam sandwich panels using full factorial method. Figure 2 below shows the flowchart that need to be follow in order to conduct the research.

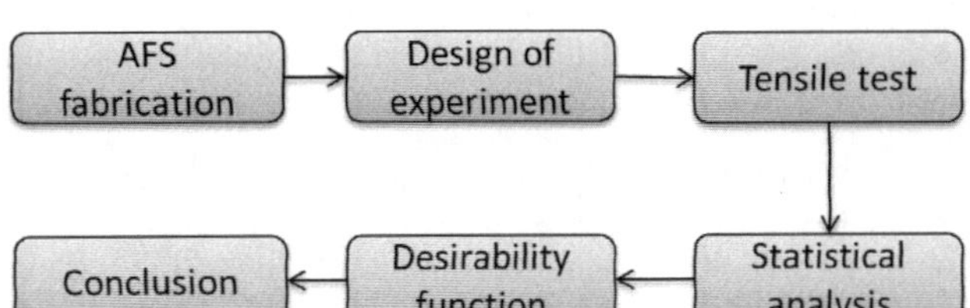

Figure 2. Steps of the research

The AFS panels were prepared by attaching the open-cell aluminum foam core with aluminum skin sheets using adhesive bonding of epoxy resin mixed with hardener with ratio 2:1. The dog-bone specimens of aluminum sheets and foam core were shaped with an electrical discharge machine (wire cut). The shape and ratio of the sandwich structure were designed based on ASTM standard of C393.

Figure 3 below shows the final sandwich panels ready to be test under tension using Universal INSTRON machine with 100KN load.

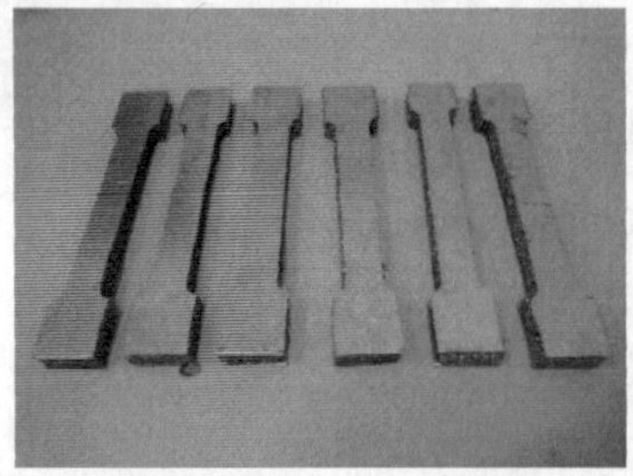

Figure 3. Tensile specimens of AFS

In order to conduct the test, design of experiment was developed in JMP statistical analysis using full factorial method for maximizing the input parameters combination. Six runs with input parameters combination of two levels of foam core thickness and three levels skins thickness have been generated as tabulated. The results from tensile test then documented in table in order to conduct the statistical analysis in JMP statistical analysis software. The AFS panels were tested under tensile test for two times using Universal INSTRON machine in order to have result that is more consistent. The relationship between input parameters with the stress/strain result was analysed.

3. Results and Discussion

The stress/strain values of tensile test were compiled in the table 1 below.

Table 1. Result compilation of tensile test of AFS sandwich

Run	Skin thickness (mm)	Core thickness (mm)	Average Stress (MPa)	Average Strain (mm/mm)
1	0.4	6.35	13.9495	0.0045
2	0.6	6.35	22.7035	0.009
3	0.8	6.35	27.7995	0.016
4	0.4	10.0	9.1935	0.005
5	0.6	10.0	15.864	0.0125
6	0.8	10.0	20.419	0.018

Based on the data above, statistical analysis was conducted for determining the relationship between input variables of skin and core thickness on stress/strain result. Table 2 and 3 below show the analysis of variance and summary of fit models for relation between input variables and stress value of AFS panels after loaded in tension.

Table 2: Analysis of variance (ANOVA)

	DF	Sum of Squares	Mean Square	F Ratio
Model	3	218.93194	72.9773	49.0431
Error	2	2.97605	1.4880	Prob > F
c. Total	5	221.90799		0.0200*

Table 3. Summary of Fit

Summary of Fit	
RSquare	0.986589
RSquare Adj	0.966472

It is found that the result for ANOVA analysis of stress, there are 0.02% chance of error that the model could be occur due to noise which can be accepted since the value of 'Prob > F' less than 0.05. Next, for summary to fit model shows the differences between RSquare and RSquare adjusted and RSquare is always increased when the factors increase while the RSquare adjusted is increased when the bad factors were removed such as slippage between the grip and sample. Based on the results obtained, the RSquare value is reasonably agree with RSquare adjusted which is 0.986589 and 0.966472 respectively and it is acceptable since the value are close to 1. While table 4 and 5 below show the ANOVA and summary of fit model analysis of strain output.

Table 4. Analysis of variance (ANOVA)

	DF	Sum of Squares	Mean Square	F Ratio
Model	3	0.00015663	0.000052	61.1220
Error	2	0.00000171	8.542e-7	Prob > F
c. Total	5	0.00015833		0.0161*

Table 5. Summary of fit

Summary of Fit	
RSquare	0.989211
RSquare Adj	0.973026

Based on tables above, it can be seen that the chance of error of the model for strain also less than 0.05 which 0.0161%. For RSquare and RSquare adjusted also reasonably agree with each other and the value are almost 1 which is 0.989211 and 0.973026 respectively. In order to find out more about the effect of skins and cores thickness on stress/strain value, the scatter diagram had been generated in JMP statistical analysis software. Figure 4 below shows the scatter diagram of relationship between input and output variables.

IOP Conf. Series: Materials Science and Engineering **290** (2017) 012083 doi:10.1088/1757-899X/290/1/012083

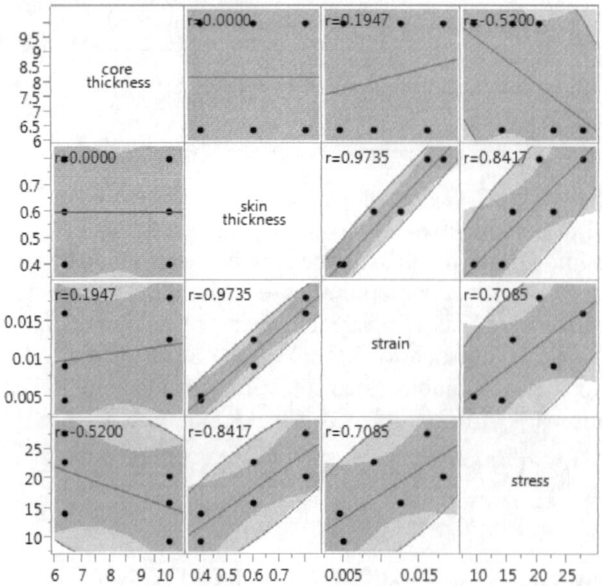

Figure 4. Scatter diagram of relationship between skins and cores thickness on stress/strain value.

Based on the figure above, it shows that skin thickness gives a high positive correlation to the stress/strain value with 84% and 97% correlation respectively. The value of correlation are contrast with the core thickness which is fairly effect the stress/strain value with value less than 50% correlation. This shows that skin thickness give more effect to stress/strain result compared to core thickness. For further analysis of relationship between cores thickness and skins thickness on stress/strain value, 3D plot surface profiler had been generated. Figure 5a and 5b below show the 3D plot surface profiler for inputs with stress and strain respectively.

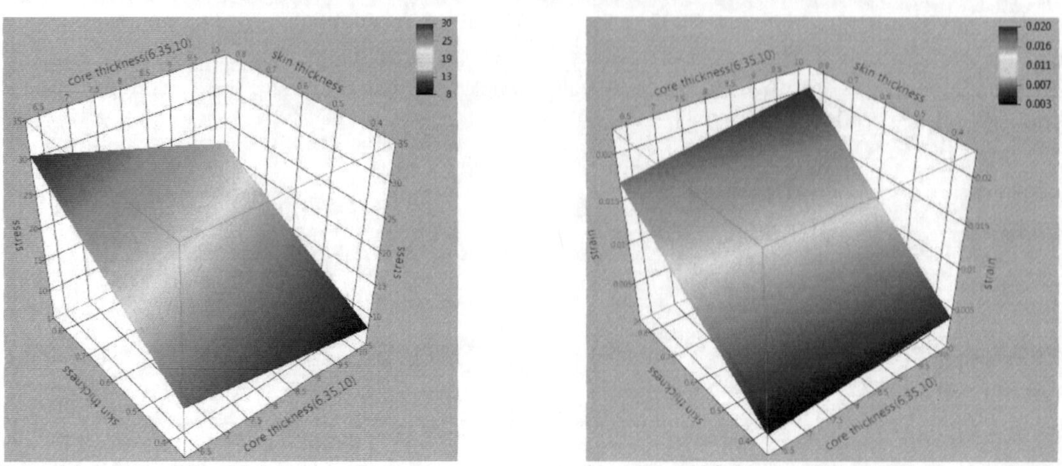

a. Core thickness, skin thickness and stress b. Core thickness, Skin thickness and strain

Figure 5. 3D plot surface profiler

Based on the figure 5a above, it can be seen that the stress value increases when the skin thickness increases and core thickness decrease. While according to figure 5b, it shows that the strain value were increasing when both skin and core thickness increase. Lastly, for finding the optimum input parameters combination that will contribute the maximize output, desirability function analysis had been used. Figure 6 below shows the prediction profiler for analysing the desirability function.

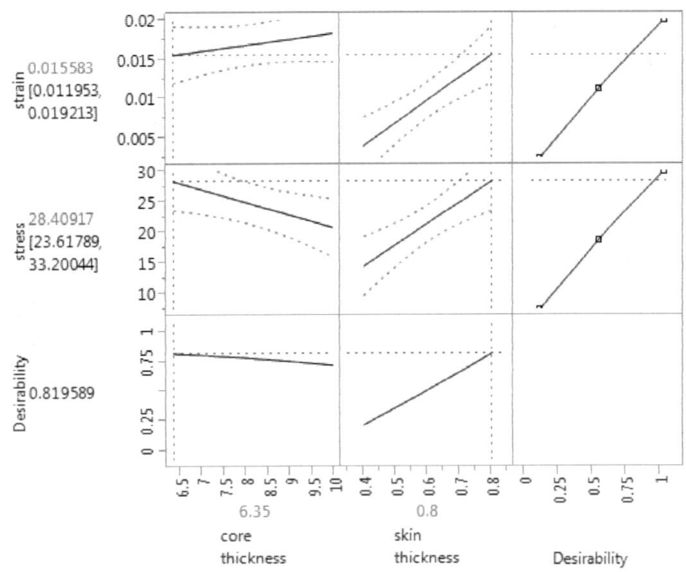

Figure 6. Prediction profiler

Based on the prediction profiler above, the desirability function analysis shows that the optimum input parameters combination for finding the maximize stress was skin thickness of 0.8 mm and core thickness of 6.35 mm with 81% desirability which means that as the skin to core ratio increase, the value of stress/strain also increase. Previous researcher also found the tensile modulus increase as the skin to core ratio increase [10].

4. Conclusion

In summary, it can be can concluded that, the tensile test was successfully conducted. The stress/strain data from tensile test were compiled and analyzed using statistical analysis. According to the ANOVA and summary of fit models, it shows that the results were acceptable with error less than 0.05. Based on scatter diagram, it confirms that skins thickness give the major effect on the stress/strain value compared to core, which fairly effect the stress/strain result. Lastly, according to 3D surface profiler plots, it show that the stress/strain value increase as the skin thickness increase and core thickness decrease which agree with the scatter diagram.

Acknowledgement

The authors would like to express their sincere appreciation to the Research Management Centre, International Islamic University Malaysia and the Ministry of Higher Education, MOHE for financially providing and supporting fund for the research under science fund with the No: FRGS 15-247-0488.

References

[1] Zu G Y, Lu R H, Li X B, Zhong Z Y, Han M B and Yao G C 2013 *Trans. Nonferrous Met. Soc. China* **23** 2491-2495
[2] Prabhu S, VK B R and Rajan N 2015 *Appl. Mech. Mater.* **766** 511-517
[3] Veale P J 2010 *Masters Theses 1911 - February 2014* (University of Massachusetts Amherst)
[4] Szyniszewski S, Smith B, Arwade S, Hajjar J and Schafer B 2012 *12th International LS-DYNA Users Conf.* (Detroit USA)

[5] Nesic S, Unruh K, Michels W and Krupp U 2012 *Steel Res. Int.* **83** 972-980

[6] Blazy J S, Marie-Louise A, Forest S, Chastel Y, Pineau A, Awade A, Grolleron G and Moussy F 2004 *Int.J. Mech. Sci.***46** 217-244

[7] Liu C, Zhang Y X and Ye L 2017 *Int. J. Impact Eng.* **100** 139-153

[8] Hangai Y, Ishii N, Koyama S, Utsunomiya T, Kuwazuru O and Yoshikawa N 2012 *Mater. Trans.* **53** 584-587

[9] Ibrahim, N. A., Al Hazza, M., Adesta, E. Y. T., Endut, N. A., & Asman, A. H. 2017. International Journal of Engineering Materials and Manufacture 2(2) 25-30

[10] Shetty R and Han C D 1978 *J. Appl. Polym. Sci.* **22** 2573-2584

IOP Conf. Series: Materials Science and Engineering **290** (2017) 012084 doi:10.1088/1757-899X/290/1/012084

Compressive Behaviour and Energy Absorption of Aluminium Foam Sandwich

N A Endut[1], M H F Al Hazza[1], A A Sidek, E T Y Adesta[1], N A Ibrahim[1]

[1] Department of Manufacturing and Materials Engineering, International Islamic University Malaysia (IIUM), Jalan Gombak, 53100 Kuala Lumpur, Malaysia

Email: muataz@iium.edu.my

Abstract. Development of materials in automotive industries plays an important role in order to retain the safety, performance and cost. Metal foams are one of the idea to evolve new material in automotive industries since it can absorb energy when it deformed and good for crash management. Recently, new technology had been introduced to replace metallic foam by using aluminium foam sandwich (AFS) due to lightweight and high energy absorption behaviour. Therefore, this paper provides reliable data that can be used to analyze the energy absorption behaviour of aluminium foam sandwich by conducting experimental work which is compression test. Six experiments of the compression test were carried out to analyze the stress-strain relationship in terms of energy absorption behavior. The effects of input variables include varying the thickness of aluminium foam core and aluminium sheets on energy absorption behavior were evaluated comprehensively. Stress-strain relationship curves was used for energy absorption of aluminium foam sandwich calculation. The result highlights that the energy absorption of aluminium foam sandwich increases from 12.74 J to 64.42 J respectively with increasing the foam and skin thickness.

1. Introduction

Recently, researchers have been shown an increased interest in production of automotive industries. Material selection plays vital role in order to meet the functional requirements of components [1]. In the past few years, metallic foams has been attracting considerable attention. One of the excellent metal foam that are used by previous researchers in automotive application is aluminium foam sandwich due to their unique properties such as low density, good energy absorption characteristics. Aluminium foam sandwich panels are good energy absorbers and lighweight which provide wide range of potential applications in automotive industries [2],[3]. Rajak et al.,[4] highlighted about to calculate energy absorbed of aluminium foam by performing compression test. For this study, the authors were investigated the effect of face sheets and core thickness due to empty mild steel sample and aluminium foam filled sample. It had been clearly shown that there were increasing trend for energy absorption of aluminium foam filled samples compared to empty mild steel samples at different strain rates. However, the study by Ansari et al., [5] were presented sandwich composites using polyurethane (PU) foam and acrylic sheet to analyze mechanical behavior by conducting compression test. The input of this research were skin(acrylic sheet) thickness, core(polyurethane) foam thickness and strain rates values whereas the output were compressive strength and deformation behaviour. The compressive stress-strain curve of ideal foam are presented in figure 1.

As shown in the figure, For the first region, it refers to collapse stress, whereas for the plateau region, total deformation energy is absorbed at constant stress value. The final region which is densification region related to increasing of stress at constant strain value. At this region, the cell walls are in close contact each other [6].

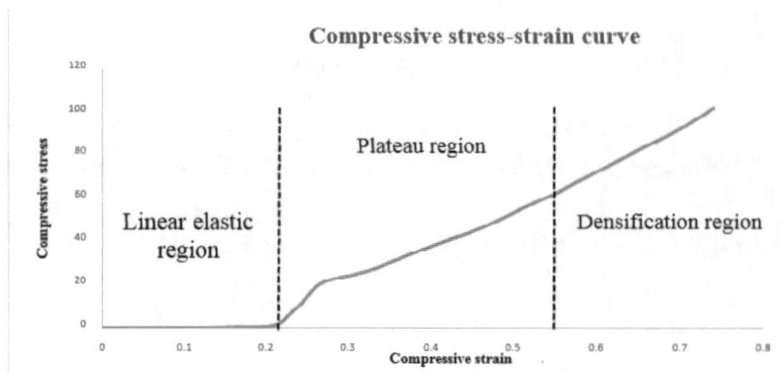

Figure 1. Compressive stress-strain curve of ideal foam [6].

Although there were many research had been done experimentally about mechanical behaviour of aluminium foam sandwich but researching into aluminium foam sandwich with respect to energy absorption characteristics is limited if any. Meanwhile, there were also some gaps that need to be fill by varying parameters and testing methods for energy absorption behaviour.

2. Experimental procedures

For this research, aluminium foam sandwich was used for conducting the experimental work. Aluminium foam were attached together with aluminium sheets using mixing of epoxy resin and hardener with a ratio of 2:1 in terms of volume 100 ml of epoxy and 55 ml of hardener to fabricate aluminium foam sandwich. According to ASTM standard of C393, the preferable ratio for designing sandwich structure should be approximately less or equal to 0.10 in order to select the suitable thickness for core and skin as shown in equation 2.0 below .

$$\frac{t}{c} \leq \sim \mathbf{0.10} \qquad (2.0)$$

t: skin thickness, c: core thickness

Based on design of experiment (DOE), Taguchi design were developed with two level of thickness of core were used which were 6.35 and 10 mm, whereas three level of thickness of aluminium sheets were 0.4, 0.6 and 0.8 mm respectively. Table 1 shows the design of experiment for the a AFS in terms of aluminum core thickness and skin thickness.

Table 1. Input parameters

Nu of run	Core thickness, mm	Skin thickness, mm
1	6.35	0.4
2	10	0.4

3	6.35	0.8
4	10	0.6
5	10	0.8
6	6.35	0.6

For compression test, Universal Testing INSTRON machine was used to analyze the compressive behavior of aluminium foam sandwich (AFS) as shown in figure 2 (a) below. This machine was set up with 100 kN load cell [6] and speed of test with 1.5 mm/min. Figure 2 (b) shows the setup of compression sample. The compression samples were compressed between two steel loading platens to 50% and the compressive stress-strain curves were generated by using Bluehill 2 software.

a) 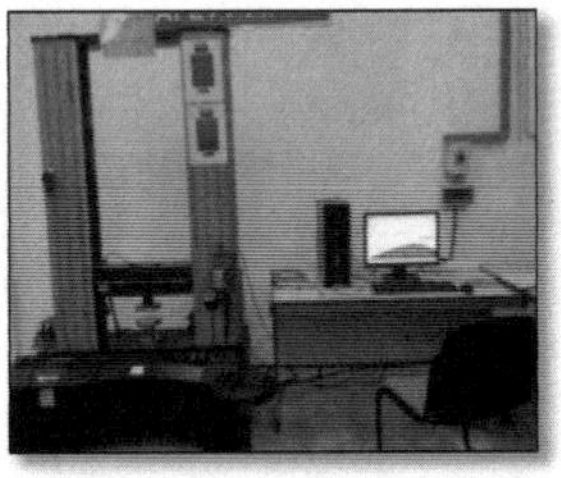 b) 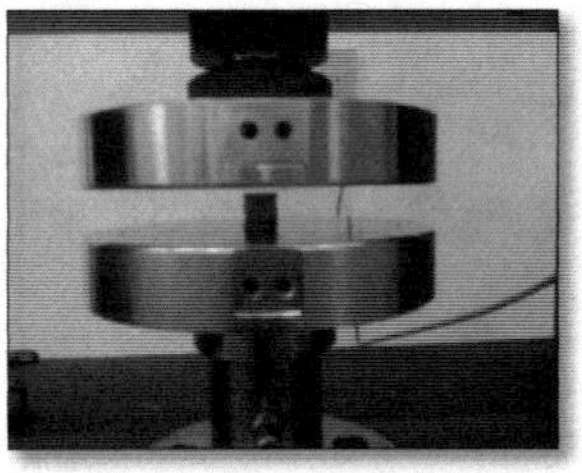

Figure 2. Compression test: (a) Universal testing machine INSTRON, (b) setup of compression sample.

3. Results and Discussion

The mechanical properties of aluminium foam sandwich (AFS) can be determined by conducting compression test. During compression test, the aluminium foam sandwich (AFS) samples will deform after undergoes some load. Energy is absorbed by aluminium foam core which approximately equal to one of AFS structure. This findings is supported by Dou et al., [7] which relate to the stress-strain relationship with relevant formulae. Based on this formulae, the energy absorbed of aluminium foam sandwich, AFS can be expressed in equation (3.0).

$$E = \int_0^{\varepsilon} \sigma(\varepsilon) d\varepsilon \tag{3.0}$$

where E is energy absorbed of aluminium foam sandwich, σ is compressive stress and ε is compressive strain. It had been clearly shown on table 2 that there were increasing trend for compressive stress, compressive strain and energy absorbed based on area under curve when increasing the thickness of aluminium foam as a core.

Figure 3 depicts results of experimental work that obtained from compression. The sample have been compressed until 50% of its initial length Based on figure 3, the alteration point between plateau stress and rupture (densification) region are clear trend. As compressive stress increases, the compressive strain also increases due to densification region. It can be found that the energy absorbed of aluminium foam sandwich follows the same trends with the increasing of compressive strain and stress

Table 2: Compression test results.

Run	C,mm	F,mm	Compressive strain,mm/mm				Compressive stress, N/mm^2				Energy absorbed , J			
			1	2	3	**Average**	1	2	3	**Average**	1	2	3	**Average**
1	6.35	0.4	0.39	0.39	0.39	0.389	91.88	91.91	93.57	92.45	12.5	11.9	13.8	12.74
2	10	0.4	1.3	1.31	1.36	1.322	153.7	153.6	153.4	153.59	29.2	27.4	34.5	30.36
3	6.35	0.8	0.6	0.6	0.6	0.598	95.2	94.97	94.92	95.03	23.1	23.4	23.2	23.22
4	10	0.6	1.26	1.24	1.24	1.249	292	290.3	291.1	291.13	37.5	39.9	38.6	38.65
5	10	0.8	1.64	1.63	1.67	1.645	321.6	317.2	318.5	319.12	62.3	66.7	64.3	64.42
6	6.35	0.6	0.45	0.47	0.45	0.458	88.04	86.78	87.21	87.34	20.7	19.4	21.4	20.51

The result was consistent with [8] which also stated that increasing the foam thickness and density of foam will increase the collapse strength and energy absorbed. The sample for run 5 with core thickness of 10mm and skin thickness of 0.8mm shows the highest value of energy absorbed. Meanwhile, compressed aluminium foam sandwich shows that breaking process of cell wall initiated from cylindrical surface to the central axis along the radius direction. This finding reflects the findings of Y-G. Li et al., [9].

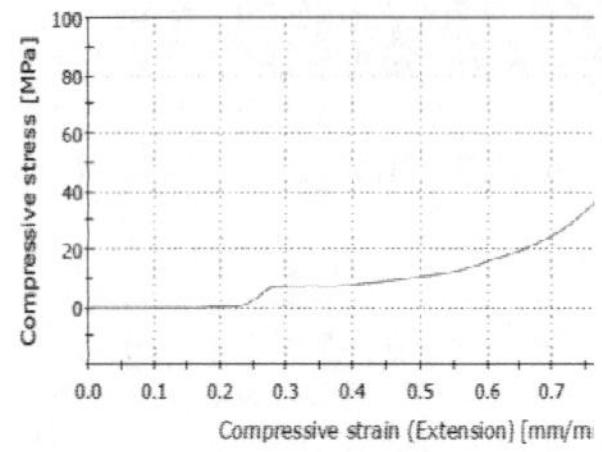

Run 1 with maximum stress=89.927 MPa, strain=0.206 mm/mm, energy absorbed=19.340 J

(a)

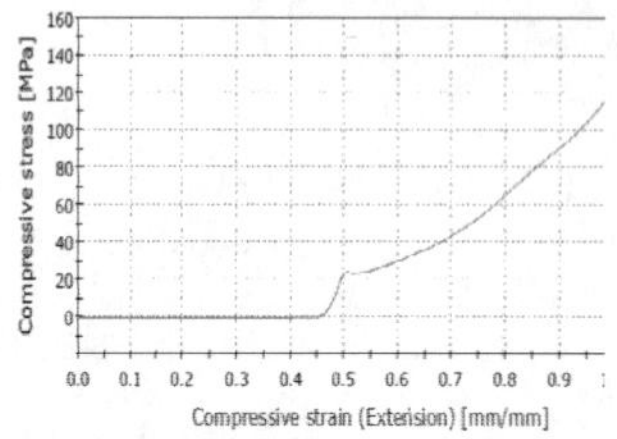

Run 2 with maximum stress=142.460 MPa, strain=1.039 mm/mm, energy absorbed=35.007 J

(b)

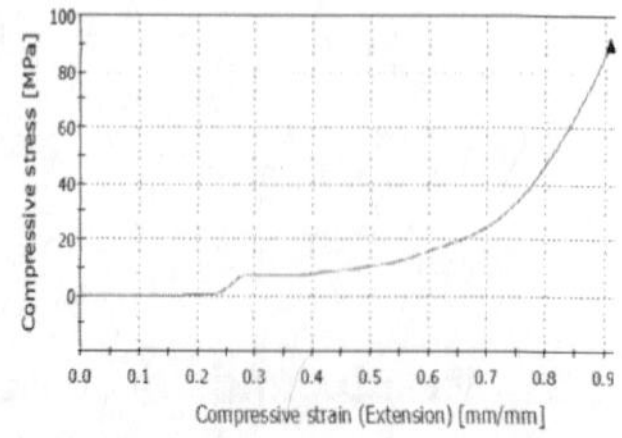

Run 3 with maximum stress=92.450MPa, strain=0.789mm/mm,energy absorbed=30.493 J

(c)

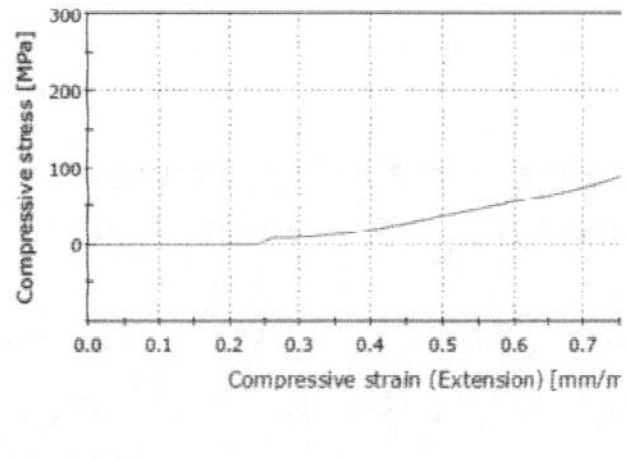

Run 4 with maximum stress=280.57 MPa, strain=1.098 mm/mm, energy absorbed=46.630 J

(d)

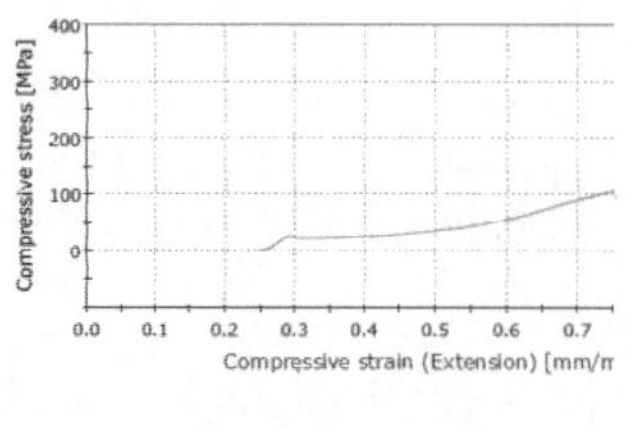

Run 5 with maximum stress, 310.850 MPa, strain=1.356 mm/mm, energy absorbed=56.505 J

(e)

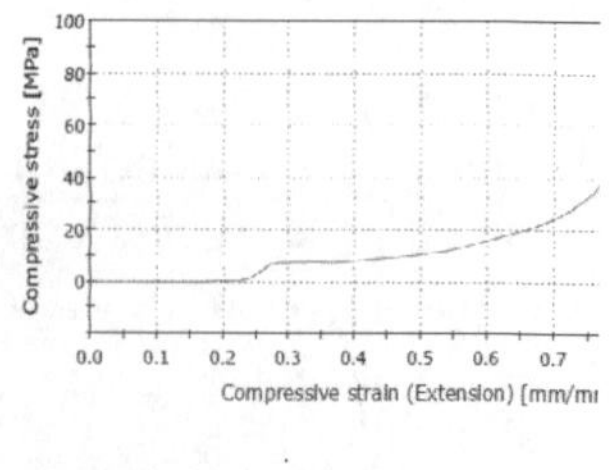

Run 6 with maximum stress, 90.346 MPa, strain=0.307 mm/mm, energy absorbed=24.398 J

(f)

Figure 3. Compressive stress-strain curves.

IOP Conf. Series: Materials Science and Engineering **290** (2017) 012084 doi:10.1088/1757-899X/290/1/012084

4. Conclusion

For this paper, samples of aluminium foam sandwich with two levels of core thickness and three levels of face thickness have been fabricated. After that, experimental work through compression test has been conducted in order to generate the stress-strain curve. Hence, this current study will help to provide the reliable experimental data that can be used to analyze energy absorption behavior of aluminium foam sandwich in automotive application. Important conclusions that can be drawn from this work are the absorbed energy in compression test increases with increasing foam and face sheet thickness of aluminium foam sandwich which is 64.42 J. Moreover, equation (3.0) provides plausible estimate for energy absorbed of aluminium foam sandwich.

Acknowledgement

The authors would like to express their sincere appreciation to the Research Management Centre, International Islamic University Malaysia and the Ministry of Higher Education,MOHE for financially providing and supporting fund for the research under science fund with the No: FRGS 15-247-0488.

References

[1] M W Andure, S C Jirapure and L P Dhamande 2012 *IJCA Proceedings on International Conference on Benchmarks in Engineering Science and Technology* pp 15–22

[2] S Yu, J Liu, M Wei, Y Luo, X Zhu and Y Liu 2009 *Materials & Design* **vol. 30** pp. 87–90

[3] D Ruan, G Lu and Y Choy 2010 *Compos. Struct.* **vol. 92** no. 9 pp. 2039–2046

[4] D K Rajak, L A Kumaraswamidhas and S Das 2014 *Procedia Materials Science* **vol. 5** pp 164–172

[5] M I Idris, T Vodenitcharova and M Hoffman 2009 *Mater. Sci. Eng. A* **vol. 517** no.1–2 pp. 37–45

[6] F Tarlochan, S Ramesh and S Harpreet 2012 *Compos. Part B Eng.* **vol. 43** no. 5, pp. 2198–2208

[7] R Dou, S Qiu and Y Ju *Mater. Trans.* 2016 **vol. 57** no. 2 pp 99–102

[8] S. B. Basturk and M Tanoglu *J. Reinf. Plast. Compos.* vol. 30 no. 18 pp. 1539–1547 2011

[9] Y G Li, Y H Wei, L F Hou, C L Guo and S Q Yang 2015 *J. Alloys Compd.* vol. 649 no. July pp. 76–81

Effects of water vapor on protectiveness of Cr_2O_3 scale at 1073 K

S K Arifin, M Hamid, A N Berahim, M H Ani

Department of Manufacturing and Materials Engineering, International Islamic University Malaysia (IIUM), Jalan Gombak, 53100 Kuala Lumpur, Malaysia

Email:mhanafi@iium.edu.my

Abstract. Fe-Cr alloy is commonly being used as boiler tube's material. It is subjected to prolonged exposure to water vapor oxidation. The ability to withstand high temperature corrosion can normally be attributed to the formation of a dense and slow growing Cr-rich-oxide scale known as chromia, Cr_2O_3 scale. However, oxidation may limit the alloy's service lifetime due to decreasing of its protectiveness capability. This paper is to presents an experimental study of thermo gravimetric and Fourier transform infrared analysis of Cr_2O_3 at 1073 K in dry and humid environment. Samples were used from commercially available Cr_2O_3 powder. It was cold-pressed into pellet shape of 12 mm diameter and 3 mm thick with hydraulic press for 40 min at 48 MPa. It then sintered at 1173 K in inert gas environment for 8 h. The samples are cooled and placed in 5 mm diameter platinum pan. It is subjected to reaction in dry and wet environment at 1073 K by applying 100%-Ar and Ar-5%H_2 gas. Each reaction period is 48 h utilizing Thermo Gravimetric Analyzer, TGA to quantify the mass changes. After the reaction, the samples then characterized with Fourier Transform Infrared Spectroscopy, FT-IR and Field Emission Electron Scanning Microscopy, FE-SEM. The TGA result shows mass decreasing ratio of Cr_2O_3 in wet (P_{H_2O} =9.5x10^5Pa) and dry environment is at a factor of 1.2 while parabolic rate at 1.4. FT-IR results confirmed that water vapor significantly broaden the peaks, thus promotes the volatilization of Cr_2O_3 in wet sample. FESEM shows mostly packed and intact in dry while in wet sample, slightly porous particle arrangement compare to dry. It is concluded that water vapor species decreased Cr_2O_3 protectiveness capability.

1. Introduction

Chromium oxide has variety of applications in various technologies such as catalyst, protective layer (corrosion and wear resistance of stainless steel), adhesion promoter, and magnetic recording media. Typically, the protective passive film on Fe–Cr stainless alloys is composed of an inner Cr_2O_3 film layer and an outer hydroxylated layer [1]. Despite their technological importance, basic understanding of Cr_2O_3 stability in any oxidation reaction involving gas adsorption is inadequate. There is strong correlation between adsorption and dissociation of water molecule on this activity of the surfaces. Thus, it is important to look into the surface reaction of Cr_2O_3 at high temperature [2]. Several studies of water on Cr_2O_3 surface have been done previously. Some were summarized in table 1.

Table 1. A summary of studies on the interaction of water with Cr_2O_3 surfaces by various methods.

Oxide	Gas	Method	Ref.
Cr_2O_3 (0001)	H_2O	Density Functional Theory (DFT+U)	[1]
Cr_2O_3	H_2O	FT-IR	[2]
Cr_2O_3	H_2O	GCMC Simulation	[3]
Cr_2O_3	H_2O	FT-IR, QNS, Dielectric Relaxation	[4]
ZnO, SnO_2, Cr_2O_3	H_2O	Physisorption Apparatus	[5]

Costa et al. [1] has studied water adsorption on (0001)-Cr–Cr_2O_3 surface done by applying the Density Functional Theory, DFT+U method. They conclude that water adsorbs with energies of adsorption between 80 and 115 kJ/mol on the surface. The driving forces for the adsorption are the formation of Cr–O coordination bonds with Cr(t_{2g})–O(2p) hybridization and the stabilization through H-bonds between adsorbate molecules in an ice-like network. They propose a representative model for a hydroxylated Cr_2O_3 surface to be used in further theoretical investigation of modeling passive films formed at the surface of stainless steels.

Kittaka et al. [3] has studied the structure of monolayer water molecules on a hydroxylated Cr_2O_3 surface using Grand Canonical Monte Carlo (GCMC) simulations. The experimental 2D phase diagram was described by the simulation, which gave the 2D critical temperature of T=319 K. They proposed structure of the monolayer of water below the 2D critical temperature is composed of a net of many hexagonal water molecules registered on the hexagonal structure of the hydroxylated surface. An analysis of Cr_2O_3-H_2O system using FT-IR and neutron scattering were done by Kuroda [4] and Morishige et al. [5]. In this system, the relaxation frequency for the dipolar relaxation of adsorbed water was found. The activation energy of this relaxation was also estimated to be 48 kJ/mol. IR spectra at 3656 cm^{-1}, 3616 cm^{-1}, 3606 cm^{-1}, 3529 cm^{-1}, 3490 cm^{-1} and 3402 cm^{-1} were found broader by increasing the temperature from 93 K until 298 K. This revealed the presence of strong hydrogen bonding among the water molecules and the restricted motion for the water molecules adsorbed.

Both Kittaka and Kuroda's investigations were done in temperature lower than 320 K. Focus has been given on sub-ambient temperature surface analysis. Meanwhile Costa's study is limited to computational simulation and temperature lower than 800 K. Kaderi et al. [2] has studied the initial stage of Fe-Cr oxidation. At T=1073 K, it was found that formation of metal oxide including Cr_2O_3 is retarded due to water vapor. IR analysis on sample surface was done to explore the possibility of metal's surface catalytic activity. It was observed that dry sample has more intense peak and tends to be narrower, while samples oxidized in wet tend to have relaxed broader peak. However, in case of Fe-Cr alloy oxidations, Fe may also oxidized alongside Cr. Study on high purity Cr_2O_3 shall give better clarity of Kaderi's findings.

The objective of this work is to study the stability of highly pure Cr_2O_3 samples in dry and wet environment at 1073 K by thermo gravimetric and FT-IR analysis.

2. Experimental Procedures

2.1 Materials

Samples were used as it commercially available Cr_2O_3 powder from Sigma-Aldrich®. It weighed at 100 mg then was poured into a 12 mm diameter cylindrical steel die. It was cold-pressed using hydraulic press at 48 MPa of pressure and stay for 40 minutes. The sample then was carefully taken out from the die and ~3 mm thick pellet was obtained. Sample then subjected to calcination process at 473 K for 2 h in oven for dehydration. Finally, it was sintered at 1173 K for 8 h in argon, Ar gas environment. The sample was let cooled to room temperature in the furnace. Final weight of the sample was measured to be 98.3 mg.

2.2 Thermo gravimetric measurement

Sample was subjected to dry and wet environment in Hitachi® STA7300 TGA. Figure 1 shows the schematic diagram of sample placement in the TGA system. Major parts of this machine are furnace, thermocouples and weighing scale. The weighing scale is an alumina insulated Pt-Rh thermocouple connected to laser weighing device. It has two sides – reference side and sample side. For reference, a blank 5 mm diameter platinum sample pan was placed on reference side and calibrated. Cr_2O_3 sample was cut and placed in another platinum pan. The pan with sample then was placed on sample side and its weight was measured and compared to reference. Sample weight was measured to be 41.87 mg. Specific area of the sample was calculated to be 2.1 mm^2/mg. Heating rate was programmed at 5 K/min to heat from 300 K to 1073 K.

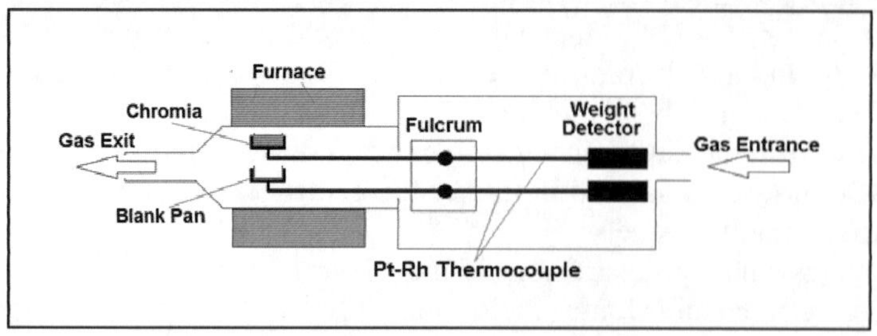

Figure 1. Schematic diagram of sample placement in the TGA system.

Before the reaction started, furnace chamber was evacuated with Ar gas for 30 minutes. After evacuation process was done, the reaction was started and set to hold temperature for 48 h when the temperature reaches 1073 K. For reaction in dry environment, Ar gas was used throughout the reaction at 100 ml/min flow rate. For wet environment, a gas mixture of Ar-5%H_2 was used with the same flow rate. The 5%H_2 is expected to react with impurity O_2 to form water vapor. Below is the possible reaction occurs to form water vapor.

$$2H_2 + O_2 \rightarrow 2H_2O \qquad (1)$$

Using Gibbs free energy and thermochemical data collected from Barin [6], water vapor partial pressure, P_{H_2O} is calculated to be 9.5x10^5 Pa. After 48 h of reaction, gas supply is shut off. Sample then was let cooled to room temperature in the furnace. Both reactions took place in the lab where no human activity within radius of >10 m to avoid vibrations and obtain mass change data within the range of ±1 μg.

2.3 FT-IR measurement

IR spectra measurement was done by using Perkin-Elmer® Spectrum 100 and utilizing the attenuated total reflection, ATR technique. Before measurement, the apparatus was purged with N_2 gas for 25 min to avoid any organic contaminators. Sample from the TGA furnace immediately taken out after being cooled to room temperature. The surface of the sample placed on top of FT-IR window. Eight sheets of tissue papers were placed on top of the sample act as a damper to prevent sample from fractured while applying force on it. Force applied on the sample was at range of 100~110 N. Each sample (control, dry and wet) IR spectra were accumulated 20 times at a resolution of 4 cm^{-1} and 32 scan time.

3. Results and Discussions

3.1 Mass change analysis.

Figure 2 shows the mass change of Cr_2O_3 after being exposed to dry and wet environment for 48 h at 1073 K. The TGA result shows decreasing of Cr_2O_3 mass in both dry and wet conditions. For dry condition, a mass linearly decreased at $1.10x10^{-5}$ mg.cm^{-2}/s was observed in the first 5.6 h. The decreased rate starts to plateau until the end of reaction time (48 hours). For wet condition, a steep mass decrease at $1.37x10^{-5}$ mg.cm^{-2}/s was observed at the first 3.1 h before its slope changed at rate of $7.36x10^{-7}$ mg.cm^{-2}/s. It is interesting to note that the Cr_2O_3 mass is continually decreasing until end of reaction time, in contrast to the observation in dry environment. The overall mass decrease for both conditions is 0.49 % in dry condition and 0.59 % in wet condition.

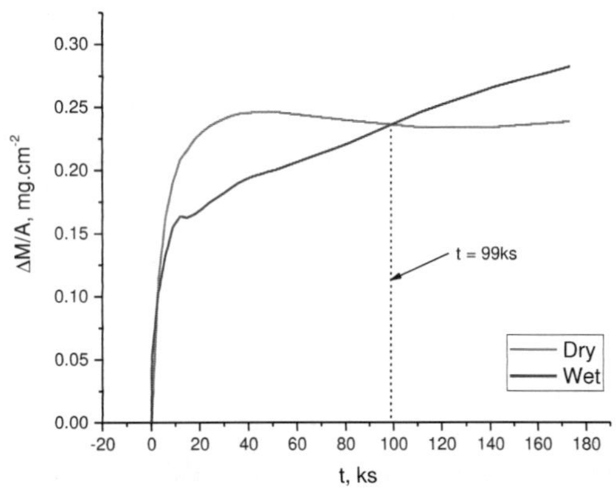

Figure 2. Mass change of Cr_2O_3 (linear plot) being exposed at 1073K for 48 h in dry and wet environment.

By ratio, the mass change different between these conditions is at 1.2. Another important observation is that at 99 ks (27.5 h), the mass change of sample in wet condition surpassed sample in dry condition. This shows that water vapor affect the stability of Cr_2O_3 at exposure time above 27.5 h.

Figure 3 shows the parabolic rate for dry and wet condition. At the end of the reaction time, k_p for dry environment is $3.29x10^{-7}$ mg^{2}cm^{-4}/s while for wet environment is $4.61x10^{-7}$ mg^{2}cm^{-4}/s. From the k_p obtained, the ratio for wet and dry condition, $k_{p,wet}/k_{p,dry}$ is calculated at 1.4. It is consistent with the previous study by Graham and Davis [7] which reported the weight loss ratio of Cr_2O_3 when exposed at 1473 K for 72 h in high oxygen partial pressure to be 1.5.

The presence of water vapor in oxidation environment greatly affects the formation of oxide scales especially on Fe-Cr alloys [8], [9]. Its presence is also increases the evaporation loss of Cr_2O_3 [10]. Significant loss of Cr_2O_3 in wet condition is due to several factors. One of the factors is volatilization. Cr_2O_3 scale protectiveness capability can be degraded by the volatilization of Cr_2O_3 [11].This problem is one of the major limitations on the high temperature use of Cr_2O_3 forming alloys which is more serious in rapidly flowing gases [12]. Asteman's [13] studies showed that H_2O reduced the ability of stainless steel to maintain its protective Cr_2O_3 scale due to volatile species. Below is the possible reaction occurs during volatilization of Cr_2O_3.

IOP Conf. Series: Materials Science and Engineering **290** (2017) 012085 doi:10.1088/1757-899X/290/1/012085

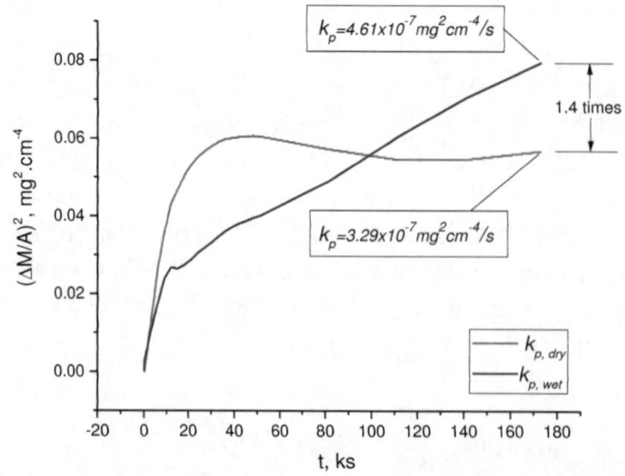

Figure 3. Oxidation kinetics (parabolic plot) k_p of Cr_2O_3 being exposed at 1073K for 48 h in dry and wet environment.

Equation (2) represents reaction that occurs in dry environment while equation (3) and (4) are in wet. In this study, the case is most likely similar to equation (4) where the oxygen partial pressure is low. Michalik et al. [14] stated that at high oxygen potentials, the addition of H_2O to the gas had negligible effect on the scaling behavior. However, at low oxygen potentials when the P_{H_2O}/P_{H_2} ratio was held constant, the oxidation rate increased with water partial pressure. It is suggested that the evaporation of volatile chromium species $CrO_2(OH)_2$ primarily occur along the grain boundaries. It accounts for scale cracking and spallation of oxide scales in the presence of water vapor [15].

$$Cr_2O_3(s) + \frac{3}{2}O_2(g) \rightarrow 2CrO_3(g) \qquad (2)$$

$$Cr_2O_3(s) + O_2(g) + H_2O(g) \rightarrow 2CrO_2(OH)_2(g) \qquad (3)$$

$$2Cr_2O_3(s) + H_2O(g) \rightarrow 2CrO_2(OH)_2(g) + \frac{1}{2}O_2(g) \qquad (4)$$

3.2 IR transmission spectra.

Figure 4 shows the IR transmission spectra of all samples; control, dry and wet. Control sample is which the Cr_2O_3 is not been exposed to high temperature reaction. The differences in IR transmission spectra are apparent by making comparisons between samples control (black), reaction in dry environment (red) and reaction in wet environment (blue).

There are 4 main range of area that needs to be given a specific attention. It is grouped and named Range 1 to 4 to represent IR spectra in the range of 3500-4000 cm^{-1}, 1800-2500 cm^{-1}, 1000-1200 cm^{-1} and 400-700 cm^{-1} respectively. Figure 5-8 are the close up view of IR spectra of all group of range.

IOP Conf. Series: Materials Science and Engineering **290** (2017) 012085 doi:10.1088/1757-899X/290/1/012085

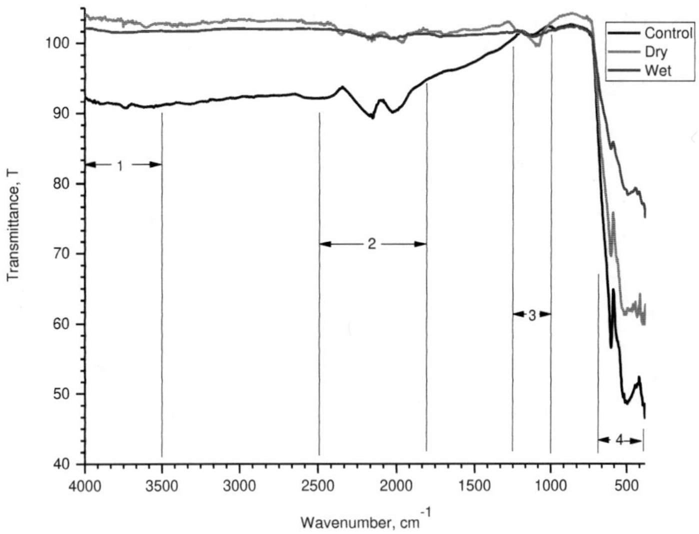

Figure 4. IR spectrum of Cr_2O_3 in dry and wet condition.

Figure 5 shows a close up view for Range 1. Two peaks appeared at 3750 cm^{-1} and 3600 cm^{-1} for control and dry samples while very broad IR spectrum were seen at wet condition sample. This effect is due to the water adsorption in the wet environment. At this region, typical of the stretching vibrations of the hydroxyls of the Cr_2O_3 surface observed at 3714 cm^{-1}, 3656 cm^{-1}, 3614 cm^{-1}, and 3408 cm^{-1} from chemisorbed hydroxyl groups on the Cr_2O_3 surface [4].

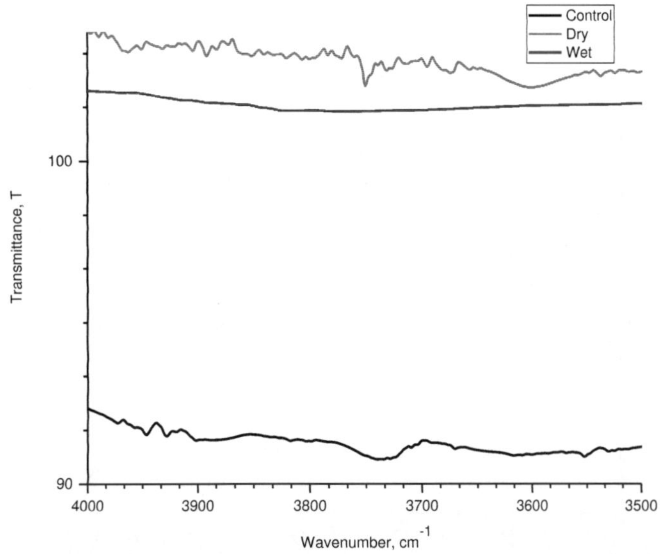

Figure 5. Close up view of IR spectra in the Range 1.

Figure 6 shows a close up view for Range 2. At 2155 cm^{-1} and 2040 cm^{-1}, intense peaks were observed at control sample while it is broader at wet sample. Kuroda [4] has studied, when Cr_2O_3

IOP Conf. Series: Materials Science and Engineering **290** (2017) 012085 doi:10.1088/1757-899X/290/1/012085

exposure to higher temperature, the peak would become broader and would disappear upon heating. This is in good agreement with this spectrum where control sample is a sample which not being exposed to high temperature environment.

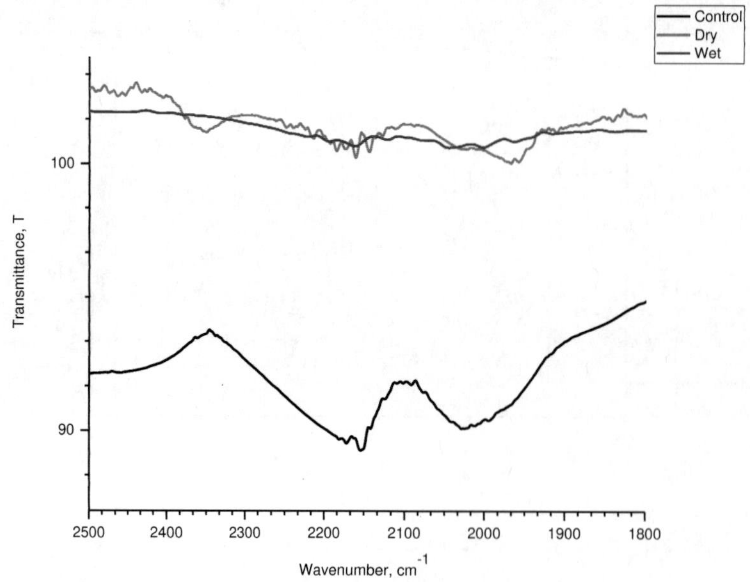

Figure 6. Close up view of IR spectra in the Range 2.

Figure 7 shows a close up view for Range 3. One range of intense peak is seen at 1080-1105 cm^{-1} for control sample and broad spectrum band for wet sample. The peak appeared to be broadened and shifted to 1121 cm^{-1}. This effect is due to the water adsorption in the wet environment.

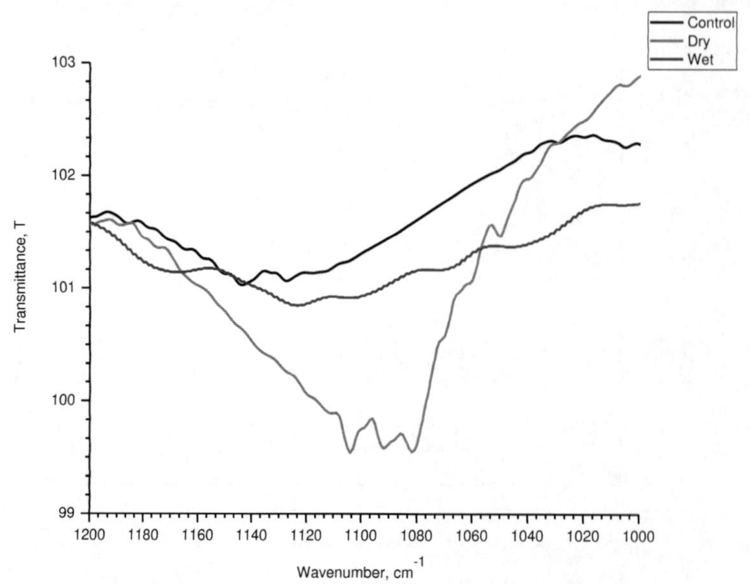

Figure 7. Close up view of IR spectra in the Range 3.

Figure 8 shows a close up view for Range 4. In this region, all three samples have very similar peaks appear at 607-608 cm^{-1}. Dry sample shows heavy fluctuations while no significant peaks at wet sample. The IR spectra is broader in wet compare to both control and dry sample wet This is due to the introduction of wet environment to the system has been shown to decrease the intensities of the bands and new broad OH bands appeared [5].

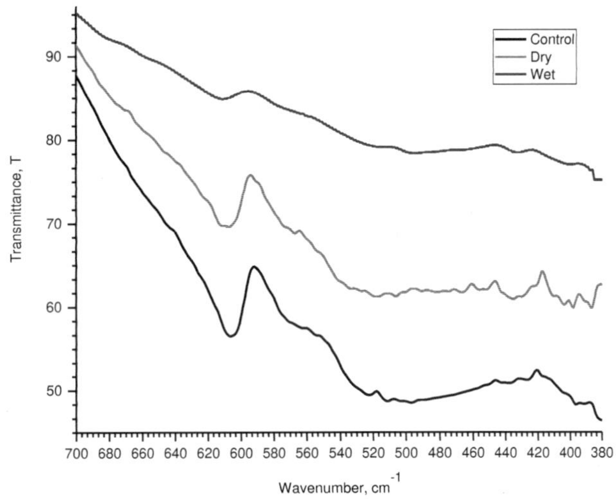

Figure 8. Close up view of IR spectra in the Range 4.

3.3 Surface Morphology

Figure 9 shows images of surface morphology of Cr_2O_3 after 1073 K for 48 h in dry and wet condition respectively. In dry sample, the surface particle arrangement is more compact and intact. Smaller grain size of Cr_2O_3 particles seen to attach with bigger particle. In wet sample, the arrangement of particle is slightly loose compare to dry sample. Smaller particle that attached to bigger particle is also lesser. This is might be due to the introduction of water vapor that increases the volatilization activity.

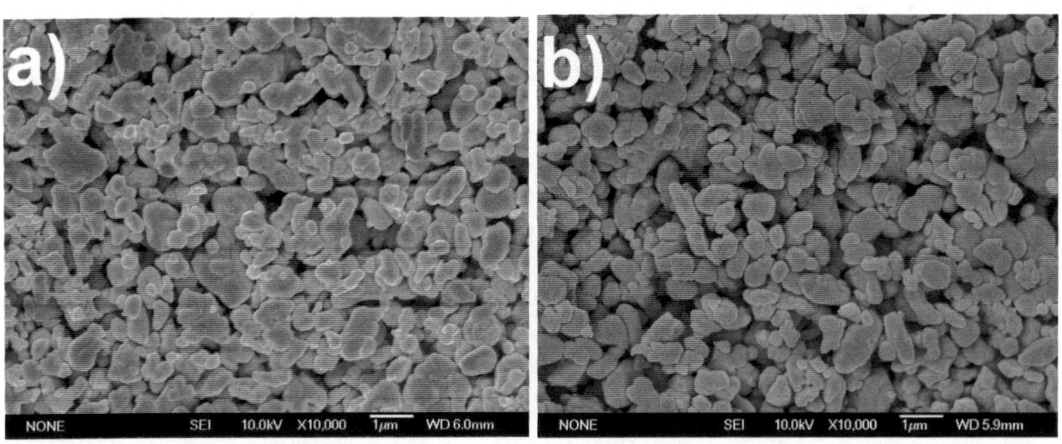

Figure 9. Surface morphology of Cr_2O_3 after 1073K for 48 h in a) dry and b) wet condition.

Hickman and Gulbransen [16] in their study has concluded that the presence of water vapor increase the volatilization of metal oxide by 30 % as compared to dry environment.

IOP Conf. Series: Materials Science and Engineering **290** (2017) 012085 doi:10.1088/1757-899X/290/1/012085

Young and Pint [17] studied to predict Cr-evaporation fluxes in the temperature range 923–1073 K. They concluded that a volatilization model, such as that also proposed by Asteman et al. [13], for $CrO_2(OH)_2$, can account for the losses observed during exposure to air and H_2O.

4. Conclusions

We have shown from thermo gravimetric measurement that initially mass loss in wet and dry environment is almost similar at exposure time <1.4 h. Later, the mass change is separated between these two environments up to 27.5 h. Above 27.5 h, the mass of Cr_2O_3 in dry condition is kept constant, while sample in wet condition increasingly losing its weight. It could be deduced that exposure time of 27.5 h is the critical time where Cr_2O_3 in wet condition loss its protectiveness based on thermodynamic calculation. $CrO_2(OH)_2$ is susceptible of volatile species. Careful analysis of FT-IR results supported that OH species somehow broaden the peaks significantly, and it is apparent proof of formation of volatile species. This study clearly shows that OH species degrade the protectiveness of Cr_2O_3 scale. Controlling it species is important to maintain protectiveness of Cr_2O_3 species.

References

[1] Costa D, Sharkas K, Islam M M, and Marcus P 2009 *Surf. Sci.* **603** 2484–2493
[2] Kaderi A, Meskon S R, Ani M H, and Othman R 2012 *Adv. Mater. Res.* **576** 619–622
[3] Kittaka S, Sugiyama H, Sumi H, and Takahara S 2005 *Adsorpt.* **11** 103–107
[4] Kuroda Y, Kittaka S, Takahara S, Yamaguchi T, and Bellissent-Funel M C 1999 *J. Phys. Chem. B* **103** 11064–11073
[5] Morishige K and Kittaka S 1981 *Surf. Sci.* **109** 291–300
[6] Barin I and Platzki G 1995 *Thermochemical Data of Pure Substances* (VCH, Weinheim)
[7] Graham H C and Davis H H 1971 *J. Am. Ceram. Soc.* **1970** 89–93
[8] Ani M H, Kodama T, UedaM, Kawamura K and Maruyama T 2009 *Mater. Trans.* **50** 2656–2663
[9] Setiawan A R, Ani M H, Ueda M and Kawamura K 2010 *ISIJ Int.* **50** 259–263
[10] Holcomb G R 2008 *Oxid. Met.* **69** 163–180
[11] Yamauchi A, Kurokawa K and Takahashi H 2003 *Oxid. Met.* **59** 517–527
[12] Hammer J E, Laney S J, Jackson R W, Coyne K and Pettit F S 2007 *Oxid. Met.* **67** 1–38
[13] Asteman H, Segerdahl K, Svensson J E and Johansson L G 2001 *Mater. Sci. Forum* **369–372** 277–286
[14] Michalik M, Tobing S L, Hansel M, Shemet V, Quadakkers W J and Young D J 2014 *Mater. Corros.* **65** 260–266
[15] Fry A, Osgerby S and Wright M 2002 *NPL Rep. MATC* **90** 1–39
[16] Hickman J W and Gulbransen E A 1947 *Trans. Am. Inst. Min. Met. Engrs.* **171** 371
[17] Young D J and Pint B A 2006 *Oxid. Met.* **66** 137–153

IOP Conf. Series: Materials Science and Engineering **290** (2017) 012086 doi:10.1088/1757-899X/290/1/012086

Study on tea leaves extract as green corrosion inhibitor of mild steel in hydrochloric acid solution

A B Hamdan, Suryanto, and F I Haider

Department of Manufacturing and Materials Engineering, International Islamic University Malaysia (IIUM), Jalan Gombak, 53100 Kuala Lumpur, Malaysia

Email: surya@iium.edu.my

Abstract. Corrosion inhibitor from extraction of plant has been considered as the most preferable and most chosen technique to prevent corrosion of metal in acidic medium because of the environmental friendly factor. In this study, black tea leaves extraction was tested as corrosion inhibitor for mild steel in 0.1M of hydrochloric acid (HCl) with the absence and presence of corrosion inhibitor. The efficiency and effectiveness of black tea as corrosion inhibitor was tested by using corrosion weight loss measurement experiment was carried out with varies parameters which with different concentration of black tea extract solution. The extraction of black tea solution was done by using aqueous solvent method. The FT-IR result shows that black tea extract containing compounds such as catechin, caffeine and tannins that act as anti-corrosive reagents and responsible to enhance the effectiveness of black tea extract as corrosion inhibitor by forming the hydrophobic thin film through absorption process. As a result of weight loss measurement, it shows that loss in weight of mild steel reduces as the concentration of inhibitor increases. The surface analysis was done on the mild steel samples by using SEM.

1. Introduction

Corrosion can be defined as degradation of materials when being in contact with corrosive environment as stated by Fontana [1]. Despite of many definitions provide by scholars, corrosion basically can be defined as interaction material with the environment. Corrosion of metal in engineering field has produced undesirable environment in industry. In which, the degradation of metal that happened in pipeline, tank and cooling system has brought many issues and problems. As a result there will be wastage of money, time, materials and manpower. Pournarazi [2] has stated that one of the important and useful techniques to protect metal from the deterioration due to corrosion is by using corrosion inhibitors. According to Otaibi [3], corrosion inhibitor is known to be the most preferable method to avoid or prevent the destruction of metal because it is low cost and it has excellent anti-corrosive reagent especially when it used in acidic environment. Although corrosion inhibitors have a great anti-corrosive properties but it is proven to bring secondary effect to the environment and human health.

Thus, there are tremendous studies and researches have been carried out to develop environmental friendly inhibitors. Research done by Bribri [4] and Loto [5] have found out that the use of environmental friendly corrosion inhibitors or known as "green inhibitors". The intensive study of green inhibitors is provoked by properties of green inhibitor which are known as environmental friendly, non-toxic, biodegradable and low cost. Considering the cost and the availability of the raw

IOP Conf. Series: Materials Science and Engineering **290** (2017) 012086 doi:10.1088/1757-899X/290/1/012086

materials, the natural plant extraction are commonly used. Research done by Fouda [6] has stated that black tea leaves extraction contains many compounds such as polyphenols, caffeine, flavonoids, tannins, volatile oil and alkaloids that will act as anti -reagents for corrosion in acidic environment

Thus in this study, the effectiveness of black tea leaves extraction as corrosion inhibitors was studied for mild steel in Hydrochloric acid solution by using weight loss measurement technique to determine the corrosion rate and the inhibitor efficiency percentage. The mild steel specimen will be further characterized by using Scanning Electron Microscope for the surface analysis.

2. Experimental procedures

2.1 Preparation of Black Tea solution extract

The black tea powder was obtained directly from the tea bag packets. The extraction process was done using the aqueous solvent method. The solution of inhibitor was prepared by boiling the 4.5 grams of black tea bag in 750ml of distilled water for 90 minutes. Then, the extract solution was left for 24 hours at room condition and filtered. Next, the solution extract of black tea was further concentrated by using Rotary Vacuum Evaporator for 4 hours. The 30 ml of extract left was collected kept in the chiller at 4℃ to keep it fresh and avoid from it denatured.

2.2 Preparation of mild steel specimens

The mild steel was cut into average dimension of 30 x 20 x 2mm. The specimens were grinded with different grade from 120 grits to 420 grits of silicon carbide abrasive paper to remove the oxide film. Then it was polished and degreased with acetone, cleaned and dried for the corrosion weight loss measurement experiment.

2.3 Preparation of test media and solution extract

The 37% AR of Hydrochloric acid (HCl) is diluted with distilled water to get specific concentration of 0.1M, 0.5M and 1M. With ratio 1:10, 25 ml of HCl was added slowly in the 250 ml of distilled water in 7 different beakers. Then tea extract of inhibitor was added by using dropper with difference amount of drops of 2, 4, 6, 8, 10 and 12 drops for the first 6 beakers. There is no tea extract added in the seventh beaker as it will use for sample reference.

2.4 Weight loss experiment

Mild steel specimens were immersed in 0.1M HCl solutions with various concentration of inhibitor which 2, 4, 6, 8, 10 and 12 drops for 3 hours at room temperature. The test specimens were immersed in the acid medium that contains different amount of tea extract concentration and without the tea extract as sample reference. For every 1 hour, the test specimens were removed, rinsed with distilled water and dried for the weight measurement. The weight loss of the test specimens were calculated by using equation in (1).

$$\text{Weight loss} = Wo - Wi \qquad \textit{Eq (1)}$$

Where Wo is the initial weight of mild steel specimen in gram and Wi is the final weight of mild steel specimens in gram after every 1 hour interval. The inhibitor efficiency (IE %) can be measured and calculated by using the equation shows in equation (2). Where, IE% is the inhibitor efficiency, R0 and R1 (in g) are the values of the weight loss observed of mild steel in the absence and presence of inhibitor respectively.

$$IE = \frac{R_0 - R_1}{R_0} X\ 100 \qquad Eq\ (2)$$

$$R = \frac{86.7x\ \Delta w}{A\ x\ T\ x\ \rho} X\ 100 \qquad Eq\ (3)$$

Where R = Corrosion rate (mpy), Δw = Weight loss (mg), A = Area of metal exposed to corrosive media (cm^2), ρ = Density of mild steel (g/cm^3) = $7.86 g/cm^3$, T = Time of immersion (h)

2.5 Surface analysis

Surface morphology of the mild steel specimen was examined by using SEM and EDX. For the surface analysis, mild steel specimens were immersed in the 0.1M HCl solution with condition of the presence and absence of black tea extract solution. The mild steel was cleaned and dried after 24 hours and 7 days immersion for the surface analysis.

3. Results and Discussion

3.1 Characterization of black tea extract

The black tea extract solution was characterized by using FT-IR to determine the presence of certain functional group in a molecule of organic and inorganic compounds. Figure 1 shows the FT-IR spectrum peaks of the tea extract solution. In this spectrum, the peak appeared at 3307 cm^{-1} corresponds to stretching mode of hydroxyl (O-H) and (N-H) amide group. While at peak 1633 cm^{-1} corresponds to the stretching mode of carbonyl group (C=O).

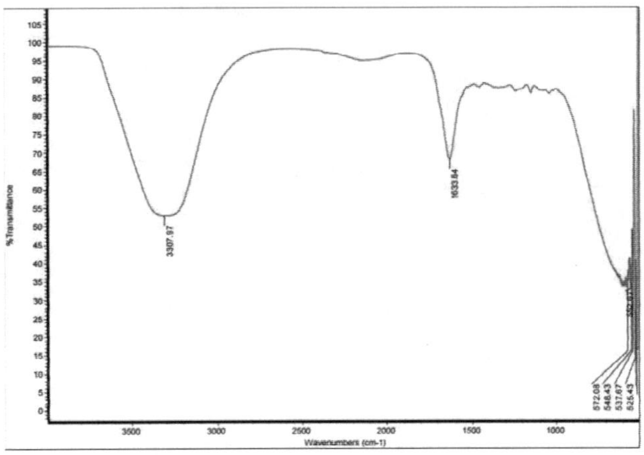

Figure 1. FT-IR spectrum of black tea extracts solution.

Therefore based on the presence of these two peaks, it indicates that the tea extract solution contains the functional group of molecular structure of main constituent of tea leaves which are catechin, caffeine and tannins which responsible as anti-corrosion reagent in black tea extract. Besides, the presence of tannin in black tea extract is proved to be an effective corrosion inhibitor. While the presence of catechin and caffeine provide high complexation affinity to metals which responsible for effective corrosion inhibitor performance .In other words, the presence of catechin and caffeine will provide strong adsorption molecules of the tea extract film on the mild steel surface. To conclude with, presence of these compounds are really important to make black tea as effective corrosion inhibitors.

3.2 Weight loss measurement

To start with, one drop of tea extract solution is approximately equal with ±0.005ml. As shown in figure 2, the effectiveness of black tea extract solution as corrosion inhibitor of mild steel in 0.1M HCl can be proven by increasing the concentration of black tea extract. It appears that, the loss in weight of mild steel reduced as the concentration of black tea inhibitor increased. Basically, the concentration of the black tea extract can be varies by increasing or reducing the number of black tea droplets in the HCl medium.

IOP Conf. Series: Materials Science and Engineering **290** (2017) 012086 doi:10.1088/1757-899X/290/1/012086

Thus, mild steel immersed in HCl with the absence of tea extract solution has the highest weight loss of mild steel which the values ranging between 13.8 mg at the beginning to 27.3 mg at the end of experiment in 3 hours. While the lowest of weight loss change was achieved with the presence of 12 drops of tea extract solution with values ranging between 4.8 mg at the beginning and 10.3 mg after 3 hours of immersion as shown in figure 2.

The curve obtained for the variation of weight loss of varies concentration of black tea extract with exposure time of 3 hours for mild steel specimen in 0.1M HCl with and without the tea extract solution. It seems that, the first highest curve represents the weight loss of mild steel specimen immersed throughout the hours with the absence of black tea inhibitor. However, mild steel immersed in 0.1M HCl with the presence of black tea extract solution shows lowest weight loss of mild steel which has the highest concentration (12 drops) of black tea solution.

It appears that, as the concentration of black tea extract is increased it will speed up the formation of thin film on the surface of mild steel as well as reduce the aggressiveness of HCl solution. Therefore undoubtedly, it is proved that the presence of black tea extract can reduce the weight loss of mild steel as the concentration of the black tea increased.

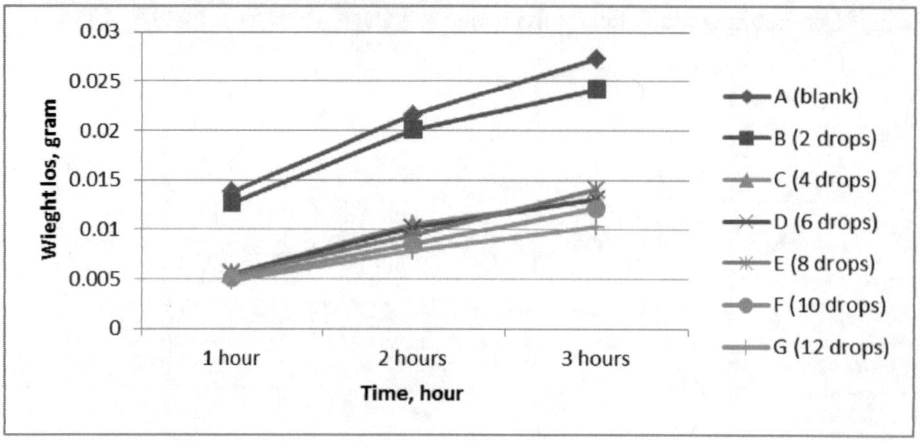

Figure 2. Variation of weight loss with exposure time for mild steel specimens immersed in 0.1M HCl with various tea leaves extract concentrations.

It appears that rate of corrosion of mild steel will drop drastically as the black tea extract solution is added in to the HCl solution as shown in figure 3. The rate of corrosion is reduced by range of 70%-90% with presence of black tea extract that act as corrosion inhibitor in HCl solution. The rate of corrosion is also reduced as the weight loss of mild steel reduced. To some extent, corrosion reagents that presence in black tea extract such as tannins, catechins and caffeine can form thin film through the adsorption process that act as barrier from the corrosive environment attack.

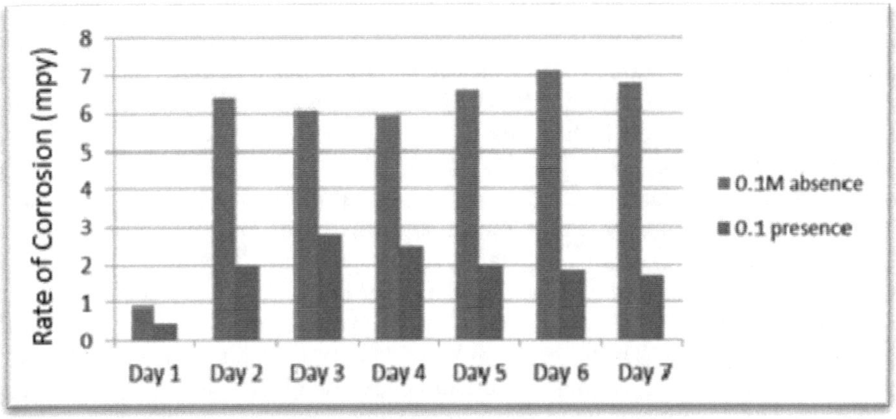

Figure 3. Corrosion rate of mild steel immersed in 0.1M HCl after 7 days with presence and absence of black tea extract solution in (mpy).

On the other hand, the percentage of inhibitor efficiency is calculated by using equation (2) and the graph of percentages obtained is recorded in figure 4.10. The percentage of inhibitor efficiency for 0.1M, 0.5M and 1M of HCl solutions are at 97%, 92% and 87% respectively. The percentage of inhibitor efficiency reduced as the concentration of HCl solutions increased and as the days passed because of high aggressiveness of corrosive environment present. In which high concentration of acid solution can reduce the effectiveness of black tea extract to slow down the rate of corrosion. On the contrary, the corrosion inhibitor is classified as good inhibitor when the inhibitor efficiency percentage is greater than 90%. Therefore, black tea extract solution performed very well and effective as corrosion inhibitor for 0.1M and 0.5M of HCl solutions at which the percentage of inhibitor efficiency is 97% and 92% respectively.

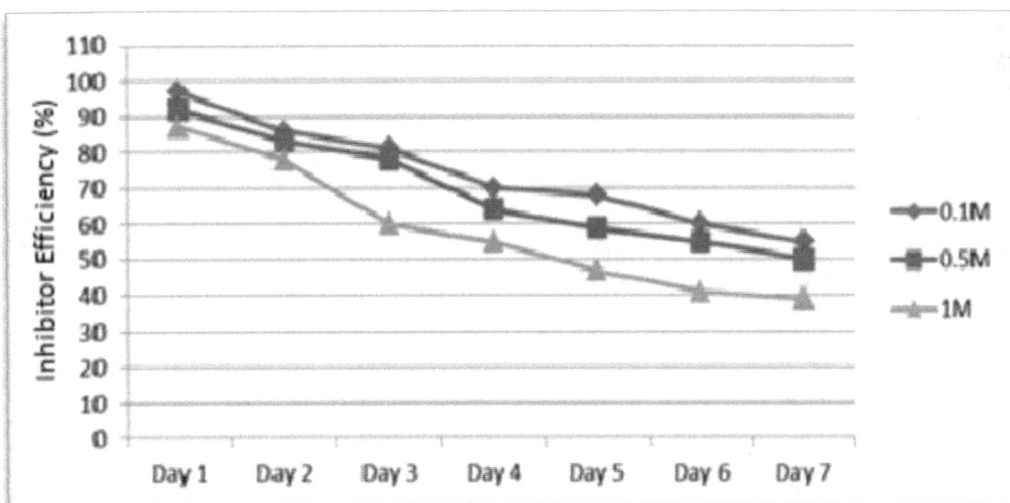

Figure 4. Inhibitor efficiency of black tea extract in HCl solution.

3.3 Surface Analysis

The Scanning Electron Microscope (SEM) images were recorded in figure 5 to observe the changed of mild steel before and after the immersion in 1M of HCl solution with the absence and presence of black tea extract. Figure 5(a) shows the scratches of mark that appeared from the grinding step before the mild steel is immersed in HCl solution. The surface of mild steel is observed to be free from any corrosion product. However, in figure 5(b) the surface of mild steel was observed with the production of corrosion product such as cavities and pits where it shows that mild steel that immersed in 0.1M HCl solution without the presence of black tea extract are exposed to corrosion activities which is degrading the metal surface.

On the other hand, the figure 5(c) shows the mild steel that immersed in 1M HCl solution with the presence of black tea extract It is proved that, the smooth surface of mild steel can be seen because of the adsorption of black tea extract molecules that covered the surface of mild steel from the exposure of corrosion attack from the aggressiveness of the environment. Anti – corrosion reagents such as tannins, catechin and caffeine were absorbed on the surface of mild steel to form hydrophobic film to prevent from the corrosion attack. These compounds are said to be stable to form the thin even in high concentration acidic medium. This shows the effectiveness of black tea extract in HCl solution in order to slow down the corrosion process of the mild steel.

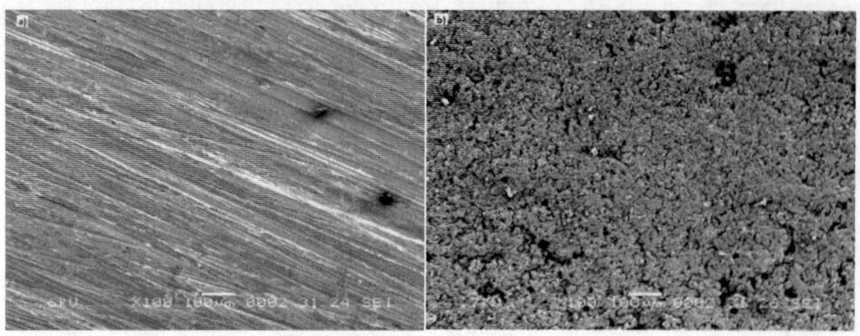

Figure 5. Mild steel surface after immersion in 0.1M HCl for 24 hours with the presence of black tea extract (left), without the presence of black tea extract (right).

4. Conclusion

The efficiency of black tea extract solution of mild steel in 0.1M HCl solution increased by increasing the black tea extract concentration. By weight loss measurement, it proved that loss in weight of mild steel reduced as the concentration of inhibitor increased. While, inhibitor efficiency increased as the concentration of HCl solution decreased. It revealed that black tea extract has very good inhibitor efficiency for 0.1M where it achieved 97%. In other words, black tea extract solution is consider as a very good corrosion inhibitor in order to slow down the rate of corrosion of mild steel in HCl solution.

Acknowledgements

The authors would like to acknowledge the financial support given from FRGS-14-117-0358 grant of International Islamic University Malaysia.

References

[1] Fontana, M.G, (2005), Corrosion Engineering, 3rd ed. New York: McGraw-Hill.
[2] Pournazari, S, Moayed, M, and Rahimizadeh, M, (2013), Admixture of benzaldehyde and benzene-1,2-diamine as a new corrosion inhibitor for mild steel in 0.5M sulphuric acid, *Corrosion Science*, 7(1)20-31.
[3] Otaibi, M. A, et all (2012), Corrosion inhibitory of some plant extracts on mild steel in acidic media, *Arabian Journal of Chemistry*, pp. 1-7.

[4] Bribri, A.E, et al, (2013), The use of Euphorbia falcata extract as eco-friendly corrosion inhibitor, *Materials Chemistry and Physics*, 141(1) 240-247.
[5] Loto, C.A., (2011), Inhibition effect of Tea (Camellia Sinensis) extract on the corrosion of mild steel in dilute sulphuric acid, *J. Mater. Environ. Sci.*, 2 (4), 335-344.
[6] Fouda, S. et all, (2014) Gibberellic acid as green corrosion inhibitor for carbon steel in hydrochloric acid solutions, *J. Mater. Environ. Sci.,* 5(1) 767-778.

IOP Conf. Series: Materials Science and Engineering **290** (2017) 012087 doi:10.1088/1757-899X/290/1/012087

Ginger extract as green corrosion inhibitor of mild steel in hydrochloric acid solution

A Fidrusli, Suryanto and M Mahmood

Department of Manufacturing and Materials Engineering, International Islamic University Malaysia (IIUM), Jalan Gombak, 53100 Kuala Lumpur, Malaysia

Email: surya@iium.edu.my

Abstract. Ginger extract as corrosion inhibitor from natural resources was studied to prevent corrosion of mild steel in acid media. Ginger rhizome was extracted to produce green corrosion inhibitor (G-1) while ginger powder bought at supermarket was also extract to form green corrosion inhibitor (G-2). Effectiveness of inhibitor in preventing corrosion process of mild steel was studied in 1.0 M of hydrochloric acid. The experiment of weight loss method and polarization technique were conducted to measure corrosion rate and inhibition efficiency of mild steel in solution containing 1.0 M of hydrochloric acid with various concentration of inhibitor at room temperature. The results showed that, the rate of corrosion dropped from 8.09 mmpy in solution containing no inhibitor to 0.72 mmpy in solution containing 150g/l inhibitor while inhibition efficiency up to 91% was obtained. The polarization curve in polarization experiments shows that the inhibition efficiency is 86% with high concentration of inhibitor. The adsorption of ginger extract on the surface of mild steel was observed by using optical microscope and the characterization analysis was done by using pH measurement method. When high concentration of green inhibitor in the acid solution is used, the pH at the surface of steel is increasing.

1. Introduction

Corrosion is a deterioration of a metal alloy under the influence of its environment [1]. Corrosion is a destructive result of chemical reaction between a metal and its environment. The environment such as humid air, steam, seawater, alkaline and acids are positive towards the corrosion process. Corrosion happens because metal wants to achieve a stable state since metals are unstable after extracted from their ores.

Green corrosion inhibitor had been studying to relate to this "greenish" theme and reduce the corrosion process as well. The plants such as rosemary leaves, garlic, tamarind, tea leaves, pomegranate juice and peels, aloe leaves, mango or orange peels, and banana peels are very useful for green inhibitor application.

Organic compound that contains oxygen, sulphide and nitrogen which is basically the active center for the absorption process on the metal surface to reduce the corrosion attack on several steel has been studied recently. Corrosion prevention is very important because many industrial companies have suffered from this activity process. Environment with pollutant will cause damage to the systems and reduce machines lifetime especially equipment that uses acid solution.

In this paper, the effect of ginger extract as a green corrosion inhibitor of mild steel in 1.0M of hydrochloric acid has been studied.

IOP Conf. Series: Materials Science and Engineering **290** (2017) 012087 doi:10.1088/1757-899X/290/1/012087

2. Experimental procedures

2.1 Preparation of Ginger extract

The ginger was washed, cut, dried and ground into a fine powder. The powder is then used to soak in distilled water with the ratio of 1g of powder with 8 ml of distilled water for 20 hours at room temperature. The extract was filtered and evaporated in the oven with the temperature 40°C for about 24 hours to produce ginger extract (G-1). Another ginger extract (G-2) was prepared from ginger powder bought at supermarket. It was macerated in the distilled water and filtered. The concentration of 50g/l, 100g/l, and 150g/l of aqueous extract ginger were prepared.

2.2 Preparation of specimens

Mild steel was cut into 6.2 x 2.1 x 0.1 cm^3. The steels were ground with emery paper of 180 to 1200 grits. Then, they were washed using distilled water and degreased with acetone.

2.3 Weight loss experiment

The weight of sample was measured using electronic balance before and after soaking in 100 ml of 1 M HCl. The samples were immersed in HCL solution for 22 hours at room temperature with the absence and presence of different concentrations of ginger extract (G-1) and also (G-2). To get the best result, the experiment was conducted twice. The weight loss of the test specimens were calculated by using equation (1).

$$Weight\ loss\ (\Delta w) = W_i - W_f \qquad Eq\ (1)$$

Where W_i is the initial weight and W_f is the final weight of mild steel specimens. Corrosion rate and inhibitor efficiency can be calculated by using the equation shows in equation (2) and equation (3), respectively.

$$R = \frac{86.7x\ \Delta w}{A\ x\ T\ x\ \rho} \qquad Eq\ (2)$$

$$IE = \frac{R_0 - R_i}{R_0}\ X\ 100 \qquad Eq\ (3)$$

Where R = Corrosion rate (*mmpy*), Δw = Weight loss (*mg*), A = Area of metal exposed to corrosive media (cm^2), ρ = Density of mild steel (g/cm^3) = $7.86 g/cm^3$, T = Time of immersion (h), R_o = Corrosion rate without inhibitor and R_i = Corrosion rate with inhibitor.

2.4 Polarization test

Polarization test uses three types of electrodes which are working electrode, counter electrode and the reference electrode with the mild steel, graphite, and the silver/silver chloride electrode respectively. These three electrodes were connected to the Autolab potentiostat. The scan rate of 0.5mV/s was set up on the setting option. The curve was set to start at -800mV until -200mV at room temperature. The inhibition efficiency can be calculated and known based on the formula:

$$IE = \frac{I^o_{corr} - I_{corr}}{I^o_{corr}}\ X\ 100 \qquad Eq\ (4)$$

2.5 Surface analysis

The surface of the mild steel was analysed by using Nikon Model Epiphot 200 optical microscope. The analysis was conducted after the experiment of weight loss was done. The samples after immersing in 1M HCl with different concentration of ginger extract were collected to be observed using the optical microscope.

3. Results and Discussion

3.1 Weight loss measurement

Experiment was conducted to study the corrosion rates and the percentage of inhibition efficiency from four samples of mild steels in the absent and presence of different concentration of ginger extract inhibitor. The concentration of 50g/l, 100g/l and 150g/l of inhibitors at room temperature were used in this experiment as shown in table 1.

Table 1. Effect of ginger extract (G-1) concentration on corrosion rate of mild steel.

Inhibitor Concentration (g/l)	Corrosion rate (mmpy)	Inhibition Efficiency (%)
0	8.09	-
50	1.04	87
100	0.96	88
150	0.72	91

The weight loss experiment also has been conducted on ginger extract that produced from the source of ginger powder product (G-2). The result is shown on the table 2.

Table 2. Effect of ginger extract (G2) concentration on the corrosion rate of mild steel.

Inhibitor Concentration (g/l)	Corrosion rate (mmpy)	Inhibition Efficiency (%)
0	6.91	-
50	0.94	86
100	0.85	88
150	0.66	90

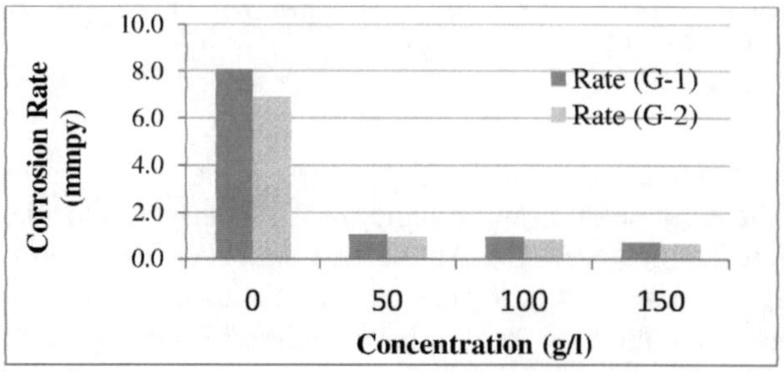

Figure 1. Corrosion Rate of mild steel in various inhibitor concentrations.

From the graph shown in this experiment, the line is increasing where the higher concentration of inhibitor used, the higher the inhibition efficiency. The inhibitor was absorbed in the steel and forming the layer on the surface of the steel. The layer which is like the coating, prevents the contact of the acid to the surface of the steel. The corrosion process will not happen when there is thick protection layer on the surface.

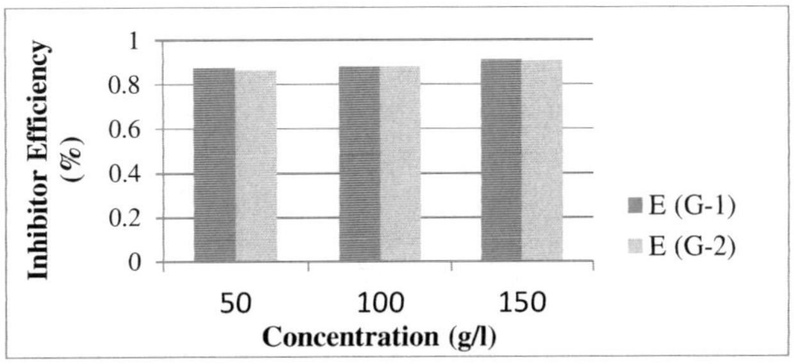

Figure 2. Inhibitor Efficiency in various inhibitor concentrations

The graph in figure 2 clearly shows the trend of the inhibition efficiency. The efficiency of inhibition of is 91% at concentration of inhibitor of 150g/l. Inhibition efficiency of 50g/l and 100g/l inhibitor is 87% and 88% respectively which showed slightly increasing in inhibition efficiency.

From the results, when comparing between G-1 and G-2, it shows that G-1 produced higher percentage in inhibition efficiency of the inhibitor that can prevent corrosion process of mild steel in 1M HCl better than G-2. But the trend from both aqueous ginger extract show same pattern which is the inhibition efficiency is increasing when the higher concentration of the extract was used.

3.2 Polarization test

The polarization curve is presented in figure 3. The data of the test used which is 1M HCl without inhibitor and 1M HCl with different concentration are summarized in the table 3. The corrosion potential (Ecorr), corrosion current density (Icorr) and corrosion rate were obtained from the curve. The inhibition efficiency was calculated using equation 4.

Table 3: Data from polarization experiment for mild steel in Ginger extract

Concentration (g/l)	Ecorr (V)	Icorr (mA/cm^2)	Rate (mmpy)	I.E (%)
-	-0.411	0.394	1.412	-
50	-0.424	0.135	0.495	65
100	-0.411	0.109	0.401	72
150	-0.409	0.054	0.195	86

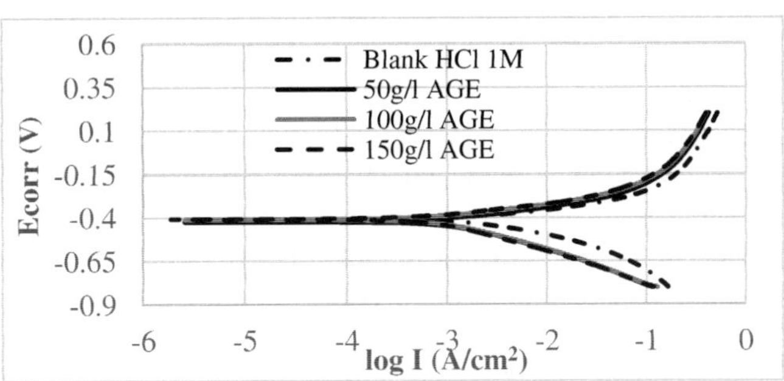

Figure 3: Polarization curve of mild steel in 1M HCl with different inhibitor concentration

From the graph in below, the colour lines are called observed polarization curve. In this curve, there is partial current for forward reaction, known as the anodic reaction and the partial current for reverse reaction known as the cathodic reaction. The both partial currents are precisely equal and

IOP Conf. Series: Materials Science and Engineering **290** (2017) 012087 doi:10.1088/1757-899X/290/1/012087

opposite. Since there is no net current flow, the potential shift from their equilibrium value with the help from the external current is called polarization. In this experiment, the value started from -0.8V and the curve shifted at the point about -0.4V until the reaction is fully utilized. Anodic reaction occurred at the curve after the observed polarization curve was shifting to the positive values. The Tafel slopes can be sketch by putting the tangent line to the both anodic and cathodic polarization curves. The data can be shown that the values for Tafel slopes decrease when the higher concentration of aqueous ginger extract was used in 1M HCl at room temperature. As the corrosion potential is increase towards positive, corrosion current density, Icorr decrease with higher concentration of aqueous ginger extract in 100ml of 1M HCL solution.

3.3 Surface analysis

The microstructure of sample was observed to identify the reaction of the inhibitor with mild steel in 1M of HCl. The microstructure of the mild steel experiences the corrosion process in the solution containing only 1 M HCl is presented in figure 4(a). The corrosion formed darker appearance than the original mild steel. When the 50g/l concentration of the inhibitor was mixed in the acid solution, the microstructure of the layer of the mild steel changed as shown in figure 4(b). It can be observed when comparing the mild steel microstructure in HCl solution with the microstructure of mild steel mixed with the 50g/l concentration of inhibitor where the corrosion process retard in certain places because the inhibitor prevents the corrosion. In figure 4(c) and (d), the brighter in color indicates the inhibitor on the surface of the mild steel where there is no corrosion process happens. Figure 4(c) and (d) show the effect of inhibitor concentration to the microstructure of the mild steel in 1M HCl. It can be said that the inhibitor was absorbed into the mild steel to produce a protective layer to prevent the corrosion process on the surface.

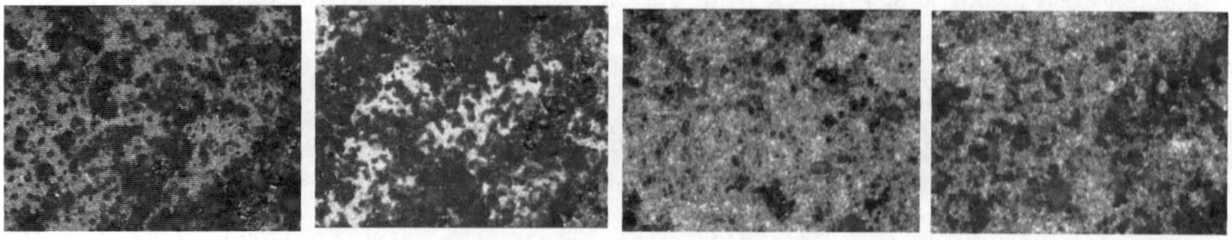

(a) (b) (c) (d)

Figure 4. (a) Mild steel in 1M HCl, (b) Mild steel in 1M HCl +50 g/l Inhibitor, (c) Mild steel in 1M HCl +100 g/l Inhibitor, (d) Mild steel in 1M HCl +150 g/l Inhibitor.

4. Conclusion

Ginger extract can be used as corrosion inhibitor for mild steel in hydrochloric acid solution. The higher concentration of inhibitor is the lower corrosion rate and the higher corrosion inhibition efficiency. Weight loss experiment shows that maximum inhibition efficiency of the high concentration of inhibitor is 91%. The surface analysis has carried out to observe the microstructure of mild steel in the absence and presence of inhibitor. Polarization experiment also shows similar results to that of weight loss experiment. It showed that the ginger extract from rhizome (G-1) and the ginger extract from powder (G-2) have same effectiveness in preventing corrosion of mild steel in hydrochloric acid solution. The efficiency both of them is about 90%.

Acknowledgements

The authors would like to acknowledge the financial support given from FRGS-14-117-0358 grant of International Islamic University Malaysia.

IOP Conf. Series: Materials Science and Engineering **290** (2017) 012087 doi:10.1088/1757-899X/290/1/012087

References

[1] F. Yousfi, M. El Azzouzi, M. Ramdani, H. Elmsellem, A. Aouniti, and N. Saidi, (n.d.), Zingiber Officinal Roscoe Extract Using as Green Corrosion inhibitor for Mild Steel.

[2] Gentil V 2003 Corrosão. 4ed. *Rio de janeiro: Editora LTC.*

[3] Hong J, Kai Z, and Li Y 2008 *Corrosion Science* **50** 865-871

[4] Ramanathan L 2014 *Corrosão e seu controle*

[5] Yaro A, Khadom A, and Wael R 2013 *Alexandria Engineering Journal* **52** 129-135

IOP Conf. Series: Materials Science and Engineering **290** (2017) 012088 doi:10.1088/1757-899X/290/1/012088

Resistive switching of Cu/Cu_2O junction fabricated using simple thermal oxidation at 423 K for memristor application

M H Ani[1], F Helmi[1], S H Herman[3], S Noh[2,*]

[1] Department of Manufacturing and Materials, Kulliyyah of Engineering, International Islamic University Malaysia (IIUM), Jalan Gombak, 53100 Kuala Lumpur, Malaysia
[2] Department of Mechanical, Kulliyyah of Engineering, International Islamic University Malaysia (IIUM), Jalan Gombak, 53100 Kuala Lumpur, Malaysia
[3] NANO-Electronic Centre (NET), Faculty of Electrical Engineering, Universiti Teknologi MARA, 40450 Shah Alam, Selangor, Malaysia

*corresponding author : syednoh@iium.edu.my

Abstract. Recently, extensive researches have been done on memristor to replace current memory storage technologies. Study on active layer of memristor mostly involving n-type semiconductor oxide such as TiO_2 and ZnO. This paper highlight a simple water vapour oxidation method at 423 K to form Cu/Cu_2O electronic junction as a new type of memristor. Cu_2O is a p-type semiconductor oxide, was used as the active layer of memristor. $Cu/Cu_2O/Au$ memristor was fabricated by thermal oxidation of copper foil, followed by sputtering of gold. Structural, morphological and memristive properties were characterized using XRD, FESEM, and current-voltage, *I-V* measurement respectively. Its memristivity was indentified by pinch hysteresis loop and measurement of high resistance state (HRS) and low resistance state (LRS) of the sample. The $Cu/Cu_2O/Au$ memristor demonstrates comparable performances to previous studies using other methods.

1. Introduction

The fourth fundamental circuit element, memristor (memory resistor) proposed by Leon Chua in 1971 is a suitable candidate to replace current memory technologies [1]. Feature size of memory device such as DRAM and FLASH has been successfully decreased while increasing its memory density. However, it will no longer be able to decrease sooner or later. Specifically, when the memories shrink down to 21nm, it will reach miniaturization limit and cannot be shrink down any further [2]. In 2008, Hewlett-Packard (HP) research laboratory has produced the first physical memristor . It has a metal-insulator-metal (MIM) structure using platinum for both of its electrode and n-type semiconductor oxide of titanium dioxide, TiO_2, as the active layer [3]. Since then researchers have use different material for the active layer and mostly consist of other n-type semiconductor oxide such as TaO and ZnO [4]. While only a few does on p-type semiconductor oxide such as SnO and CuO [5].

Cuprous oxide (Cu_2O) is a p-type semiconductor oxide with a band gap energy of 2.1 eV and it has a wide range of application such as photovoltaics, gas sensing, catalysis, dilute magnetic semiconductors and as active layer of memristor [6]. Cu_2O is cheap and fairly easy to be fabricated compared to TiO_2. N-type semiconductor oxide, for example ZnO, contain oxygen vacancies due to lack of oxygen while p-type semiconductor oxide such as Cu_2O contain copper vacancies due to excess in oxygen [7]. It is expected that copper vacancies will contributes to the formation of resistive switching in Cu_2O.

In this research, Cu_2O was prepared using simple thermal oxidation in a water vapour environment at 432 K [6, 8, 9]. This method is favourable as it can be done at low temperature and cheaper than other methods. The memristive performance of the electronic junction was compared and evaluated.

2. Experimental procedures

Thin layer of Cu_2O was produced by oxidising a copper foil in a water vapour environment at temperature of 423 K. The temperature was chosen to avoid the oxide layer from being porous. Copper foil were cut into 1cm x 1cm and polished. The sample was then sonicated using ultrasonic cleaner (Branson 3510) to remove any residue.

Figure 1 shows the furnace set up specially to oxidise copper foil in a water vapour environment. The argon gas flowrate were fix to 100 sccm. Once the heater and electric furnace has achieved their intended temperature, sample were insert and place in the middle of the furnace followed by vaccumisation of the furnace using vaccum pump to remove air trapped inside the furnace. The sample was oxidised at 423 K for 25 seconds and 1 hour.

The fabricated Cu/Cu_2O junction was analysed using XRD (PANalytical X'Pert PRO) to confirm the existance of oxide layer. The surface morphology of the junction was characterized using FESEM (JOEL JFC-6700F).

Gold was sputtered on top of Cu_2O using auto fine coater (JEOL JFC-1600) to produce a Cu/Cu_2O/Au memristor. I-V characteristics of the memristor was then measured using a 2-point probe connected to AUTOLAB Potentiostat (PGSTAT302N) at room temperature. The voltage sweep was 0V→2V→2V→0V repeated 10 cycles for each measurement.

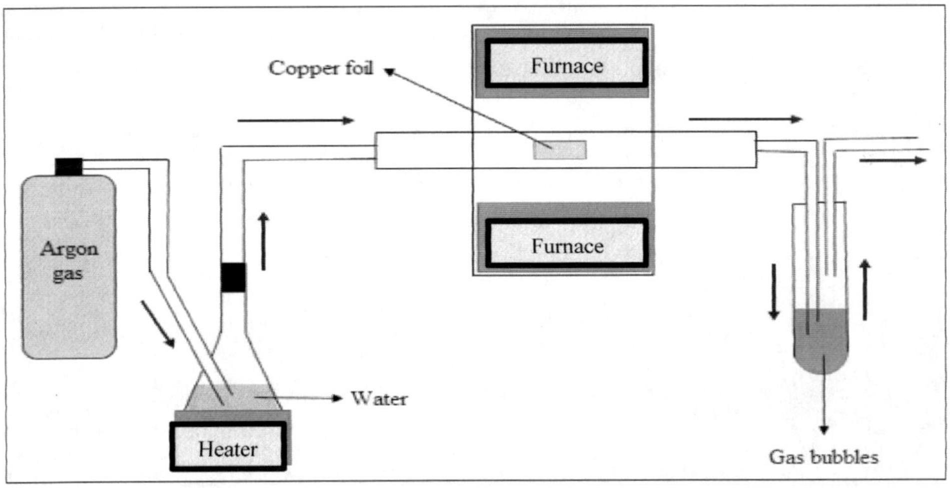

Figure 1 : Tube furnace setup

3. Results and Discussion

Figure 2 shows a diffraction peaks of copper foil after thermal oxidation in water vapour environment for 25 seconds. XRD result shows a (221) Cu_2O diffraction peaks at 66 degree angle and (321) Cu_2O diffraction peaks at 86 degree angle. Due to thin oxide layer, substrate Cu was also detected at respective peak angle with facet orientation of (111), (200) and (220). CuO diffraction peak was not detected. This is due to oxygen partial pressure during the reaction is higher than dissociation pressure of Cu_2O but lower than dissociation pressure of CuO.

IOP Conf. Series: Materials Science and Engineering **290** (2017) 012088 doi:10.1088/1757-899X/290/1/012088

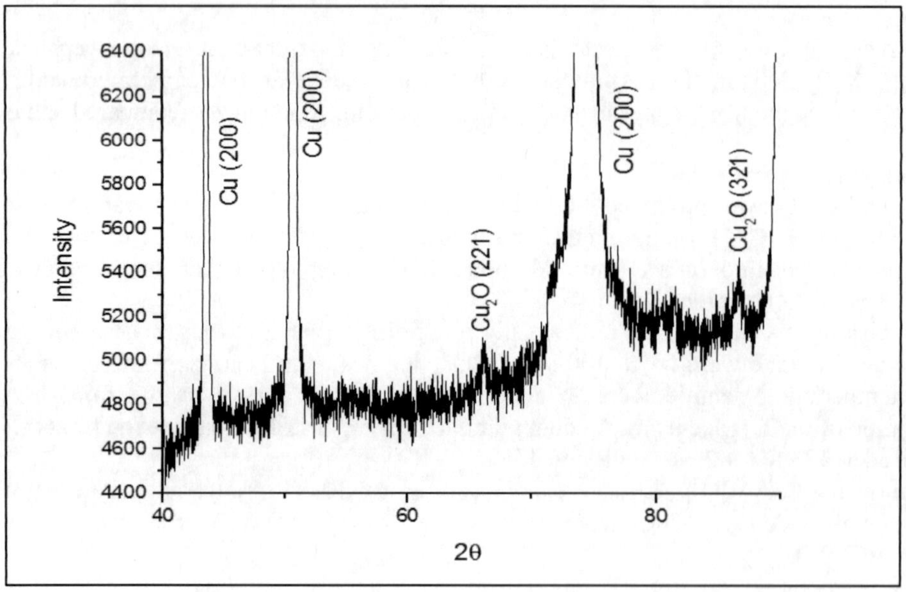

Figure 2 : Diffraction peaks of sample oxidized for 25 seconds

Figure 3 shows FESEM micrograph of the sample at various magnification. The oxide layer was uniformly formed on the surface and show almost no void / porosity. We are unable to detect any form of microstructure such as rod-like and flakes as normally observed using other deposition techniques. It is postulated that water vapor somehow facilitates to the formation of dense and even surface.

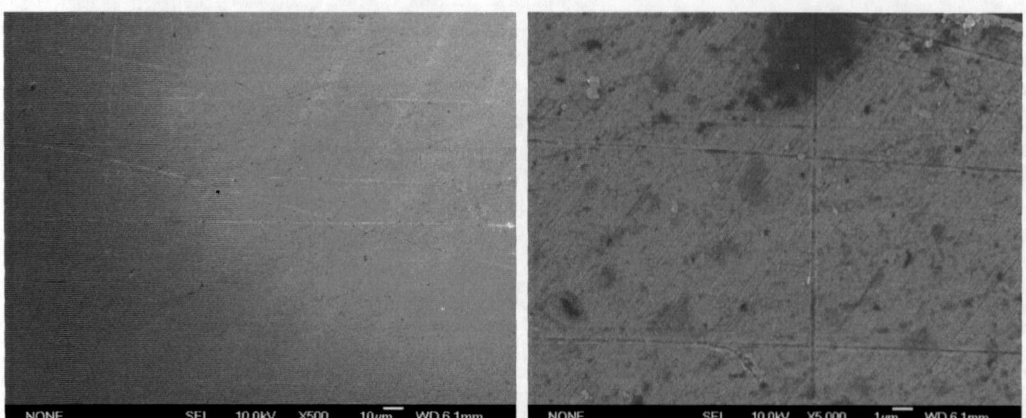

Figure 3 : FESEM image of Cu surface morphology

The *I-V* measurements of the samples show the pinched hysteresis loops as shown in Figure 4 which follows the characteristics of a memristor. Calculated HRS value for sample oxidised for 25 seconds is 15.80Ω and the LRS value is 14.85Ω, while the HRS for sample oxidised for 1 hour is 15.56Ω and the LRS is 14.74Ω. The ratio of HRS and LRS for both sample was determined to be 1.064 for sample oxidised for 25 seconds and 1.056 for sample oxidised for 1 hour. The higher the ratio of HRS and LRS the better as higher ratio means that the distinction between HRS and LRS value is bigger. A memristor required to have big hysteresis loop and a clear distinction between HRS and LRS value so that it can be read by voltage signals.

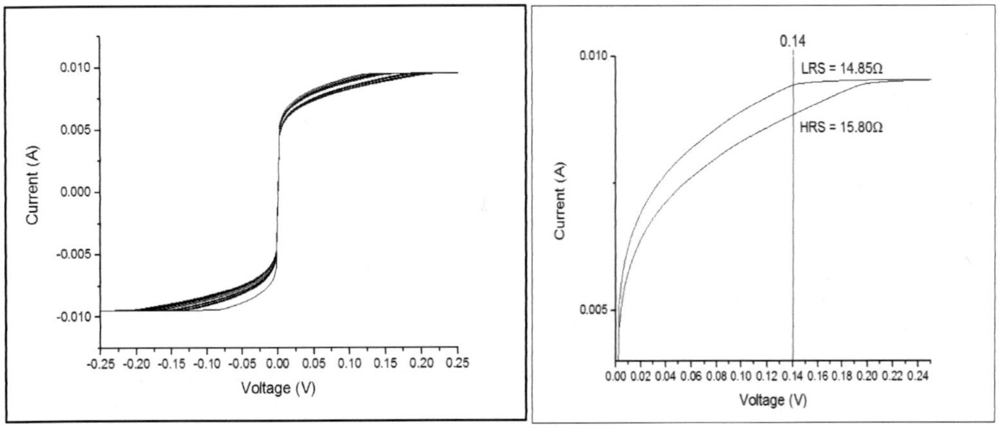

(a) (b)

Figure 4 : (a) hysteresis loop of sample oxidized in 25 seconds, (b) its' HRS and LRS value

The polarity dependence of the resistance to voltage is called bipolar resistive switching. Resistive switching behavior is where the current changes from high resistance state, HRS to low resistance state, LRS at a set voltage from '0 V to 2 V' path to '2 V to 0 V' path. The difference of HRS and LRS creates the hysteresis loop in *I-V* curve. HRS/LRS ratio of Cu/Cu_2O/Au junction shows results that is within the range of our previous findings as reported in [4], [10] and [11]. The difference between HRS and LRS in memristor occurred because of the diffusion of defect vacancy in oxide layers when polarized. The formation of metal vacancy defects in transition metal oxide produces pinched hysteresis loop in memristor. We conclude that Cu/Cu_2O/Au junction is a promising memristor, and its memristive ability relies on the existence of metal vacancy defects in Cu_2O active layer.

4. Conclusion

Cu/Cu_2O/Au memristor has successfully been fabricated by thermal oxidation of copper foil in water vapour environment, and gold sputtering technique. The I-V measurement of a Cu/Cu_2O/Au memristor show a hysteresis loop, proving its resistive switching behaviour. The thinner the oxide layer produced the better the performance of memristor and the sample produce by oxidation of copper foil for 25 seconds have the higher HRS/LRS ratio of 1.064.

Acknowledgement

This work was financially supported by International Islamic University Malaysia (IIUM).

References

[1] L O Chua: *IEEE Trans.* Circuit Theory 18 **(1971)** 507-519

[2] Jeong D S, Thomas R, Katiyar R S, Scott J F, Kohlstedt H, Petraru & Hwang **(2012)** Emerging memories: resistive switching mechanisms and current status *Reports on Progress in Physics* **75** 76502

[3] Strukov D B, Snider S, Stewart D R & Williams S **(2008)** The missing memristor found *Nature* **453** 80–3.

IOP Conf. Series: Materials Science and Engineering **290** (2017) 012088 doi:10.1088/1757-899X/290/1/012088

[4] Fauzi F B, Ani M H, Othman R, Azhar A Z A, Mohamed M A & Herman S H **(2015)**. Fabrication of flexible Au/ZnO/ITO/PET memristor using dilute electrodeposition method. *In IOP Conference Series: Materials Science and Engineering* **99**

[5] K DAquila, C Phatak, M V Holt, B D Stripe, S Tong, W I Park, S Hong, and A K Petford-Long, **(2014)**. *Appl. Phys. Lett.* **104** 242902.

[6] Yan P, Li Y, Hui Y J, Zhong, S J, Zhou Y X, Xu L, Miao X S **(2015)**. Conducting mechanisms of forming-free TiW/Cu_2O/Cu memristive devices. *Applied Physics Letters*, **107**

[7] D O Scanlon, B J Morgan, and G W Watson, **(2009)** *J. Chem. Phys.* **131** 124703

[8] A O Musa, T Akomolafe, M J Carter **(1998)** Production of cuprous oxide, a solar cell material, by thermal oxidation and a study of its physical and electrical properties. *Solar Energy Materials and Solar Cells* **51** 305–316

[9] Liang J, Kishi N, Soga T, Jimbo T & Ahmed M **(2012)**. Thin cuprous oxide films prepared by thermal oxidation of copper foils with water vapor. T*hin Solid Films* **520** 2679–2682.

[10] Yusoff M M, Ani M H and Suryanto **(2013)** *Adv. Mater. Res.* **701** 172-175

[11] Fauzi F B *et al* **(2015)**. *Materials Transactions* ***56*** 1302-1306